T0137328

Wireless Networks

Series Editor

Xuemin Sherman Shen
University of Waterloo
Waterloo, ON, Canada

The purpose of Springer's new Wireless Networks book series is to establish the state of the art and set the course for future research and development in wireless communication networks. The scope of this series includes not only all aspects of wireless networks (including cellular networks, WiFi, sensor networks, and vehicular networks), but related areas such as cloud computing and big data. The series serves as a central source of references for wireless networks research and development. It aims to publish thorough and cohesive overviews on specific topics in wireless networks, as well as works that are larger in scope than survey articles and that contain more detailed background information. The series also provides coverage of advanced and timely topics worthy of monographs, contributed volumes, textbooks and handbooks.

More information about this series at http://www.springer.com/series/14180

Jiangxing Wu

Cyberspace Mimic Defense

Generalized Robust Control and Endogenous Security

 Springer

Jiangxing Wu
National Digital Switching System Engineering
& Technological R & D Center
Zhengzhou, Henan, China

ISSN 2366-1186   ISSN 2366-1445   (electronic)
Wireless Networks
ISBN 978-3-030-29846-3   ISBN 978-3-030-29844-9   (eBook)
https://doi.org/10.1007/978-3-030-29844-9

This Springer imprint is published by the registered company Springer Nature Switzerland AG
The registered company address is: Gewerbestrasse 11, 6330 Cham, Switzerland

Preface

While human beings enter in high spirits the era of network-based digital economy and enjoy to their hearts' content the wonderful material and cultural life delivered by science and technology, they encounter the problem of cyberspace security, which haunts, like a ghost, in both the physical and virtual worlds of the Internet of Everything, constituting the "Achilles' heel" of the network information society and the digital economy. This is due to of the four theoretical and engineering security problems arising from the related original sources which are difficult to break through:

1. Loopholes: Loopholes result from the hardware and software defects that are unavoidable in the current stage of science development, though hardware and software are the bedrock of the information era.
2. Backdoors and Trojans: They are usually planted on the hardware during the making or supply process and are impossible to be eradicated due to the totally open ecosystem—the global value chain characterized by division of labor across countries, industries, and even within a product.
3. Lack of theoretical and technical means to thoroughly examine the complicated information system or control the hardware/software code configuration of devices in the foreseeable future.
4. Backdoors and loopholes polluting the cyberspace from the source. This is due to the above-mentioned causes, which lead to ineffective quality assurance and supervision during the design, production, maintenance, application, and management processes.

As the human society has been speeding up the informatization, cyber security technologies are not developing synchronously at the same level. On the contrary, the ever-increasing technology gap is forcing people to take the opportunistic trend that "the informatization is the first priority," thus opening the Pandora's box in the cyberspace. In addition, there exist too many interest temptations in cyberspace in the digital economy, alluring individuals, enterprises, entities, organizations, or even states or government organs to launch network attacks for self-interests and even take the pursuit for the unrestricted control of the fifth space and absolute freedom in cyberspace as the national strategy. The pan-cyber terrorism is a serious impediment to the continuous prosperity of the modern society, causing people to live in unprecedented anxiety and prolonging darkness.

Human beings have never given up their efforts to address the rigorous cyber security issue. With the emergence of various security technologies, e.g., intrusion detection, intrusion prevention, intrusion tolerance, and encryption and authentication, especially the introduction of big data, artificial intelligence, blockchain, and other analytic techniques and means in recent years, we no longer have to trace the sources only after the occurrence of security problems. In other words, we take proactive measures for prevention, detection, and response rather than mend the fold after a sheep is lost. For example, it is possible to find the vulnerability or suspicious functions of file codes early by software and hardware gene mapping analysis; massive problem scene data can be collected and analyzed via big data for the detection and early warning of hidden attacks; AI can be employed to optimize the state explosion problem in the vulnerability analysis process; and the tamper-resistant technology can be provided through the blockchain consensus mechanism and the timestamp-based chain relations. In addition, in order to offset the advantage of the "single attack" launched by the attacker at the static, certain, and similar vulnerabilities of the target system, we can introduce multilevel defense techniques such as dynamicity, randomness, diversity, trusted computing, and trusted custom space, in the hope of reversing the unbalanced cyber attack and defense game where the defender is all the time in an unbalanced and declining position.

Unfortunately, these defensive measures, whether they are passive models based on various a priori knowledge or behavioral feature or active models using big data intelligence analysis or randomly changing address, data, and instruction, whether they are the sandbox technology for online real-time perception or the intelligent analysis method for offline/background screening, and whether they are the behavior perception technology using trusted computing or the data tampering-resistant technology using blockchain, are, in essence, the "attached" perimeter defense technologies, irrelevant to the functions and structure of the protected object. There exists the lack of necessary feedback control mechanisms or operation just runs like a "black box." Although they have achieved good application results in preventing or reducing the availability or exploitability of security defects in the target object (without considering performance overhead), they do not perform remarkably or fail completely in suppressing "coordinated internal and external" attacks based on the hidden or built-in backdoor functions of the target object or in addressing the attacks based on "side-channel effects" and hardware construction defects (such as CPU's Meltdown and Spectre). To make matters worse, in most cases, these attached perimeter defenses cannot even guarantee the service credibility of their own security functions. For example, the bottom-line defense device for encryption and authentication cannot give any convincing proof on whether it is possible to be "bypassed" by the host system or even has any backdoors or Trojans. It even cannot provide any convincing quantitative and measurable indicators on the security of its ontological service functions when the host cannot guarantee its credibility.

It should be emphasized that in the era of network-based digital economy, it has become a national strategy for defending cyber sovereignty and protecting data resources. All major jurisdictions in the world now regard it as their national strategy to seek a prioritized position in cyberspace, sparing no effort to mobilize national resources and even exploiting market forces and trade protection regulations to compete for cyberspace rights and information control. In particular, with the help of the first mover technical advantage, the market monopoly status, and the control or influence on the design, manufacturing, supply, and maintenance sections across the industrial chain, "hidden loopholes, built-in backdoors, and implanted Trojans" will become indispensable strategic resources for the so-called active defense, which can be almost freely used without the constraints of the current legal systems, ethics, or cyber codes of conduct. They can even be combined with conventional firepower weapons to gain an overwhelming strategic edge. This means that known unknown risks or unknown unknown security threats will pervade the entire industrial chain environment like a horrible plague, polluting and poisoning the entire cyberspace. It will not only pose severe challenges to the ultimate human ideal of "an intelligent and connected world" but also fundamentally shake the basic order and the principle of good faith on which the human society depends for survival and development in the age of digital economy and network information.

In view of the fact that the underlying conditions on which the traditional perimeter defense theory relies are constantly blurring and collapsing, coupled with the promotion and application of the "Zero Trust Architecture" supporting new business models such as mobile office, the perimeter defense software/hardware facilities are not only unable to guarantee their own service reliability but also fail to effectively deal with coordinated internal and external backdoor attacks or attacks exploiting other dark features of the target object of the "Zero Trust Architecture." Both the science and industrial communities must transform the traditional cyber security concepts, mindsets, and technological development models by abandoning the illusion of pursuing utopian "sterile, virus-free" cyberspace. In the "global, open, and shared" digital economy and ecology, we strive to innovate the endogenous security theories and methods. As the software/hardware component design chain, tool chain, production chain, supply chain, and service chain cannot guarantee their credibility, we are now developing disruptive theories and techniques based on system engineering to dispel the attack theories and methods targeted at software/hardware code problems at the structural level of the systems. Without relying on (but with access to) attached security measures or means, we can endow the "structure-determined security" functions in the target system (including the defense facilities) through the innovative software/hardware structural technology.

I remember that when I was studying the variable structure high-performance computer system 11 years ago, I occasionally watched a video showing the striped octopus (also known as the mimic octopus) on an NGC program and was deeply fascinated by the unique features of the magical marine creature. While admiring the greatness of the creator, I came up with an exciting idea: Is it possible to construct a collaborative computing and processing device with a variable structure similar to the mimic function of the octopus, so that the device can change its own structure, operation mechanism, and processing scenario synergistically against different computing models, processing procedures, and resource conditions? Unlike the classic computing and processing model put forward by Von Neumann, the "structure for computer service"—a software/hardware computing method—can not only greatly improve the performance of the area-specific computing processing system but also make the parasitic backdoors in the system lose the stability of their apparent functions and characteristics due to uncertain changes in the structure. The device can, on the one hand, achieve joint optimization and coordinated management in effectiveness, performance, and security, and on the other hand, it makes it much more difficult to create effective and reliable attack chains. In my mind, the mimic computing system should be able to handle diverse, dynamic, and random processing scenarios through its active cognition and coordinated management functions. The nondeterministic relations between its task function, performance goal, and algorithm structure can just make up for the security flaws of staticity, certainty, and similarity in conventional information processing systems when addressing backdoor attacks. I have named the two types of applications based on software and hardware variable structure coordinated computing as "Mimic Structure Calculation" (MSC) and "Mimic Structure Defense" (MSD), respectively. However, the prerequisite for MSC and MSD to make coordinated variable structure reactions is the accurate perception or timely recognition of the on-site environment. Fortunately, MSC only needs to obtain the current running scene or posture data to implement structural transformation or scene migration according to the preset fitting rules, while MSD must judge whether there is a security threat. This may be acceptable when the a priori knowledge or the behavioral characteristics of the attacker are available or even the known unknown threats are perceptible, but for unknown threats lacking behavioral characteristics, there are philosophical cognition contradictions and technical feasibility challenges to conquer before a timely and reasonable judgment can be made.

Obviously, if we can propose an endogenous security function based on the system structure effect and "quantifiable design" and invent an "identification of friend or foe (IFF)" mechanism with controllable credibility, we can conditionally convert unknown unknown events into known unknown events and then into events that can be quantified and represented by probability based on such robustness control mechanisms as measurement awareness, error recognition, and feedback iteration. Namely, man-made attacks based on individual software/hardware backdoors can be normalized to general uncertain disturbances of the target object's heterogeneous redundant structure, so that mature reliability and robust control theories and methods can be used to handle traditional and nontraditional security issues in a unified manner and no longer stay in the period of thinking experiments. Cyber mimic defense is the outcome of theoretical exploration and engineering practice of this vision that "simplicity is the ultimate sophistication."

From the perspective of the defender, whether it is an incidental failure of the information system or control device or a man-made backdoor attack, it is generally an unknown event in nature, where the former is a known unknown event that can be expressed by probability in most cases, while the latter is often an unknown

unknown event that belongs to the uncertain problems and cannot be expressed by probability. Nevertheless, in the scientific sense, being "unknown" is always strongly correlated to cognitive scenarios and perceptual means. As our science and technology evolves, we may change these scenarios and means, and the "unknown" may be transformed into the "known." For instance, humans had thought that the Earth was the center of the solar system before the telescope was invented, while life and death were the will of god or devil before the invention of the microscope.

In fact, in the field of reliability engineering, the dissimilarity redundancy structure (DRS) has been able to, by means of the heterogeneous redundancy scenarios and multimode consensus mechanisms under functionally equivalent conditions (referred to herein as the "relatively correct" axioms), convert the unknown disturbances caused by uncertain physical elements or logic design defects of the random nature of a single device in the target system into an "abnormal event" that can be perceived by the multimode voting mechanism of the heterogeneous redundant architecture and obtain stability robustness and quality robustness through the heterogeneous redundant structure, which are measurable and verifiable. However, if it is directly used to deal with nonrandom man-made attacks featuring "one-way transparency" or "insider-outsider collaboration," there are security flaws in the mechanism such as staticity, certainty, and similarity, especially when the type and amount of heterogeneous redundant bodies are limited (relative to large-scale redundancy scenarios upon the blockchain consensus mechanism). In theory, an attacker mastering certain "common-mode" attack resources can still invalidate the dissimilarity redundancy mechanism for "majority ruling or 51% consensus" by using the "one fatal hit," trial and error, exclusion, and other violent attack methods. In other words, the structure of DRS does not possess stable robustness when dealing with deliberate attacks based on vulnerabilities and backdoors. Therefore, based on the dissimilarity redundancy structure, the author proposes a multi-dimensional reconfigurable "DHR" structure with a strategic decision-making, strategic scheduling, and negative feedback control mechanism, which allows the functionally equivalent "uncertain scenario" effect even in the small-scale space of heterogeneous redundancy. Under the premise of unchanged apparent service functions, any brute-force attacks, whether they are "trial and error" or "coordinated or non-coordinated," against the intra-architecture service elements will be "blocked without being perceived" or "made the attack results difficult to sustain" as long as it can be perceived by the multimode ruling segment. Changes in the ruling status or control policies will give rise to changes of variables in the feedback control functions, leading to changes in the combination of executors in the DHR architecture or changes in the executor's own structure. The basic premise of unchanged background conditions will no longer exist for trial-and-error or common-mode attacks.

It should be emphasized that the mimic defense discussed herein does not include the trial-and-error attacks that aim at an unrecoverable "downtime," or DDoS to block the target object's service chain, or cyber attacks that utilize communication protocols, procedures, specifications, and other design loopholes and backdoors.

You may easily find out that the cyber mimic defense essentially adopts a unique general robust control architecture that integrates high reliability, high credibility, and high availability. Such a defense architecture, supported by the theory of bio-mimic camouflage, can produce an uncertain effect for attackers. This endows the target object with an endogenous security function that is independent from (but can naturally converge with) the effectiveness of attached defense measures. It has seven features: First, in the small-scale space, a man-made apparent uncertain attack against the individual loopholes of the target object's heterogeneous redundant body can be converted into an event with uncertain effects at the system functional level. Second, the event with uncertain effects can be further transformed into a reliability event with controllable probability. Third, the strategic decision-making, strategic scheduling, and multi-dimensional reconfigurable feedback control mechanism can prevent any form of trial-and-error attack based on its designable, quantifiable, verifiable, and measurable endogenous security effects through mimic camouflage Fourth, the coordinated expression based on the relatively correct axiom or consensus mechanism makes it possible to offer the IFF feature with controllable credibility and without relying on the attacker's a priori information or behavioral characteristics, thereby creating a prerequisite for the application of the traditional security defense technology based on the "detection-perception-removal" mechanism. Fifth, it can normalize non-conventional security threats into robust control problems under the framework of classical reliability theories and auto-control theories, which can then be handled through mature security defense measures. Sixth, despite of the uncertain attack effects on the backdoors above the mimic domain (such as those exploiting undetected design flaws or deep-planted backdoors in network protocols), all random failure disturbances and man-made attack disturbances within the mimic domain can be managed or suppressed by the general robust control structure, and the defense effectiveness is subject to quantitative design and verifiable metrics. Last but not least, the "difficulty of dynamic multi-target coordinated attacks under the non-cooperative conditions" provided by the mimic structure will fundamentally turn over the attack theories and methods based on the defects in the hardware/software codes of the target object.

We are excited to see that as the practical network information systems and control device products based on the mimic structure get entry to various application fields in recent years, the rule-changing mimic defense principle and its endogenous security mechanism are constantly revealing its revolutionary vitality. It is expected that in the globalized ecosystem where the credibility of the component supply chain or even the industry chain of the target product cannot be guaranteed, the innovative DHR architecture can blaze a new trail to address the dilemma of hardware/software component security and credibility from the origin of the product.

The author is deeply convinced that with the rapid evolution of the "open-source, diverse, and multiple" industrial and technological ecology in cyberspace, as well as the ceaseless improvement of the mimic defense theory and the continuous innovation of the applied technologies, the mimic structure system can, by naturally integrating or accepting the existing or coming information and security technology outcomes, achieve significant nonlinear defense gains (as the addition of the relevant security elements increases the heterogeneity in the mimic brackets). The strategic landscape of cyberspace, which is "easy to attack but hard to defend," is expected to be reversed from the source of hardware/software products, and the unity of "security and openness," "superiority and maturity," and "independent controllability and security and credibility" will greatly reduce the severe negative impacts of non-tariff barriers (e.g., the reasons involving national security) on global free trade and the open industrial ecology at the engineering level. With the incremental deployment and upgrading of the new generation of information systems, industrial control devices, network infrastructure, terminal equipment, and even basic software and hardware components with the mimic structure and endogenous security functions, the basic order and behavioral code of conducts for cyberspace will be reshaped, and the sharp confrontation between the informatization development and the standardization of cyberspace security order will be relieved, or there even exists the possibility of eliminating it.

At that point, it will no longer be an impossible mission for us to reclose the "Pandora's box" in cyberspace and eradicate the "Achilles' heel" of IT products in the original sources, nor is it a wish difficult to be realized in the thinking experiments.

When the book was about to be published, the first permanently online and globally open Network Endogens Security Testbed (NEST) founded by the Purple Mountain Laboratory for Internet Communication and Security started the acceptance of online public testing on June 26 and welcomed challenges from individuals and organizations around the world. The readers of this book are also welcome to participate in the experience and challenges!

https://nest.ichunqiu.com/

Zhengzhou, Henan, China   Jiangxing Wu
July 2019

Author's Profile

Jiangxing Wu was born in Jiaxing, Zhejiang, in 1953 with his parental native place in Jinzhai, Anhui, China. He is a Professor and the Director of the China National Digital Switching System Engineering and Technological Research Center (NDSC). In 2003, he became an Academician of China Academy of Engineering through strict selection. During the period from the Eighth Five-year Plan, the Ninth Five-year Plan, the Tenth Five-year Plan to the Eleventh Five-year Plan, he served as an Expert and the Deputy Director of the Group for the Communication Technology Theme and of the Information Domain Experts Group for the National Hi-Tech Development Program (863 Program) and as the General Director of the Experts Board for such national major dedicated projects as "High-Speed Information Demo Network," "China's High-Performance Broadband Information Network (3Tnet)," "China's Next-Generation Broadcast (NGB) Television Network," and "New Concept High-Efficiency Computer System Architecture Research and Development" and was in charge of the organizing group for the dedicated projects like the "New-Generation High Trustworthy Network" and the "Reconfigurable Flexible Network." He also served as the Director of the Verification Committee of the National Mobile Communication Major Projects and as the First Deputy Director for the "National Tri-network Convergence Experts Group." In the mid-1980s of the last century, he successfully developed such core switching technologies as software-defined functions, duplicate-T digital switching network, and the hierarchical distributed control architecture. In the following decade, he presided over the successful development of China's first large capacity digital SPC switch—HJD04—with NDSC's own intellectual property rights, which promoted the growth of China's communication high-tech industry in the world. At the beginning of this century, he invented such network technologies as the full IP mobile communication, the indefinite long packet asynchronous switching network, the reconfigurable flexible network architecture, and the IPTV based on router selected broadcasting mechanisms. He also took charge of the successful development of information and communication network core equipment like the complex mobile communication system CMT based on full IP, China's first high-speed core router and the world's first large-scale tandem access router ACR. In 2010, he came up with the high-efficiency oriented mimic computing architecture (the multi-dimensional reconfigurable software and hardware cooperative computing architecture). In 2013, his high-efficiency computing prototype system based on the mimic computing came into being in the world for the first time and passed the national acceptance test, which was selected on the list of China's top ten S&T developments for the year 2013 by China's Academy of Sciences and China's Academy of Engineering. In the same year, he set up the cyberspace mimic defense theory. In 2016, the principle verification system was completed and passed the national test and assessment. In December 2017, he published his book *An Introduction to the Cyberspace Mimic Defense Principles*. On the above basis, he made further modifications and improvements and had his book reprinted, entitled *Cyberspace Mimic Defense Principles: General Robustness Control and Endogenous Security*. He won the First Prize for the National Science and Technological Progress for three times and the Second Prize for the National Science and Technological Progress for four times. He was granted the Science and Technological Progress Award and the Science and Technological Accomplishments Award from the Ho Leung Ho Lee Foundation in 1995 and 2015, respectively. In the same year, the network and switching research team headed by him was awarded the Innovation Team Prize for the National Science and Technology Progress.

Brief Introduction (Abstract)

This book, focusing on the most challenging and difficult problem of uncertain security threats in cyberspace and starting from the current technological limitations in the era, summarizes four basic security issues and three important inferences and comes up with the conjecture that the information system can successfully deal with uncertain threats from unknown sources if it possesses non-specific and specific immune functions like vertebrates. From the axiom perspective of structure-determined security, it elaborates on the formation of the concepts and theorem, original intention and vision, principles and methods, implementation basis and engineering cost, and other theories and methods which remain to be improved regarding the "cyberspace mimic defense" which can change the game rules. Various kinds of materials and contents including the system application examples, the authoritative testing reports, and the principle verification have proved both in theory and practice that the effect of indeterminacy generated by the innovative dynamic heterogeneous redundant architecture and mimic guise mechanisms enables the mimic software and hardware to possess the designable, quantifiable, verifiable, and measurable endogenous security efficacy. Without relying on a priori knowledge and behavioral characteristics of attackers and other attached defense methods except for integration, this approach can properly suppress, manage, and control in time general uncertain disturbances caused by attacks from dark functions based on software/hardware object vulnerabilities and backdoors or occasional failures within the mimic boundary and provides a "simplified and normalized" solution to the problem of conventional security reliability and unconventional cyber security threats through innovative robust control mechanisms. As a new enabling technology, it enables IT, ICT, and CPS software/hardware products to have endogenous security functions. This book has put forward the model of mimic architecture and provides a preliminary quantitative analysis and conclusions regarding the cyber reliability and anti-attack effects.

The book is designed to be used for scientists, researchers, and engineers in such areas as information technology, cybersecurity, and industrial control as well as for college faculty and postgraduates.

Preface

The human society is ushering in an era of digital economy at an unprecedented speed. The information network technology driven by the digital revolution has penetrated into every corner of the human society, creating a cyberspace which expands explosively to interconnect all things. A digital space associating both the real world and the virtual world is profoundly changing the ability of human beings to understand and transform the nature. Unfortunately, however, the security of cyberspace is increasingly becoming one of the most serious challenges in the information age or the digital economy era. It is the greediness of man and the periodical attributes in the development of science and technology that prevent the virtual world created by mankind from becoming a pure land beyond the real human society. The world today has its "Achilles' heel," for example, unscrupulously spying on personal privacy and stealing other people's sensitive information, arbitrarily trampling on the common codes of conduct of the human society and the security of cyberspace, and seeking illegitimate interests or illegal controls.

Despite the variety of cyberspace security risks, the attackers' means and goals are changing with each passing day, imposing unprecedented and far-reaching threats to human life and production. The basic technical reasons, though, can be simply summarized as the following five aspects. First, the existing scientific and technological capabilities of human beings cannot completely get rid of the loopholes caused by defects in software/hardware design. Second, the backdoor problem derived from the ecological context of economic globalization cannot be expected to be fundamentally eliminated in a certain period of time. Third, the current scientific theories and technical methods are generally not yet able to effectively check out the "dark features," such as loopholes and backdoors in the software/hardware systems. Fourth, the abovementioned reasons lead to the lack of effective safety and quality control measures for hardware/software products in terms of design, production, maintenance, and use management, where the cyber world gets severely polluted by the loopholes of technical products as the digital economy or social informatization accelerates, even heading toward annihilation. Fifth, the technical threshold for cyber attacks is relatively low in view of the defensive cost of the remedy. It seems that any individual or organization with cyber knowledge or the ability to detect and exploit the hardware/software vulnerabilities of the target system can become a "hacker" to trample on the guidelines on cyberspace morals or behavior wantonly.

With such a cost disparity in attack-defense asymmetry and such a large interest temptation, it is difficult to believe that cyberspace technology pioneers or market monopolies will not deliberately take advantage of the opportunities arising from globalization, for instance, division of labor across countries, inside an industry and even among product components, to apply strategic control methods, such as hidden loopholes, preserved backdoors, and implanted Trojans. Then, they can obtain improper or illegal benefits other than the direct product profits in the market through the user data and sensitive information under their control. As a super threat or terrorist force that can affect individuals, businesses, countries, regions, and even the global community, dark features such as cyberspace loopholes have become a strategic resource, which are not only coveted and exploited by many unscrupulous individuals, organized criminal gangs, and terrorist forces but also undoubtedly used by stakeholder governments to build up their armed forces and operations for the purpose of seeking cyberspace/information supremacy. In fact, cyberspace has long been a normalized battlefield, where all parties concerned are trying to outplay others. Nowadays, however, the cyberspace is still vulnerable to attacks and yet not resilient to defend itself.

The majority of the current active/passive defense theories and methods are based on precise threat perception and perimeter defense theory and model characterized by threat perception, cognitive decision-making, and problem removal. In fact, in the current situation where intelligent handset or terminal-based mobile offices or e-commerce have become the main application mode, as for the target object or the attached protection facilities, neither the intranet-based regional defense nor the comprehensive ID certification measures based on the "Zero Trust Architecture" can completely eliminate negative effects caused by the loopholes or backdoors. Thus, in view of the "known unknown" security risks or "unknown unknown" security threats, the perimeter defense is not only outdated at the theoretical and technological level but also unable to provide suitable engineering means in practice for quantifiable defense effects. More seriously, so far, we have not found any ideas about the new threat perception that does not rely on attack attributes or behavioral information or any new defense methods that are technically effective, economically affordable, and universally applicable. The various dynamic defense technologies represented by "Moving Target Defense" (MTD, proposed by an American) have really achieved good results in reliably disturbing or crumbling the attack chains that make use of the vulnerabilities of the target object. However, in dealing with dark features hidden in the target system or unknown attacks through the hardware/software backdoors, there still exists the problem of ineffective mechanisms. Even if the underlying defense measures and mechanisms such as encrypted authentication are used, the risks of bypass, short circuit, or reverse encryption brought by dark functions from the internal vulnerabilities/backdoors of the host object cannot be completely avoided. The WannaCry, a Windows vulnerability-based ransomware, discovered in 2017 is a typical case of reverse encryption. In fact, the technical system based on the perimeter defense theory and qualitative description has encountered more severe challenges in supporting either the new "cloud-network-terminal" application model or the zero trust security framework deployment.

Research results in biological immunology tell us that a specific antibody will be generated only upon multiple stimulations by the antigen and specific elimination can be performed only when the same antigen reinvades the body. This is very similar to the existing cyberspace defense model, and we may analogize it as "point defense." At the same time, we also notice that a variety of other organisms with different shapes, functions, and roles, including biological antigens known as scientifically harmful, coexist in the world of vertebrates. However, there is no dominant specific immunity in healthy organisms, which means the absolute majority of the invading antigens have been removed or killed by the innate non-specific selection mechanism. The magic ability obtained through the innate genetic mechanism is named non-specific immunity by biologists, and we might as well compare it to "surface defense." Biological findings also reveal that specific immunity is always based on non-specific immunity, with the latter triggering or activating the former, while the former's antibody can only be obtained through acquired effects. Besides, since there are qualitative and quantitative differences between biological individuals, no genetic evidence for specific immunity has been found to date. At this point, we know that vertebrates acquire the ability to resist the invasion of known or unknown antigens due to their point-facet and interdependent dual-immune mechanisms. What frustrates us is that humans have not created such a "non-specific immune mechanism with clean-sweep properties" in cyberspace; instead, we always try to address the task of coping with surface threats in a point defense manner. The contrast between rational expectation and harsh reality proves that "failure in blocking loopholes" is an inevitable outcome, and it is impossible to strategically get out of the dilemma of dealing with them passively.

The key factor causing this embarrassing situation is that the scientific community has not yet figured out how non-specific immunity can accurately "identify friend or foe." According to common sense, it is impossible for the biological genes, which cannot even carry the effective information generated from biological specific immunity, to possess all the antigenic information against bacteria, viruses, and chlamydia that may invade in the future. Just as the various vulnerability/attack information libraries in cyberspace based on behavioral features of the identified backdoors or Trojans, it is impossible for today's library information to include the attributes of backdoors or Trojans that may be discovered tomorrow, not to mention the information on the form of future attack characteristics. The purpose of our questioning is not to find out how the creator can endow vertebrate organisms with the non-specific selection ability to remove unknown invading antigens (the author believes that with the restraint of operational capability of the biological immune cells, the method of coarse-granule "fingerprint comparison" may be used based on their own genes and all the invading antigens not in conformity with the genes will be wiped out. As an inevitable cost, there exists a low probability of some "missing alarms, false alarms, or error alarms" in the coarse-granule fingerprint comparison.

Otherwise, vertebrate biological beings will not fall ill or suffer from cancers. And it would be unnecessary for extraordinary immune powers to exist. The comparison of own credibility and reliability is a prerequisite for the efficacy of the comparison mechanism but with an unavoidable risk.) but to know whether there is a similar identification friend or foe (IFF) mechanism in cyberspace, and whether there is a control structure that can effectively suppress general uncertain disturbances, including known unknown risks and unknown unknown threats, to obtain endogenous security effects not relying on (but naturally converging with) the effectiveness of any attached defense techniques. With such mechanisms, structures, and effects, the attack events based on vulnerability backdoors or virus Trojans can be normalized to conventional reliability issues. In accordance with the mature robust control and reliability theories and methods, the information systems or control devices can obtain both stability robustness and quality robustness to manage and control the impact of hardware/software failures and man-made attacks. In other words, it is necessary to find a single solution to address the reliability and credibility issues at both the theoretical and methodological level.

First, the four basic security problems in cyberspace are generally regarded as the restrictive conditions because the basic security problems will not change when the system host or the attached or parasitic organizational forms change or when system service functions alter. Hence, we can come up with three important conclusions: security measures may be bypassed in the target system with shared resource structure and graded operational mechanisms; attached defense cannot block the backdoor function in the target object; and defense measures based on a priori knowledge and behavior information and features cannot prevent uncertain threats from unknown vulnerabilities and backdoors in a timely manner.

Second, the challenge to be conquered is how to perceive unknown unknown threats, i.e., how to achieve the IFF function at low rates of false and missing alarms without relying on the a priori knowledge of attackers or the characteristics of attack behaviors. In fact, there is no absolute or unquestionable certainty in the philosophical sense. Being "unknown" or "uncertain" is always relative or bounded and is strongly correlated to cognitive space and perceptual means. For example, a common sense goes like this: "everyone has one shortcoming or another, but it is most improbable that they make the same mistake simultaneously in the same place when performing the same task independently" (the author calls it a "relatively correct" axiom, and the profession also has a wording of the consensus mechanism), which gives an enlightening interpretation of the cognitive relationship of "unknown or uncertain" relativity. An equivalent logic representation of the relatively correct axiom—the heterogeneous redundant structure and the multimode consensus mechanism—can transform an unknown problem scene in a single space into a perceptible scenario under the consensus mechanism in a functionally equivalent multi-dimensional heterogeneous redundant space and the uncertainty problem into a reliability problem subject to probability expression and transfer the uncertain behavior cognition based on individuals to the relative judgment of the behavior of a group (or a set of elements). In turn, the cognitive or consensus results of the majority are used as the relatively correct criteria for reliability (this is also the cornerstone of democracy in human society). It should be emphasized that as long as a relative judgment is made, there must be a "Schrödinger's cat" effect like the superposition state in quantum theory. "Right" and "wrong" always exist at the same time, while the probability is different. The successful application of a relatively correct axiom in the field of reliability engineering dates back to the 1970s, when the first dissimilarity redundancy structure was proposed in flight controller design. For a target system based on this structure under certain preconditions, even if its software/hardware components have diversely distributed random failures or statistically uncertain failures caused by unknown design defects, they can be transformed by the multimode voting mechanism into reliability events that can be expressed with probabilities, enabling us to not only enhance system reliability by improving component quality but also significantly enhance the reliability and credibility of the system through innovative structural technology. In the face of uncertain threats exploiting the backdoors of the software/hardware system (or man-made attacks lacking in a priori knowledge), the dissimilarity redundancy structure also has the same or similar effect as the IFF. Although the attack effect of uncertain threats is usually not a probability problem for heterogeneous redundant individuals, the reflection of these attacks at the group level often depends on whether the attacker can coordinately express consensus on the space-time dimension of multimode output vectors, which is a typical matter of probability. However, in a small-scale space and a certain time, a target object based on the dissimilarity redundancy structure can suppress general uncertain disturbances, including unknown man-made attacks, and has the quality robustness of designable calibration and verification metrics. However, the genetic defects of the structure, such as staticity, similarity, and certainty, mean that its own backdoors are still available to some extent, where trial and error, exclusion, common model coordination, and other attack measures often corrupt the stability robustness of the target object.

Third, if viewed from the perspective of robust control, the majority of cyberspace security incidents can be considered as general uncertain disturbances arising from attacks targeted at the backdoors or other vulnerabilities of target objects. In other words, since humans are not yet able to control or suppress the dark features of hardware/software products, the security and quality problems, which originally arise from the design or manufacturing process, are "forced to overflow" as the top security pollution in cyberspace due to "the unconquerable technical bottleneck." Therefore, where a manufacturer refuses to promise the safety and quality of its software/hardware products, or is not held accountable for the possible consequences caused thereby, seems that it has a good reason to justify its behavior by the "universal dilemma." In the era of economic and technological globalization, to restore the sacred promise of product quality and the basic order of commodity economy and fundamentally rectify the maliciously polluted cyberspace ecology, we need to create a new type of robust control structure that can effectively manage and control the trial-and-error attacks and the uncertain effect generated by the feedback control mechanism driven by the bio-mimic camouflage strategy, providing the hardware/software system with stability robustness and quality robustness against general uncertain disturbances.

Furthermore, even if we can't expect the endogenous security effects of the general robust control structure and the mimic camouflage mechanism to solve all cyberspace security problems or even all the security problems of the target object, we still expect the innovative general robust structure to naturally converge with or accept advances in existing or coming information and security technologies. Whether the technology elements introduced is static or dynamic defense, active or passive defense, the target object's defense ability should be enhanced exponentially so as to achieve the integrated economic and technological goal of "service-providing, trusted defense, and robustness control."

In order to help the readers better understand the principles of cyberspace mimic defense, the author has summarized its key theoretical points into the following: one revolving premise (unknown vulnerabilities and backdoors in cyberspace can lead to uncertain threats); one theory-based axiom (conditional awareness of uncertain threats can be provided); discovery of one mechanism (with the self-adaptable mechanism of "non-decreasing initial information entropy," uncertain threats can be stably prevented); invention of one architecture (the dynamical heterogeneous redundant architecture DHR with the general robust control performance has been invented); introduction of one mechanism (mimic guise mechanism); creation of one effect (difficult to detect accurately); achievement of one function (endogenous security function); normalization of dealing with two problems simultaneously (making it possible to provide an integrated solution to the problems of conventional reliability and non-conventional cyber security); and production of one non-linear defense gain (introduction of any security technology can exponentially promote defense effects within the architecture.)

Finally, it is necessary to complete the full-process engineering practice through the combination of theory and application, covering architecture design, common technology development, theoretical verification, application piloting, and industry-wide demonstration.

"Cyberspace mimic defense" is just what comes out from the iterative development and the unremitting exploration of the abovementioned ideas.

Commissioned by the MOST in January 2016, the STCSM organized more than 100 experts from a dozen authoritative evaluation agencies and research institutes across the country to conduct a crowd test verification and technology evaluation of the "mimic defense principle verification system." The test lasted for more than 4 months and proved that "the tested system fully meets the theoretical expectations and the theorem is universally applicable."

In December 2017, *An Introduction to Cyberspace Mimic Defense* was published by the Science Press. The book was renamed as *The Principle of Cyberspace Mimic Defense: General Robust Control and Endogenous Security* and republished after modification and supplementation in October 2018.

In January 2018, the world's first mimic domain name server was put into operation in the network of China Unicom Henan Branch; in April 2018, a variety of network devices based on the mimic structure, including web servers, routing/switching systems, cloud service platforms and firewalls, etc., was systematically deployed at the Henan-based Gianet to provide online services; in May 2018, a complete set of information and communication network equipment based on the mimic structure was selected as the target facility of the "human-machine war" in the first session of the "Cyber Power" International Mimic Defense Championship held in Nanjing, China, where it underwent high-intensity confrontational tests under new rules. The challengers came from the top 20 domestic teams and 10 world-class foreign teams. A large number of live network operation data and man-machine battle logs persuasively interpret the scientific mechanism of the endogenous security effects generated by the general robust control structure and prove the significance of the unprecedented innovation of the mimic defense technology with trinity features of high reliability, high availability, and high credibility. In May of the same year, nearly 100 domestic research institutes and industrial pioneers co-initiated the "Mimetic Technology and Industrial Innovation Alliance," embarking on a new chapter in the history of the cyber information technology and security industry.

To help readers better understand the principles of mimic defense, the book is made with 14 chapters and 2 volumes. Chapter 1 "Security Threats Oncoming from Vulnerabilities and Backdoors" is compiled by Wei Qiang, which begins with an analysis of the unavoidable backdoors, with a focus on the dilemma of backdoor/vulnerability prevention and control, pointing out that the majority of the information security incidents in cyberspace are triggered by attackers exploiting the hardware/software backdoors and vulnerabilities. The original intention of transforming the defense philosophy was put forward through perception and thinking of these details. Chapter 2 "Formal Description of Cyber Attacks" is compiled by Li Guangsong, Zeng Junjie, and Wu Chengrong. It provides an overview and attempt to summarize the formal description methods of typical network attacks for the time being and proposes a method of formal analysis of cyber attacks targeted at complex cyber environments featuring dynamic heterogeneous redundancy. Chapter 3 "A Brief Analysis of Conventional Defense Technologies" is compiled by Liu Shengli and Guang Yan. It analyzes three current cyberspace defense methods from different angles, pointing out the four problems of the conventional cyber security framework model, especially the defect in the target object and the defense system: a lack of precautions against security threats such as possible backdoors. Chapter 4 "New Defense Technologies and Ideas" and Chap. 5 "Diversity, Randomness, and Dynamicity Analysis" are compiled by Cheng Guozhen and Wu Qi. The two chapters provide a brief introduction to new security defense technologies and ideas such as trusted computing, custom trusted space, mobile target defense, and blockchain and point out the major problems concerned. They give out a basic analysis of the effects and significance of diversity, randomness, and dynamicity of basic defense methods on destroying the stability attack chain and put forward the main technical challenges. Chapter 6 "Revelation of the Heterogeneous Redundancy Architecture" is co-produced by Si Xueming, He Lei, Wang Wei, Yang Benchao, Li Guangsong, and Ren Quan, outlining the mechanisms of suppressing the impacts of uncertain faults on the reliability of the target system based on heterogeneous redundancy techniques and indicating that the heterogeneous redundancy architecture is equivalent to the logical expression of the "relatively correct" axiom and has an intrinsic attribute of transforming an uncertain problem into a controllable event of probability. The qualitative and quantitative methods are used to analyze the intrusion tolerance properties of the dissimilarity redundancy structure and the challenges of at least five aspects, assuming that the introduction of dynamicity or randomness in this structure can improve its intrusion tolerance. Chapter 7 "General Robust Control and Dynamic Heterogeneous Redundancy Architecture" is co-compiled by Liu Caixia, Si Xueming, He Lei, Wang Wei, and Ren Quan, proposing a general robust control architecture, called "dynamic heterogeneous redundancy," for the information system and proving through quantitative analysis methods that the endogenous defense mechanisms based on the architecture can, without relying on any characteristic information of the attacker, force unknown attack behaviors based on unknown backdoors of the target object to face the challenge of "dynamic multi-target coordinated attack in non-cooperating conditions." Chapter 8 "Original Intention and Vision of Cyberspace Mimic Defense" is written by Zhao Bo et al. It aims to apply the biological mimic camouflage mechanism to the feedback control loops of the dynamic heterogeneous redundancy architecture to form uncertain effects. It is expected that the attacker will be trapped in the cognitive dilemma of the defense environment (including the dark functions such as backdoors) within the mimic border, so that the cross-domain plural dynamic target coordinated attack will be much more difficult. Chapter 9 "Principles of Cyberspace Mimic Defense," Chap. 10 "Implementation of Cyberspace Mimic Defense Projects," and Chap. 11 "Bases and Costs of Cyberspace Mimic Defense" are co-compiled by He Lei, Hu Yuxiang, Li Junfei, and Ren Quan. The three chapters systematically describe the basic principles, methodologies, structures, and operating mechanisms of mimic defense, with a preliminary exploration of the engineering implementation of mimic defense, a discussion on the technical basis and application costs of mimic defense, and an outlook to some urgent scientific and technical concerns. Chapter 12 "Application Examples of the Mimic Defense Principle" is co-written by Ma Hailong, Guo Yudong, and Zhang Zheng, respectively briefing on the verification application examples of the mimic defense principle in the route switching system, the web server, and the network storage system. Chapter 13 "Testing and Evaluation of the Mimic Principle Verification System" is co-compiled by Yi Peng, Zhang Jianhui, Zhang Zheng, and Pang Jianmin, respectively introducing the verification of the mimic principle in the router scenario and the web server scenario. Chapter 14 "Application Demonstration and Current Network Testing of Mimic Defense" introduces the usage and tests of the mimic structure products, such as routers/switches, web servers, and domain name servers, in the current networks.

The readers can easily find the logic of the book: point out that the backdoors and vulnerabilities are the core of cyberspace security threats, analyze the genetic defects of existing defense theories and methods in dealing with uncertain threats, exploit the dissimilarity redundancy structure based on the relative correct axiom to get enlightenment of converting random failures to probability-controllable reliability events without a priori knowledge, propose the dynamic heterogeneous redundancy architecture based on multi-model ruling strategy scheduling and the negative feedback control of multi-dimensional dynamic reconstruction, propose to introduce a mimic camouflage mechanism on the basis of this structure to form uncertain effects from the attacker's perspective, and discover that the general robust control architecture, which is similar to the dual mechanism of non-specific and specific immunity across vertebrates, has an endogenous security function and unparalleled defense effect as well as the expected target function, which can independently deal with known unknown security risks or unknown unknown security threats through the backdoors within the mimic border, as well as the impacts of conventional uncertain disturbances, systematically expounded. The principles, methodologies, bases, and engineering costs of cyberspace mimic defense provide the online pilot application cases with principle verification and give out the testing and evaluation results of the principle verification system. In conclusion, it describes the pilot operation of several mimic structure products in the real networks and demos.

Undoubtedly, the DHR-based cyberspace mimic defense will inevitably increase the design cost, volume power consumption, and operation and maintenance overhead along with its unique technical advantages. Similar to the "cost-efficiency" rule of all security defense technologies, where "protection efficiency and defense cost are proportional to the degree of closeness to the target object," the mimic defense is no exception. However, any defense technology is costly and cannot be applied ubiquitously. That's why "deployment in the gateway and defense at the core site" becomes a golden rule in military textbooks. The preliminary application practice in information communication networks shows that the increased cost of applying the mimic defense technology is far from enough to hinder its wide application when compared to the overall life-cycle benefit of the target system. In addition, the continued progress in microelectronics, definable software, reconfigurable hardware, virtualization, and other technologies and development tools, the widespread use of open source community models, and the irreversible globalization trend have made the market price of the target product highly correlated to the application scale only but relatively decoupled from its complexity. The "breaking a butterfly on the wheel" approach and the modular integration have become the preferred mode for market-leading engineers. Moreover, with the continuous sublimation of the "green, efficiency, safety, and credibility" concept, so while pursuing higher performance and more flexible functions of information systems or control devices, people are placing more emphasis on the cost-effectiveness of applications and the credibility of services, shifting from the traditional cost and investment concept to the concept of comprehensive investment and application efficiency of the system throughout its life cycle (including security protection, etc.). As a result, the author believes that with continuous progress made in the theorem and methodology of cyberspace mimic defense, the game rules in cyberspace are about to undergo profound changes. A new generation of hardware and software products with "designable," verifiable, and quantified endogenous security functions and efficacy is on their way, and a carnival of innovation in the mimic defense technology is around the corner.

At present, the mimic defense theory has undergone the phases of logic self-consistency, principle verification, and common technology breakthroughs. The targeted application research and development are being carried out according to the relevant industry characteristics. Valuable engineering experience has been acquired, and significant progress has been made in some pilot and demonstration application projects. New theories and technologies are often incomplete, immature, or not refined. And mistakes are unavoidable. The same is undoubtedly true of this book, for some technical principles are not fully segregated from the "thought experiment" stage, so the immature and rough expressions are inevitable. In addition, the book also lists some scientific and technical problems that need urgent studying and solution in theory and practice. However, the author is convinced that any theory or technology cannot grow to its maturity only in the study or laboratory, especially the cross-domain, game-changing, and subversive theories and techniques, such as mimic defense and general robust control, which are strongly related to application scenarios, engineering implementation, hierarchical protection, industrial policies, etc., and have to undergo rigorous practical testing and extensive application before they can produce positive outcomes. As a saying goes, he who casts a brick aims to attract jade. This book is just like a brick, the publication of which is intended to "attract" better cyber security theories and solutions and maximize the outcome through collective efforts. We sincerely appreciate all forms of theoretical analyses and technical discussions on our WeChat public account (Mimic Defense) and the mimic defense website (http://mimictech.cn). And we wholeheartedly hope that the theory and basic methods of mimic defense can bring revolutionary changes to the strategic landscape of today's "easy to attack yet hard to defend" cyberspace and that the general robust control structure and its endogenous security mechanism characterized by "structure-determined security," quantifiable design, and test validation can bring about strong innovation vitality and thriving replacement demand for the new generation of IT/ICT/CPS technology and the related industries.

This book can be treated as a textbook for postgraduates major in cyber security disciplines or a reference book for the related disciplines. It also serves as an introductory guide for researchers interested in practicing innovation in mimic defense applications or intended to perfect the mimic defense theories and methods. To give the readers a full picture of the connection between the chapters thereof and make it easier for professionals to read selectively, we attach a "chapter-specific relation map" to the contents.

Zhengzhou, Henan, China   Jiangxing Wu
March 2019

The Chapter Relationships Chart

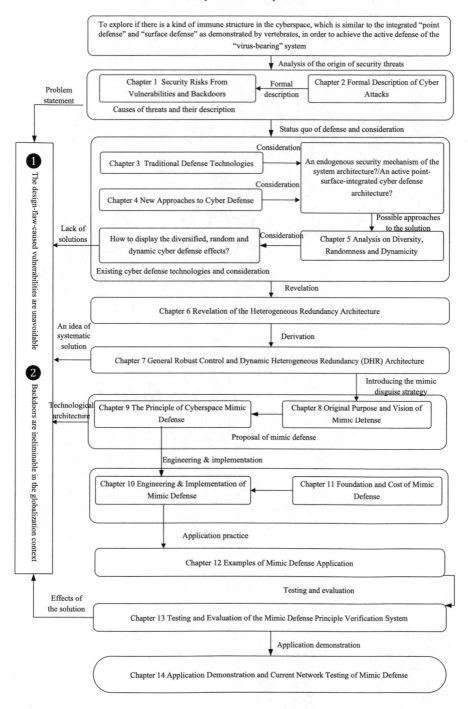

To explore if there is a kind of immune structure in the cyberspace, which is similar to the integrated "point defense" and "surface defense" as demonstrated by vertebrates, in order to achieve the active defense of the "virus-bearing" system

Analysis of the origin of security threats

Problem statement

Chapter 1 Security Risks From Vulnerabilities and Backdoors

Formal description

Chapter 2 Formal Description of Cyber Attacks

Causes of threats and their description

Status quo of defense and consideration

Consideration

Chapter 3 Traditional Defense Technologies

An endogenous security mechanism of the system architecture?/An active point-surface-integrated cyber defense architecture?

Consideration

Chapter 4 New Approaches to Cyber Defense

Lack of solutions

How to display the diversified, random and dynamic cyber defense effects?

Possible approaches to the solution

Consideration

Chapter 5 Analysis on Diversity, Randomness and Dynamicity

Existing cyber defense technologies and consideration

Revelation

Chapter 6 Revelation of the Heterogeneous Redundancy Architecture

An idea of systematic solution

Derivation

Chapter 7 General Robust Control and Dynamic Heterogeneous Redundancy (DHR) Architecture

Introducing the mimic disguise strategy

Technological architecture

Chapter 9 The Principle of Cyberspace Mimic Defense

Chapter 8 Original Purpose and Vision of Mimic Defense

Proposal of mimic defense

Engineering & implementation

Chapter 10 Engineering & Implementation of Mimic Defense

Chapter 11 Foundation and Cost of Mimic Defense

Application practice

Chapter 12 Examples of Mimic Defense Application

Testing and evaluation

Effects of the solution

Chapter 13 Testing and Evaluation of the Mimic Defense Principle Verification System

Application demonstration

Chapter 14 Application Demonstration and Current Network Testing of Mimic Defense

❶ The design-flaw-caused vulnerabilities are unavoidable

❷ Backdoors are ineliminable in the globalization context

Acknowledgments

I am very grateful to all my colleagues who have contributed to the publication of this book. In particular, I would like to express my sincere gratitude to those who directly or indirectly engaged in the writing, revising, or supplementary work. In addition to the colleagues mentioned in the preface to the reprint of this book, who were responsible for or coauthored in the relevant chapters, I also extend my heartfelt thanks to the following people: Liu Xiaolong from the compilation team of Chap. 1, who summarized the related materials of the mitigation mechanism for vulnerability exploitation, and Ma Rongkuan, Song Xiaobin, and Geng Yangyang from the same team, who collected the types of vulnerabilities and made statistical analysis of the cases; He Kang, Pan Yan, and Li Ding from the compilation team of Chap. 3, who were responsible for the collection of the related materials, and Yin Xiaokang from the same team, who was responsible for the adjustment and modification of the entire format of the chapter; Wang Tao and Lin Jian from the compilation team of Chapters 4 and 5, who collected and compiled massive information on new defense types; Liu Qinrang from the compilation team of Chap. 8, who participated in the preparation of the relevant content; and Zhang Jiexin from the compilation team of Chapters 12 and 13, who participated in the writing of the verifying application cases of the web server and the theoretic verification test of web server scenarios. I would also like to thank the organizations engaged in the writing of the related content in Chap. 14, including China Unicom and Gianet (Henan), RuneStone and TopSec (Beijing), ZTE (Shenzhen), FiberHome (Wuhan), Maipu (Chengdu), etc. In addition, Qi Jianping revised the English version of the book and compiled the abbreviations. Ji Xinsheng participated in the planning, writing texture design, and revision of the book; Zhu Yuefei, Chen Fucai, and Hu Hongchao gave valuable suggestions for the writing and the arrangements of some contents; while Chen Fucai and Hu Hongchao, together with Liu Wenyan, Huo Shumin, Liang Hao, and Peng Jianhua, participated in the review of the book.

My special thanks go to Directors General Feng Jichun and Qin Yong, Deputy Director General Yang Xianwu, and Division Heads Qiang Xiaozhe and Wen Bin of the Department of High and New Technology Development and Industrialization under the Ministry of Science and Technology (MOST); Director Shou Ziqi, Deputy

Directors Chen Kehong and Gan Pin, Division Heads Miao Wenjing and Nie Chunni, and Deputy Division Head Xiao Jing of the Science and Technology Commission of Shanghai Municipality (STCSM); Deputy Director Wang Xiujun of the Office of the Central Cybersecurity and Informatization Leading Group (CCILG); Former Deputy Director General Huang Guoyong of the PLA-GSD Department of Communications and Information Technology; etc. They have provided ever-lasting support for the research program.

I would like to sincerely thank the National High-Tech Research and Development Program (863 Program), Zhejiang Lab, the National Natural Science Foundation of China, the Chinese Academy of Engineering, and the STCSM for their long-term funding of this research work.

To conclude, I wish to wholeheartedly thank all my colleagues at the National Digital Switching System Engineering and Technological Research Center (NDSC) and my wife, Chen Hongxing, for their constant contribution to and consistent engagement in this research over the years.

Contents

Abbreviations

ABC	artificial bee colony
ACK	acknowledgment
ACL	access control list
ADR	attack disturbance rate
AnC	ASLRCache
AP	availability probabilities
API	application programming interfaces
APT	Advanced Persistent Threat
AS	attack surface
ASD	all shielding
ASIC	application-specific integrated circuit
ASLR	Address Space Layout Randomization
ASMP	asynchronous symmetric multiprocessor
ASR	address space randomization
AST	attack surface theory
ATA	average time of attack
ATD	average time of defense
AV	access vector
AWGN	additive white Gaussian noise
BGP	Border Gateway Protocol
BIOS	basic input/output system
BNF	Backus-Naur form
BV	backward verification
BVI	backward verification information
C&R	cleaning and recovery
CA	central authentication
CAC	complexity of attack chain
CAICT	China Academy of Information and Communications Technology (hereinafter referred to as the CAICT)
CDN	content delivery network
CERNET	China Education and Research Network

CERT   Computer Emergency Readiness Team
CMD   cyberspace mimic defense
CMDA   CMD Architecture
CMED   common mode events defend
CMF   common mode failure
CNCERT   National Internet Emergency Center
CNNVD   China National Vulnerability Database of Information Security
CNVD   China National Vulnerability Database
CPU   central processing unit
CR   cleaning and recover
CRTM   core root of trust for measurement
CTMC   continuous-time Markov chain
CUHB   China Unicom Henan Branch (hereinafter referred to as CUHB)
CVE   common vulnerabilities and exposures
CVSS   common vulnerability scoring system
DAPAR   Defense Advanced Research Projects Agency
DC   data center
DDN   dynamic domain name
DDoS   distributed denial of service
DEP   data execution prevention
DES   dynamically executing scheduler
DF   dark feature
DHCP   Dynamic Host Configuration Protocol
DHR   dynamic heterogeneous redundant
DHRA   dynamic heterogeneous redundancy architecture
DiffServ   differentiated services
DIL   database instruction labelling module
DMA   differential mode attack
DMF   differential mode failure
DNS   domain name system
DOS   denial of service
DP   damage potential
DP   degradation probabilities
DP   dormancy probability
DPI   deep packet inspection
DPL   deep learning
DRR   dynamic reconstruction rate
DRRV   dissimilar redundant response voter
DRS   dissimilar redundancy structure
DSA   digital signature algorithm
DSAs   domain-specific architecture collaborative computing
DSP   digital signal processing
DVSP   dissimilar virtual web server pool
ECC   elliptic curve cryptography
ED   exploiting difficulty

EK	endorsement key
EP	escape phenomena
EP	escape probability
ES	endogenous security
ESM	endogenous security mechanism
FC(S)	feedback control (system)
FCD	feedback control device
FCL	feedback control loop
FCPT	formal correctness proof techniques
FCS	feedback control system
FCSP	flaw channel scheduling policy
FE	functionally equivalent
FEO	functionally equivalent executor
FM	fault masking
FMEA	fault mode effect analysis
FP	degradation /failure probability
FP6	EU Sixth Framework Plan
FPGA	field-programmable gate array
FSM	finite state machine
FTA	fault tree analysis
FTP	File Transfer Protocol
GFC	Gianet Fast Cloud
GOT	global offset table
GPU	graphics processing unit
GRC	generalized robust control
GRCS	general robust control structure
GSPN	General Stochastic Petri Net
GUD	general uncertain disturbances
GUI	graphical user interface
HE(S)	heterogeneous executor (set)
HFE	heterogeneous functionally equivalent
HIT	heterogeneous intrusion tolerance
HMAC	hash-based message authentication code
HPC	high-performance computing
HPN	Hybrid Petri Net
HR	heterogeneous redundancy
HRS	heterogeneous redundancy system
HRWSEs	HR web service executors
HTR	hard to reproduce
HTTP	Hypertext Transfer Protocol
IA	input agent
IaaS	infrastructure as a service
ICANN	Internet Corporation for Assigned Names and Numbers
ICMP	Internet Control Message Protocol
ICS	industrial control systems

IED	intelligent electronic device
IFF	identification friend or foe
IntServ	integrated services
IoT	Internet of Things
IPS	Integrated Point-Surface
IR	isomorphism redundancy
IS	input sequence
ISR	instruction system randomization
LCS	longest common substring
LOT	low observable technology
LSDB	link state database
MADN	mimic authoritative domain name
MAFTIA	malicious and accidental fault tolerance for Internet applications
MAS	mobile attack surface
MB	mimic brackets
MC	Markov chain
MC	mimic computing
MCAP	microcore and perimeter
MCNC	Microelectronics Center of North Carolina
MCTCC	max concurrent TCP connection capacity
MD	mimic defense
MD	mimic disguise
MD	mimic display
MDL	mimic defense level
MDN	mimic domain name
MDNS	mimic domain name system
MDR	Multi-dimensional dynamic reconfiguration
MDRM	multi-dimensionality dynamic reconfiguration mechanism
MDS	metadata server
MDT	mimic defense theory
ME	management engine
MF	mimic field
MI	mimic interface
MID	multiple independents-events defend
MIIT	Ministry of Industry and Information Technology
MMR	multimode ruling
MMU	memory management unit
MOV	multimode output vector
MQ	message queue
MR	mimic ruling
MRDN	mimic recursive domain name
MRM	Markov reward model
MRP	Markov renewal process
MSC	mimic structure calculation
MSWS	mimic-structured web server

MSWVH	mimic-structured web virtual host
MTBF	mean time between failures
MTD	moving target defense
MTTF	mean time to failure
MTTFF	mean time to first failure
MV	multimode voting
NDSC	National Digital Switching System Engineering and Technological Research Center
NE	network element
NFC	negative feedback controller
NFCM	negative feedback control mechanism
NFSP	negative feedback scheduling policy
NFV	network function virtualization
NGFW	next-generation firewall
NI	non-specific immunity
NIST	National Institute of Standards and Technology
NKD	no prior knowledge of defense
NMD	narrow mimic defense
NoAH	Network of Affined Honeypots
NPU	network processing unit
NRS	non-redundant structure
NSA	National Security Agency
NSAP	non-specific awareness probability
NSF	National Science Foundation
NVD	National Vulnerability Database
O&M	operation and maintenance
OA	output agent
OASIS	organically assured and survivable information systems
ODIN	open data index name
OFC	OpenFlow controller
OFS	OpenFlow switch
OR	output ruling
ORNL	Oak Ridge National Laboratory
OS	operating system
OSA	open system architecture
OSD	object-based storage device
OSPF	open shortest path first
OSVDB	open source vulnerability database
OV	output vector
PA	protection analysis
PaaS	Platform as a Service
PAS	policy and schedule
PBFT	practical byzantine fault tolerance
PCCC	parallel concatenated convolutional code
PCON	primary controller

PCR	platform configuration register
PD	point defense
PES	parity error state
PLT	procedural linkage table
POP3	Post Office Protocol-Version3
PoS	proof-of-stake
PoW	proof-of-work
R&R	reconstruction and reorganization
RBD	reliability block diagram
RC	robust control
RCS	radar cross-section
RD	redundancy design
RDB	request dispatching and balancing module
RE	re-exploitability
RE(S)	reconfigurable executor (set)
RISOS	Research into Secure Operating System
ROP	return-oriented programming
RR	relativity ruling
RRFCSP	rapid recovery and flaw channel scheduling policy
RRSP	rapid recovery scheduling policy
RSVP	Resource Reservation Protocol
RTM	root of trust for measurement
RTR	root of trust for report
RTS	root of trust for storage
RTT	average response time
SaaS	software as a service
SAGE	Scalable Automated Guided Execution
SCADA	supervisory control and data acquisition
SDA	software-defined architecture
SDC	software-defined calculation
SDDC	software-defined data center
SDH	software-defined hardware
SDI	software-defined infrastructure
SDI	software-defined interconnection
SDL	security development life cycle
SDN	software-defined network
SDS	software-defined storage
SDX	software-defined everything
SE	super escape
SEB	symbol error rate
SEH	structured exception handling
SEM	symbiote embedded machines
SF	systematic fingerprint
SGX	Intel software guard extensions
SHA-1	Secure Hash Algorithm

SI	specific immunity
SITAR	scalable intrusion-tolerant architecture
SLA	service-level agreement
SMC	self-modifying code
SMTP	Simple Mail Transfer Protocol
SPA	stochastic process algebra
SQL	structured query language
SRK	storage root key
SSA	steady-state availability
SSAP	steady-state AP
SSE	steady-state escape
SSEP	steady-state escape probability
SSL	semi-supervised learning
SSNRP	steady-state not response probability
SSNSAP	steady-state non-specific awareness probability
SSP	stochastic scheduling policy
SSS	state or scene synchronization
SST	shortest spanning tree
ST	stealth technology
SVM	support vector machine
SVS	supplementary variable analysis
TCB	trusted computing base
TCG	Trusted Computing Group
TCM	trusted cryptography module
TCP	Transmission Control Protocol
TCS	thread control structure
TCS	TSS core service
TDD	TPM device driver
TDDL	TCG device driver library
TEE	trusted execution environment
TFTP	trivial file transfer protocol
TLS	transport layer security
TLU&F	table look-up and forwarding
TMR	triple module redundancy
TPCM	trusted platform control module
TPM	trusted platform module
TPS	transactions per second
TPU	tensor processing unit
TRA	true relatively axiom
TRON	The Real-Time Operating System Nucleus
TRR	transparent runtime randomization
TSP	TSS service provider
TSS	TCG software stack
TT	tunnel-through
TTS	tailored trustworthy spaces

UAC	user account control
UDP	User Datagram Protocol
URL	uniform resource locator
USB	unsustainable
USDOE	United States Department of Energy
VHDL	very high-speed integrated circuit hardware description language
VM	virtual machine
VMM	virtual machine monitor
VPN	virtual private network
WAF	web application firewall
WCSH	web cloud service host
XPP	eXtreme processing platform
XSS	cross-site scripting

Part I

Chapter 1
Security Risks from Vulnerabilities and Backdoors

It is well-known that Alan Mathison Turing, a mathematician, put forward an abstract computing model, Turing Machine, and provided a solution to the computability problem. Another mathematician John Von Neumann came up with the storage program architecture, solving the computational issue of how to control program directions. Later, computer scientists and engineers, one generation after another, have devoted themselves to the promotion of computational capability, the reduction of application threshold and improvement of man-machine efficiency research and practice. And progress in microelectronics and engineering has initiated a digitized, computerized, networking and intelligent era for the human society, creating a computer-controlled cyberspace which exists everywhere. Unfortunately, the security in the computer architecture has long been neglected. Meltdown and Spectre (refer to Sect. 1.1.3) hardware vulnerabilities which have been discovered recently have made people awkward and embarrassed. It seems impossible to find a remedy through pure software. Although people have paid early attention to software vulnerabilities and worked hard to solve the problem, so far no ideal solution has been found. People have to doubt whether software technology alone can overcome the obstacle. Developments in globalized industry work division and the opening of open-source code technology, supply chains or technology chains, and even industrial chains cannot guarantee the trust, which makes threats from potential backdoors even more complicated. The author believes that attack theories and approaches based on software and hardware code defects are the main cause (rather than one of the causes) for the general security threats in the current cyberspace.

1.1 Harmfulness of Vulnerabilities and Backdoors

Since the 1990s, the Internet has seen unusual exponential growth, spreading widely into various economic and social fields. Particularly, over the past decade, there has been a boom of connected devices on cyber information communication

© Springer Nature Switzerland AG 2020
J. Wu, *Cyberspace Mimic Defense*, Wireless Networks,
https://doi.org/10.1007/978-3-030-29844-9_1

infrastructure of IoT (Internet of Things) deriving from the concept of "all things connected." Many new application areas of Internet, such as the industrial control system, artificial intelligence, cloud computing/service, and mobile payment, have experienced vigorous development. On the one hand, the Internet is creating enormous value for the human society. According to Metcalfe's Law ($V = K \times N^2$), the value of a telecommunications network is proportional to the square of the number of connected users of the system; on the other, however, the security risks in cyberspace are on the rise, posing severe impacts on every aspect of the human society and its development. In other words, cyber security threats have never been so close to our life. Software and hardware vulnerabilities have become the decisive factors of cyber security. Most information security incidents are launched by attackers exploiting software and hardware vulnerabilities, as it is an established mode to attack and control the target computer or server through backdoors developed on vulnerabilities or deliberately created by software or hardware designers.

(1) Ransomware WannaCry swept the world. On May 12, 2017, a massive information security attack hit 150 countries, with 200,000 computer terminals being infected. It was related to a type of ransomware called WannaCry. This virus could spread rapidly amid infected computers, and the files and documents saved on these devices could be encrypted. Only by paying a certain amount of bitcoins (a kind of untraceable cyber currency) as a ransom can the users decrypt their documents. More than ten British hospitals, as well as FedEx and Telefonica, fell victim to the virus. Many universities and colleges in China found their networks had suffered from the attack. 20,000 CNPC gas stations got disconnected for nearly 2 days.

(2) The biggest distributed denial of services (DDoS) attack in the US history paralyzed the entire US east coast network in 2016. In October 2016, a botnet controlled by malware Mirai launched a DDoS attack against Dyn, an American DNS service provider, which led to the downtime of many websites in the US east coast, such as GitHub, Twitter, and PayPal. Users were unable to visit these sites by their domain names. After the event took place, a number of security agencies responded by tracking, analyzing, and tracing this incident and found the source of a botnet composed of cameras and other intelligent devices that launched the attack.

(3) Hillary Clinton's email controversy. The story came to light in early 2015: during 2009–2013 when Hillary Clinton served as the US Secretary of State, she handled a large number of confidential emails via her private email box and personal email server at home—an alleged breach of the *Federal Records Act*. The emails were promptly deleted just prior to the investigation. In the summer of 2016, the US Democratic National Committee, the House Ways and Means Committee, and Hillary's presidential campaign team were all attacked by hackers, with nearly 20,000 emails being disclosed via WikiLeaks. According to the emails, Hillary Clinton was suspected of tarnishing her rivals and covering up other financial crimes such as money laundering. On October 28, a

hacker named Kim Dotcom retrieved the emails deleted by Hillary Clinton, leading to the restart of an FBI investigation. The frequent outgoing of such negative news became a key cause for Hillary's defeat in the presidential campaign.

(4) Blockchain smart contract security. In April 2018, hackers successfully transferred a huge amount of BEC tokens by attacking the BEC smart contract through the BatchOverFlow vulnerability of the ETH ERC-20 smart contract, which led to the dumping of a massive quantity of BEC in the market. As a result, the value of BEC that day was almost reduced to zero, with RMB 6.4 billion being vaporized in an instant. Simply 3 days later, another smart contract—SmartMesh (SMT)—was reported to have a vulnerability. Consequently, various exchange platforms suspended the deposit, withdrawal, and trade of the SMT.

(5) TSMC incident. Around the midnight of August 3, 2018, the 12-inch-wafer fabrication works and operation headquarters of the world leading wafer foundry Taiwan Semiconductor Manufacturing Company (TSMC) in Taiwan Hsinchu Science Park were suddenly attacked by computer viruses, with all the production lines being shut down. Only a few hours later, the same happened in TSMC located in Taichung Science Park on Fab 15 and in Tainan Science Park on Fab 14. The three key manufacturing bases of TSMC in north, central, and south Taiwan stopped production due to virus attacks on almost the same day.

Compared with the past, cyber security incidents nowadays take on the following three new trends:

(1) The financial network security has aroused wide concerns. For instance, the Bangladesh Bank had 81 million US dollars stolen; Banco del Austro of Ecuador lost 12 million US dollars, and TienPhong Bank of Vietnam was also the target of attempted hacker attacks. Over the past year, hackers accessed one financial institution after another via SWIFT system vulnerabilities, with Russia being no exception either. The Central Bank of Russia lost 31 million US dollars during a hacker attack.

(2) The key infrastructure has been a new target of hackers. In retrospect of all major cyber security incidents in 2016, hackers not only stole core data but also targeted key infrastructures, such as governments, financial institutions, and energy industries.

(3) The incidents caused by hackers with political backgrounds are on the rise. It can be seen from various cyber security events in 2016 that state-backed hacker activities have been increasing. In the future, cyber security may even have an impact on national stability. Ultimately, it is up to every country to react to cyber security, a strategically important issue.

Basically, all these information security incidents have one thing in common: the information system or software itself has exploitable vulnerabilities and backdoors, which can be a severe threat to national economy, state security, and social stability.

1.1.1 Related Concepts

The academia, industry, community, and international organizations have given different definitions of "vulnerability" at different historical stages and from different perspectives. However, a broad consensus is yet to be achieved. The definitions of "vulnerability" vary from one another: either based on access control [1], on state transition [2], on security policy violation [3], or on exploitable weakness [4–6].

One definition which is easy to understand and broadly accepted [7] is that vulnerability is the defect, flaw, or bug caused by an operating entity deliberately or by accident during the process of design, realization, configuration, and operation of a software system or an information product. It exists at every level and in every link of the information system in various forms and varies with the system. Once utilized by malicious entities, the vulnerability will jeopardize the information system and further affect the normal service operation on this system, exposing both the system and information to risks. Actually, this definition well explains the concept from the perspective of software and system. However, it requires further discussion of the connotation and denotation of vulnerability in order to fully understand the concept against the backdrop of ubiquitous computing, Internet of Things (IoT), and complicated systems.

Prior to the days of computers, people talked little about "vulnerabilities." Hence, the meaning of vulnerability can be more essential and general if we try to understand it from the point of computation. Computation is considered as a process of calculation according to given rules [8]. In fact, computation aims to explore the relationship of equivalence or identity between different objects. Computers, however, are the tools that process data electronically according to a series of instructions. In this sense, we may regard a vulnerability as a "flaw" created by computation, which can be a "dark feature" of logic significance or a specific form of concrete "bug." The existence and form of vulnerabilities are directly related to the computing paradigm, information technology, system application, and attackers' ability and resources. Meanwhile, a vulnerability is also a unity of opposites in that it is both lasting and time sensitive, general and specific, as well as exploitable and repairable.

Backdoor refers to a programmed method of gaining access to a program or a system through bypassing the security control. It is a way of system access, which can not only bypass the existing system security settings but also break through enhanced security fences. Based on our understanding of vulnerability described in the preceding passages, we may well regard "backdoor" as a logic "dark feature" intentionally created during computation. Therefore, as we try to find vulnerabilities, we are also trying to find the "dark features" of computation or any "bugs" that may change the original intention of computation. For less technologically advanced countries and those dependent on the international market, backdoor is a problem that must be addressed. It is also one of the fundamental reasons why the supply chain credibility cannot be guaranteed in the open industry chains and open-source innovation chains. With the continuous improvement of hardware technologies, one single chip can integrate billions of transistors. Also as the software systems are getting more complicated, one operating system requires tens of millions of lines of codes, and some application systems realized by millions of code lines. The extension and application of new development and production tools and technologies, such as automatic generation technology for executable files, spare people from considering specific programming or coding problems, while reusable technology also helps software and hardware engineers realize the logic of standard modules, middleware, IP cores, or manufacturers' process libraries. All this has made it easier to set up and hide backdoors in the global innovation chains, production chains, supply chains, and service chains. The development of attack techniques has enabled the core backdoor technology to evolve from one relying on the exploitation of key vulnerabilities to one independent from the use of vulnerabilities.

1.1.2   Basic Topics of Research

There are four basic topics of research on vulnerability: accurate definition, reasonable classification, unpredictability, and effective elimination.

1.1.2.1   Accurate Definition of Vulnerability

The constant development of computing paradigms, information technology, and new applications leads to constant changes of connotation and denotation of vulnerability, making it difficult to define "vulnerability" accurately.

For instance, Tesler [9] presented four computing paradigms that emerged from the advent of computers to 1991—from batch processing, time-sharing, and desktop system to the network. Since the 1990s, there have appeared new paradigms such as cloud computing, molecular computing, and quantum computing. Under different conditions (single user, multiple users, and multiple tenants) and in different computing environments (distributed computing and centralized computing), vulnerabilities may vary.

Moreover, different systems have different security needs, so their vulnerability identification standards are different, too. Those identified as vulnerabilities in some systems may not be necessarily identified as the same in other systems, while even the same vulnerabilities may have different impacts on different systems. Therefore, it is also hard to define the degree of harmfulness of vulnerabilities in one single manner. The interference of human factors may particularly turn a vulnerability into an uncertainty. Some weaknesses in software and hardware design may or may not become vulnerabilities for attackers, depending on the resources and abilities of the latter.

As time changes, our understanding of vulnerability is also changing—from the earliest one based on access control to the current one based on a whole process including system security modeling, system design, implementation, and internal control. Thanks to the development of information technology, we will gain a deeper knowledge of vulnerability and even develop a more accurate definition thereof.

1.1.2.2   Reasonable Classification of Vulnerabilities

Carl Linnaeus, a Swedish scientist, once said, "...objects are distinguished and known by classifying them methodically and giving them appropriate names. Therefore, classification and name-giving will be the foundation of our science." Given the ubiquity, quantity, and variety of vulnerabilities in different information systems, it is necessary for us to research into the classification methods of vulnerabilities so as to better understand and manage them. The classification of vulnerabilities refers to the categorization and storage of vulnerabilities in line with their causes, forms, and outcomes for ease of reference, retrieval, and use. Unfortunately, as we have not developed a full understanding of the nature of vulnerability, it is rather difficult to classify vulnerabilities exhaustively and exclusively.

In early studies, the classification of vulnerabilities aimed to eliminate the coding errors in operating systems. Therefore, it focused more on the cause of vulnerabilities, which, as we see it today, has certain limitations and cannot fully reflect the essence of vulnerability. These classification methods include RISOS (Research Into Secure Operating System) [10] and PA (Protection Analysis) [11]. For example, Aslam [12] from the COAST Laboratory of Purdue University proposed a cause-based fault classification method for the UNIX operating system. Taking a step further, researchers became aware of the vulnerability life cycle. Meanwhile, cyber attacks began to pose severe threats to system security. As a result, researchers at this stage started to focus on the harmfulness and impact of vulnerabilities to weave it into the basis of classification. This part of research includes Neumannn [13] presented a classification method based on risk sources; Cohen [14] studied the classification of vulnerabilities based on system attacks; and Krsul et al. [15] proposed an impact-based vulnerability classification method. As they understood more of vulnerabilities, researchers tried to integrate their knowledge of the relation between vulnerability and information system, the attributes of vulnerabilities, as well as the exploitation and repair of vulnerabilities with the classification methods in order to describe the nature and interdependence of different vulnerabilities more accurately. In this sense, Landwher [16] established three classification models according to the origin, time of formation, and location of vulnerabilities; Bishop [17] from the University of California in Davis brought forward a six-dimensional method to classify vulnerabilities on the basis of the causes, time, way of use, scope, number of vulnerability exploitation components, and code flaws; Du et al. [18] argued that the life cycle of a vulnerability can be defined as a process of "introduction-destruction-repair" so as to classify vulnerabilities based on their causes, direct impact, and repair methods; Jiwnani et al. [19] put forward another way of classification based on the cause, location, and impact of vulnerabilities.

As vulnerabilities are growingly becoming a security problem with extensive influence, specialized vulnerability management agencies have emerged to tackle the issue, and vulnerability database management organizations have also come into being. They have formulated strict standards for the naming and classification of vulnerabilities. The US National Vulnerability Database (NVD) provides a list of Common Vulnerabilities and Exposures (CVE). The China National Vulnerability Database of Information Security (CNNVD), China National Vulnerability Database

(CNVD), Open Source Vulnerability Database (OSVDB), BugTraq, Secunia, and many other commercial vulnerability databases have their own classification methods. China has issued the related rules and regulations on information security technology security vulnerabilities, such as "Information Security Technology—Vulnerability Identification and Description Specification" (GB/T 28458–2012), but the rules on vulnerability classification have not yet been released. It shows that there remains a considerable difficulty for people to reach a consensus in this regard.

Unfortunately, none of these classification methods have been widely recognized, as the theoretical research and practice analysis of vulnerability characteristics, development trends, differences and internal relations, as well as the role of human factors in the formation and exploitation of vulnerabilities, are still inadequate.

1.1.2.3   Unpredictability of Vulnerabilities

The unpredictability of vulnerabilities encapsulates the "4W" questions, i.e., we may not know *when* and *where* the vulnerabilities will be discovered, *who* will discover them, and *what* vulnerabilities will be discovered. Unknown vulnerabilities include both the known ones under unknown categories and unknown ones under known categories. Today, humans can neither predict the emergence of new vulnerability categories nor obtain an exhaustive knowledge of particular categories of vulnerability.

The categories of vulnerability have increased, from the earliest password questions, buffer overflow, SQL (structured query language) injection, XSS (cross-site scripting), and race condition to complicated vulnerability combinations. Vulnerabilities exist in various software, modules, and firmware. Some of them even involve the combination of system and firmware. Meanwhile, the causes and mechanisms of vulnerabilities are becoming increasingly sophisticated. Some "geeks" have been trying every effort to look for undiscovered new vulnerabilities rather than to pursue the known ones.

According to the security report released by Symantec [20], the world's largest security software provider, released in 2015, 54 unknown vulnerabilities were discovered and deployed by hackers, far more than 24 vulnerabilities in 2014 and 23 in 2013. In 2007, the figure was 15, ranking the fourth over the past decade. The rapidly increasing exploitation of unknown software vulnerabilities shows that cybercrimes and cyber spying activities are getting more advanced technologically. The "zero-day vulnerabilities" or unknown ones gathered by Symantec included those discovered via the trace left by top hackers and those released publicly and recognized by software developers. In particular, secret vulnerabilities in computer programs were paid great attention to by criminal gangs, law enforcement agencies, and spies, as software developers will not release any patch if there is no alert sent to them. In 2015, the Hacking Team's files were disclosed on the Internet, in which there were six "zero-day vulnerabilities" that could be readily used by criminals. In the development and release of patches, software developers may either make the vulnerabilities known to all or disclose them when releasing the patches.

1.1.2.4 Elimination of Vulnerabilities

There is now a consensus that software security vulnerabilities will be a long-standing and unavoidable problem as the software systems are getting complicated. Among the many reasons for it, one has to do with the special characteristics and traditional mindset of the software industry. Software development aims to match software with hardware in order to realize a particular function. Therefore, software security will be taken care of only after the software functions are realized. Moreover, with faster software upgrading and greater competition, it is more important to seize the market by introducing usable software ahead of others than focusing on the security issue in the first place. People have gradually realized that there will be bugs in software development, and these bugs will further affect the software stability and functionality. As more software systems and devices are getting connected to the Internet, bugs will be likely to become a security issue. In the beginning, people used to be quite optimistic, trying to guarantee the security of software in the way of theorem proving. However, it is by no means a once-and-for-all task. Then, people started to classify the bugs, while researching on the methods to eliminate particular vulnerabilities. Disappointingly, analytical models based on program analysis, such as abstract interpretation and static symbolic execution, went bottle-necked. Major problems included low accuracy of interprocedural analysis and high difficulty of orientation analysis. Since the 1980s, model checking technology has been well applied in the checking of timing problems, but the challenge of state explosion followed. So far, it has been a key barrier for the development of model checking technology. In the late 1990s, the industrial community began to adopt fuzzing testing to facilitate the discovery of vulnerabilities. Fuzzing testing is a method to trigger the internal bugs of a program by random construction of samples.

Under the Moore's Law and the convergence of software and hardware design, the complexity of hardware systems is on the rise. Similar to software systems, the design flaws and potential security vulnerabilities in hardware systems will be long-standing and difficult to repair. Vulnerabilities such as Meltdown and Spectre are such cases. These two vulnerabilities existed in Intel CPUs produced over the past 5 years.

1.1.3 Threats and Impacts

1.1.3.1 Broad Security Threats

Theoretically, all information systems or devices have vulnerabilities in their design, realization, and configuration. In a word, vulnerabilities are ubiquitous. In recent years, the related studies on vulnerabilities are extending to closed systems, such as CPU chips and industrial control systems. Compared with software and system security problems, those hidden in CPUs will have a more severe impact. The security of industrial control systems—key national information infrastructure—is even related to national interest, people's livelihood, and social stability.

1. CPU vulnerabilities under heated debate

With the exposure of Intel's CPU vulnerabilities that almost affected computers around the globe, the research on CPU vulnerabilities has drawn wide attention from the security community. Since the first CPU vulnerability was detected in 1994, the harmfulness of CPU vulnerabilities has evolved from denial of service to information stealing, which inflicts greater damages and escalates the skills of vulnerability exploitation.

The first CPU vulnerability "Pentium FDIV bug [21]" was found in Intel CPUs by Professor Thomas at University of Lynchburg in 1994. In order to raise the computing speed, Intel recorded the entire multiplication table in the CPU. However, there were five recording bugs in 2048 multiplying digits, which led to errors of special number crunching. The later "CPU F00F" vulnerability [22] was also discovered in Intel CPUs and had an impact on all those CPUs based on P5 microarchitecture. After that, AMD CPUs were also reported to have TLB vulnerabilities. In the affected Phenom CPUs, TLB would cause reading page errors and further led to denial of service (e.g., system crash).

In 2017, ME (Management Engine) vulnerability [23] was discovered in Intel CPUs, which became the turning point of CPU vulnerability evolvement. ME is a low-power built-in subsystem in Intel CPUs to facilitate remote computer management. The original purpose of such a design is to realize remote maintenance of computers. But the vulnerability enabled attackers to control the computers via the ME backdoor. Among the high-risk CPU vulnerabilities, Meltdown [24] and Spectre [25] were two good examples. They were jointly discovered by Google Project Zero, Cyberus, and foreign universities in January 2018. They had an impact on all Intel CPUs produced after 1995. Meltdown can "melt down" the hardware isolation boundary between the user mode and the operating system kernel mode. Taking advantage of this, an attacker may utilize this vulnerability to break through the limit of system privileges to read the memory information in the system kernel and cause data leakage. Similar to Meltdown, Spectre could destroy the isolation between different applications. An attacker may utilize the CPU predictive execution mechanism to attack the system, control a variable, or register the target application through malware, and then steal the private data which should have been isolated. According to a Financial Times report on August 15, Intel disclosed the latest CPU vulnerability "L1 TF," nicknamed "Foreshadow," which may enable hackers to access the memory data. It was discovered by Catholic University of Louvain and a team composed by the staff of University of Michigan and University of Adelaide, respectively. The Computer Emergency Readiness Team (CERT) of the US government warned on August 14 that an attacker could take advantage of this vulnerability to obtain sensitive data, such as secret keys and passwords. According to experts, it was more difficult to utilize Foreshadow than other average vulnerabilities. Table 1.1 shows some typical CPU vulnerabilities discovered in recent years.

The causes of CPU vulnerabilities include both design logic problems and realization problems. Meltdown, Spectre, and ME are the ones resulted from inadequate consideration of potential security hazards in CPUs' function design, leading to

Table 1.1 Typical CPU vulnerabilities

Manufacturer	Vulnerability	Type	Description	Affected products
Intel	CVE-2012-0217	Local elevated privilege	Vulnerability in the sysret command	Intel CPUs produced prior to 2012
	Memory sinkhole	Local elevated privilege	Installing rootkit under the "system management mode" of CPU	Intel x86 CPUs produced between 1997 and 2010
ARM	CVE-2015-4421	Local elevated privilege	Obtaining root and exploiting the vulnerability through ret2user technology	Huawei Kirin series
AMD	CNVD-2013-14,812	Service rejection	Microcode fails to process the related commands and leads to service rejection	AMD CPU 16 h 00 h-0Fh
Broadcom	CVE-2017-6975	Code execution	Attackers in the same WiFi network may exploit the vulnerability to execute malicious code via the Broadcom WiFi chip (SoC) used by the device	iPhone5–7, Google Nexus 5, 6/6P, and Samsung Galaxy S7, S7 Edge, S6 Edge
Intel	CVE-2017-5689	Privilege escalation	Loading any program and reading and writing files remotely	Intel management firmware, including 6.x, 7.x, 8.x, 9.x, 10.x, 11.x, 11.5, and 11.6
Intel	CVE-2017-5754	Unauthorized access	Low-privileged users may visit the kernel and access the low-level information of the local operating system	All Intel CPUs produced after 1995 (except Itanium and Atom CPUs manufactured before 2013)
Intel/AMD	CVE-2017-5753/ CVE-2017-5715	Information leakage	Utilizing Spectre to break through the user isolation and steal the data of other users in a cloud service scenario	All Intel CPUs produced after 1995 (except Itanium and Atom CPUs manufactured before 2013), AMD, ARM, and NVDIA chip products

privilege isolation failure and unauthorized access. Realization problems, however, are caused by potential security risks in CPU's realization details, such as FDIV and F00F vulnerabilities.

As the CPU itself is the core at the bottom layer of a computer, its vulnerabilities are deeply hidden and exert considerable harms and broad damages.

Hiding deep: CPU is the bottom-level terminal that carries out the computation, with its internal structure being transparent to the upper level. The CPU vulnerabilities are hiding so deep that they are hard to be captured by non-CPU security professionals.

Considerable harmfulness: The privileged level of a CPU is even lower than the operating system. Thus attackers may obtain higher privileges through CPU vulnerabilities than operating system ones. In the age of "cloud," CPU vulnerabilities could enable harmful operations, such as virtual machine escape and breaking through virtual machine isolation.

Broad damages: As CPU is a basic computing component, it is a necessary part in the computing devices, such as IoT devices, PCs, servers, and embedded devices. One CPU vulnerability will usually affect many devices, causing cross-industrial and cross-domain damages.

2. Industrial control system vulnerabilities in focus

In August 2016, a well-known US dotcom called FireEye released a report, which said that over the previous 15 years, there had been more than 1500 ICS (industrial control systems) vulnerabilities being discovered around the world. What's worse, nearly one third of them have not been repaired [26].

Attacks on ICS have been widely reported in recent years. A most typical incident is the outbreak of Stuxnet in 2010. As the first worm virus targeting at ICS, Stuxnet could exploit the vulnerability of SIMATIC WinCC/Step 7, a control system manufactured by Siemens, to infect the Supervisory Control and Data Acquisition (SCADA) system. Aiming at paralyzing the Bushehr Nuclear Power Plant in Iran, Stuxnet utilized seven latest vulnerabilities in the products of Microsoft and Siemens to attack the target, finally leading to the delay of start-up of the nuclear plant. Similar incidents include the Industroyer virus attack on Ukraine's power grid control system in December 2016. As a result, Kiev, the capital of Ukraine, suffered a power outage for over an hour, with millions of households being trapped in power failure and the power facilities seriously damaged. Attackers exploited the vulnerability CVE-2015-5374 in SIPROTEC to disable services and responses.

The ICS vulnerabilities cover many areas. They are highly diversified, harmful, and increasing in number. So far, CNVD has collected more than 1610 ICS vulnerabilities, many of which are from communication protocols, operating systems, applications, and field control level devices, including buffer overflow, hard-coded certificate, bypassing of authentication, and cross-site scripting. Figure 1.1 shows the vulnerabilities collected by CNVD between 2010 and June 2018. Since the outbreak of Stuxnet in 2010, ICS vulnerabilities have seen an exponential increase. Those collected by CNVD grew sharply from 32 in 2010 to 381 in 2017.

The connection of all things has become a major development trend. More and more individuals will be connected to the same network, not only including smart wearable devices and smart home appliances but also involving many other areas, such as business and trade, energy and traffic, social causes, and urban management. According to Gartner's statistics [27], there will be 25 billion devices being connected to the Internet by around 2020. The ratio of number of connected devices to that of humans will exceed 3:1.

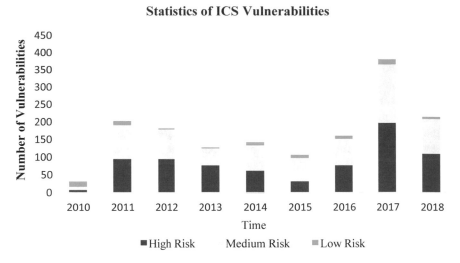

Fig. 1.1 Statistics of ICS vulnerabilities between 2010 and June 2018

Today, the IT community is eagerly running after "software definition," from software-defined function, software-defined computing (SDC), software-defined hardware (SDH), software-defined network (SDN), software-defined architecture (SDA), software-defined storage (SDS), software-defined Internet (SDI) to software-defined data centers (SDDC) and software-defined infrastructure (SDI). Many products and technologies are labeled as "software-defined" ones. Some have even put forward the idea of "software-defined anything" (SDX), "software-defined world," and "everything is software." "Software defined" has almost become the synonym of the state-of-the-art technology. In the future, more and more computing, storage, transmission, and exchange—and even the entire IT infrastructure—will be software defined. Simultaneously, there will be all kinds of risks brought about by potential vulnerabilities, with far greater security threats.

1. Homogeneity leads to similar system environments

Years of research has warned that the homogenization or unicity of software and hardware will cause huge security risks. However, most information systems today are still using a relatively static-fixed architecture and a similar operational mechanism. Under this condition, any attack that is effective on one system can be easily and quickly applied to all the other systems with the same environments or similar configuration.

The cyberspace is dominated by the Turing-Von Neumann Architecture, a processing architecture, which, coupled with the laws of market economy and monopoly, has further led to the lack of diversity of architectures in the cyber space, and the result is serious homogenization of cyber technology and system architecture. At the same time, the staticity, predictability, transparency, and similarity of the information system architecture mechanism and operating mechanism have formed the largest "security black hole," as the exploitation of

Table 1.2 Web browsers built on open-source kernels

Chinese web browsers	Open-source kernels
Cheetah safe browser	V1.0~4.2: Trident+Webkit; V4.3: Trident+Blink
360 Safe browser	V1.0~5.0: Trident; V6.0: Trident+Webkit; V7.0: Trident+Blink
Maxthon	V1.x, 2.x: IE kernel; V3.x: IE + Webkit dual kernels
TheWorld	IE kernel initially and Chrome+IEkernel in 2013
Sogou Explorer	V1.x: Trident; V2.0 and after: Trident+Webkit
UC Browser	Blinkkernel +Tridentkernel

"vulnerabilities or backdoors" will cause continuous security threats. In a globalized world, the cyber eco-environment is extremely vulnerable. The information devices are particularly sensitive to unknown technological defects or vulnerabilities and backdoors. With no prior knowledge, the current defense system based on accurate perceptions will be unable to address the threats of "unknown vulnerabilities." Possibly, even the blockchain technology based on mass user "consensus mechanism" has not taken into consideration the industrial or market fragmentation factors. For instance, the Wintel alliance has seized over 80% of the desktop terminals and operating systems, and Google's Android system has occupied over 70% of the mobile terminal market. If the vulnerabilities and backdoors of these software and hardware products are exploited along with the global positioning and timing services, it is fairly possible that one can break through the "51% consensus threshold."

2. Code reuse leads to genetic similarity

Code reuse includes at least three aspects, i.e., (1) integrating third-party libraries; (2) using open source codes; and (3) inheriting historical codes.

First disclosed in April 2014, Heartbleed is a program bug of the encrypted library OpenSSL. The library is broadly used to realize the Transport Layer Security (TLS) protocol. Many large portals in China, including Alibaba, Tencent, and Baidu, were all affected.

Also, many Chinese web browsers, such as 360, Maxthon, TheWorld, UC, and Sogou, use open-source kernels. If vulnerabilities are detected in the kernels, these web browsers will also be exposed to security risks. Table 1.2 shows the related statistics.

The backward compatibility of operating system codes will lead to a broadened impact of one vulnerability code snippet, from vulnerability MS1–011 of Windows to Shellshock of Linux. The impacts of these vulnerabilities are extensive and lasting.

MS11–011: There has been a local privilege elevation vulnerability in Windows since 1992. It can enable hackers to obtain ultimate control over a system so as to easily destroy and disable any security software, including antivirus software, firewalls, proactive defense software, sandbox, and system restoration. Meanwhile, it can also bypass the UAC (user account control) of Windows Vista/Windows 7 or elevate the privilege on the server site to control the entire web server, posting a direct threat to the information security of governments, enterprises, cyber cafes, and PC users. Lurking for as long as 18 years, the vulnerability may affect various versions of Windows operating systems, including Windows NT 4.0, Windows 2000, Windows XP, Windows 2003, Windows Vista, Windows 7, and Windows Server 2008.

Shellshock: Bash was reported to have a remote code execution vulnerability on September 24, 2014. Hiding for over a decade, the vulnerability can affect current mainstream Linux and Mac OS X operating systems, including but not limited to Red Hat, CentOS, Ubuntu, Debian, Fedora, Amazon Linux, and OS X 10.10. It can execute the attack code scripts by simulating the value of environment variables so as to affect various applications which interact with Bash, including HTTP (HyperText Transfer Protocol), OpenSSH, and DHCP (Dynamic Host Configuration Protocol). It can also generate a severe impact on the security of cyber infrastructure, including but not limited to the network devices, network security devices, cloud, and big data centers. Particularly, the broad distribution and existence of Bash may lead to a "long tail effect" in the elimination of vulnerability. It can be easily utilized to write the worm code for automatic transmission and give rise to the development of the botnet.

1.2   Inevitability of Vulnerabilities and Backdoors

In the age of digital economy, the increase of data is under explosive development. On the one hand, people's life and work are growingly connected with software programs and information devices, including various networking software and control systems. With the development of smart devices and IoT, software and information devices have become omnipresent in our life. On the other hand, the development of cyber and information technologies in the age of cloud computing/big data, the national economy and people's livelihood are also increasingly dependent on the key information infrastructure, information service systems, and relevant data resources. It is no exaggeration that any damage to the cyber infrastructure, service systems, and data resources will consequently affect the political, economic, and military operating mechanisms of a country and even affect national strategic issues, such as national sovereignty in the cyber space and effective control over the data resources.

As the software and hardware products are becoming diverse and endowed with more complicated smart functions, they are growing in large numbers and are found to have various design flaws. For hackers, these flaws could be vulnerabilities, through which they can attack the information systems. Meanwhile, as the information systems are becoming more and more important, backdoors and vulnerabilities are frequently planted in the devices or systems for political, economic, and military purposes, resulting in a mushrooming of APT (Advanced Persistent

Threat) attacks. Vulnerabilities and backdoors bring about huge challenges for the maintenance of cyber security. In retrospect of the development of information technology and social informatization, human beings have not yet found the way to completely eradicate and control the design flaws or vulnerabilities of the information systems. At the same time, the cyber attack based on vulnerabilities and backdoors has become a new battlefield for political, economic, and military interests. Therefore, as the software and hardware products vary with more complicated functions and more open supply chains, design flaws, vulnerabilities, and backdoors will also increase.

As far as one system is concerned, its vulnerabilities will emerge one after another, rather than being discovered once and for all, because their discovery requires constant enhancement of people's security awareness and technical capacity. In the same way, different people have different technical capacities, resources, and security awareness; the vulnerability exposure is also limited. In contrast to vulnerabilities which are unintended, backdoors are intentionally designed hidden functions (especially the hardware backdoors). Often, backdoors are well disguised, with clear targets and multiple triggers. Apart from activation and function upgrading via the input/output channel of the host systems, backdoors can also take advantage of nonconventional side channels, such as sound waves, light waves, and electromagnetic waves, to penetrate the physical isolation and achieve secret interaction of internal and external information. These coordinated attacks will usually turn the security protection system into an empty shell and even cause permanent damages to the host systems in just "one shot." Backdoors are much more difficult to resist and far more destructive than vulnerabilities. In a word, unknown vulnerabilities and backdoors will always be the "Sword of Damocles" over the cyber security.

1.2.1   Unavoidable Vulnerabilities and Backdoors

Vulnerabilities and backdoors are discovered from time to time. According to the statistics, the number of vulnerabilities is proportionately related to the quantity of codes. As the systems become more sophisticated and the quantity of codes increases, the vulnerabilities will inevitably increase in number. When writing their codes, software and hardware engineers may have different understandings of the rules of design, with cognitive deviation during cooperation—not to mention that each engineer has his/her own coding habits. Hence, all kinds of vulnerabilities, including the logic, configuration, and coding vulnerabilities, will be carried into the final application system. In the context of economic globalization and industrial specialization, the integrated innovation or manufacturing has become a common mode of production organization. The design, tool, production, support, and service chains are drawn even longer. Therefore, untrusted supply chains or those with uncertain credibility pose great challenges to security control and provide more chances for the pre-embedding of vulnerabilities and backdoors. So the vulnerabilities and backdoors are unavoidable from the perspective of either technology or the game of interest.

1.2.1.1 The Contradiction Between Complexity and Verifiability

With the development of modern software and hardware industries, testing has become a key link to ensure product quality. In some well-known large software and system companies, the cost of testing has even surpassed that of development. However, no one dares to say that every logic defect in software and hardware products can be eliminated, theoretically or engineeringly.

This is because, on the one hand, the quantity and complexity of codes are constantly increasing, which leads to a sharp rise in the demand for code verification. Unfortunately, our current capacity of verification is still inadequate to address such complexity; on the other hand, we are unable to verify unknown and unpredictable security risks. The more complicated the codes are, the more likely unpredictable vulnerabilities will be. This is exactly how the contradiction between complexity and verifiability is formed. In addition, the design of our verification rule is theoretically incomplete, and with the rapid increase of codes, the global implementation of the verification rule will often be hard to achieve.

1. Code complexity adds to the possibility of vulnerability existence

The number of vulnerabilities in software and hardware is related to the number of code lines. Generally, the larger the quantity of codes is, the more complicated the product functions are, and the more vulnerabilities there will be [28]. First, a considerable increase in the number of code lines will directly lead to a more complicated code structure and more potential vulnerabilities. Second, the logic relations in complicated codes are much more intricate than those in simple codes, and it is more likely to discover logic and configuration vulnerabilities in complicated codes. For instance, in web browsers, operating system kernels, CPUs, and complicated ASICs, race condition problems often arise in the processing of complex synchronous and asynchronous events. Lastly, it is a greater challenge for developers to harness sophisticated codes in complicated or even larger programming development projects. As the developers are different in terms of their personal ability and work experience, vulnerabilities could be inevitable in the structure design, algorithm analysis, and coding.

2. The capacity of coding verification is limited

For now, there are three major problems in terms of coding verification. First, it is difficult to build accurate models and develop standardized description for all kinds of vulnerabilities. The capacity in the generalization, extraction, modeling, and matching of vulnerability characteristic elements is yet to be improved. Second, when analyzing the computability of problems and implementing scalable algorithms, the current code program analytics may encounter challenges, such as "path explosion" and "state space explosion" in the process of path traversal and state search. Given this, it has become rather difficult to achieve a high-code coverage rate and check every analytic target in testing. Third, as the information systems have become even more complicated, new problems resulted from complex intersystem connections, dynamic configuration management, as well as system evolvement and derivation may pose greater challenges to the discovery, analysis, and verification of system vulnerabilities. Under the combined effect of the three factors, the "scissors difference" between growing complexity and verifiability will be more prominent and the search and discovery of vulnerabilities and backdoors be more difficult.

1.2.1.2 Challenges in Supply Chain Management

In a world of economic globalization and fine division of labor, "design, manufacturing, production, maintenance, and upgrading" have been collectively deemed as a complete supply chain. Each link on the supply chain can be regarded as a "plank" to form the finished products. However, the security of finished products would face risks due to one of these "planks." With rapid globalization and increasingly refined division of labor, the product supply chain will also be longer. As a result, the management of supply chain is getting more and more important and difficult. In the meantime, as the supply chain consists of a multitude of links, it naturally has become the easiest target of attacks, such as social engineering attacks, vulnerability attacks, and pre-embedding of backdoors. The Xcode event of Apple Inc. is an example. As a version of Apple's Xcode complier was pre-embedded with backdoors, which left all programs produced by this complier being embedded with vulnerabilities. The vulnerability developers can then readily control the systems running the programs. Therefore, how to ensure the credibility of commercial components from untrusted sources in the global markets has become a tricky problem.

1. The importance of supply chain security

Since the provision and launch of more products and services of national strategic significance rely on the Internet, national security and cyber security are becoming increasingly inseparable. As Chinese President Xi Jinping once said, "Without cyber security, there will be no national security." Therefore, the security of various network products and their supply chains, particularly that of network infrastructure and its supply chain, plays a crucial role in ensuring cyber security and even national security.

For example, by tapping its leading role in the IT supply chain and market monopoly, the United States has managed to deploy a worldwide monitoring system in its "Quantum" project. According to the information leaked by Edward Snowden, the United States has utilized its core competitive advantages in chip manufacturing, network equipment, and cyber technology to pre-embed vulnerabilities and backdoors in its export electronic information products or control and cover the spreading of vulnerability data for global network intrusion and intelligence gathering so as to reinforce its global presence and monitoring capacity.

It shows that if a country can control more links of the product supply chain, particularly the irreplaceable core ones, it can have strategic dominance, including network dominance.

2. The difficulty in supply chain management

The supply chain is difficult to be managed mainly because it is too open and involves too many links to adopt a closed model of accurate management and control. Today, in our globalized world with more refined division of labor, the supply chain will only be longer and more influential, and its manageability can be a greater challenge for all.

(1) The contradiction between supply chain risks and globalization. The supply chain risks mainly stem from the fact that there are too many links covering a too wide area. That makes the supply chain an easy target of cyber attacks. With more uncontrollable factors, the supply chain sees more risks. For this reason, to ensure the supply chain security, we need to shorten the supply chain links and narrow down the affected areas. Theoretically, only a closed supply chain can be less exposed to attacks. However, globalization is now an unavoidable trend. Meanwhile, with the continuous development of science and technology and gradual accumulation of knowledge, it is hard for people to fully grasp all of them. Thus, specialization and refined division of labor have become necessary, which further leads to the uncertainty of an open, diversified, coordinated, and trusted supply chain.

(2) The contradiction between supply chain risks and infrastructure security. In its "National Strategy for Global Supply Chain Security," the US government pointed out that information technology and network development constitute the key cause for supply chain risks. The US National Institute of Standards and Technology (NIST) also argued in "Best Practices in Cyber Supply Chain Risk Management" that infrastructure security is an important target of supply chain risk management.

The development of electronic information technology and cyber technology has removed the geographic barrier of people's social networking activities. Wherever you are, you can easily interact with each other in real time. Also, with the development of the Internet, the entire process of commodity researches and developments, production, transportation, storage, marketing, utilization, and maintenance will increasingly depend on the network. At the same time, the emergence of the industrial Internet, from the market demand analysis, product design, processing design, process planning and organization, as well as quality control, will see revolutionary changes thanks to the introduction of cyber elements. Therefore, as the foundation of nearly all modern social activities, cyber infrastructure needs to be highly controllable and credible. However, cyber infrastructure itself is an information product. Similar to other products, its security is subject to the supply chain credibility throughout the whole process, including the design, production, sales, and service links. It is barely convincing that cyber infrastructure built on untrusted software and hardware can guarantee the security of upper-level application.

Under the current technical conditions, the controllability of supply chain risks and security of cyber infrastructure is essentially contradictory.

1.2.1.3 Inadequacy of Current Theories and Engineering Techniques

From the epistemological perspective, vulnerabilities seem to be an inherent part of an information system and entangled with it throughout the entire life cycle. Scientific theories and engineering techniques are currently unable to identify vulnerabilities for lack of systematic theories and sound engineering basis. Vulnerabilities are often found accidently, since they cannot be discovered in total in advance and be dealt with in time.

It is impossible to evaluate the probability of software vulnerabilities. After years of study of nearly 13,000 programs, Humphrey W. S. [29] from Carnegie Mellon University believes that a professional code programmer may produce 100–150 bugs in every 1000 lines of codes, so it can be easily calculated that there should have been no fewer than 160,000 bugs in the Windows NT 4.0 operating system, which contains 1.6 million lines of codes. Many of the bugs could be too tiny to have any impact, but still thousands of them would cause severe security risks.

1. The predicament of program proof

Program verification and analysis based on formal methods is an important means to ensure the correctness and credibility of software. In contrast to software testing, program verification based on theorem proving is complete with strict syntax and semantics. It has long been an issue for the academia to prove the security of software by means of program analysis so as to base the credibility and correctness of a program proof system totally on strict mathematical logic. However, what's tricky is that many problems have been proved to be halting problems, such as typical points to analysis and alias analysis. As the software size and functions are getting larger and more complicated, both the theories and methods for verification of program correctness have become neither adequate to analyze their completeness nor able to prove the security of designated software, leaving the theorem proving in a predicament. Likewise, program proof for hardware code also faces the same or similar problems.

Apart from security proving, two other problems are intractable as well, i.e., how to prove the correctness of bug fixes in patched programs and how to prove the security consistency between the source codes and the executable compiled codes. For instance, the patch MS11–010 released by Microsoft in 2011 was used to fix patch MS10–011 released in 2010, because the latter had not completely fixed the vulnerability and would constitute a logic condition to trigger the vulnerability again.

2. The limitation of software and hardware testing

As an analytical method to test the security of a program, software and hardware testing can effectively discover all sorts of code bugs. Nevertheless, its inadequacy lies in its incomplete coverage of all bugs. First, it is impossible to take into consideration all possible input values and their combinations and all different preconditions of testing to carry out exhaustive testing. In the actual testing process, exhaustive testing will often generate an enormous number of test cases. Usually, each test is simply a sample. Therefore, we must control the workload of tests according to the risk level and priority. Second, program testing can barely cover all the codes. Meanwhile, along with the growth of code quantity and rising complication of the functional structure, the code paths will see an exponential increase. Under the constraint of limited computation resources and time conditions, the good performance of program testing can hardly be ensured. Even traditional white-box testing cannot guarantee that all problems on a code path can be thoroughly tested. As a result, the uncovered part of codes turns out to be a "testing blind zone." For the covered part of codes, the testing process only tests the covered code path under certain conditions whether through common fuzz testing or via symbolic execution. It is by no means complete testing. Last but not least, testing can prove the existence rather than nonexistence of bugs.

3. The bewilderment in safe coding

The safe coding standards are formulated to reinforce code security, lower the probability of vulnerability emergence, and provide guidance on safe engineering. As one of the best security practices, the standards have undoubtedly promoted the improvement of coding security. However, many problems also arise in its actual application. First, as a way to inherit the experience in safe coding, it takes time for the safe coding standards to summarize, enrich, and optimize themselves. And it will be delayed to template new problems, resulting in a time lag for extension, which will lead to the introduction of similar problems into massive coding practice during a long period. Second, programmers are on different skill levels and may have had inadequate training. For instance, some standards have stricter stipulations on the code logic and time sequence, requiring the programmers to know well the standards and to use them proficiently. Finally, in the Internet era, industry competition leads to rapid upgrading of software and hardware. Generally, development teams focus more on function development than on safe coding. This approach probably entails a rising number of code security risks and quality risks.

4. The constraint of automation

The current static analysis tools, including automatic debugging products such as Coverity and Fortify also contain a toolkit for fuzzing (e.g., Peach). One major advantage of these automatic tools is that the automatic program analysis can save a lot of manual work in code auditing. Today, automatic analysis is mainly based on pattern matching, with quite many false alarms and report failures. However, the discovery of vulnerabilities depends heavily on the accumulation of experience. So most automatic tools can only be used as auxiliary tools for the discovery of vulnerabilities. At the same time, it is a challenge for automatic tools to strike a balance between man-machine interaction and automatic analysis accuracy. Moreover, the discovery of vulnerabilities often relies on the handling capacity under the exhaustible algorithm.

Even if the financial circumstances permit, the time spent on handling vulnerabilities could be totally intolerable.

To sum up, the current vulnerability discovery theories and engineering techniques are flawed in that they cannot provide a thorough solution to software vulnerabilities, not to mention that they usually are used for analysis, discovery, and handling of known vulnerabilities. In terms of unknown vulnerabilities, the current theories and engineering techniques are either barely cost-effective or totally invalid. Hence, we should not count on the current thoughts of passive defense based on feature extraction. Instead, we must vigorously develop active defense theories and technologies to endow our information systems with an endogenous attribute of security defense and change the status quo by increasing the difficulty in exploiting vulnerabilities rather than utterly wiping out vulnerabilities and backdoors.

1.2.2 Contingency of Vulnerability Emergence

As it was previously mentioned, vulnerabilities are introduced into a system due to either the limitation of the programmers' thoughts and their coding habits or the intentional embedding of vulnerabilities by some stakeholders. Therefore, the emergence of vulnerabilities is contingent. Although vulnerabilities are discovered from time to time, it is contingent as to when and how each vulnerability is discovered. The reasons rest with the limitation of people's knowledge about vulnerabilities in a certain historical period as well as the technical ability to examine the completeness of complicated codes.

1.2.2.1 Contingent Time of Discovery

As Fig. 1.2 shows, each vulnerability or backdoor has its own life cycle. Since the day it was introduced into a system, the time of its discovery, disclosure, fix, and elimination has been under the influence of multiple factors, including the

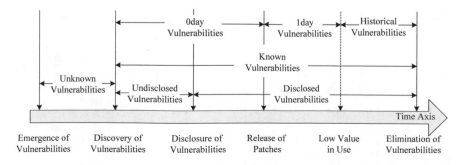

Fig. 1.2 Name of vulnerabilities at different time periods

development of theoretical methods, the maturing of technical tools, and the skill level of people handling the vulnerability. It is contingent in one way or another. During the constant evolution of vulnerability categories—from code vulnerability, logic vulnerability to composite vulnerability—there have been many interesting historical turning points. New vulnerabilities and categories are often discovered in a flash of genius, while the emergence of new defensive mechanisms will further propel that of new countermeasures and vulnerability categories.

Sometimes, vulnerabilities would "hide" deep in a program for over a decade, but could simply be exposed by chance, or discovered along with the improvement of analytical tools in the analysis of "deeper" paths or captured by someone in a line of codes neglected by all the other people. From Table 1.3 we may see that some vulnerabilities have been existing for nearly 20 years before they are discovered. They are "old" in that the discovery of such vulnerabilities is highly contingent.

Table 1.3 A time line of "senior" vulnerabilities

Vulnerability	Number	Category	Time of discovery	Time of existence	Impacts and harms
LZO	CVE-2014-4608	Buffer overflow	July 2014	20 years	Remote attackers may use this vulnerability to cause the denial of service (memory corruption)
Senior vulnerability	MS11–011	Privilege escalation	February 2011	19 years	Gaining the root-level access to a system
Local privilege elevation	MS10–048	Privilege escalation	January 2010	17 years	Gaining the root-level access to a system
MY power "explosive database" vulnerability		SQL injection	March 2012	10 years	Leading to the outflow of database address and leakage of web user privacy
Shellshock	CVE-2014-6271	Operating system command injection	September 2014	10 years	Remote attackers may use customized environment variables to execute any code
Dirty Cow	CVE-2016-5195	Competitive condition	October 2016	9 years	Malicious users may exploit this vulnerability to obtain the root-level access to a system
Phoenix Talon	CVE-2017-8890, CVE-2017-9075, CVE-2017-9076, CVE-2017-9077	Remote code execution	May 2017	11 years	Attackers may use the vulnerability to launch DOS attacks and even execute codes remotely under certain conditions. The TCP, DCCP, SCTP, as well as IPv4 and IPv6 at the network layer are all subject to its impact

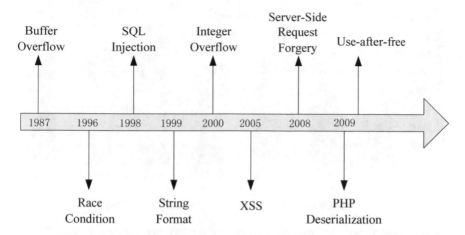

Fig. 1.3 The time of emergence of some important vulnerabilities

The temporal characteristic of vulnerability emergence reflects the process through which people learn about vulnerabilities. A vulnerability may be discovered along with the appearance of some mistakes or be proved by a theory, that is to say, the ultimate discovery of a vulnerability is often contingent. Figure 1.3 shows the time sequence in which people discover nine types of common vulnerabilities on the web and in binary codes and also their contingent time of emergence. Among them, the buffer overflow [30] was discovered at the earliest time in the 1980s. After the 1990s, race condition vulnerabilities [31], SQL injection vulnerabilities [32], and format string vulnerabilities [33] were discovered successively. In the twenty-first century, new types of vulnerabilities emerged one after another, such as the integer overflow [34], XSS [35], Server-Side Request Forgery (SSRF) [36], use-after-free [37], and PHP deserialization [38].

1.2.2.2 Contingent Form of Emergence

In terms of the vulnerability distribution statistics, as well as the time and place of discovery, the emergence of vulnerabilities is in most cases highly contingent. The distribution of vulnerabilities is influenced by multiple factors, such as the hot research topics, release of new products, technical breakthroughs, product trend, and financial interest. The statistics of vulnerability changes in quantity, category, and size is irregular. Figure 1.4 shows the vulnerabilities collected by CNNVD between January 2010 and December 2016. The total number is 46,771. We can see that there are quite many statistical changes in the quantity of vulnerabilities discovered each year and their risk levels.

In addition, we discussed the 4W features of vulnerabilities in Sect. 1.1.2. Based on that, how and when the vulnerabilities will be disclosed is also highly contingent. There are a number of ways in which vulnerabilities can be exposed, including analysis of captured attack samples, APT attacks exposure, release by the researchers

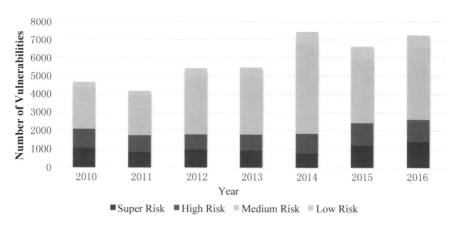

Fig. 1.4 Distribution of vulnerabilities collected by CNNVD between 2010 and 2016

on their own (often known as irresponsible disclosure), bug fix announced by manufacturers, and exposure in PWN competitions (e.g., 0-day vulnerability displayed in a PWN competition). Therefore, the exposure of a vulnerability could be highly unpredictable.

1.2.3   The Temporal and Spatial Characteristic of Cognition

Cognition is a process of knowledge accumulation, which refers to not only the accumulation of individual knowledge under restrictive conditions but also that of knowledge of all human beings throughout the history. As Isaac Newton said, the reason why we have seen further is because we stand on the shoulders of giants. The knowledge of each individual is limited to some degree, especially in the current age of data, information, and knowledge explosion. No one can master all knowledge. Hence, the human society needs a stratified knowledge structure with different specialized domains so that they can carry out continuous in-depth research in various areas and constantly improve people's knowledge of the natural world. It is such a cognitive process that has determined the temporal and spatial characteristic of vulnerabilities.

The systems which are deemed safe today may not be necessarily safe tomorrow; the systems that "I" consider safe are not necessarily safe in "his" eyes; and a system safe in environment A might not be safe in environment B. This is the temporal and spatial differences which result from different perspectives of vulnerabilities.

1.2.3.1   From Quantitative Changes to Qualitative Changes

Vulnerabilities are always there, but their discovery often takes place at different time, for they cannot be found out until knowledge has been accumulated to a certain degree. For example, the world-shaking vulnerability Rowhammer [39] in 2015 was initially only a phenomenon discovered by the researchers of integrated circuits at Carnegie Mellon University. As we know, the repetitive use of machine code instructions CLFLUSH or Cache Line Flush may clear the cache, along with forced data reading and updating. However, the researchers found that if someone used this technique to force repetitive memory reading and capacity charge, it would cause massive bit-flipping, or the so-called Rowhammering. Project Zero, a security research team under Google, carried out a further analysis of this problem and found that malware may exploit Rowhammering to run unauthorized codes. To address the problem, Project Zero designed a method to reorient the CPU, run codes at a wrong memory address, and then use Rowhammering to change the memory mapping of an operating system. Moreover, four researchers from VU University Amsterdam have found a way to combine Rowhammering in Windows 10 with a newly discovered method of de-duplication to successfully launch an attack. In this way, even if a system is fully patched and running various security-reinforcing software, attackers may still control the operating system. Therefore, hardware vendors are now trying to introduce the functions to prevent or reduce Rowhammering into the DDR4 architecture. However, the research carried out by Google and Third I/O shows that even DDR4 cannot be immune to Rowhammering.

With the extensive application of information systems and the increase of certain types of vulnerabilities, more attention will be paid to the issue. Thus, we will discover new problems in these systems or conduct more in-depth research on the vulnerabilities, triggering a technological revolution that leads to growing popularity of attacks through these vulnerabilities. For instance, as cloud computing has become more widely used, some people are trying to find vulnerabilities in virtual host penetration and allocation algorithm [40]. The array-index-out-of-bounds exception is another example. In the 1970s, people did not check the index-out-of-bounds problem, because it was commonly considered as a data integrity error of C programming language. Under the given computing conditions, automatic checking of this problem will reduce the effectiveness of program execution. Taking into account the cost-effectiveness, people would usually leave the problem to the programmers to resolve, not realizing that the error would cause severe security risks. Until the 1980s when the Morris Worm broke out, people began to realize the severity of the problem. It exactly proves that the stack buffer overflow vulnerabilities have always been in the program, and their discovery is restricted by people's knowledge of them. In this case, only when people become able to understand how the stack overflow was caused and what harms it would bring about can the vulnerability of stack overflow be exposed in a real sense.

1.2.3.2   Absolute and Relative Interdependence and Conversion

The relativity of conversion between vulnerabilities and bugs: in software and hardware engineering, we often mention the concept of code bugs. It is generally believed that only the security-related bugs will be considered as vulnerabilities. However, whether a bug is related to security or not depends on both the runtime environment and the subjective judgement. In other words, a bug may be classified as a bug under certain conditions and as a vulnerability under some other conditions.

In addition, there is no one vulnerability which can be absolutely exploited, nor one which cannot be utilized. Even if the existence of vulnerabilities is absolute, we can still technically reduce the exploitation of them, which in fact has already been an important research topic of active defense. Similarly, although some vulnerabilities haven't been exploited yet, they may be utilized by attackers sometime in the future with the development of technologies. Some vulnerabilities cannot be used in a given system environment, but they could possibly be exploited in other environments.

What's more, some vulnerabilities cannot be used separately but can be exploitable when they are combined. For instance, most memory-destructive vulnerabilities will be invalid under the ASLR (address space layout randomization) protective mechanism. However, when combined with chip implementation, they could be well exploited. For example, in early 2017, VUSec [41], an Amsterdam-based cyber security research team, designed a JavaScript program that can easily bypass the ASLR protection of 22 CPUs manufactured by Intel, AMD, and NVDIA. The attack showed by VUSec was launched via a vulnerability existing in the interaction between the chip and the memory. There is a component in every chip called memory management unit (MMU) which is especially used to map the address of stored programs in the computer memory. In order to track these addresses, the MMU will continuously check a directory called "page table." Usually, the page table is stored in the processor cache so that the most frequently accessed data can be readily called by the computing core to a small block of memory. Therefore, a piece of malicious JavaScript codes running on a web page can also be written into that block of cache. Most importantly, it can check the working speed of the MMU. Through close monitoring of the MMU, the JavaScript codes can find their own address. In this way, VUSec created a side channel attack program called ASLRCache (AnC) to detect the page table address.

1.2.3.3   The Unity of Specificity and Generality

Vulnerabilities exist in particular environments, so it is not a scientific way to discuss vulnerabilities without referring to the environments. As we know, each vulnerability emerges from a given environment and condition. For instance, in an isolated system, a remote overflow vulnerability will not have any impact on the system for there is not such an environment. That is why a secure system in environment A is not necessarily secure when being put in environment B. The pattern of a vulnerability stems from the specific system or software codes. It is a common problem of the same type or pattern of vulnerabilities.

At the same time, vulnerabilities are also general. For instance, those existing in one version of the Windows operating system also exist in all the application systems that run this version of Windows. In the same or similar operating systems, there are the same or similar exploitable conditions. Nonetheless, the existence and exploitability of vulnerabilities are not always consistent. Differences in operating environments may render vulnerabilities unexploitable, but the unexploitability of vulnerabilities does not mean that they are nonexistent. Without certain conditions, a bug may not become a vulnerability, while a vulnerability will probably be unusable if it has lost its dependent environment. Therefore, the unity of such specificity and generality of vulnerabilities constitutes a key foundation to ensure the effectiveness of an endogenous defense mechanism which does not rely on feature extraction.

1.3   The Challenge of Defense Against Vulnerabilities and Backdoors

1.3.1   Major Channels for Advanced Persistent Threat (APT) Attacks

APT, or "Advanced Persistent Threat," is a kind of intelligent cyber attack. Under an APT attack, an organization or an institution may use advanced computer network and social engineering tools to launch persistent attacks on specific and highly valuable data. It is a term coined by a US Air Force information security analyst. Generally, an APT has the following three features:

Advanced: attackers are expert hackers who can independently develop attack tools or discover vulnerabilities and reach the intended target through multiple attack methods and tools.

Persistent penetration: attackers will persistently penetrate into the target of attack and maximize the effect of persistent attacks without being discovered.

Threat: it is an attack coordinated and directed by a human organizer. The attack team will set a specific target, for its members are well trained, highly organized, and fully funded, with strong political or financial motives.

Table 1.4 shows the use of major vulnerabilities in some APT attacks.

We can see that 0 day vulnerability is involved in every APT attack. Some of the attacks also include Nday vulnerabilities. It shows that vulnerabilities have become an important tool for APT attacks, because they are playing a crucial role in cyber attack and defense and are the key lever to break the asymmetry in the game between the attackers and defenders.

1.3.2   Uncertain Unknown Threats

A known threat refers to a detectable attack with obvious features and labels. Certainly, all known threats are once unknown in the beginning, including XSS, brute force, misconfiguration, ransomware, watering hole, phishing, SQL injection, and DDoS.

Table 1.4 APT events and vulnerabilities

No.	APT event	Use of vulnerabilities	Time of occurrence	Impacts
1	APT-C-09	CVE-2013-3906 CVE-2014-4414 CVE-2017-8570	November 2009 March 2018 or earlier	Mainly stealing data from the science, education, and government areas
2	BITTER	CVE-2012-0158	November 2013	Stealing intelligence from a government ministry and large centrally owned energy enterprise
3	Patchwork	CVE-2014-4114	2014	Stealing intelligence from military and political agencies
4	Equation drug	CVE-2016-6366 CVE-2016-6367	2016	Controlled IPs and domain names distributed in 49 countries, mainly in the Asian-Pacific region
5	DX-APT1	CVE-2015-1641 CVE-2012-0507 CVE-2013-0640	March 2015 or earlier	Stealing the address books and related military and diplomatic documents from some embassies
6	MONSOON	CVE-2015-1641	December 2015	Stealing a lot of military intelligence from China and some South Asian countries
7	Petya ransomware	CVE-2017-0199	June 2017	Encrypting documents and even leading to system crash

Here, we may borrow an economic concept to understand an unknown threat. US economist Frank Knight saw the relation between risk and uncertainty in this way: risk is a kind of uncertainty, with its probability distribution known to all. People can speculate on future possibility based on the past. However, uncertainty means the ignorance of people, as people cannot predict a future event that has not yet happened. It is brand new, one and only, and unprecedented. So, we call a predictable unknown threat "risk" and an unpredictable unknown threat "uncertainty."

In email-based APT threats, attackers will control the target through information gathering and scanning in the early phase and by exploiting Flash and Excel vulnerabilities at a later time. In the entire attack process, early information gathering and scanning are known threats, for which the industrial and academic communities have put forward a range of defensive schemes. In the later use of vulnerabilities, the 0 day vulnerabilities and backdoors are unknown threats that are difficult to detect and defend. But we can still model and prevent such threats.

However, unknown unknown threats often refer to backdoors and unknown kinds of vulnerabilities. For instance, according the Reuters, the files disclosed by Edward Snowden show that with the approval of NIST, the US National Security Agency (NSA) and encryption company RSA signed an agreement valued more than US$ 10 million. NSA requires the placement of backdoors in the widely used encryption technology in mobile terminals so that NSA may generate a Bsafe backdoor program through random digits to easily crack all kinds of data encryption. If that is the case, the backdoors placed by NSA can be regarded as unknown unknown threats, because it is totally impossible to predict any backdoors embedded in an algorithm which has been proved to be safe mathematically.

Vulnerabilities and backdoors, regarded as an asymmetrical cyber deterrent force, are strategic resources persistently pursued, deliberately designed, and carefully reserved by the countries with advanced cyber technologies. These countries will surely turn their scientific research advantage in the cyber area into a technological and industrial edge and then create a vendor advantage of "embedding backdoors" and "hiding vulnerability" through technology export, product supply, channel distribution, service provision, and market monopoly in order to further obtain the strategic advantage of "one-way information transparency and absolute freedom of action" in the cyberspace.

Moreover, with the robust development of open-source communities, the open-source software has often become the target of backdoor embedding. OpenBSD, OpenSSL, and similar open-source software have come under question. Take vulnerability Heartbleed as an example, it is hidden too deep to be discovered. Meanwhile, based on the current development of software and hardware, it is highly possible to embed vulnerabilities in hardware and firmware through either available resources or technologies. For instance, Dr. Skorobogatov [42] from Cambridge University found that chip ProASIC3 had a backdoor in May 2012. The development of integrated circuit processing technique enables one single chip to easily hold tens of millions or billions of transistors. It will at most take up tens of thousands of or millions of transistors to embed an intelligent backdoor with highly complicated functions, without having any perceptible impact on the nominal function and performance of the target, making it highly impossible to detect a backdoor in the chip. The latest research published on *Science* on October 7, 2016 said that the gate length of a transistor made of CNTs and molybdenum disulfide was only 1 nm (1/50,000 of the human hair diameter), breaking the industry limit of not shorter than 5 nm. It means that the embedding of backdoors in chips will be easier and more covert, and in turn, the detection and discovery of backdoors will be more difficult. Even a chip is so hard a nut to crack, not to mention the modules, components, parts, systems, or bigger networks, where the hardware backdoors are more chilling to the bone.

1.3.3   Limited Effect of Traditional "Containment and Repair"

In the past, the defensive measures against vulnerabilities and backdoors are characterized by "containment and repair."

1.3.3.1   Reduce the Introduction of Vulnerabilities into Software Development, but Oversights Are Inevitable

In order to reduce the introduction of vulnerabilities into the development stage as much as possible, Microsoft has designed the Security Development Lifecycle (SDL) management mode to guide the software development process. SDL is a process of security assurance that applies security and privacy principles to every stage of development. Since 2004, SDL has been a company-wide mandatory scheme implemented by Microsoft. However, the execution, operation, and maintenance cost of SDL are so high that only giant companies are still using it.

Apart from SDL, some lightweight tools have been developed to help reduce the occurrence of vulnerabilities in coding, including compiler-based source analysis tools, such as Clang Static Analyzer. As a static analytical tool embedded in the open-source compiler front-end Clang for C, C++, and Objective-C source code, it can detect conventional security vulnerabilities (e.g., buffer overflow and format string). In addition, gcc compiler supports the option of—Wformat—security, which can be used to detect format string vulnerabilities in the source codes. Compilers running these plug-in options will send out a warning once they detect a vulnerability and notify developers of the location of the suspected vulnerability. Later, the developers may search, locate, analyze, and confirm the suspicious codes before fixing and removing them.

Surely, no matter which of the above measures is taken, such "containment" of vulnerabilities during software development can effectively eliminate part of the problems, but may still leave out many problems, "the escaped fish."

1.3.3.2   Discovering Vulnerabilities in the Testing Phase, but New Ones Are Emerging

Vulnerability discovery has developed from manual discovery, fuzz testing, and symbolic execution to intelligent mining. In the early days, vulnerability discovery mainly relied on manual reverse analysis, which is not only time- and energy-consuming but also unscalable. Later, people utilized fuzz testing to test programs. Fuzz testing can generate samples through random variation, which can raise the level of automation in vulnerability mining. It was a testing method commonly accepted by the industrial world. Well-known fuzz testing tools include SPIKE, Peach Fuzzer, and AFL. To address the need for higher code coverage rates and deal with the inherent blindness of fuzz testing, security researchers then proposed the technology of symbolic execution [43–46]. Under symbolic execution, people can simulate the execution of a program by symbolic expressions. When there are conditional branches, they must gather the constraint conditions and get the corresponding input information for two branches so as to define the relation between the input and the path, effectively raising the code coverage rate in the testing process. SAGE (Scalable Automated Guided Execution) [47] is a typical symbolic execution tool. By using SAGE, Microsoft found a lot of vulnerabilities in its own products. Recently, with the development of machine learning, intelligent tools have been a hot research topic. For example, Yamaguchi [48] from Germany designed the mode of automatic retrieving of vulnerabilities through machine learning to address taint propagation vulnerabilities. To raise the code coverage rate, Microsoft proposed a method to strengthen fuzz testing through deep learning. The test result showed that, compared with AFL, the testing coverage rate of parsers for ELF and PNG increased by 10%.

Although the vendors and security researchers have been committed to improving their ability to discover vulnerabilities and have actively checked possible omissions, they still have to fix and repair many vulnerabilities submitted by the third parties. Recent years have witnessed many cases of targeted attacks exploiting vulnerabilities, an evidence of prevalence of 0 days.

1.3.3.3   Exploit Mitigation Measures Keep Improving, but the Confrontation Never Stops

As it is impossible to eradicate vulnerabilities, security personnel then try to take mitigation measures to increase the difficulty in vulnerability exploitation. Vulnerability exploitation and mitigation technologies have been a hot topic of vulnerability attack and defense research. Yet in this area, the status quo is that "while the priest climbs a foot, the devil climbs ten." That's to say, the emergence of an mitigation tactic always gives rise to a bypassing technology.

Take the attack-defense game model of the Windows stack protection technology as an example. As Table 1.5 shows, GS protection v1.0 was introduced into the earlier Visual Studio 2002. Under this protection mechanism, a safe cookie was inserted into the function prologue, and then the cookie was checked in epilogue. If any inconsistency was discovered, the program execution would then be terminated. In 2003, Litchfield [49] put forward the method of structured exception handing (SEH) to cover the bypass. Then, GS protection v1.1 was introduced in Visual Studio 2003, adding the new protection mechanism SafeSEH to the software. In 2010, Berre brought forward the method of faking an SEH chain to bypass the SafeSEH protection. In order to prevent the execution of ShellCode in a stack, the data execution prevention (DEP) [50] protection was introduced as well. If the executed code is located in a non-executable memory page, it will throw an exception and terminate the process. Then, to bypass the DEP protection, the return oriented programming (ROP) [51] technology emerged to realize the function of ShellCode by utilizing existing code snippets. What's more, Windows also introduced the ASLR mechanism to prevent an attacker from locating the ROP gadgets accurately by randomizing the DLL loading base address. Correspondingly, the attackers designed a way of information disclosure by leaking the DLL loading base address before the precise positioning of ROP gadgets.

In the Pwn2Own hacker competitions over recent years, hackers can still bypass the protection mechanism and have the root-level access in the latest versions of Windows, MacOS, and Ubuntu systems, which have been constantly reinforced by

Table 1.5 Attack-defense game model of the Windows platform stack protection

Vulnerability mitigation measures	Bypassing technology
GS cookie protection	Covering SEH handler
SafeSEH protection	Faking SEH chains
DEP protection	ROP technology
ASLR technology	Information leakage

vulnerability mitigation measures. It shows that such measures have increased the difficulty in vulnerability exploitation but still cannot prevent it completely.

1.3.3.4   The Careful Designing of White List Detection Mechanisms for System Protection Fails to Prevent Bypassing from Taking Place from Time to Time

If a system is broken in, security personnel will still protect the key system resources and reduce losses through feature-based detection mechanisms, such as a white list. For instance, since the launch of Vista, Windows has introduced the User Access Control (UAC) mechanism, under which, when a malicious program tries to change the computer settings, Windows will pop up a dialogue box for the users to confirm the operation. So even if a piece of malicious code breaks into a system, it can only have access to the current user permission, thus being confined within the scope of current user privilege in order to reduce the harms. The UAC protection of Window 7 has adopted a white list mechanism to choose the trusted Microsoft applications. However, attackers can still utilize a program in the white list to bypass the UAC protection. Specifically, they can inject codes into the explorer white list process, create an IfileOperation object, use the object to copy a file under the system directory, and then utilize the dll hijacking vulnerability in this white list process to successfully bypass the UAC protection.

Some security protection software, such as Bit9, has also adopted the white list strategy—only allowing the applications in the white list to run on a terminal, and others will be disabled. Nonetheless, Bit9 itself could be under attack. For instance, Metasploit once released a DLL injection attack load against Bit9 Parity 6.0.x. Moreover, Bit9 and other whitelist vendors were once attacked, during which their digital certificates were bypassed by the hackers.

1.4   Inspirations and Reflection

After going through the technological development from feature search based on pattern matching, fuzz testing based on data structure, and traversal search based on exhaustive methods, we have been trapped in the predicament of improving the path search capacity enhancement and sample space structure optimization in order to address the vulnerabilities caused by code flaws, to say nothing of solving more complicated problems such as logic flaws and backdoor setting. The attitude of prudently "doubting everything" has enabled us to realize more clearly the severe fact that "vulnerabilities and backdoors are pervasive and unavoidable." As the credibility of the software and hardware component supply chain cannot be guaranteed, the industrial eco-environment is "contaminated," and design flaws can be utilized for a vicious purpose, it has become a highly challenging task for us to depict the system risks and ensure the security and credibility of the system.

1.4.1   Building a System Based on "Contamination"

In the globalization era, open and cooperative innovation chains and industrial chains are becoming the basic mode of human technological development and modern production activities. It is almost impossible for a single country to make the supply chains completely independent, controllable, secure, and credible; meanwhile, there have been no effective theoretical or technical solutions to address the vulnerabilities arising from the software and hardware design flaws, and it is also against the objective laws of human cognition and scientific development for the attempt to fundamentally eradicate these problems, which means that, technically or economically, we cannot guarantee that the cyberspace will be free of vulnerabilities. Therefore, a natural step to be taken is to change both the scenario under question and our mentality to build a security "sand castle," while alleviating the "known unknown" risks and "unknown unknown" threats. To do this, people need to jump out of the conventional "better-late-than-never" mindset of bug fixing and defense, so that the security of information devices should no longer depend excessively on its components and parts or on the "independent controllability" and security of individual software and hardware design, production, operation, and management. The aim is to enable the information system to tolerate "contaminated" components to some extent or under certain restrictive conditions and show impressive robustness and stability amid random failures and cyber attacks.

1.4.2   From Component Credibility to Structure Security

As mentioned earlier, vulnerabilities and backdoors are difficult to be eradicated. The sheer pursuit of independent controllability, security, and credibility cannot help address the unknown threats. Therefore, people must change their mindset of "precisely perceiving and eliminating the threats." Today, as the credibility of components is not guaranteed, we need to address the problem of component supply chain incredibility and unknown cyber attacks through the endogenous effect of mechanisms and system innovation, while fusing current security technological results with practice to significantly improve the cyber security defense capacity. What needs to be taken into special consideration is that technologies and products shall take place in a global, open environment, rely less on closed links and confidentiality means, and can technically ensure the effective implementation of security management systems.

1.4.3   From Reducing Exploitability to Destroying Accessibility

Often, security technologies focus on the resource-sharing mechanisms in the single-entity processing space or virtual multi-space. It is committed to alleviating the exploitability of vulnerabilities, trying to change the vulnerable conditions of such bugs. While turning static processing into dynamic processing, turning single-entity space into virtual multi-space, and adding strict performance state verification to the processing that lacks process validation, it can also reduce high-risk vulnerabilities into low-risk or unavailable ones by lowering the accessibility through destroying the attacks. However, it has been proved that attackers can bypass all these measures, particularly when the vulnerabilities and backdoors are combined together or have been penetrated by Trojan virus; these protective measures will be barely effective. For instance, the dynamic instruction address and data address can hardly play a role in cyber attacks under the coordinated impact of inside and outside factors. One of the key reasons why the exploitability mitigation measures take little effect lies in the lack of independence in the perception of threats or environment virtualization under the single-space shared resource processing mechanism. As a result, the flaws or bugs are hard to be discovered and avoided. Another reason is that attackers can possibly bypass the protection if there is no physical isolation or defensive lines. Once they break through the security protection, the system is likely to collapse.

1.4.4 Transforming the Problematic Scenarios

Based on the discussion above, we suggest to change our problem-solving mindset, which is limited to the single-process space-shared resource mechanism. By doing this, we can achieve the expected goal of transforming the problematic scenarios and overcoming the differential-mode interference through the heterogeneous redundant structural effect of reliable technology. First, the target system has the attributes of heterogeneous redundant space in the functional equivalence conditions so that the interruptions caused by independent vulnerabilities and backdoors in any space will not deprive the system of its stable functions. Second, even if the attackers may utilize the heterogeneous vulnerabilities in each single space, they can hardly launch coordinated attacks from multiple spaces under the multimode voting mechanism and affect the robustness of the system functions. Third, the effectiveness of defense is not initially determined by information perception of threat features. It only depends on the endogenous security mechanisms of the structural effect. Fourth, it integrates the defensive and service functions of the target system. Most fundamentally, it is imperative to shift the focus of defense from simply destroying the visibility or accessibility of vulnerabilities, to increasing the difficulty in forming temporal and spatial consistency under multi-space attacks. In other words, the current scenario of "independent attack on a single target" can be transformed into that of "coordinated attack on multiple targets under noncooperative conditions" to generate an effect of "common mode rejection" in similar reliable technology.

It should be noted that as the connotation of vulnerability is rich and broad, we do not discuss the following two types of problems in this book: (1) the vulnerabilities caused by particular functions of the target system and (2) the same design flaws (in engineering implementation) caused by the conventional thinking or limited knowledge of humans.

References

1. Dorothy, E.: Cryptography and Data Security. Addison-Wesley, Reading (1982)
2. Bishop, M., Bailey, D.: A Critical Analysis of Vulnerability Taxonomies. University of California, Davis, Sacramento (1996)
3. Krsul, I.V.: Software Vulnerability Analysis. Purdue University, West Lofayette (1998)
4. Shirey, R.: Internet Security Glossary. USA: IETF (2007)
5. Stoneburner, G., Goguen, A., Feringa, A., et al.: Risk Management Guide for Information Technology Systems. National Institute of Standards and Technology. U.S. Government Publishing Office, Washington, DC (2002)
6. Gattiker, U.E.: The Information Security Dictionary. Springer, New York (2004)
7. Shizhong, W., Tao, G., Guowei, D.: Software Vulnerability Analysis Technology. The Science Press, Beijing (2014)
8. Wikipedia. Computation. https://en.wikipedia.org/wiki/Computation (12 Dec 2016)
9. Tesler, L.G.: Networked computing in the 1990s. Sci. Am. **265**(3), 86–93 (1991)
10. Abbott, R.P., Chin, J.S., Donnelley, J.E., et al.: Security Analysis and Enhancements of Computer Operating Systems (1976)
11. Bisbey, R., Hollingsworth, D.: Protection Analysis Project Final Report. Southern California University Information Sciences Institute (1978)
12. Aslam, T.: A Taxonomy of Security Faults in the UNIX Operating System. Purdue University, West Lafayette (1995)
13. Neumann, P.G.: Computer-Related Risks. Addison-Wesley, Reading (1995)
14. Cohen, F.B.: Information system attacks: a preliminary classification scheme. Comput. Secur. **16**(1), 26–49 (1997)
15. Krsul, I., Spafford, E., Tripunitara, M., et al.: Computer Vulnerability Analysis. Coast Laboratory (1998)
16. Landwehr, C.E.: A taxonomy of computer program security flaws. ACM Comput. Surv. **26**(3), 211–254 (1994)
17. Bishop, M.: A Taxonomy of UNIX System and Network Vulnerabilities. University of California, Davis (1995)
18. Du, W., Mathur, A.P.: Categorization of software errors that led to security breaches. In: The 21st National Information Systems Security Conference, pp. 392–407 (1998)
19. Jiwnani, K., Zelkowitz, M.: Susceptibility matrix: a new aid to software auditing. IEEE Secur. Priv. **2**(2), 16–21 (2004)
20. Symantec: 2018 Internet security threat report. http://www.199it.com/archives/486129.html (12 Dec 2016)
21. https://en.wikipedia.org/wiki/Pentium_FDIV_bug (12 Dec 2016)
22. https://en.wikipedia.org/wiki/Pentium_F00F_bug (12 Dec 2016)
23. https://www.intel.com/content/www/us/en/security-center/advisory/intel-sa-00075.html (12 Dec 2016)
24. Lipp, M., Schwarz, M., Gruss, D., et al.: Meltdown. arXiv preprint arXiv:1801.01207 (2018)
25. Kocher, P., Genkin, D., Gruss, D., et al.: Spectre attacks: exploiting speculative execution. arXiv preprint arXiv:1801.01203 (2018)
26. https://www.fireeye.com/blog/threat-research/2016/08/overload-critical-lessons-from-15-years-of-ics-vulnerabilities.html
27. Sohu. The power behind 25 billion connected devices by 2020. http://mt.sohu.com/20170109/n478196610.shtml (12 Dec 2016)
28. Chujiang, N., Xianfeng, Z., Kai, C.: A model of predicting the number of micro vulnerabilities. J. Comput. Res. Dev. **48**(7), 1279–1287 (2011)
29. Humphrey, W.S.: Personal software process (PSP). In: Encyclopedia of Software Engineering (2002)
30. Pincus, J., Baker, B.: Beyond stack smashing: recent advances in exploiting buffer overruns. IEEE Secur. Priv. **2**(4), 20–27 (2004)

31. Bishop, M., Dilger, M.: Checking for race conditions in file accesses. Comput. Syst. **9**(2), 131–152 (1996)
32. Halfond, W.G., Viegas, J., Orso, A.: A classification of SQL-injection attacks and countermeasures. Proc. IEEE Int. Symp. Secure Softw. Eng. **1**, 13–15 (2006)
33. Shankar, U., Talwar, K., Foster, J.S., et al.: Detecting format string vulnerabilities with type qualifiers. In: USENIX Security Symposium, pp. 201–220 (2001)
34. Wang, T., Wei, T., Lin, Z., et al.: IntScope: automatically detecting integer overflow vulnerability in x86 binary using symbolic execution. In: Network and Distributed System Security Symposium, NDSS 2009, San Diego, CA, DBLP, 2009
35. Duchene, F., Groz, R., Rawat, S., et al.: XSS vulnerability detection using model inference assisted evolutionary fuzzing. In: 2012 IEEE Fifth International Conference on Software Testing, Verification and Validation (ICST), pp. 815–817 (2012)
36. DeralHeiland. Invest in security-ERPScan. http://www.doc88.com/p-6631236362751.html (12 Dec 2016)
37. Feist, J., Mounier, L., Potet, M.L.: Statically detecting use after free on binary code. J. Comput. Virol. Hack. Tech. **10**(3), 211–217 (2014)
38. CVE. Zend Framework Zend_Log_Writer_Mail quasi-shutdown function authority permit and access control vulnerability. http://cve.scap.org.cn/CVE-2009-4417.html (12 Dec 2016)
39. Kim, Y., Daly, R., Kim, J., et al.: Flipping bits in memory without accessing them. ACM SIGARCH Comput. Archit. News. **42**(3), 361–372 (2014)
40. Ristenpart, T., Tromer, E., Shacham, H., et al.: Hey, you, get off of my cloud: exploring information leakage in third-party compute clouds. In: ACM Conference on Computer and Communications Security, CCS 2009, Chicago, IL, pp. 199–212 (2009)
41. Gras, B., Razavi, K., Bosman, E., et al.: ASLR on the line: practical cache attacks on the MMU. In: NDSS (2017)
42. Skorobogatov, S., Woods, C.: Breakthrough silicon scanning discovers backdoor in military chip. In: International Workshop on Cryptographic Hardware and Embedded Systems, Springer, Berlin, pp. 23–40 (2012)
43. Boyer, R.S., Elspas, B., Levitt, K.N.: SELECT: a formal system for testing and debugging programs by symbolic execution. ACM SIGPLAN Not. **10**(6), 234–245 (1975)
44. Clarke, L.A.: A program testing system. In: Proceedings of the 1976 Annual Conference, ACM, pp. 488–491 (1976)
45. de Moura, L., Rner, N.: Satisfiability modulo theories: introduction and applications. Commun. ACM. **54**(9), 69–77 (2011)
46. Cadar, C., Engler, D.: Execution generated test cases: how to make systems code crash itself. Lect. Notes Comput. Sci. **3639**, 902–916 (2005)
47. Godefroid, P., Levin, M.Y., Molnar, D.: SAGE: Whitebox fuzzing for security testing. Queue. **10**(1), 20 (2012)
48. Yamaguchi, F., Maier, A., Gascon, H., et al.: Automatic inference of search patterns for taint-style vulnerabilities. In: 2015 IEEE Symposium on Security and Privacy (SP), pp. 797–812 (2015)
49. Litchfield, D.: Defeating the stack based buffer overflow prevention mechanism of Microsoft Windows 2003 server. http://blackhat.com (12 Dec 2016)
50. Andersen, S., Abella, V.: Data execution prevention: changes to functionality in Microsoft Windows XP Service Pack 2, Part 3: Memory protection technologies. http://technet.microsoft.com/en-us/library/bb457155.aspx (12 Dec 2016)
51. Checkoway, S., Davi, L., Dmitrienko, A., et al.: Return-oriented programming without returns. In: Proceedings of the 17th ACM Conference on Computer and Communications Security, ACM, pp. 559–572 (2010)

Chapter 2
Formal Description of Cyber Attacks

With the ever-evolving cyber attacks as well as defense technologies, the attack behaviors are characterized by uncertainty, complexity, and diversity, and the attack operations are becoming large-scale, synergistic, and multilevel. To study any cyber attacks, it is necessary to establish an objective and scientific and descriptive methodology for accurate feature analysis so as to come up with some general laws based on which the overall defense strategies will be put forward. So far, there is not a universal scientific theoretical model for us to depict the cyber attack behaviors, and the existing theoretical models are proposed for specific scenarios or certain attack categories. The scientific description of cyber attack behaviors is the premise and basis for analyzing the theory of cyber attacks and establishing a general theory of cyber defense. This chapter is an overview or an attempted summary of the existing formal description methods of mainstream network attacks and proposes some preliminary suggestions for the formal analysis of cyber attacks against the complicated dynamic heterogeneous redundant (DHR) network environments. The content herein, though not directly applied in the following chapters, is of guiding significance and reference value for the research on the theory of cyber attacks based on vulnerabilities or backdoors, the formulation of cyberspace defense strategies, and the design of cyber attack defense mechanisms.

Typical attack modeling methods in the early times include attack trees, attack graphs, attack net, etc., which mainly describe the repeated penetration process of an attacker from the network terminal to the network node and then from one node to another and finally attack the target network devices. In any case, each penetration in the process is manifested as a successful attack on some network subjects—vulnerable routers or terminal devices that the attacker deliberately filters out from the entire network system. Targeted at relatively simple scenarios, the methods mentioned above are insufficient to describe complex attack behaviors in a large-scale cyber environment and also impossible to quantitatively analyze the process or defense effects of cyber attacks. The attack surface (AS) [1, 2] is an evaluation theory that has emerged in recent years, which evaluates cyber attack and defense capabilities scientifically. Unlike the usual analytics of cyber attacks based on

© Springer Nature Switzerland AG 2020   39
J. Wu, *Cyberspace Mimic Defense*, Wireless Networks,
https://doi.org/10.1007/978-3-030-29844-9_2

backdoors or hardware/software vulnerabilities, the AS theory characterizes potential attacks through all possible interfaces, channels, and untrusted data items entering the system from outside and gives a measurement of the AS backed by empirical data. However, in order to improve the security of the target object, the defender will consciously deploy and run networks and systems that are changing dynamically, so that uncertain changes may occur to the resources available on the AS, thereby transferring the AS. In this case, the static AS theory will no longer apply. The mobile attack surface (MAS) [3] is an extension of the AS theory, aiming to establish a description and measurement method for the cyber attack process upon the transferred AS. In response to complex cyber attacks such as APT, we often need to introduce elements, such as dynamics, heterogeneity, and redundancy, as well as closed-loop feedback control mechanisms, in the target environment to enhance system robustness or defense capability. At this point, it is difficult to describe and measure such a target environment by either the AS or MAS theory, where new descriptions and metrics need to be created. The last section of this chapter proposes a modeling idea and method for studying cyber attacks from two aspects: the atomic attack behavior and the combined attack process, revealing the mechanism of the DHR closed-loop control environment for network attacks. By classifying and modeling the mainstream atomic attack behaviors, it describes and measures the vulnerabilities and backdoors which cyber attack behaviors target at, the preconditions for such attacks, the knowledge input, and the probability of gains after a successful attack.

2.1 Formal Description Methods of Conventional Cyber Attacks

The academic community has long expected to scientifically describe cyber attacks so as to grasp the law and find effective defense measures. There are many methods for cyber attack modeling, including the typical attack tree, the attack graph, the attack net, etc. These methods are detailed in literature [4]. The introduction of the relevant contents in this section refers much to the original text in literature [4], which are not labeled here one by one for your reading convenience.

2.1.1 Attack Tree

The concept of the attack tree was proposed by Schneier [5], which is a modeling method that uses a tree hierarchy to describe the target and sub-goal of cyber attacks. The attack tree model focuses on the security status of the system and can describe the full set of events causing security failures of the system.

The attack tree has a multilayer structure, comprising the root node, leaf nodes, and sub-nodes. The subordinate level of the root node is a sub-node or a leaf node. The subordinate level of the sub-node is also a sub-node or a leaf node. In the hierarchy, the root node of the tree represents the ultimate attack target, the leaf nodes represent various available attack methods, and the intermediate nodes between the root node and the leaf nodes represent the sub-goals of the attack. The correlation between the sub-nodes of the same parent node may be one of the three: "or," "and," and "sequence and." The "or" correlation means that the implementation of any sub-node target can lead to the implementation of the parent node target; the "and" correlation means that only the implementation of all sub-node targets can lead to the implementation of the parent node target; and the "sequence and" correlation means that only the sequential acquisition of all sub-node targets can lead to the implementation of the parent node target. Figure 2.1 shows the ties between the attack tree hierarchy and the nodes.

The attack tree model provides a convenient, target-oriented approach to describe multistage cyber attacks. The attack tree identifies the sub-goals to achieve the final goal in the simplest form. All attack nodes can be divided into "and"/"or" sequences, from which the relevant and unrelated attack conditions can be found. The successful attack probability of each attack node can be reflected by weight/value settings. Using the attack tree to describe a cyber attack can be divided into three phases: first, generate an attack tree targeted at a dedicated network system and attack type; then, assign a certain value to each node on the tree; and finally, derive qualitative or quantitative security metrics through calculation.

There are many methods for formal description of the attack tree. Here we introduce the BNF (Backus-Naur Form), a system specification language, to describe

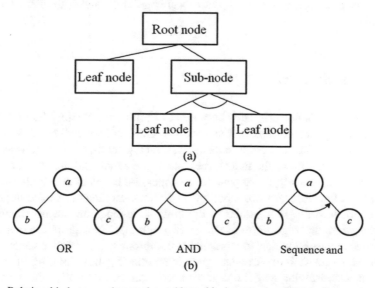

Fig 2.1 Relationship between the attack tree hierarchical structure and the nodes

the attack tree. Established by Garshol [6], this system specification describes the syntax of a given language by introducing formal symbols. The BNF system specification is a hierarchical type description language, where each type can inherit the attributes defined by another type by extending keywords. The attributes of the parent type will be passed to the sub-type by default, and they can be redefined at the same time. A type name consists of an identifier and a namespace. A namespace is primarily used to assist in organizing and classifying entity categories. The elements of a namespace are similar to Java in syntax structure. The description of the system specification follows certain naming conventions, usually defining the standard names of entity categories such as the computer and application. There are cases where a defined type is used to define an actual system, network, or host. The subject part of the entity declaration determines the attributes that the entity defines in the type declaration. The entity itself can also be used as an attribute for the system and subsystem modeling.

A node in the attack tree can be represented by a description template defined by a unique identifier and a series of system attributes. The description template includes description properties, preconditions, sub-goals, and post-conditions. The description properties are primarily used to describe arbitrary attack features. The preconditions consist of the system environment and configuration properties, determining whether the attack can be successfully implemented and mainly including local system variables and parameter instructions. Unlike preconditions, sub-goals refer to several targets before system invasion. That is to say, an attacker should collect as many beneficial components of the preconditions as possible to execute the specific attack procedure, while for sub-goals, the attacker must achieve each sub-goal before accomplishing the final goal. Post-conditions refer to the final state in the system/environment after the attack event. The state of the network and the host may change, or it may appear as a change in the survivability of the node attacked.

2.1.2 Attack Graph

The attack graph model, originally proposed by Phillips and Swiler [7] in 1998, uses the graph theory to describe cyber attacks. The graph contains all possible paths of the attacker from the start point to its attack target, providing a way of visualizing the process of attacks. In the model, the state of the attack and the attack action of the attacker, respectively, correspond to the nodes and edges of the graph. Based on the configuration information, topology, and attack method set of the target network, a structure diagram similar to the finite state machine can be obtained by using the generation algorithms of the attack graph. An attack graph is essentially a set of attack plans by which an attacker can compromise a particular security attribute of the target system. An attack plan is an attack path consisting of multiple atomic attacks. Relying on the attack plan, the attacker can realize the attack purpose step by step from the initial state of the attack. The construction of an attack graph also considers the logical and time sequence between various attack actions. The attack graph approach not only describes the attacker's behavior pattern for the defender to take defenses but also helps the attacker by providing optimized attack strategies.

The state transitions (edges) in the attack graph correspond to various atomic attacks. If we analyze the network system as a whole, an atomic attack event has random factors, which is a probability event affected by the difficulty of exploiting the vulnerability, the amount of a priori knowledge, and the credibility of the scan results. The total success probability of the attack path depends on the probability of the attacker successfully implementing atomic attacks. The value of atomic attack probability can be obtained through comprehensive evaluation by experts, and when it is assigned to the edge in the attack graph, the most vulnerable attack sequence can be found by calculating the total success probability of each attack sequence.

The construction of the attack graph is a key part of the model. In early research, each attack graph was completed by manual analysis. With the complexity of network topology and the increasing vulnerabilities of the target object, it has become infeasible to rely on manually constructed attack graphs. Automatic generation of attack graphs by model detectors have become the mainstream method, the effect of which is restricted by the expression ability of the model detector. The commonly used model checking tools include SMV, NuSMV, SPIN, etc. [8–12]. In the specific construction, the cyber attack event model can be abstracted via the input languages of these model checking tools, with the corresponding security attributes described by the calculation tree logic and the attack path map obtained from the output of the model verification tools. 2.1.3 Attack Net.

The attack net [13] is a model built for attackers and is essentially a special type of an attack map. It consists of an attack state set, an attack method set, and the node correlations. The attack net model provides decision-making services for the attacker, who can generate a complete attack plan as needed. The attack net can be represented using a graphical method similar to Petri net [14], called the attack net map. A typical attack net map is shown in Fig. 2.2, where the circle represents the attack state and the square represents the attack method. The attack net map reflects the logical correlations between all kinds of attack methods. As for a given attack net, each state represented by a circle in the attack net map represents the attack state reached in the attack process, and all the state nodes constitute a state set. A transition node represented by a square in the attack net map represents the attack method, the input node of the transition node represents the enabled state of the attack method, while the output node represents the new state the attack method attains. Each transition node in the attack net map corresponds to an actual case of an attack method. The set of directed arcs in the model represents the correlation between the attack method and the attack state.

A cyber attack behavior is a constantly alternating process: state →attack method→state. The implementation of the attack method and the change of the attack state constitute the process of a network attack. The attack net model embodies the preconditions for the successful implementation of various network attack methods and the logical correlations between them. The attack net model mainly

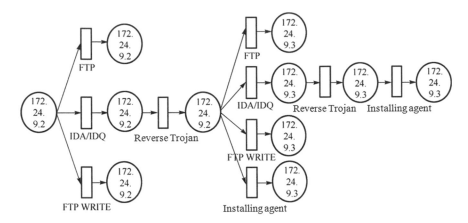

Fig. 2.2 An actual case of an attack net map

describes the logical and timing relationships of various feasible attack methods, highlighting the process characteristics of network attacks. The runtime mark distribution of the attack net model indicates the dynamic attack process.

Since the attack net is an attack model based on Petri net, the relevant Petri net theories can be used to analyze the attack net. The Petri net-based attack net model has become the most suitable model for describing cooperative attacks. The basic construction of Petri net is quite simple, but it has a rigorous mathematical background and easy-to-understand graphical features. Petri net consists of three structures: places, transitions, and flow relations [15]. Places are represented by a circle, indicating the state or conditions upon system migration. Transitions are represented by entries or squares to describe events that may change the state of the system. The correlation between places and transitions is represented by a collection of flow relations, which is a unidirectional connection used to connect places and transitions. Two identical structures cannot be connected.

The attack net model can guide attackers to launch cyber attacks and help them with pre-attack deduction and attack plan selection. Based on this model, the relevant analysis on the rollout of cyber attacks can be performed, including accessibility analysis, attack scenario analysis, and timing analysis of the attack method.

2.1.3 Analysis of Several Attack Models

At this stage, cyber attacks are featured by uncertainty, complexity, and diversity, where attack conducts are turning large-scale, cooperative, and multilevel. Therefore, the requirements for defense and maintenance efforts are getting higher and higher. There is a need for correct understanding and description of the attack behavior, followed by the establishment of an effective formal model for qualitative or quantitative analysis. Any analytical model for the attack tree, attack graph, or the attack net has its own characteristics and advantages, but they all targeted at relatively simple scenarios, hardly applicable to current large-scale and complex cyber attacks.

The beauty of the attack tree model is that it is simple and explicit and easy to understand and has good practicability. But it is not perfect: For example, both the attack conduct and the result are represented by nodes, which may cause confusion given the lack of further distinction and may fail to effectively handle frequent modification and extension of the nodes. Compared with the attack tree model, the Petri net-based attack net model distinguishes between the attack conduct and the result, solves the problem of node extension, and supports node addition without changing the original structure. The transition of the attack net can also express the logic expressed by the nodes in the attack tree. The graph representation based on Petri net is more suitable for visually displaying the loophole and its cause. Token coloring added to the complex Petri net improves the model description ability. However, it is easy for an attack net graph to outgrow the size of paper. Both the attack tree and the attack net focus on the correlation between the various attack behaviors involved in the attack process, while the attack behavior is not distinguished from its impact on the state of the target system, and the damage of the attack and the security requirements of the system are not integrated. Security warnings, though, are often closely tied to the security demand of the target system. Therefore, neither the attack tree nor the attack net should be directly applied to the attack checking and early warning systems.

The attack graph approach can describe the attacker's behavior pattern for the defender and help with defense and can also provide an attacker with optimized attack strategies. Nevertheless, an attack graph requires a large number of accurate input parameters, which is often unrealistic in application. Besides, the model's description of the attack method is too simple to illustrate the complex logical ties between the attack methods [11]. In addition, the generation of an attack graph itself is a technical bottleneck. With the expansion of the network scale and the rapid increase of network loopholes, it is necessary to develop a special model checking tool to automatically analyze the network state.

2.2   The AS Theory

In order to scientifically evaluate the relationship between cyber attacks and system resources, some scholars have proposed the attack surface (AS) theory [1, 2], which quantitatively portrays cyber attacks. Some system-specific attacks (such as those that exploit buffer overflows) are implemented by sending data to the system from the operating environment. There are also system-specific attacks (such as symbolic link attacks) that occur when the system sends data to the operating environment. In both types of attacks, an attacker enters the system through a system channel (such as a socket) and invokes a system method (or program) to send data items to or receive data items from the system. An attacker can also indirectly transfer data to and from the system using shared persistent data items (such as files). As a result, an attacker can attack the system by exploiting the methods and channels of the system and the data items in the operating environment. These methods, channels, and data items are used as system resources herein, based on which the AS of the system is defined. In a nutshell, a system's AS is a subset of the system resources that an attacker can invoke to launch attacks. Its size indicates the extent to which an attacker may damage the system and the effort the attacker needs to make for such damage. In general, the larger the AS, the less secure the system will be, and the more difficult it is to guarantee its security. Therefore, security risks in a system can be mitigated by reducing its AS.

2.2.1 The AS Model

The following model illustrates the AS theory. Assume that there is a system set S, a user U, and a data storage area D. For a given system $s \in S$, define its environment as three agents $E_s = \langle U, D, T \rangle$, where $T = S \backslash \{s\}$ is the system set S excluding s. The system s interacts with the environment E_s. Figure 2.3 depicts the system s and its environment $E_s = \langle U, D, |s_1, s_2| \rangle$. For example, s can be a web server, with s_1 and s_2, respectively, being an application server and a directory server.

Each system s includes a set of communication channels, namely, the path through which a user U or any system $s_1 \in T$ communicates with s. Specific examples of the channels include the Transmission Control Protocol (TCP)/User Datagram Protocol (UDP) sockets and named pipes. The user U and the data storage area D are modeled as the I/O automaton, and they are global to the systems in S.

Assume that the resources and features of the system remain unchanged and the target system is always reachable for the attacker. From the attacker's perspective, many attacks on the hardware/software system require data to be sent to or received from the system. Therefore, the entry and exit points of the system serve as the attack basis. An entry point of the system can be thought of as an input method

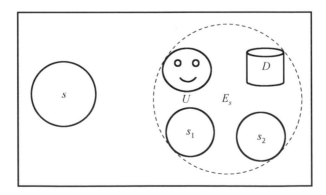

Fig. 2.3 System s and its system environment E_s

through which data can enter the system from the external environment. Accordingly, an exit point can be regarded as a way for data to be sent out from the system to the external environment. An attacker can attack the system by sending data into the system (or receiving data from the system) through its entry and exit points, channels, and untrusted data items. Data storage area D contains permanent and temporary data items, where files, cookies, and records in the database belong to permanent data items. User U can indirectly send (or receive) data to s using permanent data items, which constitute another basis for attacking s. If an entry point of s can read data to a permanent data item d in data area D or an exit point of s can write data to d, then d is called an untrusted data item. These factors, all belonging to the AS, are a subset of the relevant resources. However, not all resources belong to the AS, and different resources contribute differently to the AS metrics, depending on the likelihood of the resources to be used for the attack.

The system's methods, channels, and data items are all represented as system resources, so that the system's AS can be defined by system resources (Fig. 2.4). The AS of the system is a subset of the system resources that an attacker can invoke to launch attacks. An attacker can use the set M of entry and exit points to send data to or retrieve data from the system through channel set C and untrusted data item set I, thereby attacking the system. Therefore, M, C, and I are subsets of the AS-related resources. For a given system s and its environment, the AS of s can be defined as three agents <M, C, I> [16].

In order to measure the AS of a system, it is necessary to identify the relevant resources and determine the contribution of each resource to the AS metrics. If an attacker can exploit a certain resource to attack the system, the resource will be viewed as part of the AS. The contribution of a resource to the AS metrics is manifested by its likelihood of being used for the attack.

When a system s and its environment E_s are given, the AS of s is the triple <M, C, I>, where M is the set of entry and exit points of s, C is the set of channels of s, and I is the set of untrusted data items of s.

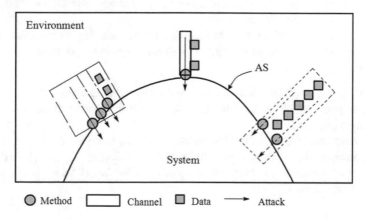

Fig. 2.4 The AS model

The ratio of damage potential to the attack cost can be used to estimate the effect of a resource on the AS of the system, where damage potential refers to the degree of damage caused by an attacker using the resource to attack the system, while the attack cost refers to the effort the attacker makes to obtain the necessary access to the resources for the attack [16]. Therefore, it is necessary to separately evaluate the damage potential-attack cost ratio of each method $m \in M$, each channel $c \in C$, and each data item $d \in I$, respectively, $der_m(m)$, $der_c(c)$, and $der_d(d)$, where der represents a function of the corresponding ratio of damage potential to the attack cost, thereby quantitatively evaluating the AS of the system. The AS metrics for the system s shall be determined by the triple, $<\sum_{m \in M} der_m(m), \sum_{c \in C} der_c(c), \sum_{d \in I} der_d(d)>$.

The size of AS metrics does not indicate the density of vulnerabilities in the system, while few vulnerabilities do not mean small-sized AS metrics. Larger AS metrics indicate that an attacker exploiting the inherent vulnerabilities can cause greater damage to the system at a lower cost. In this sense, a smaller AS is a wise choice to reduce security risks.

2.2.2   Defects in the AS Theory

The moving target defense (MTD) technology [17] embodies new concepts and techniques of the game rules for cyberspace security, aiming to break off from the passive defense scenario in cyberspace by introducing uncertainty, deploying and running randomized networks and systems, and significantly increasing attack costs. Specifically, the approach includes the adoption of multiple types of dynamic and randomized technologies at the platform, network, software, and data levels to enhance the uncertainty of system targets and destroying the preconditions for attackers to implement attacks. These mechanisms achieve the transfer of the AS by ceaselessly changing the resources of the AS or changing their roles. If the resources that the attacker relies on have disappeared or changed, the attack method that worked before will no longer function.

The existing AS metrics are not fully applicable to assessing the transfer (dynamic changes) of the AS, where new checking systems need to be introduced for the following reasons:

(1) The current checking metrics are based on the premise that "the AS remains unchanged," which is inconsistent with the actual situation of the dynamically changing system.
(2) The reachability of an attack has become an issue of uncertainty, as an attacker cannot control the attack packet to be correctly directed to the selected target. This goes against the previous assumption that the target AS is always accessible to the attacker.

2.3   The MAS

Intuitively, a defender can try changing system resources or modifying their contributions to reduce the AS for improved system security. However, not all modifications can reduce the AS. Modifications to the system resources or their contributions can lead to the transfer of the AS, and the dynamic AS can no longer be analyzed by the original AS theory. For example, if an attack is dependent on a resource removed (modified), the attack that succeeded in the past may no longer be feasible. However, in this transfer process, some new resources may have been added to the AS, which will make new attacks against the system possible. That is to say, the attacker needs to either pay a higher price to maintain the original attack capability or try a new attack method on the transferred AS. The mobile attack surface (MAS) theory [3] attempts to analyze and quantify the transferred AS. Although it is not yet possible to accurately predict the probability of successful attacks on the MAS, there is no doubt that the MAS can significantly increase the difficulty of launching any successful attack.

2.3.1   Definition and Nature of the MAS

When a system s and its environment E are given, the AS of s is three agents $\langle M, C, I \rangle$, and the set of resources belonging to the AS of s is represented by $R_s = M \cup C \cup I$. Similarly, with two resources of s given, respectively, r_1 and r_2, $r_1 \succ r_2$ means that r_1 contributes more to the AS than r_2 does. If R_0, the AS of s is modified to obtain a new AS R_n, and then it indicates that the contribution of resource r to R_0 is r_0 and that to R_n is r_n. Based on the above assumptions, the AS transfer can be defined [3].

Definition 2.1 Given a system s and its environment E, assume the original AS of s is R_0 and its new AS is R_n; if there is at least one resource r that satisfies $r \in (R_0 \backslash R_n)$ or $r \in (R_0 \cap R_n) \wedge (r_0 \succ r_n)$, then the AS of s has shifted.

Where the AS of s shifts, some of the attacks implemented on the original AS will not work on the new AS. Assume that the system and the environment use the I/O automaton modeling; a parallel combination of models can be established for the interaction between s and its environment, i.e., $s \| E$. Since an intruder attacks the system by sending data to or receiving input from the system, any input/output action invoking the combination $s \| E$ and including s is a potential attack on s. Suppose that attacks (s, R) represent all potential sets of attacks targeted at s, where R is the AS of s. In the I/O automaton model, if the AS of s is transferred to R_n from R_0, some of the potential attacks on R_0 will fail on R_n given the same attacker and environment. For instance, if a resource r is removed from the AS or the contribution of r to the AS is reduced during the transfer process, r will not be reused in new attacks targeted at s (Fig. 2.5).

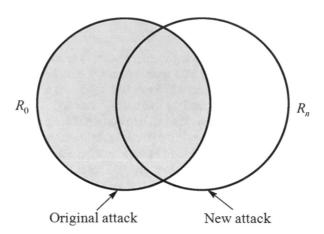

Fig. 2.5 AS transfer

Theorem 2.1 Given a system s and its environment E, if the AS of s is transferred to the new attack surface R_n from R_0, then $attacks(s, R_0) \backslash attacks(s, R_n) \neq \varnothing$.

The above explanation of the MAS is just a qualitative one. The transfer of the AS is quantified as follows.

Definition 2.2 Given a system s and its environment E, assume that the original AS of s is R_0 and its new AS is R_n, and then the transfer amount of the AS ΔAS is $\mid R_0 \setminus R_n \mid + \mid \{r : (r \in R_0 \cap R_n) \wedge (r_0 \succ r_n)\} \mid$.

In Definition 2.2, $\mid R_0 \backslash R_n \mid$ represents the number of resources belonging to the original AS but removed by the new AS; $\mid \{r : (r \in R_0 \cap R_n) \wedge (r_0 \succ r_n)\} \mid$ represents the number of resources making greater contribution to the original AS over the new AS. If $\Delta AS > 0$, it indicates that the AS of s has been transferred to R_n from R_0.

2.3.2 MAS Implementation Methods

A defender can modify the AS in three different ways, but only two of them involve the transfer of the AS.

(1) The defender may move the AS or reduce the AS metrics by disabling or modifying the features of the system (Scenario A). Disabling system features will reduce the number of entry and exit points, channels, and data items, thereby changing the number of resources belonging to the AS. This helps to reduce the ratio between the damage potential of the resources belonging to the AS and the attack cost.
(2) The defender moves the AS by enabling new features and disabling old features. Disabling certain features can remove the corresponding resources from

the AS, which may lead to one of the three scenarios: reduction (Scenario B), remaining unchanged (Scenario C), or increasing (Scenario D). Enabling features refer to adding resources to the AS to increase the AS metrics; disabling features will reduce resources in the AS, thus shrinking the AS. Therefore, the overall change in the metrics may be negative, zero, or positive.

(3) The defender may modify the AS by enabling new features, which will add new resources to the AS to increase attack surface metrics. However, the AS has not shifted because the original AS still exists and all attacks implemented in the past are still implementable (Scenario E). The defender can enlarge the AS metrics by increasing the ratio between the damage potential of existing resources and the attack cost rather than transferring the AS. These scenarios are summarized in Table 2.1.

From the perspective of protection, the defender's preference for these scenarios is as follows: A > B > C > D > E. Scenario A outperforms Scenario B because Scenario B adds new resources to the AS, which may bring new attacks to the system. Scenario D adds the AS metrics, but it may be attractive in moving target defense, especially when there is little increase in metrics and a large shift of the AS.

2.3.3 Limitations of the MAS

In order to cope with complex cyber attacks, such as APT, and to improve the security and reliability of the target system, the defender may need to construct a dynamic, heterogeneous, and redundant (DHR) network environment to protect the target object. The practice of cyber security protection shows that the introduction of the DHR mechanisms can usually significantly improve the security and robustness of the system. Unfortunately, the MAS theory fails to give a reasonable explanation of the security protection effectiveness of the DHR mechanism.

If the system has reconstructable, reorganizable, or reconfigurable features and functionally equivalent hardware/software redundancy resources and adopts a dynamic heterogeneous "fault-tolerant" or "intrusion-tolerant" structure, the original rules, channels, and data of the system will remain unchanged in this scenario, while the overall AS and available resources will actually increase. The application practice in "fault tolerance and intrusion tolerance" also proves that the heterogeneous redundant architecture has an outstanding defense effect against attacks based

Table 2.1 Different scenarios for modifying and transferring the AS

Scenario	Features	AS shift	AS measurements
A	Disabled	Yes	Decrease
B	Enabled and disabled	Yes	Decrease
C	Enabled and disabled	Yes	No change
D	Enabled and disabled	Yes	Increase
E	Enabled	Yes	Increase

on unknown vulnerabilities, which is obviously contrary to the statement of "reducing the AS is conducive to security" in the AS or MAS theory. This suggests that a new method needs to be introduced to describe such object scenarios.

2.4   New Methods of Formal Description of Cyber Attacks

The practice of cyber security defense shows that the dynamic heterogeneous redundancy mechanism has become an important measure to improve the security and robustness of the target system. In order to characterize cyber attacks in the DHR environment, this section proposes a new strategy as well as a new method for the formal analysis of cyber attacks from two aspects, respectively, the atomic attack behavior and the combined attack process, and describes the preconditions, knowledge input, and probability of successful attacks.

2.4.1   Cyber Attack Process

Although different scholars and theories describe the host attack process in the network environment in different ways, the process can be roughly described as one that consists of several stages, some stages of which may form an iterative subprocess [18] shown in Fig. 2.6.

A large number of cyber security events indicate that attackers mainly employ known network vulnerabilities and often exploit different host vulnerabilities for hopping attacks. A path with the least resistance is usually selected to launch attacks, difficult for defenders to detect. The defender expects the subjects of the network to greatly reduce or even block such paths, so that high-balanced security features can be exhibited across the subjects. It is a challenge to quantitatively describe this feature and theoretically prove that the HDR anti-attack mechanisms are difficult, but if the upper bound of certain quantitative security indicators can be given, it will be meaningful in engineering practice.

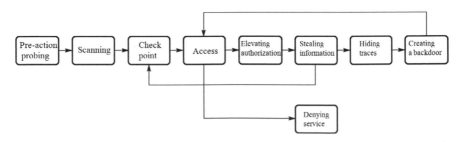

Fig. 2.6 Description of the host attack process

In order to solve the upper bounds of these quantitative security indicators, reasonable assumptions can be made on the staticity of conventional networks and devices regarding the device subject and its properties, as well as the capabilities of attackers.

(1) A network subject is always active. In other words, we tend to think that the time an intruder takes to complete the attack task is much shorter than the standby time or the intruder always chooses to launch attacks when a vulnerable service or application is enabled. In fact, this assumption is sometimes not true. For example, the target machine may not be turned on at night, and even if it is turned on, the service or application on it may not be active.
(2) A network subject can always be located. Before a substantial cyber attack is initiated, the attacker may spend much more time positioning or penetrating the target subject.
(3) The state of the network subject will not change. Due to the staticity, certainty, and similarity of conventional information systems, the target information that an attacker needs to exploit often remains unchanged during a certain period of time.
(4) The attacker has an asymmetric resource advantage. Conventional information systems suffering from cyber attacks reflect the asymmetry between attack and defense. While the attacker knows some vulnerabilities and has the ability to exploit them, the defender often knows nothing about them. Also, there may be a considerable time lag between the attacker and the defender regarding the mastery of the vulnerabilities.

Cyber attacks mainly plague network systems in the security attributes such as confidentiality, integrity, authentication, non-repudiation, and availability. The hazards are divided into two categories, respectively, targeting at information security attributes and network availability. Because the occurrence of hazards is often hysteretic and the systems are not equally sensitive to these hazards, we assume that an attacker should be considered as successful upon any intrusion in the target host. Therefore, we just need to take into account at most the first five stages of the attack process. The damages are then simplified and classified as follows according to the different attributes:

(1) The intrusion attack simplification, as shown in Fig. 2.7.
(2) Denial of service attack simplification, as shown in Fig. 2.8.

Fig. 2.7 Intrusion attack simplification

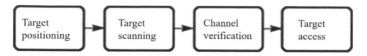

Fig. 2.8 Denial of service attack simplification

A denial of service attack usually exploits an agent host to launch an attack. An agent host is selected randomly and is regarded as a random attack target of the attacker. The acquisition of agent hosts is implemented by the intrusion into the attack chain. In a conventional network, a denial of service attack is initiated generally after enough agent hosts have been accumulated, so we think that the acquisition of agent hosts is completed during the attack preparation stage. In the various models and theories of cyber attack analysis, the probability of success of network attacks is a key indicator. Assume that authorization escalation can always be obtained automatically during the denial of service attack process or the probability of successful permission escalation is 1; the attack process analysis is then reduced to the analysis of intrusion attack simplification without affecting the solution of the upper bound of the quantitative security indicator.

2.4.2   Formal Description of the Attack Graph

Drawing on the ontology definition, a conceptual description of the concepts in a certain field and their correlations [19, 20], we extract the attack behavior atomic ontology of cyber attacks (referred to as the atomic attack) and apply all kinds of atomic ontology to each attack process atomic ontology (referred to as the attack stage) to form an attack graph. The attack graph is actually a directed graph of the attack behavior atomic ontology. For the sake of description, two atomic attacks are initiated, respectively, on the front and back of the attack graph, i.e., the start of the attack and the end of the attack, as shown in Fig. 2.9.

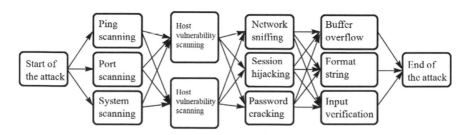

Fig. 2.9 Diagram of the attack graph

Suppose that the formal process of a complete cyber attack has n attack stages (n represents the number of attack process atomic ontology), which is recorded as a sequence:

$$\langle Step_0, Step_1, Step_2, Step_3, \cdots, Step_n, Step_{n+1} \rangle$$

Stage $Step_i$ ($0 \leq i \leq n + 1$) has a total of N_i types of atomic attacks (attack behavior atomic ontology), constituting the atomic attack set $PASet_i$ in stage $Step_i$:

$$PASet_i = \left\{ A_i^1, A_i^2, \cdots, A_i^{N_i} \right\}$$

$Step_0$ and $Step_{n+1}$ have only one type of atomic attack each, respectively, the start of the attack and the end of the attack, namely,
$N_0 = 1$, $PASet_0 = \left\{ A_0^1 = Start \right\}$, and $N_{n+1} = 1$, $PASet_{n+1} = \left\{ A_{n+1}^1 = End \right\}$.

The attack graph $G = \langle V, E, p \rangle$ is organized as follows:

(1) $V = \bigcup\limits_{i=0}^{n+1} PASet_i$.

(2) $E = \left\{ e_k^{ij} = < A_k^i, A_{k+1}^j > \mid 0 \leq k \leq n, 1 \leq i \leq N_k, 1 \leq j \leq N_{k+1} \right\}$.

(3) $p : E \rightarrow R, e_k^{ij} \mapsto p\left(e_k^{ij} \right)$.

The function $p\left(e_k^{ij} \right)$ is used to measure the conditional success probability of an atomic attack A_{k+1}^j after the atomic attack A_k^i is successfully implemented. In fact, the function is a weighting function of the edge e_k^{ij}. For example, the weight value can also be used to measure the attack cost (such as time) of a successful atomic attack A_{k+1}^j after the atomic attack A_k^i is successfully implemented. The edge set E is defined here as a complete $n + 1$ multigraph, while it is actually a subgraph of a complete $n + 1$ multigraph; any subgraph can be converted to a complete $n + 1$ multigraph through a reasonably weighed nonexistent edge (e.g., the probability is 0, or the cost is infinite).

2.4.3 Formal Description of an Attack Chain

In a complete formal process group covering n stages of cyber attacks:

$$path = \langle AC_1, AC_2, AC_3, \cdots, AC_n \rangle$$

whereas the normal attack stage AC_i ($1 \leq i \leq n$) is completed via some atomic attacks. Generally speaking, we decide that the host system always responds to atomic attacks sequentially. Assume the sequence of atomic attacks is AC_i^j ($1 \leq j \leq N_i$).

Definition 2.3 An attack chain is a sequence of atomic attacks:

$$\mathrm{ac}_{11} \to \mathrm{ac}_{12} \to \cdots \to \mathrm{ac}_{1N_1} \to \mathrm{ac}_{21} \to \cdots \to \mathrm{ac}_{2N_2} \to \cdots \to \mathrm{ac}_{n1} \to \cdots \to \mathrm{ac}_{nN_n}$$

or $\left\langle \mathrm{ac}_{11}, \mathrm{ac}_{12}, \cdots, \mathrm{ac}_{1N_1}, \mathrm{ac}_{21}, \cdots, \mathrm{ac}_{2N_2}, \cdots, \mathrm{ac}_{n1}, \cdots, \mathrm{ac}_{nN_n} \right\rangle$,

whereas ac_{ij} $(1 \leq j \leq N_i)$ corresponds to the attack stage AC_i $(1 \leq i \leq n)$.

For an attack chain $\left\langle \mathrm{ac}_{11}, \mathrm{ac}_{12}, \cdots, \mathrm{ac}_{1N_1}, \mathrm{ac}_{21}, \cdots, \mathrm{ac}_{2N_2}, \cdots, \mathrm{ac}_{n1}, \cdots, \mathrm{ac}_{nN_n} \right\rangle$, a probabilistic approach is used to measure its impact on system security. The success probability of the atomic attack ac_{ij} $(1 \leq i \leq n, 1 \leq j \leq N_i)$ is recorded as

$$P\left(\mathrm{ac}_{ij} \,\middle|\, \mathrm{ac}_{11}, \mathrm{ac}_{12}, \cdots, \mathrm{ac}_{ij-1}\right)$$

Suppose that A_0 is an empty set $\varnothing$ and $\mathrm{ac}_{i0} = \mathrm{ac}_{i-1N_{i-1}}$; then the success probability of the attack chain shall be

$$P = \prod_{i=1}^{n} \prod_{j=1}^{N_i} P\left(\mathrm{ac}_{ij} \,\middle|\, \mathrm{ac}_{11}, \mathrm{ac}_{12}, \cdots, \mathrm{ac}_{ij-1}\right)$$

If the success probability of an attack chain is equal to 0, the attack chain is claimed to be blocked. Obviously, an attack chain

$$\left\langle \mathrm{ac}_{11}, \mathrm{ac}_{12}, \cdots, \mathrm{ac}_{1N_1}, \mathrm{ac}_{21}, \cdots, \mathrm{ac}_{2N_2}, \cdots, \mathrm{ac}_{n1}, \cdots, \mathrm{ac}_{nN_n} \right\rangle$$

is blocked if and only if there exist $1 \leq i \leq n$ and $1 \leq j \leq N_i$ to make $P(\mathrm{ac}_{ij} | \mathrm{ac}_{11}, \mathrm{ac}_{12}, \cdots, \mathrm{ac}_{ij-1}) = 0$. Furthermore, if we can find the smallest set of system factors that make all known attack chains blocked, we can get satisfactory defense effects with less defense efforts.

2.4.4 Vulnerability Analysis of Cyber Attack Chains

In conventional systems with a static structure, the attack chain model provides a good qualitative and quantitative analysis basis for the defender to successively defend against atomic attacks and break the attack chain, but it does not consider the complex dependencies between cyber attacks and the environment as well as the operating mechanisms of the target object.

2.4.4.1   Conditions for the Successful Implementation of Atomic Attacks

According to the formal description of a cyber attack chain, an attack chain is a sequence of atomic attacks. If this attack chain is to be successfully implemented, all the atomic attacks forming it have to be successfully implemented sequentially. Therefore, the successful implementation of atomic attacks is the basis for the success of the attack chain. An atomic attack is usually an attack that is performed on a specific target object for a specific purpose. The name "atomic attack" mainly indicates that it is regarded as an attack step that cannot be split or subdivided. For an atomic attack to be successfully implemented, a series of preconditions must be met. The preconditions for different types of atomic attacks may vary, while the common preconditions include but are not limited to the following:

1) The attacker needs to understand the relevant vulnerabilities.

An atomic attack is usually implemented against a certain vulnerability of the target object. The most common vulnerabilities include loopholes and pre-embedded backdoors in the information system, misconfigurations made by users of the target object, and so on and so forth. An attacker needs to be aware of the knowledge associated with these vulnerabilities to select the tool and method for a targeted attack. Vulnerability-related knowledge includes the meaning and principle of the vulnerability, the conditions that trigger the vulnerability, the methods and tools to exploit the vulnerability, the effect of successful exploitation of the vulnerability, etc. However, atomic attacks in the stages of target positioning, target scanning, channel verification, etc. may not necessarily require prior knowledge of the related vulnerability, because atomic attacks at these stages mainly aim at discovering the vulnerabilities of the target object and other related knowledge. Pre-attack preparations such as vulnerability mining, vulnerability collection/purchase, vulnerability-oriented proof of concepts, etc. are primarily intended to accumulate such preconditions so that they can be basically put in place when a specific atomic attack is implemented.

2) The attacker has the tools and the execution environment to implement atomic attacks.

There are generally specialized attack tools for certain vulnerabilities, the most common of which being the program *Exploit* that exploits a specific vulnerability. The execution of these attack tools and methods requires a corresponding system environment. However, some atomic attacks need to run in a dedicated environment. For example, some local attack tools need to run in the *Shell* of the target host's operating system upon certain permissions. In this case, other atomic attacks may be required to create such conditions in advance, or there should be a backdoor previously embedded. There are also some attacks that need to hide their origins. The relevant execution environment is on the "broiler chicken," which should be controlled by other atomic attacks in advance.

3) The attacker has mastered the knowledge/information on the target of the attack.

Attack tools and methods for specific vulnerabilities will be effective in the implementation process only when they are combined with the knowledge on the target of the attack. Just as a callable "function program" needs to be assigned with parameters, the knowledge on the target object is an input parameter for attack tools and methods. For remote attacks, the most basic knowledge includes the IP address and port number of the attack target, the type and version number of the operating system, the account and password with basic access permissions, the specific function entry or the memory address of data on the target host, the directory where a specific file is located in the file system on the target host, and so on. Only by getting such attack-related knowledge can an attacker invoke the related tools or use the related methods for targeted attacks. Many application systems today apply an isomorphic design, where the knowledge on these targets is the same (e.g., the library function entry address, the executable file path, and the default username password, etc.), which makes it easier for the attacker to master the related knowledge. Sometimes, unintended random changes in certain elements in the attack target also help to eliminate the preconditions.

4) Access to the attack target from the attack source.

When launching an atomic attack, an intruder needs to have a reachable access path to the attack target from the attack source, which means that the attacker can connect or "contact" the attack target in some way. For remote attacks, the most basic access path is a reachable network channel to the attack target from the attack source. For instance, the access to the target's IP address, port number, uniform resource locator (URL), other marks, etc. is not blocked and filtered out by security mechanisms in cyberspace such as firewalls and the web application firewall (WAF), and the access is usually sustainable for a certain period of time. There are some atomic attacks that require more access paths, such as a remote login with a certain authorization, a remote access to a database, a connection that allows the packet to follow a specific route to reach the target, an access to a file in the "read"/"write" mode, etc.

5) The operating mechanism of the network environment and the target object allows the vulnerability to be triggered or utilized.

Even if the target object is vulnerable, triggering this vulnerability still requires the network environment and the target object to meet certain specific conditions, including the current state of the system (such as specific combinations of in-memory data, variable values, etc.) and special input from the outside world. A vulnerability cannot be triggered or exploited as long as the conditions that trigger the vulnerability are blocked. For example, an external input check processing mechanism (such as WAF) filters the extra-long input, so that the extra-long data which pass in through the WEB request variable cannot reach the function where the vulnerability resides, nor can it trigger the vulnerability, though the defender has not necessarily gained an in-depth insight into the vulnerabilities of the system and the specific attack features. If there is a vulnerability (e.g., a known vulnerability) in the system, as long as the vulnerability cannot be triggered or exploited through certain settings of the system or its peripheral components, the preconditions for the relevant atomic attacks can be eliminated.

6) The target environment allows the expected attack function to be executed once the vulnerability is triggered.

Even if the vulnerability is successfully triggered, it does not mean that the ultimate goal of atomic attacks can be achieved, for triggering or exploiting a vulnerability is likely to cause a module/process of the system to enter an unexpected operating state at a certain time, such as a program exiting erroneously or entering an "infinite loop." In usual cases, an attacker is more willing to steal the relevant information in secret, obtain a foothold on the target object (such as Shell), rewrite system data as expected, and enhance the related operation permissions. Only attacks like denial of service (DoS) can set the ultimate goal to cause system service anomalies or interruptions. Therefore, an attacker or attack tool manufacturer often needs to precisely design the information as well as excitation sequence able to pass through the AS entry point, so that the system can perform the relevant functions according to the attacker's expected target when the vulnerability of the system is triggered; otherwise it is vulnerable, or it cannot achieve the purpose of performing the expected attack function even when the vulnerability is triggered.

7) The network environment allows the attacker to control and access the attack results at any time.

Some atomic attacks achieve their goals the moment the attacks are launched. For example, an attack may cause the system to return the information that the attacker wants to obtain via the AS exit point. Some atomic attacks are designed to gain long-term control over the target object and create conditions for subsequent atomic attacks. For the latter, successful attacks also include establishing an ability to control and access the outcome of the attack so as to support subsequent atomic attacks or sustaining the ability for a certain period of time.

The preconditions required for the successful implementation of an atomic attack (ac_{ij}) can be integrated into the logical expression:

$$\text{Premise}\left(ac_{ij}\right) = \wedge_{k=1}^{t_{ij}} \left(\text{Pre}_{ij-k1} \vee \text{Pre}_{ij-k2} \vee \cdots \vee \text{Pre}_{ij-ks_k}\right)$$

where all variables are Boolean variables and Premise(ac_{ij}) = 1 indicates that the preconditions for the successful implementation of an atomic attack are met. To make Premise(ac_{ij}) = 1, then at least one of the preconditions in each group $\left(\text{Pre}_{ij-k1} \vee \text{Pre}_{ij-k2} \vee \cdots \vee \text{Pre}_{ij-ks_k}\right)$ is met; otherwise the attack cannot be implemented. In fact, even if these preconditions are all satisfied, it does not mean that the intended purpose of the attack can be fully achieved in the actual attack process, which always has certain variables.

2.4.4.2   The Conditions on Which the Successfully Completed Attack Chain Depends

Under the attack chain model, the atomic attack sequence represented by an attack chain is arranged according to the implementation time of atomic attacks. Further analysis shows that the relation between atomic attacks has dependencies in addition to purely chronological relationships. In general, the attack results obtained by a previous atomic attack provide support for the preconditions for subsequent atomic attacks, while a purely chronological relationship can certainly exist herein. For example, the dependencies of a series of atomic attacks are shown in Fig. 2.10.

Assume that the attack results of the atomic attack ac_{11} can provide support for the precondition Pre_{21-11} of the atomic attack ac_{21}, the results of the atomic attack ac_{12} can provide support for the precondition Pre_{21-12} of the atomic attack ac_{21}, the results of the atomic attack ac_{13} can provide support for the precondition Pre_{21-21} of the atomic attack ac_{21}, and so on. Assume that the relation between the preconditions of ac_{21} and ac_{31} can be expressed as

$$Premise(ac_{21}) = (Pre_{21-11} \vee Pre_{21-12}) \wedge Pre_{21-21}$$
$$Premise(ac_{31}) = (Pre_{31-11}) \wedge (Pre_{31-21})$$

According to Fig. 2.10, there can be two attack chains, namely:

$$\langle ac_{11}, ac_{13}, ac_{21}, ac_{22}, ac_{31} \rangle$$

$$\langle ac_{12}, ac_{13}, ac_{21}, ac_{22}, ac_{31} \rangle$$

The successful execution of these two attack chains means that all of the atomic attacks are successfully executed.

According to Fig. 2.10, there is no dependency between ac_{11}, ac_{12}, and ac_{13} or between ac_{21} and ac_{22}. This sequence can be reversed (or executed in parallel). However, there is a dependency between ac_{11}, ac_{12}, ac_{13}, and ac_{21}. This sequence

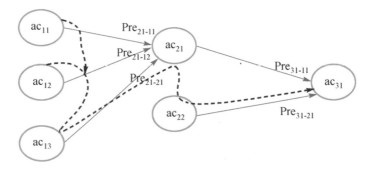

Fig. 2.10 Illustration of the dependencies between atomic attacks

cannot be reversed. In the above attack chain, the relation between the preconditions of the subsequent atomic attack ac_{21} supported by the attack results of ac_{11} and ac_{12} is "or," so it requires the success of only one of the two atomic attacks. In the actual attack, if ac_{11} is successful, ac_{12} can be skipped; similarly, if ac_{12} is executed first and succeeds, ac_{11} can be skipped. However, if the first executed atomic attack ac_{11} is unsuccessful, the precondition Pre_{21-11} of the atomic attack ac_{21} will not be supported, so the first attack chain is broken, and the second attack chain needs to be executed instead, that is, the atomic attack ac_{12} will be executed to provide support for the precondition Pre_{21-12}. The remaining atomic attacks are all designed to provide necessary support for the basic conditions required for subsequent atomic attacks. As a result, the actual execution sequence may be as follows:

$$\langle ac_{11}, ac_{12}, ac_{13}, ac_{21}, ac_{22}, ac_{31} \rangle$$

This attack sequence contains the steps to jump to another attack chain when execution of ac_{11} fails.

In Fig. 2.10, atomic attacks that can replace each other, such as ac_{11} and ac_{12}, form an "alternative atomic attack set" (if there is only one atomic attack in an "alternative atomic attack set," it indicates that the atomic attack has no other alternative atomic attack). Some atomic attacks that cannot be replaced by others constitute an "alternative atomic attack set." Then the above two attack chains can be integrated into the following sequence to increase the semantics of the alternative atomic attack:

$$\langle \{ac_{11}, ac_{12}\}, \{ac_{13}\}, \{ac_{21}\}, \{ac_{22}\}, \{ac_{31}\} \rangle$$

The successful execution of the attack process represented by the above sequence refers to the successful execution of at least one atomic attack in all "alternative atomic attack sets" from the beginning to the end of the attack. For each atomic attack to be successfully executed, the precondition $Premise(ac_{ij}) = 1$ must be met. Each $Premise(ac_{ij})$ is determined by a Boolean expression. Among the basic variables Pre_{ij-kl} constituting the expression, some are supported by the attack outcome obtained by an atomic attack upon success, and some are provided by the external environment or satisfied by the attacker per se (such as an awareness of a vulnerability). Among all Pre_{ij-kl}, excluding those supported by atomic attacks in the attack chain, the remaining preconditions can be viewed as external preconditions provided by the external environment or the attacker for the execution of this attack chain, some of which cannot be replaced by other preconditions and become the necessary preconditions for the attack chain. The attack will not succeed as long as these necessary preconditions are not met.

Among the preconditions of a subsequent atomic attack, if a precondition is provided by the outcome obtained in the previous atomic attacks, when the previous atomic attacks are all successfully implemented according to the dependencies, we still need to meet the following requirements in view of the connection between the atomic attacks:

1) Preconditions for the previous atomic attack outcome to effectively support subsequent atomic attacks.

In the attack chain, the role of a previous atomic attack is to create conditions for subsequent atomic attacks. However, the successful implementation of previous atomic attacks does not necessarily equate to the corresponding preconditions for subsequent atomic attacks being satisfied. For example, if previous atomic attacks can only narrow the possible value range of an account password to a certain interval (such as 100 passwords), the precondition for subsequent atomic attacks is to prepare three attempts of the password, which must contain the correct password (i.e., only three attempts are permitted). In this case, although the outcome of the previous atomic attack partially supports the satisfaction of the corresponding conditions for the subsequent atomic attack, such support is not sufficient. In addition, in actual attacks, even if an atomic attack is successfully implemented, the expected outcome may not be necessarily achieved 100%. For the subsequent atomic attacks to be successfully executed, the attack outcome of the previous atomic attack should fully support the relevant preconditions for the subsequent atomic attacks. The results obtained by the previous atomic attack have to be accurate and effective, i.e.:

$$P\left(\text{Pre}_{ij-kl} == 1 \middle| \text{Result}\left(\text{ac}_{mn}\right) == \text{success, Any ac}_{mn} \text{ that supports pre}_{ij-kl}\right) \approx 1$$

Or simply expressed as

$$P\left(\text{pre}_{ij-kl} == 1 \middle| \left\{\text{ac}_{mn,}\right\}, \forall \text{ac}_{mn} \text{ that supports pre}_{ij-kl}\right) \approx 1$$

2) The outcome of the previous atomic attack is persistently available for a period of time.

In most cases, the successful execution of an attack chain takes a certain period of time. Therefore, the outcome of the previous atomic attack needs to remain available for a period of time, at least until the end of the subsequent atomic attacks relying on the attack outcome. Being persistently available means that the impact of the atomic attack on the target object can last for a period of time and that the knowledge acquired in the atomic attack is valid/available for a period of time.

3) The impact of the network environment and the operating mechanism of the target object on cyber attacks has not destroyed the external preconditions required for the attack chain.

The atomic attacks and attack chains cannot be successfully implemented until they meet certain preconditions. Some of the necessary preconditions are provided by the external environment for the attack chain. Both the network environment factors and the operating mechanism of the target object can disturb or destroy the preconditions required for a cyber attack. Therefore, the attack chain is actually very fragile. The goal of an attack will not be achieved as long as the conditions for one section are not met in the sophisticated attack method.

However, at present, there is a high probability for the successful implementation of atomic attacks and attack chains against certain specific vulnerabilities. The main cause is that the deployment of current target objects is mostly static, isomorphic (or homogeneous), and certain. As long as the preconditions required for atomic attacks are all satisfied, the staticity and certainty of the system then guarantee the high success rate of the attacks, that is, for the atomic attack ac_{ij}:

$$P\Big(\mathrm{Result}\big(ac_{ij}\big) = \mathrm{success|Premise}\big(ac_{ij}\big) = 1\Big) \approx 1$$

or simply expressed as

$$P\Big(ac_{ij}\,\mathrm{|Premise}\big(ac_{ij}\big) = 1\Big) \approx 1$$

When combined with the formula

$$P\Big(\mathrm{pre}_{ij-kl} == 1, \big\{ac_{mn}\big\}, \forall ac_{mn}\,\text{that supports}\,\mathrm{pre}_{ij-kl}\Big) \approx 1$$

It can be concluded that

$$P(ac_{ij}\,|\,\big(\big\{ac_{mn,}\big\}, \forall ac_{mn}\,\text{that supports}\,ac_{ij}\big)\,\text{and}$$

$$\big(\mathrm{pre}_{ij-kl} = 1,\ kl\,\text{is an external condition}\big) \approx 1$$

In other words, as long as the previous atomic attacks are all completed and the external conditions can be satisfied at the same time, the subsequent atomic attacks can basically succeed (with a high probability).

Isomorphism (homogeneity) determines that if a system's configuration (the operating system version, basic settings, etc.) can satisfy the majority of the conditions required for an attack, other systems adopting the same configuration, which are basically isomorphic to the system (perhaps they differ only in the IP address, but it is also easy to obtain), can therefore satisfy the majority of the conditions of this attack. In this way, as long as a veteran attacker designs a tool that can successfully implement an attack, this tool will also apply to attacks on other identical or similar systems in most cases and will be a general tool for a primary intruder who can successfully implement attacks if he or she knows how to use it. The tool may, in most cases, launch attacks automatically. This partially explains why many malicious worm codes can spread widely.

In addition to the successful implementation of each relevant atomic attack, the successful implementation of an attack chain also requires several extra conditions. However, when the information systems are static, certain, and isomorphic, these conditions are not difficult to be satisfied; they are even to be satisfied naturally. Obviously, as long as the network environment and the target object are dynamic, random, and diversified/heterogeneous to some extent, the basic conditions required for a successful network attack are easily smashed, and though there may be some atomic attacks, they will have their preconditions met or will suffer lower probability of success even if the conditions are barely met. The support for subsequent atomic attacks provided by previous atomic attacks in the attack chain will be weakened, so the probability of successfully implementing the entire attack chain will also tend to zero, namely:

$$\exists i, j, P\left(\text{Premise}\left(\text{ac}_{ij}\right) = 1\right) \to 0; P\left(\text{ac}_{ij}\right) = \to 0$$

$\exists i, j, P(\text{ac}_{ij} | \text{Premise}(\text{ac}_{ij}) = 1) \approx 1$ will no longer be tenable or tends to 0
$\exists i, j, P(\text{ac}_{ij} | \text{Premise}(\text{ac}_{ij}) = 1) \approx 1$ does not hold or tends to 0
$\exists i, j, P(\text{Pre}_{ij-kl} = = 1 | \{\text{ac}_{mn}\}, \forall \text{ac}_{mn}$ that supports $\text{Pre}_{ij-kl}) \approx 1$ will no longer be tenable or tends to 0
$\exists i, j, P(\text{Pre}_{ij-kl} = = 1 | \{\text{ac}_{mn}\}, \forall \text{ac}_{mn}$ that supports $\text{Pre}_{ij-kl}) \approx 1$ is impossible or tends to 0
leading to

$$P = \prod_{i=1}^{n} \prod_{j=1}^{N_i} P\left(\text{ac}_{ij} \middle| \text{ac}_{11}, \text{ac}_{12}, \cdots, \text{ac}_{ij-1}\right) \to 0$$

The impacts of the network environment featuring dynamicity, randomness, and diversity on the conditions on which a cyber attack depends include the following: the attacker will lose some of the potentially lurking resources and operating conditions; the attacker cannot ceaselessly "align to" the attack target, making the source-to-target path no longer stable and reliable; and the attacker finds it hard to constantly control and access the attack outcome. If the target object itself is also injected with characteristics such as dynamicity, randomness, diversity, etc., even if the conditions on which the attack depends are satisfied, the success of an attack is no longer a certain event: the attacker cannot obtain the required accurate information about the target object, the triggering or exploitation of vulnerability is no longer a certain event, and even if the vulnerability is triggered, the expected function cannot be performed on a certain basis. The success of previous atomic attacks may not be able to effectively support subsequent atomic attacks, the time duration for exploiting the outcome of previous atomic attacks is significantly shortened, and the coordination or cooperation between atomic attacks is difficult to establish and maintain.

In short, the network environment and the system operation mechanism are strongly correlated with the success of cyber attacks. In the target systems with dynamic, heterogeneous, and redundant features, cyber attacks are forced to face the difficulty of coordinated attacks of heterogeneous, diverse, and dynamic targets under non-cooperative conditions. The attack chain becomes extremely vulnerable in this defense scenario, where a slight change in the underlying conditions or operating laws can disrupt or destroy the spatiotemporal consistency of the results of a coordinated attack, so that any successful attack will become least probable or uncertain.

References

1. Manadhata, P.K.: An Attack Surface Metric. Carnegie Mellon University, Pittsburgh (2008)
2. Howard, M., Pincus, J., Wing, J.M.: Measuring relative attack surfaces. In: Computer Security in the 21st Century, pp. 109–137. Springer, New York (2005)
3. Manadhata, P.K.: Game Theoretic Approaches to Attack Surface Shifting. Moving Target Defense II —Application of Game Theory and Adversarial Modeling, pp. 1–13. Springer, New York (2013)
4. Ming, X., Weidong, B., Yongjie, W.: Introduction to the Evaluation of Cyber Attack Effects, pp. 96–120. NUDT Press, Changsha (2007)
5. Schneier, B.: Attack trees: Modeling security threats. Dr.Dobb's Journal. **24**(12), 21–29 (1999)
6. Garshol, L.M.: BNF and EBNF: What are they and how do they work. http://www.garshol. priv.no/download/text/bnf.html. [2016-12-13].
7. Phillips, C, Swiler, L.P.: A graph-based system for network-vulnerability analysis. In: Proceedings of the 1998 Workshop on New Security Paradigms. Charlottesville, pp. 71–79 (1998)
8. Ramakrishnan, C., Sekar, R.: Model-based analysis of configuration vulnerabilities. J. Comput. Secur. **10**(1–2), 189–209 (2002)
9. Ritchey, R, Ammann, P:. Using model checking to analyze network vulnerabilities. In: Proceedings of the IEEE Symposium on Security and Privacy. Washington, pp. 156–165 (2000)
10. Sheyner, O, Haines, J, Jha, S, et al.: Automated generation and analysis of attack graphs. In: Proceedings of IEEE Symposium on Security and Privacy, Oakland, pp. 273–284 (2002)
11. Sheyner, O.: Scenario Graphs and Attack Graphs. Carnegie Mellon University, Pittsburgh (2004)
12. Zhang, T., Mingzeng, H., Xiaochun, Y., et al.: Research on Generation Methods of the Cyber Attack Graph. High Technol. Lett. **16**(4), 348–352 (2006)
13. Mcdermott, J.P.: Attack net penetration testing. In: Proceedings of the 2000 New Security Paradigms Workshop, Ballycotton, pp. 15–22 (2000)
14. Petri, C.A., Kommunikation, M.I.T.: Automaten. Bonn University, Bonn (1962)
15. Yuan, C.: Principles and applicaztions of Petri Net. Publishing House of Electronics Industry, Beijing (2005)
16. Lin, Y., Quan, Y.: Dynamic empowerment of cyberspace defense. Posts & Telecom Press, Beijing (2016)
17. Jajodias, S., Ghosh, A.K., Swarup, V., et al.: Moving target defense: creating asymmetric uncertainty for cyber threats. Springer, New York (2011)
18. Zhao, J., Yunchun, Z., Hongsong, C., et al.: Hacking exposed — network security secrets & solutions, 7th edn. Tsinghua University Press, Beijing (2013)
19. Guarion, N.: Formal ontology, concept analysis and knowledge representation. Int. J. Hum. Comput. Stud. **43**(5–6), 625–640 (1995)
20. Zhihong, D., Shiwei, T., Ming, Z., et al.: Ontology Research Review. J. Peking Univ. Nat. Sci. **38**(5), 730–738 (2002)

Chapter 3
Conventional Defense Technologies

3.1 Static Defense Technology

3.1.1 Overview of Static Defense Technology

From the perspective of technology, the current cyberspace defense methods fall into three categories: the first category focuses on the protection of information by strengthening the system with such technologies as firewall, encryption and decryption, data authentication, and access control. They offer basic protection for normal network access, legitimate user identification and rights management, and the security of confidential data. The second category includes mainly intrusion detection and other technologies, such as vulnerability detection, data authentication, traffic analysis, and log auditing. They aim to perceive attacks in real time and initiate immediate defenses according to the known features of an attack. This category relies on dynamic monitoring and alarm system enabled by feature scanning, pattern matching, data comprehensive analysis, and other methods to block or eliminate threats. The third category is network spoofing, represented by honeypot and honeynet. Their basic approach is, before any attack is performed, to actively construct special preset monitoring and sensing environments, serving as "traps," to lure potential attackers to enter for the purpose of carrying out analysis of possible attack moves and gathering information necessary for cracking down on, tracing back, or countering the attacks.

The abovementioned categories are all static, exterior protection methods, independent of the structural and functional design of the protection object. This precise defense technology based on a priori knowledge can be first deployed on a scale in a network and later enhanced to provide more security. Such defense idea conforms to the progressive cognition of cyberspace security. It is proved effective in defending against the attacks with known features or fixed patterns but is found invalid when confronting attacks based on unknown vulnerabilities and backdoors, complex joint attacks of a multi-model nature, or those coming from within. This reveals

© Springer Nature Switzerland AG 2020
J. Wu, *Cyberspace Mimic Defense*, Wireless Networks,
https://doi.org/10.1007/978-3-030-29844-9_3

the "genetic defect" of the traditional defense methods, which has been proven repeatedly by every disclosure of big unknown vulnerabilities and typical APT attacks.

3.1.2 Analysis of Static Defense Technology

Currently, there are many static defense technologies to protect the cyberspace [1], such as firewall, intrusion detection systems, intrusion prevention systems, antivirus gateways, virtual private network (VPN), vulnerability scanning, and audit forensics systems, among which firewall, intrusion detection, intrusion prevention, and vulnerability scanning are the most extensively adopted. The following is a brief analysis of the four methods.

3.1.2.1 Firewall Technology

Firewall, by definition, is a security barrier between internal and the external networks, preventing illegal users from accessing internal resources and stopping such illegal and vicious access behavior from causing damage to the internal network [2]. By applying access control strategies, firewall checks all the network traffic and intercept packets that do not comply with security policies. Its basic function is to receive or discard data packets in network transmission according to the security rules, to shield internal network information and operation, to filter malicious information, and to protect the internal network from unauthorized users.

The default of firewall's access control is usually set to either Allow or Reject. By "Allow," the firewall gives access to all visits except those which meet the preset rejection rules, and by "Reject," it refuses all requests except those which meet the preset admission rules. In comparison, the default Allow strategy is easier to be configured and allows most traffic to pass, whereas the default Reject strategy enjoys a higher degree of security by only allowing minority traffic to pass.

Firewalls fall into several different types, for example, packet filters, application proxy filter, stateful inspection filters, web application filters, etc [3]. The characteristics of the commonly used firewalls are come as follows:

Packet Filtering Firewalls [4] Situating in the IP layer of TCP/IP, the packet filtering firewalls check every data packet according to the defined filtering rules to determine whether it matches the rules or not and then decide to transfer or disregard it. Filtering rules examine the data in sequence until there is a match. If there is no match according to the rules, the data packet will be dealt with according to the default rule, which should be "Forbid." Filtering rules are customized to the header information of the data package, which includes IP source address, IP destination address, transmission protocol (TCP, UDP), Internet Control Message Protocol (ICMP), TCP/UDP destination port, ICMP message type, and ACK (acknowledgement) in the TCP header, etc. Therefore, packet filtering firewalls can only perform filtering function based on IP address and port number. They are not effective as to the application protocols unfit for packet filtering. Besides, they normally operate at the network layer to check only the IP and TCP headers and cannot completely prevent address spoofing.

Proxy Firewalls Working at the application layer, proxy firewalls mainly rely on the proxy technology to prevent direct communication between internal and external networks to hide the internal network. Every application needs to install and configurate different application proxies, such as HTTP to visit websites, File Transfer Protocol (FTP) for file transfer, and Simple Mail Transfer Protocol (SMTP)/ Post Office Protocol-Version 3 (POP3) to support email, etc. Proxy firewalls receive data from an interface, check its credibility according to pre-defined rules, and pass the data to another interface if the result is positive. The proxy technology does not allow direct dialogue between internal and external networks, thus leaving them completely independent from each other. The advantage of this technology is that it can not only include the functions of packet filtering firewalls but also act as an agent to process and filter data contents, carry out authentication of user identity, and also perform other more detailed control functions in the application protocol. Thus, in theory, a whole set of security strategies can be realized through the proxy technology by following a certain rule and taking certain measures, and a higher level of filtering can be done by applying filtering rules to the proxy, making its security function more extensive than the packet filtering firewalls. However, the proxy technology demands configuration modification on the client-side, checking and scanning of the content of data packets according to protocols like HTTP, and also needs the proxy to transfer requests or responses, losing the advantage of transparency compared with the packet filtering method.

Status Inspection Firewalls They add state inspection to the conventional packet filters. Also working at the network layer and processing the data packets based on quintuple information, the difference is that a status inspection filter can make decisions by inspecting sessions. When processing a packet, the state inspection firewall will first save the session in it and determine whether it is allowed through the wall. Since the data packet will be decomposed into smaller data frames by the network device (such as a router) during the transmission process, status inspection firewalls first need to reorganize these small IP data frames into complete data packets before making decisions. However, when designing a status inspection firewall, it is impossible to protect applications, thus making them extremely vulnerable to threats.

Deep Packet Inspection (DPI) Firewalls The abovementioned situation gives rise to the deep packet inspection (DPI) firewalls [5], which, based on the DPI technology and supported by packet filters and status inspection, are also able to carry out in-depth analysis of TCP or UDP packets, thereby being capable of defending the applications in complex networks and improving the performance of the firewalls and the security and stability of the internal network.

Web Application Firewalls (WAFs)   WAF is a new defense technology tailored to protect web applications by a series of security strategies for HTTP/HTTPS. Unlike conventional firewalls, WAF works at the application layer and is able to solve problems concerning web application, such as SQL injection, XML injection, and XSS attacks. WAF protects websites by carrying out content detection and verification on various types of requests from web application users to ensure that they are safe and authorized and blocking the unauthorized ones. More importantly, artificial intelligence is incorporated into the latest WAF solutions. For example, Imperva's WAF product provides not only protection but also automatic learning function for the web application page. That is, it records and analyzes the visiting models of frequently visited pages of certain websites and then defines normal visiting and using behavior models for those pages on the websites to identify suspicious intrusion behaviors. Another example is Citrix's WAF product, which establishes several user behavior patterns through the two-way traffic analysis to detect suspicious intrusion behaviors. In China, Chaitin Tech's WAF product "SafeLine" adopts a threat recognition technology based on intelligent semantic analysis, improving its recognition ability and big data processing ability to cope with a sudden flow of traffic. Moreover, it lowers management and maintenance difficulties by eliminating the security risks caused by improper configuration or management of regulations.

In a word, all firewalls need a priori knowledge to set proper rules, and all of them are defenseless to unknown attacks based on unknown vulnerabilities and backdoors or protocol defects. Failure to timely upgrade the rule repository or security strategy will lead to dramatic decline in the effect of firewall defense. Thus, robust background analysis and real-time upgrading become the very essential elements of an effective firewall, which are exactly what the emerging "data-driven security" technology trying to enhance through the use of cloud computing and big data.

3.1.2.2   Intrusion Detection Technology

The intrusion detection technology [6] is to spot and issue real-time alarms on behaviors violating security policies or attack signs by monitoring cyberspace or system resources. As a supplement to the firewall, the technology is considered to be the second gate behind the firewall. It collects and analyzes information about network behavior, security logs, audit data, key points in the computer system, and other available information to check if there are any behaviors violating the security policies or there are potential attacks in the system or network. As a kind of proactive protection, intrusion detection provides real-time defense against internal and external attacks or misoperations [7] and tries to intercept and respond to intrusions before they harm the network. Intrusion detection is realized by the following moves: monitoring; analyzing user and system activities; system construction and vulnerability auditing; identifying and responding to known attack patterns and issuing alarms in real time; statistical analysis of abnormal behavior patterns;

assessing data integrity of critical systems and files; performing audit trail of operating systems; and identifying users' violations of security policies.

The advantage of intrusion detection is that it extends the system administrator's abilities in security management (e.g., security audit, monitoring, offense identification, and response) and improves the integrity of information security infrastructure. A good intrusion detection system can not only keep the administrator informed in time about any change in the system (program, files, hardware, etc.) but also give guidance to him on the strategy-making. More importantly, the system shall support simple management settings so that nonprofessionals can find it easier to manage the cyber security. In addition, the intrusion detection coverage should vary with changes in threats, system architecture, and security requirements. Once an invasion is detected, the intrusion detection system shall respond promptly to record events and send alarms, etc.

Intrusion detection is an effective approach to perceive or identify offenses. It is placed in parallel to the network to check all the layers of information carried in all data packets transmitted, for which comes its capability to identify threatening visits [8]. This system is carried out in many different manners, each of which has its merits and shortcomings: the host-based methods detect attacks by analyzing the system audit trail and system log in a single computer system; the network-based ones do this by capturing and analyzing data packets in key network segments or exchange sessions; the application-based ones are to analyze the events sent out by an application or the transaction log files of an application; the anomaly detection decides whether current event is normal according to the existing normal behavior patterns; the misuse detection is to check whether the current event is suspicious of attacks based on existing attack history. Other methods include those based on protocol, neural network, statistical analysis, expert system, etc., each of which has its advantages and disadvantages.

The advent of big data witnesses more intelligent and complex cyber intrusions with constantly changing ways of attacks. For those attacks, conventional anomaly detection methods prove inadequate, leading to the emergence of many new and more intelligent algorithms, such as the intrusion detection system based on Markov model [9], the support vector machine (SVM) model based on artificial bee colony (ABC) algorithm [10], the SVM model based on autoencoder [11], and the model based on deep learning and semi-supervised learning (SSL) [12], to name just a few.

The intrusion detection system functions to prevent or mitigate threats by analyzing features of an attack and detect quickly and comprehensively whether it is a probing attack, a denial-of-service attack, a buffer overflow attack, an email attack, or a browser attack and then take countermeasures. In addition, the data monitoring, active scanning, network auditing, and statistical analysis functions of the intrusion detection system can be used to further help monitor network faults and improve its management.

However, the intrusion detection system also has its inherent structural weaknesses. For example, the increase in traffic and more complex attack methods put high pressure on the adaptability of security strategies [13]. If the attack feature database fails to be updated in time, the detection accuracy which relies on pattern matching will be directly undermined. Other shortcomings of this method include high false-positive rates and false-negative rates, excessively voluminous logs, and too many alarms. In this way, the administrator is overwhelmed in the sea of log information and is occupied with dealing with alarms (both true and false), in which important data would easily get unnoticed and real attacks would be overlooked.

Intrusion detection is essentially a passive defense. Most detection systems can only detect known and defined attacks and can only record the alarms without actively taking effective measures to stop such intrusions. With the emergence of new methods of attacks, an intrusion may have multiple variants, and people find it hard to accurately define and describe all the attack behaviors. However, if the intrusion detection system fails to precisely extract the attack features and update the target description database in time, it will not accurately perceive and judge the intrusion behavior. This is the Achilles' heel of this system. To make it even worse, an attack launched against the target based on undisclosed vulnerabilities or backdoors is considered an unknown attack by the intrusion detection system, and the lack of a priori knowledge about the attack will, in theory, make it impossible for the system to detect such attacks.

3.1.2.3   Intrusion Prevention Technology

Intrusion prevention technology [14], usually placed in the channel between the internal and external networks, receives traffic through a port and then, after making sure there is no suspicious behavior or content, transfers the traffic through another port into the inner network. The problematic packets and all subsequent ones from the same source will be eliminated. As it integrates both the detection function of the detection system and access control of the firewall system, the intrusion prevention system can provide better defense for the cyberspace than the single detection system. So it is increasingly popular in this field.

Focusing on risk control, the intrusion prevention system can detect and defend against clearly defined attacks or malicious behaviors that may harm the network and data, significantly reducing the overhead in dealing with abnormalities of the protection object. Unlike the intrusion detection that only gives out alarms when malicious traffic is being transferred or has been transferred, the intrusion prevention is designed to detect intrusions and attack traffic earlier than the target and intersect them before they reach the inner network.

The intrusion prevention system falls into three categories according to its working principle[15]:

1. Host-based intrusion prevention

The host-based intrusion prevention system prevents attacks through an installed software agent program on the host/server. Relying on customized security policies and analytics learning mechanisms, it prevents the intrusions which attempt to take control of the operating system by blocking buffer overflows, changing login passwords, overwriting dynamic link library, and other attempts to manipulate operations. In this way, the host security will be improved.

2. Network-based intrusion prevention

The network-based intrusion prevention protects the network by checking the traffic that flows through. Connected online, it can remove the whole network session once an intrusion is detected, instead of just resetting the session. In the meantime, the real-time online working mode demands quite a high performance. Thus, the network-based intrusion prevention is normally designed as a network device like a switch, which is able to provide a linear throughput rate and multiple ports.

Feature matching, protocol analysis, and anomaly detection are among the latest major mature technologies in this field that the network-based intrusion prevention system draws on. Feature matching is the most widely applied technology, with advantages of high accuracy and fast speed. The state-based feature matching not only detects the characteristics of attacks but also checks the current network session state to avoid spoofing attacks. Protocol analysis is a relatively new intrusion detection technology. By exploiting the high orderliness of network protocols based on a full understanding of their working principles and being aided by high-speed packet capture, it can quickly detect or extract attack features to identify suspicious or abnormal visits.

3. Application intrusion prevention

Application intrusion prevention is a special case of network-based intrusion prevention. It extends the host-based prevention into a network device located in front of the application server. Configured on the network link of application data, application intrusion prevention is designed to be a high-performance device, directly detecting or intercepting packets to ensure that users comply with the pre-described security policies. In theory, it works independently of object host/server operating systems.

Compared with conventional firewalls and intrusion detection, the intrusion prevention indeed proves more advantageous, for it can check traffic in a more granular way and respond to security issues in time and is able to prevent attacks at all levels. Although it can strengthen defense and block attacks to a certain extent, the intrusion prevention system still cannot solve the problem of false positives and false negatives. The current prevention system only integrates the access control of firewalls and the detection function of the intrusion detection system, and it also brings about new problems, such as single points of failure, performance bottlenecks, false positives, and false negatives.

Moreover, the intrusion prevention system has to be embedded in the network as required by its working principle, causing single points of failure. Even if there are no such failures, there are still performance bottlenecks for the target object or network, which may increase service latency and reduce efficiency, especially when there is a need to synchronize the rapidly growing traffic and load with an enormous amount of detection feature database. At this point, the prevention system is challenged both in technology and engineering to support such a response rate. The majority of high-end prevention product suppliers use customized hardware field-programmable gate arrays (FPGAs), network processors, and application-specific integrated circuits (ASIC) chips to maintain and improve their efficiency. Another problem is the high rate of false positives and false negatives, caused by unavoidable inaccuracies in feature extraction, diversity and complexity of applications, and attacker's high behavior hiding technique. The high rate of false positives and false negatives is a problem that cannot be avoided by the intrusion prevention system, and sometimes even legitimate traffic can be deliberately intercepted.

It is worth noting that the firewall, intrusion detection, and intrusion prevention technologies have a common feature of precise perception and recognition of attack behaviors, supported by a priori knowledge including a rich and updated feature database and an effective security policy algorithm library. They are all passive protection models affiliated to the target object and cannot perform proactive defense against attacks based on unknown vulnerabilities or backdoors or those lacking in historical information.

3.1.2.4 Vulnerability Scanning Technology

Vulnerability scanning is a cyber security technology based on the vulnerability database to check the vulnerability of designated remote or local computer systems. When integrated with the firewall, intrusion detection, and intrusion prevention technologies, the vulnerability scanning will effectively improve cyber security. Through vulnerability scanning, administrators will be informed of the security setting and the condition of application services, so they can timely find out vulnerabilities, objectively evaluate the network risks, patch the vulnerabilities, correct wrong settings, and take preventive measures before attacks happen [16]. Compared with the passive defense of firewalls, vulnerability scanning is a relatively proactive approach, nipping risks in the bud. Figure 3.1 shows its working principle.

The scanner simulates attacks and checks all the known vulnerability in the target object, for example, a workstation, a server, a switch, database, or an application, and generates analysis reports for the administrator to respond. In the network security system, scanners are characterized by cost-effectiveness, good and quick effects, independence of network operation, and easy installation and running.

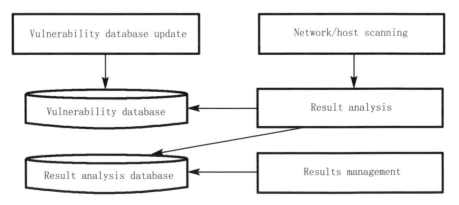

Fig. 3.1 Principle of vulnerability scanning

Vulnerability scanning works mainly in two ways: (a) gathering information about ports at work and services that are active on them through port scanning and then matching the information with the vulnerability database provided by the system to see if there is a vulnerability satisfying the matching conditions in the host and (b) launching offensive vulnerability scanning of the target host to test the weak passwords by simulating a malicious attack. If the simulated attack is a success, then there are vulnerabilities in the host system [17].

The vulnerability scanners are divided into three types according to their scanning mode [18].

(1) *Network vulnerability scanner.* It imitates attacks by sending packets from the network side and checks vulnerabilities of the host-based network operating system/services and application against the responses of the host when receiving packets.

Its main functions include the scanning of ports, backdoors, password cracking, applications, blocked services, system security, analysis reports formulation, and the upgrading of the security knowledge base.

(2) *Host security vulnerability scanner:* It aims to carry out deeper scanning of the security issues inside operating systems such as UNIX, Linux, and Windows. It compensates the network scanner in that the latter only works from outside of the network.

The host security vulnerability scanner has a client/server software architecture, a unified console program, and agent applications deployed in important operating systems. When it is in use, the console sends commands to each agent to start scanning, receives results from them, and generates a security vulnerability report.

(3) *Database security vulnerability scanner:* It targets database servers scanning, similar to that of a network scanner. It mainly checks unsafe or expired passwords, detects login attacks, closes accounts that are unused for a long time, and tracks user activities during the login period.

It is generally believed that through the use of firewalls, intrusion detection and prevention, and vulnerability scanning technologies, the network or host is immune to outside disturbances or attacks. Sadly, it's not true. Those traditional static defense measures have innate drawbacks as networks become increasingly complex and attacks appear in increasingly varied forms.

(1) The first major drawback of static defenses is their strong dependence on a priori knowledge (rules repository, feature database). The firewall, intrusion detection, intrusion prevention, and vulnerability scanning rely more on a priori knowledge in formulating their protection policies and attack features. They are protections of a passive nature. When new types of attacks appear, the continuance of protection depends heavily on the expertise of personnel and the capability of producers to guarantee services. This is quite demanding, and the protection is quite uncertain. It's hard to maintain a robust and continuous defense if perception and analysis are not clear enough to form clear rules.

Databases are completely passively formulated, like vaccinations which, if not syndrome specific, will be defenseless in facing new types of threats.

(2) The second drawback is the difficulty of guaranteeing timely defense. To be effective, the defense system has to be able to identify new features and create new rules, which, again, depends highly on whether the security analysts can timely form a prior knowledge based on new threats and whether they can send the knowledge timely to the defense target by updating the feature library and rule library. In this industrialized chain of technology, for cost and technology reasons, a seamless closed loop is hard to realize, the loop being a process from new threats identified to analysis and new features found to new defense rules formed. In practice, due to various factors and limitations, the response chain is too long, and it's impossible to deliver defense upon discovery.

(3) The third drawback is its inability to defend against internal intrusions, especially so when it comes to the border defense of the abovementioned defense technologies. For example, the firewall is designed to detect packets from outside to inside and is difficult to detect those the other way around. However, in reality, when launching APT attacks, the intruders most probably bypass the border protection through spear phishing or water hole to go directly inside and steal information through legal protocols. The border protection methods also fall short to defend the internal horizontal attacks.

(4) Its own weakness might be taken advantage of. As stated in Chap. 1, static defense products (e.g., firewalls, intrusion detection, and prevention systems), consisting of software and hardware components, came out under the current technological and manufactural circumstances, inevitably have their vulnerabilities or backdoors. Once the intruder spots those security flaws, any border defense could be bypassed, taken over or deceived, and be turned into invalid Maginot line or even serving as a guide to attacks, just as Snowden did to NSA and the Equation Group attacks.

To sum up, conventional static defense technologies focus on adding reinforcement to protect the target system, normally using risk assessment predicated on the ability to detect, analyze, and extract the features of threats through precise perception and defense. The static defense is essentially a hindsight technology, which learns its lessons through attacks and uses them for future preventions. It yields positive effects at the early stage of cyber security, but its flaws are obvious. Given the serious asymmetry of information on both sides, attackers can easily get what they want about the target systems, and the target objects are in a disadvantageous state due to their staticity, similarity, and certainty, passively defending against backdoor attacks, for example, 0 day or N day, or other new types of attacks.

3.2 Honeypot

Network spoofing, a technology complementary to conventional defenses, contributes to building a multilayered cyber security system [18]. By attracting intruders and significantly consuming their resources to lower the threat, the spoofing system gathers information before real attacks happen to win the time to deploy defense and strengthen security. Honeypot is one form of the network spoofing.

Honeypot, as a security resource, has no other functions than to lure attackers into illegal use of it [19]. Essentially a technology to deceive attackers, honeypot deploys hosts, network services, or information as baits to induce attacks, thus enabling the capture and analysis of attacking behaviors, for example, tools and tactics and intentions and motives. In this way, the defense systems could be informed of the threats they are facing and to strengthen their defense by technology or administration.

As a proactive protection method, a honeypot focuses on how to construct a flaw-less spoofing environment (e.g., controlled or software-simulated network or host) to trick intruders into attacking it and to issue alarms to protect the targets [20]. Honeypots, like bacterial culture dishes used in medicine, enable people to observe intruder behaviors in a monitored environment and to take records of their activity patterns, their severity, motives, tools, and tactics of intrusions.

Typical applications of honeypots include network intrusion and malicious code detection, malicious code sample capturing, security threat tracking and analysis, and extraction of attack features.

How to build a monitoring environment not easily perceived by attackers is the problem a honeypot faces in dealing with the quickly evolving cracking technology. Another problem it faces is that it can only be used as a temporal "isolation room" rather than a routinely operational carrier because a virtual honeypot environment hampers the performance of applications. Regarding this feature, intruders could wait until they are closed or actively detect the difference between a honeypot and the true network environment. Once a honeypot is spotted, intruders could stop their attack and wait to act after they leave the isolation room.

3.2.1 Network Intrusion and Malicious Code Detection

As an active security defense technology, a honeypot is initially used to assist intrusion detection to find malicious codes and attackers. There are many kinds of honeypot products with the following as examples. Based on the dynamic honeypot concept, Kuwatly et al. [21] dynamically configured Honeyd virtual honeypots by integrating active detection and passive identification tools and built an adaptive honeypot in the changing network environment to detect intrusions. Artail et al. [22] proposed a hybrid architecture, using low-interaction virtual honeypots to simulate services and operating systems and direct the malicious traffic to the high-interaction real-service honeypots to enhance monitoring, analysis, and control of intruding. Anagnostakis et al.[23] innovated shadow honeypot and combined it with anomaly intrusion detection. The latter is responsible to monitor all traffic flow and channel suspicious traffic to the honeypot. Shadow honeypot shares all internal states with its protected systems. Attacks will be detected by the shadow honeypot before it reaches the protected system, while legitimate traffic identified by the anomaly detection will be responded transparently by the system after verified by honeypot.

The honeypot is effective in detecting malicious codes which could actively spread, such as network worms. To detect worms in early stages of spreading in LANs (local area networks), Dagon et al. [24] implemented a script-driven HoneyStat system which covers a large amount of IP addresses. According to the spreading features of LAN worms, HoneyStat issues alarms in three different categories: those on memory operation, disk write, and network operation. Then, it can detect the outbreak of 0-day worms in a fast and accurate manner based on its auto-generated alarm data and correlation analysis. SweetBait system [25] is used in the NoAH (Network of Affined Honeypots) project funded by the European Union's Sixth Framework Programme (FP6). Integrating the low-interaction SweetPot with the high-interaction Argos [26], it can detect worms in real time, generate detection signatures automatically, and build a worm response mechanism through distributed deployment and signature sharing. It was proven effective in the online detection and analysis during the outbreak of Conficker worm at the end of 2008 [27]. In addition to traditional types of malicious codes such as network worms, researchers also use the honeypot to detect and analyze malicious web attacks against client software such as browsers. The HoneyMonkey system [28] detects and discovers malicious pages which carry out infiltration attacks against browser vulnerabilities by using high-interaction HoneyClient and operating systems and browser software with different patch levels. Google's Safe Browsing project, by using a combination of machine learning and high-interaction HoneyClient, detected more than three million uniform resource locator (URL) links from voluminous web pages crawled by search engines and did a systematical analysis of malicious web pages.

3.2.2 Capturing Samples of Malicious Codes

Latest honeypots develop the ability to automatically capture samples of malicious codes after detection. Nepenthes [29] is the earliest open-source tool to perform this. With flexible expandability, Nepenthes can cover 18-segment monitoring with a single server. It once captured more than 5.5 million penetration attacks and generated 1.5 million malicious code samples within 33 hours, which, after de-duplicated by MD5 value, left 408 samples. Those data prove that Nepenthes is effective in capturing active malicious codes. Wang [30] and other people proposed the ThingPot, used in the Internet of Things (IoT) platform, to help identify attacks to its devices by analyzing mobile phone data. The devices deployed have discovered five samples of smart device attacks. HoneyBow performs auto-capturing of malicious code samples based on the high-interaction honeypot. For example, it captures 296 code samples per day during the 9-month monitoring period in the matrix distributed system, performing better than Nepenthes. Capturing effect will be much improved if the advantages in the two systems could be combined in one. WebPatrol [31] develops a method of automatic capturing which combines low-interaction honeypot and "proxy-like" caching and replay technology to defend against client-attacking malicious web Trojan scenes. It stores distribution across multiple websites and dynamically generates Trojan scene webpages with multiple steps and paths for more comprehensive collection and storage. It also supports a reply of attack scenes for offline analysis. WebPatrol has captured 26,498 malicious web Trojan attacks from 1248 hacked websites in China Education and Research Network (CERNET) within 5 months. The honeypot is widely recognized and deployed in the anti-virtual industry for its ability to capture malicious code samples, for example, the world-renowned antivirus vendor Symantec.

3.2.3   Tracking and Analysis of Security Threats

Honeypot's detection and capturing of security threat data provide support for tracking and analyzing specific types of security threats such as botnets and spams. Botnet monitoring and tracing are a hotspot in the honeypot application for in-depth analysis of threats. The basic process comes as follows. At first, it captures the bots which actively spread through the Internet and then monitors and analyzes the bots in a controlled environment or sandbox to get information about their connection to the botnet commands and control server and to track the botnet through the Sybil node. Finally, when information is adequate, honeypot can further exploit the sinkhole shutdown, takeover, and other active containing methods [32]. Freiling et al. [33] are among the earliest to use the honeypot technology for botnet tracking, on which Rajab and others improve its method, tracking from multiple angles at the same time. For instance, they develop a distributed collecting system to capture bots, an IRC tracking tool to observe the actual botnets from within, and the DNS cache probing which could assess botnet spreading footprint. Through correlation analysis of data captured from multiple angles, they give insight to some behavior and structural features of Botnets. Jianwei Zhuge, a researcher on network science, revealed phenomenon characteristics after conducting long and comprehensive researches [34] on IRC botnets using matrix distributed honeynet [35]. Stone-Gross et al. [36] take over Torpig botnet by pre-registering the Dynamic Domain Name (DDN), and thus track 180,000 IP addresses of bot hosts and collect 70GB sensitive information. Accordingly, the fact that botnets can be contained actively by tracking and takeover is verified. To curb spams, Project Honey Pot project makes use of over 5000 honeypots installed by website administrators to monitor more than 250,000 spam phishing addresses and conducts large-scale tracking analysis [37] of behaviors of collecting email addresses and spamming. Steding-Jessen et al. [38] used the low-interaction honeypot to study spammers' abuse of open agents.

3.2.4   Extraction of Attack Features

Honeypot is a reliable source of attack features for many reasons. First of all, the threat data captured by it is pure and in a relatively smaller amount, which usually does not contain normal traffic. Secondly, it is able to detect threats such as network probing, penetration attacks, or worms in the early stages as long as the honeypot system covers a small amount of IP addresses. Researchers have worked out a variety of approaches to extract attack features from honeypot data, for example, Honeycomb, Nemean, SweetBait/Argos, and Honeycyber. Honeycomb [39] is the earliest announced research to automatize the feature extraction. As an extension module of Honeyd, it connects with attack links to conduct a one-to-one longest common substring matching with the saved network connection load of the same target port. A candidate feature is generated if a match is identified that exceeds the minimum length of common substring. All the candidate features will join the existing feature pool to generate an updated attack feature database. Honeycomb introduces a basic extracting method but fails to consider the similar semantic information in the application layer protocol, so the same content inside the head of the protocol may lead to extraction of invalid features in many cases. Noticing this, Nemean system [40] proposes a method of extracting attack features with a semantic perception ability by taking as input the original packets captured by virtual honeypot and Windows 2000 Server physical honeypot. The procedure comes as follows: firstly, data packets are converted into a shortest spanning tree (SST)—semi-structured network session tree—through the data abstract module. Then, the feature extraction module applies the MSG multilevel feature generalization algorithm to cluster network sessions and generalize them into semantically sensitive features based on the finite state machine (FSM). Last but not least, those features will be converted to the regulated format of detection system features and put into practice [31]. SweetBait/Argos is aimed at the honeypot which could be released actively. Firstly, it uses the dynamic taint analysis technology to detect penetration attacks and trace back to the specific location of the tainted data in the network session traffic, which could cause the EIP instruction register to be maliciously controlled. In the automatic feature extraction, it supports algorithms such as the longest common substring (LCS) and the penetration attack key string detection (CREST), in which the latter could extract attack features that are more concise and more accurate and also could fight better against the multiple forms of network attacks. HoneyCyber [41] utilizes the double honeynet deployment structure to capture the inflow sessions and outflow sessions of worms in different forms and automatically extracts features through the obvious data gained through analyzing the main content in different cases. In the lab environment, it could achieve zero false positive and low false negatives on worms artificially made various in forms, while it's not verified in practical traffic circumstances against polymorphic worm cases. To combat a large number of complex problems demanding strong expertise, Fraunholz et al. [42] proposed a strategy of configuring, deploying, and maintaining the dynamic honeypot based on machine learning.

3.2.5   Limitations of Honeypot

Unlike encryption, access control, and firewalls whose functions are clearly defined in each part of the overall defense system, honeypot takes the confrontational thinking and approach into every chain of the defense system. Therefore, without a fixed connotation or denotation, it always changes, making it difficult to formulate a theoretical basis as access control could but, rather, relying much on confrontational strategies and tactics. Unfortunately, it means that the existing honeypot cannot be backed up by scientific theories, which prevents it from forming a generalized engineering form with standard specifications. It only serves as a solution tailored to the high-security demand of enterprise clients, rather than a common and universal product form despite the efforts made by Symantec and other vendors to promote honeypot. As a technology that directly confronts attacks, current honeypot has the following drawbacks, putting it on a disadvantaged position in the game of technology confrontation.

(1) The honeypot environment fails to solve the contradiction between fidelity and controllability. A high-interaction honeypot, almost identical with the operation system environment, has a better camouflage and spoofing but is in an unfavorable condition due to its insufficiency to detect attacks, maintenance cost, and other costs. A low-interaction honeypot with high controllability and low maintenance cost has an inadequate fidelity, thus being easier to be identified by attackers and becomes defenseless to new threats.
(2) The existing honeypot technologies, primarily aiming at normal or general threats with large-scale impacts, is weak at deception and detection in dealing with Advanced Persistent Threats. Honeypots have to be highly customizable, adaptive to dynamic environments, and highly concealed. It needs to be able to hide in the actual operating system without easy perception.
(3) In the game of core technology, the honeypot is not quick enough to catch up with the fast development of attack technologies and lags behind the study of new threats on new platforms such as supervisory control and data acquisition (SCADA) and intelligent mobile platform.
(4) The conventional passive defense chain consists of perception, recognition, decision, and execution and follows a technical model of capturing threats, analyzing them, extracting threat features, forming defense rules, and finally deploying defense resources to deal with all types of threats. Honeypot is among the most effective in the perception process and also in analyzing and extracting features. However, the perception technologies of attackers keep evolving in the heightened confrontation. For example, malicious codes will remain silent or destruct itself once a honeypot environment is identified using the defects of honeypot, stopping attacks to avoid exposing themselves, thus jeopardizing the effectiveness of the primary chain of passive defense.

3.3   Collaborative Defense

Collaborative defense combines different security tools with their individual adaptability and expertise to improve the overall effect in order to form a holistic defense to guard against ever-appearing new attack methods. The collaboration of products is, in essence, a security mechanism of information communication. With the development of information security, an increasing number of users realize that it's wrong to see security as an isolated issue and that no single product can guarantee their company's network security. The overall security that customers want requires a tightly connected defense system. Collaboration across tools as an inevitable trend would have wider and wider application, just imagine it.

Collaborative cyber security defense is a system to minimize losses by ensuring the security of the local and host network as well as detecting ongoing attacks and preventing them. It has the following characteristics:

(a) It is a multilayer defense system with its components deployed according to their features and functions, working independently as well as interactively. Attackers must break through all defense layers to harm the system.
(b) The defense ability improves dynamically.
(c) The components can interact and form a cooperative relationship. For example, after intrusion detection informs attacks to defense system, the latter chooses a strategy to respond, such as notifying the firewall to cut down the intrusion connection or use attacking programs to counterattack the hacker.
(d) It combines automatic response with artificial intelligence, thus improving the ability to fight against hackers and reducing the workload of administrators.

3.3.1   Collaborative Defense Between Intrusion Detection and Firewall

By monitoring, limiting, and changing the trespassing data flow, the firewall offers protection and reliable security services to sources by shielding the interior information, structure, or operation as much as possible from the exterior. However, if you want to fully use a firewall to protect your network from connecting the Internet, you have to combine the firewall with other security measures so that they will work complimentarily with one another.

The coordination between intrusion detection and a firewall is the most common form of intrusion prevention at present. As soon as an attempt is identified, the detection system will inform the firewall to block the attacker's IP address or port. Information security products have a tendency to converge and be centralized [43]. However, there are still weaknesses in their collaboration. Firstly, it complicates the system by using two defense products, and failure of either will lead to the failure of the whole system. Plus, it costs more to maintain the two products. Secondly, there is not a widely accepted standard for coordination of the two products, and most vendors just follow their own practices. Producers of intrusion detection devices only interact with the firewalls they recognize, whereas firewall producers have different open interfaces. There is no universal standard that allows all intrusion detection products to interact with all firewalls. In this way, users have to consider the interactivity of different products when making purchasing decisions, thus limiting the options of products users could choose from.

3.3.2 Collaborative Defense Between Intrusion Prevention and Firewall Systems

The coordinated defense between intrusion prevention and firewall systems is considered as early as the designing phase of the intrusion prevention system, which could largely reduce the pressure on the prevention system. That means, when the firewall is deployed, a large amount of illegal traffic will be shielded, and the efficiency of prevention system can be thus improved.

The collaborative defense between intrusion prevention systems and firewalls is to build an overall defense by transferring policy rules to and from each other. Firewalls fall into network firewalls and host firewalls according to their configuration location. The prevention system passes the rules concerning security events with the highest blocking rate to the network-based firewall so that the latter will be able to prevent such events, alleviating the pressure on the intrusion prevention system. The rules passed on to the firewall vary dynamically with the prevention system and are subject to an effective time period. In other words, they will become invalid upon expiry. When the network-based firewall is using the dynamically opened port protocol, it must send rules to the prevention system for the communication traffic to pass. The prevention system has to remove these rules after the communication is finished. Similarly, a firewall on a host can identify which strategy prevents the most packets, and therefore it can use those rules to stop the largest number of attacks or unnecessary data flow. Generally, such rules also have the same effect on other hosts in the intranet. Likewise, if those strategic rules are passed on to the prevention system, the packets will be intercepted before entering the intranet, reducing the pressure on the host firewall, as shown in Fig. 3.2.

However, a major problem facing the collaboration of these two totally different defense mechanisms lies in the inability to apply all the rules of the firewall to the prevention system and vice versa. Another problem is that different host firewalls operate in different environments and its semantic difference would prevent direct dialogue in the rules exchange between host firewalls.

In response to the above problems, the newly issued next-generation firewall (NGFW) [44] makes use of the integrated intrusion prevention system, visual

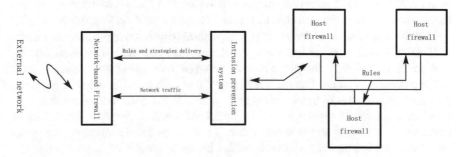

Fig. 3.2 Coordinated defense between the intrusion prevention and firewall

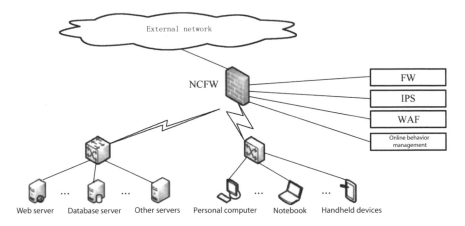

Fig. 3.3 Topological structure of NGFW

application identification, and intelligent firewalls to provide users with integrated application layer protection. Figure 3.3 shows the topological structure of the NGFW.

By integrating or coordinating different systems, NGFW is able to perform functions, such as user identification, malware detection, application control, intrusion prevention, and program visualization, and is able to identify and control applications on any port, significantly improving the protection of existing firewalls.

3.3.3 Collaborative Defense Between the Intrusion Prevention System and Intrusion Detection System

The intrusion prevention system itself has a detection function but is not always efficient. The main reason is that this system is bound with the target; thus the efficiency of the former will influence the performance of the latter, limiting the size of the knowledge base used by the detection engine. As a result, it is impossible for the intrusion prevention system to detect all the security events [45]. The point of using collaboratively both the intrusion prevention system and the intrusion detection system is to complement the prevention system with a stronger detecting ability enabled by multiple intrusion detection systems and perform prevention to the greatest extent.

This collaboration of the intrusion prevention system and intrusion detection system is mainly realized through the passing along of rules from the detection system to the prevention system. The prevention system is located at the network gateway, monitoring and controlling the inflow and outflow of data, while the detection systems are deployed at several important nodes. At a certain interval, the detection system will, according to the security events spotted by the detection system, pass the detection rules on security events with a higher frequency of occurrence to the prevention system for the latter to defend against such events at the gateway. The prevention system will update the knowledge base of its detection engine against the detection rules sent by the detection systems. For example, it can store the rules inactive for a long time as cache and prioritize those which occur more frequently. The active and dynamic update of rules helps to detect more attack events and minimizes influence to the performance of the target. The prevention system can also transfer rules to the detection systems to verify if the rules are effective in preventing threats. If no attack can be detected after a rule is applied to multiple detection systems, it would be safe to say this rule is invalid. The right thing to do is to give feedback to the prevention system, delete the rule, and cut the overhead of invalid operation of the system.

You shall bear in mind the compatibility of rules, so the rules need to be abstracted to a certain degree before they are transmitted across the detection and prevention systems. In addition, the key of a successful collaboration lies on the detection rate and correctness of the detection system [46].

3.3.4 Collaborative Defense Between Intrusion Prevention and Vulnerability Scanning Systems

The vulnerability scanning system prevents intrusions by detecting the vulnerability or the incorrect settings of a system, which are the targets of most attacks. The purpose of the scanning system is to find and fix them before an attack occurs. However, it happens when the system is unable to fix some of those vulnerabilities. At this time, it is advisable to collaborate with the intrusion prevention system to prevent such vulnerabilities from being attacked, for the latter can block the network traffic of applications containing such vulnerabilities or prevent the system from invoking such applications.

This collaboration starts with the vulnerability scanning system. After scanning, a list of applications with vulnerabilities is generated and converted into vulnerability rules to be transmitted to the prevention system. Then the prevention system converts the rules and adds them to its knowledge base, which could be used to detect and defend relevant vulnerabilities. The prevention system will prevent attacks by blocking the invocation of the vulnerable applications. However, it can restore the normal functions of the applications after the vulnerabilities are fixed.

The key to the intrusion prevention is the provision of a rule which describes the vulnerable applications and of the necessary information to take actions. The prevention system can use this rule to arrange responses, scanning, system checks, vulnerability, and vulnerable application detection.

3.3.5 Collaborative Defense Between the Intrusion Prevention System and Honeypot

Honeypot, based on the virtual honeypot technology [47], can be deployed behind the intrusion prevention system. It constructs virtual hosts to attract hackers and can work as a subsystem within the intrusion prevention system or work independently.

In this kind of collaboration, honeypot helps the prevention system expand its scope of security defense and make up for the deficiencies of its false positives and false negatives. It also helps in collecting hackers' behaviors or intrusion evidence.

Moreover, the prevention system is weak at detecting new types of attacks, where a honeypot can be a powerful complement to improve security levels of the target.

Honeypots, acting as the substitute of the target host, simulate one or more vulnerable hosts or servers equipped with a file system to make them deceptive. They are designed to imitate the appearance and content of key systems so as to determine intruders' intention and to record their behaviors without exposing or damaging the actual system.

The honeypot system lures intruders by simulating the operating system, service processes, protocols, application vulnerabilities, and fragilities, as shown in Fig. 3.4.

A honeypot keeps detailed log records of the intruders' behaviors and tracks the behaviors at different levels to facilitate its analysis. It can send alarms and log information to the administration console, based on which the intrusion detection system upgrades its knowledge base and modifies its prevention strategy to better fend off intrusions. Furthermore, the honeypot can also be set at the active response mode so that, once an illegal activity is detected, the connection will be cut off.

In short, the collaborative defense is more multilayered and intelligent than a single system. Focusing on known attacks, it monitors dynamically and gives alarms based on technologies, such as feature scanning, pattern matching, data analysis, etc., and blocks or eliminates threats by emergency response, manual or automatic. However, in practice, security products are deployed on different platforms, so customers have to deploy several different systems for collaboration purpose. It's a waste of resources and internet space. The collaboration between different products is surely a challenge to their deployment.

The main drawback of the collaborative defense model, similar to the static defense, is that it falls short to defend against unknown vulnerabilities and complex network attacks. The detection-oriented collaboration defense is unable to detect complicated attacks, especially the covert attacks which are aimed at an open network. They are mostly carried out through social networks against network

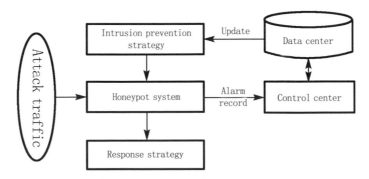

Fig. 3.4 Collaborative Defense of the intrusion prevention system and honeypot system

maintenance and administrating personnel. Besides, it's not hard for hackers to take advantage of the potential vulnerabilities and backdoors within the system to open covert channels to avoid detection.

3.4 Intrusion Tolerance Technology

As the static defense means are inadequate to meet the demands of networks for security, the intrusion tolerance technology is introduced, which allows the system to have some vulnerabilities and presumes that some attacks against system components can come through. Instead of trying to stop every single attack, intrusion tolerance triggers a mechanism which can prevent system failure and ensure at a certain predictable probability the security and dependability of the system [48].

In essence, intrusion tolerance is the executive part of the passive defense system and the last line of network defense. Fault tolerance is its core algorithm in the execution. It improves security by tolerating some degree of failure in the defense system. Since it still requires a priori knowledge from the detection process, intrusion tolerance is also a method of passive defense.

3.4.1 Technical Principles of Intrusion Tolerance

The idea of intrusion tolerance was proposed by Fraga and Powell as early as in 1985. For it is impossible to predict all unknown attacks nor to prevent all attacks against potential vulnerabilities, it is necessary to work out a robust system which can still perform specific functions even under attacks.

In 1991, Deswarte, Blain, and Fabre developed the first distributed computing system with intrusion tolerance. In January 2000, Europe launched the MAFTIA research project (malicious and accidental fault tolerance for Internet applications) to systematically study this method and to establish a reliable large-scale distributed application. The main results of MAFTIA are as follows:

(1) The intrusion-tolerant structural framework and conceptual model are defined, which makes up for the reliability and security gaps.
(2) A set of mechanisms and protocols for an intrusion tolerance system have been developed. They include a set of modular and scalable communication middleware security protocols, a large-scale, intrusion-tolerant distributed intrusion detection system, and the distributed authentication service of intrusion tolerance.
(3) A formal verification and evaluation method for some components of MAFTIA middleware has been introduced. In 2003, the US DARPA (Defense Advanced Research Projects Agency) launched a new research program named OASIS (Organically Assured and Survivable Information Systems) to study intrusion tolerance. It focuses on the system's ability to maintain its availability and security

under attack and on the methods to evaluate such kinds of ability. The funded programs include SITAR, ITTC, COCA, ITUA, etc. China also carried out some researches on intrusion tolerance in 2001.

3.4.1.1 Theoretical Model

1. Security objectives

Intrusion tolerance divides the security objective of a system into three categories: confidentiality, integrity, and availability. It is made according to the function and characteristics of the system and the purpose of network attacks.

Confidentiality: The system needs to make sure that some of its specific information and the confidential data contained in its service are safe from invaders.
Integrity: The system needs to make sure that its internal data and those in the provided services are not deleted or tampered with by attackers.
Availability: The system needs to make sure the service is usable and continuous.

2. System fault models

There are mainly two reasons why the system or its components are attacked. The first is its internal vulnerabilities, essentially defects in requirements, specifications, design, or configuration, for example, unsafe password or coding failure which causes a stack overflow. The second is the external attacks. Attackers carry out a malicious operation on system vulnerabilities, for example, port scanning, uploading Trojans, privilege escalation, or DoS attacks. A successful intrusion can cause an error in the system state and can crash the system. To apply traditional fault tolerance to the intrusion tolerance, it is necessary to abstract attacks, intrusions, or system vulnerabilities into a system fault. The process of a system from being attacked to becoming invalid involves those sequence of events: fault-error-failure. To infer a mechanism for blockage and tolerance, it is necessary to model system faults, dealing with vulnerability, attack, and intrusion as three elements.

Vulnerability: A potential defect in the system that can be triggered or exploited by attackers.
Attack: An act launched by an attacker against a certain vulnerability, with the purpose of exploiting this vulnerability to cause damage to the system.
Intrusion: When an attack triggers a vulnerability and causes damage to the system's security target, an intrusion is completed.

Figure 3.5 is the theoretical model of intrusion tolerance.

3.4.1.2 Mechanisms and Strategies

1. Intrusion tolerance mechanism

Intrusion tolerance is essentially a survival skill of the system. Based on security requirements, the system should be able to (a) block and prevent attacks; (b) to

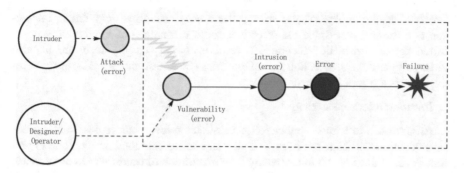

Fig. 3.5 Theoretical model of intrusion tolerance

detect them and assess the damage caused; (c) to maintain or recover critical data, service, or all of its service after an attack. To realize those objectives, certain security mechanisms should be in place. The main types of mechanisms are as follows:

(1) Security communication mechanism. It's necessary to maintain a secure and reliable communication by preventing or blocking attacks such as eavesdropping, masquerading, and denial of service. An intrusion-tolerant one usually adopts techniques such as encryption, authentication, message filtering, or classic fault-tolerant communication.

(2) Intrusion detection mechanism. It detects possible attacks, intrusions, or vulnerabilities and gives response by monitoring and analyzing network events. It should be combined with vulnerability analysis and attack forecasting to predict possible errors in order to identify the cause of the security threat. It can also combine with auditing to make a record of system behaviors and security events to perform post-analysis.

(3) Intrusion containment mechanism. It increases the difficulty and cost of attack attempts by resource redundancy and the variety of designing and also isolates damaged components through separation or resource reconfiguration to limit or prevent further spread of intrusion.

(4) Error handling mechanism. It includes error detection and error recovery, with the purpose of preventing catastrophic system failures.

Error detection includes system integrity check and log auditing. It serves multiple purposes: (a) to stop further spread of errors, (b) to trigger the error recovery mechanism, and (c) to trigger the fault handling mechanism. The error recovery mechanism is to recover the system from error condition to maintain or recover key data or services, or all of its services, if possible. It includes (a) forward recovery (the system moves on to a functioning situation), (b) backward recovery (the system moves back to the previously functioning situation), and (c) error shielding. Error shielding means that system redundancy is used to shield errors for the system to provide normal service. Its tools include component or system redundancy, threshold cryptography, system voting operations, Byzantine negotiation, interaction consistency, etc. Since the effectiveness of error detection relies on a priori knowledge or information accuracy, its result is not only largely uncertain, but also causes bigger delays, affecting the effectiveness of error recovery. Therefore, error shielding is a priority for error recovery.

2. Intrusion tolerance strategy

An intrusion tolerance application necessitates relevant strategies, that is, what kind of action will be taken in case of an attack to tolerate the intrusion and to prevent system failure. Tolerance strategy is an integration of traditional fault tolerance technologies and security strategies. It is decided through measuring the cost of intrusion and the cost the protected system could afford on condition of operation types, quality, and technology available. Once the intrusion tolerance strategy is defined, the intrusion tolerance mechanism is also defined, based on which an intrusion tolerance system will be designed. More specifically, an intrusion tolerance strategy has the following aspects.

(1) Fault avoidance and tolerance. Fault avoidance aims to rule out faults in the system designing, configuring, and operating processes. For it is impractical and impossible to remove all vulnerabilities, the fault tolerance approach of offsetting the negative impact of system failure is more economical than fault avoidance. Therefore, when designing an intrusion tolerance system, it's advisable to maintain a balance between fault avoidance and fault tolerance, though, in some important systems, fault avoidance could be the main goal to pursue.

(2) Confidentiality strategy. When the primary strategic goal is to maintain the confidentiality of the data, intrusion tolerance need not reveal any useful information when some of the unauthorized data are exposed. This service can be realized by error shielding methods, such as threshold cryptosystem or quorum scheme.

(3) Reconfiguration strategy. It can assess the damage caused by an ongoing intrusion on the system component or subsystems and reconfigure system resources or survive accordingly. This strategy relies on the detection process to decide whether the components are unharmed or faulted when there are errors detected so that it can use unharmed ones to replace the faulted ones or reconfigure the system. Its main purpose is to guarantee the usability or integrity of services, such as transaction databases or web services. It also needs to be noted that since all sources and services will be reorganized, the system will be temporarily out of function.

(4) Recovery operation. Assume that (a) t_i is the amount of time it takes for a system to fail; (b) t is the time it takes from system failure to recovery, and it's an acceptable amount of time off for the users; (c) the system is recoverable; and (d) t_c is the duration of a specific attack and $t_c < t_i + t$. If a system satisfies the above four assumptions, it could launch the recovery action when it falls under

attack. In a distributed environment, recovery can be realized through a secure negotiation protocol.

(5) Failure-proof. When some components are attacked, the system's function may be compromised. If something unbearable happens, the system is in danger of losing control and becomes highly insecure. At this time, it's necessary to have emergency measures (e.g., to shut down the system) to prevent the system from further damage. This failure-proof strategy is often used in systems with highly important tasks and is a supplement to other strategies.

3.4.2   Two Typical Intrusion Tolerance Systems

3.4.2.1   Scalable Intrusion-Tolerant Architecture

SITAR (scalable intrusion-tolerant architecture), launched by the research institute MCNC (Microelectronics Center of North Carolina) and Duke University, is an intrusion tolerance project aiming to protect COTS servers. Sponsored by the DARPA, it is a part of the DARPA OASIS (Organically Assured and Survivable Information Systems) program (1999–2003). Its structure is shown in Fig. 3.6.

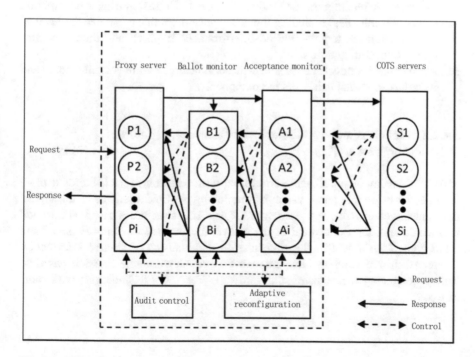

Fig. 3.6 SITAR system construction

SITAR is a distributed service architecture of intrusion tolerance based on COTS servers. Its research motives are twofold. Firstly, there are no security precautions that can guarantee that the system is free from being penetrated. Secondly, key applications should be able to provide the lowest-level services even under attack or partial damage. Therefore, SITAR shall mainly ensure the continuity of service. Instead of trying to determine whether the system failure is caused by a malicious attack or it is accidental, it focuses on the fact that the system must handle and withstand the effects of negative disturbances. The SITAR system utilizes redundancy and diversity as the basic building blocks of its structure and defines five key components:

(a) The proxy server carries out the service strategy designated by the intrusion tolerance strategy, and the service strategy will decide on which COTS servers the requests should be forwarded to and how to present the final results.
(b) The acceptance monitor checks the validity of the requests and responses and then selectively forwards them with the directions of the results to the ballot monitor. The ballot monitor checks the security problems of the COTS servers and generates the intrusion trigger information.
(c) The ballot monitor acts as the representative for the COTS server to resolve conflicts. The final response is decided by the majority rule or Byzantine protocol, and its implementation will depend on the current level of security threat detected.
(d) The adaptive reconfiguration module receives the intrusion trigger information from other modules (including the acceptance monitor); evaluates the level of threats, tolerance targets, and cost/performance impacts; and then generates new system configurations.
(e) Audit control conducts diagnostic tests periodically. It verifies audit records and identifies abnormal behaviors in components.

3.4.2.2 Malicious and Accidental Fault Tolerance for Internet Applications [75]

MAFTIA, an important EU-supported long-term research project, builds a more complex intrusion tolerance system by employing multiple tolerance technologies, distributed system skills, and security strategies. This structure is based on trust and trustworthiness. The conceptual model and architecture are shown in Figs. 3.7 and 3.8. Figure 3.9 shows that the transmission and analysis of control information between distributed architectures are realized through wormhole channels based on trusted computing mechanisms and hardware processing technologies with roots of trust.

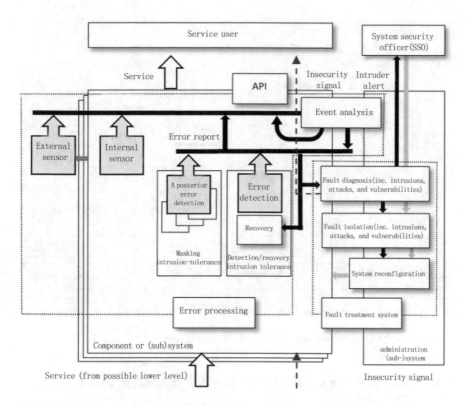

Fig. 3.7 MAFTIA conceptual model and architecture

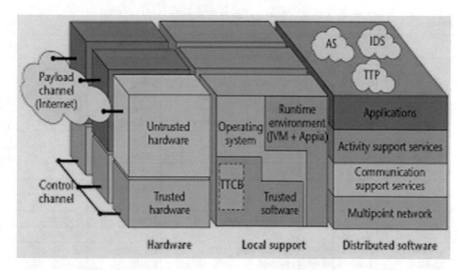

Fig. 3.8 MAFTIA architecture

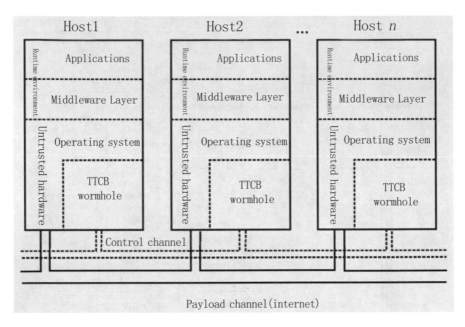

Fig. 3.9 MAFTIA architecture and TTCB wormhole mode [74]

3.4.3 Comparison of Web Intrusion Tolerance Architectures (Table 3.1)

Table 3.1 Comparison of web intrusion tolerance architectures [73]

Parameters	Generic intrusion tolerance architecture	Scalable intrusion-tolerant architecture for distributed services	Self-cleaning intrusion tolerance architecture	Intrusion-tolerant architecture using adaptability and diverse replication	Malicious and accidental fault tolerance for Internet applications
Dependability characteristics					
Availability	Yes	Yes	Yes	Yes	Yes
Confidentiality	No	Yes	No	Yes	Yes
Integrity	Yes	Yes	Yes	Yes	Yes
Design criterion					
Diversity	Required	Required	Optional	Required	Required
Redundancy	Required	Required	Required	Required	Required
Servicing of request	By randomly selected servers	By randomly selected servers	Online virtual server	By active nodes	Distributed servers
Configuration manager	Distributed	Centralized	Centralized	Centralized	Distributed

Parameters	Generic intrusion tolerance architecture	Scalable intrusion-tolerant architecture for distributed services	Self-cleaning intrusion tolerance architecture	Intrusion-tolerant architecture using adaptability and diverse replication	Malicious and accidental fault tolerance for Internet applications
Intrusion detection mechanism	SNORT, EMERALD, SNORT, EMERALD, agreement protocol, challenge/response protocol, runtime verification	Acceptance monitor, proxy servers, audit control	No	Acceptance test and vote	Wormhole
Research issue addressed	Tolerating accidental faults and deliberate faults in public access web servers	Continuity of operation and scalability of the architecture	Reducing a server's exposure time	Survivability of critical services	High resilience in case of arbitrary failure and high performance in case of controlled failures
Research challenge	Silver bullet attack (an attack that would be successful on all servers), an attack that aims at the leader elected	Cost and complexity	High response time, Malicious alteration of session information of the servers	Sacrifice noncritical services, timing overhead	Trusted components, which the entire architecture relies on, will resist only well-defined set of attacks

3.4.4   Differences Between Intrusion Tolerance and Fault Tolerance

In summary, both fault tolerance and intrusion tolerance attempt to enable the system to continuously provide acceptable services under abnormal conditions. Therefore, some key fault-tolerant mechanisms can be used for intrusion tolerance, including (a) redundancy backup, diversity, reconfiguration, etc. and (b) independence of redundant components (diversity of design), for example, by adopting different operating systems, the general model could have fewer vulnerabilities or less likely to be attacked. The error detection, damage assessment, error recovery, error processing, and continual service of fault tolerance are deployed in different phases, which can also be a reference for the designing and actualizing an intrusion-tolerant system.

However, there are still differences between the two tolerance systems. The difficulty to integrate fault tolerance in intrusion tolerance owes to the following points:

(1) Fault tolerance takes into prior consideration possible accidental or random or permanent errors in its designing and implementing phases. It has made reasonable assumptions about nearly all predictable errors. While intrusion is almost all maliciously motivated and intrusion behaviors are mostly hidden or deliberately contagious, which cannot be described by random events or mathematical tools and is difficult to predict.

(2) Intrusion behavior refers to external attacks against vulnerabilities of system components; such deliberate triggering behavior is seldom considered by the traditional tolerance system triggered by random faults. In addition, intrusion behavior in the general sense includes internal and external coordinated attacks initiated by malicious codes on backdoors, hardware, and software in the system, which the fault-tolerant systems usually do not consider.

(3) The current fault-tolerant technology is generally supported by a priori knowledge, including historical statistics of relevant hardware and software components, a large number of trial and experimental data, proven failure model, etc. However, it's difficult to generate an engineerable error pattern when facing complex internal and external collaborated attacks, especially those unrecognizable and unexpected attacks based on unknown vulnerabilities or backdoors.

(4) Deliberate attacks are usually repeatable. Intruders attack all homogeneous redundant components using the same method until the fault tolerance system is unable to handle. In particular, the target object will be forced by countless attacks of the same kind to reconfigure repeatedly and fruitlessly until it loses its robustness or even fail to function at all. Of course, a heterogeneous redundant component configuration mode can greatly improve this situation, which will be discussed later in the book.

Because of their inner connectivity, intrusion tolerance is widely used in recent years in areas where fault tolerance was conventionally applied, such as the industrial control system and wireless sensors. In 2017, Wu Chuankun et al. [49] proposed that the redundant configuration of main control computers enable the system to continue running normally even if a certain number of the controlling computers have been attacked. Wu offers different security architectures and analyses according to different abilities of attackers and also a reliability analysis of the intrusion tolerance system. Analyses show that the newly designed intrusion-tolerant architecture is highly secure and helpful to increase system reliability. In 2018, Lee et al. [50] proposed an intrusion tolerance-based quantitative assessment method for the industrial control system, which is used for security assessment and real-time security defense of the digitalized instrument console and information system of nuclear power plants. In terms of wireless sensors, Li Fan et al. [51] noticed that the existing sensor network only considered the performance of network intrusion but not of network QoS and put forward a detection mechanism based on performance feedback. This mechanism marks the abnormal nodes, selects reliable nodes as a cluster head, and designs the PFITP intrusion tolerance protocol. PFITP protocol can not only tolerate common types of attacks but also help optimize network performance continuously, so it can improve the robustness and practicability of the sensor network.

3.5 Sandbox Acting as an Isolation Defense

3.5.1 Overview of Sandbox

Sandbox is an isolated environment in which program behaviors are restricted according to a certain security strategy. The sandbox-based security mechanism can monitor the program behaviors and limit its actions of carrying out the operations violating the security strategy.

The idea of sandbox defense comes from the software error isolation technology, which uses software methods to restrict untrusted modules from causing damage to the software. The main idea is to isolate, that is, to ensure the robustness of the software by isolating untrusted modules from the software system. In cyber defense, sandbox is a technology to defend the network by building an isolated environment for analyzing and processing of untrusted resources and by restricting their possible malicious acts.

Malicious codes must have sufficient operating authority to reach their goals of intrusion, spreading, and activation. For example, they must obtain the permission to read and write the target if they intend to infect the target files, to cause leaks or to modify target content. Therefore, obtaining the operation permission is the prerequisite for malicious codes to cause damage [52].

Sandbox is a software security technology to mitigate this threat. By running a program in the sandbox, the program's operation of the system, such as registry of files, will be virtually redirected. All such operations are virtual, and the real registry and files are unmodified, ensuring that the user's real system environment won't get affected.

Therefore, the core of sandbox is to build an environment for applications to run within limits, especially the programs with tentative dependability, to limit possible damages they may cause to the system.

The theoretical basis of sandbox is the access control mechanism, whose task is to manage users' access authority while allowing them to maximumly share the system resources, so as to prevent any unauthorized tampering or abuse of information. After identity authentication, a legal user is granted limited access to the object. Any unauthorized visit request will be refused so that the system can operate in accordance with the planned security order. The application-oriented access control sets rules according to the program's functional and security demands. In short, using access control rules to limit the program's resource access cannot only meet normal access demands of programs but also do it on a security basis. The core method of sandbox technology is application-oriented access control.

According to the approaches of access control, sandbox at present is mainly of two types: virtualization-based sandbox and rule-based sandbox.

A virtualization-based sandbox (Fig. 3.10) builds a closed operating environment for untrusted software for it to function normally and safely. This type of sandbox ensures that the untrusted software won't affect the host. According to the virtualization level, this book divides the virtualization-based sandbox into system-level and container-level sandboxing.

The system-level sandbox uses hardware-layer virtualization to provide a complete operating environment for untrusted software. Related researches include WindowBox, VMware, VirtualBox, QEMU, etc. In comparison, the container-level sandbox uses the more lightweight virtualization, adding a virtual layer between the operating system and the application to virtualize the user space. Solaris Zones, Virtuozzo Containers, and Free VPS have conducted researches on this.

The rule-based sandbox (Fig. 3.11) uses the access control rules to limit a program's behavior. It consists mainly of an access control rule engine, a program monitor, etc. The program monitor transforms and submits the monitored behaviors to the access control rule engine for it to decide whether the requests will be permitted or not. Unlike the virtualization-based sandbox, the rule-based sandbox

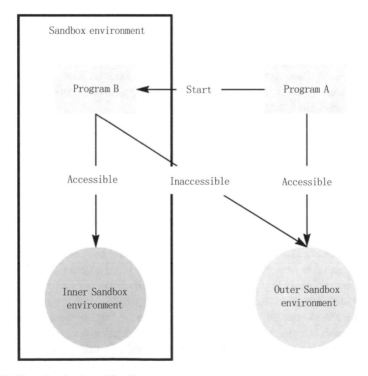

Fig. 3.10 Virtualization-based Sandbox

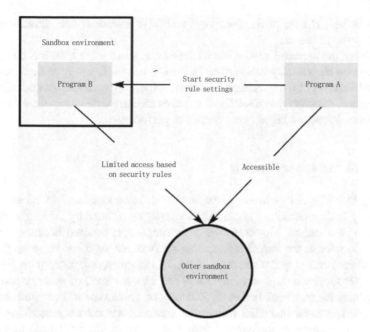

Fig. 3.11 Rule-based sandbox

does not require replication of system resources, thus reducing the impact of redundant resources on system performance. Besides, the rule-based sandbox makes resource sharing more convenient for different applications. Relevant research projects include Nucleus TRON (The Real-time Operating System Nucleus) and AppArmor [53, 54].

3.5.2   Theoretical Principles of Sandbox

There are three types of the actualization of sandbox: application layer, kernel layer, and hybrid sandboxes.

3.5.2.1   Application Layer Sandbox

This kind of sandbox runs on the user layer of the operating system. Usually written in a high-level language, it has the same access authority as ordinary user applications do. All the services it needs are provided by the operating system, and its basic function of isolation is realized by relevant software technology. The implementation mechanism of such sandbox has the following advantages: easy realization, requiring not too much knowledge on the part of programmers about the OS kernel; simple utilization; high stability; abundant user interfaces; and suitability for PC users.

However, the mechanism also has its defects: low in security, for it is completely dependent on the security mechanism of the OS kernel. It's easy to be bypassed if the vulnerabilities of the operating system are exploited by malicious codes; poor in performance, for this type of sandbox, requires an extra detection code to be added in the normal process, largely comprising its performance.

3.5.2.2 Kernel Layer Sandbox

Residing in the kernel's address space, the kernel layer sandbox has the same level of security as the operating system and can easily be isolated by the hardware-level protection mechanism. Moreover, the basic function is ensured because the functional code runs at the kernel layer, avoiding frequent switches between the user layer and the kernel layer while monitoring. In summary, compared with the application layer sandbox, the kernel layer one has a higher level of security and better performance. Its drawback lies in its difficulty of development. Programmers have to be experts versed in kernel principles. Besides, it's not easy to transplant or deploy in the non-open operating system. For example, it's difficult to apply such kind of sandbox to non-open-source systems such as Windows. Lastly, having the same security authority as the operating system, the sandbox will put the whole system at risk if there are vulnerabilities in the sandbox itself.

3.5.2.3 Hybrid Sandbox

Hybrid sandbox combines the above two types. In a hybrid sandbox, the kernel code provides isolation support and relevant executive mechanism, while the rest part of the system will be realized in the application layer. In this way, it can achieve a higher level of security isolation and a higher level of sandbox system stability, and, moreover, it's easier to design and develop [55].

3.5.3 Status Quo of Sandbox Defense Technology

Recent years have witnessed a rapid growth of the sandbox technology due to the increase of APT attacks. Sandbox is widely used in browsers (e.g., IE, Firefox, and Chrome), document readers (e.g., Adobe Reader), and other applications which analyze and process network resources. The sandbox system applicable on Windows platform is mainly business oriented, but there are also open-source free versions for personal or academic use. More academic-oriented contents can be found in the sandbox based on the Linux platform.

CWSandbox is used for malicious code analysis on Windows platforms. It works by executing malicious code samples in a virtual environment, monitoring all system calls, and automatically generating detailed reports for simplifying and automating malicious code analysis. The core is API hooking and DLL injection techniques. The downside of such technologies is that they need to execute malicious codes in a virtual environment and the modification done to the system cannot be rolled back. Also, API hooking, monitoring commonly used Windows APIs, are easily bypassed by new types of shelled malicious codes.

Sandboxie allows user applications easy to get infected by malicious codes, such as browsers, to run in a sandbox environment, and the changes made during the running process can be deleted afterward. It can be used to erase traces of Internet surfing and program use and also to restore favorites, home pages, or registry. Sandboxie provides a completely isolated operating space from the host. Even the files downloaded in sandboxing will be deleted along with the emptying of the sandbox. As a business software, Sandboxie is normally used to ensure a safe and secure online surfing experience. It's not open-source and not for free. Therefore, its internal details remain yet unknown.

Similar to CWSandbox, Cuckoo is a sandbox system for the Windows platform. It also works by running malicious codes in a virtual computer to perform related behavioral analysis and recording the sequence of system calls of malicious codes. Because Cuckoo also performs detection by monitoring normal Windows APIs, there are also possibilities of being bypassed for some novel shelled malicious codes. For example, the source codes leaked by Hacking Team in 2015 revealed a mechanism to bypass Cuckoo.

Vx32, a versatile user-level sandbox, isolates the memory access path of malicious codes through the x86 CPU segment registers. A variety of common applications can run in this system, for example, self-extracting compressed files, scalable public key infrastructure, and user-level operating systems. The advantages of Vx32 are low-performance loss, significant role in portable scenarios, and non-modification of the Linux OS kernel. However, as Vx32 sandbox runs at the user level, it is easily bypassed by kernel-level malicious codes to carry out illegal activities.

Rule-based sandbox [56] uses access control rules to limit a program's behavior. It consists mainly of an access control rule engine, a program monitor, etc. The program monitor transforms and submits monitored behaviors to the access control rule engine for it to decide whether the requests will be permitted or not. Unlike the virtualization-based sandbox, the rule-based sandbox does not require replication of system resources, thus reducing the impact of redundant resources on system performance. Besides, the rule-based sandbox makes resource sharing more convenient for different applications. Berman and others [53] designed TRON, the first rule-based sandbox system that designates resources into different control domains. In control domains, it uses a unique string to mark each resource and sets access rights. Users use command to assign an application to the control domain, and the sandbox system enforces restrictions on program behaviors by access control rules in the domain. AppArmor [54], designed by Cowan and his team, is similar to TRON and uses a white list to define accessible resources in the sandbox.

Although sandbox can prevent malicious code invasion to a certain degree, the sandbox escape technology develops rapidly as well. For example, virtual environment perception, long-term latency, and hidden traces may help attackers to escape.

The main drawbacks of the sandbox technology are as follows:

(1) It only monitors commonly seen application interfaces of the operating system, making it easier for special malicious codes to bypass such sandboxes to attack local or host computing environments.
(2) There is no reliable recovery or rollback mechanism after a malicious code is detected, endangering to a certain degree the integrity of the user's application data.
(3) There are multiple functional modules and a large number of codes in the existing sandbox system, as well as the unavoidable designing defects or even backdoors imported by open-source software. These pose a big challenge to the security, stability, and maintainability of the sandbox system.

3.6 Computer Immune Technology

3.6.1 Overview of Immune Technology

When the term "computer virus" was mentioned by Adelman in 1987, computer experts began to draw an analogy between computer security problems and biological processes [57, 58]. Virus is characterized by self-replication and transmission, and the immunity system aims to inhibit its spreading. The correlation between the immune system and computer security was proposed in *Computer immunology* [59]. The natural immune system protects animals from dangerous foreign pathogens (including bacteria, viruses, parasites, toxins, etc.), acting as the computer security system. Though living organisms and computer systems are significantly different, they share dramatic similarities that can guide the computer experts to improve computer security. To this end, the computer security experts are trying to design a computer immune system according to the biological immune system, and the mapping of their relationship is shown in Table 3.2.

The natural immune system [60] has three layers of protection, with the outermost, the skin, as the first protection, the innate immune system as the second, and the adaptive (or acquired) immune system as the third. Once pathogens enter a body, the immune system will be activated and the scavenger cells, such as macrophages, will clean some of the pathogens. Then the adaptive immune system will be activated. As a complex system, it includes a variety of immune cells and immune molecules, which can produce corresponding immune cells and antigens to fight against and finally eliminate the invading pathogens.

In *Principles of a Computer Immune System*, Somayaji [61] and others point out the basic differences between the two immune systems: (a) the computer system is an electronic system made of digital signals rather than cells or molecules; (b) it's

Table 3.2 Mapping of natural and computer immune system

Antigen	Computer virus
	Network intrusion
	Other targets to be tested
Binding of antibody and antigen	Pattern matching
Self-tolerance	Negative selection algorithm
Memory cell	Memory detector
Cell cloning	Copy detector
Antigen detection/response	Identification/response to non-self-bit strings

impossible to reconfigure the signal system of the natural immune system; and (c) the natural immune system makes sure the biological body survives, while the computer system needs to do more. In this book, they propose and design a set of principles the computer immune system needs to abide by:

(a) The system should be distributed so that failure at one single point won't affect the whole immune system and can increase the robustness of the system.
(b) It should be multilayered for better security.
(c) It should be diverse to prevent the virus from spreading.
(d) It should be manageable to control the system.
(e) It should have some self-governance to manage and maintain itself.
(f) It should be adaptive, being able to learn by itself to detect new viruses, identify new invasions, and record the signature of previous attack.
(g) There should be no security layer; system components should be able to protect each other instead of being separated by setting security code.
(h) Coverage should be dynamic; the system checks the system in a dynamic way rather than through an enough large coverage.
(i) It should be able to identify behaviors and detect system invocations.
(j) It should be able to perform anomaly detection to find out invasions or unknown safety issues.
(k) The detection should not be complete. The immune system needs to be changeable, just like macrophages, which can detect an extensive range of pathogens, but will not be that effective when fighting specific pathogens.
(l) It should be able to join the game. Viruses can develop resistance to the antibody. Likewise, a computer virus will also evolve with the help of hackers.

3.6.2   Artificial Immune System Status

The concept of artificial immune system, as a branch of computational intelligence, was proposed by DasGupta [62]. Artificial immune system can be applied in many scenarios, for example, anomaly detection/pattern recognition, computer security, adaptive control, and error detection [63]. Figure 3.12 shows the system's steps to solve an actual problem. To apply an immune model to solve a specific problem in

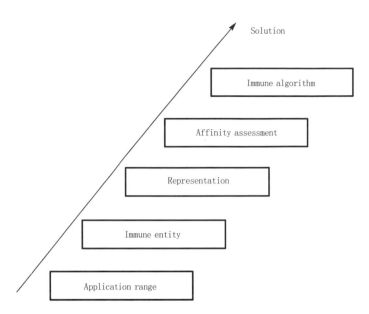

Fig. 3.12 The steps for the artificial immune system to solve actual problems

a particular area, an immune algorithm should be selected according to the type of the problem. Then questions such as the elements involved, how to represent those elements, and how to model them into an entity should be decided. Next, an appropriate affinity (distance) should be settled to determine a matching rule. Finally, choose an immune algorithm that generates a suitable set of entities to provide a suitable solution to the problem.

Reference [64] proposes the main steps to achieve the immune system: (a) to find a previously unknown computer virus on the user's computer; (b) to collect the virus sample and send it to the central host; (c) to automatically analyze the virus and get a method to remove the virus on any host; (d) to send the clearing method back to the user computer and write it in the antivirus database; and (e) to spread the method to other areas to clear viruses.

Reference [65] proposes a malicious code detection technology based on the artificial immune system. As an important application field of the artificial immune system, malicious code detection is an important research direction of many immunologists and information security experts at home and abroad, and several of such models have been introduced. They have basically the same logic as borrowed from the natural immune system: to extract features of malicious codes and encode them into antigens, to generate relevant antibody by artificial immune system algorithms and models such as negative selection, to calculate the distance between the antigen and the antibody using methods like Euclidean distance or r-Contiguous bits matching, and finally to realize accurate identification of malicious codes. The correspondence between the biological immune system and the malicious code detection system is shown in Table 3.3.

Table 3.3 Correspondence between biological immune system and malicious code detection system

Biological immune system	Malicious code detection system
Autologous cell	Normal file
Antigen (bacteria, pathogen)	Malicious code file
Lymphocyte/antibody	Detector or signature codes
Binding of antigen-antibody	Detection process pattern matching
Vaccination	Feature library update
Antigen clearance	Malicious code removal
Autoimmune disease	Normal document false positive
Tumor, cancer, or other diseases	Malicious code false negative

Reference [66] proposes and analyzes the application and future of artificial immune system in the intrusion detection system. The characteristics of the artificial immune system, namely, diversity, distributiveness, dynamics, self-adaptation, self-identification, learning, memory, etc., can effectively make up for the shortcomings of the traditional intrusion detection system and thus boost its intelligence and accuracy.

Reference [67] proposes an efficient proactive artificial immune system for abnormal detection and prevention, as shown in Fig. 3.13. Through the combination of the idea of the artificial immune system and proxy, a defense system has been developed, which is a self-configuring, adaptive, and synergistic system with the ability to detect abnormalities.

In Reference [68], the multiagent technology is integrated into the evolving process of traditional artificial immune network. The neighborhood-clonal selection algorithm is introduced, and the operating process moves from the local to the whole, making possible a more comprehensive emulation of the natural evolving model of the immune system. Moreover, more competition and cooperation between antibodies are introduced into the process, improving the dynamic analysis ability of the network.

The natural immune system ensures merely survival, while the computer security system has five tasks: confidentiality, reliability, availability, accountability, and accuracy. Immune technology also has its limitations [69]: a certain immune method is only effective for a certain type of viruses, and it can only protect files that are running stable and in no need of modification. The sheer complexity and size of the immune system forbids even the immunologists from fully understanding or describing this immune phenomenon. There are not many research results for the artificial immune system to borrow from; thus it faces problems in modeling and algorithms. Besides, the immune system is not flawless in itself. To build a defense system (the optimal feasible solution set) and to obtain initial antibodies (solution samples), a lot of calculations [70] need to be done.

Although the artificial immune system has already been applied to many fields, it is still difficult to establish a unified theoretical basis for it [71]. Like the natural immune system, the artificial immune system has the problem of false positive.

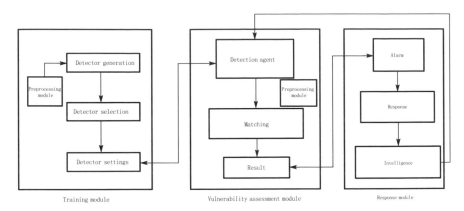

Fig. 3.13 An efficient proactive artificial immune system

Removal of normal files further leads to "autoimmune diseases" or inability to fully identify malicious files and codes, thus hampering the efficiency of detection. A single artificial immune system cannot achieve the effect of its natural immune system. Without the natural protection of innate immunity in the latter, the artificial immune system is more vulnerable than the natural immune system.

The artificial immune system is still essentially a passive defense system based on perception. An antigen is generated from the extraction and encoding of the features of known malicious codes, so it still follows the mindset of "draw lessons from the past for future immunity" and hasn't overcome its inherent structural flaws, such as staticity, similarity, and certainty of the target object. It's still difficult for the artificial immune system to get the upper hand in combating vulnerabilities, backdoors, Trojan virus, or new types of attacks or threats.

3.7   Review of Conventional Defense Methods

The problems of traditional cyber security framework are as follows:

(1) Cyber security technologies, such as firewalls, are able to fend off the majority of attacks. However, inner vulnerabilities and pre-implanted backdoors make it easy for attackers to penetrate the static cyber security defense shield. In addition, the effectiveness of passive defenses such as firewalls is often strongly related to the experience and expertise, game strategies, and confrontational skills of security officers.
(2) The defense capabilities of conventional cyber security components are fixed, while attackers keep evolving their skills. It's possible that, at the initial phase of installation, the defense components are able to fend off hackers' attacks. However, over time, hackers' ability will eventually exceed the defensive capabilities.

(3) The traditional cyber security defense or detection capabilities cannot upgrade dynamically but can only be upgraded manually or periodically. The effectiveness, continuity, and timeliness of the defense capabilities depend largely on the expertise of security personnel and the ability of manufacturers to guarantee their service. It's hard to meet all requirements, so great uncertainty exists. Take the WannaCry ransomware attack in May 2017 as an example. The WannaCry ransomware took advantage of the SMB vulnerability. Although the SMB vulnerability and the official patch were released by Microsoft (security bulletin MS17-010) 2 months earlier, there were still more than 300,000 host computers falling victim to WannaCry when the virus broke out on May 12, recording a loss of more than eight billion US dollars.

(4) The information that a single security component can obtain is quite limited and is not enough to detect complex attacks. Even if attacks were detected, it is unable to make an effective response. The defense of any single security component is limited in the face of the increasingly popular, distributed, or collaborative attacks. To achieve interaction or coordination among components, you will face many difficulties and challenges to realize rigorous and effective detection and protection.

(5) In a self-contained processing space with shared resources, it is not only an engineering problem but also a philosophical dilemma to prove the security of software or hardware components.

It should be noted that the abovementioned conventional defense systems have a prominent feature that both the target object and the defense system lack an endogenous security attribute and that there are no precautionary measures in place to defend against possible vulnerabilities or backdoor attacks. The defense functions, mainly external or additional or embedded, are externally attached protection means, meaning that the protection method and the protected objects are independent in structure and function. This piling up of defense components makes it difficult to realize reliable protection in an intensive and optimized manner. The defense system thus lacks a domain for organic security mechanism, and the protection interfaces cannot dynamically merge with each other. In addition, the singularity and limitations of the processing and cognitive space and the transparency required in providing services make real-time perception, attack behavior identification, and accurate extraction of attack feature insurmountable engineering challenges. The defense system turns into a dilemma of quantitatively proving its effectiveness in defending against undisclosed vulnerability attacks, complex and multimodal joint attacks, and attacks based on inner backdoors or resident Trojan viruses.

In summary, the conventional defense measures are unable to deal with unknown attacks based on unknown vulnerabilities or backdoors due to the structural weaknesses (caused by certainty, staticity, similarity, and sharing of resources) of the target system, the unavoidability of vulnerabilities, and backdoors of hardware and software. Recent security events disclosed at home and abroad and their serious consequences have fully revealed the generic defects of conventional cyber security defense theories and technologies: being powerless to attacks based on unknown vulnerabilities, to attacks through potential backdoors, and to the increasingly complex and intelligent penetrating cyber intrusions. The situation is getting worse when hacking techniques keep improving themselves, for example, techniques in digging vulnerabilities and making use of them, in presetting backdoors and activating them, in better concealment of APT attacks, etc.

Conventional defense focuses on external security reinforcement and the detection and eliminating of known threats. It seldom considers the possibility to change its defense logic or the technology pattern based on precise perception. Although researchers have made much progress in finding vulnerabilities and detecting backdoors, there is still a big gap to fill before achieving the goal of eliminating vulnerabilities and backdoors. Therefore, it is necessary to introduce systematic engineering in cyber security defense, that is, by changing the scenario of the problems, to move the solving space from the component layer to the structure layer, to extend from the single or homogeneous environment to the heterogeneous redundant one, to turn the operational mechanism from a regular one to one with an apparent uncertainty attribute, to weaken the role of accurate acquisition of attack features and accurate removal of problems in the defense system, to upgrade the availability design from dealing with random failure to the general robust control stage of dealing with even artificial attacks, to generate endogenous security defense effect through the innovative architectural technology, and to effectively maintain the certainty of guarding against attacks by lessening to the maximum degree the availability of system vulnerabilities and blocking the internal and external collaborative functions of backdoors. Only in this way can we achieve a better control of the defense effects and fundamentally change the current gross imbalance cost between attackers and defenders.

The good news is that the industry has witnessed efforts to try to introduce organic security measures in information architecture, operating system, and compiler designing from the perspective of integrated designing. Among the successful cases are the Harvard architecture which separates codes from data, the randomization of memory addresses in the operating system, and the randomization cookie of compiler-based stack protection. A convincing example is Windows and Office series software, which are now more difficult for hackers to take advantage of their vulnerabilities, and the attacks are less general and have a less chance of success. Nevertheless, these endogenous defense measures are only applicable to certain operating systems and compilers, and they are basically measures for mainstream buffer overflow vulnerabilities. Up to now, the theory about structure-based design of endogenous security is still in its infancy, and its technologies and methods are far from becoming systematic. Despite this, it can be expected that the structure-derived endogenous security defense technology is going to be one of the most important research directions in cyber security.

References

1. Nie Yuanming, Qiu Ping: Network Information Security Technology. Tsinghua University Press, Beijing (2001)
2. Desai N.: Intrusion prevention systems: the next step in the evolution of IDS. https://www.symantec.com/connect/articles/intrusion-prevention-systems-next-step-evolution-ids. [2017-07-05]
3. Goncalves, M.: Firewalls: A Complete Guide. Translated by Kong Qiulin, Song Shumin, Zhu Zhiqiang, et al. Mechanical Industry Press, Beijing (2000)
4. He Haibin: Technology and application of firewall based on packet filtering of linux. J. UEST (Univ. Electron. Sci. Technol.) China. **33**(1), 75–78 (2004)
5. Lu Qi, Huang Zhiping, Lu Jiaqi: System design of firewall based on deep packet inspection. Comput. Sci. **44**(S2), 334–337 (2017)
6. Top Layor Networks: Beyond IDS: essentials of network intrusion prevention. http://www.toplayer.com/bitpipe/IPS_Whitepaper_112602.pdf. [2017-07-05]
7. Rusty Russell: Linux 2.4 packet filtering HOWTO. http://www.netfilter.org/documentation/HOWTO/packet-filtering-HOWTO.html. [2017-07-05]
8. Tang Zhengjun: Introduction to Intrusion Detection Technology. Mechanical Industry Press, Beijing (2004)
9. Han Hongguang, Zhou Gaiyun: Network intrusion detection based on Makov chain state transition probability matrix. Control. Eng. **24**(3), 698–704 (2017)
10. Xie Weizeng: Artificial bee colony algorithm optimization network intrusion detection of support vector machine. Microcomput. Appl. **33**(1), 71–73 (2017)
11. Gao Ni, Gao Ling, He Yiyue, Wang Hai: A lightweight intrusion detection model based on feature reduction of self-encoding network. Chin. J. Electron. **45**(3), 730–739 (2017)
12. Wang Shengzhu, Li Yongzhong: Intrusion detection algorithm based on deep learning and semi-supervised learning. Inf. Technol. **2017**(1), 101–104
13. Fan Chengfeng, Lin Dong: Network Information Security and PGP Encryption. Tsinghua University Press, Beijing (1998)
14. Lindstrom P.: Understanding intrusion prevention. http://www.networkassociates.com/us/_local/promos/_media/wp_spire.pdf. [2017-07-05]
15. Welteh.: Pablo Neira Ayuso Netfilter/Iptables Project. http://www.netfilter.org [2017-07-05]
16. Yan Tihua, Zhang Fan: Network Administrator Tutorial. Tsinghua University Press, Beijing (2006)
17. Zhang Yuqing: Security Scanning Technology. Tsinghua University Press, Beijing (2004)
18. Zhao Wei, Liu Yun: Analysis and Design of Network Information System: Theory, Method and Case. Tsinghua University Press, Beijing (2005)
19. Spitzner, L.: Honeypots: Tracking Hackers. Addison-Wesley, Reading (2003)
20. Perdisci, R., Dagon, D., Lee, W., et al.: Misleading worm signature generators using deliberate noise injection. 2006 IEEE Symposium on Security and Privacy (S&P'06), 2006, pp 15–31
21. Kuwatly I, Sraj M, Al Masri Z, et al.: A dynamic honeypot design for intrusion detection. IEEE/ACS International Conference on Pervasive Services, 2004, pp 95–104
22. Artail, H., Safa, H., Sraj, M., et al.: A hybrid honeypot framework for improving intrusion detection systems in protecting organizational networks. Comput. Secur. **25**(4), 274–288 (2006)
23. Anagnostakis, K.G., Sidiroglou, S., Akritidis, P., et al.: Detecting targeted attacks using shadow honeypots. Proceedings of the 14th Conference on USENIX Security Symposium, Berkeley, 2005
24. Dagon D, Qin X, Gu G, et al.: Honeystat: local worm detection using honeypots. International Workshop on Recent Advances in Intrusion Detection, 2004: 39–58
25. Portokalidis, G., Bos, H.: SweetBait: zero-hour worm detection and containment using low- and high-interaction honeypots. Comput. Netw. **51**(5), 1256–1274 (2007)

26. Portokalidis, G., Slowinska, A., Bos, H.: Argos: an emulator for fingerprinting zero-day attacks for advertised honeypots with automatic signature generation. ACM SIGOPS Oper. Syst. Rev. **40**(4), 15–27 (2006)
27. Kohlrausch, J.: Experiences with the NoAH honeynet testbed to detect new Internet worms. The Fifth International Conference on IT Security Incident Management and IT Forensics, 2009, pp 13–26
28. Wang, Y.M., Beck, D., Jiang, X., et al.: Automated web patrol with strider honeymonkeys. Proceedings of the 2006 Network and Distributed System Security Symposium, 2006, pp 35–49
29. Baecher, P., Koetter, M., Holz, T., et al.: The nepenthes platform: an efficient approach to collect malware. International Workshop on Recent Advances in Intrusion Detection, 2006, pp 165–184
30. Wang, M.: Understanding Security Flaws of IoT Protocols through Honeypot Technologies: ThingPot-an IoT platform honeypot. Delft University of Technology, Delft (2017)
31. Chen, K.Z., Gu, G., Zhuge, J., et al.: WebPatrol: automated collection and replay of web-based malware scenarios. Proceedings of the Sixth ACM Symposium on Information, Computer, and Communications Security, 2011, pp 186–195
32. Zhuge Jianwei, Tang Yong, Han Xinhui, et al.: Progress in research and application of honeypot technology. J. Softw. **24**(4), 825–842 (2013)
33. Freiling, F.C., Holz, T., Wicherski, G.: Botnet tracking: exploring a root-cause methodology to prevent distributed denial-of-service attacks. European Symposium on Research in Computer Security, 2005, pp 319–335
34. Han, X., Guo, J., Zhou, Y., et al.: Investigation on the botnets activities. J. China Inst. Commun. **28**(12), 167 (2007)
35. Zhuge, J., Holz, T., Han, X., et al.: Characterizing the IRC-based Botnet Phenomenon. Universität Mannheim/Institut für Informatik, Mannheim (2007)
36. Stone-Gross, B., Cova, M., Cavallaro, L., et al.: Your botnet is my botnet: analysis of a botnet takeover. Proceedings of the 16th ACM Conference on computer and Communications Security, 2009, pp 635–647
37. Prince, M.B., Dahl, B.M., Holloway, L., et al.: Understanding how spammers steal your e-mail address: an analysis of the first six months of data from project honey pot. CEAS, 2005
38. Steding-Jessen, K., Vijaykumar, N.L., Montes, A.: Using low-interaction honeypots to study the abuse of open proxies to send spam. INFOCOMP J. Comput. Sci. **7**(1), 44–52 (2008)
39. Kreibich, C., Crowcroft, J.: Honeycomb: creating intrusion detection signatures using honeypots. ACM SIGCOMM Comput. Commun. Rev. **34**(1), 51–56 (2004)
40. Yegneswaran, V., Giffin, J.T., Barford, P., et al.: An architecture for generating semantic aware signatures. USENIX Security, 2005 pp 34–43
41. Mohammed, M.M.Z.E, Chan, H.A., Ventura N.: Honeycyber: automated signature generation for zero-day polymorphic worms. MILCOM 2008 IEEE Military Communications Conference, 2008, pp 1–6
42. Fraunholz, D., Zimmermann, M., Schotten, H.D.: An adaptive honeypot configuration, deployment and maintenance strategy. 2017 19th International Conference on Advanced Communication Technology (ICACT), 2017, pp 53–57
43. Deng Shichao: Design and Delivery of Intrusion Prevention System Based on Data Mining. Guilin University of Electronic Technology, Guilin (2005)
44. Chen Zhizhong: Analysis of next generation firewall (NGFW) features. Netw. Secur. Technol. Appl. **2017**(10), 21–22 (2017)
45. Ierace, N., Urrutia, C., Bassett, R.: Intrusion prevention systems. Ubiquity, 2005.
46. Cummings, J.: From intrusion detection to intrusion prevention. http://www.nwfusion.com/buzz/2002/intruder.html. [2017-07-05]
47. Provos, N.: Developments of honeyd virtual honeypot. http://www.honeyd.org. [2017-05-27]
48. Plato, A.: What is an intrusion prevention system. http://www.anition.com/corp/papers/ips_defined.pdf. [2017-07-05]

49. Wu Chuankun, Zhang Lei: A security defense architecture for industrial control systems with intrusion tolerance. National Network Security Level Protection Technology Conference, Nanjing, 2017
50. Lee, C., Yim, H.B., Seong, P.H.: Development of a quantitative method for evaluating the efficacy of cyber security controls in NPPs based on intrusion tolerant concept [J]. Ann. Nucl. Energy. **112**, 646–654 (2018)
51. Li Fan: Research on Intrusion Tolerance of Wireless Sensor Network. Southeast University, Nanjing (2016)
52. Cohen, F.: Computer Viruses. University of Southern California, Los Angeles (1985)
53. Berman, A., Bourassa, V., Selberg, E.: TRON: process-specific file protection for the UNIX operation system. Proceedings of USENIX Winter, Manhattan, USA, 1995, pp 165–175
54. Canonica: AppArmor: linux application security framework. https://Launchpad.net/apparmor. [2014-04-16]
55. LeVasseur, J., Uhlig, V.: A sledgehammer approach to reuse of legacy device drivers. Proceedings of the 11th Workshop on ACM SIGOPS European Workshop, Leuven, Belgium, 2004, pp 240–253
56. Zhao Xu, Chen Danmin, Yan Xuexiong, et al.: On sandbox technology research. Henan Computer Society Academic Conference and Henan Computer Conference, Anyang, 2014
57. Cohen, F.: Computer viruses: theory and experiments. Comput. Secur. **6**(1), 22–35 (1987)
58. Lamont, G.B., Marmelstein, R.E., van Veldhuizen, D.A.: A distributed architecture for a self-adaptive computer virus immune system. In: New Ideas in Optimization, pp. 167–184. McGraw-Hill, Maidenhead (1999)
59. Forrest, S., Hofmeyr, S.A., Somayaji, A.: Computer immunology. Commun. ACM. **40**(10), 88–96 (1997)
60. Janeway, C.A., Travers, P., Walport, M., et al.: Immunobiology: The Immune System in Health and Disease. Current Biology, Singapore (1997)
61. Somayaji, A., Hofmeyr, S., Forrest, S.: Principles of a computer immune system. Proceedings of the 1997 Workshop on New Security Paradigms, ACM, 1998, pp 75–82
62. DasGupta, D.: An overview of artificial immune systems and their applications. In: Artificial Immune Systems and Their Applications, pp. 3–21. Springer, Berlin (1993)
63. DasGupta, D.: Advances in artificial immune systems. IEEE Comput. Intell. Mag. **1**(4), 40–49 (2006)
64. Kephart, J., Sorkin, G., Swimmer, M., et al.: Blueprint for a computer immune system. In: Artificial Immune Systems and Their Applications, pp. 242–261 (1999)
65. Lu Tianliang: On Malicious Code Detection Technology Based on Artificial Immune System. Beijing University of Posts and Telecommunications, Beijing (2013)
66. Yang Chao: Application of artificial immune system in intrusion detection system. Inf. Commun. **2015**(1), 6–7 (2015)
67. Saurabh, P., Verma, B.: An efficient proactive artificial immune system based anomaly detection and prevention system. Expert Syst. Appl. **60**, 311–320 (2016)
68. Hong Ming, Liu Peizhong, Luo Yanmin: A data classification method for artificial immune network based on multi-agent strategy [J]. Appl. Res. Comput. **34**(1), 151–155 (2017)
69. Chen Lijun: New approaches to immunity from computer viruses. J. Peking Univ. Nat. Sci. Ed. **34**(5), 581–587 (1998)
70. Jiao Licheng, Du Haifeng: Development and prospect of artificial immune system. Chin. J. Electron. **31**(10), 1540–1548 (2003)
71. Mo Hongwei, Zuo Xingquan, Bi Xiaojun: Research advances in artificial immune system. J. Intell. Syst. **4**(1), 21–29 (2009)

Chapter 4
New Approaches to Cyber Defense

As noted in Chap. 3, conventional cyberspace defense technologies focus on bolstering the target system externally and trying to test, discover, and eliminate known threats. In spite of the substantial and fruitful researches in the field of vulnerability detection and backdoor testing, we are still far from the ideal goal of rooting out vulnerabilities and backdoors. Both the academia and the industry have realized that the conventional static or coordinated approaches of defense are not well adept at handling Advanced Persistent Threat (APT). In response, many advanced countries have launched research programs based on new defense approaches (such as moving target defense, or MTD) [1] in an attempt to change the rules of the game. The nature of cyberspace is such that it is easy to attack but difficult to defend. By increasing the dynamism, randomness, and redundancy of the system or network against external attacks, the new defensive technologies render attackers unable to effectively sustain the cognitive advantage over the target system and the available resources over time and space, making it difficult to accumulate information, duplicate the mode of attack, recreate the effect, or continue the method of attack, which significantly increases the cost of the attack and hopefully would change the rules of the game.

4.1 New Developments in Cyber Defense Technologies

In Chap. 3, we briefly introduced the traditional cyberspace defense concepts and technologies. In a nutshell, traditional defense technologies are external security technologies that are attached on top of or in between the targets that are constantly under passive protection. As cyber attacks become more intelligent and coordinated, however, they can easily breach the traditional perimeters of defense or override the defense mechanism. Therefore, it is not only necessary but inevitable to develop new, game-changing defense approaches.

© Springer Nature Switzerland AG 2020
J. Wu, *Cyberspace Mimic Defense*, Wireless Networks,
https://doi.org/10.1007/978-3-030-29844-9_4

The new types of cyber defense effectively guard against cyber attacks at all stages by utilizing the endogenous security mechanisms of the system. It should not rely too heavily on obtaining and recognizing a priori knowledge, such as code of attack or characteristics of behaviors, nor should it intend to eliminate vulnerabilities, backdoors, and Trojans in real time, which conventional technologies tend to focus on. Instead, the new types of defense aim to create an operating environment with an endogenous uncertainty effect, using basic defense approaches based on dynamism, redundancy, and heterogeneity, which alters the staticity, certainty, and similarity of the system so as to decrease the success rate of attacks via vulnerabilities, destroy or interrupt the coordination of backdoors, and block or interfere with the reachability of attacks, thereby significantly increasing the difficulty and cost of attacks.

In essence, the new cyber defense methodology aims to introduce an endogenous element of security through innovation at the target system's structural level and of the operating mechanism. In recent years, as the security sector and the industry are becoming more aware that unknown vulnerabilities and backdoors are among the most crucial issues of cyber security threats, they have started to introduce certain security mechanisms in the design of system structures, operating systems, as well as network structures. Different from the traditional, external approaches, these new and active defense mechanisms—including trusted computing technology, tailored trustworthy space (TTS), and moving target defense (MTD)—could effectively increase the difficulty for attackers to take advantage of vulnerabilities or backdoors. In order to lower the possibility of target vulnerabilities being used and increase the robustness of the system, the author believes that the new defense technologies should focus on the following aspects in system design:

(1) Adaptability design [2]. Adaptability refers to the reconstruction capability of the system to dynamically adjust configurations or operating parameters in response to external events. Firstly, when developing a system, the designers shall plan ahead a path for the execution of external events or build a fault mode for the system. Secondly, when the system is running, they shall establish a mode of perceiving external events through machine learning and trigger its adaptive reconstruction mechanism at proper times. Some examples of adaptability design include on-demand resource scaling, system diversity design (see Chap. 5), and decreasing attack surface [3]. In 2014, Gartner, an IT research and consulting company, claimed that the adaptive security architecture design is the next-generation security system against advanced attacks that exploit unknown vulnerabilities [4]. In 2016, adaptive security architecture was again listed by Gartner as one of the top ten most noteworthy strategic technologies of the year [5].

(2) Redundancy design. It is generally believed that redundancy design is one of the most important means to increase system robustness or resilience. Gene redundancy [6], for example, improves the adaptability of species to the environment. In trusted engineering, redundancy design [7] has always been an effective way to protect key subsystems or components. Information theory, on

the other hand, proves theoretically that redundancy can enhance code robustness [8]. Redundancy of a network system means that multiple resources are deployed for the same network function, so that in case of main system failure, services could be promptly transferred to other backup resources. Redundancy is also the most prominent technological characteristic of new cyber defense technologies and the prerequisite for operating mechanisms such as diversity, dynamism, or randomness. Redundancy is the central element of the new cyber defense strategies, such as MTD and TTS, and it could significantly improve the overall resilience of the target system.

(3) Fault tolerance design [9]. Fault tolerance is the means through which the system tolerates faults to achieve reliability. There are generally three types of fault tolerance in system design: hardware, software, and system fault tolerances. Hardware fault tolerance includes redundancy of communication channels, processors, memories, and power supplies. Software fault tolerance includes structural designs, exception handling mechanisms, error authentication mechanisms, multimode operation, and ruling mechanisms. System fault tolerance uses unit-level heterogeneous redundancy and multimode ruling mechanisms to compensate for any operation errors caused by random physical failures or design errors.

(4) Mitigation mechanism design. Mitigation system can automatically respond to faults or support manual fault handling. Mitigation policy refers to an established standard process or execution plan that is used to guide the system or the administrator through fault handling in case of fault or attack. Common forms of mitigation policies include automatic system quarantine and segregation, redundant communication channel activation, and even counterattack strategy.

(5) Survivability design. In ecology, survivability means the ability of some life-forms to survive more successfully than others when faced with unknown changes of physical conditions, such as flood, disease, war, or climate change [10]. In engineering, survivability means the ability of the system, subsystem, device, process, or program to continue functioning against natural or artificial interference. In cyberspace, the survivability of a network means that, under the circumstances of (unknown) attacks, faults, or accidents, the system can still perform tasks [11]. Survivability is considered part of resilience [12] or robustness. We believe that survivability should be considered an important indicator of the new defense system, so that when under known or unknown attacks, the target system could maintain its normal operation and maintenance indicators as much as possible or maintain corresponding level of services by smooth downgrading.

(6) Recoverability design. Recoverability means that the system can rapidly and effectively recover its operation in the event of service interruptions. Specific strategies include automatic switching of hot backup components, dynamic embedding of code backup components, and diagnosing, cleaning, and recovering of fault components.

In fact, before the concept of new cyber defense emerged, the abovementioned six design approaches have already been used to varying degrees in cyber defense technology research and system design, without forming a systematic new defense theory. The past decade saw increasing attention to the research and practice of new defense strategies as more efforts were made to change the asymmetry or imbalance of the cost of attack versus defense. In order to change the "rules of the game," academics in China and overseas have proposed a variety of defense theories and technologies. This chapter will focus on the design thinking behind four typical defense strategies or security technologies, namely, trusted computing, TTS, MTD, and blockchain. In Sect. 4.6, we will discuss the available new defense technologies.

4.2 Trusted Computing

There are different understandings of the concept of "trusted." For example, in the *Common Criteria for Information Technology Security Evaluation*, the International Organization for Standardization and the International Electrotechnical Commission (ISO/IEC) point out from an evaluation perspective that a trusted component, operation, or process is one whose behavior is predictable under almost any operating condition and which is highly resistant to subversion by application software, viruses, and a given level of physical interference [13]. Trusted Computing Group (TCG), on the other hand, gives the definition from a behavioral perspective, stating that an entity can be trusted if it always behaves in the expected manner for the intended purpose [14]. We can conclude from these different definitions that trusted computing means an entity has a subjective anticipation toward whether another entity could reach the expected goal in an accurate, non-disruptive manner. When an entity always reaches the goal in expected ways, it is trusted. Practice shows that trusted computing could guarantee that the system is secure and trusted when all the behaviors of the target are known or expected.

4.2.1 Basic Thinking Behind Trusted Computing

The mainstream view is that the security of hardware system and operating system is the foundation of information system security, whereas passwords and cyber security are the key technologies. Security issues of information system can be effectively solved only in a comprehensive way. Therefore, trusted computing proposes that in order to improve security of the information system, we must start from the basic layer of hardware, such as chips, mainboard, hardware structure, basic input output system (BIOS), and operating system, and combine them with database, network, and application in the design process. In fact, the clue of trusted computing can be found in security design of earlier information systems. For example, its document encryption technology could be to a large extent mitigate invasions by virus; its data backup and restoration mechanism could effectively improve the reliability and usability of the system; and its access control technology can effectively improve system security. The thinking behind such systems is the prototype of trusted computing technology and can be regarded as the early attempts of exploration of trusted computing.

The basic thinking of trusted computing borrows the social management experience and introduces its successful management methods into the computer security system. History tells us that any stable society must have a root of trust as well as a "chain of trust" based on that root. The governing of a nation starts from the root of trust, with the upper level of officials checking the integrity of the lower level and passing trust, then trust can be extended into the whole governing system, and the "Transmitted Trust" is a way by which the reliable national governance system can be established.

Similarly, the common approach to trusted computing is that we need to first of all build a root of trust as the basis and starting point of trust. From that root, a chain of trust can be established, where each level measures and trusts the next level, so that the trust can be extended to the entire system, ensuring that the entire computing environment is trusted [15]. Trusted computing needs to establish a reporting mechanism to notify the properties of the system in real time, while on the other hand, it also needs to provide protection for the reporting mechanism and the content of the report.

4.2.2 Technological Approaches of Trusted Computing

Today, the technological roadmap of trusted computing includes root of trust, trust measurement model and chain of trust, and trusted computing platform [16].

4.2.2.1 Root of Trust

Root of trust is the foundation of trusted computing system. TCG believes that a trusted computing platform must contain three roots of trust: root of trust for measurement (RTM), root of trust for storage (RTS), and root of trust for report (RTR).

1. RTM

The beginning of a chain of trust is called RTM, or core root of trust for measurement. RTM refers to the initial root for measurement or the beginning of the chain of trust, and it is the code that is executed first when booting the platform. It can credibly measure the platform configurations defined by any user [15]. If the RTM is not secure, then the credibility of all following measurements cannot be guaranteed. In PCs, RTM is defined as the first portion of codes to run first, which can be interpreted as an extended BIOS. Such portion of codes is embedded in BIOS chips. No access interface to these codes is allowed nor can it be edited or refreshed.

2. RTS

RTS refers to the storage root key (SRK). In the trusted platform module chips, RTS is the memory of the platform configuration register (RCR) and SRK. It is the basis for trusted storage on the trusted computing platform and can protect all stored root key and data, directly and indirectly. Considering the security of the root key and the cost benefit of the chips, trusted computing platforms usually have strict rules on the storage area and scope of use. SRK is always stored in nonvolatile memory so as to ensure its safety and guarantee its credibility.

3. RTR

In some sense, RTR means the endorsement key (EK). In the trusted platform module chips, RTR includes the PCR and EK and is the basis for platform integrity report. RTR is unique and is responsible for creating the identity for a platform in order to achieve proof of identity and integrity report. It is also responsible for protecting report value and verifying the accuracy of stored data.

4.2.2.2 Trust Measurement Model and Chain of Trust

TCG believes that the condition for establishing a trusted environment is that the integrity of system configuration is not compromised when it's being transferred from the manufacturer to the user. In other words, as long as the device is provided by a legitimate manufacturer and has not been tampered with, it can be trusted. Integrity refers to the extent to which the data or resources can be trusted, which include the content and source of the integrity information as well as the source of integrity data. TCG adopts the integrity mechanism to verify the content and source of system configuration. In order to build a trusted computing environment, TCG starts from the basic level of the system and proposed the key technology of chain of trust based on hardware.

The chain of trust is a mechanism for transferring trust and extending the perimeter of trust. It is based on and starts with the root of trust. Each level measures the next and trusts the next, thereby extending trust to the whole system, creating a trusted computing environment [15].

Generally speaking, creating a chain of trust consists of three parts, namely, measuring the trustworthiness of the platform, storing the measured value, and providing reports in case of visitor inquiry.

1. Measurement

The chain of trust is measured by checking the integrity of the data, which in turn is measured by using hash function to test whether the integrity of the data is compromised. For correct system resource data, its hash value is calculated and stored in secure memory in advance. When the system boots, the hash value of the system resource data will be recalculated and compared to the correct value. If the two values do not match, then we know that the integrity of the data has been compromised.

2. Storage

To save the storage space for trust measurements, an extended approach is adopted to calculate the hash value, where it is recalculated by linking the present value to the new one, and the result is stored in the PCR as the newly measured value of integrity.

It should be noted that the hash value stored in the PCR and the log stored in the disk are linked and attest to each other. The PCR is located inside the TPM chips and is more secure, whereas the log on the disk is less secure. But due to this attestation relationship, any tamper with the log will be detected immediately through the value in the PCR [15].

3. Report

The next piece of the puzzle, after measuring and restoring, is to provide reports upon inquiry so that the visitor can judge the credible state of the platform. The content of the report includes the PCR value and the log. There is also a function called platform remote attestation, which ensures the security of the content by using encryption, digital signature, and authentication technologies.

Figure 4.1 illustrates how a chain of trust works. When a system is powered, the core root of trust for measurement (CRTM) measures the integrity of the BIOS first, which usually means calculating the hash value of the current BIOS code and comparing it with the expected value. When the two figures match, it means that the BIOS has not been changed and is trustworthy. Otherwise, it means that the BIOS has been attacked and its integrity has been compromised. If the BIOS is deemed trustworthy, then the scope of trust will be extended from CRTM to CRTM+BIOS. Next, CRTM+BIOS will go on to measure OSLoader, or operating system loader, which includes the master boot sector, the operating system boot sector, and so on. If OSLoader is trusted, then the borderline of trust will expand further to CRTM+BIOS+OSLoader, while the system will start to load and boot the operating system (OS). If the OS is trusted, then the perimeter of trust expands to include OS (CRTM+BIOS+OSLoader+OS). After the OS boots, it will measure the integrity of the applications, including the pre-run static measurement and the dynamic measurement when running. If the applications are trusted, then the perimeter of trust expands further to cover all the components (CRTM+BIOS+OSLoader+OS+ Applications). When the chain of trust is authenticated, the operating system will load and run the applications [15].

Fig. 4.1 Working process of a chain of trust [15]

4.2.2.3 Trusted Computing Platform (TCP)

The trusted computing platform (TCP) is an entity that sends information to users and receives information from the latter. Trusted platforms are those that have security chip architectures embedded, whereas TCP [15] is a software and hardware entity that can provide trusted computing services and ensure the reliability of the system. The basic idea of TCP is to build a root of trust by using TPM—the core technology of trusted computing—and to build a chain of trust using that root of trust as the starting point and with the help of trusted software. When building the chain of trust, the system expands the bottom-layer trusted relationship to the entire computer system, thereby ensuring that the whole system is trusted.

TPM is the root of trust of the TCP and also one of the key technologies of trusted computing. Compared with ordinary computers, the biggest difference is that the trusted computer has a security module—TPM—embedded in its mainboard. TPM is equipped with the function to perform most security services the TCP needs and provides the basic security services for the platform. At the same time, TPM is also the hardware root of trust of TCP and the starting point of trust for the entire platform. As such, TPM is strictly protected. It has the ability to physically prevent attacks, tampering, and probing, and therefore it can guarantee that TPM and its data will not be attacked. TCG stipulates that authorization is required to run any TPM orders that might affect security, lead to privacy leaks, or expose secrets of the platform. The main structure of TPM is shown in Fig. 4.2.

As Fig. 4.2 shows, the I/O unit manages the coding and decoding of bus protocol and sends messages to different units [16]. The crypto coprocessor is used to cipher, decipher, sign, and verify the signature. TPM uses the RSA asymmetric cryptographic algorithm but also allows elliptic curve cryptography (ECC) or digital signature algorithm (DSA). Calculation of Hash-based message authentication code (HMAC) is achieved using the HMAC engine and follows the RFC 2104 standards. The secure hash algorithm (SHA-1) engine calculates the SHA-1 hash value. Nonvolatile memory is used to store the embedded operating system and file system, as well as important data such as keys, licenses, and identifiers. The key

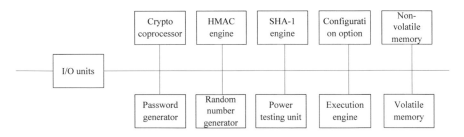

Fig. 4.2 Hardware structure of TPM

generator unit is used to generate RSA key pairs. The random number generator is the built-in source of randomness for TPM. The power testing unit manages the power supply status of TPM and the platform. The execution engine includes the central processing unit (CPU) and embedded software, and it executes received orders by running software. Volatile memory is mainly used for internal operations of TPM.

A trusted computing platform also requires the TCG software stack (TSS) [16] to provide the TPM interface for applications. The purpose of designing TSS is to provide robust software support for trusted hardware platforms and TPM synchronized access. There are three layers to TSS, with TCG device driver library (TDDL) at the bottom, TSS core service (TCS) in the middle, and TSS service provider (TSP) on the top. TDDL and TCS are system processes, and TSP and top-tier applications are user processes. The TSS structure is shown in Fig. 4.3.

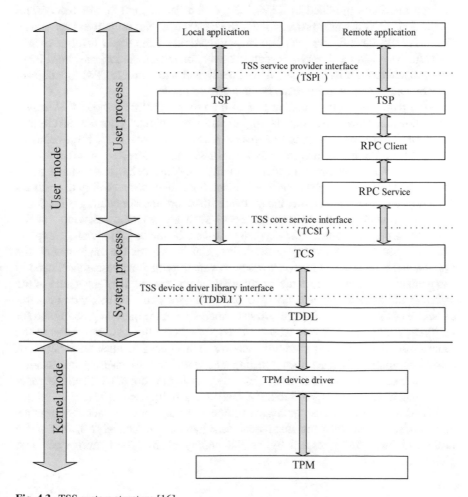

Fig. 4.3 TSS system structure [16]

TDDL [16] is an intermediate module between TCS and TPM device driver (TDD). It is the transitional phase between user status and kernel status and provides an open interface for user mode. It does not serialize TPM orders, nor does it manage the interaction between the thread and TPM. TCS is the system process under the user mode and provides a set of standard platform service interfaces for users. Usually in the form of services, TCS provides services for multiple TSPs and communicates directly with TPM through TDDL. TCS can provide not only all original functions of TPM but also other core function services such as audit management, measurement event management (managing of PCR access and event log input), key and license management (storing and managing keys and licenses related to the platform), context management (to allow TPM thread access), and parameter block generation (responsible for serialization, synchronization, and processing of TPM orders). The applications at a higher level can also directly use TPM functions through interfaces provided by TCS. TSP, which lies at the topmost tier of TSS protocol stack, is the user process under the user mode. Not only does it provide a rich array of object-facing interface for applications at the upper level of trusted platform system but also services such as context management and password functions, so that the applications can more directly and easily utilize TPM functions to construct necessary security features the platform needs.

To sum up, the work process of trusted computing platforms shows that trusted computing emphasizes the predictability of entity behaviors. Under normal circumstances, an entity operates according to an expected behavioral track. If the behavior deviates from the track, then the trustworthiness will be affected. In other words, trusted computing requires knowledge of the expected behavior or status of an entity, rather than viewing the entity as a black box. Take the credibility of software dynamic behavior, for example, the software trustworthiness modeling is the foundation for researching the trustworthiness of software dynamic behavior, and the key is to "properly" describe the expected software behavior [15]. Accurate expression of expectations is the key to modeling, and it involves the environment, the system, and their interaction. To express software expectations, researchers need to carry out profound analysis of and acquire a deep understanding of the nature of the software and its expected behaviors. To learn about the expected behavior, property or nature of the software through software analysis is an important prerequisite for verifying trustworthiness. But who is there to guarantee the trustworthiness of the root of trust itself? Nobody knows the answer to this question when the function of trusted computing changes from boosting trustworthiness to ensuring information security, because users do not believe that the provider of root of trust and the transfer of trust which is managed through society are two different issues.

In addition, if the member or target of a chain of trust contains backdoor or malicious code, can the trust transfer mechanism handle it effectively? This is not a farfetched issue today, especially as the supply chain grows more open and globalized.

4.2.3   New Developments in Trusted Computing

Recent years saw a surge of research findings in trusted computing. In this section, we will briefly introduce a number of typical new trends in trusted technologies, including trusted computing 3.0, Trusted Cloud, and SGX (Intel Software Guard Extensions) architecture [17].

4.2.3.1   Trusted Computing 3.0

Trusted computing research is growing progressively, and its development has gone through several stages. The initial trusted computing 1.0 came from computer reliability, and it is a security protective measure based on fault tolerance, with fault elimination and redundant backups as its main approaches. TPM 1.0 launched by TCG marks the beginning of trusted computing 2.0, which mainly uses hardware chips as the root of trust, using trusted measurement, trusted storage, and trusted report to protect computers. The limitation, though, is that it does not approach security issues from a structural level of the computer system, so it is difficult to engage in proactive defense. Reference literature No. 18 proposes a trusted computing 3.0 strategy that is based on a "computing model of active immune." Based on innovation to platform password options, the document proposes a trusted cryptography module (TCM). It also proposes a trusted platform control module (TPCM) as an autonomous, controllable trusted node where the root of trust is embedded, so that it is activated before CPU in order to run BIOS verification. By embedding the trusted measurement node inside the trusted platform mainboard, a dual node of CPU + TPCM on the host computer is formed. As a result, the chain of trust will start to build the moment the system is powered on. The document proposes a framework of trusted computing supporting software which is a dual system structure of host software system + trusted software base. It also proposes a trusted connection framework that is three-tiered and based on tri-element peer architecture, which increases the overall trustworthiness, security, and manageability of the network connection.

Based upon active immune-based defense approach and the strategic thinking behind trusted computing 3.0, the document [18] further proposes an active defense strategy, which, "based on active immune trusted computing and with access control at its core, establishes an active, three-tiered protective framework under the support of trusted security management center," as shown in Fig. 4.4, the active immune trusted computing technology lies at the core of the strategy, while secure computing environment, secure area perimeters, and secure communication networks form an active system of in-depth defense, which centers around a security management center to coordinate the protection, response, and audit mechanisms at different levels of the defense system.

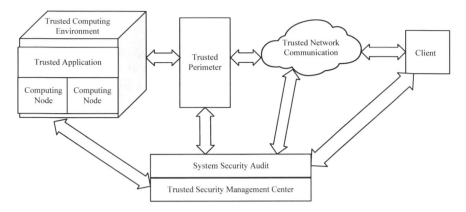

Fig. 4.4 Three-tier protection, active defense framework of active immune [18]

Table 4.1 Defense features of trusted computing 3.0 [19]

Category	Features
Theoretical foundation	Complicated computing; trusted verification
Applicability	Applicable to servers, memory system, client, embedded systems
Security strength	Strong/being able to defend against unknown virus, attacks on unknown vulnerabilities, smart cognition
Protective goal	Unify management of trustworthiness of data processing and system service resources supported by platform strategy
Technological approach	Password as the gene; active recognition, active measurement and active encrypted storage
Area of protection and prevention	Source of action; automatic management of network platform
Cost	Low; enabling trusted nodes within a multi-core processor
Difficulty	Easy to implement; applicable to establishing new systems and adjustment of legacy systems
Impact on business	Do not need to revise original application; real-time protection achieved by developing strategies; keep impact on business function under 3 percent

Reference [19] gives a more comprehensive summary and projection of trusted computing 3.0, stating that its main feature is system immunity and its target of protection is virtual dynamic chain of nodes. The defense features of trusted computing 3.0 (Table 4.1) makes it especially suitable to provide security guarantee for important productive information systems, as it can provide active immune protection to the information system based on a "host + trusted" dual-node trusted immune architecture

It is worth mentioning that trusted computing 3.0 does not touch upon the trustworthiness of the root of trust itself or the issue of self-verification nor does it seem to contribute to solving the problem of supply chain backdoors.

4.2.3.2   Trusted Cloud

In recent years, the trusted computing research findings inspired some researchers to propose architecture designs for trusted cloud [18]. Trusted cloud architecture is a distributed trusted system consisting of root of trust, trusted hardware, and trusted basic software at different nodes—such as cloud security management center, host computer, virtual machine, and cloud perimeter devices—through trusted connection. It supports the security of cloud environment and provides trusted services to cloud users. Generally speaking, trusted cloud architecture needs to be connected to a trusted third party, which provides trusted service recognized jointly by cloud provider and user and performs trusted supervision over cloud environment. Figure 4.5 explains the security framework of a trusted cloud computing system.

In the trusted cloud architecture, due to the varying security mechanism and trust functions of different nodes, the trust functions executed by trusted basic software also vary. Such trust functions coordinate with each other to provide overall trust support functions for the cloud environment. The functions of the security components are as follows:

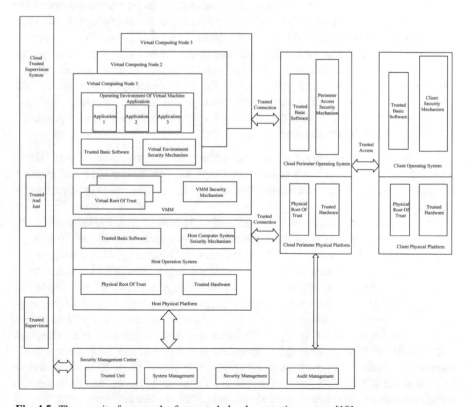

Fig. 4.5 The security framework of a trusted cloud computing system [18]

(1) Security management center. The security management center runs cloud security management applications, including a variety of mechanisms for trusted unit system managements, security management, and audit management. As the management center of the trusted cloud architecture, the trusted basic software is able to monitor security management behaviors and connect with other trusted basic software on nodes of different hosts, thereby achieving security in a systematic way.

(2) Cloud perimeter device. The cloud perimeter devices run a perimeter security access mechanism, which couples with the trusted basic software to provide services, such as trust recognition and verification, and to ensure that the perimeter security access mechanism can be trusted.

(3) Hosts. The trusted basic software support mechanism for hosts needs to both ensure the security of the host machine and the virtual machine monitor and provide virtual root of trust service for the virtual machine at the same time. The active monitoring mechanism of the host security mechanism acts like a trusted server for the cloud. It can receive and localize trusted management policies from the cloud security management center and provide trusted services for the virtual environment based on the trusted policies.

(4) Virtual machine. The trusted basic software on the virtual machine supports its own trust security mechanism and actively monitors the operating environment of cloud applications on the virtual machine. The trusted basic software on the virtual machine, host, and security management center makes up a client-proxy server management center, a triple distributed and trusted cloud architecture.

(5) Trusted third party. Third parties are parties that are recognized by both cloud provider and users, for instance, cloud computing supervision authorities and evaluation centers. The trusted third parties provide attestation and supervision functions for the cloud architecture.

Trusted client. The trusted basic software can be installed on cloud clients, where a trusted computing base can be constructed as well. A client that has been installed the trusted basic software and constructed a trusted computing base is considered a trusted client.

The trusted cloud architecture provides cloud services with trusted computing service functions and security mechanisms in a systematic way, thereby solving a series of security issues caused by an open cloud environment [20]. But because the cloud is a very complicated system and contains a large quantity of status information, if we are to use the trusted computing technology to verify all statuses in the cloud, it would lead to tremendous computing costs and severe damage to the function of the cloud. In addition, since a lot of service modules in the cloud need frequent expansion or upgrading, especially because modules from third-party software and modules of binary executable documents are often not transparent, their trustworthiness cannot be guaranteed, and therefore it is very difficult to establish a root of trust and chain of trust that are completely secure and trustworthy. Even though some researchers have proposed the idea of trusted cloud, due to the abovementioned issues, it is not optimistic about the prospect of real-life application of trusted cloud architecture, especially universal application.

4.2.3.3 SGX Architecture

The concept and principles of SGX were raised by Intel at an ISCA workshop in 2013, and the first generation of CPU that supports SGX technology was introduced in October 2015. SGX is considered a similar technology that rivals the ARM TrustZone. The difference between the two is that TrustZone requires a secure OS and therefore only allows one secure zone, while SGX allows multiple parallel secure regions by guarding only the hardware of the operation environment of the codes without altering the existing Windows development ecosystem, so when compared to the former, it is more advanced.

As Fig. 4.6a shows, SGX can provide a trusted space on the computing platform to ensure the confidentiality and completeness of key user codes and data. SGX is a new type of software architecture as it reduces the trusted computing base (TCB) to include only the CPU and the secure application and supports the exclusion of untrusted, complicated OS and virtual machine monitors (VMMs) from the security perimeter.

Figure 4.6b illustrates a relatively advanced method of key encryption in SGX where a new key (e.g., Seal key or Report key) is generated using the key generation algorithm that combines SGX version key, device key, and user key, and the keys are used to encrypt the code and data of the application that needs to be loaded.

Figure 4.7 demonstrates how an enclave is created [17]. (1) Create a user application. (2) While creating an SGX user application, use the relevant key and SGX key to protect the application and run SGX applications through Hash. (3) When running the application, first load the code and date onto the SGX Loader before they are loaded to the enclave. (4) Loader notifies SGX driver (which is actually a micro kernel OS) to create an enclave in the memory, at which point the tasks in memory would be handed over to the SGX driver. (5) SGX driver creates an enclave in the user's virtual memory according to the size set by the application. The memory segment and page operation in this region are identical to those in a normal operating system. (6) Use the key certificate to decrypt the applications and data that

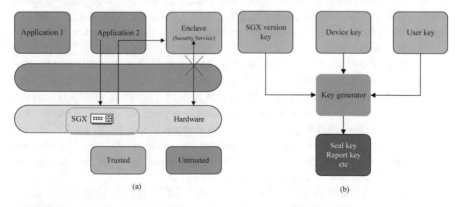

Fig. 4.6 Principles of SGX. (**a**) Schematic diagram of SGX architecture. (**b**) Diagram of SGX key generation

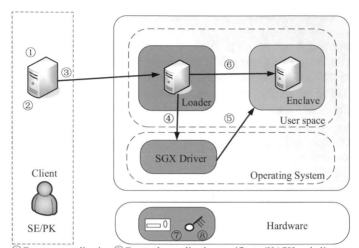

①Create an application ②Create the application certificate (HASH and client
PK) ③Load the application onto the loader ④Create an enclaveAllocate the
enclave page ⑤Load the test application ⑥Verify the completeness of the
certificate and enclave ⑦Generate the enclave key

Fig. 4.7   Process of creating an enclave

need to be loaded in the form of EPC (Enclave Page Cache). (7) Prove that the
decrypted applications and data are trusted through the SGX hash and load them
into the enclave. (8) Start the enclave initialization application, forbid the continuation of EPC loading and verification, generate and encrypt the enclave key, and save
it in the thread control structure (TCS) of the enclave as its tag, until now the enclave
is completely protected by SGX.

The SGX architecture does not recognize and shield all malware on the platform.
Instead, it follows the idea of trusted computing to seal all secure operations of
legitimate software in an independent storage space, so as to protect them against
malware attacks. None of the other software could access this independent space,
whether it has authorization or not. In other words, once the software and data are
placed within the independent storage space, not even the operating system nor the
VMM could affect that code and data. The security perimeter of the independent
storage space contains only the CPU and the space itself; therefore it could be seen
as a trusted execution environment (TEE) or trusted space. This is slightly different
from trusted zone (TZ) which usually divides the CPU into two shielded environments—the secure world and the normal world—which communicate with each
other through self-modifying code (SMC). In comparison, in SGX, a CPU can run
multiple independent and secure storage spaces and is able to run them concurrently,
although similar results can be achieved by carrying out multiple shielded secure
tasks within the secure spaces of a trusted zone [21].

Software developers responded positively to SGX. A typical application architecture is Haven of Microsoft [22], which is the first of its kind to achieve isolated

operation of unmodified legacy applications on a commercial OS and hardware. Haven leverages the hardware protection of Intel SGX to defend against privileged codes and physical attacks like memory probes, protecting a variety of service systems or software, such as SQL Server and Apache Web, from a malicious host [23].

SGX could effectively improve the sandbox mechanism and also substantially improve the security of hypervisor, for instance, by eliminating memory analysis attacks. However, security risks still remain for SGX.

Risk 1: how to guarantee the trustworthiness of the CPU, one of the two security perimeters, in particular, how to prove that the hardware functions in the controlled independent storage space are free from vulnerability. It still did not solve the issue of trustworthiness of the root of trust itself.

Risk 2: if the shielded applications carry design flaws or malicious codes, then they could avoid security monitoring under the protection of the shield.

Risk 3: once the malware enters the enclave, the entire function of SGX would be used by malware developers. In 2017, five researchers from Graz University of Technology in Austria published a paper, claiming that the malware they developed could defeat SGX. They built an SGX enclave that coexisted with other SGX but was embedded with malicious codes, which enabled them to access confidential information stored in other enclaves through side-channel attacks. What is worse, such attacks were protected by SGX mechanisms.

4.3 Tailored Trustworthy Spaces

To avoid the security threats brought about by the homogenization of cyberspace, it is necessary to adopt the idea of Tailored Trustworthy Spaces (TTS) in order to increase differentiation of applications, certain opaque elements, or privacy. But customization does not mean isolation. The basic components, units, and systems should remain open, with only certain tailored features being kept private to strengthen the defense line and increase the threshold of attacks. The author believes that communication, computing, and security are the three fundamental technologies in trusted cyberspace. Recent years have seen quite a number of practical technologies in the three areas, which address different needs. For instance, Integrated Services (IntServ) [24] address the need of operators to provide differentiated services. Cloud computing mainly targets the on-demand services. Heterogeneous computing helps accelerate different computing components in different applications, while the trusted computing technology mainly addresses the need to strengthen reliability and the security need of cyberspace. Even though these technologies grew out of different commercial needs, their wide deployment provides the physical or technological foundation for adopting a TTS strategy. Next, we will discuss the foundations and main features of TTS.

4.3.1 Preconditions

4.3.1.1 Communication

It has always been the main goal of Internet operators to provide differentiated services. Although the primary motive is to provide segmented services and differentiated resources, it also forms the base for the level of diversity and redundancy required by the new type of defense technologies. At the beginning, the Internet provides the best-effort services, so users are equal. However, the development of new IP network applications is exposing technical flaws of the traditional IP network service models, one of which is the lack of service differentiation. As a result, IntServ was born, where all the intermediate systems and resources would explicitly provide preordered services for communication flows. IntServ uses Resource Reservation Protocol (RSVP) [25] to reserve network resources in order to ensure that the network could satisfy the specific needs of the communication flow. RSVP is required on any network devices along the route between the two ends. Here is how it works: before transmitting the data, the beginning node would request a specific service from the network, provide the flow specs to all the intermediate nodes, and request services that meet the bandwidth and delay requirements. Upon confirmation from the network, the data will be transmitted. Later on, differentiated services, or DiffServ [26], are proposed to meet the extendibility, robustness, and simplicity requirements of network applications. DiffServ categorizes communication flows into groups of varying efficiency according to service requests, so that some communication flows are prioritized. These network resource allocation technologies paved the way for TTS—a new type of defense strategy.

4.3.1.2 Computing

As the new computing platforms or models, such as heterogeneous computing, customizable computing, and cloud computing, become a research hotspot, they provide the physical foundation for TTS.

Heterogeneous computing is a computing method that uses different types of processing units, such as CPU, NPU (network processing unit), TPU (tensor processing unit), FPGA (field-programmable gate array), and ASIC (Application-Specific Integrated Circuits), to accelerate the operation of applications. It has two important directions:

1. It fully utilizes the diversity of processing units and pairs the most appropriate processing unit with the tasks.

The general-purpose processor (GPP), by and large, has different efficacies when handling different workloads. That is why there are many specialized or specific processors, such as DSP (digital signal processing), GPU (graphic processing unit), NPU (network processing unit), and FPGA, which are more adaptable and flexible.

If the diversity of processing units could be utilized in the right way at the right time, then the processing capability and efficacy of the information system would be greatly improved. There are many researches and practices in this field. One typical example is the design of ASMP (asynchronous symmetric multiprocessor), as represented by Qualcomm Snapdragon mobile processors. IBM, Intel, AMD, and National University of Defense Technologies in China have generalized the mobile computing via applying CPU + GPU or CPU + FPGA to GPPs or high-functioning computers. They are planning to move on to the more advanced structure by combining CPU, GPU, and FPGA.

2. It makes it easier for developers to use a multiprocessor environment.

The GPP has developed many advanced programming languages and application tools, whereas the GUDA helped to popularize the GPU. Even hardware programing languages for FPGA, such as Verilog and VHD (Very-High-Speed Integrated Circuit Hardware Description Language), are also becoming more popular and common. Especially with the rise of heterogeneous computing, people are exploring the possibilities of integrating all the different programming languages in order to use them in a diverse processing environment. For instance, a Java-based integrated language already exists for data processing.

Heterogeneous parallel computing in distributed environment has been extended academically. In 2013, Chinese scientists proposed the idea of mimic computing (MC, see Chap. 9), which in principle has more advanced application forms and more flexible structural features than heterogeneous computing.

While cyber attacks strongly depend on the level of regularity of the target environment, the inherent randomness, dynamism, uncertainty, and functional equivalence of the heterogeneous computing application system, as well as the introduced non-public characteristic, could have an equivalent impact on that dependency. But mimic computing focuses more on the diversity of algorithms under the functional equivalence; therefore the seemingly disordered choice of algorithms could create a stronger "fog of defense."

In 1974, Dennard et al. [27] came up with the famous Dennard's scaling law, which states that, as transistors get smaller, their power density stays constant, so that the power use stays in proportion with area. However, this law failed at the beginning of the twenty-first century, and improving computing efficacy became the focus of academia and the industry. Given the huge gap of efficacy between GPPs and ASICs, customizable computing [28] aims to design a load-adaptive computing architecture that combines the two, namely, a computing architecture that customizes computing models based on user-specific load type. Researches into customizable architecture of high-efficacy and energy-efficiency include customizable processor core and accelerator, customizable on-chip memory, and interconnect optimization. While the intention of customizable computing is to increase efficacy, it also introduces heterogeneity at the computing level. Therefore it not only meets the needs of different users for specific computing models but ensures that the target computing structure or processing environment is no longer static or certain.

Cloud computing [29] is seen as the fourth IT industrial revolution following the mainframe computer, personal computer, and the Internet. As a new computing model, cloud computing is an Internet-based IT environment that provides on-demand measurable computing resources, such as computer networks, servers, memory, application, and services. The five major characteristics of cloud computing as summarized by the National Institute of Standards and Technology of the USA (NIST) are on-demand self-service, broad network access, resource pooling, rapid elasticity, and measured service [30]. Cloud computing offers a new model of IT infrastructure and platform service. It can provide on-demand customizable services to users based on a uniform heterogeneous resource pool, thereby allocating resources according to need.

There are three features of cloud computing:

(1) Low cost. Cloud computing provides users with scalable resources through resource virtualization and supports the simultaneous running of different applications. It also improves the overall resource utilization through resource planning and dynamic migration across virtual machines, physical machines, and data centers, for the purpose of lowering the cost.
(2) High reliability. Cloud computing is highly reliable because it can store multiple copies of user data, so damage to any physical machine would not lead to data loss. Its multiple data centers also help prepare for disasters.
(3) High return on security measures. Because the targets are concentrated, the security measures in theory can be shared to the maximum degree, and the high volume of services also means effective amortization of security investment. These advantages of cloud computing are supports for customizable applications at the infrastructure level.

4.3.1.3   Security

In addition to communication and computing, the trusted computing technology provides the foundation for TTS from the security perspective, as the technology provides security augmentation to TTS and effectively guarantees the security of TTS. The technology has been detailed in Sect. 4.2.

4.3.1.4   Summary

As described above, the technological advancement in communication, computing, and security have prepared the foundation for developing TTS capacity. In fact, researches into on-demand cybersecurity through customizable mechanisms are gradually receiving more attention in recent years, such as customizing network subspace based on users' needs, or cloud-based dynamic secure service function chain, which provide customized secure service chains to cloud tenants or enterprise networks by deploying security devices in the cloud. Such researches usually achieve customized transmission based on redundant network resources or establish secure spaces based on special scenarios. Their purpose is to further explore the technological path and economic way of customization of key components or tailored architecture differentiation against the backdrop of the popular Internet environment and open-source software.

4.3.2   Tailored Trustworthy Spaces (TTS)

Tailored Trustworthy Spaces (TTS) aim to create a flexible, distributed trustworthy environment to support activities in the network and back the multi-dimension management capabilities, including confidentiality, anonymity, data and system integrity, provenance, availability, and performance. There are three layers to the purpose of TTS.

(1) Achieve trusted computing in a non-trusted environment.
(2) Develop a general-purpose framework to provide different types of network behaviors and events with a variety of trustworthy space policies and trusted services in specific context.
(3) Develop rules of trustworthiness, measurable indicators, and flexible and trustworthy consultation tools; deploy decision support capabilities and trusted analysis that can execute notifications.

In the real world, people switch from space to space, such as home, school, workplace, supermarket, clinic, bank, or theater. Each space comes with specific functionalities and rules of behavior, and people have to obey the rules in order to enjoy the services provided in these spaces. For instance, people can watch movies at the theater, but they must not speak loudly. In other words, specific behavior and restrictions apply only to a specific space.

The cyberspace is a virtual space created by mankind. A lot of activities are happening in this space, e.g., chatting, video, shopping, and gaming, for instance. Logically, the areas for such activities are virtual subspaces. Therefore we can see cyberspace as a flexible, distributed, and trusted environment that can provide customized support for functions, policies, and trusted needs when faced with different variable threats. Users choose different subspaces for different activities in order to acquire different trust dimensions, such as trustworthiness, anonymity, data integrity, performance, and availability. They can also create a new environment through negotiation with each other and customize the agreed-upon features and time.

At present, the research progress in TTS is mainly focused on features, trust negotiation, operation set, and privacy.

4.3.2.1   Features Research

Features research in the field of TTS focuses on space description, how to translate advanced management needs into execution policy, how to define customized requests, and how to translate the customized requests into executable rules. The US

National Science Foundation (NSF) funded a research program at Carnegie Mellon University on definition and execution of privacy policy semantics. Take healthcare in cyberspace, for example, the reasonable requirement for keeping healthcare records private is different from the traditional secure access to computers. First of all, the protection policy requires not only the protection of privacy of such information when it is in use but also restrictions on future access to data, so as to avoid leaking sensitive information. Secondly, the policy may change according to status. Therefore, we need to explore how to establish appropriate strategies and corresponding execution mechanisms.

4.3.2.2   Trust Negotiation

The field of trust negotiation mainly researches the policy-based framework, methods, and technologies for building trust between different system components. The policy must be clear and unambiguous, and consist of dynamic, artificially intelligible, and machine-readable orders. This requires that the trust level of specific security properties be adjustable, for instance, building anonymous, lower-level or high trust level TTS. In the future, dynamical TTS would also require customization of spaces that address different threat scenarios.

4.3.2.3   Set of Operations

Dynamic TTS contains a large quantity of necessary instructions or operations, such as joining, dynamic tailoring, splitting, merging, and dismantling. Such operations could support the customization function of trusted spaces. In 2012, the US National Science Foundation funded a security and trustworthy cyberspace program, which focused on researches in two aspects: fundamental technology that makes systems customizable and the development of TTS applications that target at specific environment [31]. The former mainly studies how cyber defense systems can adaptively learn "normal" behaviors, while the latter deals with achieving a trustworthy, reliable communication mechanism through untrustworthy nodes.

4.3.2.4   Privacy

To design a reasonable model for privacy protection in an open cyberspace has many challenges in terms of rapid model innovation and technology advancement, to name a few. The US government is heavily supporting researches on technologies to protect the privacy in cyberspace. As the framework for tailored cyberspace, TTS implements granular control on the environmental features and sets anticipated security and privacy goals. Players are able to establish a trusted interaction context by customizing environmental features and developing policies for the data and activities in the TTS. The ability of customization directly supports the desired conditions of privacy.

4.4 Mobile Target Defense

Moving target defense (MTD) is a set of theories and methods that target the current information system defense flaws to make the system more dynamic, random, and diverse. Its central idea is to build a dynamic, heterogeneous, uncertain cyberspace target environment so as to increase the difficulty for attacks, using the randomness and unpredictable nature of the system to defend against cyber attacks.

MTD technology can be applied at multiple levels like the network, platform, operating system, software, and data. Some examples include IP address hopping, port hopping, dynamic routing, and IPSec Protocol channel, randomization of network & host identity, and of executing code, address space, instructions, and of the form of data storage. Figure 4.8 illustrates the structure of MTD technology.

MTD uses manageable means to change the static, certain, and similar parts or links within the system so as to increase the difficulty of attacks that are based on unknown system vulnerabilities. The attacker must be familiar with the dynamic nature of the target system so as to establish a chain of attack using all the system vulnerabilities. While the attacker is trying to learn about the dynamic nature to establish attacks using methods such as exhaustive parameters, the system will have changed its environment again, making the attacker's knowledge about the cyberspace outdated, thus undermining its attack. MTD is a type of external or shell technique that limits the target of protection to external attacks based on unknown vulnerabilities. Its central idea is to actively add noises in a controllable way to relatively static contents or positions such as order program, data files, memory address, and even ports. It can be understood as a general encryption technology that targets at key environmental parameters to achieve anticipated effects of defense. In a

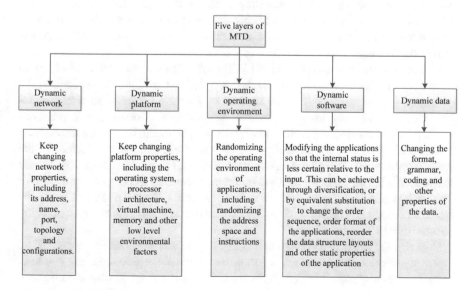

Fig. 4.8 MTD structure

broad sense, it is to introduce broad but controllable dynamism and uncertainty to the network, platform, and system, thereby increasing the difficulty of attacks that are unauthorized and based on unknown vulnerabilities. It should be stressed that in theory, MTD is not effective in defending against coordinated attacks or authorized attacks based on backdoors or Trojans.

As MTD works by introducing randomness, dynamism, and diversity in the system, and dynamizes system parameters to decrease the attack surface, the pursuit of speed of change and sufficiency of entropy space is the key to a strong defense. Recent research findings and events have given us clues and examples to breaking the MTD, which means we must reflect on the pros and cons of MTD and magnify its advantages while eliminating flaws and disadvantages.

4.4.1 MTD Mechanism

4.4.1.1 Randomization

The randomization mechanism of MTD technology randomly scrambles or encrypts important executable programs or data so that they are decrypted before being executed, thereby defending against injection tempering or scan attacks from the outside. This mechanism adds uncertainty to the system and turns the defined and certain system parameters into the form of probabilities, which defies malicious control and manipulation.

The most successful MTD technology is probably the Address Space Layout Randomization, or ASLR, which is widely adopted by commercial operating systems, including most modern operating systems such as Mac OS X, Ubuntu, Windows Vista, and Windows 7. Its effective defense mechanism is that by randomizing memory address, the location-based attacks would be unable to locate the correct code or data segment and therefore disabled.

Instruction set randomization (ISR) [32] uses ambiguous target instructions to defend against code injection attacks. By generating random byte sequences, the system runs an exclusive OR on the random bytes that correspond to each instruction when loading code texts. In practice, the program is loaded into a simulator which recovers the original instruction by running an exclusive OR on the instruction and the random bytes. Before being executed, the code injected to the application/program will go through an exclusive OR against the random bytes used during loading, but the code injected by the attacker into the target application or program's vulnerabilities cannot be recovered into a normal instruction, so the attack will not only fail to achieve its expected effect but run the risk of being detected due to instruction error.

Data storage can also be randomized [33]. The idea is to run an Exclusive OR on a memory pointer and a random key in order to prevent attacks caused by pointer error. The basic approach is that Exclusive OR is required both when inputting the pointer value into the register and when accessing the pointer. Without knowing the

Exclusive OR key, it is impossible to launch an effectively attack by tempering with the pointer. There is another data randomization approach that disguises data in a specific way based on the type of memory. If the attacker tries to determine the memory area related to a specific target through static analysis of the program, his attempt to write in an external target in a specific memory area will be stopped because of the randomization disguise.

Although the randomization mechanism is capable of preventing many types of attacks, as mentioned before, due to limited information entropy, it is still subject to violent attacks and probe attacks. Besides, any randomization mechanism requires a trusted CPU function. Unfortunately, a 2017 report on the "Attack Technology That Breaks ASLR" (https://www.vusec.net/projects/anc) details how to break the ASLR through interaction between the memory management unit (MMU) and the page tables.

4.4.1.2   Diversification Mechanism

The approach of the diversification mechanism is to change the similarity of target systems, so that attackers will not be able to simply duplicate successful attacks onto similar targets. It helps create an ecosystem with a certain degree of resistance, hoping to automatically generate applications or system variants that could change the system performance. The changes are intended to protect the basic semantics of the original application input in a normal way. But when input in a malicious way, the results will be different.

Researchers commonly use the principle of compiler optimization to increase the level of differentiation between executable files (length, partial algorithm structure, operation patterns, etc.) while maintaining the function by adjusting parameters to generate executable files with multiple variants during the process of compiling source application into executable application. What's more, the diversification mechanism is widely used in other aspects. Reference literature [34] shows that risks of homogeneous vulnerabilities being used by attackers can be lowered through the deployment heterogeneous components equivalent in function. However, it should not be expected that, in principle, diversified compilation can eliminate backdoor or malicious codes in the source file.

4.4.1.3   Dynamic Mechanism

Randomization and diversification bring uncertainty to the target system, but, as keys need to be changed regularly, long-running server processes also need to be randomized and diversified repeatedly to ensure the security of the system. To this end, MTD introduced a dynamic mechanism to change its static nature, which further improves the level of security by breaking the limitation of static randomization and diversification.

If the attack surface changes fast enough, then the dynamic defense can still protect the system more effectively than a static system, even with low entropy or when faced with probe attacks. But it can be anticipated that the dynamic mechanism will bring unneglectable additional loads to lower its usability. As the frequency of change increases, the performance of the system will drop dramatically. Just as the one-time pad is too idealistic for an idea, it is difficult for the system to change each time. Therefore, the cost efficiency of dynamic mechanism will be an area of focus. In recent years, some researchers have tried to apply the game theory to this field, using the balance optimization model on the system's usability, security, and cost, in order to design the best dynamic defense strategy [35–37]. Since the model is rather simple at this time, a satisfying option of optimal dynamic mechanism is still not available.

4.4.1.4   Symbiotic Mechanism

The Symbiotic Embedded Machines (SEM) [38] is a defense mechanism based on the host machine. The defense coordination of symbionts is a natural phenomenon which usually refers to the short- or long-term dependency between different populations. Such symbiotic relationship strengthens the chance of survival or adaptation of one or more species. When two or more organisms respond to an emergency, the results are usually mutually beneficial. In the field of information technology, the SEM can be seen as a digital life-form or organism that closely coexists and forms coordinated defense mechanisms with any executable files. It takes computing resources from the host while at the same time protects the host from being attacked. What's more, the SEM is diverse by nature and can provide natural protection against any direct attacks on the host's defense system on the basis of defense coordination.

In principle, a coordinated defense entity is formed by the SEM and the host application. Every instance of the host applications is embedded with an SEM which is autonomous and unique. The SEM can stay anywhere in the application regardless of the location of the application in the system stack. The SEM is injected into the host in different ways and randomly generates code variants through a polymorphic engine. The combination of SEM and the host creates a unique executable application, forming a redundant moving target to change the static, certain, and similar nature of the system.

4.4.2   Roadmap and Challenges of MTD

There are three milestones in the MTD development plan: creation, evaluation/analysis, and deployment. And each stage has its short-term, midterm, and long-term goals. As Fig. 4.9 shows, it is now at the creation, evaluation, and deployment stage of the midterm goal.

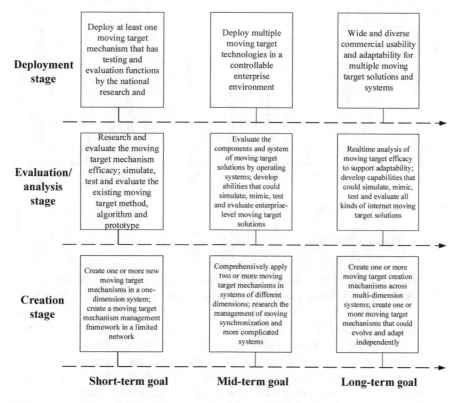

Fig. 4.9 Short-term, midterm, and long-term goals of MTD

Propelled by a series of strategic planning, the US government, companies, and academia have joined forces, formed a clear direction for developing this technology, and rolled out a number of research programs. The MTD technology is quickly gaining momentum. Figure 4.10 shows some of these programs.

MTD still faces many challenges, such as additional overhead that affects service performance, security problems caused by isomorphism, and the issues on how to manage dynamic network and system architecture, as well as how to scientifically proof the efficiency of MTD. But the most deadly challenge, without doubt, is that it is unable to handle backdoor-based unknown attacks.

4.5   Blockchain

This section briefly introduces the concepts, features, key technologies, or algorithms of blockchain and also discusses its advantages and disadvantages.

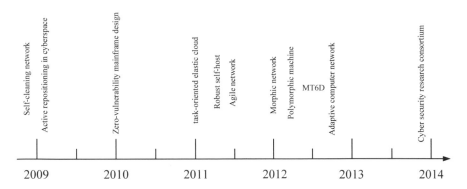

Fig. 4.10 Milestones of MTD development

4.5.1 Basic Concept

Blockchain is a data exchange, processing, and storage technology involving multiple parties. It integrates different technologies including the modern cryptography, consensus mechanism, P2P protocol, distributed architecture, authentication, and smart contract. It stores data using time sequence-based chain of blocks. The consensus mechanism guarantees consistency of data across different nodes, while cryptography ensures the security of data storage and transmission. Using automated scripts to create smart contracts enables automatic data judgment and processing, solving the common problems in a centralized model, namely, problems like low security, poor reliability, and high cost. In other words, blockchain is a distributed database that is consistent and achieves Byzantine fault tolerance. In terms of data structure, it is a chain of data blocks based on time sequence. From the perspective of node topology, all of its nodes are redundant backups, while in terms of operation, it manages accounts using the cryptography-based public and private key management system.

From the participants' perspective, there are three types of blockchains: public blockchains, consortium blockchains, and private blockchains.

1. Public Blockchain

Public blockchains are open to the public. Users can participate anonymously without registering and have access to the network and blockchains without authorization. The blocks on a public blockchain can be read by anyone in the world, who can send a transaction on a public blockchain, and can participate in the consensus process online. The public blockchain is a completely decentralized blockchain in the real sense. Cryptography ensures that the transaction cannot be revised, while encrypted authentication and economic incentives (namely, encrypted digital economy) help build consensus among strangers in the network, thereby forming a completely decentralized credit mechanism. The public blockchain combines economic incentives with encrypted digital authentication, using proof of work (PoW) or proof-of-stake (PoS) as its consensus mechanism. A user's influence on the consensus directly depends on their proportion of resources in the network. Bitcoin and Ethereum are both typical examples of public blockchains and are generally applied to scenarios such as virtual currency, e-commerce, and Internet finance (B2C, C2B, C2C).

2. Consortium Blockchain

The consortium blockchain is a type of blockchain jointly managed by several organizations. Each organization runs one or more nodes, and only members of the consortium are allowed to read or write the data or send transactions, and they all participate in recording transaction data. R3, a blockchain consortium participated by many banks, and Hyperledger, program backed by Linux Foundation, are both consortium blockchains. The consortium blockchain requires registration and authorization. The consensus process is controlled by pre-selected nodes and is applicable to B2B, such as transactions, settlement, or clearance between organizations. The openness of a consortium blockchain can be decided according to specific scenarios. Due to relatively few nodes engaged in consensus forming, consortium blockchains usually do not use the PoW mining mechanism, but rather use PoS or consensus algorithms such as PBFT (Practical Byzantine Fault Tolerance) or RAFT. The consortium blockchain has transaction confirmation times and transaction volumes per second different from a public blockchain, and its requirements on security and performance are also higher. A consortium blockchain network is jointly maintained by member organizations and accessed through the member organization's gateway node. A consortium blockchain platform should provide security management functions, such as membership management, authentication, authorization, monitoring, and auditing.

3. Private Blockchain

The private blockchain is only open to a private organization, which has the authority to decide who has access to read and write on a blockchain or participate in record keeping. The blockchain can be open to the public or restricted to varying degrees. It's usually applied in corporations or internal departments of the government to database management or data auditing, for example. A private blockchain provides a secure, traceable, unalterable, and autorun computing platform while preventing security attacks from both internal and external sources. Compared with centralized database, the private blockchain could prevent any single node inside the organization from deliberately omitting or altering data. Even in case of error, the source can be identified quickly. Therefore many large financial organizations today prefer to use the private blockchain technology.

4.5.2   Core Technologies

Blockchain is the combination of many innovative technologies: cryptography, consensus mechanism, and smart contract.

1. Cryptography

In order to ensure the security and integrity of information on the blockchain, there are a lot of modern cryptographic technologies in the definition and structure of blocks and blockchains, such as cryptographic hash function and elliptic curve public key cryptography. At the same time, these cryptographic technologies are also used to design PoW-based consensus algorithms and to identify users. A blockchain does not usually save the raw data or transaction records directly, but rather saves its hash value. For instance, the bitcoin blockchain uses a double SHA256 hash function, which means, the raw data of any length are run through SHA256 hash function twice and converted to a 256-bit binary number for storage and identification. The blockchain uses the public key system for digital signature, and the bitcoin blockchain uses the elliptic curve public key system.

2. Consensus mechanism

In a distributed system, there are multiple host machines that form a network cluster through asynchronous communication. Status replication between host machines is needed in order to ensure that a status consensus is formed. However, in an asynchronous system, there may be a faulty host that cannot communicate, the performance of a host machine may drop, or the network may become congested, causing the spread of wrong information in the system. Therefore it is necessary to define a fault tolerance agreement in a network deemed unrealizable by default, in order to ensure all host machines can reach a secure and reliable status consensus. The blockchain structure is a distributed structure. The three types of blockchains (public, consortium, and private) correspond with the three distributed structures: decentralized distributed system, partially decentralized distributed system, and weak center distributed system. To ensure the consistency and accuracy of data, blockchain adopts the consensus algorithm in distributed systems to select a ledger node and ensure accurate and consistent consensus over the ledger data throughout the network.

The consensus algorithm for distributed systems was invented in the 1980s. It was intended to solve the Byzantine generals problem, including a variety of consensus agreements or algorithms: "status machine" Byzantine agreement, Practical Byzantine Fault Tolerance agreement, and RAFT, of which the Practical Byzantine Fault Tolerance agreement and RAFT are consensus algorithms for consortium and private blockchains. The consensus mechanism of public blockchains is usually PoW and PoS, which will not be elaborated here.

3. Smart contract

The smart contract is a type of contract that uses a computer language instead of a legal language. It can be autorun by a computer system. If the blockchain is a database, then smart contract is the application layer that enables blockchain technology to be applied to real life. The contract in the traditional sense usually does not have any direct connection with the computer code used to execute the content of a contract. Paper contracts are mostly used for archiving purposes, while software can execute contract provisions written in computer codes. The potential benefit of smart contract is that it can help lower the cost of contract signing, execution, and supervision. Therefore, for a lot of low-value transactions, smart contract can greatly reduce the cost of manpower.

4.5.3 Analysis of Blockchain Security

Based on Byzantine fault tolerance technologies, a blockchain can solve the problem of reliable information transmission on unreliable networks. Independent of authentication and management through a centralized node, it can prevent risks of data leakage or authentication failure caused by attacks on the central node. The algorithm and data structure on which a blockchain is based have three advantages over conventional cybersecurity defense approaches.

(1) Decentralized trust mechanism. The traditional network uses central authentication (CA) for user identification, and therefore the security of the system relies entirely on the centralized CA center and the relevant staff. If the CA is under attack, then all user data can be stolen or altered. But a blockchain uses consensus mechanism and does not rely on any third-party trust platform, and the written-in data need to be recognized by the majority of nodes in order to be recorded. Therefore, attackers need to control at least 51 percent of the nodes so as to forge or alter data. This will significantly increase the cost and difficulty of attacks.
(2) Significantly increased cost of data manipulation. A blockchain adopts a time-stamped chain of blocks to store data, which adds the dimension of time, making the data record verifiable and traceable. Any changes to any information in any particular block will cause the data on all the blocks beyond that point to be changed, which greatly increases the difficulty of data manipulation.
(3) Distributed denial of service, or DDoS. The nodes on the blockchain are dispersed. Each of the nodes contains the complete blockchain information and can authenticate the validity of data on other nodes. Therefore DDoS attacks on blockchains are more difficult. Even if one node is broken, the rest of the nodes can maintain the functioning of the entire blockchain system.

Given the above advantages, it is worth researching to apply a blockchain in the field of cybersecurity. ODIN (Open Data Index Name) [39] can be taken as an example, where the blockchain is to replace the DNS.

As researches and applications of the blockchain technology grow, it becomes clear that a blockchain also faces the same risk of vulnerability and backdoor attacks like any information system. More specifically, the risk lies in three aspects. Firstly, even though the blockchain combines different cryptographic technologies to ensure security, as a high-density program, there may be unknown vulnerabilities to the realization of crypto algorithms, which threatens the foundation of blockchain technology. Secondly, there exist the security risks of consensus mechanisms.

A blockchain is the application of Byzantine fault tolerance technology and designs and adopts a lot of consensus algorithm mechanisms, such as PoW, PoS, or DPoS, but it requires more rigorous scientific proof over time to determine whether these mechanisms could truly bring and guarantee security. Lastly, when applying the blockchain technology, the foundation of its irreversibility and unforgeability is the private key, which is generated and kept by the users. But once the hackers were able to obtain the private key, they will be able to access all the data.

It should be emphasized that the software and hardware products currently taking up the majority of market shares have vulnerabilities and backdoors, which pose great challenges to the secure application of blockchain technology. For instance, x86CPU and Windows operating systems account for at least 80 percent of the PC market, Android by Google takes up over 80 percent of the smart phone market, while operating systems like Linux also take up the lion's share of the server market. But none of the available mechanisms—consensus, cryptography, or timestamp—seem capable enough to block attacks on the vulnerabilities and backdoors of these products.

Generally speaking, from the security perspective, blockchain faces challenges in terms of algorithm realization, consensus mechanism, and application. Hackers can take advantage of its security vulnerabilities and design flaws to launch attacks.

4.6 Zero Trust Security Model

An inherent problem with the information system is that people are too generous with their trust of the Internet. Too many devices have free access through default connection, which makes it possible for people to share any information anywhere and anytime. This is the cause for the Internet's rapid development, but it is also the crux of the issue because "if you believe everything, you stand no chance of protecting anything." What's more, the IT- or ICT-based modern corporate management and production organization industry has evolved to a stage where some enterprise applications are located in the headquarters and some in the cloud, connecting employees, partners, and customers all over the world. The traditional firewall-based perimeter security protection can no longer support such applications that blur the boundary of intranet and Internet. Perimeter rules of microsegmentation and fine granularity are needed to determine whether user/host/application requests for software/hardware resources within specific areas of the company can be trusted. In other words, we need to shift from the regional defense model to the garrison defense model, establish a new security framework that is based on network logical perimeter or sensitive resource protection, and use comprehensive identity verification as a means to enable strict process management over people, terminals, and systems, such as identity verification, access control, real-time tracking, behavior fitting, and channel encryption, so as to adapt to the shift toward the "cloud-network-terminal" model. This is the idea behind Zero Trust Security Model [39]. It should be stressed that the "Zero Trust Architecture" security framework is not so much a technical solution but a network deployment plan of resources, because it does not involve any specific security technologies.

4.6.1   Basic Concept

The Zero Trust Security Model has been popularized in the industry ever since it was created in 2010 by John Kindervag at Forrester Research Inc [40]. However, there has been very little research on Zero Trust Security—perhaps because in essence it is nothing more than a new network usage and protection deployment model and does not involve any new science and technological issues. In 2017, Evan Gilman and Doug Barth wrote *Zero Trust Networks: Building Secure Systems in Untrusted Networks*, in which they described in detail the history and principles of the Zero Trust Model [41], providing important reference materials, which are for anyone who wants to learn about this model.

The creation of the Zero Trust Model is to address the flaws of the traditional perimeter security model. It is based on the following five assertions:

1. The network is always insecure.
2. External and internal threats always exist.
3. The location of the network is not enough to determine its trustworthiness.
4. Every device, user, and network flow should be verified or authenticated.
5. All the policies need to be updated in a dynamic way, taking full advantage of the diverse nature of data.

Figure 4.11 is a Zero Trust Model Architecture based on those assertions.

The system it supports is called a control plane, while the rest is called a data plane. The control plane is responsible for configuration and coordination. A user initiates an access request to privileged software/hardware resources through the control plane, which, in response, verifies and authenticates the accessed device and user and makes decisions based on the nature of the request. After the request is approved, the control plane will dynamically configure the data plane so as to receive data from the customer and establish encrypted channels between the request initiator and the resources to ensure security of follow-up communications. A control plane usually includes the following content, as shown in Fig. 4.12:

- Identity provider: to track user and user information
- Device directory: to maintain the list of devices that have access to internal resources and corresponding device information (such as device type and integrity)
- Policy evaluation service: to determine whether a user or device meets the tactic requirements proposed by the security manager
- Access proxy: use the abovementioned signals to authorize or deny access requests

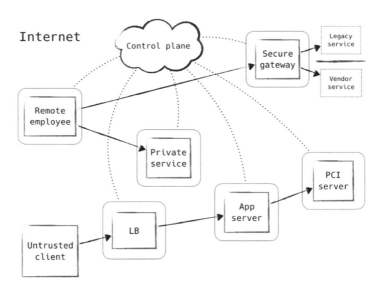

Fig. 4.11 A general Zero Trust Model Architecture

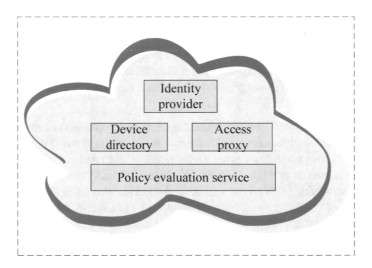

Fig. 4.12 Main content of a control plane

4.6.2 Forrrester's Zero Trust Security Framework

Created by a Forrester analyst John Kindervag in 2010, the Zero Trust Network Architecture is characterized by segmentation, parallelization, and centralization. Segmentation means that the network is divided into segments that are easier to manage, parallelization means establishing multiple exchange centers that are parallel to each other, and centralization means centralized control from a single control platform. Based on these requirements, the Forrester Security Framework proposed

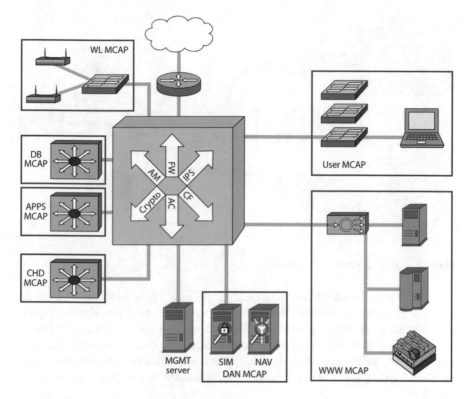

Fig. 4.13 Forrester Zero Trust Security Framework

the Zero Trust Network Architectural component as the microcore and perimeter (MCAP). Figure 4.13 is the Forrester Zero Trust Security Framework.

- The integrated "segmentation gateway" is used as the core of the network; the segmentation gateway defines the overall policies and has multiple high-speed interfaces.
- Parallel and secure divisions are created; each exchange area is connected to an interface called microcore and perimeter (MCAP), such as User MCAP and WWW MCAP in Fig. 4.13.
- Centralized management: MGMT server.
- DAN, data acquisition center monitoring network: making it easier to retrieve network data (data pack, system log, or SNMP notifications) to one location for examination and analysis.

4.6.3 Google's Solution

Google was undoubtedly the first company to embrace the Zero Trust Security Framework and benefited from its technical application. BeyondCorp is a Zero Trust Security framework modeled by Google, which introduces a Zero Trust

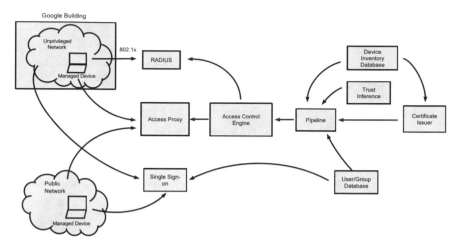

Fig 4.14   Google BeyondCorp framework

Architecture to the client-server model. By shifting the access control from the network perimeter to individual devices and users, BeyondCorp allows employees to work more securely from any location without the need for a traditional VPN. BeyondCorp consists of many cooperating components to ensure that only appropriately authenticated devices and users are authorized to access the requisite enterprise applications. Figure 4.14 illustrates its structure and information flow.

The following are the principles of Google BeyondCorp framework:

4.6.3.1   Principles of Identifying Security Devices

1. Component 1: Device Inventory Database

BeyondCorp uses the concept of a "managed device," meaning a device that is managed by the enterprise. Only managed devices can access corporate network and applications. Google keeps track of changes made to the device and makes that information available to other parts of BeyondCorp. Because Google has multiple inventory databases, a meta-inventory database is used to amalgamate and normalize device information from these multiple sources.

2. Component 2: Device Identity

All the managed devices need to have a unique ID. One way to accomplish this unique identification is to use a device certificate. The certificate is stored on a hardware or software trusted platform module (TPM) or a qualified certificate store. A device qualification process validates the effectiveness of the certificate store, and only a device deemed sufficiently secure can be classed as a managed device.

4.6.3.2 Principles for Identifying Users' Security

1. Component 3: User and Group Database

This database is used to store and track the information of all users and their group memberships.

2. Component 4: Single Sign-On System

A single sign-on (SSO) system is a centralized user authentication portal that validates credentials for users requesting access to enterprise resources. After successful verification, the SSO system generates short-lived tokens that can be used as part of the authorization process for specific resources.

4.6.3.3 Removing Trust from the Network

1. Component 5: Deployment of an Unprivileged Network

An unprivileged network closely resembles an external network, though it is within a private address space. The unprivileged network only connects to the Internet, limited infrastructure services (e.g., DNS, DHCP, and NTP), and configuration management systems, such as Puppet. All client devices are assigned to this network. There is a strictly managed ACL (access control list) between this network and other parts of the Google's network.

2. Component 6: 802.1x Authentication on Wired and Wireless Network Access

Google uses RADIUS servers to dynamically assign devices to an appropriate network based on 802.1x authentication. Rather than relying on the switch/port static configuration, it informs the switch of the appropriate VLAN assignment for the authenticated device. Managed devices provide their certificate as part of this 802.1x authentication and are assigned to the unprivileged network, while unrecognized and unmanaged devices on the corporate network are assigned to a candidate or guest network.

4.6.3.4 Externalizing Applications and Workflows

(1) Component 7: Internet-Facing Access Proxy

All enterprise applications at Google are exposed to external and internal clients via an Internet-facing access proxy. The proxy enforces encryption between the client and the application. It is configured for each application and provides common features, such as global reachability, load balancing, access control checks, application health checks, and denial of service protection. After the access control checks are completed, the proxy will delegate requests to the appropriate back-end application.

(2) Component 8: Public DNS Entries

All of Google's enterprise applications are open to the public and are registered in public DNS.

4.6.3.5 Implementing Inventory-Based Access Control

1. Trust Inference for Devices and Users

The level of access given to a single user and/or a single device may change over time. By interrogating multiple data sources, we are able to dynamically infer the level of trust to assign to a device or user.

2. Access Control Engine

The Access Control Engine provides service-level authorization to enterprise applications based on every request. The authorization decision makes assertions about the user, the groups to which the user belongs, the device certificate, and artifacts of the device from the Device Inventory Database.

3. Pipeline into the Access Control Engine

The pipeline dynamically extracts information useful for access decisions and feeds it to the Access Control Engine.

A number of companies have established a trust system based on Zero Trust, with Google BeyondCorp leading the way. Some companies started out with similar concepts as Zero Trust Security when designing their own products. Zero Trust Model undoubtedly has an important impact on cyber defense. However, the Zero Trust Model is not able to resolve the unknown threat problem, such as unknown vulnerabilities and backdoors of the target or Trojans. Whether the control components or control plane has a reliable and trustworthy security control function is one of the most challenging issues facing the Zero Trust Security framework. In this sense, it is similar to trusted computing, which must face the problem of how to prove if the root of trust is trusted or not.

4.7 Reflections on New Cyber Defense Technologies

The new cyber defense technologies are now widely studied in the field of information security. In trusted computing, the secure mainframes using autonomous and controllable TCM security chips are being mass produced. In the field of TTS and MTD, the theoretical studies of trusted cloud and elastic cloud have received broad attention. New cyber defense technologies no doubt have great potential. In this section, we will analyze the advantages and disadvantages of the mainstream new cyber defense technologies, discuss the challenges they are facing or will face in the future, and reflect on the ramifications for cybersecurity.

There are a number of issues concerning the new cyber defense technologies.

(1) Although trusted computing and TTS greatly enhanced the security of information systems, technically speaking "trusted" does not necessarily means "trustworthy" to the users. Trusted technologies are flawed in the following five aspects:

    (i) It is difficult to guarantee the security of the root of trust. The root of trust is the security foundation of trusted computing. Only when the security of the root of trust is guaranteed can the trusted computing have any security values. At the moment, the root of trust is mainly designed through the root of trust for measurement, root of trust for storage, and root of trust for report. The process is relatively complicated. There are a few tough issues to address, for instance, how to make sure that the design, development, and usage management processes are completely autonomous and controllable; how to avoid design flaws or backdoors; and how to prove that the design of the root of trust and even the manufacturer are trustworthy.

    (ii) It is difficult to accurately set status check points for the chain of trust. In trusted computing, the status of target systems must be checked when building a chain of trust in order to measure the integrity of its data. However the internal status of a complex system is too complicated to run through each and every one of them. It is impossible to test the rationality of each status or behavior, nor is it possible to conclude that the system is harmless simply on the basis of reasonable behavior of all the parts. If a check point is set for each unit of the system, then the cost would go up significantly and directly affect the service performance of the target system. What's more, how to properly set up status check points of the chain of trust in order to avoid situations where trusted units are tricked or bypassed is a problem that must also be solved when designing the chain of trust.

    (iii) Black box problem. Trusted technologies in theory require that the behavior of the target under protection is completely knowable and predictable. This is impossible in a lot of cases for a number of reasons. Firstly, with the current engineering technologies, no software/hardware manufacturer can guarantee that its product is free from design flaws. Secondly, in the age of globalization and open industrial chain, no manufacturer can make the product entirely on its own, and therefore it is impossible to ensure that the supply chain or production chain is not tempered with. Thirdly, the open-source model is the mainstream of technological development. But if someone is to take advantage of the open-source community and embed a backdoor code on purpose, there is no dependable ways to detect that code. Fourthly, software/hardware manufacturing depends heavily on generalization of technologies. It is common practice to use black box software modules or black box IP core in the system or to directly integrate software/hardware parts or components or subsystems from other sources. It is almost impossible to fully grasp all the functions of the integrated software/hardware codes. The fifth reason is that "status or behavior know-able" does not necessarily mean that its purpose is reasonable. And even if the rationality of behavior can be partially proven, it does not guarantee the rationality of the overall behavior. In practice, the problem of "status explosion" still exists. Therefore it is not convincing that, as long as the device is provided by a legitimate producer and is not tempered with, it must be trustworthy. In addition, because it is impossible to completely grasp all the software/hardware code status of the target, some status not included in the chain of trust but might be normal may trigger "false alarm." It is a long and tiresome task to eliminate false alarms in trusted computing systems. On the other hand, there may be situations where a system needs frequent updates or new services need to be frequently loaded. It remains questionable whether trusted computing is applicable in those situations.

(iv) Compatibility. Trusted computing requires that all statuses of the chain of trust are transparent so that their integrity can be checked. However, this process may very well conflict with other security defense technologies. For instance, when randomization technologies are combined with trusted technologies, if the system tries to access an address that has been randomized, then errors might occur when performing integrity verification on the chain of trust, because the address information has been changed, and the normal call might be halted, leading to instabilities. From both technological and commercial perspectives, it is impossible to include all the software/hardware code designs and elements in the world into the chain of trust management, which severely limits the timeliness, convenience, and cost efficiency of function expansion of trusted computing, thus limiting its scope of application.

(v) It guards against external threats rather than the internal ones. Even if there are unknown vulnerabilities or design flaws in components or elements included in the chain of trust, the principles of trusted computing dictate that any attempt to attack by uploading Trojans through the vulnerabilities will be detected. This means that trusted computing can not only effectively deal with any attacks based on known or unknown vulnerabilities but also detect any random malfunction or errors in the target. However, if there are backdoor-type software/hardware codes in the chain of trust elements or components, and that the verification process does not detect any harm, then the trust computing-based defense mechanism is useless. It is comparable to the nonspecific immune system of our body. Although it can non-selectively clean up any invasion antigens, it can neither eliminate nor detect and warn against the cancer cells.

(2) The MTD technology can take advantage of the time, space, and physical environment of the target to protect it. Instead of establishing a perfect system that has zero vulnerabilities or zero flaws to defend against attacks, MTD admits the reality of target systems carrying vulnerabilities and takes diverse, ever-changing mechanisms and strategies to prevent the vulnerabilities from being reliably or effectively exploited. This defense approach significantly increases

the difficulty and cost for attackers, turning the game to the advantage of the defenders. However, there are four main problems with MTD.

(i) Lacking in systematic technology. First of all, dynamism, diversification, and randomization technologies are elementary defense technologies, not unique to MTD. Secondly, MTD is simply a stack of these technologies, lacking in any systematic, comprehensive defense effect. Lastly, it applies dynamism, diversification, and randomization technologies non-discriminatorily, from small targets (instructions, addresses, and data) to larger targets (systems, platforms, and networks). Its generalization leads to many problems in practice, such as performance loss and low-cost efficiency. In fact, under the mechanism of shared resources in the same processing environment, attackers can use the MMU (memory management unit) vulnerabilities to bypass ASLR (Address Space Layout Randomization), use the side-channel effect of CPU caches to circumvent the check points, or even directly take control of the operating system and ban dynamism and randomization operations. These are some of the challenges MTD faces. Unfortunately, like trusted computing, the MTD mechanism is ineffective against threats such as software/hardware backdoors or embedded Trojans.

(ii) Security hard to measure. Using dynamism, randomization, and diversification technologies, MTD changes the static, similar, and certain nature of the target system, thereby increasing the cost of attacks and making the results of vulnerability-based attacks less certain. But it also makes the level of complexity difficult to measure—not only for the attackers but for the defenders as well. This creates obstacles for defenders to run a playback analysis of the attacks and organize targeted defense mechanisms accordingly. As a security technology product, how to test and measure its security level is also an issue that must be clearly addressed.

(iii) Impact on target system performance. The approach of randomization—be it of instructions, addresses, or data—or increasing dynamism, will inevitably sacrifice the processing capability. The more uncertain the defense behavior, the higher the cost to the system, because the defense functions and the system's service functions are realized in the same resource-sharing environment. For example, randomization of instructions is basically a scrambling process of the executable code entering into the memory. Decoding is needed when running the code. The effectiveness of defense is connected to the level of complexity of scrambling algorithms as well as the expense of system processing. Without the specialized decoding hardware, the normal performance of the system will be dramatically reduced to an unbearable level. Increase in dynamism will increase the level of uncertainty of the system, but the more dynamic the system, the higher the cost of processing. Especially when it is necessary to rapidly switch environments, the cost is unacceptably huge. Increased level of diversity can also hike up the cost in terms of design, manufacturing, energy consumption, and maintenance.

(iv) Difficult to defend against unknown unknown risks. Before establishing a wall of defense, MTD needs to build a model based on known risk information. For instance, the IPv6 address randomization of MT6D technology is developed on the assumption of the "known" IP attack. Likewise, instruction and address randomization are predicated on the SQL injection attack or buffer-overflow attack. Therefore, MTD requires knowledge of overall characteristics of external attacks or mechanisms of vulnerability-based attacks. Only by then can it change key parts of the system in a dynamic, random, and diverse way—without considering backdoors—to lower "known unknown risks" or increase the level of uncertainty to attacks. However, it is not effective in dealing with the "unknown unknown threats" based on unknown vulnerabilities or Trojans inside the targets.

Below is a summary of the similarities and differences of the three types of defense technologies, as shown in Table 4.2.

Trusted computing cannot treat the target as a black box, but rather must operate under the assumption that all target behaviors are knowable. It can detect any attempt at external attacks using vulnerabilities in the chain of trust elements and monitor any internal malfunction or faults. However, it cannot detect attacks from malicious codes that have passed behavior check but have unknown intentions. While MTD can make it more difficult to use vulnerabilities or cause the results of attacks to be less certain without relying too much on the features of attacks, it is unable to offer any quantified measures of protection levels. MTD is also theoretically ineffective against attacks based on backdoors or malicious codes. The Zero Trust Security framework on the other hand puts more emphasis on the reliability and trustworthiness of key applications or services. It does not have much to do with the security of element system or control devices. Is there a defense technology or system architecture that can do all of the above? Namely, a technology that, like a nonspecific immune system, non-selectively eliminates any invasion antigens while having specific immunity against known antigens; a technology that can defend against both known unknown risks and unknown unknown threats; a technology that is capable of both key point and overall defense; a technology that can provide reliable, available service functions and performance and guarantee the trustworthiness of the service itself—all without relying on a priori knowledge of the attacks or characteristics of behaviors. We will try to find the answer to this question in the next few chapters of the book.

Table 4.2 Comparison of three types of defense technologies

Technology	Characteristics of attacks	Target treated as a black box	Known unknown risks	Unknown vulnerabilities/ backdoors	Effect
Trusted computing	Not needed	No	Yes	Yes/No	Certain
TTS	Not needed	No	Yes	Yes/No	Certain
MTD	Needed	Yes	Yes	Yes/No	Uncertain

Both the traditional and the new types of defense technologies mainly focus on attacks that target vulnerabilities from the front and defend against breach through the main channel. As attack technologies advance and develop, there are more cases of attacks through the side channels. This has become an area of focus in the field of cyber security. Side-channel attacks take advantage of the characteristics (flaws) of the computer system to obtain information. Such attacks have nothing to do with the shortcomings or flaws of the algorithms or protocols (such as software vulnerabilities). To be more specific, when the computer system is processing information, it also generates "side information," a kind of information that is obtained from paths other than the main channel and unrelated to target information. That is the flaw of the computer system. The typical side information includes runtime, energy consumption, electromagnetic leakage, and acoustic-optical information. There is not an effective universal solution to the side-channel attacks at the moment. As side-channel attacks depend on the corresponding relationship between information leaked from side channels and the target data, there are accordingly two types of defense approaches or strategies. One approach is to cut off information source so as to eliminate or decrease the side information; the other approach is to target the corresponding relationship and eliminate the connection between information leak and target data, so that the leaked information has nothing to do with the target data.

References

1. The White House: Trustworthy Cyberspace—Strategic plan for the federal cybersecurity research and development program. https://www.whitehouse.gov/sites/default/files/micro-sites/ostp/fed_cybersecurity_rd_strategic_plan_2011.pdf. [2017-04-16]
2. Widi: Adaptation (computer science). https://en.wikipedia.org/wiki/Adaptation_%28computer_science%29. [2017-04-16]
3. Manadhata P.: An attack surface metric. https://reports-archive.adm.cs.cmu.edu/anon/2008/CMU-CS-08-152.pdf. [2017-04-16]
4. Designing an adaptive security architecture for protection from advanced attacks. https://www.gartner.com/doc/2665515/designing-adaptive-security-architecture-protection. [2017-04-16]
5. Gartner: Gartner identifies the top 10 strategic technology trends for 2016. http://www.gartner.com/newsroom/id/3143521. [2017-04-16]
6. Wiki: Gene redundancy. https://en.wikipedia.org/wiki/Gene_redundancy. [2017-04-16]
7. Wiki: Redundancy (engineering). https://en.wikipedia.org/wiki/Redundancy_%28engineering%29. [2017-04-16]
8. Wiki: Redundancy information theory. https://en.wikipedia.org/wiki/Redundancy_%28information_theory%29. [2017-04-16]
9. Avizienis, A.: Towards systematic design of fault-tolerant systems. IEEE Trans. Comput. **30**(4), 51–58 (1997)
10. Wiki: Survivability. https://en.wikipedia.org/wiki/Survivability. [2017-04-16]
11. Ellison, R.J., Fisher, D.A., Linger, R.C., et al.: Survivable network systems: an emerging discipline. In: Carnegie-Mellon Software Engineering Institute Technical Report CMU/SEI-97-TR-013 (1997)
12. Mohammad, A.J., Hutchison, D., Sterbenz, J.P.G.: Poster: Towards quantifying metrics for resilient and survivable networks. In: The 14th IEEE International Conference on Network Protocols (ICNP 2006), pp. 17–18. Santa Barbara, CA, USA (2006)

13. Common Criteria Project Sponsoring Organization: Common Criteria for Information Technology Security Evaluation. In: ISO/IEC International Standard 15408 Version 2.1. Genevese: Common Criteria Project Sponsoring Organization (1999)
14. Trusted Computing Group: TCG specification architecture overview. https://www.trustedcomputinggroup.org. [2017-04-16]
15. Zhang, H., Bo, Z.: Trusted Computing. Wuhan University Press, Wuhan (2011)
16. Trusted Computing Group: TCG Specifications. https://www.trustedcomputting-group.org/specs. [2017-04-16]
17. Shih, M.W., Kumar, M., Kim, T., et al.: S-NFV: securing NFV states by using SGX. In: ACM International Workshop on Security in Software Defined Networks & Network Function Virtualization, pp. 45–48 (2016)
18. Shen, C., Zhang, D., Jiqiang, L., et al.: Trusted 3.0 strategy: revolutionary development of trusted computing. Eng. Sci. **18**(6), 53–57 (2016)
19. Changxiang, S.: Developing a solid cyber security defense line with trusted computing 3.0. Inf. Commun. Technol. **3**(3), 290–298 (2017)
20. Changxiang, S.: Accelerating the development of trusted computing in the principle of autonomous innovation. Comput. Secur. (6), 2–4 (2006)
21. Mckeen, F., Alexandrovich, I., Anati, I., et al.: Intel Software Guard Extensions (Intel SGX) support for dynamic memory management inside an enclave. In: The Hardware and Architectural Support for Security and Privacy, pp. 1–9 (2016)
22. Baumann, A., Peinado, M., Hunt, G.: Shielding applications from an untrusted cloud with Haven. ACM Trans. Comput. Syst. **33**(3), 1–26 (2015)
23. Mckeen, F., Alexandrovich, I., Berenzon, A., et al.: Innovative instructions and software model for isolated execution. In: International Workshop on Hardware & Architectural support for Security & Privacy, p. 10 (2013)
24. RFC 2998: A framework for integrated services operation over diffserv networks. http://www.rfc-base.org/txt/rfc-2998.txt. [2017-04-06]
25. RFC 2205: Resource ReSerVation Protocol (RSVP)—Version 1 functional specification. https://tools.ietf.org/html/rfc2205. [2017-04-06]
26. RFC 2475: An architecture for differentiated services. https://www.rfc-editor.org/rfc/rfc2475.txt. [2017-04-06]
27. Dennard, R.H., Gaensslen, F., Yu, H.N., et al.: Design of ion-implanted MOSFET's with very small physical dimensions. IEEE J. Solid State Circuits. **9**(5), 256–268 (1974)
28. Chen, Y., Cong, J., Gill, M., et al.: Customizable Computing, p. 118. Morgan & Claypool (2015)
29. Wiki: Cloud computing. https://en.wikipedia.org/wiki/Cloud_computing. [2017-04-06]
30. Mell P.: The NIST definition of cloud computing. http://nvlpubs.nist.gov/nistpubs/Legacy/SP/nistspecialpublication800-145.pdf. [2017-04-06]
31. Epstein J.: Secure and trustworthy cyberspace (SaTC) program. https://www.nsf.gov/pubs/2015/nsf15575/nsf15575.htm?WT.mc_id=USNSF_25&WT.mc_ev=click. [2017-04-16]
32. Lu, K., Song, C., Lee, B., et al.: ASLR-Guard: Stopping address space leakage for code reuse attacks. In: ACM Sigsac Conference on Computer and Communications Security, pp. 280–291 (2015)
33. Wang, Q., Wang, C., Li, J., et al.: Enabling public verifiability and data dynamics for storage security in cloud computing. In: European Conference on Research in Computer Security, pp. 355–370. Springer (2009)
34. Wang, L., Zhang, M., Jajodia, S., et al.: Modeling Network Diversity for Evaluating the Robustness of Networks Against Zero-Day Attacks, pp. 494–511. Springer, Berlin (2014)
35. Ahn, G.J., Ahn, G.J., Ahn, G.J., et al.: A Game Theoretic approach to strategy generation for moving target defense in web applications. In: Conference on Autonomous Agents and Multiagent Systems. International Foundation for Autonomous Agents and Multiagent Systems, pp. 178–186 (2017)

36. Lei, C., Ma, D.H., Zhang, H.Q.: Optimal strategy selection for moving target defense based on Markov game. IEEE Access. (99), 1 (2017)
37. Zangeneh, V., Shajari, M.: A cost-sensitive move selection strategy for moving target defense. Comput. Secur. **75**, 72–91 (2018)
38. Cui, A., Stolfo, S.J.: Defending embedded systems with software symbiotes. In: Recent Advances in Intrusion Detection, pp. 358–377. Springer, Berlin (2011)
39. ODIN.: http://www.ppkpub.org/
40. Zero Trust Network Architecture with John Kindervag-Video. https://www.paloatonetworks.com
41. Evan Gilman, Evan Gilman, et al. "Zero Trust Networks: Building Secure Systems in Untrusted Networks" O'Reilly Media, 2017.06

Chapter 5
Analysis on Diversity, Randomness, and Dynameicity

As mentioned earlier, moving target defense (MTD) [1] adopts multiple security technologies featuring diversity, dynamicity, and randomness, aiming to change the passive cyber defense situation by greatly increasing the cost of attack and vulnerability exploitation through the deployment and operation of networks, platforms, systems, devices, and even components that are subject to apparent uncertainty and random dynamics. In fact, such basic defense methods or technical elements as diversity, randomness, redundancy, and dynamicity are not exclusively for MTD or certain security defense systems. They have been widely used in all aspects of the related fields, and the purpose is nothing but how to endow the target system with security attributes, such as diversity, randomness, and dynamicity, to build a survivable, recoverable, and fault-tolerant self-adaptive system in the harsh context of an elusive operating environment, asymmetrical threats, or uncertain failures. For example, the biodiversity mechanism in natural communities guarantees the stability of the ecosystems, while the dynamic multipath forwarding mechanism in communication networks guarantees the anti-interception of data transmission. Various encryption technologies are applying pseudo-random properties or methods, and the redundancy technology is a "talisman" in the field of reliability. This chapter will elaborate on the concepts, characteristics, and applications of diversity, randomness, and dynamicity technologies themselves and analyze the possible engineering challenges for introducing these underlying defense technologies into information systems, with an attempt to comb the ties among the three technologies and propose the relevant arguments.

© Springer Nature Switzerland AG 2020   159
J. Wu, *Cyberspace Mimic Defense*, Wireless Networks,
https://doi.org/10.1007/978-3-030-29844-9_5

5.1 Diversity

5.1.1 Overview

The concept of diversity comes from the theory of biodiversity. In this theory, the environment for living organisms is constantly changing. The fluctuation of climate, the increase and decrease of the number of predators, the outbreak of infectious diseases, and the intrusion of alien species will have an impact, greater or lesser, on the continuation of biological populations. In general, biodiversity mainly has a three-layer connotation: biological population diversity, genetic diversity, and ecosystemic diversity (using quantitative measures for diversity indicators [2]). In the biological world, each species is composed of several populations formed through the aggregation of individual members. The diversity of biological populations refers to the abundance of biological species in a certain region, which determines the stability of biological populations in the region to adapt to the environment. Genetic diversity results from the variation of genetic structures within or between populations due to mutations, natural selection, or other reasons across the populations concerned. Facing the pressure of survival, populations must ensure that they contain multiple types of genes. Rich genetic diversity means that there are phenotypes adaptive to various kinds of environments. When a population has a greater genetic diversity, it has a bigger chance to breed individuals adaptive to specific environmental changes. Namely, diversity increases the suitability or fitness of the population and enables it to cope with the ever-changing environmental conditions. For instance, to adapt to the topographical conditions such as grassland, plateau, or desert, livestock like cattle and sheep must constantly change the genetics of their population in consistency with environmental changes so as to improve the survival of their population in the environment.

Diversity exists not only in the natural world but also in people's daily life. There are many ideas of diversity applied to areas ranging from common daily necessities to the founding of national systems, all reflecting the wisdom of diversity. However, the trend is sadly opposite in cyberspace: in order to simplify system architecture and operation and maintenance costs, people usually instantiate one function as a single protocol standard or homogeneous hardware/software for implementation. For example, Internet addressing is still dominated by the single IP addressing. Although homogenization improves the interoperability of the system and reduces the cost of production, operation, and maintenance, it has become a serious potential vulnerability in the entire cyberspace. The homogenization of system components leads to the explicit expression of internal laws, where the handling process and operating mechanisms of programs are also detectable, making it easy for attackers to exploit these invariant laws and implicit defects to construct attack chains and implement universal attacks. A typical case is Microsoft's vast portfolio of products, which includes a homogenous software hosting environment (the operating system), as well as application software (Word, Outlook, PowerPoint, etc.). Once a loophole in a running system is successfully exploited by an attacker, such successful experience can affect thousands of network users subject to the same operating environment. Fortunately, in the current Internet, the Internet-wide service provision has not been affected by the malicious exploitation of a single vulnerability. This is because the Internet in nature originates from a heterogeneous network environment, where attacks against Outlook clients' vulnerabilities will not affect the UNIX mail client users.

Therefore, diversity is an effective means to ensure the functionality of cyberspace services or network elements and terminals to resist attacks. An increasingly eye-catching hotspot observed by security researchers in academic and industrial circles is how to introduce the diversity mechanism of biological populations into the cyber defense system. Through the diversity design and implementation of the defense target, an object whose laws are difficult to predict or detect is presented to the attacker, making it difficult for the latter to achieve the desired deliberate conduct. Thus, the security of the target system is guaranteed. The core of diversity design is to express a target entity in a diverse manner through multiple variants to complicate the attacks, without affecting the spatial relations between executors or without excessively increasing the cost of implementation while ensuring the functional equivalence of executors. As for a high-reliability system, functionally equivalent components that require redundant configurations must differ in one or more aspects to avoid as much as possible the common mode or homomorphic failures caused by common vulnerabilities of the redundant components.

The difficulty of cognitive detection of unknown backdoor vulnerabilities in a modern information system is measured by the degree of diversification of its hardware/software components. The author believes that the diversity application in an information system mainly includes two aspects, namely, the executor diversity and the execution space diversity. The executors refer to hardware/software components, devices, and other physical functional units in the information system, such as heterogeneous CPUs, different types of operating systems, databases, and functional software. The execution space refers to the running environment isolated from the host executors, such as different virtual environments, different execution platforms, and so on. The diverse executors and execution space jointly constitute the diversity of the information system, and the two concepts will be introduced, respectively, in subsequent chapters.

5.1.2 Diversity of the Executors

With the measurement indicators of biodiversity, researchers can design and measure the diversity of information systems. Over the past few years, as the role of diversity in active cyber defense draws more and more attention, the quantitative measurement of the diversity of executors has become a new research direction. The author will illustrate executor diversity by two examples as follows: controller diversity and network path diversity in the software-defined networking (SDN).

5.1.2.1   Executor Diversity in Network Operating Systems

In recent years, the software-defined networking (SDN) has become a buzz word among technology researchers. In the SDN architecture, the core component of the control layer is the controller, which can control several switches logically for fast data forwarding and convenient and flexible service management.

However, there are many types of controllers on the market, and their specifications, features, and even possible backdoors are not the same. How to select an appropriate controller is a hot-debated issue among researchers. Therefore, some researchers begin to quantitatively describe the diversity of controllers in SDN. Reference [3] first uses performance as the analytical basis for comparison of the three controllers, NOX, Beacon, and Maestro, but the model built therein is relatively simple, and the controllers analyzed are relatively obsolete. On this basis, Reference [4] quantitatively describes the diversity of controllers in SDN. As shown in Table 5.1, the literature analyzes five mainstream controllers, POX, Ryu, Trema, Floodlight, and OpenDaylight, and gives a quantitative analysis of the diversity design of them through the modeling of 12 features: interface, virtualization, graphical user interface (GUI), REST API, open source, documentation, language support, modularity, platform support, TLS, OpenFlow support, and OpenStack support.

5.1.2.2   Executor Diversity in the Path

With the wide-range application of network technology and the constant increase of emerging services, cyber survivability has become a more prominent issue. Researchers are now focusing on the study of how to ensure low-latency, non-disruptive transmission of network services and how to ensure the security of network transmission paths and the diversity of heterogeneous paths.

A heterogeneous path [5] refers to a transmission path that includes multiple types of executors such as network components and transmission protocols, etc. Variety and heterogeneity are the most important features of this type of path. Figure 5.1 shows a diagram of typical heterogeneous network forwarding paths.

Figure 5.1 contains a total of six paths and four protocols, as shown in Table 5.2.

Take <http,0,1> as an example, where http represents the protocol type of the transport stream, 0 represents the source address, 1 represents the destination address, and <http,0,1> means that there is data sent to address 1 from address 0 according to http.

As far as the path length alone is concerned, the lengths of paths 3, 4, 5, and 6 are longer than that of paths 1 and 2, but this does not mean that paths 3, 4, 5, and 6 are safer than paths 1 and 2. If the number of protocols on a path is considered, there are only two component types on path 4, which is outnumbered by the types of protocols on path 1. If an attacker has mastered the vulnerabilities of one type of protocols on the path, he can directly launch attacks on all similar protocols on the path. Therefore, with regard to the resource diversity and the path length, path 5, whose length is 4 and protocol number is also 4, is a relatively safe path in Fig. 5.1.

Table 5.1 Feature comparison of five mainstream controllers

	POX	Ryu	Trema	Floodlight	OpenDaylight
Interface	OpenFlow	OpenFlow + OVSDB JSON	OpenFlow	OpenFlow + Java and REST	OpenFlow + REST and Java RPC
Virtualization	Mininet and Open-Switch	Mininet and Open-Switch	Built-in emulation virtual tool	Mininet and Open-Switch	Mininet and Open-Switch
GUI	Y = Yes	Y	N = No	Y	Y
REST API	N	Y	N	Y	Y
Open source	Y	Y	Y	Y	Y
Documentation	P = Poor	M = Medium	M	G = Good	G
Language support	Python	Python	C/Ruby	Any Java+ language that utilizes REST	Java
Modularity	M	M	M	H = High	M
Platform support	Linux, Mac OS, Windows	Linux	Linux	Linux, Mac OS, Windows	Linux
TLS	Y	Y	Y	Y	Y
OpenFlow support	OF v1.0	OF v1.0, v2.0, v3.0	OF v1.0	OF v1.0	OF v1.0
OpenStack support	N	S = Strong	W = Weak	M	M

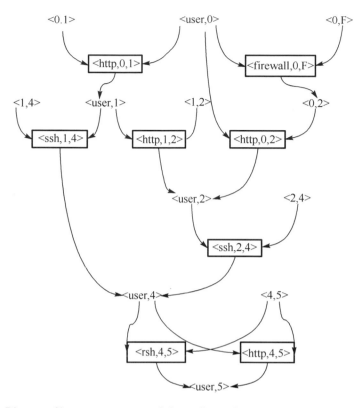

Fig. 5.1 Diagram of heterogeneous network forwarding paths

Table 5.2 Forwarding paths

Attack path	Number of steps	Number of resources
1. <http,0,1>→<ssh,1,4>→<rsh,4,5>	3	3
2. <http,0,1>→<ssh,1,4>→<http,4,5>	3	2
3. <http,0,1>→<http,1,2>→<ssh,2,4>→<rsh,4,5>	4	3
4. <http,0,1>→<http,1,2>→<ssh,2,4>→<http,4,5>	4	2
5. <firewall,0,F>→<http,0,2>→<ssh,2,4>→<rsh,4,5>	4	4
6. <firewall,0,F>→<http,0,2>→<ssh,2,4>→<http,4,5>	4	3

In recent years, there have been many research results on heterogeneous path diversity. For example, from the perspective of network overheads and transmission performance, Newell et al. [6] designed a secure transmission mechanism for maximizing the heterogeneous diversity paths. From the perspective of prevention against zero-day attacks, Borbor et al. [7] designed a 0 day security metric based on the features of 0 day attacks to measure the heterogeneity of services flowing through the transmission path. This method proves the 0 day defense ability of heterogeneous paths. Reference [5] published in 2016 defines the diversity indicators by the Shannon-Wiener index and analyzes the paths in the network. It first assumes a given set of resource types R in a network, where $|R|$ is called the abundance of network resources. This definition describing the diversity of network resources has an obvious drawback: the association of each type of resource is not considered. For example, the two sets $\{r_1, r_1, r_2, r_2\}$ and $\{r_1, r_2, r_2, r_2\}$ have the same resource abundance, but they may differ in the diversity level. To this end, the paper in Reference [5] groups the systems at the same diversity level and defines a new diversity indicator. In the design of the indicator, the paper considers both the resource type and the resource length, which assumes that, in the network G, the host set is $H = \{h_1, h_2, \cdots, h_n\}$, the resource type set is $R = \{r_1, r_2, \cdots, r_m\}$, and the resource mapping is $\mathrm{res}(\cdot) : H \rightarrow 2^R$ (where 2^R represents the power set of R), and when the total number of resource instances is $t = \sum_{i=1}^{n} |\mathrm{res}(h_i)|$, the relative frequency of each type of resource is $p_j = |\{h_i : r_j \in \mathrm{res}(h_i)\}|/t$ $(1 \leq i \leq n, 1 \leq j \leq m)$. At this time, the network diversity can be expressed as $d_1 = \dfrac{r(G)}{t}$, where $r(G)$ represents the network diversity of the resources, and the calculation formula of $r(G)$ is

$$r(G) = \frac{1}{\prod_{1}^{n} p_i^{p_i}}$$

The experimental results of this paper show that this new diversity indicator can be used to qualitatively measure the network diversity, which helps to greatly improve cyber security.

5.1.3 Diversity of the Execution Space

At present, the majority of researchers are focused on the study of executor diversity—only one aspect of the information system. In practical application, the diversity of execution space is equally important. Executors of the same type may present different effects when the execution space is different. In recent years, many researchers have studied the diversity of the execution environment. Similar to Sect. 5.1.2, the author still briefs on the diversity of the execution space through two cases: controller diversity and network path diversity in a software-defined network.

5.1.3.1 Execution Space Diversity in Network Operating Systems

As described in Sect. 5.1.2, in the SDN architecture, the controller is a core component of the control layer, while centralized control may face single point failures, that is, once a single centralized controller fails or gets attacked externally, the entire SDN will face huge risks.

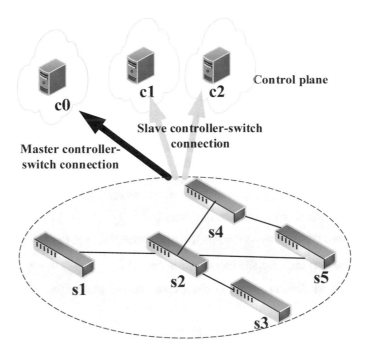

Fig. 5.2 Diagram of the SDN's tripple backup strategy

Some researchers have attemped to solve the single point failures of SDN controllers via the execution space diversity technology. Backup technology is a common execution space diversity technology. By designing the backup or slave controller, it places the master and slave controllers in different execution spaces and designs an effective controller backup migration strategy to solve the single point failures, such as SDN controller failures [8–10]. Figure 5.2 is a diagram of the tripple backup strategy for a simple SDN controller. It contains three controllers, C_0, C_1, and C_2, of which C_0 is the master controller mainly responsible for functions, such as SDN data forwarding and rule circulation, while C_1 and C_2 are slave controllers; they only read information in the network in day-to-day operation instead of generating rules and operating on the network. However, once C_0 fails or halts upon an attack by an external attacker, C_1 and C_2 will, according to the allocation algorithm, automatically replace C_0 to become the new master controller responsible for network data forwarding and rule circulation.

The SDN architecture based on the backup strategy can solve single point failures of the controller to some extent through multiple backup controllers, but this method can only empower the controller to block attacks; it cannot guarantee cyber security when an attacker takes advantage of the vulnerability of a certain controller to forge the controller's commands or even directly modify its rules, and then launch attacks. Therefore, on the basis of the SDN architecture of the backup strategy, some researchers draw on the Byzantine defense technology and propose an SDN architecture with stronger fault tolerance, namely, the SDN architecture based on the Byzantine mechanism [11–14].

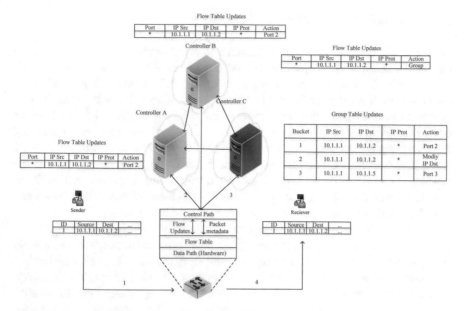

Fig. 5.3 Diagram of the Byzantine controller ruling model

Figure 5.3 shows a simple Byzantine controller ruling model, where the sign $*$ represents the wildcard; A, B, and C in the figure are three different types of controllers. Differing in the architecture, design philosophy and execution space, they can be considered as diverse controllers; even if an attacker masters the vulnerability of one controller, he is unlikely to hijack all controllers simultaneously. For example, assume that Controller C is hijacked by the attacker, since there are two normal working controllers A and B, the entire network will make a correct ruling according to the consistent output of A and B, thereby ensuring the security of the system. For example, forwarding rule $10.1.1.1 \rightarrow 10.1.1.2$ is embedded in all the controllers A, B, and C. Once an attacker controls C and modifies the forwarding rule to $10.1.1.1 \rightarrow 10.1.1.x(x! = 2)$, the ruler will judge according to the respective rules of the three controllers A, B, and C to ensure the correctness of the rules. The specific ruling method is as follows:

(1) Assume that the sender only communicates with the master controller and considers that Controller A is the master controller. Once A receives the sender's request, it immediately sends a request to other slave controllers (B and C).
(2) Once Controllers B and C receive the request sent by Controller A, they will send a ruling message to other controllers to rule all the messages, with a reply message returned to the recipient after the ruling is completed.
(3) Once the receiver receives $f + 1$ coincidence messages (f represents the number of failed controllers, which is 1 in this example), the consensus message will be regarded as the ruling result to complete the ruling.

The SDN architecture based on the Byzantine mechanism can guarantee that as long as the total number of controllers is greater than $3f + 1$, the ruling at the control layer will remain consistent even if there are "f failed controllers" in the network.

5.1.3.2 Execution Space Diversity in the Path

Similar to the network operating system, there is also execution space diversity in the path. Backup technology can also be applied to the study of path diversity. Backup path diversity is a common path diversity defense technology, which can effectively improve the survivability of the network. Survivability is the ability of a system to recover from failures and meet the demand of given key missions to provide ongoing services in the face of failures and uncertain accidents. Figure 5.4 shows a simple double backup path. Normally, the data is transmitted on path 1 through two switches, respectively, S_1 and S_2, and the transmission path is the sender $\rightarrow S_1 \rightarrow S_2 \rightarrow$ the receiver. However, once a problem occurs to S_1 or S_2 on path 1, the data cannot be transmitted normally, and the system will use the backup path 2 for data transmission according to the preset allocation rule.

At present, the research on backup path diversity mainly includes two aspects: protection and recovery/restoration [15]. The protection mechanism usually allocates resources in a preplanned way and uses the node/link backup method to address failures, such as the protection strategies 1 + 1 and 1:1. The recovery/ restoration mechanism usually refers to the use of existing backup paths to establish new available paths dynamically according to the failure scenario to bypass the failed points. The protection mechanism requires strict physical backups, and the deployment cost can be expensive; besides, the protection mechanism cannot protect the layers above the physical layer from failures, which is not suitable for the IP-centered Internet. The recovery/restoration mechanism features greater flexibility and optimized resource utilization [15], so it can handle failures in a larger scope and at lower cost. Currently, the majority of backup path diversity methods adopt the path recovery mechanism. When the network underlying topology ensures that the connectivity and the capacity meet the requirements, the path recovery mechanism

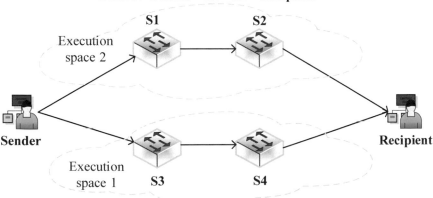

Fig. 5.4 Diagram of backup paths

is also able to handle the underlying fault self-adaptively. The current path recovery mechanism methods are mainly dedicated to the path establishment model, the resource usage model, the recovery scope, and the recovery domain [16–19].

5.1.4 Differences Between Diversity and Pluralism

The author argues that an important difference between diversity and pluralism is that they are functionally equivalent but not homological. Diversity usually refers to multiple manifestations of the same entity, also known as multiple variants of the same entity, which share many identical elements. The biological mimicry in the natural world is a classic representation of this form. While there are equivalent functions among plural entities, there are few or even no homologous components across identical elements. Diversity and pluralism are ultimately manifested as the heterogeneity difference in the component implementation algorithms. For an attacker, it is far more difficult to spot the same vulnerability in plural entities than seek a homologous vulnerability among multiple variants.

Based on the NIST National Vulnerability Database (NVD), Reference [20] analyzes the vulnerability data spotted in the released versions of 10 operating systems over 18 years to statistically study how the number of vulnerabilities spanning across different operating systems. The results of the study show an extremely small number of public vulnerabilities occurring in different operating systems. As shown in Table 5.3, the operating systems mainly include two major families: the UNIX-like system series for the open source community and the Microsoft Windows series. Among them, the UNIX-like systems and Windows systems show the pluralized development of operating systems, and the diversity of operating systems is mainly reflected in that UNIX-like systems gave rise to OpenBSD, NetBSD, FreeBSD, Solaris, Debian, Ubuntu, Red Hat, etc., while Windows systems gave birth to Windows 2000, Windows 2003, and Windows 2008.

Table 5.3 Statistics of public vulnerabilities in ten major operating systems

	OpenBSD	NetBSD	FreeBSD	Solaris	Debian	Ubuntu	Red Hat	Windows 2000	Windows 2003	Windows 2008	
OpenBSD	—	33	43	9	2	3	10	3	2	1	
NetBSD	4	—	36	9	4	0	6	3	2	1	
FreeBSD	4	6	—	12	4	2	12	4	3	1	
Solaris	1	1	2	—	2	2	8	8	7	0	
Debian	0	0	0	0	—	14	52	1	1	0	2000~2009
Ubuntu	0	0	0	0	0	—	27	1	1	0	
Red Hat	0	0	0	2	0	0	—	2	2	0	
Windows 2000	0	0	0	1	0	0	0	—	216	42	
Windows 2003	0	0	0	1	0	0	0	49	—	53	
Windows 2008	0	0	0	1	0	0	0	38	229	—	

2010~2011 年

Table 5.3 shows the historical vulnerability statistics of the operating systems installed with a large number of applications of the same functionality. The upper right data triangle shows the comparison in the number of public vulnerabilities of ten types of operating systems from 2000 to 2009, while the lower left data triangle represents the statistics of the public vulnerabilities of the ten types of operating systems from 2010 to 2011. Take OpenBSD-Red Hat as an example; the number of public vulnerabilities is ten in 2000–2009, which turns zero in 2010–2011. It can be seen from the data in Table 5.3 that there is a relatively big number of public vulnerabilities in operating systems from the same series or origin, while the number of public vulnerabilities between different series of operating systems is relatively small or even zero in a short period.

5.2 Randomness

5.2.1 Overview

Randomness is a form of contingency and is the uncertainty exhibited by all events in a set of events with a certain probability. A random event or state can be measured by the probability that it may occur, which reflects the likelihood that the event or state will occur. All random events have the following characteristics:

(1) An event can be repeated under essentially the same conditions, like multiple shots to the same target by the artillery. If there is only a single accidental process, the repeatability of which cannot be identified, so it cannot be called a random event.

(2) Under essentially the same conditions, an event may be manifested in a variety of ways, and it is impossible to predetermine the particular way in which it occurs. For example, no matter how we control the firing conditions of the artillery, the position of the impact point cannot be predicted without error. If the event has only one single possible process, it will never be a random event.

(3) It is possible to predict all the possibilities of the event's occurrence in all ways or to predict the probability of its occurrence in a certain way, namely, the frequency of its occurrence during the repetition process. This is like that the normal distribution of the impact points after a large number of shots, with each impact point subject to a certain probability within a certain range. A phenomenon without a certain probability when it occurs repeatedly is not a random event in a unified process.

The above characteristics indicate that a random event is a phenomenon and process between an inevitable event and an impossible event. The introduction of the randomness philosophy into the target object in cyberspace can change the internal information expression state of the target object and endow unpredictability in the target object's defense behavior, thus increasing the cost of attackers and reducing the reliability of their attacks.

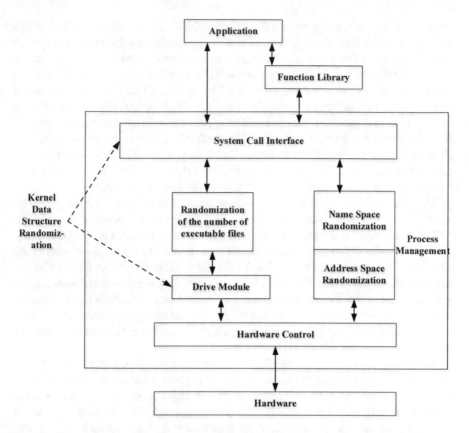

Fig. 5.5 Overall plan for operating system randomization

The use of randomness in information systems can significantly improve system security and reduce the deliberate jeopardy based on unknown vulnerabilities. For example, the randomness of the operating system can be realized by modifying the existing Linux kernel source code. The randomization of the operating system can improve the difficulty of detecting and exploiting vulnerabilities in the operating system and software layers. The overall plan of the operating system randomness is shown in Fig. 5.5 and mainly includes three key technologies: address space randomization, instruction system randomization, and kernel data randomization.

5.2.2   Address Space Randomization

Address space randomization (ASR) is a common randomization method that attempts to turn a defined address distribution into a random one, blocking an attacker from relocating a control stream or reading data from a particular fragment using a known memory address. The basic idea of ASR is to randomize the location information of the running application in memory, thereby invalidating the attacks relying on that information. It is usually done by modifying the operating system kernel, and sometimes needs additional support provided in the application. At present, there are mainly three mechanisms for specific ASR: (1) randomize the stack base address or the entry address of the global library function, or add a random offset for each stack frame (the offset in a single compilation process can be a fixed value or a variable); (2) randomize the position of the global variable and the offset allocated for the local variable in the stack frame; (3) assign an unpredictable position to each new stack frame, randomly for instance, rather than assign it to the next connected unit. The specific randomization timing may be the period of program compilation, of process loading, or of program execution.

The ASR technique is a relatively mature applicable defense technology so far, mainly deployed in the form of Address Space Layout Randomization (ASLR). ASLR is a security protection technology targeted at buffer overflow. By randomizing the layout of the linear area such as stacks and shared library maps, it ensures the security of the address, making it harder for the attacker to predict the destination address while preventing the attacker from directly positioning the attack code location so as to block overflow attacks. Research shows that ASLR can effectively reduce successful buffer overflow attacks. So, the technology has been widely used in some mainstream operating systems, such as Linux, FreeBSD and Windows, as well as some specific mobile phone systems.

PaX [21] is the most typical application sample of the ASR layout technology. By patching the Linux kernel, it randomizes the stack base address, the main program base address and the loading address of the shared library. The implementation method is as follows: 24 bits (4~27) of the stack base address and 16 bits (12~27) of the main program address (image, static data area and heap base address), as well as of the shared library load address, are randomized when the process is loaded. This ASR method based on the randomized segment base address can effectively improve the security of Linux systems. In addition to PaX, there are other ASR applications. Transparent runtime randomization (TRR) implements ASR by modifying the process loader [22]. Similar to PaX, it modifies the stack base address, the heap base address, and the loading address of the shared library when the process is loaded. But unlike PaX, TRR adds randomization to the global offset table (GOT), which moves the GOT to a new location when the process is loaded and uses binary code modifiers to modify the procedural linkage table (PLT) in the code segment, which prevents attacks that relocate the GOT function pointer to a malicious code.

The advantage of the ASR technology is that it can randomize the address space distribution of the process stack, the code/data segment of the program, and the like, thereby reducing the success probability of attackers and effectively enhancing the security of the system. Besides, characterized by address randomization, the ASR technology helps to effectively reduce the system's exposure to unknown vulnerabilities. Even if an attacker launches an attack on an address via a vulnerability of the system, the control stream of the program cannot jump to a specified location. Thus, the attack code cannot be executed, causing a failed attack. However, the ASR technology also has the following defects:

(1) Randomization can only be limited to a part of the memory space rather than be applied to full memory. Then, the intruder can attack the vulnerabilities of the program by exploiting the nonrandomized part of the memory.

(2) It has not been able to resist any attack against an open memory (make a memory object public) address.

(3) The relative address in a program segment remains unchanged. Generally, in the randomization operation, only the reference address of the memory segment is randomized, while the relative ubiety of all parts in the memory segment remains unchanged. This design provides convenience for attackers.

(4) The address randomization technology assumes that the attacker is ignorant of the contents of the memory.

(5) The available entropy is limited.

5.2.3 Instruction System Randomization

Instruction system randomization (ISR) [23] is a method of blocking code injection attacks through a fuzzy target instruction set. Its basic philosophy is to randomize specific instructions in the applications. Random operations can occur at the operating system layer, the application layer or the hardware layer. ISR is usually implemented in three ways:

(1) At the time of compiling, the machine code of the executable program is XOR-encrypted, where a specific register is used to store the randomization key used for encryption, with the machine code XOR-decrypted when the program interprets the instructions.

(2) Block encryption is used instead of XOR operation. For example, the AES encryption algorithm is adopted to encrypt the program code with an encryption interval of a 128-bit block for instruction set randomization.

(3) Randomization can also be implemented by different keys during the user-controlled program installation process so that the entire software stack is subject to instruction set randomization, thereby avoiding the execution of unauthorized binary codes and script programs.

With the instruction set randomization technology, it is difficult for an attacker to detect the instruction set of the attack target, and the code injected into the target application's vulnerability will not generate the expected attack effect. Figure 5.6 shows a typical ISR method. For the source code stored in the hard disk, the instructions can be randomized during the compilation process. When the program is compiled, the system randomly generates a series of keys from the key space to encrypt each instruction or instruction operation code in the file. Once encrypted, all the instructions in the program can only exist in the form of ciphertext in the memory, and the system will not read the decryption key from the key space to decrypt the corresponding ciphertext until the program executes the instruction [23].

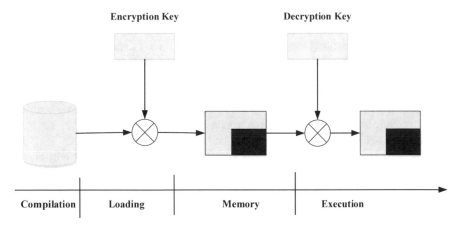

Fig. 5.6 ISR methods

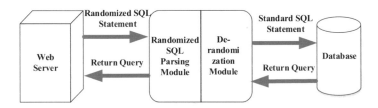

Fig. 5.7 Defense system model featuring SQL instruction set randomization

ISR is used across many defense systems. One example is the defense system model featuring SQL instruction set randomization [24]. As shown in Fig. 5.7, the system consists of a randomized SQL parsing module and a de-randomization module. The randomized parsing module is responsible for parsing the randomized SQL statement containing the user input data: if the interpretation succeeds, it is passed to the de-randomization module and restored to a standard SQL statement for execution by the database; if the interpretation fails, it is determined as containing unrandomized SQL keyword(s), indicating that the statement contains the code injected by the attacker, where abnormality handling is performed. The system workflow is described below:

(1) Perform MD5 encryption on the KEY to generate a key.
(2) Add the key to the SQL keyword, and assemble it with the user input data into a randomized SQL statement.
(3) Analyze the assembled SQL statement through the randomized parsing program.
(4) An injection is affirmed upon an interpretation error, with the user IP address recorded, and the attack code subject to security review.
(5) If the interpretation succeeds, the instruction will be de-randomized and passed to the database for execution.

Similar to ASR, instruction randomization also exploits the randomization technology to reduce threats based on unknown vulnerabilities or backdoors of the system, thus effectively improving system security. However, instruction randomization still faces many limitations now:

(1) The system execution overhead is considerable. The randomization of the instruction set makes it difficult for an attacker to predict the specific operational model of the program. At the operating system, the application software or even the hardware level, instruction set randomization can prevent the attacker from predicting the operation of the application (such as an attack laying a jump instruction in the program area to acquire the control power). An example of this technology is to encrypt the instruction when the program file is loaded and to decrypt it prior to execution. However, without trusted hardware as a guarantee, the system operation purely relying on software suffers remarkable performance loss. Besides, it involves potential limitations on advanced applications (e.g., dynamic links are not supported).
(2) In the absence of mainstream hardware architecture and specialized component support, it is difficult for simple XOR-encrypted instructions (because complex encryption can cause huge processing overhead) to defend against brute force by the attacker, who can overcome randomized defense by developing higher-level programs unaffected by randomization. For example, randomized web server's instruction set and address space layout cannot alleviate SQL injection attacks because it is targeted at the logic vulnerabilities of high-level applications. In fact, a randomized instruction set guarding against code injection attacks is not immune to memory leak attacks.

5.2.4 Kernel Data Randomization

The basic philosophy of kernel data randomization [25] is to change the way in which data is stored in memory. It usually performs XOR operation of a memory storage pointer with a random key to complete the encryption so as to prevent errors caused by pointer attacks.

At present, the technology has generated a lot of research results. PointGuard [25] uses a specific encryption algorithm to encrypt the pointer or address data stored in the memory. If an intruder directly attacks the data encrypted in the memory, the pointer tampered with by the attacker is likely to be one that contains an invalid value due to the lack of information on the encrypted key. Similarly, Cadar et al. [26] and Bhatkar et al. [27] are also based on registers. Before the program reads and writes, the XOR operation is conducted on the data in the memory by means of a random mask sequence. When the program reads and writes the data tampered within the memory, it will produce unpredictable results. Since the specific control stream modification strategy the attacker uses need not be considered in the data randomization method, which simply encrypts the data, the technology and address randomization are complementary to each other.

Figure 5.8a shows a simple data attack and a data randomization defense system. Suppose there is a pointer datum in the system memory with a value of 0x1234. When the CPU parses the pointer, it will get the data at the address of position 0x1234 in the memory. However, when an attacker tampers with the pointer by buffer overflow or a formatted string, the value of the pointer datum is changed to 0x1340. At this time, de-referencing according to the pointer datum will cause the execution path of the instruction addressing program to be directly or indirectly hijacked by the attacker.

However, in a system with data randomization, such as PointGuard shown in Fig. 5.8b, the pointer data in the memory exists in the form of ciphertext. For example, 0x1234 exists in the memory as the value of 0x7629. Before referring to the data of the pointer datum, the CPU needs to decrypt it first, namely, decrypting 0x7629 to 0x1234 and storing it in a register, and then access the 0x1234 position according to the normal process. If an attacker can still tamper with the specified datum according to the normal process by overriding 0x7629 with 0x1340, then, after decrypting the pointer datum, the CPU will get an unpredictable value. At this point, if the CPU accesses the memory according to this uncertain value, it will cause the program to crash to a large extent.

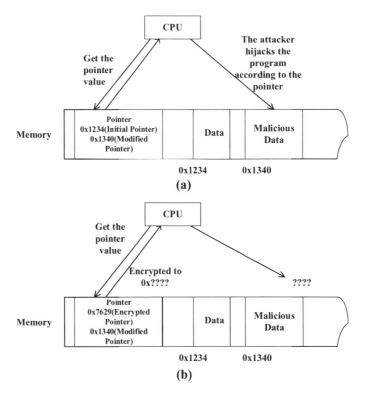

Fig. 5.8 Data randomization attack and the defense system model. (**a**) Data attack (**b**) Data randomization defense

5.2.5 Cost of Introduction

In general, the key parts of a network have high requirements for system reliability. For example, in the SDN architecture, the controller is the core component of the control layer and can logically and centrally control the switch network. However, once a single point failure occurs in single centralized control, the entire SDN will face huge security risk. As can be seen from the foregoing description, the security performance in the SDN scenario can be improved by increasing the diversity of the network control systems.

Therefore, by increasing the diversity at the hardware and software levels and integrating hardware and software resources through certain strategies, we can achieve higher system reliability. Of course, the introduction of diversity will inevitably increase the corresponding costs. First, diversity adds to the complexity of the system. In order to achieve a high level of diversity, a given service function shall be implemented in completely different ways. Next, higher system complexity will result in increased maintenance costs. Therefore, this section summarizes the costs of diversity into the following aspects:

5.2.5.1 Different Software and Hardware Versions Require Different Expert Teams to Design and Maintain

For hardware products with the same function, their quality control varies with the implementation algorithms, raw materials, and process flows of different vendors. If the server uses network cards, motherboards, power systems, etc., from different vendors for increased diversity, it will inevitably increase the difficulty or cost of production debugging, engineering deployment, and maintenance of hardware.

Under the multi-version system architecture (Fig. 5.9), a service runs multiple independent, functionally identical programs simultaneously, and their output results will be collected and tested by a voter to avoid possible single point failures of a single software system. However, a multi-version system needs to run multiple different and functionally identical versions of a program simultaneously, which

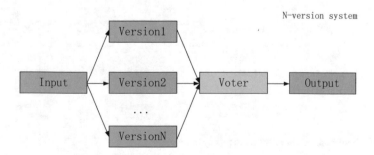

Fig. 5.9 Diagram of a multi-version system

inevitably increases the difficulty of development, deployment, and maintenance. While considering how to realize the functional requirements of the application software, application designers should also take into account the diversity of software (programs), which will double the workload of application system development and increases the complexity of the system.

When the theory of diversity system is applied to engineering practice, the great differences in the problems encountered in different engineering fields make it impossible to solve all the problems in a unified framework. The models and assumptions used in theoretical research differ much from that used in engineering practice, and this is also the main bottleneck for the diversity system theory applied to engineering practice.

System diversity not only puts higher demands on developers but also poses a much higher threshold to system administrators. The shortcomings of system isomorphism have been discussed in detail above, indicating the advantages of heterogeneous and diverse systems. It can be seen that system administrators who only master a single technology (skill) no longer meet the demand for managing a diversity system, so the future system administrators must have a more comprehensive professional knowledge base.

Program diversity usually takes the following four ways, as shown in Fig. 5.10. The first is to exert influence on the program during the compilation process from the source code to the executable file. In this way, the executable files obtained from each compilation are somewhat different from each other. The second is to modify the compiled program body per se, and the modifications will be solidified into the executable files of the program, with its impact showing each time when it is loaded. The third is to change the image in the program memory during the loading of the program body into the memory, so that it shows different behaviors or features in the current process, and the last is more complicated, where the program in running can change its own features on a regular basis or at any time to make them present diversity at different times. The four methods and the tools used to complete the conversions in these four phases are shown in Fig. 5.10.

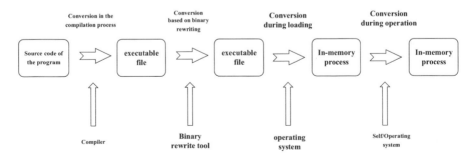

Fig. 5.10 Four methods of program diversity conversion

5.2.5.2 The Cost Will Inevitably Increase if a Multi-version Service System Is Constructed

Due to the differences in structure and communication methods, the cost of design and deployment of a multi-version service system is necessarily higher than that of a single-version one. Taking the SDN network as an example, the SDN Byzantine model can increase the reliability of the system and address single point failures in the network via a variety of heterogeneous redundant controllers, but the fact that multiple controllers serve one switch increases the service cost of the system, reducing the utilization of resources to some extent.

As shown in Fig. 5.11, each SDN switch in the SDN Byzantine defense system is managed by at least three controllers. The SDN switch will only work normally when more than half of the SDN controllers send correct control messages. Obviously, the mechanism will reduce the throughput of the SDN network and increase the communication latency.

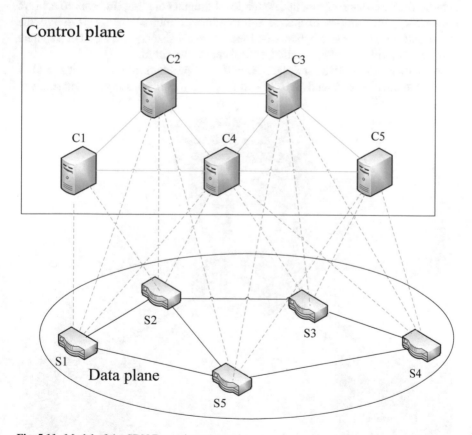

Fig. 5.11 Model of the SDN Byzantine system

5.2.5.3   Introduction of Diversity Makes Multi-version Synchronized Updating a New Challenge

Large-scale diversity processing increases system and management complexity, which mainly introduces two different problems, respectively, (1) the cost of establishing a diversity system and (2) the challenges of continuously managing a diversity system. The construction cost of a diversity system has been specified above and will not be repeated herein. This section will be a specific analysis of the challenges stemming from multi-version synchronized updating.

In order to increase the reliability of the system, synergy is required across multi-version executors. Then how to ensure multi-version update synchronization, or specifically, how to ensure consistency in multi-version updates, has become a new challenge.

Consistency refers to the fact that, in the absence of central coordination and control or global communication, several entities or processors distributed in space achieve the same state or output through local mutual coupling. In order to achieve consistency, the entities exchange information of mutual interest through the communication or sensing network and seek state consistency via a certain separate coupling algorithm, which is called a consistency algorithm.

As shown in Fig. 5.12, first, let's simplify the problem. Suppose that A adds a record Message_A and sends it to M and B adds a record Message_B and sends to

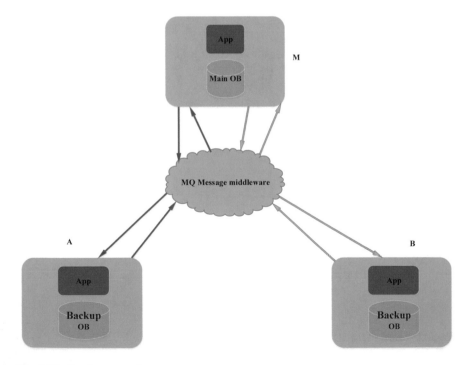

Fig. 5.12 Consistency updates

M; both are forwarded through the message queue (MQ) server. Then the system M has two more pieces of information after receiving the messages, and then group-sends the added messages to A and B. A and B then get their respective missing data and update the data, thus completing the synchronization of a datum.

The above process follows a completely reasonable logic in normal situations, but the following issues still need to be considered:

(1) How to ensure that A receives the message sent by M in the process of A → M, and vice versa.
(2) Assume that A sends multiple update requests simultaneously, how to ensure the orderly processing.
(3) How to ensure complete consistency of data between A and B after the data updates.

These problems involve consistency, an issue widely discussed across distributed systems. A diversity system is also a kind of distributed system, so it needs to address such problems. Of course, diversity systems have their own particularities: different executors are heterogeneous, and multi-version update synchronization faces greater challenges. Conventional consistency algorithms, such as Paxos [13], must be improved appropriately if they are to be applied to the diversity systems.

The system overhead arising from randomization cannot be ignored. Randomization is the process of making something random, which is not arbitrary but subject to a specific probability distribution. Randomization increases the security of a system by making the system more uncertain to attackers, usually by adding encryption and decryption modules, which is essentially an increase in space. For example, instruction randomization and data randomization are usually implemented via encryption. If the encryption algorithm is too simple, the expected protection effect is hardly achieved, but if the processing is too complicated, the system can hardly stand the loss of performance. Similarly, if the randomization operations are too few, the target object will not be well-hidden, but if the operations are too frequent, the system overhead will rocket up.

5.3 Dynamicity

5.3.1 Overview

Dynamicity refers to an attribute of the system that changes over time, also called "time-variance" in some areas. A time-varying system is characterized by the fact that its output is not only relevant to its input but also to the moment when the input is loaded. From the perspective of forward analysis, this essential feature increases the complexity of system analysis and research; from the perspective of defense, this feature is bound to increase the cost and difficulty of the attackers.

In common sense, cyber attacks based on static system architecture are much easier than defense. In particular, attacks featuring "one-way transparency and internal and external coordination" implemented on the basis of known vulnerabilities and pre-verified methods hardly miss. As the functionality of the system continues to increase, its complexity also rises sharply. The defects in the design and implementation of management and defense become serious engineering challenges, for an attacker can easily intrude into the system by detecting an exploitable vulnerability and launching an attack action, the expected effect of which is certain and can be planned. A possible defense method is to dynamically change the system's structure or operational mechanism under the resource (or time) redundancy configurations to create an uncertain defense scenario for attackers, to randomly use the system's redundant components or reconstructable, reconfigurable or virtualized scenarios to increase the uncertainty of the defensive conduct or deployment, or to use dynamicity (combined with diversity or pluralism) to maximize the implementation complexity based on coordinated attacks. In particular, it is necessary to turn over the mechanism-based vulnerability of static system defense. Even if it is unable to completely resist or invalidate all attacks, the system can be reconstructed to improve its resilience and flexibility or reduce the sustainable usability of the attack results. The theoretical basis of the Elastic Cloud program of the Defense Advanced Research Projects Agency (DAPAR) is that "an attacker will fail to figure out an ever-changing target system and carry out an attack." Therefore, the essence of dynamic defense is to disrupt as much as possible the attack chain and degrade the sustainable usability of the attack results with uncertainty.

Just as animals move fast or run to escape their predators, people use frequency hopping to improve the anti-interference ability of communication systems, dynamically adjusting the operating parameters or scene configurations of the target system according to a certain time law to reduce the vulnerability to attacks due to static parameters or configurations.

The effectiveness of dynamic defense depends on the entropy of the target system. The larger the entropy, the higher the apparent uncertainty of the target system, and the lower the reliability of attacks. Therefore, any effort to increase dynamicity, randomness, and redundancy will produce the following effects:

(1) It can reverse the asymmetry advantage of attackers and minimize the impact of attack actions on the key capabilities of the target object.
(2) It enables the system to resist unintentional, deliberate, or targeted attacks, so that the service functionality presented by the target system is resilient enough to withstand initial or consequent damage.
(3) It has the characteristics of segregation, isolation, and closure. For example, uncertain or uncontrollable parts are segregated from a trusted system to narrow down the attack surface, or critical resources are isolated from non-critical areas to reduce defense costs, or technologies such as reconstruction, reconfiguration, and virtualization are applied to make the defense scenarios of the target object uncertain and non-sustainable. The chances for an attacker to identify the vulnerability of a target object are reduced to protect the target object from

long-term effects of APT attacks and the like. APT is a form of long-term persistent cyber attack against specific targets using advanced attack means. The principle of APT attack is more advanced than that of other forms of attack and is mainly reflected in the need for precise collection of the business processes and target systems of the attack object prior to the launch of an attack. When collecting that information, the attacker will actively mine the vulnerabilities of the attack target's credit systems and applications, use these vulnerabilities to build the network needed for the attack, and finally attack via 0 day vulnerabilities. Latency and persistency are the biggest threats posed by APT attacks, mainly characterized by the following:

(1) Latency: These new types of attacks and threats may last for more than a year (or above) in the user's environment, where they constantly collect all kinds of information until something critical is detected. The purpose of the hackers launching APT attacks is not to profit in a short period of time but to use a "controlled host" as a springboard and continue to search until they can completely handle the targeted people, things, and objects. In this sense, the APT attack is essentially a model of "malicious commercial espionage" threat.

(2) Persistency: APT attacks are characterized by persistency, which can last even several years and remain undetectable to the corporate management. In the meantime, this persistency is reflected in various attacks constantly tried by the attacker and the long-term stagnation following the penetration into the network.

In short, APT is trying every way to bypass conventional code-based security solutions (e.g., antivirus software, firewalls, and intrusion prevention systems) and lurking in the system for a longer period of time, making its existence difficult for conventional defense systems to detect.

The dynamic defense technology has a good reason to launch an "arms race" with the attackers, for it can irregularly transfer the attack surface, preventing the attacker from acquiring the defender's information accurately and from ensuring the reachability of the attack packet. Thus, attackers will find it difficult to implement an effective attack. This is a big edge of dynamic defense.

Indeed, dynamicity also faces some challenges in design and implementation, including resource redundancy configuration, randomness cost, and performance cost.

5.3.1.1 Resource Redundancy Configuration

The prerequisite for introducing dynamics is the need to configure redundant resources. Whether it is a space resource or a time resource, a processing resource or a storage resource, and a physical resource or a logical resource, it requires redundancy configuration. In a sense, the redundancy of resources determines the range of dynamic changes, as well as the capacity or resilience/flexibility of the system in fault/intrusion tolerance.

Redundancy refers to the probability of introducing additional resources to reduce unpredictable events damaging the system. Commonly used redundancy methods include hardware redundancy, software redundancy, information redundancy, and time redundancy. Hardware redundancy increases the reliability of the system at the expense of additional system hardware resources and is divided into static redundancy and dynamic redundancy. Software redundancy improves reliability at the expense of increased program size and complexity, as well as higher resource overhead and performance loss, commonly achieved through N-text and recovery block methods. Information redundancy checks potential data deviation by attaching an additional error detection/correction code to the data and corrects the deviation if any. The commonly used error detection codes include the parity code, the cyclic code, and the fixed ratio transmission code, while the commonly used error correction codes include the Hamming code and the cyclic code. Time redundancy diagnoses whether the system has a permanent fault by attaching an additional execution time, and meanwhile offsets the impact of transient faults. Time redundancy has two basic modes: command re-execution and program rewind.

Resource redundancy configuration technology applied to the target object's architecture design not only helps to significantly improve the endogenous robustness of the system and the resilience of service provision but also provides the means and scenario support for the attack-defense game in the space-time dimension and maximizes the defense outcome.

5.3.1.2   Cost of Randomness

Dynamic defense with a random connotation enables the defense conduct to show greater uncertainty and unpredictability, which is extremely challenging for an external attacker attempting to build the attack chain based on a certain defense system, as well as for an internal penetrator attempting to achieve coordinated internal and external interaction. The conscious introduction of randomness into dynamicity can significantly improve the defender's game advantage in the attack-defense confrontation. We know that an attacker usually needs to implement different attack strategies according to the response of the dynamic defense system, which constitutes a non-cooperative incomplete information dynamic game posture. This non-cooperative decision-making process based on attack-defense confrontation has the following behavioral characteristics:

(1) Unlimited repetition. The attacker and defender play the game one round after another, each gaining their respective returns. Neither side knows when the confrontation will end. In each round, both sides must make decisions based on their own strategies so that they can get the best returns in an effective period of time.
(2) Reason. The attacker wants to choose the most effective way to attack the defender, while the defender always wants to defend the attacker in the best manner. The principle of the attack-defense game is the minimization of one's own cost loss or the maximization of one's returns.

(3) Non-cooperation. The attacker and defender are confrontational and non-cooperative. It can be known from the Nash theorem that there must be a Nash equilibrium for this cyber attack-defense process of limited strategies.
(4) Incomplete information. This refers to all knowledge related to attack and defense, covering the vulnerability of the information network systems or target audiences, the capabilities of the attacker and defender, the past attack and defense actions and results, and the role of the external environment. In the cyber attack and defense, the confronting decision-making parties only know their own expected objectives and do not know their opponent's targets. It is difficult for one side to know all the strategies adopted by the other side, and the information about which both sides are concerned and its value are likely to be different.
(5) Interdependence. The strategies and returns of the attacker and defender are interdependent. "While the priest climbs a post, the devil climbs ten." There is neither an absolute secure defense system nor a permanently effective attack means. The returns of the intrusions depend not only on the attacker's strategy but also on the defender's vulnerabilities and available defense measures. Similarly, the defense effect depends not only on the defender's strategy but also on the attacker's judgment and decision on the target object.

Obviously, the cyber attack and defense behaviors comply with the basic characteristics of the game theorem. Both parties expect to safeguard their own information while acquiring the information of their rivals, aiming at maximizing the returns and thus forming an infinitely repetitive, non-cooperative dynamic game posture with incomplete information, where the ability and level of applying randomization is one of the key factors determining who can gain a dominant position.

5.3.1.3 Cost of Effectiveness

In addition to the cost of redundant configuration of resources, the introduction of dynamicity may also sacrifice the performance and effectiveness of the target system, especially dynamic migration or scheduling within a complex system or between heterogeneous redundant executors, which often involves the relocation of large amounts of data or the synchronization or reconstruction of the internal state, the translocation between heterogeneous modules, or even the reconstruction or restart of the operating environment, all causing effectiveness loss of the system in varying degrees. In some scenarios, the degradation of system performance and effectiveness may be intolerable.

5.3.2 Dynamic Defense Technology

As shown in Table 5.4, in this section, the dynamic technology is classified into five categories: dynamic network, dynamic platform, dynamic operating environment, dynamic software, and dynamic data.

Table 5.4 Classification of dynamic technology

Serial No.	Category	Typical technology
1	Dynamic network	IP address hopping/port hopping
2	Dynamic platform	N-variant system based on compiled diversity technology
3	Dynamic operating environment	Address space randomization, instruction set randomization
4	Dynamic software	Software implementation diversity
5	Dynamic data	Randomized encrypted storage of data and programs

The key technology randomization in the dynamic operating environment has been described in Sect. 5.2 and will not be reiterated hereafter.

5.3.2.1 Dynamic Network

Through detection, an attacker can obtain the attributes of a network, including its address, port, protocol, and logical topology, to launch cyber attacks. In order to increase the cognitive difficulty of the target object and reduce the probability of a successful cyber attack, it is necessary to constantly change the attributes of the network. The basic idea of a dynamic network is to increase the cost of attackers by dynamically modifying the network attributes, attempting to prevent or disturb the attacker's detection and attack actions by frequently changing the address and port or even the logical topology of the network, preventing the attacker from spotting or exploiting the target object's vulnerabilities and invalidating the conventional scanning means. On the one hand, the dynamic network can prevent an attacker from discovering the exploitable vulnerability, and on the other hand, the historical scan data obtained by the attacker can be invalidated during the attack phase. The following describes the dynamic network technology by taking the dynamics of the network address as an example.

Randomized processing of the information on the packet header of the network makes it difficult for an attacker to identify the communicating parties, the services being used, and the location of the key systems in the network. Specifically, this includes the source and destination MAC addresses, the source and destination IP addresses, the TOS field of the IP address, the source and destination TCP ports, the TCP sequence number, the TCP window length, and the source and destination UDP ports. As a prelude to cyber attacks, cyber sniffing [28] poses a great threat to cyber security. As a result, IP hopping technology is widely used for defense purpose, with its implementation and verification architecture shown in Fig. 5.13. The core consists of a DNS server, an authentication server, N IP routers, and an IP hopping manager. A legitimate user will identify the IP address of the server through the DNS server and the authentication server and will then perform normal transmissions. A malicious user accesses the network server according to the

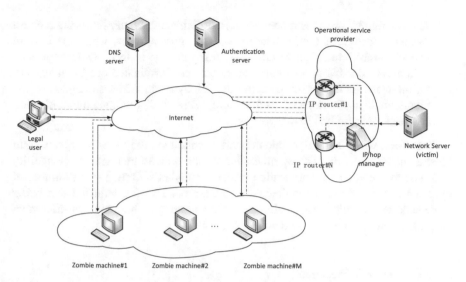

Fig. 5.13 Diagram of the IP hopping architecture

previously detected IP address, while the IP hopping manager periodically converts the access router so that the IP address obtained by the malicious user becomes unavailable, preventing the "steady" attack implementation.

Dynamic network technology based on network address has undergone broad-based development [29–35]. This technology effectively adds to the difficulty for an attacker to steal valid message information to map the network and launch an attack, but it is opaque to users, and the mechanism is designed to protect a group of static nodes deployed behind the centralized gateway, responsible for the execution address translation of all messages travelling between the protected network and the external network through an interface. For example, it defends against worm attacks by adjusting the frequency of IP address updates in the dynamic allocation of network addresses. The mechanism needs to configure a dynamic host configuration protocol (DHCP) server to terminate the DHCP lease at different time intervals for address randomization. The address alternation can be performed when the node is restarted, or implemented based on the setting of the timer. Since this mechanism must be deployed in a network with dynamic addresses, the deployment cost is relatively high. In addition, as the mechanism implements address randomization at the LAN level, the unpredictability it can provide is relatively low.

Although the dynamic network technology can bring about certain defense effects, it may affect the normal operation of the system. In addition, the application scenarios of the dynamic network technology itself are also limited by conventional network technologies.

(1) This type of defense is unlikely to work on services that must reside on a fixed network address or on network services that must remain otherwise reachable (e.g., network servers and routers.).

(2) The existing C/S service model network architecture strictly limits the random dynamic change of the IP address and the port number, while the current network architecture fails to afford the rapid change of the network topology, with a considerable part of upper-layer services invalidated due to changes in the network hierarchy. In fact, the use of dynamic fuzzy strategies on the hierarchical static network architecture without disrupting communication connections is limited in many ways.

Therefore, when using dynamic network technology, the network service shall be at a known network location, otherwise it shall be with guaranteed reachability and support common communication protocol standards. Only a small number of dynamic cyber defense technologies have been used so far, with the rest being achievable but not universally adopted. To illustrate, port hopping, a well-known technology though, has not been widely deployed.

5.3.2.2   Dynamic Platform

The basic philosophy of the dynamic platform [36] is to dynamically change the properties of the computing platform to disrupt attacks relying on specific platform structures and operational mechanisms. Properties that can be changed include the operating system, the processor architecture, the virtual machine instance, the storage system, the communication channel, the interconnected architecture, and other underlying environmental elements. In the cloud computing field, this technology serves to migrate applications between virtual machines or execute the same application in parallel in the context of multiple different structures.

As shown in Fig. 5.14, Reference [37] provides a proactive defense scheme through the diversity of services achieved by the rotating of multiple heterogeneous executors. The same WordPress service is deployed with different versions of Linux virtual machines sharing the storage data. The system provides an outbound IP interface, sets up an in-house IP resource pool, and assigns an IP address to each virtual machine. These IP addresses are either live or spare. All the live hosts will run normally, while each host using a spare address will be subject to security check. If an attack is detected, the host will no longer be used. The IP addresses of all the hosts are rotating, and the host previously in the live state will be rotated to the spare state for and be checked.

The advantages of dynamic platform technology are remarkable, for it can provide certain defense capabilities in each attack phase, which are mainly reflected in the approach, development, and continuity phases. Defense in the approach phase is achieved by changing the defense system's attributes or behaviors exposed to the attacker. If an attacker needs to base his construction of an attack chain on multiple security vulnerabilities from a variety of platforms of the defender, he will face a sticky issue of coordination of attack in the development phase. Besides, even if an attack is successful, its attack effect is unlikely to be reproduced due to the dynamic migration behavior of the defender, thus losing the value of sustainable exploitation.

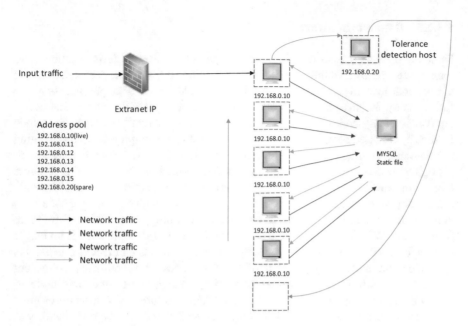

Fig. 5.14 Operation sheet of a dynamic platform

Of course, the dynamic platform technology also has obvious technical flaws:

(1) The biggest concern is the loss of performance and the challenge of implementation cost. Besides, cross-platform migration is a stressful high-cost, low-performance task. At present, the bottleneck is the lack of a common system as well as mechanism for platform-independent application state transition.

(2) It is difficult to extract the state as a platform-independent format. It is a daunting task to keep the application state unchanged or maintain synchronization across platforms during the dynamic platform migration process. The emergence of cloud computing has helped to troubleshoot the problem, but it has also created a new one: suppose there is an attacker, and if the platform migration and the state transition cannot be performed in real time, the attacker will be able to continue with the attack.

(3) Dynamic platform technology will inevitably increase the attack surface (AS) of the protected system. For example, not only new software/hardware components need to be added to control and manage the migration process, but also new vulnerabilities may exist in these new components, creating new targets of attack.

(4) The application of dynamic platform technology is based on the conditions for platform diversity. In fact, due to the limitations of diversity and compatibility and cross-platform technology, the engineering implementation of dynamic platform technology is by no means an easy mission.

5.3.2.3 Dynamic Software

Dynamic software means that the internal state of the software during runtime is no longer determined certainly by the input stimulus (the random parameters in the application system environment need to be added) on the premise that the functionality is not affected. Instead, it is implemented through semantically equivalent binary code conversion programs or different versions of execution code formed by changed compiler parameters. Specifically, it is implemented by changing the internal running rules or processes determined by the source program or changing the execution instruction sequence and form, dynamically storing resource allocation scheme or randomly invoking the redundant function modules. The main philosophy of dynamic software technology is to dynamically change the code of an application, including modifying the order, grouping, and style of program instructions. With this modification, the internal state of the application is no longer certain for the input when the existing functionality remains unaffected. Diversity is generated via mutual replacement of the sequences of functionally equivalent program instructions, as shown in Table 5.5, changing the order and style of instructions, rearranging the layout of internal data structures, or altering the original static nature of the application. This natural conversion reduces the exploitability of vulnerabilities at the specific instruction layer, where an attacker has to guess which software variant is being used. Two functionally equivalent instruction sequences are given below, as shown in Table 5.5.

The ChameleonSoft [38] system is a typical application of dynamic software technology, and its system architecture is shown in Fig. 5.15, which illustrates ChameleonSoft's core concepts: function, behavior, and organizational structure. Its function is described by "task" and defines different roles through tasks; its behavior is the dynamic performance of the system, which is the "variable" of the system; and its organizational structure is the infrastructure that supports the variables. In the abstract sense, a system consists of "elements," where each element plays a different role; from the perspective of instantiation, the instantiation of multiple elements is "unit," and the unit encapsulates "resources," which in operation appear as different variables; the elements, units, and resources are connected by a uniform communication bus.

In practice, the system relies on software diversity to encrypt software behaviors so as to build a dynamic system. Its design philosophy is to decouple the functional roles of software from the behavioral roles in operation, design a combinable,

Table 5.5 Comparison between two functionally equivalent instruction sequences	Sequence 1	Sequence 2
	xor eax, eax	mov eax, 0x0
	shl ebx, 0x3	mul ebx, 0x8
	pop edx	Ret
	jmp edx	

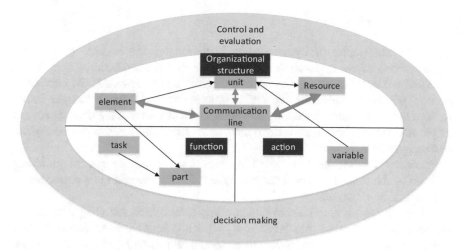

Fig. 5.15 Diagram of the core components of ChameleonSoft

programmable online compilation block with endogenous resilience to separate the logic, state, and physical resources, and meanwhile utilize the code variant with the same functions but different behaviors. The ChameleonSoft system can suppress chaos by encrypting execution behaviors at a reasonable cost. Since there are multiple variants, it is difficult for an attacker to attack the system through code injection and code reuse.

So far, dynamic software technology has not been widely used. It is still limited to academic research [39–43], hindered by two problems:

(1) Binary code conversion and simulation involve the performance of upper-layer services. There are some uncertainties in the impact on the functions, especially the static translation. As a result, it is unlikely to be used in large scale, and there may be other unexpected side effects. The main shortcoming of dynamic software technology lies in the difficulty to ensure that the converted software is functionally equivalent to the original software. Heavyweight binary conversion and simulation incurs significant overhead and lacks scalability, which may produce other unexpected edge effects even if the desired effect is achieved.

(2) Although the compilation-based approach has a relatively high credibility, many software uses compilation measures that optimize software performance. If only the software compiled with semantic equivalence is adopted, the loss or change of optimized processing details may greatly affect the performance and lead to differences between execution variants or versions of the source program. In addition, the compilation-based approach requires the source code, which inevitably touches the trading rules of existing commercial software (the mainstream is the trading of executable files rather than source codes).

5.3.2.4   Dynamic Data

The main philosophy of dynamic data is to change the internal or external representation of the application data to ensure that the semantic content is not modified and that unauthorized use or intervention can be prevented. The internal or external representation of the application data is changed through changes in the form of the data, syntax, encoding, and other attributes, so that the data penetrated by the attacker will be detected due to incorrect format, and the data stolen may not be displayed in an intuitive form.

When the data appears to be in an inappropriate format, the penetration attack initiated by the attacker can be detected. Table 5.6 shows two different performance formats for the same data.

Microsoft has proposed a data dynamism/randomization technology to defend against attacks that exploit memory errors [44]. It applies XOR operation to the data in the memory by means of a random mask sequence, increasing the difficulty of the attacker using a "memory error," and a successful attack becomes a probability event. This technology splits the instruction operand into equivalent categories based on static analysis. For example, if two instruction operands correspond to the same object in one execution, then the two operands are classified into the same category. Then, a random mask is assigned to the category, and the XOR operation is performed on the corresponding operand with the mask to generate the dynamic data. Therefore, an attacker who violates the results of static analysis will get unpredictable results, and will be unable to normally exploit the memory error vulnerability.

Similar to dynamic software technology, the dynamic data technology attempts to disrupt the development and launch stages of attacks. The development stage of an attack is impaired because it is difficult to generate a suitable payload for a plurality of different data presentations, making it hard to successfully attack a vulnerability of a specific data format. The dynamic data technology has not yet actually deployed and applied, and the existing research is focused on memory encryption or randomization of specific data. The dynamic data technology has the following limitations:

(1) Due to the lack of diversity in data encoding methods, the majority of standard binary formats only support one recognized representation, with a confined development space. The new binary format is not encouraged unless new requirements are fulfilled or significantly improved over the original method.
(2) Besides, each additional data representation format requires a corresponding parsing and check code. New formats of data may also lack compatibility. The

Table 5.6 Formats of data representation

Format 1	Format 2
<Age = 24;	<Gender = F;
Gender = female;	Age = 11,000;
ID = 159,874;	Salary = $65 K;
Salary = $65,000; >	ID = 00159874; >

creation of new data formats can lead to loss of generality, so it requires the application to add a new usable code instead of using a tested standard library.

(3) Any additional processing of the data format during program execution will reduce the performance of the system, and managing the diverse data formats simultaneously will also affect the application efficiency.

(4) The use of enhanced defense measures in a recognized data format is usually easier to realize than using a diverse form of data protection, because the former only requires the data in use to be in an explicit form rather than an implicit form as long as the timeliness of the encryption algorithm is guaranteed.

(5) The diverse presentations of data categories will inevitably lead to the expansion of the AS.

(6) Although encryption can effectively protect the internal data state of the application, given the lack of a universally pragmatic encryption scheme, all data decryption needs to be returned to the original form before any processing, which is likely to bring new vulnerabilities.

5.3.3 Dynamicity Challenges

The dynamic diverse defense technology is dedicated to three attributes: comprehensiveness, timeliness, and unpredictability. It emphasizes that defense should be all-inclusive and comprehensive, able to effectively avoid attacks; that the system response should be timely, without providing predictable defense models to the opponent; and that dynamicity and randomness should be in place, without providing the attacker with reproducible attack results. This will force the attacker to constantly chase the attack target, weaken the attacker's advantage in time and space, and make the defender more flexible in the face of persistent advanced threats, thus effectively preventing or even eliminating the impact of cyber attacks. Nevertheless, dynamization is also challenged in four aspects:

(1) The service mechanisms and functions presented by the network, platform, or operating environment are difficult to be dynamized. Since the access mode, service interface, service function, standard specification, and even user custom of service facilities must have long-term stability, some open ports and device addresses are mostly used as basic identifiers for cyber addressing, with extremely limited variable space. Given that the inheritability of software is an essential constraint of cyberspace technology advancement, any idea, approach, or method on dynamicity is hard to be accepted in practical applications.

(2) We have to consider the performance and effectiveness loss due to dynamization under the same processing environment or shared resources. In particular, the cost of dynamization upon complex systems and large-scale network facilities may grow nonlinearly with changes in the depth and breadth, while the range of dynamic change (i.e., the size of the entropy space) and the complexity and rate of the change determine the defense outcome of such dynamization.

(3) How to ensure the security and credibility of the dynamic control and scheduling section in the same "virus/bacteria-bearing" environment. Unknown security threats such as vulnerabilities and backdoors are fundamental problems that cannot be avoided by any security systems, posing a particularly serious challenge for countries with backward technology and high market dependence. If dynamization is based on the "backdoored" hardware/software devices or components, it is difficult to trust the effectiveness of the dynamic defense results. To make matters worse, for any given device, modern science and technology has not been able to prove the existence or nonexistence of vulnerabilities. This requires the effectiveness of dynamization to be based on the components "without absolute loyalty/reliability," which will again resort to the paradox of the underlying cyber security issues.

(4) Dynamicity, randomness, and diversity require a bearer architecture that can perform iterative or additive effects. From the perspective of engineering implementation, none of the measures listed in the previous chapters is systematic and comprehensive, as most of them are means of partial transformation and unable to dodge the inheritance problem of existing software/hardware achievements (e.g., the cases where there are only executable code files). A defender seeks a systematic or architected solution that can affect all stages of an attack chain in all directions, where any defensive elements added can reduce the reliability of the attack nonlinearly, with the defense effect of "one hair is pulled, the whole body will be moved."

Dynamicity, diversity, and randomness enable the defender to display strong uncertainty, so it is a promising active defense technology, which can make the vulnerabilities of information service devices less exploitable to attackers. The effectiveness of defense depends not only on the number of performance characteristics of dynamic changes but also on the size of the entropy in the randomization process and the degree of diversity. The defensive effect of dynamic diversity can be improved through a randomized system process environment or a "cleaning operation" based on a trigger event. However, research data [16] indicate that for many scenarios, the benefits of dynamic randomization do not meet the expectations. Dynamic randomization does not provide benefits for evasive and proxy attacks, and it at most doubles the attack difficulty for entropy reduction attacks. For other types of attacks, dynamic diversity defense may provide more significant advantages. And more in-depth research and practice are needed to achieve the expected high cost-effective results.

5.4   Case of OS Diversity Analysis

We know that one of the main advantages of an intrusion-tolerant system is that it can guarantee compliant service or performance to a certain extent upon attacks and intrusions. Its safety gain is strongly correlated with the diversity of faults generated by its components. Fault diversity that can be observed in actual deployment depends on the diversity of the system components. On the basis of the OS-related vulnerability data in the NVD, in their *Analysis of OS Diversity for Intrusion Tolerance*, Garcia and Bessani et al. [20] studied the vulnerabilities discovered in 11 operating systems over 18 years and analyzed the statistics of these vulnerabilities simultaneously appearing in various operating systems. They found out that the figure was very small (with a minimum of 0) for several combinations of OSs. Analysis shows that choosing the right combination of operating systems can eliminate or greatly reduce the number of common vulnerabilities that occur simultaneously in heterogeneous redundant systems.

The author believes that although Sect. 5.1.4 outlines some of the research outcomes of OS diversity, it is necessary to stress on the main work of, and the valuable conclusions derived from, the paper *Analysis of OS Diversity for Intrusion Tolerance*, because it provides critical reference value for the OS diversity selection and the related heterogeneity design of the mimic defense system in the subsequent chapters.

5.4.1 Statistical Analysis Data Based on the NVD

Maintained by the NIST, the NVD is a standardized vulnerability management database in support of automated vulnerability management, security measurement, and compliance investigation. As shown in Table 5.7, Garcia and Bessani et al. selected 2563 OS-related vulnerabilities from more than 44,000 vulnerabilities in the NVD, with a time span of 18 years (1994–2011).

The results of the analysis of 2270 valid vulnerabilities are summarized in Table 5.8 and then assigned to each OS component category. As shown in this table, all OS products have a certain number of vulnerabilities in each component category except for the driver. Therefore, if some software components can be removed or cut

Table 5.7 Distribution of OS vulnerabilities in the NVD

(OS)	(Valid)	(Unknown)	(Unspecified)	(Disputed)	(Duplicate)
OpenBSD	153	1	1	1	0
NetBSD	143	0	1	2	0
FreeBSD	279	0	0	2	0
OpenSolaris	31	0	52	0	0
Solaris	426	40	145	0	3
Debian	213	3	1	0	0
Ubuntu	90	2	1	0	0
Red Hat	444	13	8	1	1
Windows2000	495	7	28	5	5
Windows2003	56	5	34	3	5
Windows2008	334	0	8	0	3
#distinct vulns.	2270	63	210	8	12

Table 5.8 Categorized analysis of the OS vulnerabilities

OS	(Driver)	(Kernel)	(Sys. Soft.)	(Application)	(Total)
OpenBSD	2	76	37	38	153
NetBSD	9	64	39	31	143
FreeBSD	4	153	61	61	279
OpenSolaris	0	15	9	7	31
Solaris	2	155	120	149	426
Debian	1	25	39	148	213
Ubuntu	2	22	8	58	90
Red Hat	5	94	108	237	444
Windows2000	3	146	135	211	495
Windows2003	2	171	96	291	560
Windows2008	0	123	36	175	334
Total	1.00%	33.50%	22.50%	42.90%	

off from the OS distribution, the security gain can be enhanced accordingly. In the BSD/Solaris series, vulnerabilities mainly appear in the kernel, while in Linux/Windows systems, the application vulnerabilities are more prevalent. It can be seen that default distributions in Windows/Linux typically contain more preinstalled (or bundled) applications compared to BSD/Solaris. As a result, the number of application vulnerabilities preinstalled in Windows/Linux has a higher reported value.

The last row of Table 5.8 shows the percentage of each category in the total dataset. It should be noted that drivers account for a small percentage of the released OS vulnerabilities, which is a bit surprising, for device drivers make up a large portion of the OS code, usually subject to closed development and likely to contain more programming flaws. In fact, previous research has shown that driver errors are the main cause of some OS crashes. However, software errors do not necessarily translate into security vulnerabilities. If a design flaw/error is to be exploited as vulnerability, an attacker must also have the ability to enforce certain conditions or means to activate the flaw/error. It is also not easy to effectively exploit device driver design flaws, since they are at the bottom of the system.

5.4.2   Common OS Vulnerabilities

The paper analyzes the shared vulnerabilities across various "OS combinations" during the period 1994–2011. Three highly general server settings are considered, as they provide the relevant support results directly from NVD data:

(1) Fat server: The server contains the majority of the packages for a given OS, so it can be used for local or remotely connected users to run various application services.
(2) Thin server: A platform that does not contain any applications (except for replication services). As the application-related vulnerabilities have been largely eliminated, the server security risks are significantly reduced.

(3) Isolated thin server: The configuration of the server is similar to that of a thin server, but it is placed in a physically isolated environment, where remote login is disabled, so it is only exposed to security risks implemented by malicious packets received through the local network.

Table 5.9 shows the shared OS vulnerabilities for each combination. In all cases, the number of vulnerabilities shared between two OSs is greatly reduced when compared to global vulnerabilities. Even when the fat server configuration is taken into account, we can find combinations of OSs that have no common flaws (e.g., NetBSD-Ubuntu) and very few common vulnerabilities (e.g., BSD and Windows). As expected, OSs from the same family have more common vulnerabilities due to software components and reusable applications (e.g., Debian-Red Hat or Windows 2000-Windows 2003). Compared to a fat server, a thin server shows improvements in several OS combinations, but there are often some common pitfalls. In contrast, the use of an isolated thin server has a greater impact on security guarantee, as it greatly reduces the number of common vulnerabilities—the number of zero-vulnerability pairs increases to 21 from 11. In general, this means that a large portion of common vulnerabilities are local (i.e., not remotely exploitable) and/or from applications available in both OSs. Table 5.9 vulnerabilities (1994–2011): fat servers, all vulnerabilities; thin servers, no application vulnerabilities; and isolated thin servers, no application vulnerabilities, and only local exploitation. The columns $v(A)$ and $v(B)$ show the total number of vulnerabilities collected by OS A and OS B, respectively, while $v(A, B)$ is the number of vulnerabilities affecting both A and B.

Figure 5.16 depicts the number of common vulnerabilities that exist simultaneously across OSs of different numbers (using isolated thin servers). The figure shows that, as the number of OSs increases, the number of common vulnerabilities decreases, and they usually occur to systems of the same family with a common ancestry, which means a larger portion of the code base is reused. As it can be observed in the previous tables, this is especially true for Windows and BSD series.

Table 5.10 lists in more detail the vulnerabilities that can be exploited in (4–6) combinations of large OSs. The first three errors have a considerable impact because they allow a remote opponent to run arbitrary commands on the local system using a high-authority account. They occur in a wide range of login services (telnet and rlogin) or basic system functions, so several products in the BSD and Solaris series are affected. The NVD entry for the vulnerability CVE-2001-0554 also has external references to the Debian and Red Hat websites, indicating that the systems concerned may encounter similar (or identical) problems. Vulnerability CVE-2008–1447 occurs to a large number of systems because it is caused by a bug in the DNS BIND implementation. Since BIND is a popular service, more OSs may be affected. The above vulnerabilities show that from a fault-tolerant perspective, it is unwise to run the same server software everywhere, and different servers should be selected for the purpose.

Overall, the above results are encouraging, because for a long time (about 18 years) there are few vulnerabilities in many OSs, of which a large portion stems from the TCP/IP stack software, and in most cases due to shared components.

Table 5.9 Common OS vulnerabilities for each combination

Operating systems	Fat server			Thin server			Isolated thin server		
Pairs (A-B)	v(A)	v(B)	v(A, B)	v(A)	v(B)	v(A, B)	v(A)	v(B)	v(A, B)
OpenBSD-NetBSD	153	143	45	115	112	34	62	46	17
OpenBSD-FreeBSD	279	57		218	49		90	33	
OpenBSD-OpenSolaris	31	1		24	1		6	0	
OpenBSD-Solaris	426	13		277	10		108	6	
OpenBSD-Debian	213	2		65	2		28	0	
OpenBSD-Ubuntu	90	3		32	1		10	0	
OpenBSD-Red Hat	444	11		207	6		69	4	
OpenBSD-Win2000	495	3		284	3		183	3	
OpenBSD-Win2003	560	2		269	2		138	2	
OpenBSD-Win2008	334	1		159	1		56	1	
NetBSD-FreeBSD	143	279	54	112	218	41	46	90	25
NetBSD-OpenSolaris	31	0		24	0		6	0	
NetBSD-Solaris	426	18		277	14		108	10	
NetBSD-Debian	213	5		65	4		28	4	
NetBSD-Ubuntu	90	0		32	0		10	0	
NetBSD-Red Hat	444	12		207	9		69	6	
NetBSD-Win2000	495	3		284	3		183	3	
NetBSD-Win2003	560	1		269	1		138	1	
NetBSD-Win2008	334	1		159	1		56	1	
FreeBSD-OpenSolaris	279	31	0	218	24	0	90	6	0
FreeBSD-Solaris	426	23		277	15		108	8	
FreeBSD-Debian	213	7		65	4		28	1	
FreeBSD-Ubuntu	90	3		32	3		10	0	
FreeBSD-Red Hat	444	20		207	13		69	5	
FreeBSD-Win2000	495	4		284	4		183	4	
FreeBSD-Win2003	560	2		269	2		138	2	
FreeBSD-Win2008	334	1		159	1		56	1	
OpenSolaris-Solaris	31	426	27	24	277	22	6108	6	
OpenSolaris-Debian	213	1		65	1		28	0	
OpenSolaris-Ubuntu	90	1		32	1		10	0	
OpenSolaris-Red Hat	444	1		207	1		69	0	
OpenSolaris-Win2000	495	0		284	0		183	0	
OpenSolaris-Win2003	560	0		269	0		138	0	
OpenSolaris-Win2008	334	0		159	0		56	0	
Solaris-Debian	426	213	4	277	65	4	108	28	2
Solaris-Ubuntu	90	2		32	2		10	0	
Solaris-Red Hat	444	17		207	10		69	6	
Solaris-Win2000	495	10		284	3		183	3	
Solaris-Win2003	560	8		269	1		138	1	
Solaris-Win2008	334	1		159	0		56	0	
Debian-Ubuntu	213	90	15	65	32	6	28	10	2

(continued)

Table 5.9   (continued)

Operating systems	Fat server			Thin server			Isolated thin server		
Debian-Red Hat	444	66		207	28		69	13	
Debian-Win2000	495	1		284	1		183	1	
Debian-Win2003	560	1		269	1		138	1	
Debian-Win2008	334	0		159	0		56	0	
Ubuntu-Red Hat	90	444	28	32	207	8	10	69	1
Ubuntu-Win2000	495	1		284	1		183	1	
Ubuntu-Win2003	560	1		269	1		138	1	
Ubuntu-Win2008	334	0		159	0		56	0	
Red Hat-Win2000	444	495	2	207	284	1	69	183	1
Red Hat-Win2003	560	2		269	1		138	1	
Red Hat-Win2008	334	0		159	0		56	0	
Win2000-Win2003	495	560	265	284	269	120	183	138	85
Win2000-Win2008	334	80		159	29		56	16	
Win2003-Win2008	560	334	282	269	159	125	138	56	40

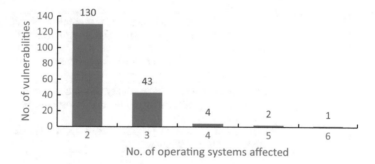

Fig. 5.16 Number of common vulnerabilities simultaneously existing in n different OSs (isolated thin server)

Table 5.11 shows the number of vulnerabilities that are common to the OSs in a fat server environment on a yearly basis, indicating a large number of OS combinations with zero common vulnerability over the years (45% of non-empty units are zero). This is especially true for OSs that belong to different families, such as between Debian and Solaris or between Debian and Windows 2008. Years with zero entries even appear on OS combinations with very high vulnerability counts, for sometimes errors are aggregated in some years rather than evenly distributed (e.g., Debian-Red Hat). In addition to the OSs in Windows systems that typically share a variety of software, Debian-Red Hat had a maximum of 20 vulnerabilities in 2005. For the rest of the data, the number is much smaller, with only three other OS combinations exceeding 10 (Ubuntu-Red Hat in 2005, Debian-Red Hat in 2001, and OpenBSD-FreeBSD in 2002). The last row of the table shows the average value, standard deviation and maximum number of vulnerabilities each year. As shown by

Table 5.10 Vulnerabilities affecting more than four OSs

Serial No. of CVE	Number of affected OSs	Description
CVE-1999-0046	4	A RLOGIN buffer overflow vulnerability that gains the administrator authority OSs affected: NetBSD, FreeBSD, Solaris and Debian
CVE-2001-0554	4	A TELNETD buffer overflow vulnerability that allows remote attackers to run arbitrary commands OSs affected: OpenBSD, NetBSD, FreeBSD and Solaris
CVE-2003-0466	4	A Kernel function fb_realpath() vulnerability that gains the administrator authority OSs affected: OpenBSD, NetBSD, FreeBSD and Solaris
CVE-2005-0356	4	A TCP implementation vulnerability that implements a denial of service attack by forging a packet with an extra-long timer OSs affected: OpenBSD, FreeBSD, Windows2000 and Windows2003
CVE-2008–1447	5	A DNS software BIND 8/9 vulnerability that leads to cache poisoning attacks OSs affected: Debian, Ubuntu, Red Hat, Windows2000 and Windows2003
CVE-2001-1244	5	A TCP implementation vulnerability that implements a denial of service attack by setting a packet that exceeds the maximum fragment size OSs affected: OpenBSD, NetBSD, FreeBSD, Solaris and Windows2000
CVE-2008-4609	6	A TCP implementation vulnerability that implements a denial of service attack by abnormally setting multiple vectors of the TCP state table OSs affected: OpenBSD, NetBSD, FreeBSD, Windows2000, Windows2003 and Windows2008

the standard deviation, there are very few OSs with massive common defects (Windows series), and the number of other common vulnerabilities is decreasing. In fact, if the impact of Windows 2008 is removed from the table, the average number of common vulnerabilities has decreased over the past 3 years. From the perspective of deploying an intrusion-tolerant system, there are several OS combinations with hardly any shared vulnerabilities at reasonable intervals (several years). In addition, in some cases, these vulnerabilities will decrease as the system matures. Therefore, it should be feasible to choose a configuration of four or more OSs as they are highly resilient to intrusions.

5.4.3 Conclusions

Potential security benefits can be obtained by using multiple OSs in heterogeneous redundant intrusion-tolerant systems:

Table 5.11 Number of common vulnerabilities in OS combinations (using fat servers) from 2000 to 2011

OS combination	Number of vulnerabilities 2000–2011
OpenBSD-NetBSD	5 7 5 2 2 0 3 2 7 0 0 4
OpenBSD-FreeBSD	2 6 11 6 6 2 2 1 7 0 0 4
OpenBSD-Solaris	0 2 3 2 1 0 0 1 0 0 0 1
OpenBSD-Debian	0 0 0 0 0 0 0 1 1 0 0 0
OpenBSD-Ubuntu	- - - - 0 0 0 3 0 0 0 0
OpenBSD-Red Hat	3 0 1 0 3 0 0 3 0 0 0 0
OpenBSD-Win2000	0 1 0 0 0 1 0 0 1 0 0 -
OpenBSD-Win2003	- - - 0 0 1 0 0 1 0 0 0
OpenBSD-Win2008	- - - - - - - - 1 0 0 0
NetBSD-FreeBSD	4 6 6 5 1 0 3 1 9 1 2 4
NetBSD-Solaris	0 2 0 4 0 0 2 0 0 1 0 1
NetBSD-Debian	0 2 2 0 0 0 0 0 0 0 0 0
NetBSD-Ubuntu	- - - - 0 0 0 0 0 0 0 0
NetBSD-Red Hat	2 2 2 0 0 0 0 0 0 0 0 0
NetBSD-Win2000	0 1 0 1 0 0 0 0 1 0 0 -
NetBSD-Win2003	- - - 0 0 1 0 0 1 0 0 0
NetBSD-Win2008	- - - - - - - - 1 0 0 0
FreeBSD-Solaris	0 2 2 4 0 1 0 1 1 1 0 2
	2000 01 02 03 04 05 06 07 08 09 102,011
FreeBSD-Debian	1 1 2 0 0 0 0 0 0 0 0 0
FreeBSD-Ubuntu	- - - - 0 2 0 0 0 0 0 0
FreeBSD-Red Hat	2 2 2 0 3 2 0 0 1 0 0 0
FreeBSD-Win2000	0 1 0 1 0 1 0 0 1 0 0 -
FreeBSD-Win2003	- - - 1 0 1 0 0 1 0 0 0
FreeBSD-Win2008	- - - - - - - - 1 0 0 0
Solaris-Debian	1 0 0 0 0 0 0 1 0 0 0 0
Solaris-Ubuntu	- - - - 0 1 0 1 0 0 0 0
Solaris-Red Hat	1 0 1 1 0 1 0 1 3 0 2 0
Solaris-Win2000	0 1 0 1 0 1 0 5 0 0 1 -
Solaris-Win2003	- - - 0 0 1 0 5 1 0 1 0
Solaris-Win2008	- - - - - - - - 0 0 1 0
Debian-Ubuntu	- - - - 0 8 0 2 2 2 0 0
Debian-Red Hat	8 11 5 0 2 2 0 0 4 2 0 0 0
Debian-Win2000	0 0 0 0 0 0 0 0 1 0 0 -
Debian-Win2003	- - - 0 0 0 0 0 1 0 0 0
Debian-Win2008	- - - - - - - - 0 0 0 0
Ubuntu-Red Hat	- - - - 3 16 1 5 2 0 0 0
Ubuntu-Win2000	- - - - 0 0 0 0 1 0 0 -
Ubuntu-Win2003	- - - - 0 0 0 0 1 0 0 0
Ubuntu-Win2008	- - - - - - - - 0 0 0 0
Red Hat-Win2000	0 0 0 1 0 0 0 0 1 0 0 -
Red Hat-Win2003	- - - 0 0 0 0 0 2 0 0 0

(continued)

Table 5.11 (continued)

OS combination	Number of vulnerabilities 2000–2011
Red Hat-Win2008	- - - - - - - - 0 0 0 0
Win2000-Win2003	- - - 10 23 44 29 47 44 19 49 -
Win2000-Win2008	- - - - - - - 1 25 16 38 -
Win2003-Win2008	- - - - - - - 2 30 21 94,135
Mean	1.4 2.2 2.0 1.4 1.2 2.9 1.1 2.3 3.4 1.4 4.2 4.1
Std Dev	2.1 2.9 2.8 2.4 4.0 8.2 4.9 7.6 8.5 4.7 16.5 22.2
Max	8 11 11 10 23 44 29 47 44 21 94,135

(1) The higher the diversity or dissimilarity of the configured OSs, the greater the security gain that can be obtained, and only a few vulnerabilities can affect multiple OSs simultaneously.

(2) The number of vulnerabilities that can affect more than one OS simultaneously depends on the diversity of the configuration. The OSs of the same family share a large number of vulnerabilities, while the number of common vulnerabilities shared by OSs of different families (e.g., BSD-Windows) is small, which is zero in most cases.

(3) By removing applications (e.g., thin servers) and restricting remote access to the system, we can greatly reduce the number of common vulnerabilities among OSs (by 76%).

(4) In view of the previous studies (on non-security-related crashes and failures), we decide that the OS pair failure is largely caused by the driver error. Nevertheless, the author has found that driver vulnerabilities only account for a small portion (less than 1%) in this case.

(5) Three strategies are proposed to use NVD data in selecting the most diverse OS combination, depending on: All common vulnerabilities are of equal importance; the more recent common vulnerabilities are more important (i.e., the number of recent vulnerabilities shared across OSs is minimized); and the less frequently the vulnerability reporting is, the better we think the case is (for the operator will have more emergency response time). The analysis also proves that a combination of four OSs is the best combination among all the three configuration strategies adopted for intrusion tolerance.

5.5 Chapter Summary

Diversified technology can double the security gain, yet it remains a static defense means in nature, which can't block the trial-and-error or exclusion attacks. Randomized and dynamized defense technologies can change the static properties of a system or architecture and construct an uncertain defense scenario but also expose the defender to complex economic and technical challenges while significantly increasing the cost of attacker for detection and attack.

Since any attack is based on a certain attack chain, any successful attack should be reproducible or have a high repeatability. For this sake, a reliable attack chain naturally calls for an accurate grasp of the real posture information of the defender, including the overall architecture, operation mechanism, resource allocation, network topology, function distribution, and flaws (vulnerabilities). However, the reliability of the attack depends, in a sense, on the invariance or staticity of these underlying data. If the defender can make these (or part of these) data exhibit external or internal uncertainty or dynamicity without affecting (or less affecting) their own functionality and performance, the stability or effectiveness of the attack chain will undergo an earthquake.

The elements related to the target system are diversified, randomized, and dynamized, so that attackers cannot understand or predict the defense deployment and behavior of the target system, where the MAS can be formed. With the uncertainty of the system, the reliability of the attack chain will be deeply affected. Even if an attacker can achieve a successful attack, changes in the relevant elements of the system can still invalidate the subsequent attacks, so that the sustainable use value of the attack results is deteriorated or the attack experience is difficult to be replicated, thus achieving the defense purpose. Despite of this temptation, the practice of MTD [1] shows that it is difficult to obtain architecture-based integrated defense benefits by discretely using the defense methods, such as diversification, randomization, and dynamization. On the contrary, it only increases the performance overhead of the target object. The high cost of related design and use also creates substantial obstacles to the promotion and application. The non-systematic technology implementation even affects the stability of the system in terms of compatibility and function provision, which urgently requires innovation in the pursuit of more advanced and economical application forms.

In short, from the perspective of a defender, diversity enhances the complexity of the target environment in the space dimension, dynamicity increases the uncertain expression of the defense behavior in the time dimension, while randomness extends the defender's advantage in the space-time dimension. Therefore, it can be foreseen that the basic development characteristics of new defense technologies will take diversity as the basis of positive defense while dynamization and randomization as the strategy and means for active defense.

As a bold speculation, there should be some kind of (possibly not unique) synergy mechanism based on the architectural effect among diversity, dynamicity and randomness, which can be used to maximize the defense outcome technically and economically. However, what is the physical or logical form of this architecture? How shall the synergy mechanism of the three factors be presented through the architecture? What implications will the changes in the "cyberspace game rules" bring about? What are the engineering and economical challenges that may emerge? These issues will be discussed in depth in the subsequent chapters.

References

1. Jajodia, S., Ghosh, A.K., Swarup, V., et al.: Moving Target Defense: Creating Asymmetric Uncertainty for Cyber Threats. Springer, Berlin (2011)
2. Diversity index.: https://en.wikipedia.org/wiki/Diversity_index. [2017-04-20]
3. Tootoonchian, A., Gorbunov, S., Ganjali, Y., et al.: On controller performance in software-defined networks. In: Hot-ICE'12 Proceedings of the 2nd USENIX Conference on Hot Topics in Management of Internet, Cloud, and Enterprise Networks and Services, pp. 10–15 (2012)
4. Khondoker, R., Zaalouk, A., Marx, R., et al.: Feature-based comparison and selection of Software Defined Networking (SDN) controllers. In: Computer Applications and Information Systems, IEEE, pp. 1–7 (2014)
5. Guo, H., Wang, X., Chang, K., et al.: Exploiting path diversity for thwarting pollution attacks in named data networking. IEEE Trans. Inf. Forensics Secur. **11**(9), 2017–2090 (2016)
6. Newell, A., Obenshain, D., Tantillo, T., et al.: Increasing network resiliency by optimally assigning diverse variants to routing nodes. In: IEEE/IFIP International Conference on Dependable Systems and Networks, pp. 1–12 (2013)
7. Borbor, D., Wang, L., Jajodia, S., et al.: Diversifying network services under cost constraints for better resilience against unknown attacks. In: Data and Applications Security and Privacy XXX. Springer, Berlin (2016)
8. Yao, G., Bi, J., Li, Y., et al.: On the capacitated controller placement problem in software defined networks. IEEE Commun. Lett. **18**(8), 1339–1342 (2014)
9. Ros, F.J., Ruiz, P.M.: Five nines of southbound reliability in software-defined networks. In: ACM SIGCOMM Workshop on Hot Topics in Software Defined Networking, pp. 31–36 (2014)
10. Wenbo, W., Binqiang, W., Feiyu, C., et al.: A controller hot backup and election algorithm in software-defined network. Acta Electron. Sin. **44**(4), 913–919 (2016)
11. Castro, M.: Practical byzantine fault tolerance and proactive recovery. In: Symposium on Operating Systems Design and Implementation, pp. 173–186 (1999)
12. Veronese, G.S., Correia, M., Bessani, A.N., et al.: Efficient byzantine fault-tolerance. IEEE Trans. Comput. **62**(1), 16–30 (2013)
13. Li, H., Li, P., Guo, S., et al.: Byzantine-resilient secure software-defined networks with multiple controllers. In: IEEE International Conference on Communications, pp. 695–700 (2014)
14. Eldefrawy, K., Kaczmarek, T.: Byzantine fault tolerant software-defined networking (SDN) controllers. In: IEEE Computer Society International Conference on Computers, Software & Applications, pp. 208–213 (2016)
15. Iannaccone, G., Chuah, C.N., Bhattacharyya, S., et al.: Feasibility of IP restoration in a tier 1 backbone. IEEE Netw. **18**(2), 13–19 (2004)
16. Cholda, P., Mykkeltveit, A., Helvik, B.E., et al.: A survey of resilience differentiation frameworks in communication networks. IEEE Commun. Surv. Tutorials. **9**(4), 32–55 (2007)
17. Haider, A., Harris, R.: Recovery techniques in next generation networks. IEEE Commun. Surv. Tutorials. **9**(3), 2–17 (2007)
18. Saradhi, C.V., Gurusarny, M., Zhou, L.: Differentiated QoS for survivable WDM optical networks. IEEE Commun. Mag. **42**(5), S8–S14 (2004)
19. Fumagalli, A., Valcarenghi, L.: IP restoration vs. WDM protection: is there an optimal choice. IEEE Netw. **14**(6), 34–41 (2000)
20. Garcia, M., Bessani, A., Gashi, I., et al.: Analysis of OS diversity for intrusion tolerance. Softw. Pract Experience. **10**, 1–36 (2012)
21. PaX: PaX Team. http://pax.gmecurity.net. [2017-04-20]
22. Xu, J., Kalbarczyk, Z., Iyer, R.K.: Transparent runtime randomization for security. In: Proceedings of the IEEE Symposium on Reliable Distributed Systems, pp. 260–269 (2003)
23. Huiyu, C.: Research on Structural Body Randomization Technology, pp. 10–14. Nanjing University, Nanjing (2012)

24. Yuan, L., Huawei, J.: Research on SQL injection defense technology based on instruction set randomization. Comput. Digit. Eng. **37**(1), 96–99 (2009)
25. Cowan, C., Beattie, S., Johansen, J., et al.: Pointguard TM: protecting pointers from buffer overflow vulnerabilities. In: Conference on Usenix Security Symposium, p. 7 (2003)
26. Cadar, C., Akritidis, P., Costa, M., et al.: Data randomization. Microsoft Res. **10**(1), 1–14 (2008)
27. Bhatkar, S., Sekar, R.: Data space randomization. In: Detection of Intrusions and Malware, and Vulnerability Assessment, pp. 1–22. Springer, Berlin (2008)
28. Krylov, V., Kravtsov, K.: IP fast hopping protocol design. In: Central and Eastern European Software Engineering Conference in Russia, pp. 11–20 (2014)
29. Cheng, G., Shikang, C., Hong, W., et al.: Research on address hopping technology based on the SDN architecture. Commun. Technol. **48**(4), 430–434 (2015)
30. Krylov, V., Kravtsov, K., Sokolova, E., et al.: SDI defense against DDoS attacks based on IP fast hopping method. In: IEEE Conference on Science and Technology, pp. 1–5, (2014)
31. Trovato, K.: IP hopping for secure data transfer: EP, EP1446932 (2004)
32. Shawcross, C.B.A.: Method and system for protection of internet sites against denial of service attacks through use of an IP multicast address hopping technique: US, US 6880090 B1 (2005)
33. Katz, N.A., Moore, V.S.: Secure communication overlay using IP address hopping: US, US7216359 (2007)
34. Bin, L., Zheng, Z., Daofu, G., et al.: A secure communication method for IP address hopping based on the SDN architecture: China, CN105429957A (2016)
35. Fang, L., Yunlan, W.: Research on IP address frequency hopping technology based on task optimization. China Comp. Commun. **2016**(14), 53–54
36. Petkac, M., Badger, L.: Security agility in response to intrusion detection. In: Computer Security Applications, pp. 11–20 (2000)
37. Cox, B., Evans, D., Filipi, A., et al.: N-variant systems: a secretless framework for security through diversity. In: Conference on Usenix Security Symposium, p. 9 (2006)
38. Azab, M., Hassan, R., Eltoweissy, M.: ChameleonSoft: a moving target defense system. In: International Conference on Collaborative Computing: Networking, Applications and Worksharing, pp. 241–250 (2011)
39. O'Donnell, A.J., Sethu, H.: On achieving software diversity for improved network security using distributed coloring algorithms. In: ACM Conference on Computer and Communications Security, CCS 2004, Washington, pp. 121–131 (2004)
40. Petkac, M., Badger, L., Morrison, W.: Security agility for dynamic execution environments. In: DARPA Information Survivability Conference and Exposition, vol. 1, pp. 377–390 (2000)
41. Roeder, T., Schneider, F.B.: Proactive obfuscation. ACM Trans. Comput. Syst. **28**(2), 1973–1991 (2010)
42. Chang, D., Hines, S., West, P., et al.: Program differentiation. In: The Workshop on Interaction Between Compilers and Computer Architecture, p. 9 (2010)
43. Salamat, B., Gal, A., Franz, M.: Reverse stack execution in a multi-variant execution environment. In: Workshop on Compiler and Architectural Techniques for Application Reliability Security (2012)
44. Ammann, P.E., Knight, J.C.: Data diversity: an approach to software fault tolerance. IEEE Trans. Comput. **37**(4), 418–425 (1988)

Chapter 6
Revelation of the Heterogeneous Redundancy Architecture

It can be seen from the foregoing chapters that the biggest security threat in cyberspace is uncertain attacks based on the unknown vulnerabilities of the target object, and the biggest defense concern lies in the ability of reliably addressing uncertain attacks based on unknown factors in the absence of a priori knowledge. The theoretical challenge is how to quantify the effectiveness of defense against uncertain attacks. In order to change the current situation where cyberspace cannot effectively prevent uncertain threats, the following scientific issues must be tackled:

(1) Whether it can perceive an uncertain threat without relying on the attacker's characteristic information.
(2) If the answer is affirmative, what are the theoretical basis and prerequisites.
(3) Whether this perceptual mechanism can create a measurable ability to resist uncertain attacks.
(4) How to ensure that this capability has stability robustness and quality robustness.

Fortunately, the author has found out in the field of reliability that there are theoretical and technical issues identical or similar to this, and the related theorems and methods have the significance of "advice from others to enlighten our own mind."

6.1 Introduction

With the development of industrial technology and the increasing emphasis on safe production, people grow more and more demanding on reliable industrial controls. One of the effective ways to improve system reliability is the adoption of redundancy design (RD), which can effectively enhance the reliability of the entire machine and the system by configuring redundant resources.

Redundancy technology, also known as reserve technology, is a means of improving system reliability by using the parallel model of the system, namely, by adding

© Springer Nature Switzerland AG 2020
J. Wu, *Cyberspace Mimic Defense*, Wireless Networks,
https://doi.org/10.1007/978-3-030-29844-9_6

redundant functionally equivalent components and synchronizing their operation through relevant redundant control logic so that the functionality of the system applications is guaranteed at multiple levels. The purpose of the technology is to make the system less susceptible to local or sporadic faults in operation, and the faulty components can be repaired online in time. Reasonable RD can significantly improve the reliability and availability of the system, effectively avoiding economic losses arising from shutdown or equipment damage caused by random system failures.

In order to effectively control randomly occurring physical faults, or uncertainties resulting from design defects, the redundancy technology has been widely applied. However, in practice, it has been found that in the isomorphism redundancy (IR) mode, the components with completely identical parameters (circuits, structures, materials, processes, and so on), when used in parallel, are easy to fail simultaneously under certain conditions at the impact of a single, specific fault, which is called common mode failure (CMF) and can greatly affect the reliability of the isomorphism redundancy system (IRS).To avoid the CMF as much as possible, the heterogeneous redundancy (HR) mode has been developed, which applies in parallel multiple heterogeneous components with equivalent functionality or performance to effectively reduce the probability of CMF occurrence. However, due to the limitations of design, manufacturing, process, materials, etc., different heterogeneous components may be subject to identical design defects, or have coincident failure factors to some extent, and thus cannot completely eliminate CMF occurrence. If a heterogeneous redundancy system (HRS) with independent fault factors can be constructed, where the design defects in the parallel heterogeneous components have non-coincident properties, the CMF can be avoided to a considerable extent. So, people have developed the dissimilar redundancy structure (DRS). Unfortunately, with the underdeveloped dissimilarity design theory as well as technological tools, it is difficult for any practical DRS to achieve an ideal dissimilarity. Namely, there is no guarantee that the redundant executors will be "absolutely dissimilar."

As we all know, the current security or anti-aggression of cyber information systems (including the dedicated defense equipment) is not yet measured by any effective quantitative analytics. The reason is that a cyber attack against the information system is usually not a random event, and its effect is often certain due to the staticity, certainty, similarity, and "one-way transparency" of the target object, which is statistically expressed as Boolean rather than probability, so it is difficult to describe quantitative analysis of security by way of the reliability theory and random process tools. Fortunately, the author finds that for the DRS architecture with a majority voting mechanism, its mechanism to curb uncertain differential mode failures (DMFs) has a similar or identical effect on certain or uncertain attacks based on the vulnerabilities of the heterogeneous redundancy entities (HREs). Similarly, the probability size and distribution pattern of the attack effect can be changed by adjusting the number of redundant configurations of the executors, the semantic abundance of output vectors and the ruling strategy, as well as by changing the dissimilarity of the implementation algorithm or structure of the executors. In other words, when the target system adopts the DRS, both the unknown attack events targeted at individual executors and the random failures arising from the executor components per se can be normalized to reliability issues that can be expressed by probability at the system level, making it possible to analyze anti-aggression that could not be measured before via a mature reliability theory and methodology. In this chapter, two typical information system architectures are involved, respectively, non-redundancy structure (NRS) and DRS (dissimilar redundancy structure), as well as certain and uncertain attacks, where a unified model is established on the basis of the generalized random Petri net theory to analyze the relevant anti-aggression so as to derive the steady-state available probability, steady-state escape probability, and steady-state non-specific perceptual probability.

6.2 Addressing the Challenge of Uncertain Failures

6.2.1 Proposal of the Problem

We all know that the biggest challenge for reliability lies in the uncertain faults or failures of a complex system, including two aspects: uncertain faults caused by physical failures of passive or active devices/components and uncertain failures caused by software/hardware design defects. Although the mechanism and degree of impact of the faults vary widely, the common feature is that the time, location, and trait of the faults are uncertain. Therefore, in addition to enhancing the reliability of the component itself at the material, process, and manufacturing levels, it is necessary to find a new way to improve the reliability gain of the system from its architecture. This gain should be like a triangle in the Euclidean space, the geometric stability of which is directly derived from the shape formation where the sum of the three internal angles totals 180°. As shown in Fig. 6.1, the arrangement of carbon atoms determines the hardness of graphite and diamond.

We know that, in general, the normal functionality of a given system is design-delivered and even the model perturbation threshold range is usually also certain. Engineering experience and statistics also reveal a rule that "physical failures of multiple identical functives rarely occur in exactly the same failure outcome at the same time and location," which makes it possible to perceive and mask the sporadic failures or differential mode faults under the software/hardware redundant configuration and majority voting mechanism. However, the homomorphic (common mode) faults caused by design flaws of software/hardware functives do not follow the above statistical law, so the corresponding IRS has no definite significance for such failures. To solve this problem, it is necessary to resort to new findings or another construction mechanism.

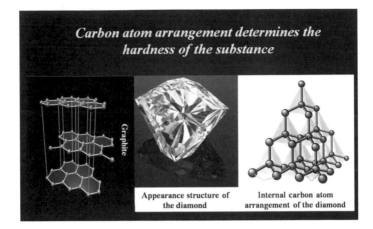

Fig. 6.1 Different arrangements of carbon atoms

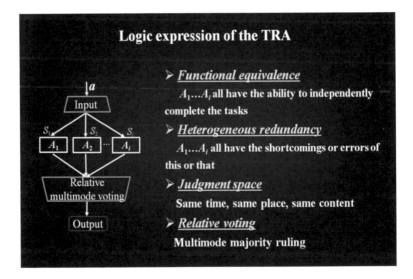

Fig. 6.2 Logic expression of the TRA

6.2.2 Enlightenment from TRA

We know that "everyone has shortcomings of one or another, but the chance is few for the majority of people to make the exactly same mistake in the same place and at the same time when accomplishing the same task independently," which is both a well-known consensus and an axiom, called by the author as a "true relatively" axiom (TRA), or a "consensus mechanism" in the blockchain technology. Figure 6.2 shows the logical expression of the TRA.

There are four preconditions for the establishment of a TRA: (1) $A_1 \sim A_i$ have the ability to complete a given task independently; (2) it is a high probability event that A_i can complete the task correctly; (3) there is not any coordination or cooperation relationship between A_i; and (4) it is possible for a multimode majority voting result to be incorrect. The reason to emphasize that the probability for an individual member to complete the task independently will generally outnumber that to fail in the task is just as that we cannot expect each member of the jury to be a moral saint but will never allow a mentally retarded patient or a misbehaving person to act as a juror. It should be stressed that when A_i in a TRA is not big enough, it is impossible to rule out that the result of majority action can be wrong in both logical and practical senses or to completely eliminate the chance of affecting a true relatively meaning resulting from "collusive cheating, bribery and canvassing" and other misconducts that make voters lose their independence. Despite this, there is no better mechanism to challenge the status of its cornerstone in the democratic society's fair system.

From the original definition of TRA we can get the following four useful conclusions:

Conclusion 1: From the perspective of individual members of A_i, although there are some uncertainties due to the diverse shortcomings or errors, it is rare for the group to make exactly the same mistake at the same time and on the same occasion when completing the same task independently. At this point, the problem of uncertainty is converted into a problem that can be perceived by majority voting.

Conclusion 2: The relative voting result has an attribute of "superposition state." That is to say, the result of a majority voting has both certainty and uncertainty attributes, which can be expressed by probability. Certainty means that in general situations, the "correct" result of a majority voting is a high probability event. Uncertainty means that correct and incorrect voting results exist simultaneously, though the probability differs as the "false" result of the voting is often a small probability event.

Conclusion 3: The greater the "probability difference" of the relative voting result is, the more likely it is relatively true. The core factors affecting the probability difference include the number of people involved in completing the task, the individual competency differences of the participating members, the detail level of the task completion indicators, the selection of the index content, the decision-making strategy, etc.

Conclusion 4: Transforming the problem scene changes the nature of the problem. The TRA switches the problem scenario to a multidimensional space from a single space, to functionally equivalent plural processing scenes from homogenous processing scenes, to relative judgment of a group from subjective perception of an individual, and to focusing on the global picture from focusing on local impact. Uncertain events that cannot be clarified in the original problem scenario become problems that can be perceived or positioned by using TRAs in the new cognitive scenario.

In engineering practice, we are more concerned with the logical expression of TRAs in the general sense. In theory, this form can conditionally (in functionally equivalent conditions) convert the problem of uncertain failures of HR individuals into an event of reliability with probabilistic attributes at the system level. If it is assumed that there are not identical hardware/software design flaws in the functional equivalent executors of HRS (i.e., the dissimilarity is infinite), the CMF will be theoretically impossible, and thus multimode voting is able to perceive any differential mode fault causing inconsistent executor output vectors. However, in practice, it is often impossible to guarantee that functional equivalent executors are completely heterogeneous. In other words, there is no guarantee on the nonexistence of "other potential equivalent functions" beyond the expected equivalent functions. Therefore, the voting result, such as multimode output vectors are consistent or most of them are the same, cannot give an absolute judgment on whether the target system is normal or not. As described in Conclusion 2, the relative voting result has a superposition state effect similar to that in quantum mechanics, where both right and wrong may coexist. The only certainty is that, when the majority ruling conditions are met, there is a large probability for completely consistent or mostly identical multimode output vectors to be artificially identified as "normal," otherwise identified as "abnormal" and attributed to a small or even a minimum probability event. In addition, according to Conclusion 3, there are multiple ways to influence the degree of the relative truth (i.e., the relative difference between large probability and small probability events in the superposition state), such as by adjusting the number of elements in the set A to increase redundancy, or enhancing the dissimilarity between heterogeneous members in the set, or increasing the semantic and content abundance of the multimode output vectors, and the fineness of the target, to strengthen the complexity of the voting strategy.

Therefore, it is not difficult to infer that, with the assistance of TRA, we can conditionally transform the uncertain failure problem of the component into a reliability problem that can be expressed by probability at the system level, given the constraints of functional equivalence, HR, and relativity. This constitutes a theoretical feasibility for solutions based on the construction layer, which is the original intention to improve reliability via the heterogeneous redundancy architecture (HRA).

6.2.3 Formal Description of TRA

It is assumed that there is a set of executors $E = \{E_i\}_{i=1}^{n}$ and a set of input vectors $I = \{I_j\}_{j=1}^{m}$ in the I[P]O system; when the executor E_i correctly responds to the input vector I_j, the resulting output vector is denoted as R_{ij}, where R_{ij} is the only correct output vector; when the executor E_i responds to the input vector I_j incorrectly, the output vector will be a particular output vector that belongs to the set of output vectors W_{ij}, where W_{ij} is the set of incorrect output vectors produced by the

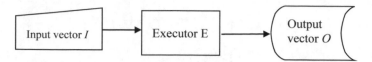

Fig. 6.3 An I[P]O System

executor E_i. As shown in Fig. 6.3, for the input vector set I, the I[P]O system will correspondingly obtain an output vector set $O = \left\{R_{ij} \cup W_{ij}\right\}_{i=1,j=1}^{n,m}$.

According to the prerequisite ① of TRA axiom, it can be seen that the case where any two executors E_i and E_k generate an incorrect output vector for a specific input vector I_j is independent of each other. Further, it can be seen from the prerequisite ③ of TRA axiom that the cases in which the executors E_i and E_k generate the same incorrect output vector are also independent of each other. Concisely, for any i, $k = 1, 2, \cdots, n$, and $i \neq k$, if the incorrect output vector $\delta \in W_{ij} \cap W_{kj}$ is satisfied, the following equation holds:

$$P\left\{E_i\left(I_j\right) = \delta, E_k\left(I_j\right) = \delta\right\} = P\left\{E_i\left(I_j\right) = \delta\right\} \cdot P\left\{E_k\left(I_j\right) = \delta\right\} \quad (6.1)$$

According to the prerequisite ② of the TRA axiom, "incorrect accomplishment of tasks is an event with small probability." Given a probability $\alpha(0 < \alpha \leq 0.05)$, for any executor E_i, $i = 1, 2, \cdots, n$ and a specific input vector I_j, $j = 1, 2, \cdots, m$, the following equation holds:

$$P\left\{E_i\left(I_j\right) \in W_{ij}\right\} < \alpha,$$

$$P\left\{E_i\left(I_j\right) \in W_{ij}\right\} > 0, \ i = 1, 2, \cdots, n, \ j = 1, 2, \cdots, m \quad (6.2)$$

For the input vector I_j, it is assumed that the incorrect output vector sets W_{1j}, W_{2j}, $\cdots$, W_{nj} generated by E_1, E_2, $\cdots$, E_n belonging to the executors set E have a common incorrect output vector set ω_j, $j = 1, 2, \cdots, m$. Concisely, there exists a ω_j that satisfies the following equation:

$$\omega_j = \bigcap_{i=1}^{n} W_{ij}$$

The number of output vectors in the incorrect output vector set W_{ij} is denoted as $card\ (W_{ij})$ and $card\ (W_{ij}) \neq 0$. The number of output vector in set ω_j is denoted as $card\ (\omega_j)$ and $card\ (\omega_j) \leq card\ (W_{ij})$. τ_j is an element in the set ω_j. It is assumed that for any one of the executor $E_i \in E$, the probability of occurrence of each element in the incorrect output vector set W_{ij} is equal in response to the input vector I_j. According to Eqs. (6.1) and (6.2), the probability that "most people committing exactly the same mistake simultaneously in the same place" is presented as follows:

$$
\begin{aligned}
P_j &= \sum_{\tau_j \in \omega_j} P\left\{E_1\left(I_j\right)=\tau_j,\cdots,E_n\left(I_j\right)=\tau_j\right\} \\
&= \sum_{\tau_j \in \omega_j} \prod_{i=1}^{n} P\left\{E_i\left(I_j\right)=\tau_j\right\} \\
&= \sum_{\tau_j \in \omega_j} \prod_{i=1}^{n} \frac{P\left\{E_i\left(I_j\right)\in W_{ij}\right\}}{card\left(W_{ij}\right)} \\
&= \prod_{i=1}^{n} \frac{P\left\{E_i\left(I_j\right)\in W_{ij}\right\}}{card\left(W_{ij}\right)} \cdot card\left(\omega_j\right) \\
&= \prod_{i=1}^{n} P\left\{E_i\left(I_j\right)\in W_{ij}\right\} \cdot \frac{card\left(\omega_j\right)}{card\left(W_{ij}\right)} \\
&< \prod_{i=1}^{n} \frac{card\left(\omega_j\right)}{card\left(W_{ij}\right)} \cdot \alpha^n
\end{aligned}
\tag{6.3}
$$

Due to the unpredictability of the types of incorrect output vectors that may be generated by the executors, and the heterogeneity between the executors, the larger the incorrect output vector set W_{ij} holds, the smaller the probability P_j that all the executors produce the same incorrect output simultaneously is acquired. Therefore, the common mode escape in the system rarely occurs. Considering real scenarios for an I[P]O system in which a user operation can be represented as an input sequence L with s steps. L consists of a finite number of input vectors in the set of input vectors I, marked as $L = \left(I_{l_1}, I_{l_2}, \cdots, I_{l_s}\right)$. According to Eq. (6.3), the probability of common mode escape aroused by inappropriate behavior is presented below:

$$
P_L = \prod_{t=1}^{s} P_{l_t} = \prod_{t=1}^{s} \prod_{i=1}^{n} P\left\{E_i\left(I_{l_t}\right)\in W_{il_t}\right\} \cdot \frac{card\left(\omega_{l_t}\right)}{card\left(W_{il_t}\right)}
$$

$$
< \left(\max\left\{P_{l_1}, P_{l_2}, \cdots, P_{l_s}\right\}\right)^s
$$

6.3   The Role of Redundancy and Heterogeneous Redundancy

6.3.1   Redundancy and Fault Tolerance

Based on redundant processing, fault-tolerant processing increases system resources in exchange for improvement in system reliability, so that the system can tolerate random faults, able to continue the specified algorithm even in the event of an error or failure, also known as fault masking technology. For example, in a dual fault-tolerant system, when one server fails, the other server can take over to ensure the normal operation of the system.

In fact, at a given level of the error detection capability, the redundant resources of different scales and natures can have the fault-tolerant functionality of different capabilities or degrees. For example, hardware redundancy supports random fault tolerance through the repeated setup and usage of hardware, while software redundancy prevents design flaws that may exist in the same software through multiple different software modules of the same function or N versions of technology and through the heterogeneity of the implementation algorithm. Typically, these two modes are often used in combination for higher fault tolerance.

Another expression of fault tolerance is error correction. For instance, Shannon's channel coding theory is usually used to ensure undistorted information carried on the noisy channel. The core concept is to add appropriate redundant bits to the information to be transmitted and perform channel coding according to the corresponding algebraic algorithm and then decode it on the output end of a given noise model transmission channel so as to obtain information transmission service with a guaranteed quality (with a controllable SER). In other words, error correction coding based on redundant bits can correct a certain number of information bits interfered by a given noise model, and the probability of success satisfies the expected value requirement. By analogy, if an attack based on vulnerabilities is considered as a specific continuous or discontinuous interference, and the target system is regarded as a transmission channel that allows discontinuous noise injection, the attack success can be regarded as the SER, the value of which is strongly relevant to the number of redundant bits and the error correction coding algorithm. Intuitively, if an appropriate analogy model can be established, the channel coding method can give a quantitative reference conclusion on the target object's anti-attack capability (to be further discussed in subsequent chapters).

Undoubtedly, one of the prerequisites for fault tolerance is to configure redundant resources and establish a system architecture to carry these resources. Therefore, fault tolerance is generally provided together with system services on the same architecture, and it is usually viewed as an endogenous or symbiotic function. When random errors, differential faults, or failures in resource processing are spotted in the detection section, the system's endogenous fault tolerance function can minimize its adverse effects on the services and can ensure the system to continue its service provision with normal or derated performance under certain probability conditions. Obviously, redundancy is the core element of fault tolerance.

It should be noted that conventional detection technology can only perceive the "known and even known unknown" failures within the perturbation range of the model, while the TRA-based multimode output vector voting method can perceive the "unknown unknown" failures. In the subsequent chapters, we will repeatedly use this detection method irrelevant to a priori knowledge.

6.3.2   Endogenous Functions and Structural Effects

There is no strict unified definition of the so-called endogenous functions. They generally refer to explicit functions or possibly existing derivative/semi-derivative secondary functions or implicit defects determined by the model system design. The multiple semantics of the concept "design" and the limitation of the times regarding the design theory and tools often make design-generated functions unable to stand the historical test. It is currently impossible to ensure the functional purity of the complex system both in theory and methodology. How to strictly constrain the multiple semantics of the given functions is still a challenge. For example, the exploitable vulnerabilities are the dark functions of the endogenous defects in some functions or functional combination. Besides, structural effects are usually based on the implementation structure or algorithms inseparable from the model system. Generally, it is impossible to pass the external-attached structure or algorithms. For example, high robustness constructed on the basis of a redundant model is an endogenous effect, whose reliability gain originates from the structural effects of the model and is not or weakly correlated with the material, process, and producer factors of the component. Let's look at graphite and diamond, their elements are all carbon atoms, but arranged differently, so they differ in hardness. You can see that the physical hardness of the material depends entirely on their respective structural effects.

6.3.3   Redundancy and Situational Awareness

Situational awareness refers to the acquisition, understanding, and presentation of the security elements that can cause changes in a system's posture, as well as to the prediction of their future development trends, in a complex system environment. If you can gain an advantage in situational awareness, you will win a positive defense standing against cyber security threats. However, in a single processing space or non-redundant conditions, especially when adopting a resource sharing mechanism and a hierarchical functional structure (e.g., classic computing system based on Von Neumann structure), we usually find it impossible to directly judge whether the processing result of a complex system is correct or not, for the reliability is not self-evident. In other words, in the above scenario, what is normal and what is abnormal are not judged upon irrefutable evidence when there is no support of a priori knowledge. Even if the encryption algorithm is rigorously proven in theory, the credibility of the encryption function is not assured when the credibility of the host system cannot be guaranteed. In fact, the biggest challenge of trusted computing (see Chap. 5) is that the credibility of the trusted root cannot be self-certified, just like a trusted operating system, which may still lack credibility for lack of trusted CPU support. Recently, it has been reported that researchers from College of William & Mary (WM), Carnegie Mellon University(CMU), University of California, Riverside (UCR), and Binghamton University exploited the "branch prediction" vulnerability of the CPU to launch the BranchScope attack and retrieved the contents stored in the SGX security zone of Intel x86 processors (see Sect. 4.2.3 for details). This shows that SGX exposed to the "branch prediction" vulnerability is still unable to deal with the BranchScope attack even if it is meticulous designed.

According to the attach surface (AS) theory [1], given that the channels, data, and rules available to the attacker remain unchanged, the introduction of redundant structures and related resources into the target system can mitigate the damage caused by potential vulnerabilities. In fact, by means of redundant resources or redundant mechanisms, the comparison between the results of multiple or multi-channel calculations helps to identify computational anomalies or failures arising from random failures of physical devices (provided that the computing devices meet the basic reliability and availability requirements) by performing repeated calculations on the same computational component in the time dimension or performing the same calculation on different computational components in the space dimension. Of course, reconfiguration of time or space resources alone is insufficient, and it is still necessary to use the same calculation/function as the main line for associated processing in the time and space dimensions for effective use of the redundant processing means to perceive abnormalities or to get an idea on whether it is relatively true or false.

Undoubtedly, the redundant structure and multimode voting mechanism can explicitly express (can perceive the result) the problem scenarios regarding randomness or uncertainty and can significantly enable the target system to perceive or detect uncertain events without, or independent from, the a priori knowledge.

6.3.4   From Isomorphism to Heterogeneity

Redundant executors fall in two modes: isomorphic redundancy and heterogeneous redundancy, classified by the isomorphism/heterogeneity of the structure or algorithm. In general, the isomorphism redundancy mode (IRM) can solve problems regarding fault tolerance and abnormality perception under the differential mode failure (DMF) conditions. However, in the case of CMF across all redundant executors at the same time and the same place in the same condition, it is impossible to detect or perceive the fault only by relying on the contrastive monitoring or cross-judgment mechanism between the executors. Unlike the IRM, the heterogeneous redundancy mode (HRM) not only deals with simultaneously physical DMFs randomly occurring in operation but also dodges or curbs DMFs possibly caused by software/hardware design defects.

6.3.4.1   Isomorphic Redundancy

According to the spatiotemporal nature of random physical faults, the error detection mechanism using isomorphic redundancy processing space or the execution result from the same algorithm is used for majority judgment, where calculation errors caused by physical hardware failures can be processed in a fault-tolerant manner at the system level, generating a reliability gain higher than the physical failure rate of the component, provided that any randomly generated physical fault is a DMF.

It is impossible for isomorphic redundant processing space to effectively deal with random failures caused by software/hardware design defects. This is because all the hardware and software within the redundant space have identical design functionality as well as identical or highly similar performance, which, together with any possible design flaws, are reflected as they are on the relevant components in the redundant space. Given the same input stimulus, the computational processing of the components in the functionally equivalent isomorphic redundant space shall be consistent, including consistent correct results (normally a high probability event) and identical error results (normally a small probability event), and when the same design flaw causes a CMF or a homomorphic fault, isomorphic redundancy will lose the structure-based fault tolerance function because it is difficult to make a correct judgment.

6.3.4.2   Heterogeneous Redundancy

With the well-known consensus that "the same problem usually has multiple solutions, and the same function often has multiple implementation structures," the use of a TRA based on heterogeneous redundancy can dodge the problem that isomorphic redundancy cannot cope with the unknown design flaws. Due to the dissimilarity in the philosophy, approach, instrument, and prerequisite for solving problems and implementing functions, the differences in practice, experience, and cultural and educational backgrounds between the implementers or engineering design teams will lead to differences in problem-solving algorithm selection or method creation, in structural design to satisfy the functionality requirements, and even in the specific implementation, where they may even follow completely different technology trajectories. This makes it possible to ensure one thing: given that the executors are functionally equivalent with a large probability of correct computational results, the probability of mostly or entirely identical faults occurring to the heterogeneous redundancy space will stay below the designed threshold even if their design defects or faults are different. It is necessary to satisfy the following assumptions.

(1) The function F has n functionally equivalent processing spaces, which are heterogeneous.
(2) The problem W has n solution algorithms, producing consistent results although they differ from one to another.
(3) Each processing space contains one algorithm, and the correct result produced is a high probability event.
(4) All algorithms have identical input stimulus.
(5) The output vector of each spatial algorithm will be judged under the majority rule, with the mostly identical output results used as the evidence for the normal state.

Therefore, heterogeneous redundancy can perceive not only an abnormal output caused by a random failure but also one caused by an unknown hardware/software design defect. This belongs to its endogenous characteristics. The theoretical and engineering meaning is that, when certain conditions are met, the heterogeneous redundancy architecture can transform the uncertainty problem into a risk management one subject to conditional perception and probability expression.

As shown in Fig. 6.4, in order to achieve a certain function, there are three functionally equivalent heterogeneous processing structures. The processing spaces are S_1, S_2, and S_3, where $S_1 \neq S_2 \neq S_3$, and the algorithms in different processing spaces are not identical, corresponding to $k \neq q \neq f$. At this time, for the same input stimulus i, the output vectors (OVs) of the three spatial algorithms are obtained as $k(i)$, $q(i)$, and $f(i)$, and then the voter is used for majority ruling, with the obtained completely consistent or mostly identical output results as the normal state (theoretically, there is a negative conclusion of a small probability), including the position information of the abnormally output vector space (based on which the important information about historical performance can be formed).

6.3.4.3   Appropriate Functional Intersections

Theoretically, only by intersecting given set functions rather than others, the heterogeneous redundant entities can ensure that the design flaws of redundant executors will not cause CMFs. However, it is often difficult to rule out the intersection of undesired functions in engineering implementation, since a given functional intersection is likely to be merely a subset of the maximum functional intersection among the heterogeneous redundant entities. So far, there is neither a theoretical nor a practical total solution the way of letting the maximum functional intersection

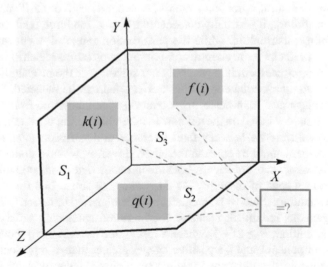

Fig. 6.4 Illustration of heterogeneous multimode redundancy processing

equal to the given function intersection, and the pursuit of a feasible approach still lingers on theory as it switches toward practice. In engineering terms, non-homology is generally used to approximately express the dissimilarity.

6.3.5 Relationship Between Fault Tolerance and Intrusion Tolerance

Although the heterogeneous redundancy architecture originally intends to avoid CMFs or homomorphic faults caused by a potential common problem across iso-morphic systems, as well as to provide error tolerance for DMFs, we know from Sect. 3.4.2 that intrusion-tolerant systems and fault-tolerant systems are different after all. It is not easy to well integrate fault-tolerant techniques in the intrusion-tolerant techniques, although fault-tolerant mechanisms generally have endogenous intrusion-tolerant properties. The classic concept of intrusion tolerance has two core elements, respectively, intrusion detection and error tolerance. The formula is tolerance = intrusion detection + heterogeneous redundancy. Similarly, the concept of fault tolerance has a similar formula, namely, fault tolerance = failure detection + heterogeneous (isomorphic) redundancy. Intuitively, as long as it can be proved that the intrusion detection function is equivalent to the fault detection function and both have heterogeneous redundant resources, it can be concluded that intrusion tolerance is equal to fault tolerance. We know that either intrusion detection or error detection is based on a priori knowledge, and the object detected usually belongs to a known unknown problem that cannot be detected without omissions or errors, while an unknown problem cannot be detected effectively or even cannot be identi-fied. This is a common attribute for both scenarios. It should be pointed out that, in the traditional reliability problem scenario, whether a DMF or CMF, or a sporadic or permanent failure, it is usually considered to occur randomly and satisfy a cer-tain probability distribution, while the perturbation range of a known unknown model is also predictable. In an intrusion-tolerant application scenario, as far as the target system is concerned, it is generally not suitable for the probabilistic expres-sion of even an attack behavior or attack effect that can be detected, because the "one-way transparent" man-made attack featuring "collaboration from within with forces from outside" based on the unknown vulnerability is an uncertain event. This is the biggest difference between fault tolerance and intrusion tolerance. In this sense, the former cannot be equal to the latter. However, we also notice that, when an unknown unknown event occurs, both the intrusion detection and fault detection are actually in a state of disability. At this point, the above two formulas can be expressed as fault tolerance = heterogeneous redundancy = intrusion tolerance, and the heterogeneous redundant resource configuration shall only satisfy one error correction condition: $N \geq 2f + 1$ (where N represents the number of heterogeneous redundant components, and f, a positive integer greater than 1, represents the num-ber of components that allow abnormalities or failures).Under the same stimulus, if a majority voting or consensus mechanism (such as the construction described in Sect. 6.5) is used, we can find the inconsistency between the output vectors of execution components without relying on a priori knowledge. Supplemented by certain functions, it is not difficult to "mask" f outstanding output vectors without the need to pay special attention to the causes and properties of the output vector differences. In other words, whether it is an uncertain intrusion behavior or a random failure/fault disturbance, as long as it can affect an output vector of the heterogeneous execution component and can be perceived or positioned by the multiple voting section, there is no essential difference at this time between a heterogeneous redundant fault-tolerant system and a heterogeneous redundant intrusion-tolerant device. In this sense, it is not difficult to understand why each heterogeneous redundant fault-tolerant system with a majority voting or consensus mechanism usually has a certain degree of intrusion tolerance. This will be further discussed in-depth in subsequent chapters of the book.

6.4 Voting and Ruling

6.4.1 Majority Voting and Consensus Mechanism

In principle, the majority voting or consensus mechanism itself can only judge whether the scenarios are identical by majority or minority, or completely consistent or totally disparate, rather than distinguish between what is normal and abnormal, and what's more, it may not directly indicate the source of such abnormalities, or to put it another way, it is a state of "knowing the symptoms rather than the root cause." Due to the equivalence requirements for the heterogeneous redundant executors on a given set of functions, and the assumption that it is a high probability event for the multimode output vectors to be consistent under given operating conditions, we often, in engineering practice, set the majority or identical voting result as the relatively true result.

However, we know that the result is actually a "superposition state," i.e., there is a small probability of misjudging an abnormality to be normal, which we call an "escape phenomena" (EP). This is because it is impossible to theoretically and practically determine whether the processing spaces and algorithms are completely independent and absolutely dissimilar, so it is impossible to eliminate the existence of identical yet erroneous output vectors beyond the expected or given function. Other factors that influence and control escape are discussed in subsequent chapters.

It should be highlighted that if the maximum function intersection of heterogeneous executors is greater than the given function intersection, the overlap is defined as a collection of "dark features" (DFs). Assume that the user or target system fails to effectively suppress or constrain the effects of these DFs, the TRA is unlikely to hold true in this scenario.

A highly useful conclusion can be obtained from the above analysis: "The HRA-based majority voting can be applied to the highly credible abnormality sensing component in the "relatively true application scenario," where its credibility is strongly correlated with the number of functionally equivalent redundant components, the semantic expression space of output vectors, the voting fineness of multi-mode vectors, the voting strategy, and the dissimilarity between the redundant executors.

The complexity of the majority voting mechanism is strongly correlated with the specific application scenario. Even the consistent ruling of majority output vectors based on standard interfaces and normative protocols faces different technical challenges in engineering implementation, which can be costly sometimes. The classic triple module redundancy (TMR) voting logic is shown in Fig. 6.5.

In Fig. 6.5, A, B, and C can be either multivariate vectors, or they can be time-based functions, where And/Or operations are operational expressions in a logical sense.

6.4.2   Multimode Ruling

The redundant executor output voting technology is one of the technical cores of a fault-tolerant system. In an isomorphic redundant (IR) system, a large number voting or a consistent voting algorithm is usually adopted. Heterogeneous redundancy often requires a multimode ruling algorithm due to the possible disparity allowed. For example, the *sin* function has multiple implementation algorithms with different precisions, which may be considered as equivalent when approximated to n decimal places in engineering. The simple large number voting alone cannot accommodate this judgment that allows biased or incorrect results. Moreover, the large number voting algorithm is obviously not applicable to scenarios with the same semantics and different grammars. It is necessary to mention the role and meaning of iterative ruling. We know that a physical fault is usually a time-relevant function, where even man-made attacks follow a strict timetable. If possible, a given input stimulus can be subject to the iterative ruling based on temporal redundancy. That is, in a given

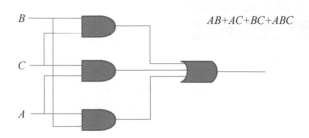

Fig. 6.5 Diagram of classic TMR voting

time, the heterogeneous redundant executor responds n times or repeatedly performs the same input excitation, and then the iterative ruling is applied to the ruling results of n times. This approach has the following benefits: confirm the nature of a random failure, simplify the post-processing flow after the sporadic failure, verify the start of the exception processing flow after the steady-state failure, or select the highest credible output vector as the system output via iterative ruling. It should be emphasized that if credibility statistics can be done on the historical performance of each executor, the process where these parameters are referred to in iterative ruling will be equivalent to constructing a negative feedback mechanism in the space-time dimension, endowing the single majority ruling with an iterative meaning strongly correlated with the nature of time and space. In addition, in view of the complex confrontation requirements of the attack-defense game, the engineering implementation often requires the weighting parameters in ruling on the basis of large number voting, including but not limited to the current security posture, the system resource status, the attack frequency, the historical performances of all executors, etc., or diversified alternative voting strategies available for different scenarios, which the author calls multimode ruling (MR).The available MR methods include plural voting, maximum likelihood voting (MLV), weighted voting, mask voting, large number voting based on historical information, fast voting, monitored voting, byzantine generals voting, and advanced ruling based on artificial intelligence. It should be emphasized that MR features not only a variety of ruling algorithms but also temporal/spatial iterative convergence mechanisms that support the corresponding algorithms.

6.5 Dissimilar Redundancy Structure

As mentioned earlier, the non-redundant architecture (NRA) is prone to single-point failures and is not immune to attacks based on vulnerabilities due to its inherent properties such as staticity, similarity, and certainty. Although a series of additional defense techniques and methods have been developed, represented by intrusion detection, intrusion prevention, and intrusion isolation, the solution to the problem still rests on active/passive defense backed by perceptions of the attacker's a priori knowledge or behavioral characteristics. It is therefore difficult to effectively resist the uncertain attacks initiated by exploiting unknown hardware/software vulnerabilities, and it lacks necessary means to handle advanced persistent threats (APTs) and other persistent penetration cyber attacks.

The fault-tolerant and survivable technologies have been developed successively to improve the reliability of the NRA. Among them, the fault-tolerant system is based on the isomorphic /heterogeneous redundancy architecture, including Similar Redundancy Structure (SRS) (primary backup, $M + N$ backup) and Dissimilar Redundancy Structure (DRS), which can effectively address the reliability problems caused by random hardware/software failures so as to significantly improve the robustness of the system.

1. Redundancy technology: For some applications with high security requirements, such as flight controllers, high-speed rail traffic systems, nuclear power plant operating systems, etc. special measures should be taken in the software/hardware architecture to improve the reliability of the computing system [2]. Redundancy technology is usually adopted to deal with the random physical hardware failures, that is, where the same program runs on N identical hardware environments, which is conducive to higher hardware reliability of the target system [3, 4].However, where errors such as hardware/software design or specification understanding occur, resource repetition alone cannot achieve the purpose of fault tolerance, since these errors will result in identical faults (CMFs or homomorphic faults) on the same operational object or work unit at the same time. In principle, it is not effective to detect and isolate such errors using cross-monitoring across the various redundancies, and the system itself may mistake the redundant function as normal due to identical output results across the processing environments, which may lead to catastrophic consequences.

2. Dissimilar Redundancy Structure

This dissimilar redundancy structure (DRS) is adopted by the US-based Boeing on the flight controller of its B777 passenger aircraft, as shown in Fig 6.6b/c/d, with an abstract model shown in Fig. 6.6a. Obviously, it is almost indistinguishable from the logical expression of TRA.

Note: $A_i(i = 1,2,\ldots,n)$ represents functionally equivalent heterogeneous executors (FEHEs) and provides normal functionality as a high probability event; $S_i(i = 1,2,\ldots,n)$ indicates possible design flaws in the corresponding executors. For input x, the operation result of each executor is output via the multimode voting.

The theoretical basis of its fault-tolerant function is well-known: "the probability of CMFs caused by common design flaws in an independently developed system is low." In order to ensure the independent design and development of the heterogeneous software and hardware, the companies concerned have adopted extremely strict engineering and technical management means and methods; selected completely independent R&D teams differing in educational, engineering, and even cultural backgrounds; and applied different development languages, development tools, and processing environments, expecting that the dissimilarity among the redundancy functives is managerially and technically as large as possible to avoid or curb the occurrence of CMFs [5, 6].Obviously, this kind of enhanced heterogeneous design relies primarily on the effectiveness of technical management and process supervision to fulfill its core mission of ensuring that the unknown design defects in the functionally equivalent independent components have non-coincident properties and the cross-judgment mechanism is applied to detect, isolate, and locate the faults, as well as to restructure available resources where possible. Engineering practice shows that a system designed on the basis of the DRS enjoys exponential improvement in reliability levels compared to non-redundant systems or other redundantly constructed systems. At the same time, it also indicates that the design principles and methods of DRS are not only effective in suppressing the common defects and errors of the target object but also effective in the common

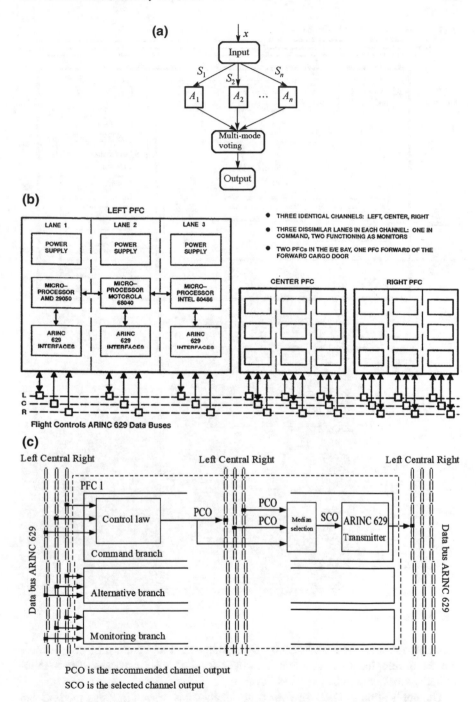

Fig. 6.6 (**a**) Dissimilar redundancy structure. (**b**) The 3∗3 dissimilar redundant flight control computer structure of the B777 passenger aircraft. (**c**) Diagram of the generation process of key output commands of B777's master flight control computer [17]. (**d**) Illustration of B777's actuator control electronic device voting. (**e**) The GSPN model for erroneous behaviors of the flight control computer

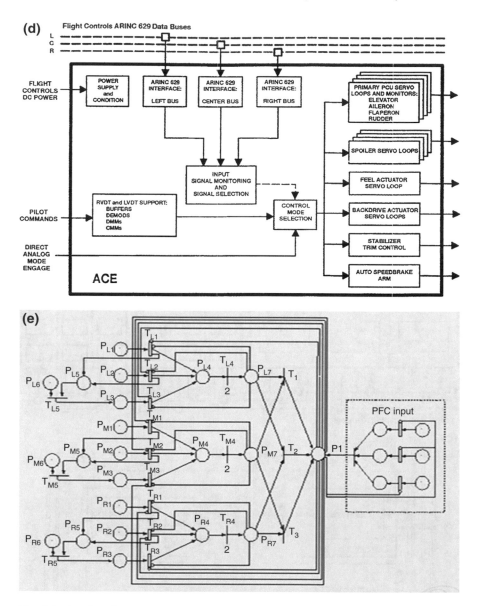

Fig. 6.6 (continued)

problems that may be introduced by the management and control of design tools and the development environment. However, its fault tolerance capability is theoretically considered to be $f \leq (N - 1)/2$.

The application of DRS for over four decades has proved that this method can effectively detect, locate, and isolate "known unknown" or "unknown unknown" failures. If a certain pre-cleaning or recovery mechanism is added, the target object system can become highly reliable. Data show that the failure rate of the flight control system of the Boeing 777 aircraft can be lower than 10^{-11} [16], while that of the F16 fighter is lower than 10^{-8}.It also proves that DRS can greatly reduce the time and cost of product verification and confirmation, facilitating the product's time to market. Figure 6.6b–d shows the block diagrams of B777's dissimilar redundant flight control system [16], and Fig. 6.6e is a Petri net-based fault analysis model.

Unfortunately, the current dissimilarity design has not yet reached the scientific quantitative degree, that is, it cannot be precisely guaranteed that the given function intersection is exactly equal to the maximum function intersection of the system. Once the latter is larger than the former, undesired DFs will be inevitable (this is true in most cases). Quite a few theoretical and technical problems are to be studied in- depth in this area. Therefore, in practical applications, it can only be achieved "non-homologously" through rigorous engineering management. In particular, the design, manufacture, and maintenance cost of the DRS system are a serious problems, which usually hinder the scope of its universal application.

3. Survivable Technologies

A research team at the CMU-SEI (Software Engineering Institute, Carnegie Mellon University) defines survivability [3] as the capability of a system to fulfill its assignments and external service requests, in a timely manner, in the presence of illegal external attacks, in-house failures, or accidents that have some reversible/ irreversible effects on the system itself. Current technologies for developing survivable systems include adaptive reconfiguration; diversity redundancy; real-time intrusion monitoring, detection, and response; intrusion tolerance; performance and function tradeoff for security; enhanced kernel, access control, and isolation control; etc. Fault-tolerant and survivable defenses are based on the comprehensiveness and timeliness of attack feature perception, so they can neither maintain the system's stability robustness nor quality robustness against uncertain attacks initiated by exploiting the unknown hardware/software vulnerabilities of the target object.

6.5.1 Analysis of the Intrusion Tolerance Properties of the DRS

The DRS is a masterpiece structural technology used to improve reliability. Its core mechanism is to convert the problem of uncertain failures of a branch circuit into a reliable event with a controllable system failure probability via the physical HRS. This gives us an important revelation as to whether we can rely solely on heterogeneous redundant resources and the output vector majority voting mechanism to convert the uncertain threat into a perceivable reliability event with a controllable probability, without relying on failure or fault detection conditions (in fact, there is no a priori knowledge or detection able to serve as a reference for unknown unknown threats). Although the rational cognition tells us that uncertain failures are completely different from uncertain threats in mechanism, the former has a random nature, while the latter is purely deliberate and artificial. It seems that they are not identical. In Sect. 6.3.5, it is pointed out that a heterogeneous fault-tolerant mechanism generally has endogenous intrusion-tolerant properties, while the DRS belongs to a typical heterogeneous fault-tolerant structure. Therefore, it is beneficial to explore its intrusion-tolerant properties.

As described in the previous sections, since the intrusion detection system (IDS) usually requires the support of a priori knowledge, it cannot detect the uncertain threats for it "does not know" the behavior and feature information of the treats. Many researchers have tried to analyze how DRS responds, in case of IDS disability, to the uncertain attacks against its endogenous intrusion-tolerant mechanism, and whether it can produce a quantitative analytical conclusion. However, this thorny problem cannot be studied by using probabilistic tools since it involves deliberate man-made attacks rather than random events. Researchers keep exploring the randomness factors of the inherent operating state of the architecture, trying to find out evidence on random factors affecting man-made attacks in the DRS, for instance, the trigger conditions of intrusion detection, the scenario conditions of state transition, the dynamic mechanism of the operating environment, the difference in the operational response latency, etc. They expect to normalize the attacks based on unknown vulnerabilities into classic reliability perturbations, so that quantitative analysis can be achieved with relevant mathematical tools. Unfortunately, no convincing proof has been observed so far.

According to the author, although in theory each functionally equivalent heterogeneous executor in the DRS may have known or unknown backdoors or even viruses, Trojans, etc., the dissimilar design and environmental conditions make their availability different, and the trigger mechanism like vulnerability or the approach to internal and external connectivity and the content of information will usually vary with the host or environmental factors. In theory, as long as the functionally equivalent heterogeneous redundant components are completely independent from one another, namely, there is no correlation or coordination between them, it is difficult to simultaneously apply identical or different attack means to all the executors and produce identical abnormal output vectors. In other words, when the model "input-process-output" (hereinafter referred to as I[P]O) is satisfied, as long as the output response vectors of the heterogeneous executors (instead of the alarm information of the general error detection section, such as the parity state, etc.) as the target of multimode voting, no attack, regardless of its category, can achieve the attack escape state until "identical errors occur to the majority or all of the heterogeneous executor output vectors." Otherwise, any individual differential mode attack is perceivable and can be expressed as a probability. This is like a precise shooting competition. Not only the number of target rings is assessed but also the distribution of the impact points should be concentrated in a sufficiently small area. In extreme cases, it may be necessary to check how many bullets can pass through the same bullet hole. For an outstanding shooter, the ability to reach the target is not a probability event, but there is definitely a probability that the impact points are distributed in a given area or that the majority or even all of the bullets can pass through the same bullet hole. This example shows that the more precise the target ruling is, the stronger the synergistic consistency of operation will be, and the lower probability the target system will have regarding the attack-and-escape state. Due to the DRS's inherent majority voting or consensus judgment mechanism, the certain attack results for a specific executor's vulnerabilities can hardly affect the output vectors of other executors participating in the voting, so they can only be perceived as DMFs and corrected or shielded by the fault-tolerant mechanism, unless the attacker can organize a coordinated attack across the heterogeneous executors and obtain mostly or completely identical output vectors, or finally achieve the attack effect $f > (N - 1)/2$ through a "trial and error or blind attack" approach. Ideally, as long as there is not any dark feature between the heterogeneous executors other than the given functions, the probability of a CMF or a homomorphic error should approximate zero.

Since "absolute dissimilarity" cannot be guaranteed in engineering implementation, it is impossible to completely avoid the possibility that n lines of output vectors have mostly or completely identical faults at the same time (for theoretically they are difficult to avoid as long as there is a dark feature intersection), even if the heterogeneous redundancy and multimode voting mechanism are used. After all, it is a small probability event. However, this probability has a strong correlation with the semantic abundance of output vectors (l), the heterogeneity between executors (h), the configuration number of redundant executors (n), and the multimode voting strategy (v), so it is possible to appropriately select these parameters to achieve the engineering objective with a controllable probability. Obviously, this is also one of the important inferences of the TRA.

In the ideal DRS configuration, there should be no other interactable or synergistic mechanism between n HREs except the same input stimulus. This is the first premise to ensure that the multimode consensus voting makes it difficult to achieve exponential cooperative attack, because for any attack based on known or unknown vulnerabilities, it is difficult to guarantee that on the DRS architecture "without cooperation" or under de-synergization, the multimodal output vectors of the service set n will make identical mistakes in time, content, and even behavior. Moreover, the larger the n is, the greater the inter-executor heterogeneity, the higher the time consistency requirement, the richer the semantics included in the output vector, the smaller the probability $\varepsilon(\varepsilon(l, h, n, t, \ldots)$ abbreviated ε)of fully or mostly identical errors, and the more difficult it is to achieve a cooperative attack (similar to the principle of using a consensus mechanism to ensure the reliability of data block operations in blockchain technology). For a given DRS system, the heterogeneity of executors (h) and the configuration number of redundant executors (n) are usually determined, while the semantic abundance of output vectors (l) can be regarded as the content complexity of the output vectors and the two-dimensional space of their bit length. Thus, ε is actually the escape probability in the three-dimensional ruling space that consists of the length and content of each output vector and the output time of each heterogeneous executor, which is also equivalent to the probability of false output vectors of n executors just falling on the same error point in the three-dimensional ruling space. Obviously, this is a small probability event, just like a sniper to shoot all bullets through the same bullet hole.

In addition, at any stage of the attack chain, any scenario involving the interaction of the I[P]O interface is subject to the consensus mechanism for voting. Once an inconsistency (including misjudgment) occurs, the following measures may be taken to immediately block the current attack process:

(1) Any executor judged abnormal will be immediately put offline for cleaning or for initialization at the corresponding level.
(2) The executor judged abnormal will be subject to a state rewind operation.
(3) The standby will take over from the abnormal executor.

Undoubtedly, to meet the ultimate requirements, an attack being manipulated internally and externally for multiple times must ensure that all operations involving majority voting sessions can create a cooperative escape under the non-cooperative conditions, and there shall be "no trial and errors or mistakes" throughout the process, which will force the attacker to face exponential challenges.

Therefore, the author believes that the fault-tolerant nature of the DRS not only ensures the robustness or resilience of system functionality but also provides a considerable degree of intrusion tolerance. This trait is derived from the "difficulty of a cooperative attack on diverse targets non-cooperatively in a heterogeneous redundant space." In other words, the difficulty can be expressed as the probability of identical or mostly consistent errors simultaneously occurring to multimodal output vectors within the structure. Thus, this intrusion-tolerant capability is "endogenous," determined by the DRS architecture and independent of any a priori knowledge and behavioral characteristics of attackers. This intrusion-tolerant mechanism of the DRS is somewhat similar to the nonspecific immunity (NI) mechanism of spinal cord animals [6, 7], whose immunity stems from the built-in biological genetic mechanism, which features "indiscriminative killing" against all invading antigens without relying on any a priori knowledge of viral/bacterial antigens (the author speculates that this immunity may be obtained by means similar to "fingerprint comparison or identification friend or foe (IFF)").However, there is another problem: the "fingerprint information" used for comparison will inevitably be missed or misjudged, making the acquired specific immunity an indispensable complement. Logically, this intrusion tolerance of the DRS has nothing to do with the content of majority-ruling comparison or match. Any attack, known or unknown, will be detected as long as it can be reflected at the output vector level.

Although the DRS architecture has an intrusion-tolerant gene that does not rely on attackers' a priori knowledge and behavioral characteristics, it is still static and certain in nature, similar to its operating mechanism, which, in a strict sense, is still characterized by "the detectability of the defense environment and the predictability of the defense behaviors." Assume that an attacker possesses or masters the information on the scenarios within the DRS structure or the resources needed to carry out an attack, then he will be able to capture or control the relevant executors one by one through trial and error or exclusion, resulting in a crash scenario of $f > (N-1)/2$ or exploiting homologous vulnerabilities or another dark feature to attack and escape without being detected by the majority-ruling mechanism. In other words, the intrusion tolerance of the DRS does not have any "time invariance" or stability robustness. Moreover, once the attack and escape are achieved, the attacker's experience and knowledge can be inherited and replicated, and the attack effect can be accurately reproduced on the target system at any time. At this point, DRS has lost all the intrusion tolerance ability, just like a decrypted password. In addition, to ensure the basic heterogeneous redundant properties of the DRS, all components shall be required to be strictly independent in the spatial sense, and the intersection of the given functions of these components shall be strictly limited, i.e., there should be no other identical function intersections except for the equivalent ones. Unfortunately, the existing dissimilarity design tools have not yet been able to provide quantifiable implementation capabilities at both theoretical and technical levels. Currently, they rely mainly on rigorous engineering management or experience-based orientation, which are creating monumental impediments to DRS promotion and application.

In short, the essence of DRS intrusion-tolerant ability is to turn the difficulty of attacking static targets under the shared resource mechanism in a single space into the difficulty of a cooperative attack against static diverse executors under the noncooperative conditions through the DRS consensus mechanism for multimodal output vectors. This results in five important insights: first, the probability of a successful attack on the target system is reduced nonlinearly; second, it is no longer necessary to distinguish between known and unknown attacks or between external and internal attacks; third, the DRS can appropriately lower the reliability threshold for individual parts or components, because the system safety gain has endogenous effects from the structural level, saving rigid requirements for reliability of the software/hardware components, so it is unlike the non-redundant or other redundant systems that are strongly correlated with the effectiveness of additional defensive or protective measures; fourth, the staticity and certainty of DRS allow attackers to use unknown vulnerabilities or other functions for trial-and-error standby tactics to attack and escape, and its intrusion tolerance is neither "time invariant" nor "information-entropy undiminished"; and fifth, once an attacker controls the majority of the executors, the DRS intrusion tolerance will become immediately invalid, exposing the system "to risks."

6.5.2 Summary of the Endogenous Security Effects of the DRS

From the above analysis, the endogenous security effects or features of the DRS can be summarized as follows:

(1) For DM attacks/failures or sporadic failures, as long as the number of actuators at an error state is $f \leq (n - 1)/2$, the security effects of the DRS are certain.
(2) DRS raises the threshold for attacks, for either known or unknown intrusions must be able to cooperatively attack multiple heterogeneous executors.
(3) The multimodal output vector voting mechanism requires known or unknown intrusions to produce a cooperative attack effect that fully matches and is consistent in time and space with the voting strategy.

(4) The endogenous intrusion tolerance effect of the DRS is not premised on any a priori knowledge, such as the attacker's information or behavioral characteristics.

(5) As the semantic or information abundance of output vectors and the voting fineness increase, the success rate of cooperative attacks on diverse targets will be subject to exponential attenuation.

(6) The sustainability of intrusion tolerance effects is related to whether the executors have a real-time cleaning and recovery mechanism and the dynamic removal capability for problem avoidance.

(7) Once the attacker escapes after successful attack, the status can be maintained or safeguarded.

It is not difficult to infer that increasing the redundancy n can improve the security of the system nonlinearly, that increasing the inter-executor dissimilarity h and the output vector information abundance l can reduce the probability of consistent errors exponentially (herein the information abundance refers to the semantic complexity and bit length of the output vectors of heterogeneous executors), and that the multimodal voting mechanism v forces the attacker to solve the problem of achieving cooperative and consistent attacks on multiple targets of different categories under the non-cooperative conditions. Obviously, DRS brings about a fundamental change in attack difficulty and cost compared to the conventional static, certain and similar systems, showing a trend of exponential increase. Just as the geometric stability of a triangle in the Euclidean space depends on the sum of the three interior angles, which is equal to $180°$, the intrusion tolerance effects of DRS are also derived from its multi-dimensional redundant structure as well as mechanism based on multimodal voting. The additional detection and defense factors, although unnecessary, can enhance the dissimilarity between executors in a functionally equivalent manner, that is, DRS can naturally accept or inherit the heritage of, or the latest progress in, the security technology in respect of construction and mechanism.

However, the intrusion tolerance effects of DRS are still subject to the genetic defects of the structure itself in theory and practice, including staticity, certainty, and similarity. As a result, the system can not only be disintegrated by internal backdoor functions or vulnerability-based trial-and-error cooperative attacks that undermine the effectiveness of its intrusion tolerance but may even facilitate the attacker to use the "systematic fingerprint" to achieve "tunnel-through" of sensitive information. See Sect. 6.5.4 for details.

6.5.3   Hierarchical Effect of Heterogeneous Redundancy

A complex information system is often composed of hierarchical hardware and software components. For example, a core router generally consists of four planes, as shown in Fig. 6.7. The lower a plane is located, the more harmful the vulnerabilities will be, evidenced by the vulnerabilities in the basic architecture of the CPU,

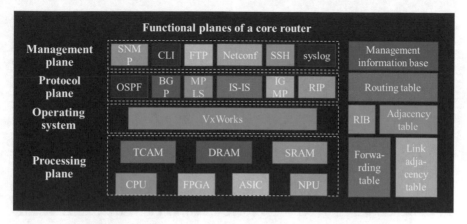

Fig. 6.7 Functional planes of a core router

such as Meltdown and Spectre discovered by the Google security team in June 2017. On January 3, 2018, Intel announced a list of affected processor products, which may even be traced back to the CPU products launched from 1995 onward. Afterward, other major CPU chip manufacturers around the world successively announced their findings: on January 4, IBM announced the potential impact of the vulnerabilities on its POWER series of processors; on January 5, ARM admitted that its chip had similar security vulnerabilities, possibly affecting Android OS devices; AMD issued an official statement acknowledging the existence of similar security vulnerabilities, while Qualcomm claimed to be fixing the related vulnerabilities without specifying the affected chips. This security incident in the IC field has affected almost all hardware devices, including laptops, desktops, smartphones, tablets, and Internet servers. More seriously, a clever exploitation of the "Meltdown" vulnerability can penetrate various security defensive measures in Windows, Linux, Mac OS, and other operating systems, while a flexible use of "Spectre" vulnerability helps to penetrate the operating system kernel's self-protection layer without being noticed in access to kernel space data from the user running space. Some researchers say that at least 30% of CPU performance shall be lost for fixing the vulnerabilities. In any case, even if there were no loophole like "Meltdown" or "Spectre," a high-performing modern CPU usually has millions or even tens of millions lines of hardware codes, where a design flaw is certainly not a rare occurrence. In particular, in an era where the global division of labor has been widely practiced in the IC chip design, processing, and packaging, the introduction of reusable IP kernels in the design is a basic development model. According to statistics, there must be at least 70–80 IP kernels in a high-performing CPU, and at least 20–30 of them need to be outsourced, which is true even to a powerful large company. Therefore, "embedded backdoors" cannot be completely avoided throughout the design chain in both theory and practice.

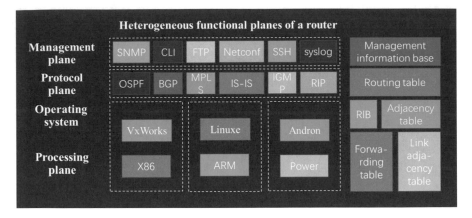

Fig. 6.8 Heterogeneous functional planes of a router

The above events indicate that if the underlying hardware has a backdoor, any security measures of the operating system will become unreliable. Similarly, if the operating system has a backdoor, no matter how many defense technologies are applied to the application software or the underlying hardware, it is difficult to achieve the desired security effect. In fact, the security measures taken within the same environment of hierarchical architecture and shared resources have a very limited defense effect on the unknown vulnerabilities, for these security measures can neither prove their credibility nor eliminate the security threats from the hardware. In addition, strong correlations at all planes result in a scenario where a loophole at any plane can affect the system security. Therefore, the DRS hierarchical heterogeneous redundancy deployment seems to be the only choice to address this security challenge, as shown in Fig. 6.8.

However, an unavoidable cost problem is that the DRS intrusion tolerance not only requires that there are no identical vulnerabilities in the functionally equivalent heterogeneous executors but also expects that the planes bearing the heterogeneous executors shall avoid applying homologous software/hardware components. Obviously, it is often too challenging to meet such demanding heterogeneous deployment requirements in engineering practice.

6.5.4 Systematic Fingerprint and Tunnel-Through

For an I[P]O digital system, when the set of functions relevant to P $[P_1, P2...P_i,$ where $P_i = P_j]$ remains unchanged, varying input stimulus information and sequence relationships can theoretically produce all response information of P at the output, which we call systematic fingerprint (ST), including such information as normal response, abnormal response, latency and state, and the correlated cognitions of the time and logic of such information.Tunnel-through (TT) refers to a scenario where an attacker obtains the sensitive information S in the redundant component A_i or the control authority of the redundant component by running the embedded malicious code and then modulates the response latency or the normal/abnormal state of the output vector, etc., of A_i, without changing any information bit or input/output compliance response of the output vector, while using the consistency requirements of the multimodal voting algorithm or the allowed window time, the sensitive information scan tunnel-through to the outside of the voter via the I[P]O mechanism as the voter's output response time or result is affected by the changes in A_i's output vector latency or state. For instance, there is a DRS system with three heterogeneous redundant components A_1, A_2, and A_3 respectively, if an attacker finds out that the range of the normal output response latency of the system is about 5 ms and the voter's time window is about 30 ms through the Trojan program implanted in A_i. Then, if it is agreed that the bit "1"is represented by the output vector with a response latency of 20 ms received from outside of the voter, while the bit "0" by the output vector has a response latency of 5 ms, the sensitive information S with a length of M bits sent in the TT form will take at most $M + n$ times (n may include the information bits required for synchronization and verification) to request a response. At this time, even if only Executor A_i is controlled in the current operating environment, as long as a strictly consistent voting strategy is adopted and other redundant executors are assumed to be working properly, the virus Trojan in A_i can merely rely on itself to regulate the working state of the voter. Similarly, if all the redundant executors except A_i are working normally, the virus Trojan in A_i can enable the sensitive information S to pass through the voter via $M + n$ times of request response as long as it can control the normality of the A_i output vector. For example, suppose the attacker sets that when the bit is "1" in the transmission of S, A_i will output the abnormal vector information so that the voter will not have any output, and when the bit is "0," A_i will respond normally. At this time, if other redundant executors have identical outputs, a normal output vector can be obtained from outside of the voter. Thus, the sensitive information S can be "demodulated" according to the voter output state upon $M + n$ times of request response. Although this "side channel" attack is not suitable for venues involving the transmission of large amounts of information, it is sufficient to covertly send sensitive information such as passwords and keys.

It should be emphasized that the latter case of the above TT examples is a security defect inherent in the DRS, as well as a reproducible problem. The former case is a special form of side channel attack, where it is difficult to achieve the goal of natural immunity by merely relying on a single voting strategy of multimodal output vector.

6.5.5 Robust Control and General Uncertain Disturbances

The intrusion tolerance property of the DRS has been discussed above. This section considers the question asked at the beginning of this chapter from the perspective of robust control, namely, to normalize the problem of various uncertain failures caused by design defects of the target system into the problem of uncertain disturbances that can be handled through control and architectural technologies so as to provide the system with stability robustness and quality robustness in respect of service and control functions. Of course, the uncertain disturbances referred to herein include those faced by conventional control systems, such as physical uncertain failures of various components and uncertain faults caused by hardware/software design defects, as well as man-made disturbances imposed on the vulnerabilities of the target system (except for TT or downtime-induced attacks), which are collectively summarized as "general" uncertain disturbances (relative to physical disturbances in conventional control systems).

To address the former uncertain disturbances, many effective suppression techniques or mechanisms have been developed, such as redundant backup, code error correcting, distributed deployment, and other reliability designs. However, for the latter uncertain disturbances, there are not yet suitable coping techniques or methods to guarantee the robustness of the target system in a stable manner.

With the ever-evolving information technology, industrial Internet, and digital economy, the control systems are turning from centralized mode into distributed one, from intranet connection to the Internet connection, from closed development to open source innovation, from automation to intelligence, and from the closed intranet deployment to the open "cloud network-terminal" mobile office. The traditional perimeter defense based on attached security facilities such as firewalls is "blurred," which highlights the new robust control challenges arising from cyber attacks such as system backdoors and malicious code injections that are different from uncertain disturbances in a narrow sense. The analysis results of massive cyber security incidents show that the current security situation in cyberspace is largely due to the lack of theories and methods for network elements or hardware/software devices to shoot the new problems regarding robust control.

The concept of robustness is a statistical term, which became popular in the study of control theory in the early 1970s, used to characterize the insensitivity of a control system to feature/parameter disturbances. Generally, the so-called robustness refers to the ability to maintain the stability or dynamicity of a system to some extent when the control target varies within a certain range [8]. When there are uncertain factors in the system, such as model perturbations or random interferences, the control theories and methods that can maintain satisfactory functional quality are called robust control. The concept of stability margin can reflect the ability of the system to resist model perturbations. Robust control methods are suitable for applications prioritizing stability and reliability. According to different definitions of robust performance, it can be divided into two categories: stability robustness and performance/quality robustness. The former refers to the degree to which a control system maintains asymptotic stability when its model parameters vary greatly or when its structure changes. The latter refers to the ability of a system's quality indicators to remain within a certain allowable range under the disturbances of uncertain factors.

A feedback control system (FCS) with robustness means that it has the properties to keep the stability, asymptotic adjustment, and dynamic features constant under a type of uncertain conditions, that is, the FCS shall be able to withstand the effects of such uncertainties [9]. In general, the robustness of the control system runs through the contents of stability, asymptotic adjustment, and dynamic features: Robust stability refers to the ability to guarantee the stability of the FCS, given the impact of a set of uncertain factors. Robust asymptotic adjustment refers to the function to realize asymptotic adjustment of the FCS given the impact of a set of uncertain factors. Robust dynamicity, usually called the sensitivity feature, requires that the dynamicity to be unaffected by uncertain factors. Among them, robust asymptotic adjustment and robust dynamicity reflect the robust performance requirements of the control system.

Robust control has special and general definitions. Special robust control minimizes the sensitivity of the controller to model uncertainty (including external disturbances and parameter disturbances, etc.) to maintain the original performance of the system, while general robust control refers to all control algorithms using a certain controller to deal with systems that contain uncertain factors. It is important to point out that due to the "known unknown security risks of, or unknown unknown security threats to" the target object, its action mechanism and dynamic processes are known, and the scope of impact can be estimated (e.g., the attacker's goal is to acquire sensitive information and modify or delete key data, etc.), so even if the target object's internal random disturbances or uncertain threat disturbances are both included in the scope of robust control, it does not violate the constraints set out by robust control.

The formal description of the robust concept given in Reference [8] is: assume that the mathematical model of the object belongs to a set F, examine some features of the feedback system, such as internal stability, and specify a controller K, if each object in the set F can maintain such feature, the controller is said to be robust to this feature. Understanding a robust control system in the traditional sense requires a description of the uncertainty of the object. It is precisely due to the existence of uncertainty that no physical system so far can be described by accurate mathematical models. The basic form of a typical object model commonly used in cybernetics is: $y = (P + \Delta)u + n$, where y is the output, P is the transfer function of the nominal object, and the uncertainty of the model is presented in two forms:

- n——Unknown noise or interference
- Δ——Perturbations of unknown objects

An input u may produce an output set due to the existence of uncertainties.

It should be emphasized that robust control requires the following constraints:

(1) The dynamicity of the control process is known, and the variation range of uncertain factors can be estimated.
(2) The uncertain factors can be measured and compared with the expected value. The comparison error can be used to correct or regulate the response of the control system.

Although the uncertain disturbances based on the vulnerabilities, etc. meet the robust control constraints, i.e., "the principle is known, the process properties are known", and "the possible range of influence can be estimated," the difficulty is how to measure general uncertain disturbances and obtain the comparison error so as to regulate the response of the control system. What error-regulating algorithms are needed for the control system to reduce its response time? What kind of ability in problem avoidance or abnormality rewind should the target object have to make error elimination possible? How to ensure the stability robustness and quality robustness of the construction function given the deliberate attacks based on unknown vulnerabilities? Therefore, the measurement and perception, error acquisition, and deviation correction of general uncertain disturbances (including man-made disturbances) have become new challenges for robust control.

In general, constrained by $f \leq \dfrac{n-1}{2}$, DRS is highly resistant to random DM disturbances (see Sect. 9.5 for details). However, for uncertain threats based on unknown vulnerabilities built in the structure, such as anti-attack analysis detailed in Sect. 6.7, the defense function does not have stability robustness. There are four reasons for qualitative analysis. First, due to the staticity, certainty, and similarity of the DRS, any attacker possessing some attack resources (such as a controllable 0 day vulnerability or Trojan) can reproduce the expected attack result on a given executor; second, if this capability makes the number of executors at an abnormal state exceed $f \leq \dfrac{n-1}{2}$, all the DRS functions will collapse; third, the majority ruling algorithm cannot perceive the existence of a CM attack-and-escape event; and fourth, when the executors are universally "virus/bacteria bearing," the validity of a deviation correction operation cannot be confirmed, and uncertainty is often bounded, while being unknown is often a concept of relativity stemming from the limitations of perceptible spaces or cognitive means, like what is told in the story of An Elephant and Blind Men. From the perspective of reliability, we can estimate the uncertain faults caused by the random physical failures of the components or by the software/hardware design defects of the system. Therefore, the reliability issue can be appropriately transformed into a system robust control issue due to uncertain disturbance factors in the context of TRA's logical representation. As described in Sect. 6.5, the DRS structure and its operating mechanism are equivalent to TRA's logical representation. When the uncertain faults caused by random physical failures of the components or software/hardware design defects of the system are small probability events, the DSR architecture can transform the issue of uncertain failures of the components into a probable issue of stability robustness at the system level in a scenario of functional equivalence, heterogeneous redundancy, and relative ruling. At this point, the relative perception premise of "an uncertain failure" is: (a) different failure scenarios occurring to the components are large probability events; (b) mostly or fully identical failure scenarios occurring to the components are small probability events. The classic engineering case where the DRS helped to improve the robustness of flight control systems dates back to the launch of Boeing's B-777 passenger aircraft and GE's F-16 fighter. Under the constraints where the dynamicity of the process is known and the variation range of the component's uncertain factors is predictable, the researchers and engineers have achieved the stability as well as quality robustness at the system level via the structural technology, which keep the failure rate of the flight control system of the two types of aircraft below 10^{-11} and 10^{-8}, respectively.

If we do not consider what happens after TT and escape after attack, in a typical DRS structure, no matter what kind of vulnerabilities and Trojans exist in the elements of a heterogeneous redundant executor set and whatever their behavior characteristics are, as long as they cannot make the output vectors produce mostly or fully identical errors, they will be detected and blocked by the multimodal voting mechanism. Therefore, TRA's logical representation is equally applicable to the perception of uncertain threat disturbances. Thus, whether it is a certain or uncertain fault, or an individual known unknown risk or an unknown unknown threat, once reflected to the output vector level, it can be normalized to a probability-controllable reliability event by the TRA-equivalent structure.

Obviously, the robust control of DRS is unable to stably curb the general uncertain disturbances.

6.6 Anti-attack Modeling

The concept of Petri net (PN) is proposed by German scholar Petri in the 1960s [10]. PN is applied at the logical level to model and analyze dynamic systems of discrete events, used to describe the relations between processes or components in a system, including order, concurrency, confliction, and synchronization. After decades of evolution, great progress has been made in the research on the classic PN and the extended PN model, with a general network theory established with a rigid mathematical foundation and multiple abstract levels, which has been applied in many fields, such as communication protocols, system performance evaluation, automatic control, flexible manufacturing systems, distributed databases, and decision-making modeling. PN has gradually grown into a "universal language" for the related disciplines.

In a PN, the place-transition-arc connection is used to represent the static functionality and structure of the system. Meanwhile, the stochastic Petri net (SPN) model contains an execution control mechanism that describes the dynamic behaviors of the system through the transition of the implementation and the change of token distribution in the position. As long as the given conditions or constraints are met, the model will automatically perform state transitions. The deduction process under this causal relationship reflects the dynamic behavioral feature of the system.

The models and methods presented in this section can be used to analyze the robustness, availability, and anti-attack of the information system architecture, mainly by using the PN theory, and the isomorphism between the reachable graph of the general stochastic PN (GSPN) modeland the continuous-time Markov chain (CTMC), to obtain the quantitative analysis conclusion of the target system's anti-attack and reliability.

Firstly, based on the GSPN, a typical anti-attack model of the information system architecture is established for the non-redundant structure and DRS. Next, the isomorphism between the reachable graph of the GSPN model and the CTMC is employed to quantitatively analyze the steady-state probability of the system. In order to establish a universal GSPN-based analysis method for the typical non-redundant, dissimilar redundancy information system architecture, three anti-attack indicators (steady-state availability probability, steady-state escape probability, and steady-state non-specific awareness probability), and three types of parameters (failure rate λ, output vector dissimilarity parameter σ, and recovery rate μ) are used for normalized modeling and analysis of anti-attack of the target architecture. The relevant conclusions and methods can provide meaningful guidance on the design of a robust information system that is highly reliable, credible, and available.

6.6.1 The GSPN Model

In a PN model, the dynamic changes marked in the position can indicate different states of the system. If a position expresses a condition, then it can either include or exclude a token. When a token appears at the position, the condition is true; otherwise, it is false; if a position expresses a state, then the number of tokens in the position sets out the state. The implementing rules for dynamic behaviors in the PN model are as follows:

(1) When the input position of a transition (connected to the transition by an arc, the direction of which points to the transition) includes at least one token, the transition may be implemented, making it an implementable transition.
(2) The implementation of an implementable transition will result in a token cancelled from each of its input positions and a token added to each of its output positions (connected to this transition by the arc, the direction of which points to the transition).
(3) When the weight of an arc is greater than 1, the implementation of the transition requires to include at least the number of tokens equal to the connected arc weight in all input positions of the transition and will generate the corresponding number of tokens in all output positions based on the connected arc weights.

According to its basic structure and the implementation rules for its dynamic behaviors, a PN system is defined as follows:

(1) (S,T,F) is a network, where S is position/state, T is transition, and F is arc.
(2) $K: S \rightarrow N + \cup \{\infty\}$ is a position capacity function.
(3) $W: F \rightarrow N+$ is an arc weight function.
(4) $M_0: S \rightarrow N$ is an initial state satisfying $\forall s \in S, M_0(s) \leq K(s)$.

Over the past few years, the basic form of PN has evolved into various PNs with different features and forms from different sides and perspectives. The general stochastic Petri net (GSPN) is an extension of PN, which can simplify the system state space and is thus widely used for analysis of a system's reliability, availability, and security [11]. The transitions in GSPN are divided into two subsets: the immediate transition set and the timed transition set. The implementation delay of an immediate transition is zero, while the latency parameter of a timed transition is a random variable exponentially distributed. The hybrid Petri net (HPN) is developed on the basis of the discrete Petri net, and its position and transition are divided into two categories, namely, continuous and discrete, to characterize continuous variable processes and discrete event processes.

The definition of the GSPN is: $GSPN = (S, T, F, K, W, M_0, \Lambda)$, where (S, T, F, K, W, M_0) is a PN system, and Λ is the average implementation rate. An inhibitor arc is allowed in F, where the implementable condition of the position to which the inhibitor arc is connected becomes unimplementable. The transitions in the GSPN are divided into two subsets: the timed transition set $T_t = \{t_1, t_2, \ldots, t_k\}$, and the immediate transition set $T_s = \{t_{k+1}, t_{k+2}, \ldots, t_n\}$, where the implementation delay of immediate transitions is 0, while the average implementation rate set related to timed transitions is $\Lambda = (\lambda_1, \lambda_2, \ldots, \lambda_k)$. When a mark M contains multiple immediate transitions, its selection probability needs to be identified.

For a given GSPN model, the reachable set $P(N, M_0)$ can be obtained by the following steps under the initial state identifier M_0.

(1) Determine the distribution of the initial identifier M_0 to solve the correlation matrix D.
(2) Under the identifier distribution $M_k(k = 0, 1, 2, \ldots)$, identify the enabling transition sequence X, and solve the new state identifier of the system M_{k+1} from the matrix equation $M_{k+1} = M_k + XD$.
(3) Determine whether the state identifier M_{k+1} is the enabling transition: if not, the system reaches the steady state; otherwise, take $k = k + 1$, and then return to Step (2).

6.6.2 Anti-attack Considerations

People always expect a sufficiently low probability of structural failures of an information system to meet its application and service requirements, i.e., the availability of the architecture should reach the required level. The author believes that the anti-attack and reliability of an ideal information system architecture should be normalized in design, and its behaviors and results should be expected by solving its model, so that the performance can be quantified, the level can be calibrated, the effect can be verified, the behavioral state can be monitored, the behavioral outcome can be measured, and the abnormal behaviors can be controlled.

For defenders, since the non-redundant structure cannot transform an uncertainty/certainty attack into a probability issue, it can only be assumed as a reliability issue to satisfy the premise of random failure analysis. The anti-attack of the non-redundant structure generated from this assumption is only the most ideal case, namely, the upper limit, which may far outnumber its actual anti-attack performance.

Being intrusion tolerant to some extent, the DRS can transform an uncertainty attack into a system-level probability event within a certain time range as long as the attack is random, so it can be normalized into, and handled as, a reliability issue during this period. However, for a trial-and-error attack based on known DM vulnerabilities or a cooperative attack based on CM vulnerabilities, it can no longer be regarded as a random event given the static and certain structure. The anti-attack of the DRS obtained under the above assumption only represents the maximum value, namely, the best case of its anti-attack.

Vulnerability refers to any feature of an information system that can be used as an attack premise, where those caused by backdoor vulnerabilities constitute an important reason for ubiquitous security threats across cyberspace. There are many conceptual descriptions of vulnerability, among which an authoritative one defines [12] it from the perspective of state transition of a computer system. A computer system consists of a series of states that describe the current configuration of the entities that make up the system. The system computes via state transitions. Beginning with a given initial state, using a set of state transitions, all states that can be reached are ultimately divided into two state types defined by the security policy: authorized and unauthorized. Then, a vulnerable state refers to a state in which the authorized state can be used and transformed to finally reach the unauthorized state; a compromised state refers to a state that can be reached by the above transformation method; a successful attack refers to a sequence of transformations beginning with an authorized state and ending with a compromised state, while vulnerability refers to the specific features that distinguish a vulnerable state from a non-vulnerable state.

Attacks can lead to failures, errors, or degradations of the information system via transformations that can eventually reach the vulnerable state. Attackers can take advantage of the vulnerability awash in the system to achieve their goals, such as obtaining system permissions and key user data or paralyzing the target system. According to the attacker's ability, we divide attacks into two types: general attacks and special attacks. Among them, a general attack refers to an attack where the attack ak at the time T has no synergistic cumulative effect on the attack a_{k+n} at the time T + X, such as the common DMAs and CMAs; a special attack refers to a super powerful attack where the attack a_k at the time T has a cumulative cooperative effect on the attack a_{k+n} at the time $T + X$, for example, an attack that can cause the target executors to halt one by one or a standby cooperative attack that cannot be perceived via multimodal output vectors under the majority ruling conditions.

Downtime attacks and standby cooperative attacks that belong to special attacks are specified as follows:

1. Downtime attack

Assume that an attacker possessing sufficient resources understands and grasps the "killer" vulnerabilities of all (or most) functionally equivalent (FE) executors or all variants within the defense community. These vulnerabilities can be exploited to cause the target executors to stop normal services. However, this hypothetical situation requires the scheduling of various resources and measures, including those in the social engineering sense, and thus suggests extremely high attack costs. Needless to say, ordinary hackers are not in a position to perform well, and this goal is tough even for a powerful hacker organization, or a professional agency with government background, or even an action planned by a nation.

2. Standby cooperative attack

Assume that an attacker grasps the "super" vulnerabilities of all executors of the target object and can exploit these vulnerabilities to obtain the supreme control over the executors one by one without touching their output vectors. In this scenario, the attacker only needs to employ the shared input channel and the normal protocol message to send the pre-designed cooperative attack command. Then the controlled executors will send the prepared outbound output vectors to achieve a stable attack and escape. Although this hypothesis may hold true theoretically, it is still very challenging to put it into practice, for there are too many uncertainties in the pursuit of this goal.

We classify the faults caused by attacks into two categories, respectively, degradation/failure faults and escape faults. Each redundancy of the information system is called an executor or a channel. In respect of a degradation/failure fault, there is usually only one fixed implementation scenario in the non-redundant system: when a single executor fails, the system will also fail, i.e., a failure fault occurs; for a triple-redundancy DRS system, it will fail only when two or more executors fail simultaneously, i.e., failure faults occur; in order to simplify the analysis and verify the anti-attack of the architecture itself, the surface defense measures (e.g., abnormal fault detection, such as intrusion detection and firewalls, as well as encrypted authentication) are not added to the executors of the abovementioned systems, the factors of fault perception and defense mechanisms (such as fault detection rate, false alarm rate, and missed alarm rate) are not considered, and the software and hardware devices of the system are not segregated, while their mutual influence not considered. To facilitate quantitative analysis, we normalize factors such as the attacker/defender capabilities, network vulnerability, and network environment into attack/defense time costs. The attack time cost refers to the average time of attack (ATA) required by the attacker to successfully launch an attack, and the defense time cost refers to the average time of defense (ATD) required by the defender to get recovered after successfully defending against the attack. When the attack disturbance arrival conforms to the negative exponential distribution, the probability of a successful attack can be expressed as $F(t) = P\{T < t\} = 1 - e^{-\lambda t}$, where λ is the transfer rate of the attack resulting in an abnormal output response. Thus, the expected time to complete the attack is $1/\lambda$, and the average time for the abnormal executor recovery is $1/\mu$. The attack scenarios can be divided into the following categories based on the attack time cost: in a weak attack scenario, it takes a relative longer time (assuming an average of 10 hours) to successfully break through the defense of a single executor; in a medium attack scenario, it takes a relative shorter time (assuming an average of 10 minutes) to successfully attack a single executor;

and in a strong attack scenario, it takes a very short time (assuming an average of 10 seconds) to successfully attack a single executor.

This section assumes that, if the TRA holds true, the threats based on the vulnerabilities of the target object can be normalized to a reliability problem by the defense system's structural effect. It also assumes that the success and recovery time of an executor-oriented attack are subject to the exponential distribution, so that the structural anti-attack analysis can be performed via the GSPN model. It can be seen from the above assumptions that for a non-redundant structure, the actual effect of the attack threat is much more severe than the assumed scenario when the conventional defense mechanism is not taken into account, because the experience can be constantly copied or inherited as long as an attack succeeds. In the worst case, the continuous availability of the non-redundant system may approach zero. For DRS, even if the average attack success time is subject to the exponential distribution, the actual anti-attack effect may be much lower than the expected value: given the staticity and certainty of the system's architecture and operating mechanism, once an attacker finds the way of breaking through the lines, it is possible to use this method without limitations to achieve a time-irrelevant attack effect. At this time, the DRS will completely lose its anti-attack performance.

6.6.3 Anti-attack Modeling

Anti-attack is the ability of the system to continuously provide effective services and recover all services within a specified time period in the event of a failure led by an external attack. The assumptions about the anti-attack model are as follows:

Assumption 1 The target system of the non-redundant architecture is a one-redundancy static system (referred to as the non-redundant system, NRS), the executors have neither attached specific fault perception and defense methods such as intrusion detection and firewalls nor a fault recovery mechanism; the system stops its service when the executors fail; the attack success time is subject to a negative exponential distribution.

Assumption 2 The DRS target system is a triple-redundancy static heterogeneous redundancy system (HRS) (the abstract model of which is shown in Fig. 6.6a). The executors do not have specific fault perception and defense methods, such as intrusion detection and firewalls, and the multimode ruling mechanism can detect the executors which failed due to general attacks and halt attacks (as the output vectors differ from the majority of the executors) and can recover single executor abnormalities in triple-redundancy scenarios; as long as most of the executors are not in the fault state at the same time, the system will always be able to provide degraded services, with the time for an executor to successfully attack and recover subject to a negative exponential distribution.

The anti-attack GSPN models for different systems are defined as follows:

Definition 6.1 The GSPN model for NRS attack failures

$$\text{GSPN}_N = \left(S_N, T_N, F_N, K_N, W_N, M_{N0}, \Lambda_N\right)$$

where $S_N = \{P_{N1}, P_{N2}, P_{N3}\}$; $T_N = \{T_{N1}, T_{N2}\}$; F_N is the arc set of the model; $K_N = \{1,1,1\}$ defines the capacity of each element in the S_N; W_N is the set of arc weights, where the weight of each arc is 1; $M_{N0} = \{1,0,0\}$ defines the initial state of the model; and $\Lambda_N = \{\lambda_{N1}, \lambda_{N1}\}$ defines the average implementation rate set relevant to the time transition.

Definition 6.2 The GSPN model for DRS attack failures

$$\text{GSPN}_D = \left(S_D, T_D, F_D, K_D, W_D, M_{N0}, \Lambda_D\right)$$

where $S_D = \{P_{D1}, P_{D2}, \ldots, P_{D24}\}$; $T_D = \{T_{D1}, T_{D2}, \ldots, T_{D33}\}$; F_D is the arc set of the model; $K_D = \{1,1,\ldots,1\}$ defines the capacity of each element in the S_D; W_D is the set of arc weights, where the weight of each arc is 1; $M_{D0} = \{1,1,1,0,\ldots,0\}$ defines the initial state of the model; and $\Lambda_D = \{\lambda_{D1}, \lambda_{D2}, \ldots, \lambda_{D6}\}$ defines the average implementation rate set relevant to the time transition.

The probability indicators used to evaluate the system's anti-attack include availability probabilities, escape probabilities, non-specific awareness probabilities, dormancy probabilities, and degradation probabilities. The definitions of the probabilities are as follows:

Definition 6.3 Availability probabilities (AP): The probabilities for the system to be in the state of normal service. In a NRS, AP refers to the probabilities that all executors of the system are in a vulnerability dormant state; in a triple-redundancy DRS, AP refers to the probabilities that all executors of the system are in a vulnerability dormant state or a single executor is in a fault state.

Definition 6.4 Escape probabilities (EP): The probabilities for the system to be in the state of outputting erroneous output vectors. The reason is that the attacker controls the majority of the executors, resulting in consistent yet erroneous output vectors after the system ruling.

Definition 6.5 Non-specific awareness probabilities (NSAP): The probabilities for the system to spot part of the executors having an inconsistent output vector state with other executors.

Definition 6.6 Dormancy probabilities (DP): The probabilities for a system's vulnerabilities not exploited by an attacker and thus in a dormant state.

Definition 6.7 Failure or degradation probabilities (FP/DP): In terms of anti-attack, for the NRS and DRS, FP refers to the probabilities for the system to be in a null output state caused by some or all failures of the executors or the inconsistent output vectors of all the executors. DP refers to the probabilities for the system to enable the relevant re-ruling strategy to generate the output vectors with reduced confidence levels when all the output vectors of the executors are inconsistent.

6.7   Anti-aggression Analysis

6.7.1   Anti-general Attack Analysis

6.7.1.1   Non-redundant System

Based on the definition of the GSPN, the GSPN model for the non-redundant system (a single-redundancy static system) under the general attack conditions is shown in Fig. 6.9. There are three states and two transitions. P_1 contains a token to indicate system dormancy, that is, a single executor has vulnerabilities not yet exploited by an attacker; P_2 contains a token to indicate system degradation; P_3 contains a token to indicate system escape; T_1 represents a post-attack degradation of an executor transited from dormancy; and T_2 represents a post-attack escape of an executor transited from dormancy.

A NRS system is a static system the executor redundancy number of which is one. Assume that the attack type is a general attack, where the executors do not have specific fault perception and defense methods such as intrusion detection and firewalls, and do not afford fault recovery.

Since the GSPN model reachable set and the CTMC are isomorphic, the isomorphic method can be used to analyze the target system's anti-attack. The CTMC model of a non-redundant system's attack failures is shown in Fig. 6.10. When

Fig. 6.9 The GSPN model for the non-redundant system under the general attack conditions

Fig. 6.10 The CTMC model of a non-redundant system's attack failures

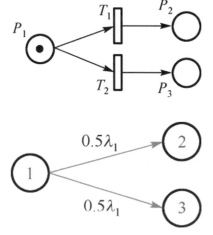

Table 6.1 Steady states of the CTMC model for non-redundant system attack failures

Serial No. of the state	Meaning
1	The executor is running normally, and the system is in a dormancy state
2	One executor is attacked, generating a null output failure, and the system is in a degradation state
3	One executor is attacked, generating an output vector error failure, and the system is in an escape state

Table 6.2 Parameters of the CTMC model for non-redundant system general attack failures

Parameters	Value	Meaning
$\lambda_1(h)$	0.1/6/360	Assume that the average time of a single executor failure due to a general attack (output vector abnormalities) is 10 hours/10 minutes/10 seconds

attacked, the system's sole executor may degrade or escape, with the same probability assumed to both scenarios. In order to unify the preconditions, it is assumed that the attack success time of the executor is subject to the exponential distribution. The actual situation is far more serious for the non-redundant system than the assumption (because the attacker can constantly exploit a successful attack once it is spotted), that is, the upper limit of the anti-attack of the non-redundant system (i.e., the best case of defense capabilities) can be obtained under the assumption.

The steady states of the CTMC model are shown in Table 6.1. The meanings of the parameters of the CTMC model are shown in Table 6.2, where λ_1 represents the average failure time of a single executor due to a general attack, classified into the weak attack scenario, the medium attack scenario, and the strong attack scenario.

The GSPN model reachable set and the Markov chain isomorphism are applied to find the steady-state probabilities of the GSPN. Assume that the reachable set of the GSPN model is R, which can be divided into two sets according to its characteristics, respectively, M_T and M_V. M_T represents the explicit state, where an immediate transition cannot be enabled; the M_V represents the implicit state, where an immediate transition is enabled. In the process of system state transition, implicit states do not take time, so they can be eliminated from the reachable set R, and their impact on the system is transferred to the explicit states for consideration. All the states are rearranged, with all implicit states in front, and explicit states in the back. The transition probabilities for the system to transfer to M_V from M_T and for M_T itself are recorded as P^{TV} and P^{TT}, respectively. The transition probability matrix between the system's explicit states is:

$$U = P^{TT} + P^{TV}\left(I - P^{VV}\right)^{-1} P^{VT} \tag{6.1}$$

Via U, we can construct a transfer rate matrix of CTMC, also called a transfer density matrix, defined as

$$q_{ij} = \begin{cases} \lim\limits_{\Delta t \to 0} \dfrac{u_{ij}(\Delta t)}{\Delta t}, & i \neq j \\[3mm] \lim\limits_{\Delta t \to 0} \dfrac{u_{ij}(\Delta t)-1}{\Delta t}, & i = j \end{cases} \tag{6.2}$$

Then, q_{ij} refers to the transfer rate to the explicit state M_j from the explicit state M_i, where $i, j \in [1,1]$, $1 = M_T$. The Q matrix is a matrix with q_{ij} as an element. The probability vector $P(t) = (p_1(t), p_2(t),\ldots, p_l(t))$, where $p_i(t)$ is the immediate probability when the system is in the explicit state M_i, and then the following differential equation holds:

$$\begin{cases} P'(t) = P(t)Q \\ P(0) = \left(p_1(0), p_2(0),\cdots, p_l(0)\right) \end{cases} \tag{6.3}$$

The above calculation steps are the same for non-redundant systems, dissimilar redundant systems, and mimic defense systems.

Based on the non-redundant system's attack failure CTMC model, the state transition equation can be obtained:

$$\begin{bmatrix} \dot{P}_1(t) \\ \dot{P}_2(t) \\ \dot{P}_3(t) \end{bmatrix} = \begin{bmatrix} -\lambda_1 & 0 & 0 \\ 0.5\lambda_1 & 0 & 0 \\ 0.5\lambda_1 & 0 & 0 \end{bmatrix} \begin{bmatrix} P_1(t) \\ P_2(t) \\ P_3(t) \end{bmatrix} \tag{6.4}$$

If M_{N0} is known, we can calculate the steady-state probabilities of each state of the NRS system's CTMC model:

$$\begin{cases} P_{M_0} = 0 \\ P_{M_1} = 0.5 \\ P_{M_2} = 0.5 \end{cases} \tag{6.5}$$

In the GSPN model, according to the implementation rules of time transitions and the no-latency feature of immediate transitions, Table 6.3 gives the reachable set of states and gives the steady-state probabilities of each state mark according to the relevant parameters of the transition process, with P_u, P_k, and P_s, respectively, representing the steady-state probabilities of the weak attack, medium attack, and strong attack scenarios.

Based on the definition of system state probabilities, the steady-state probabilities of the non-redundant system can be calculated as follows:

(1) Steady-state AP = $P(M_0) = 0$, a weak/medium/strong attack scenario.

Table 6.3 Reachable state set of the non-redundant system's attack failure CTMC model

Mark	Steady-state probabilities P_u, P_k and P_s	P_1	P_2	P_3	State
M_0	0	1	0	0	Tangible
M_1	0.5	0	1	0	Tangible
M_2	0.5	0	0	1	Tangible

(2) Steady-state EP = $P(M_2)$ = 0.5, a weak/medium/strong attack scenario.
(3) Steady-state NSAP = 0, a weak/medium/strong attack scenario.
(4) Steady-state FP = $P(M_1)$ = 0.5, a weak/medium/strong attack scenario.

The state probabilities of the non-redundant system change with the runtime, as shown in Fig. 6.11. The simulation results show that in the case of random attack, the attack on the non-redundant system will always succeed after a period of time, and the system's AP(t) will quickly tend to 0, and the FP(t) and EP(t) will be identical and will quickly approach 0.5, while the NSAP(t) is always 0. The non-redundant systems do not have constant anti-attack capabilities when not endowed with specific awareness and defense (e.g., intrusion detection or firewalls) measures.

The relationship between the non-redundant system's state probabilities and λ_1 is shown in Fig. 6.12. The simulation results show that the non-redundant system will only end with the state of degradation/escape when attacked. The steady-state DP and the steady-state EP (both valued at 0.5) are independent from the average time of a single executor failure λ_1.

When specific awareness and defense measures are not considered for a non-redundant system, a general attack will result in system degradation or attack and escape after a short period of time, and the system cannot afford any non-specific awareness of general attacks. In summary, the non-redundant systems do not have sustained resistance to general attacks, and they are not robust, which is consistent with our intuitive feelings.

6.7.1.2 Dissimilar Redundant System

Based on the definition of the GSPN$_D$, the GSPN model for the DRS system (such as a triple-redundancy DRS system) under the general attack conditions is shown in Fig. 6.13. There are 24 states and 33 transitions. P_1, P_2, and P_3 contain a token each to indicate that the executor is in a dormancy state, that is, all executors have vulnerabilities not yet exploited by an attacker; P_4, P_5, and P_6 contain a token each to indicate a post-attack executor failure; P_7, P_8, and P_9 contain a token each to indicate two simultaneously failed relevant executors with abnormal output vectors; P_{10}, P_{12}, and P_{14} contain a token each to indicate two failed executors with consistent abnormal output vectors; P_{11}, P_{13}, and P_{15} contain a token each to indicate two failed executors with inconsistent abnormal output vectors; P_{16}, P_{17}, P_{18}, and P_{19} contain a token each to indicate three simultaneously failed executors with abnormal output vectors; P_{22} contains a token to indicate three failed executors with consistent

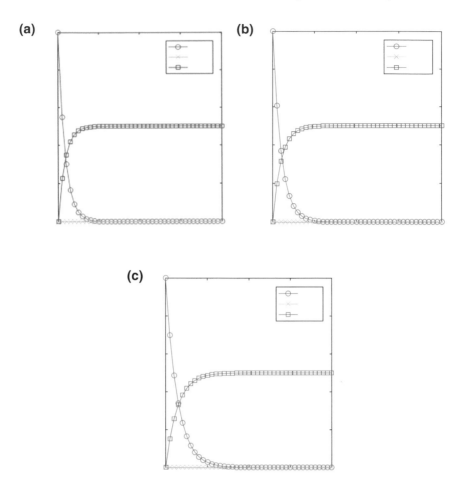

Fig. 6.11 Changes in the non-redundant system's state probabilities with runtime. (**a**) Medium attack scenario. (**b**) Week attack scenario. (**c**) Strong attack scenario

abnormal output vectors; P_{23} contains a token to indicate three failed executors with abnormal output vectors, with two of them having consistent output vectors; and P_{24} contains a token to indicate three failed executors with abnormal and totally inconsistent output vectors.

T_1, T_2, and T_3 indicate that the executor is faulty due to the attack; T_4, T_5, and T_6 indicate that the executor is restored to a dormancy state due to the changed attack/ defense scenario; T_7, T_8, and T_9 indicate that the two executors simultaneously enter the fault state; T_{10}, T_{12}, and T_{14} indicate that the abnormal output vectors of the two faulty executors enter the consistent state with the selection probability σ; T_{11}, T_{13}, and T_{15} indicate that the abnormal output vectors of the two faulty executors enter the inconsistent state with the selection probability $(1-\sigma)$; T_{16}, T_{17}, T_{18}, and T_{19}

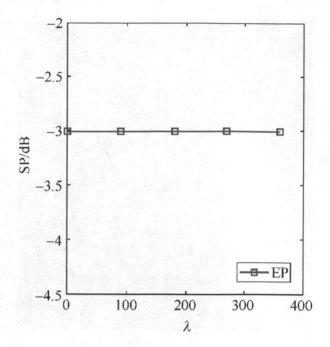

Fig. 6.12 Changes in the non-redundant system's state probabilities with parameter λ

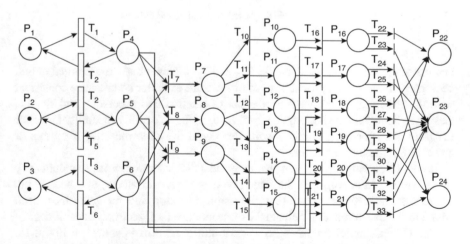

Fig. 6.13 GSPN model for the DRS system's general attack failures

indicate the simultaneous fault of three executors; T_{22}, T_{26}, and T_{30} indicate that the abnormal output vectors of the third executor enter a state consistent with the same abnormal output vectors of the first two executors with the selection probability σ; T_{23}, T_{27}, and T_{31} indicate that the abnormal output vectors, as the selection probability $(1-\sigma)$, of the third executor enter a state inconsistent with the same abnormal output

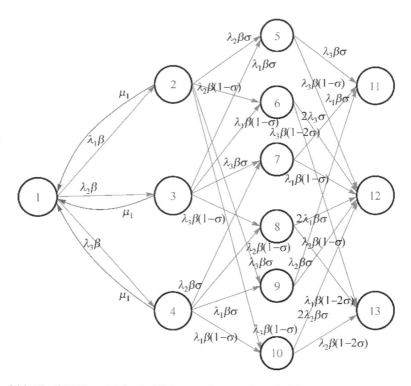

Fig. 6.14 The CTMC model for the NRS system's general attack failures

vectors of the first two executors; T_{24}, T_{28}, and T_{32} indicate that the abnormal output vectors, as the selection probability 2σ, of the third executor enter a state consistent with the abnormal output vectors of either of the first two executors; and T_{25}, T_{29}, and T_{33} indicate that the abnormal output vectors, the selection probability $(1-2\sigma)$, of the third executor enter a state inconsistent with the abnormal output vectors of either of the first two executors.

It should be emphasized that the analyzed DRS system is a triple-redundancy DRS system. Assuming that the attack type is a general attack, and no specific awareness and defense means (e.g., intrusion detection and firewalls) have been added to the executors, the individual executors have fault recovery capabilities.

The CTMC model for the NRS system attack failures is shown in Fig. 6.14. Assume that the success and recovery time of an executor-oriented attack are subject to the exponential distribution, then for a DRS system, the actual situation is far more serious than the assumption (due to its static structure, the attacker can constantly copy the attack experience once an attack succeeds). Under this assumption, the upper limit of anti-attack of the DRS system (i.e., the best case of defense capabilities) can be obtained. The steady state of the CTMC model is shown in Table 6.4.

Table 6.4 Steady state of the CTMC model for the DRS system's general attack failures

Serial No. of state	Meaning
1	All the executors are running normally and in the dormancy state
2	Post-attack failure of Executor 1
3	Post-attack failure of Executor 2
4	Post-attack failure of Executor 3
5	Executors 1 and 2 failed, and their abnormal output vectors are consistent
6	Executors 1 and 2 failed, and their abnormal output vectors are inconsistent
7	Executors 2 and 3 failed, and their abnormal output vectors are consistent
8	Executors 2 and 3 failed simultaneously, and their abnormal output vectors are inconsistent
9	Executors 1 and 3 failed, and their abnormal output vectors are consistent
10	Executors 1 and 3 failed, and their abnormal output vectors are inconsistent
11	The three executors failed, and their abnormal output vectors are consistent
12	The three executors failed, and two of them have consistent abnormal output vectors
13	The three executors failed, and they do not have consistent abnormal output vectors

Table 6.5 Parameters of the CTMC model for the DRS system's general attack failures

Parameter	Value	Meaning
$\lambda_1(h)$	0.1/6/360	Assume that Executor 1 is attacked, and the average time of its abnormal output vector is 10 hours/10 minutes/10 seconds
$\lambda_2(h)$	0.1/6/360	Assume that executor 2 is attacked, and the average time of its abnormal output vector is 10 hours/10 minutes/10 seconds
$\lambda_3(h)$	0.1/6/360	Assume that executor 3 is attacked, and the average time of its abnormal output vector is 10 hours/10 minutes/10 seconds
σ	1.0×10^{-4}	Assume that, when attacked, two faulty executors show the uncertainty in the consistency of output vectors
$\mu_1(h)$	60	Assume that one executor's output vectors are abnormal, the average time for the executor to recover to the dormancy state is 0.001 seconds when the attack/defense scenario changes

The meanings of the parameters of the CTMC model are shown in Table 6.5. $\lambda_1 \sim \lambda_3$ represents the average time of a fault caused by a general attack. It is assumed that the scenarios are classified into three: weak, medium, and strong attacks. The average time of an executor fault caused by a general attack is assumed to be 10 hours/10 minutes/10 seconds, respectively, where μ_1 indicates that 1 executor produces abnormal output vectors due to general attacks. When the attack scenario changes, the average time for the executor to recover to the dormancy state is assumed to be 0.001 seconds. For general attacks, the smaller the μ_1 is, the faster the recovery rate of the executor and the better the system's anti-attack effect will be. By definition, a general attack refers to an attack where the attack a_k at the time T

has no cumulative cooperative effect on the attack a_{k+n} at the time $T + X$. Therefore, when the next general attack arrives, along with the changes in the attack/defense scenario, the executor will go directly to the dormancy state from the fault state. Assume that the maximum request processing capacity of the system is M times/second, then the average recovery time of the executor represented by μ_1 should be $1/M$ second for general attacks, and the maximum processing capacity of a normal information system can easily exceed 100,000 times/second. That is, μ_1 should be much smaller than 1.0×10^{-5} seconds. Here, we assume that μ_1 is 1.0×10^{-3} seconds.

σ represents the proportion of consistent output vectors of two faulty executors after being attacked. Reference [13] is a file based on the National Vulnerability Database (NVD), which analyzes the vulnerabilities of 11 operating systems in 18 years and finds that the operating systems from the same family (such as Windows 2003 and Windows 2008) share a relatively large number of CM vulnerabilities, which is a rare occurrence (0 in many cases) for operating systems from different families (such as between BSD and Windows).For this reason, if two executors are sufficiently heterogeneous, in the event of general attacks, their proportion σ of consistent abnormal output vectors can be set as a reasonable small value of 1.0×10^{-4}. In the development of specific products, the engineering management method similar to that used in the development of flight control systems can be employed to ensure that the actual value of this parameter is much smaller than 1.0×10^{-4}.

Where, M_{D0} is known, the stability probability of each state can be figured out:

$$
\begin{cases}
P_{M0} = 0 \\
P_{M1} = 0 \\
P_{M2} = 0 \\
P_{M3} = 0 \\
P_{M4} = 0 \\
P_{M5} = 0 \\
P_{M6} = 0 \\
P_{M7} = 0 \\
P_{M8} = 0 \\
P_{M9} = 0 \\
P_{M10} = 0 \\
P_{M11} = 3\sigma - 3\sigma^2 \\
P_{M12} = 1 - 3\sigma + 2\sigma^2
\end{cases}
\tag{6.7}
$$

In the GSPN model, according to the implementation rules of time transitions and the no-latency feature of immediate transitions, Table 6.6 gives the reachable set of states and gives the steady-state probabilities of each state mark according to the relevant parameters of the transition process, with P_u, P_k, and P_s, respectively, representing the steady-state probabilities of the weak attack, medium attack, and strong attack scenarios.

Table 6.6 Reachable state set of the DRS system's general attack failure CTMC model

Mark	Steady-state probability P_u	Steady-state probability P_k	Steady-state probability P_s
Steady-state probabilities			
M_0	0.000000	0.000000	0.000000
M_1	0.000000	0.000000	0.000000
M_2	0.000000	0.000000	0.000000
M_3	0.000000	0.000000	0.000000
M_4	0.000000	0.000000	0.000000
M_5	0.000000	0.000000	0.000000
M_6	0.000000	0.000000	0.000000
M_7	0.000000	0.000000	0.000000
M_8	0.000000	0.000000	0.000000
M_9	0.000000	0.000000	0.000000
M_{10}	10^{-8}	10^{-8}	10^{-8}
M_{11}	2.999700×10^{-4}	2.999700×10^{-4}	2.999700×10^{-4}
M_{12}	9.996999×10^{-1}	9.996999×10^{-1}	9.996999×10^{-1}

Tangible state	P_1	P_2	P_3	P_4	P_5	P_6	P_{10}	P_{11}	P_{12}	P_{13}	P_{14}	P_{15}	P_{22}	P_{23}	P_{24}
M_0	1	1	1	0	0	0	0	0	0	0	0	0	0	0	0
M_1	0	1	1	1	0	0	0	0	0	0	0	0	0	0	0
M_2	1	0	1	0	1	0	0	0	0	0	0	0	0	0	0
M_3	1	1	0	0	0	1	0	0	0	0	0	0	0	0	0
M_4	0	0	1	0	0	0	1	0	0	0	0	0	0	0	0
M_5	0	0	1	0	0	0	0	1	0	0	0	0	0	0	0
M_6	1	0	0	0	0	0	0	0	1	0	0	0	0	0	0
M_7	1	0	0	0	0	0	0	0	0	1	0	0	0	0	0
M_8	0	1	0	0	0	0	0	0	0	0	1	0	0	0	0
M_9	0	1	0	0	0	0	0	0	0	0	0	1	0	0	0
M_{10}	0	0	0	0	0	0	0	0	0	0	0	0	1	0	0
M_{11}	0	0	0	0	0	0	0	0	0	0	0	0	0	1	0
	0	0	0	0	0	0	0	0	0	0	0	0	0	0	1

According to the definition of the system state probabilities, each steady-state probability of the DRS system can be calculated:

(1) Availability probabilities (AP)

$$\text{AP} = P(M_0) + P(M_1) + P(M_2) + P(M_3) = \begin{cases} 0, & \textit{Weak attack scenario} \\ 0, & \textit{Medium attack scenario} \\ 0, & \textit{Strong attack scenario} \end{cases} \quad (6.8)$$

(2) Escape probabilities (EP)

$$EP = P(M_4) + P(M_6) + P(M_8) + P(M_{10}) + P(M_{11})$$

$$= \begin{cases} 2.999700 \times 10^{-4}, & \textit{Weak attack scenario} \\ 2.999700 \times 10^{-4}, & \textit{Medium attack scenario} \\ 2.999700 \times 10^{-4}, & \textit{Strong attack scenario} \end{cases} \tag{6.9}$$

(3) Non-specific awareness probabilities (NSAP)

$$NSAP = 1 - P(M_0) - P(M_{10}) = \begin{cases} 9.999999 \times 10^{-4}, & \textit{Weak attack scenario} \\ 9.999999 \times 10^{-4}, & \textit{Medium attack scenario} \\ 9.999999 \times 10^{-4}, & \textit{Strong attack scenario} \end{cases}$$

The state probabilities of the DRS system vary with the runtime, as shown in Fig. 6.15. The simulation results show that under the general attack scenarios, the multimode ruling mechanism of the DRS system will bring about higher NSAP and lower EP, but its static structural characteristics will make its AP tend to 0 after a certain time (Depending on the attack intensity, the time ranges from tens of hours to tens of thousands of hours). Namely, when there is no specific awareness and defense means such as intrusion detection or firewalls, the attack will make the DRS system unavailable for a certain period of time.

The steady-state system probabilities of a DRS system are correlated only to the parameter σ, and the correlation is shown in Fig. 6.16. The simulation results show that the uncertainty σ of consistent abnormal output vectors occurring to the two executors after the attack has a great impact on the escape probabilities and the non-specific awareness probabilities. The bigger the heterogeneity between the executors, the smaller σ is, the lower the steady-state escape probabilities, and the higher the steady-state non-specific awareness probabilities. When there is no CMF with identical abnormal output vectors between the executors, equivalent to that σ is equal to 0 (the attack causes the abnormal output vectors of any two faulty executors to be different or the attack will not cause the two executors to fail at the same time), its escape probabilities shall be 0, and the non-specific awareness probabilities shall be 1.

The DRS system has a higher steady-state non-specific awareness probabilities and a lower steady-state EP, but its steady-state availability probability is 0. The DRS system's anti-attack ability is relatively sensitive and accurate but is far from being persistent, stable, and robust. The system state probabilities of the DRS system are greatly affected by the parameter σ. The bigger the heterogeneity between the executors, the smaller σ is, the lower the steady-state escape probabilities, and the higher the steady-state non-specific awareness probabilities. Therefore, to improve the anti-attack of DRS systems, we should try to choose executors with the largest heterogeneity.

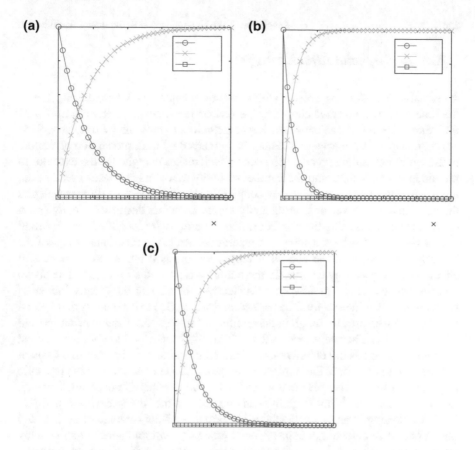

Fig. 6.15 The DRS system's state probabilities varying with the runtime. (**a**) Medium attack scenario. (**b**) Week attack scenario. (**c**) Strong attack scenario

Fig. 6.16 The DRS system's steady-state probabilities varying with the parameter σ

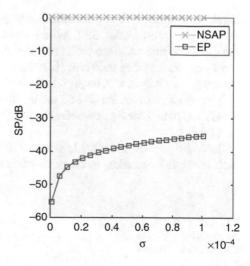

6.7.2   Anti-special Attack Analysis

A special attack refers to an attack with superpower capability, where the attack a_k at the time T has a cumulative cooperative effect on the attack a_{k+n} at the time $T + X$, including attacks that can cause the target executors to halt one by one, as well as standby cooperative attacks that cannot be perceived by multimodal output vectors in the multimode ruling process. A special attack makes the time for the executor to remain in the fault state, and the number of faulty executors increases greatly (i.e., the attack effect can be cumulatively cooperative in the space-time dimension), and the abnormal output vector heterogeneity of the executors decreases greatly (for a downtime attack, the functionally equivalent heterogeneity length of the abnormal output vectors is 1 bit; for a standby cooperative attack, the functionally equivalent heterogeneity length of the abnormal output vectors is 0 bit, where a successful attack will not cause any changes in the output vectors of the executor). The above two reasons lead to the greatly reduced steady-state AP and the greatly increased steady-state EP of the system. The background we need to understand is that special attacks are based on a thorough understanding of the real-time status of online and offline executors. They have mastered the "killer" or "super" vulnerabilities for most executors, as well as the effective use of the related attack chains. On this basis, it will be able to overcome the bottleneck of "cooperative attack to diverse dynamic targets under non-cooperative conditions" while addressing all uncertainty factors. Obviously, this "super attack" ability can only exist in the "thought experiment."

The following is an analysis of the performance of non-redundant system and DRS systems in defending against special attacks (downtime attacks and standby cooperative attacks) on the assumption that the attacker has the above capabilities.

6.7.2.1   Non-redundant System

The CTMC model for a non-redundant system that is faulty due to two special types of attack is shown in Fig. 6.17. When attacked, the sole executor may degrade or escape. P_1 contains a token indicating that the system is in a fault dormancy state, and P_2 contains a token indicating that the system is in a fault state. T_1 indicates that the system is faulty due to a special attack.

The steady states of the CTMC model are shown in Table 6.7. The parameters of the CTMC model for the non-redundant system's special attack faults are shown in Table 6.8.

Table 6.9 gives the reachable state set and gives the steady-state probability of each state mark according to the parameters relevant to the transition process, with

Fig. 6.17 CTMC model for special attack faults of non-redundant systems

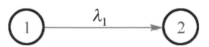

Table 6.7 Steady states of the CTMC model for the non-redundant system's special attack fault

Serial No. of state	Meaning
1	Each executor is in normal operation and in a dormancy state
2	1 executor fails after downtime attack or escapes after a standby cooperative attack

Table 6.8 Parameters of the CTMC model for the non-redundant system's special attack faults

Parameter	Value	Meaning
$\lambda_1(h)$	6/360	Assume that the MTBF of the executor after a special attack is 10 minutes/10 seconds

Table 6.9 Reachable state set of the CTMC model for the non-redundant system's special attack faults

Mark	Steady-state probability P_k	Steady-state probability P_s	P_1	P_2	State
M_0	0	0	1	0	Tangible
M_1	1	1	0	1	Tangible

P_k and P_s representing the steady-state probabilities of medium attack and strong attack scenarios, respectively.

According to the definition of the system state probabilities, the AP, EP, and NSAP of the mimic defense system in the steady state can be calculated:

(1) AP = $P(M_0)$ = 0, a medium/strong attack scenario.
(2) EP = $P(M_1)$ = 1, a medium/strong attack scenario.
(3) NSAP = 0, a medium/strong attack scenario. For non-redundant systems, the steady-state DP for halt attacks and the steady-state EP for standby cooperative attacks are both 1, and the NSAP is zero. Therefore, non-redundant systems are basically not resistant to constant special attacks.

6.7.2.2   Dissimilar Redundant System

The GSPN model for special attack faults of a dissimilar redundant system is shown in Fig. 6.18. There are 7 states and 9 transitions. P_1, P_2 and P_3 contain a token each to indicate that the executor is in a faulty dormancy state; P_4, P_5, and P_6 contain a token each to, respectively, indicate a special attack fault occurring to the executor; P_7 contains a token to indicate a special attack fault simultaneously occurring to two or more executors. T_1, T_2, and T_3 indicate a special attack fault occurring to the executor; T_7, T_8, and T_9 indicate the executor recovered to the fault dormancy state; and T_4, T_5, and T_6 indicate two or more executors simultaneously entering the special attack fault state. In this model, halt attacks are subject to a suppression arc, while standby cooperative attacks are not.

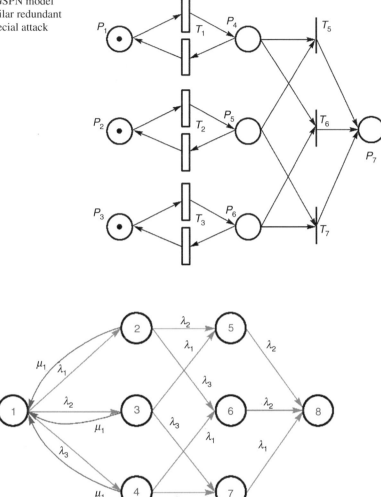

Fig. 6.18 GSPN model for a dissimilar redundant system's special attack faults

Fig. 6.19 CTMC model for a dissimilar redundant system's special attack faults

The CTMC model is shown in Fig. 6.19.

1. Downtime Attacks

The steady states of the CTMC model are shown in Table 6.10. The meanings of the parameters of the model are shown in Table 6.11.

Based on the GSPN model for special attack faults of a DRS system, the state transition equation can be obtained:

Serial No. of state	Meaning
1	All the executors in normal operation and in a dormancy state
2	Post-attack downtime of one executor
3	Post-attack downtime of two executors
4	Post-attack downtime of three executors
5	Post-attack downtime of the executors 1 and 2
6	Post-attack downtime of the executors 1 and 3
7	Post-attack downtime of the executors 2 and 3
8	Post-attack downtime of three executors

Table 6.10 Steady states of the CTMC model for a dissimilar redundant system's downtime attack faults

Table 6.11 Parameters of the CTMC model for a dissimilar redundant system's halt attack faults

Parameter	Value	Meaning
λ_1	6/360	Assume that Executor 1 is attacked, and its average downtime is 10 minutes/10 seconds
λ_2	6/360	Assume that Executor 2 is attacked, and its average downtime is 10 minutes/10 seconds
λ_3	6/360	Assume that Executor 3 is attacked, and its average downtime is 10 minutes/10 seconds
μ_1	60	Assume that the average self-recovery or reconstruction time of one executor is 1 minute

$$
\begin{bmatrix} \dot{P}_1(t) \\ \dot{P}_2(t) \\ \dot{P}_3(t) \\ \dot{P}_4(t) \\ \dot{P}_5(t) \\ \dot{P}_6(t) \\ \dot{P}_7(t) \\ \dot{P}_8(t) \end{bmatrix} = \begin{bmatrix} -\lambda_1-\lambda_2-\lambda_3 & \mu_1 & \mu_1 & \mu_1 & 0 & 0 & 0 & 0 \\ \lambda_1 & -\mu_1-\lambda_2-\lambda_3 & 0 & 0 & 0 & 0 & 0 & 0 \\ \lambda_2 & 0 & -\mu_1-\lambda_1-\lambda_3 & 0 & 0 & 0 & 0 & 0 \\ \lambda_3 & 0 & 0 & -\mu_1-\lambda_1-\lambda_2 & 0 & 0 & 0 & 0 \\ 0 & \lambda_2 & \lambda_1 & 0 & -\lambda_3 & 0 & 0 & 0 \\ 0 & \lambda_3 & 0 & \lambda_1 & 0 & -\lambda_2 & 0 & 0 \\ 0 & 0 & \lambda_3 & \lambda_2 & 0 & 0 & -\lambda_1 & 0 \\ 0 & 0 & 0 & 0 & \lambda_3 & \lambda_1 & \lambda_2 & 0 \end{bmatrix} \begin{bmatrix} P_1(t) \\ P_2(t) \\ P_3(t) \\ P_4(t) \\ P_5(t) \\ P_6(t) \\ P_7(t) \\ P_8(t) \end{bmatrix} \quad (6.10)
$$

Table 6.12 gives the reachable set of states and gives the steady-state probabilities of each state mark according to the relevant parameters of the transition process, with P_k and P_s, respectively, representing the steady-state probabilities of the medium attack and strong attack scenarios.

According to the definition of the system state probabilities, the AP, EP, and NSAP of the mimic defense system in the steady state can be calculated:

(1) AP = $P(M_0) + P(M_1) + P(M_2) + P(M_3) = 0$
(2) EP = 0
(3) NSAP = 1

Table 6.12 Reachable state set of the CTMC model for a dissimilar redundant system's downtime attack faults

Mark	Steady-state probability P_k				Steady-state probability P_s		
Steady-state probabilities							
M_0	0				0		
M_1	0				0		
M_2	0				0		
M_3	0				0		
M_4	0				0		
M_5	0				0		
M_6	0				0		
M_7	1				1		
Tangible state							
	P_1	P_2	P_3	P_4	P_5	P_6	P_7
M_0	1	1	1	0	0	0	0
M_1	0	0	0	1	0	0	0
M_2	0	0	0	0	1	0	0
M_3	0	0	0	0	0	1	0
M_4	0	0	0	1	1	0	0
M_5	0	0	0	1	0	1	0
M_6	0	0	0	0	1	1	0
M_7	0	0	0	1	1	1	1

For dissimilar redundant systems, the NSAP and the AP/EP for downtime attacks are 1 and 0, respectively. Therefore, the dissimilar redundant systems are basically not resistant to constant downtime attacks.

2. Standby Cooperative Attacks

The steady states of the CTMC model are shown in Table 6.13. The meanings of the parameters of the model are shown in Table 6.14.

Based on the GSPN model for special attack faults of a dissimilar redundant system, the state transition equation can be obtained:

$$
\begin{bmatrix} \dot{P}_1(t) \\ \dot{P}_2(t) \\ \dot{P}_3(t) \\ \dot{P}_4(t) \\ \dot{P}_5(t) \\ \dot{P}_6(t) \\ \dot{P}_7(t) \\ \dot{P}_8(t) \end{bmatrix} =
\begin{bmatrix}
-\lambda_1 - \lambda_2 - \lambda_3 & \mu_1 & \mu_1 & \mu_1 & 0 & 0 & 0 & 0 \\
\lambda_1 & -\mu_1 - \lambda_2 - \lambda_3 & 0 & 0 & 0 & 0 & 0 & 0 \\
\lambda_2 & 0 & -\mu_1 - \lambda_1 - \lambda_3 & 0 & 0 & 0 & 0 & 0 \\
\lambda_3 & 0 & 0 & -\mu_1 - \lambda_1 - \lambda_2 & 0 & 0 & 0 & 0 \\
0 & \lambda_2 & \lambda_1 & 0 & -\lambda_3 & 0 & 0 & 0 \\
0 & \lambda_3 & 0 & \lambda_1 & 0 & -\lambda_1 & 0 & 0 \\
0 & 0 & \lambda_3 & \lambda_2 & 0 & 0 & -\lambda_2 & 0 \\
0 & 0 & 0 & 0 & \lambda_3 & \lambda_1 & \lambda_2 & 0
\end{bmatrix}
\begin{bmatrix} P_1(t) \\ P_2(t) \\ P_3(t) \\ P_4(t) \\ P_5(t) \\ P_6(t) \\ P_7(t) \\ P_8(t) \end{bmatrix}
\quad (6.11)
$$

Table 6.15 gives the reachable set of states and gives the steady-state probabilities of each state mark according to the relevant parameters of the transition process,

Table 6.13 Steady states of the CTMC model for a dissimilar redundant system's standby cooperative attack faults

Serial No. of state	Meaning
1	All the executors in normal operation and in a dormancy state
2	Executor 1 is attacked, where a standby cooperative fault occurs without abnormal output vectors
3	Executor 2 is attacked, where a standby cooperative fault occurs without abnormal output vectors
4	Executor 3 is attacked, where a standby cooperative fault occurs without abnormal output vectors
5	Executors 1 and 2 are attacked, where a standby cooperative fault occurs to both without abnormal output vectors
6	Executors 1 and 3 are attacked, where a standby cooperative fault occurs to both without abnormal output vectors
7	Executors 2 and 3 are attacked, where a standby cooperative fault occurs to both without abnormal output vectors
8	The three executors are all attacked, where a standby cooperative fault occurs to them all without abnormal output vectors

Table 6.14 Parameters of the CTMC model for a dissimilar redundant system's standby cooperative attack faults

Parameter	Value	Meaning
λ_1	6/360	假
		Assume that the average standby cooperative fault time of Executor 1 when attacked is 10 minutes/10 seconds
λ_2	6/360	Assume that the average standby cooperative fault time of Executor 2 when attacked is 10 minutes/10 seconds
λ_3	6/360	Assume that the average standby cooperative fault time of Executor 3 when attacked is 10 minutes/10 seconds
μ_1	2	Assume that the average recovery cycle of an executor is 30 minutes

with P_k and P_s, respectively, representing the steady-state probabilities of the medium attack and strong attack scenarios.

According to the definition of the system state probabilities, the AP, EP, and NSAP of a DRS system in the steady state can be calculated:

(1) $AP = P(M_0) + P(M_1) + P(M_2) + P(M_3) = 0.$
(2) $EP = 1.$
(3) $NSAP = 0.$

For dissimilar redundant systems, the AP/NSAP and the EP for downtime attacks are 0 and 1, respectively. Therefore, DRS systems are basically not resistant to standby cooperative attacks.

Table 6.15 Reachable state set of the CTMC model for a dissimilar redundant system's standby cooperative attack faults

Mark	Steady-state probability P_k	Steady-state probability P_s
Steady-state probabilities		
M_0	0	0
M_1	0	0
M_2	0	0
M_3	0	0
M_4	0	0
M_5	0	0
M_6	0	0
M_7	1	1

Tangible state							
	P_1	P_2	P_3	P_4	P_5	P_6	P_7
M_0	1	1	1	0	0	0	0
M_1	0	0	0	1	0	0	0
M_2	0	0	0	0	1	0	0
M_3	0	0	0	0	0	1	0
M_4	0	0	0	1	1	0	0
M_5	0	0	0	1	0	1	0
M_6	0	0	0	0	1	1	0
M_7	0	0	0	1	1	1	1

6.7.3 Summary of the Anti-attack Analysis

The anti-general attack abilities of non-redundant systems and dissimilar redundant systems are shown in Table 6.16. Taking a strong attack scenario as an example, we can find that upon a general attack, a non-redundant system is basically not resistant if the attack is constant; the dissimilar redundant system has a higher steady-state NSAP and a lower steady-state EP, but its steady-state AP is 0, that is, the dissimilar redundant system has an anti-attack capability that is relatively sensitive and accurate yet not persistent.

As shown in Table 6.17, for special attacks, a non-redundant system is unable to resist downtime attacks and standby cooperative attacks. A dissimilar redundant system can detect all the downtime attacks but cannot escape, while the steady-state AP is 0, that is, a dissimilar redundant system is unable to persistently resist downtime attacks and standby cooperative attacks.

Table 6.16 Steady-state probabilities of system resistance to general attacks

Steady-state probabilities	Non-redundant system	Dissimilar redundant system
Weak attack scenario		
Steady-state DP(dormancy probability),$P(M_0)$	0	0
Steady-stateNSAP(non-specific awareness probability)	0	9.999999×10^{-1}
Steady-state AP(availability probability)	0	0
Steady-state EP(escape probability)	5.00000×10^{-1}	2.999700×10^{-4}
Medium attack scenario		
Steady-state probabilities	Non-redundant system	Dissimilar redundant system
Steady-state DP(dormancy probability),$P(M_0)$	0	0
Steady-stateNSAP(non-specific awareness probability)	0	9.999999×10^{-1}
Steady-state AP (availability probability)	0	0
Steady-state EP(escape probability)	5.00000×10^{-1}	2.999700×10^{-4}
Strong attack scenario		
Steady-state probabilities	Non-redundant system	Dissimilar redundant system
Steady-state DP,$P(M_0)$	0	0
Steady-stateNSAP	0	9.999999×10^{-1}
Steady-state AP	0	0
Steady-state EP	5.00000×10^{-1}	2.999700×10^{-4}

Table 6.17 Steady-state probabilities of systems against special attacks

Downtime attacks				
Steady-state probabilities	**Non-redundant system**		**Dissimilar redundant system**	
	Medium attacks	**Strong attacks**	**Medium attacks**	**Strong attacks**
Steady-state DP(dormancy probability), $P(M_0)$	0	0	0	0
Steady-state NSAP (non-specific awareness probability)	0	0	1	1
Steady-state AP (availability probability)	0	0	0	0
Steady-state EP (escape probability)	1	1	0	0
Standby cooperative attacks				
Steady-state probabilities	**NRS systems**		**DRS systems**	
	Medium attacks	**Strong attacks**	**Medium attacks**	**Strong attacks**
Steady-state DP(dormancy probability), $P(M_0)$	0	0	0	0
Steady-state NSAP (non-specific awareness probability)	0	0	0	0
Steady-state AP (availability probability)	0	0	0	0
Steady-state EP (escape probability)	1	1	1	1

6.8   Conclusion

6.8.1   Conditional Awareness of Uncertain Threats

In reference to the TRA, under the TRA constraints, the functionally equivalent, heterogeneous redundant, and multimodal vector majority ruling mechanism can be used to perceive the DM output vectors caused by unknown attacks (e.g., those targeted at unknown vulnerabilities of the executors) at the structural level. In other words, based on the logical structure of the TRA, uncertain threats to the individual elements can be transformed into the differences between the multimodal output vectors at the structural level, which can be measured and perceived. Although this kind of perceptual mechanism lacks stability robustness, with a risk of being disintegrated by trial and error or standby cooperative attacks, it still points out a new way of uncertain threat perception that does not depend on the a priori knowledge and behavior characteristics of the attacker. It also provides the necessary conditions for implementing the robust control functionality based on error elimination. However, due to the existence of standby/CM attacks, the majority rule mechanism cannot dodge the problem of "identification blind zones," which often requires iterative processing by diversified ruling algorithms and backward verification mechanisms in support of further screening.

6.8.2   New Connotations of General Robust Control

The robust control mechanism based on "measurement and perception, error recognition and feedback iteration" can keep the model perturbations within a given desired threshold when the model structure is certain and the disturbance range is known. Therefore, when the robust control mechanism and the related control functions are introduced into the heterogeneous redundancy structure (HRS) (such as a DRS architecture) equivalent to the TRA, it is not difficult to give the latter an endogenous security feature of uncertainty, largely solving the problem that the robust control functionality of a classical HRS lacks stability in response to trial and error or cooperative attacks. The author incorporates this kind of control functionality to stably suppress general uncertain disturbances into the scope of general robust control, trying to propose innovative theoretical methods and institutional mechanisms for normalized solutions to the traditional uncertain disturbances and nontraditional uncertain threats by expanding the connotation and extension of classical general robust control. Therefore, when general robust control is mentioned in the subsequent sections of this book, unless otherwise stated, it refers to the concept of general robust control that covers two types of uncertain disturbance factors with the new, extended connotation.

6.8.3 DRS Intrusion Tolerance Defect

It is easy to see from the foregoing content in this section that the DRS architecture itself is transparent to the provision of system service functionality, i.e., the increase or decrease in system services is independent from the architecture. This intensive architecture attribute, which integrates service provision, reliability, guarantee, and intrusion tolerance, has a special potential for application scenarios with cyber service reliability and security assurance requirements, a significant price/performance edge throughout the life cycle, and better technical and economic benefits over the traditional deployment of segregated services and security systems. But above all, it is necessary to address the problem of nonpersistent intrusion tolerance or stability robustness. Secondly, the engineering implementation complexity should be simplified.

Despite its inherent ability of intrusion tolerance, the DRS is at least challenged in six aspects in response to incremental, persistent, and cooperative intrusion threats such as APT.

(1) To ensure the basic heterogeneous redundant properties of the DRS, it is required that all components are strictly independent in the spatial sense and that the intersection of the given functions of these components is strictly limited, i.e., there should be no other identical function intersections except for the equivalent ones. That is to say, the properties shall be guaranteed by a very complicated and costly dissimilarity design. Unfortunately, the existing dissimilarity design is not yet able to provide quantifiable implementation capabilities at both theoretical and technical levels. Currently, we rely mainly on rigorous engineering management or experience-based orientation, which has created obstacles to the universal promotion and application.

(2) The classical heterogeneous redundancy system is still static and certain and so is its operating mechanism. In a strict sense, it still has a "detectable defense environment with predictable defense behaviors." Assume that an attacker possesses or masters the information on the target or the resources needed to carry out the attack, then the attacker will be able to expose the heterogeneous redundant executors to one by one or trial-and-error attack in sequence, causing with ease majority or consistency errors to achieve escape under the multimodal voting mechanism. More seriously, once the attack succeeds, its experience and knowledge are inheritable to accurately reproduce the attack effect, and the value of exploitation can be planned, posing severe security threats to the privacy, integrity, and effectiveness of the information system.

(3) Fault tolerance and intrusion tolerance are subject to different mathematical models. The former is premised on the detectability of faults, conditioned on the random occurrence of faults, and based on mathematical analysis tools such as Markov chain. As for the latter, since deliberate attacks, such as those based on unknown vulnerabilities of the target object, are often hidden and internally and externally cooperative, the attack effect is usually certain and often does not

constitute a probability issue. Therefore, the dissimilar redundant fault-tolerant model is obviously not a suitable model for security analysis under the complex cyber attack conditions.

(4) The DRS invented to greatly improve reliability is premised on relatively unreliable heterogeneous executors, but it does not perceive the reliability defects that may exist in the heterogeneous executors to have an attribute of deliberate diffusion. Therefore, there is no mandatory "de-synergization" requirement in the engineering implementation, for example, the executors can have ports or links in-between them, or own common physical or logical space, or boast shared communication and synchronization mechanisms, or have even two-way interactive dialogue mechanisms. In particular, there is still the practice of copying the running environment from the online executors for mutual synchronization in the faulty executor recovery algorithm, which will facilitate the deliberate construction of a cooperative attack scenario.

(5) The voting mechanism designed to overcome Byzantine failures requires a complicated negotiation mechanism (even an introduced appropriate encryption mechanism) to be set up across the heterogeneous executors while assuming that all the executors are credible before being attacked. In the context of the cyberspace globalization, especially when the vulnerabilities (even Trojans) in the hardware/software products taking more than half of the market share cannot be reliably removed, the premise of this assumption is difficult to hold true. This also explains why the current blockchain consensus mechanism cannot reliably function in the lack of credibility of the host system.

(6) The DRS has a "structure-determined security" attribute, which can normalize known or unknown personalized attacks and certain or uncertain faults into probability-controllable reliability events by exploiting the equivalent logical structure of the TRA. However, the DRS, as a non-closed-loop control structure, fails to change the certainty and staticity defects of the target system, and is thus unable to deal with "co-cheating" or "trial-and-error" attacks, failing to impose stable robust control over uncertain threats.

In summary, if an attacker has the resources and capabilities to cover the static environment of the DRS, he can be theoretically better position in terms of detection, awareness, and cooperative attacks based on the DRS defense defects. For example, assume that an attacker has mastered the vulnerabilities or implanted backdoors in the majority of the executors and can initiate a cooperative attack based on some trigger mechanism (e.g., triggering multiple "downtime" events within a period shorter than the heterogeneous executor switching time, or outputting pre-stored, identical error results based on a unified timing mechanism, or uniformly outputting the same message after t seconds upon receiving some sort of input stimulus data at a time, etc.), it can still penetrate the block of the multimodal voting process and result in an FE CMF or escape event.

Taking a triple-redundancy DRS system as an example, assume that an attacker has mastered vulnerabilities A and B in two of the components and is able to construct two Trojans exploiting these two vulnerabilities. Due to the staticity and certainty of the DRS system, the attacker always has relatively sufficient opportunities to implant the two Trojans into the corresponding executors. The attacker only needs to design a common external stimulus signal K in the implanted Trojans. When the Trojan of the corresponding executor receives the signal K, it simultaneously transmits outbound the message M, which has been long stolen and is in a pending state. Since the message M in heterogeneous executors is exactly the same, it cannot be intercepted by the multimodal voting mechanism. Similarly, if the attacker is able to implant backdoor traps and Trojans, etc. in multiple components in advance, the above attack can be achieved more conveniently. Using this technique skillfully, the attacker is fully able to get the desired information or expected attack results from a DRS system. Nevertheless, it is a daunting challenge to design, implant, and exploit the related attack code, even involving social engineering measures in event of some custom executors.

In any case, in addition to that the DRS system cannot effectively deal with the "TT" attack introduced in Sect. 6.5.4., its staticity and certainty make the system's internal operating environment, and voting mode still perceptible or recognizable, including the unknown vulnerabilities that exist in the components and the malicious features that are hidden therein in some way, which are often attack reachable [14]. In other words, although the DRS operating environment has increased the difficulty of attack based on the output vector modification to the extent that multiple targets need to be cooperatively and consistently attacked under the non-cooperative conditions, it still belongs to the target object category where the attack task can be planned and the attack effect can be expected [15].

6.8.4 DRS Transformation Proposals

We know that the original intention of the DRS does not emphasize the dynamic and random properties of reconfigurable or software-definable HR resources. Therefore, in the DRS environment, the introduction of dynamicity may change the staticity in the allocation of system resources, the introduction of randomness may change the certainty of the operating mechanism, the introduction of diversity may change the similarity between the voting strategies, and the introduction of robust control mechanisms may achieve general dynamicity of the iterative convergence in a defense scenario. In Chap. 7 of this book, we can see that the systematic introduction of these basic defense elements significantly enhances the DSR architecture's ability to resist known or unknown attacks, and in particular, it has a potential to fundamentally transform the DRS architecture to solve its genetic defect of inability to provide stable intrusion tolerance and the TT problem that cannot be solved before. Owing to the dynamic, diverse, and stochastic factors and robust control mechanisms, the endogenous security effects of the target object may not only bring about intractable challenges to cooperative attacks under the non-cooperative conditions but also make it more uncertain for the tactical effects of affecting the privacy, integrity, and effectiveness of the target object through cyber attacks. It may even directly undermine the effectiveness of attack theories and methods based on software/hardware code defects and may hinder cyber attacks from acting as a means of combat missions that can be planed and utilized at the tactical level with evaluable combat results.

As far as cost-effectiveness is concerned, simple stacking of basic defense elements, including dynamicity, randomness, and diversity, cannot shoot the problem expected to be solved, but will significantly reduce the economic service performance of the target object (see related descriptions in Chap. 4 and Chap. 5). In recent years, influenced by the US-launched thinking of moving target defense (MTD), the industry has rolled out some active defense technologies (partially with smart functions to some extent) that reduce the availability of the target object's vulnerabilities by way of dynamicity and randomness and has indeed achieved some good effects, for they can particularly reduce the severity threat based on unknown vulnerabilities, though they are completely ineffective on the internal-external cooperative attacks implemented on the backdoors of the system's internal hardware/software resources or on the embedded malicious codes. In addition, the credibility and encrypted authentication of their dynamic control or scheduling process are similar to traditional security technologies such as virus screening, neither self-proven nor self-protected in the same processing space and through a shared resource mechanism (the hacked ASLR, a Windows address randomization measure, is a well-known example). More seriously, their engineering implementation is opaque to the target object's software/hardware code function and performance, and the related components or modules have to be "inserted"; the effectiveness of their defense is at the expense of a large amount of consumption of the target object's processing resources and service performance, even leading to a seriously confrontation between the security and the cost-effectiveness of the service. Technically, we are still trapped in the status quo where the security status of the defense object cannot be designed and evaluated in a quantitative manner.

The core of the DRS transformation philosophy is to find a structure and mechanism to aggregate the dynamic, diverse, and random defense elements into a defense environment applicable both externally and internally, given that there are universal "DM" vulnerabilities in the assumed redundant executors. The closed-loop feedback control structure based on the voting mechanism is a good choice for "stringing the pearls into a necklace." The principle of robust control tells that the closed-loop feedback mechanism can easily incorporate dynamic, random, and diverse elements into a perceptual, gradual, and convergent dynamic process, which helps to control the effects of general uncertain disturbances within the expectable scope. Apparently, once the feedback loop of robust control enters a steady state, it cannot be reactivated until it detects or senses a new uncertain disturbance or receives an external command. Unlike the anti-jamming frequency hopping communication, it does not need to strive for wider frequency range, more frequency points, and higher hop speed. Massive engineering practices have shown that the ubiquitous, high-frequency, or unstructured use of dynamic, random, and diverse defense means is not a wise choice for defending the security of complex information systems.

References

1. Jajodia, S., Ghosh, A.K., Swarup, V., et al.: Moving Target Defense: Creating Asymmetric Uncertainty for Cyber Threats. Translated by Yang Lin. National Defense Industry Press, Beijing (2014)
2. Huang, C.Y., Lyu, M.R., Kuo, S.Y.: A unified scheme of some nonhomogenous poisson process models for software reliability estimation. IEEE Trans. Softw. Eng. **29**(3), 261–269 (2003)
3. Rubira-Calsavara, C.M.F., Stoud, R.J.: Forward and Backward Error Recovery in C++. University of Newcastle upon Tyne, New Castle (1993)
4. Carreira, J., Madeira, H., Silva, J.G.: Xception: a technique for the experimental evaluation of dependability in modem computers. IEEE Trans. Softw. Eng. **24**(2), 125–136 (1998)
5. Hongwei, Z., Wei, H., Deyuan, G.: Analysis of the dissimilar redundancy computer system and the corresponding reliability. J. Harbin Inst. Technol. **40**(3), 492–494 (2008)
6. Xudong, Q., Zongji, C. Reliability analysis of dissimilar redundancy flight control computers based on Petri net. Control and Decision. **20**(10), 1173–1176 (2005)
7. Baidu Encyclopedia/Immunity. http://baike.baidu.com/item/%E5%85%8D%E7%96%AB/82 5313?fr= aladdin. [2017-03-19]
8. Zhongke, S., Fangxiang, W., Bei, W., et al.: Robust Control Theories, pp. 2–16. National Defense Industry Press, Beijing (2003)
9. Doyle, J.C., Zhou, K.M.: Essentials of Robust Control. Pearson, New York (1997)
10. Petri, C.A.: Communication with Automata. Rome Air Development Center, Rome (1966)
11. Caselli, S., Conte, G., Marenzoni, P.: Parallel state space exploration for GSPN models. Appl. Theory Petri Nets. 181–200 (1995)
12. Bishop, M., Bailey, D.: A Critical Analysis of Vulnerability Taxonomies. University of California at Davis, Sacramentto (1996)
13. Garcia, M., Bessani, A., Gashi, I., et al.: Analysis of OS diversity for intrusion tolerance. Soft0 Pactice Exp. 1–36 (2012)
14. Fengshun, Y.: Research on Data Self-destruct Mechanisms in the Cloud Computing Environment. Central South University, Changsha (2011)
15. Gu, J.F., Zhang, L.L.: Data, DIKW, big data and data science. Procedia Comp. Sci. **31**, 814–821 (2014)
16. Yeh, Y.C.: Triple-triple redundant 777 primary flight computer. IEEE Aerosp. Appl. Conf. Proc. **1**, 293–307 (1996). https://doi.org/10.1109/AERO.1996.495891
17. Zongji, C., Xudong, Q., Gao, J.: Dissimilar redundancy flight control computers [J]. Acta Aeronautica Et Astronautica Sinica. **03**, 320–327 (2005)

Chapter 7
DHR Architecture

As we have learned from Chap. 6, DRS has an inherent anti-attack attribute (i.e., intrusion-tolerant attribute) in addition to the conventional robust control attribute, where the "one-way transparent attack from within and without" based on the backdoor of static target vulnerabilities under a single-space shared resource mechanism can be upgraded in respect of complexity to the cooperative attack stage based on the backdoor of static multi-target vulnerabilities. Thus, deterministic attacks targeted at each executor, in a given multimodal OV space, will be forced by the MV mechanism into a probability event strongly correlated with the redundancy size, executor heterogeneity, and OV complexity. However, the attack resistance of DRS is affected by the following conditions and factors: ① Suppose that the heterogeneity of the DHR redundant executors is infinite in size, that is, there are no existing dark function intersections; ② the number of executors in the abnormal state within the architecture must meet the condition of $f \leq (N-1)/2$; ③ external and internal coordinated attacks aiming at backdoors or malicious codes within the architecture are not taken into account; ④ majority selection ruling algorithms have judgement-blind areas against the multiple or concerted attacks and escapes; ⑤ there are no post-processing mechanisms against the executors with abnormal output vectors except hang-up/cleaning. What is worse is that the operation environment of all the executors in the DRS structure and the exploitable conditions of vulnerabilities and backdoor are statically determined and the parallel deployment of the executors usually does not change the accessibility of attack surface. Therefore, theoretically speaking, attackers can reach two aims through an unrestricted trial-and-error approach: the first aim is to break through the executors with exploitable vulnerabilities and backdoors continuously or one by one so that the number of abnormal executors which happen concurrently in the architecture is larger than $f = (N-1)/2$; the second aim is that the attacker may use the dark functions existing in the executors to launch standby coordinated attacks or tunnel breakthrough (please refer to Sect. 6.5.3) so as to use the judgement-blind areas of

© Springer Nature Switzerland AG 2020
J. Wu, *Cyberspace Mimic Defense*, Wireless Networks,
https://doi.org/10.1007/978-3-030-29844-9_7

the majority selection ruling mechanisms for attack and escape. And the attack experience can be inherited. Attack methods can be copied, and the attack effects have afterward-exploitability value. That is, DRS staticity, certainty, and similarity have serious genetic defects in the security area. As a result, it lacks the capability of maintaining nondecreasing information entropy against general uncertain disturbances and therefore does not possess such features as stable robust control and quality robust control against cyber attacks. This chapter focuses on how to use the general robust control technology to change the DRS "structural genes." Dynamic convergent and iterative diversified defense scenarios will be used to replace excessively rigid and costly heterogeneous designing to obtain the effects with the measurable designing and verifiable structure so that it is possible to stably suppress general uncertain disturbances including attacks aiming at unknown vulnerabilities and backdoors.

7.1 Dynamic Heterogeneous Redundant Architecture

The author has found out that from the perspective of the defender's initial information entropy, both sides of attacks and defenses focus their games on the increase, decrease, and maintaining of the information entropy. The attack-tolerant attribute of DRS lacks the longtime stability. That is, when precisely oriented attempt attacks or trial-and-error attacks go on persistently, the initial information entropy in the architecture will decrease due to lack of self-maintaining mechanisms until the initial information entropy drops to the level when attacks can reliably play an expected role. It is easy to infer that if the mechanism of nondecreasing initial information entropy is introduced or (added) into DRS (or the information entropy can be held in balance), robustness can be added to its features. For example, if such conventional defense elements as dynamicness, randomness, diversity, reconfiguration, encrypted authentication, or intrusion detection can be taken into the DRS, architecture and strategy ruling, control law feedback, and iterative and convergent robustness control mechanisms are also introduced[1], it can theoretically change the DRS staticity, certainty, and similarity and other genetic defects in the operating environment. These control architecture and operation mechanism after the change of the genetic engineering are featured by non-decreasing entropy or entropy addition. Therefore it possesses stable robustness and quality robustness of quantifiable designing and verifiable measurement both in attack-tolerant or error-tolerant features. This innovative construction is called DHRA (dynamic heterogeneous redundancy architecture). The abstract model of DHRA is shown in Fig. 7.1.

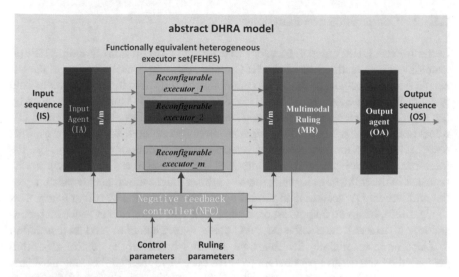

Fig. 7.1 Abstract DHRS model

7.1.1   Basic Principles of DHRA

7.1.1.1   Assumed Conditions

It is necessary to determine the hypothetical conditions before explaining the DHR principles: ① Dissimilar DRS is used as the basic structure. ② All the software/hardware in the structure allow the existence of heterogeneous vulnerabilities. Except the feedback control loop, all the components allow the existence of backdoors or malicious codes with different functions and application approaches. ③ Intrusive attacks include external attacks based on vulnerabilities inside the structure but also the internal and external coordinated attacks launched from the dark functions based on embedded backdoors inside the structure. ④ There are no interactive communication links and connections between the reconfigurative or reconstructable executors which are spatially or physically independent. ⑤ The feedback control mechanism always attempts to make the ruler enter the expected output state by changing the current operating environment or the ruling algorithms or architectural change in the executors if the multimode ruler output state is not in the expected state. ⑥ In given attack conditions, so long as the ruler can perceive the difference between output vectors of the reconfigurable executors, iterative process of diversified scenarios within the architecture can converge as long as the ruler can perceive the difference between the output vectors of the reconfigurable executors. ⑦ Feedback control response time or diversified scenario iterative convergence time is far less than the attack duration.

7.1.1.2   Composition and Functions

In the abstract DHRA model, IA needs to distribute IS to the corresponding HFE(s) according to the instructions of the NFC; there is a large probability that in the RE set, an executor subject to the input stimulus should be able to work normally and generate independently a multimodal OV that can provide the functions designed by the model system and satisfy the given semantics and syntax. The multimodal OV refers to a generalized one. In the input excitation condition, any direct or indirect response information generated by executors or variation information in the intra-executor processing and environmental state can serve as OV. This will not be repeated in the following chapters; the MR arbiter judges the content consistency of the multimodal OV according to the ruling strategy generated by ruling parameters or algorithm, forms an output response sequence, and activates NFC once an inconsistency is detected; once activated, NFC determines whether to send an instruction to IA to replace (migrate) the abnormal executor based on the control algorithm generated by the control parameters (control law) or to instruct the "output abnormality" executor to perform online/offline C&R operations (including triggering the related background processing features) or to conduct diversified operations (reorganization, reconstruction, and reconfiguration) targeted at the suspected problematic executor under the FE conditions based on soft/hard components. The activation process will not pause until the OV inconsistency disappears at the MR segment or the occurrence frequency of this case is lower than a given threshold. This iterative dynamic convergent process includes the conscious change of arbitrating algorithms and verifies the relative correctness of prior judgement in a multi-dimensional approach because the DHR mechanism believes that even if the majority executors whose volume of output vectors is identical may be in the attack-and-escape state in small probability, they need to be cleaned offline through strategic selection like the problematic executors or change their defense scenarios in a diversified manner through recombination/reconfiguration so as to find out whether there exists an attack-and-escape state by verification although the multimode OV still complies with the results of the current ruling algorithm (otherwise, attack-and-escape events will become common). It is obvious that DHRA is isomorphic with DRS in the metastable state, and they both have the same reliability and anti-attack (intrusion-tolerant) attributes (please refer to Sect. 9.5). However, when the multimode OV is inconsistent, DHRA exhibits a negative FC characteristic that can converge iteratively to a given state of the MR arbiter or a certain threshold range. In other words, for whatever reason, when perceiving a difference between multimodal OVs, DHRA attempts to strategically schedule the elements in the current executor set or to perform C&R operations on the executor itself or to implement the multi-dimensional reconstruction or other mechanisms under FE conditions so as to eliminate this difference in a gradual manner. Obviously, the dynamic, diverse, and random operations automatically introduced in this process are convergent. It should be emphasized that DHR can also trigger the above negative feedback loop according to mandatory scheduling instructions formed by internal and external control parameters and generate corresponding scheduling operations in order to disturb or destroy the relative static defense environment, which is caused by the attacker using the judgement-blind zone of MR arbiter's "superposition state" (when the OV is completely consistent or completely output-free), aiming at a hidden or standby cooperative escape performed by the virus Trojan embedded in the executor, such as the tunnel-through problem mentioned in Sect. 6.5.4. External or internal control parameters may originate from a random function generator. Operational control parameters may also be taken from the hash value formed by the indeterminate state information inside the target system, such as the number of current active processes or CPU seizure or average network traffic. It is easy for readers to find that DHR architecture not only inherits the original dominant attributes of DRS and avoids its inherent genetic defects but also significantly enhances the reliability and anti-attack of DRS through the GRC function introduced by the negative feedback mechanism; it also upgrades the attack complexity from DRS's "synchronous coordinated attack on static heterogeneous multi-target under non-cooperative conditions" to "synchronous coordinated attack on dynamic heterogeneous multi-target under non-cooperative conditions"; the difficulty of cooperative attack rising in a nonlinear manner makes it hard for external attackers or internal penetrators to find ways to escape after attack by means of trial and error or exclusion or tunnel-through. In particular, its organic "uncertain" effect from the attacker's perspective means that the trial-and-error attack on DF intersection has only one chance to succeed. Otherwise, once MR perceives that the OV becomes abnormal, the negative feedback loop will be activated, the current service environment or the construction scenario of the suspected problematic executor (including the DF intersection) will be changed, and the target object scenario information or the initial result of the attack obtained at time t is likely to lose its usable or inheritable value at time $(t + x)$. In other words, when DHR shields any OV error, it will change the current operating environment in the iterative and convergent approach based on the given policies, which makes it impossible for the trial-and-error tactics to assess the attack effects and to guarantee that the preconditions for the trial-and-error attack remain unchanged.

As described in the previous chapters, dynamics and randomness are the top disrupters or destroyers of the attacks requiring multiple participants as well as consistency or synchronization (especially under the non-cooperative conditions or for a lack of synchronization mechanisms), therefore, the introduction of underlying defense elements based on the closed-loop feedback robust control mechanism into DRS, such as dynamic, randomness and diversity/plurality, can effectively solve the problem of majority-selection blind zone and overcome the genetic defects in response to trial-and-error, standby cooperation or tunnel-through attacks, and meanwhile achieve the goal of normalized non-specific surface defense and specific point defense. In other words, DHRS can act as a jack to disintegrate or disturb "cooperative attacks on dynamic heterogeneous multiple targets under non-cooperative conditions" and significantly reduce the escape probability of the majority selection ruling algorithm in the DRSMV section. In fact, it is far from easy for an attacker to escape using the static DF intersection in DRS (excluding the use of system fingerprint for tunnel-through, which should be a special case), let alone the chance to produce a majority of consistent and escape-enabling OVs in DHRA with an uncertain DF intersection, as the difficulty between the two attacks differs at an exponential level.

7.1.1.3 Core Mechanism

DHR is based on DRS, and such robust control mechanisms as "measurement awareness, error identification, and feedback iteration," policy scheduling, multi-dimensional dynamic reconfiguration, and diversified scenario change based on policy arbitrating are introduced into DHR, thus making the general uncertainty range of the model converge within the expected threshold when the model architecture is certain and the range of disturbance is known so as to economically achieve the general robust control goal with non-decreasing initial information entropy within the architecture.

There are three goals which policy arbitrating is designed to achieve. The first goal is to measure and identify the general uncertainty disturbance exhibited by OV. The second goal is to improve the creditability of OV selection operation through iterative application of various arbitrating policies. The third goal is to activate the negative feedback control mechanism once OV is found to be inconsistent.

The feedback mechanism has four functions. The first one is to implement backward verification operation and "problematic executor" location in accordance with the current arbitrating results and related setup information as well as historical statistical data. The second one is to execute strategic scheduling of proper executors online or set up defense scenarios correspondingly through executors or defense scenarios and attack snapshots. The third one is to instruct the offline executors/ defense scenarios to make statistical analysis and routine security examination and perform operation like cleaning and restoration, reconfiguration, recombination and re-matching, and waiting for synchronization. The fourth one is to record and analyze the iterative convergence (or error removal) process.

The expected overall target of the iterative convergence aims to enable the DHR architecture to possess basic robust control functions like "measurement awareness, error identification, and iterative convergence." It is composed of three sub-goals: the first one is to reduce the response time of the model iterative convergence because this parameter is strongly related with the duration of decreasing initial information entropy, defense degradation, or availability degradation within the model; the second one is to obviously reduce the engineering implementation requirement for ideal heterogeneity through diversified defense scenarios because iterative change of problematic scenarios is much easier than heterogeneous designing of defense scenarios; and the third one is to increase the uncertainty effects to the maximum extent within the model and to avoid the stable attack-and-escape state in mechanisms.

The multi-dimensional dynamic reconfiguration includes four goals. The first one is to crumble attacks originating from memory injection, blind coordination, or complicated state transfer mechanisms through cleaning and restoration (C&R) and rollback and rebooting (R&R); the second one is to utilize virtual technologies in multicore and mass-core operational environments to economically strengthen diversified defense scenarios of the executors; the third one is to make use of highly developed software-/hardware-definable and reconfigurable technologies to promote the diversification of executors; and the fourth one is to properly introduce conventional technologies to obtain defense gains at an exponential level.

7.1.1.4   Robust Control and Problem Avoidance

In terms of implementation, DHR is based on the strategy judging and robustness control mechanisms, which, when the service function remains unchanged, can implement strategic scheduling or processing operations on the target executor or construction scenarios, including dynamic replacement (migration), C&R, and start-up of the related background processing mechanism (virus detection and elimination, backdoor scanning and repairing, etc.) or subject the suspected problematic executor to component-level structural and algorithmic changes, including reconstruction, recombination, rebuilding, redefinition, virtualization, etc., under FE conditions, so that the target object will have a variety of diverse, dynamic, and random construction scenarios (including DF intersections) and feature progressive convergence. Eventually, OV inconsistency in the MV segment tends to be minimized, so that the known or known-unknown security risk or even unknown-unknown security threat can be suppressed by using the normalized robust control function (certainly the random DMF/CMF also included). This not only embodies the ability of managing and controlling generalized uncertain disturbances but also demonstrates the "combined point-and-surface integrated defense" attribute.

The DHR architecture is featured by the function of problem avoidance rather than problem zeroing. This function does not attempt to deal with any attack event with one perfect defense scenario. It only aims at obtaining the effect of "fighting against fire with water," thus enabling small-scale diversified resource space to reach "the engineering-expected goal of non-decreasing entropy" in an equivalent approach. This lays the engineering technological foundation in the economic application of the DHR general robust control mechanism. In addition, DHR can also change the dark functional intersection of the current service set. Even if it is faced with "toxic and virus-containing" biological environments, we can still use more relaxed diversified mechanisms to mitigate the rigid engineering requirement for heterogeneous designing in the DRS. This is of important engineering implementation significance in simplifying the difficulty of implementation, reducing the design cost, and making full use of highly cost-effective COTS-level general software and hardware components whose creditability cannot be guaranteed in the supply chain.

7.1.1.5   Iterative Convergence

The above statement is valid when the feedback control response time or the time for iterative convergence of diversified defense scenarios is shorter than the attack duration. But if attack duration is shorter than the feedback control response time, three occurrences may happen: it is difficult to judge whether defense scenarios changed by feedback control are effective against this attack and is also impossible to confirm whether the target of non-decreasing initial information entropy has been reached in the structural environment; if the change of defense scenarios cannot be completed before another attack is launched, the service environment at this time must be in the state of decreasing initial information entropy or availability degradation; iterative change operation of defense scenarios has not reached the level of convergence during the attack duration. In other words, no effective defense scenario has been found against this round of attack. The following causes may account for this. ① There do exist defense scenarios, but the expected scenario matching has not been completed during the attack. ② There does not exist any effective defense scenario, and iterative operation cannot achieve the expected convergence effect. Obviously, accelerations of the feedback control and iterative convergence time are very important for the improvement of anti-attack operation and maintaining the non-decreasing initial information entropy and the availability feature within the architecture (please refer to Sect. 9.5).

7.1.2   Goals and Effects of DHR

As mentioned above, DHR is a DRS-based GRC architecture technology, which features reconfigurable (software-definable) executors, follows the control mechanism of MR negative feedback, and uses generalized dynamics as the means to change the structural environment, coupled with a "trinity" organic attribute: highly reliable, available, and credible. As simplicity is the ultimate sophistication, the underlying philosophy of the above is to, under the prerequisite of guaranteeing the FE target object service set, import MR-based strategic scheduling and MDRNFCM, software redefining, and algorithm reconfiguration and give the operating environment or reconfigurable and definable executor uncertainty from the attacker's perspective so as to endow the target object with characteristics such as dynamics, randomness, and diversity that can converge in suppressing generalized uncertain disturbances. At the same time, it strictly separates the cooperative approaches between the executors or minimizes synchronization and shared mechanisms available for attackers to maximize the inhibition of MR on DF in the non-cooperative mode in the dynamic heterogeneous environment or the "point-and-surface-combined defense" feature relying on neither attacker information nor behavioral characteristics, thus significantly improving hardware/software DMF and random failure tolerance. In other words, it is intended to obtain multifaceted endogenous security functions through the DHR architecture, which can suppress the effectiveness of non-coordinating or differential-mode attack disturbances based on the target object's backdoors, virus Trojans, etc. and can keep the escape probability of cooperative attacks below the expected threshold. It not only significantly increases the uncertainty of the attacks but also fully enhances the GRC capability, including high reliability, availability, and credibility. All in all, the author hopes that the anti-attack issue can be transformed into the difficulty for cooperative attacks on dynamic heterogeneous redundancy targets in the non-cooperative conditions through the endogenous security function and simplicity-oriented structure within the architecture and can be normalized to a matter of reliability theories which can be expressed by probability, so that the corresponding cases can be handled in a consolidated way.

7.1.2.1 Killing Four Birds with One Stone

To enable DHRA to achieve the goals at four levels at the same time, we should firstly transform the deterministic attacks based on the backdoors and virus Trojans of an individual executor of the target object into an issue of model perturbation probability perceivable at the target system level. Secondly, we should normalize the OV anomaly to a shieldable or correctable DMF processing problem. Thirdly, the reliable C&R mechanisms and effective R&R means can be used to create diversified defense scenarios focusing on problem avoidance with the support of problem zeroing for economical control over attack escape probabilities. Fourthly, conventional security technologies can be introduced to obtain exponential defense gains. The correlation is that the more certain the attack effect on a single executor is, the more uncertain the cooperative attack effect on the target object will be. In other words, the greater the backdoor heterogeneity is between the executors, the more effective DHR will be in suppressing generalized uncertain disturbances, creating a nonlinear relationship, the more complex the space-time expression of OV semantics, the more specific the spatial orientation of the sample space, the richer the ruling strategy, the narrower the OV window selection time, the quicker the feedback loop response time, and the lower the probability of cooperative escape, as well as of sustainable escape. In theory, the introduction of conventional technologies can obviously improve the inter-executor heterogeneity and promote intra-executor diversification, enabling the negative feedback mechanism of iterative convergence to possess more diversified defense scenarios so as to economically achieve the engineering goal of non-decreasing entropy.

7.1.2.2 Dynamic Variability of the Apparent Structure

The dynamic variability of DHR can be achieved in a variety of ways. The most typical method is to dynamically extract relevant components from the heterogeneous component pool of FE into k executors according to a certain strategy, where k refers to an integer greater than or equal to 1 (if k is equal to 1, it will be an atypical application case of DHRA); the selection of k is correlated to the anti-attack index of the system and the cost that the system can bear(in fact, it is correlated to software cross-platform transplantability and compatibility; the dynamic variability of the apparent structure can also be achieved by strategic reconstructing, reorganizing, and rebuilding of the executor itself or by virtualization technology to change the internal resource allocation mode or the apparent operating environment of the executor or to increase the uncertainty of the attack chain by preventional or remedial cleaning and initialization of the executor. The use of conventional security defense techniques (like the fingerprint and verification function) in the executor can increase the inter-executor heterogeneity or intra-executor defense scenarios in an effective manner; systematically application of the relevant defense methods and measures can make it difficult for an attacker to effectively perceive the defense environment or evaluate the attack effect in the space-time dimension, thus failing to take advantage of a priori knowledge to reproduce historical scenarios of successful attacks. Therefore, the DHR can effectively avoid the fatal flaw of DRS in the intrusion-tolerant mechanism, and it is unlikely to end up "being fully exposed once the gate is broken" even if attack and escape happen. At this point, readers can easily get to know that the ability of DHR to resist known or unknown attacks stems from the MV mechanism of TRA. In respect of suppressing uncertain perturbations, it benefits from the uncertain uncertainty effect (non-decreasing information entropy) generated by the iterative convergence mechanism based on multimode ruling (MR) and multi-dimensional dynamic reconfiguration (or diversified scenarios). As for attack difficulty, DHR benefits from the dynamic multi-objective consistency attack dilemma under the non-cooperative conditions resulting from the "desynchronization, strategic adjudication, and dynamic convergence" mechanism. In engineering implementation, it can lay a solid application foundation as progress has been made in technologies such as diversification and virtualization and the mature application of MDR technology. In fact, any technology able to increase the redundancy, heterogeneity, and cooperative attack difficulty of the executor and the decrease of feedback control loop response time in a FE manner can be on the option list which can be naturally accepted by DHRA.

7.1.2.3   Equivalent to TRA with the Superposed-State Authentication Function

It is not difficult to understand that DHRA is equivalent to the logical expression of TRA (the truth with relative accuracy), including all the preconditions for the establishment of axioms, and thus it has the GRC attributes, including but not limited to threat perception and point/surface integrated defense capabilities independent of attacker's a priori knowledge. There is a strong correlation between the capabilities, the feedback control response time, resource sharing, and the heterogeneity (diversity) of the executor elements in the service set, the number of executor elements (redundancy), the OV semantic abundance (bit length, etc.), the complexity of MV strategy, and the potential cooperative and exploitable paths and mechanisms, which is quantitatively expressed as the effect on the probability of MR escape. As mentioned earlier, the majority or identical consistent situations in the MR state are superimposed, with both normal and abnormal possibilities, except that we identify large probability events as normal state and small probability events as abnormal state. Therefore, there is no absolute difference between the normal and abnormal states artificially determined by probability. In fact, DHR cannot directly perceive an integrated attack and escape (including the tunnel-through attack, etc.) by mechanisms. Instead, it only applies the multimodal OV ruling function to identify any non-cooperating (lone-wolf) attack and "differential-mode attack" as a result of failed precise coordination or employs random disturbances of the feedback loop to destroy the well-organized cooperative attack in the standby mode (such as tunnel-through attack, etc.). In other words, DHR can only rely on the MR-based strategic scheduling and MDR negative feedback mechanism to, in an FE manner, increase the redundancy in the executor set and the heterogeneity of the defense scenarios or the uncertainty of DF intersection and improve the difficulty of cross-domain cooperative attack nonlinearly and prevent and disrupt the cooperative attack achieved through trial and error and by taking advantage of the blind area of the multimode ruling identification. Therefore, the effect is presented in the enlarged "scissors difference" between the normal and abnormal probabilities in the MR superposition state, indirectly achieving the goal of controlling the escape probability and enhancing the confidence of "Identification Friend or Foe." This is consistent with TRA-related inferences.

It should be emphasized that, under FE conditions, the executor itself should have diversified or pluralized reconfigurable functions (including environments based on resource virtualization) that can be dynamically defined, and various task (scenario) migrations, different strategy C&R mechanisms, R&R operations, and traditional security protection techniques can be used as measures to change robust control scenarios.

7.1.2.4 Metastable Scenarios and DRS Isomorphism

The metastable scenarios of DHRA or the differential scenarios with respect to time are isomorphic with the classical DRS, so DHR can also be regarded as DRS structural sets composed of executors (function-equivalent diversified scenario processing) different in time and space or strategy dimensions. Among them, there are diverse or plural representations in the spatial dimension, as well as dynamic and static representations in the time dimension; there are both strategic scheduling based on strategic adjudication and instructional scenario replacement based on external parameters, coupled with a combination of presentations in time and space or strategy dimensions. The purpose is to increase the apparent uncertainty of the target system's defense scenarios to significantly reduce the escape probability of cooperative attack, avoid any abuse of DRS genetic defects by tunnel-through or trial-and-error attacks, and lower the overdemanding bar of engineering requirements for heterogeneity to a great extent. Please note that since a metastable scenario of DHRS is isomorphic or equivalent to the DRS structure, we can use the related theory and mathematical tools to quantitatively study DHR.

7.1.2.5 The Uncertainty Attribute

In 1927, Heisenberg, a German physicist, put forward the uncertainty principle arising from quantum mechanics. This principle points out that to precisely measure a
particle is restricted. For example, when you measure the electronic location and
momentum around an atom, you will be faced with uncertainty, which comes from
two factors: first, when you measure a thing, it is impossible to avoid disturbing the
thing, making it changing its status; second, the quantum world is not certain but
can be expressed with probability. Therefore, if you measure the state of a particle
precisely, you will be faced with deep and fundamental restriction, that is, it is
impossible to measure the location and momentum of a particle simultaneously.
This principle shows that some physical value of a microparticle (such as location
and dynamics, azimuth angle and momentum, time and energy, and so on) cannot
have the certain measurement value simultaneously. When one value is more precisely determined, the uncertainty of another value increases correspondingly.

For external attackers or internal penetrators, DHRA has an uncertainty attribute.
Once the multimode ruling detects an executor OV abnormality in a metastable
scenario (or the frequentness reaches a certain threshold), the generalized dynamization mechanism of the negative FC will be triggered, initiating operations such as
the replacement of the suspected problematic executor in the current service set, the
online cleaning of abnormal executor, or the R&R of the current problematic executor or even the entire service set. On the other hand, the feedback controller functions to accept external commands or drive parameters, and its action effect is the
same as the former. Apparently, the direct effect of DHR's negative feedback mechanism is changing the current operating environment of the target object, making
the defense scenarios and behaviors difficult to detect and lock, so that the effect of
the attack action is hard to evaluate, and the development of the attack strategy
becomes a challenge. Moreover, the attacker will find it difficult to inherit or reuse
their achievements because they have become worthless. It subverts the preconditions for tentative attacks via a trial-and-error or standby approach (when the initial
information entropy is decreasing) and accomplishes a systematic defense effect:
even if the attack and escape happen, it is impossible to hold on.

7.1.2.6 Coding Theory and Security Measurement

From the perspective of transmission reliability, DHRA is superior in error correction. As the MR segment always selects majority or identical consistency from the
multimodal OV of the current service set, or the OV that satisfies ruling policy
requirements, as the response output of the input stimulus, the effect of the abnormal OV can be effectively shielded. Moreover, the activated generalized robustness
control mechanism with closed-loop feedback attributes always strives for iteratively changed structure, algorithms, or resource configuration to eliminate or circumvent output anomalies of executors within the service set (or to control the

frequency of exceptions below a given threshold). This shielding error correction function is highly similar to the action mechanism adopting self-adaptive coding algorithm to achieve anti-transmission channel interference, especially anti-artificially additive interference, and it is also very similar to the problem scenario described in Shannon's information coding theory. As long as the appropriate mathematical physics model can be constructed, it is possible to analyze the defensive nature of DHR by means of information coding and network analytical theories and to measure the security defense performance of the target system with a measurement method and index system similar to SE (symbol error probability). Exploratory work has been done in Sect. 7.6.

7.1.2.7 Endogenous Security Mechanism and Integrated Defense

If DHR's organic mechanism of anti-artificial attack disturbance is analogous to the human immune mechanism [2], it is easy to see that the DHRA system has an organic effect very similar to the human immune system, specifically as follows:

(1) DHRA has a human-like innate immune mechanism. The latter has no specific selection when wiping out invading antigenic substances and can "kill all" the invading antigens (valid in great probability events)without relying on their characteristic data. In theory, there exists the possibility that an attacker can take advantage of the dynamic redundancy resources for the purpose of coordinated attacks and that attack and escape may happen in the surface defense function of the DHR architecture. But from the perspective of the mechanism, such an attack-and-escape state cannot hold on steadily.
(2) Similarly, DHRA also has a human-like acquired immune mechanism. When the multimodal output ruling mechanism senses that the current executor OV is "abnormal," the system automatically removes or cleans the "problematic executor or service scenario" through the GRC mechanism with the feedback function or launches the background diagnostic processing mechanism, using loophole scanning, attack trace analysis, virus detection and elimination, Trojan removal, and other traditional security techniques and, if necessary, resorting to preventive reconstruction and reorganization mechanism to achieve the transformation of the current service environment or change the use strategy for the "suspected problematic" executor, etc. The above robust control function is called "point defense." In principle, it is similar to the activation and reaction mechanism of the antibody with special antigen identification.
(3) Just as described in the above (1) and (2), point defense and surface defense can be implemented under the same control structure and operating mechanism. The author names this effect "fusion defense (FD)"or "integrated defense (ID)."

Besides, the multi-mode ruling mechanism makes any attempts to tamper information, disclose information or announce wrong information have to control the multi-mode target executors (scenarios), also have to perform the coordination of multiple executors (scenarios) in non-cooperative environment. Obviously, the difficulty of implementation is nonlinear for the attack and escape. But on the side of the defense, the endogenous mechanism can provide exponential defense gains for technological measures used to strengthen inter-executor heterogeneity or defense scenarios.

7.1.2.8   Problem Avoidance and Problem Zeroing

It must be emphasized that generalized dynamization only changes the implementation structure of the executor or the operating environment within the current service set. It does not "completely remove" the problematic component but implements the operation "problem avoidance," which completely differs from the concept of problem zeroing in the reliability domain. A backdoor is usually part of the host structure, while a virus Trojan often belongs to "add-ons," but the availability of both is strongly correlated to the certainty of the host environment. So, if we replace or restore the host or change the host structure, we can effectively affect the stability of attack chains based on backdoors. To put it in another way, as long as the cooperation tie between the attacker and the backdoor (including the Trojan) in the current running environment of the target object is destroyed or changed, the purpose of defense can be achieved. Just as it is impossible to expect zero backdoor in the system or zero new loophole in the vulnerability patch, you can't expect DHR's FC-based GRC mechanism to completely remove design flaws or malicious codes from the system. In fact, since an absolute majority of the backdoor problems are strongly correlated to the host environment, once the environment changes, the nature of the backdoors may change, too, with the risk reduced to medium or even from a high level to a low level or even turning completely unusable. Therefore, the aim of DHRGRC is not on whether it can create a "virus-free and aseptic" or failproof defense scenario but on how to try its best to "use the right method or tool to deal with the current problem." Therefore, in the case of established attack scenarios and methods, the feedback control (FC)-based generalized dynamization mechanism "avoids" the problematic scenario of the executor in the current service set rather than "repair and clean" the executor, or it is more appropriate to describe it as replacing the executor with potential problems, which is consistent with the basic definition that DHRS allows the "virus-/bacteria-bearing" executors. More importantly, turning the problem zeroing process into problem avoidance processing can greatly increase the utilization of limited defense scenarios (or resources), provided that the problematic scenario can have problems but cannot lose normal functionality. For example, the defense scenario M is invalid for A-type attacks but can effectively defend against other attacks. The defense scenario N is invalid for B-type attacks but can defend against other attacks including the A-type ones. Then, scenario N can be used to address A-type attacks, and M can be used to deal with B-type attacks. Of course, if there are DF such as backdoors non-related to the environment, that's another pair of shoes.

7.1.3 Typical DHR Architecture

The typical DHR architecture is shown in Fig. 7.2a, which embodies the philosophy of "simplicity being the ultimate sophistication."

The author calls it the MR-based policy distribution and MDR negative FC architecture. The engineering representation of its abstract model is shown in Fig. 7.2b.

The policy distribution corresponds to the IA segment. Its main function is to decide whether to hinge external input with the specified executor in the current service set according to the strategic scheduling instruction of NFC to implement the activation, suspension, and repairing of the executor or to perform other given tasks. This usually means to complete the specific connection and distribution functions

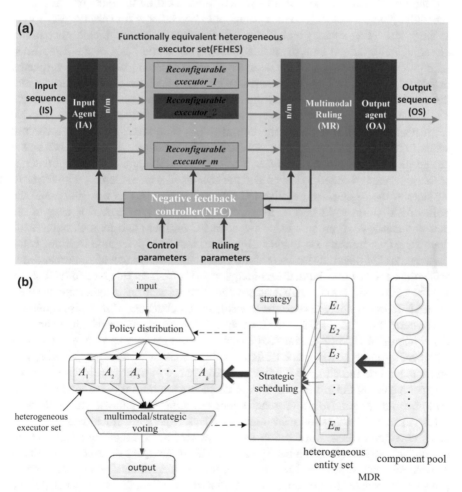

Fig. 7.2 (a) DHR architecture. (b) Typical DHR structure

with a path-and-mode-definable switch module. The HE elements in the current service set are represented by A_j ($j = 1,2,\ldots, k$), and each executor is represented by E_i ($i = 1,2,\ldots, m$) based on the multiple reconfigurable scenarios of the FE according to the strategic scheduling instruction. The construction of scenario E_i serves two purposes: (1) use standard or non-standard physical or virtual hardware/software modules in the component pool to generate m kinds of FEs via reconfigurable, reorganizable, reconstructable, redefinable, and virtualized MDR technologies based on pooled resources in order to create structure scenarios of different algorithms; (2) use different levels of initialization operations, starting or calling test/scan or other diagnostic tools and activating corresponding security functions to obtain a structure scenario with the same service functionality and a changed operating environment. The strategic scheduler is the NFC in DHRA, which, upon receiving the output state of the MR arbiter, will make at least three actions. First, the scheduler judges whether there is a compliant OV available for selection according to the given voting strategy and, if not, whether it is necessary to call other strategic adjudication algorithms for ruling; second, it, on the basis of the established scheduling policy, commands the input distribution segment, removes the suspected problematic executor from the current service set, or links the standby executor into the service set or directly assigns the suspected problematic executor to execute the new structure scenario and even replaces all executors in the current service set if needed; third, it observes the output ruling state after updating the defense scenario and decides whether to repeat the first two actions. Therefore, it forms a closed-loop control process with selective output and dynamic convergent defense scenario changes, backed by the pooled hardware/software resources, with the arbiter state as the trigger, the negative FC strategic scheduling as the center, and the policy distribution and MDR as the structure scenario changers. It is easy to see that this closed-loop process has a negative FC attribute and its metastable state depends on the information fed back by the arbiter. In a given time window, when the executor OV does not show any abnormality or the frequency of abnormalities is lower than a preset threshold, the elements in the current service set and their structure scenarios will remain stable unless the arbiter identifies another abnormality or needs to consciously change the heterogeneous elements or structure scenarios in the current executor set due to some random strategy. Obviously, any change in the elements of HES or the internal structure scenarios of a specific element will significantly increase the attack difficulty of attackers or penetrators. Whether it is an accidental perturbation or a "mistake" disturbance in a cooperative attack, once it is perceived by the arbiter, the structure scenarios within the current service set will undergo an iterative convergence change dominated by the negative feedback mechanism. This makes the inheritance of attack experience (including the previously obtained defense information and the phased outcome of the attack) a serious challenge. Therefore, the ability of the DHR architecture to respond to definite or uncertain attacks and "differential or common mode threats" is nonlinearly and positively correlated to the number of configured executors, the heterogeneity of the inter-executor structure scenarios, the complexity of multimodal OV semantics, the fineness of strategic ruling, diversified scenario iterative convergent speed, and the complexity of the attack task implementation steps. The most significant effect is that it can block or crumble the attacker's attempt to implement a cooperative attack through trial and error or make it difficult to maintain the cooperative attack in a standby mode for which the attacker strives. On the side of the attacker, the current service set presents a general uncertainty effect in the software/hardware algorithms without functional changes. It should be noted that DHR's anti-attack performance based on the uncertainty effect can be simulated not only by mathematical modeling but also by the injection test method similar to reliability verification. Namely, under the condition of "white box," a differential or common mode test sample can be injected for quantifiable verification analysis. In addition, according to the online operation record, situational information such as anti-attack confidence in different structure scenarios and defense effectiveness under different attack scenarios can be calculated. If this information is applied intelligently in MR, it can reflect the ruling effect based on time iteration. For example, when the OVs of k executors are completely inconsistent, selecting the executor output with higher confidence is functionally equivalent to the judgement of time iteration, which can, to a considerable extent, increase DHRA tolerance to attack events and perception of threat. Similarly, based on the online operation record or data collected, we can endow DHR with online or offline intelligent analysis and learning mechanism, making FC operations more targeted in replacing the executor elements in the service set or changing the executor structure scenarios, which helps to reduce the stable convergence time of the FC loop. Of course, once the analysis result is associated with the structure scenarios, it allows credibility calibration on the hardware/software components in the resource pool under different attack scenarios, providing a vital reference message and improvement basis for removing the problem or design defect. More importantly, in the problem scenario perceived by MR, the addition of a relevant data collection function will be as valuable as you know your enemy in discovering specific unknown backdoors, virus Trojans, and new types of attacks.

It can be easily inferred that DHRA will generate exactly the same or similar effect if a known or unknown attack event is replaced with a random DMF (differential mode failure) or CMF (common mode failure) event. In other words, DHRA can effectively suppress generalized uncertain disturbances such as random failures within the target object, known-unknown risks, or unknown-unknown threats. Therefore, it can be said that DHRA achieves the GRC goal of normalizing anti-attack performance and reliability.

The typical DHR structure allows a certain degree of isomorphic composition between mutually independent executors. This is because DHR introduces the generalized dynamization mechanism in terms of operating environment, operational mechanism, synchronization, scheduling strategy, parameter assignment, restructuring, and algorithm replacement and iterative convergence, while the availability of backdoors or the effectiveness of backdoor-based attacks is usually strongly correlated to the target environment factors and the operating mechanism. From another point of view, the introduction of the generalized dynamization mechanism is likely to destroy the "synchronous coordinated attack on multi-target under non-cooperative conditions," which means that the basic defense elements such as dynamics, diversity, and randomness can be systemized in DHRS, which helps to reduce the engineering difficulty of executor heterogeneity design or heterogeneity screening.

In addition, for RE (reconfigurable executor), as long as the availability or reliability parameters and recombinable diversified application scenarios of the service functionality can be guaranteed, there is no overstrict requirement in terms of credibility and security, i.e., allowing the use of software/hardware components and even the executor itself whose supply chain credibility cannot be ensured. By the same token, DHR allows (actually cannot completely avoid) the operation of "virus-/bacteria-bearing" executors within the structure, including the presence of unknown or known DFs. In short, regardless of the type of physical failure (downtime excluded) or network attacks based on target system backdoors, as long as an attack fails to realize escape representation with sustainability and spatiotemporal consistency on multimodal OV, DHRA can automatically get immunized based on its mechanism.

It should be pointed out that DHR does not impose any limitations on the functional granularity and attributes of an RE element itself except for FE diversified application scenarios (where there may also be minimum performance requirements) and MR requirements for OV semantics and syntax (including output response delay). The functionality can be physical or visual hardware or software; a module, a component, or a subsystem; or a complex system, a platform, or even a network element device. Similarly, the arbiter, IA/OA, and feedback controller can also be physical entities or virtual functives, which can be processed in either a centralized or a distributed manner (e.g., using an arbiter based on the blockchain technology). There are not too many rigid constraints on the means and methods of implementation. Thus, DHR boasts a broad application perspective and economic and technical value. It is especially suitable for building IDCs or cloud service platforms with high use requirements for reliability, availability, and credibility or information service facilities with high intelligence and sensitivity, etc.

The author believes that with the evolution of multicore parallel computing, mimetic computing, user-customizable computing, SDH (software definition hardware) and CPU + FPGA, and the in-depth application of virtualization technology, the long-standing complex and economical concerns on the implementation of typical DHRA will no longer be a sensitive issue in engineering implementation. In particular, the development of a globally open and open-source ecological environment makes it not a wild imagination any longer to use standard, cheap, and diversified COTS-level software/hardware components to form FE executors. For robust applications with high reliability, availability, and credibility requirements, typical DHRA can bring to the target system unparalleled price/performance advantage throughout the life cycle (such as guarding against deep-implanted backdoors) compared to the service system that is deployed separately between the security facilities and defense targets and the additional security safety facilities requiring real-time maintenance and upgrading.

7.1.4   Atypical DHR Architecture

In an assumed scenario with non-rigid security requirements, the target object system is always functionally recoverable from any attack, and a low probability of information leakage or service suspension is allowed within a certain time window. Then, if some kind of dynamic, diversified, and random method is adopted, the reproducibility of the attack effect will be affected, leading to the loss or reduction in the value of the attack itself regarding the planned or sustainable use. This is the protection mechanism of the atypical DHRA. This is because the target object system's functionality and service security depend not only on the reliability or availability parameters of the system itself but also on the reproducible or recurring frequency of any deliberate attack outcome and the inheritability of the attack experience.

As is shown in Fig. 7.3, the most fundamental difference between the atypical DHRS model and the typical DHRA model is that there is only one FE REA$_k$ at any time. The executor is not statically configured but is dynamically selected from HES by some strategic scheduling algorithm and acts to determine whether to clean and repair (C&R) or rebuild and reconfigure (R&R) the operating environment of the current executor based on the output state of the error detection device or the intrusion testing component, as well as the input/output guidance segments required to support this migration or reconstruction operation. The atypical and typical DHRA have a common feature that both need MDR and strategic scheduling or the generalized dynamization mechanism under the FE conditions, and both seek diversified,

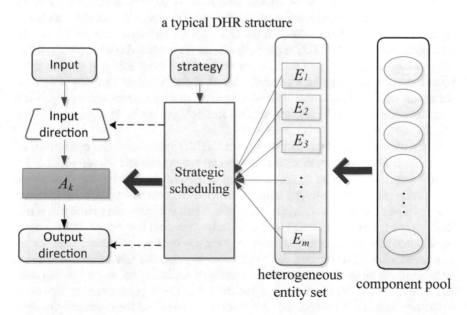

Fig. 7.3 Atypical DHR structure

dynamic, and random defense effects, and both use this mechanism to invalidate the target object's defense information that the attacker has acquired or invalidate the constructed attack chain in the desired time window. For instance, if there is a backdoor in the defense scenario of one of A_k's executors, an attacker may be able to use it when the executor in this defense scenario is dispatched to the front office, but there is no guarantee that the use will be regular or sustainable. It should be emphasized that the diversification, dynamization, and randomization of atypical DHR is usually an open-loop process. Like the mobile target defense MTD, under the FE conditions, its defense effect cannot display the quantitative relations between diversification, dynamization, and randomization. However, MTD cannot theoretically prevent and control backdoor threats because, when the protected object itself has a backdoor, the backdoor of the variant A_k will not fail due to any intelligence or illusive scheduling function. For example, when there is a backdoor with a "signing" function in the A_k source program, the attacker can perceive whether there is a preset backdoor in the current A_k variant via a pre-agreed answering method and then dynamically edit that backdoor according to the corresponding strategy to launch the desired attack. The atypical DHRA does not have such a problem.

That is because A_k in DHRA is RE of FE rather than a variant of a source code. When active MDR mechanisms are applied, they can conduct preventive C&R toward offline A_k executors, or R&R the current algorithm of standby A_k executors, or consciously and actively replace some heterogeneous resources in A_k executors. The diverse algorithms or structures actively change the available scenarios of backdoors, creating a defense advantage "where it is difficult to use a backdoor stably even if there is one."

The defense expectation of typical DHRA is to convert definite or uncertain attacks or random failures based on the backdoor of the target object into a probability controllable reliability event, so the GRC operation mechanism based on MR strategic scheduling and MDR negative feedback is adopted. As for atypical DHRA, there is neither a ruling step nor a negative feedback mechanism. It only uses the open-loop strategic scheduling and the MDR defense means and thus only constrains the backdoor exploitability, resulting in an uncertain attack effect. The difference between the two in implementation technology results in a gap in defense effectiveness, reliability, and robust control.

It should be pointed out that in atypical DHR applications, the dynamization, diversification, and randomization methods can use or schedule the resources of FE components and component pool according to the strategic mechanism or motivate the scheduling or reconstruction mechanisms of the control system by in-house uncertain parameters. For example, when the number of processes currently active in the system, memory, and CPU resource usage, network port traffic, and the number of open files are taken as the incentive parameters of the generalized dynamization mechanism, it is difficult for an attacker to grasp the law of time and space changes in the target object's defense scenario by analyzing the same type of facilities. It is easy to conclude that as long as the AS of the target system exhibits characteristics of random change, the information obtained and the resources that may be controlled by the attacker in advance, or the attack experience once effective, are insufficient to ensure continued effectiveness after the next scheduling. According to the AST, the target object's reachability premise is no longer guaranteed. Atypical DHRA can be used in application scenarios that are security demanding yet sensitive to investment costs, such as IDCs and cloud-based service platforms (IT-PaaS, ICT-PaaS, etc.), where the configuration of computing, storage, networking, and link resources are often heterogeneous, redundant, and usually of FE feature, while the pooling processing and virtualization calling mechanism and strategy based on heterogeneous resources must have a certain degree of dynamic and random nature, so it is extremely easy to damage the stability or reliability of the attack chain. With the help of the atypical DHRS, any conscious use of the operations such as in-house virtualization environment and seemingly irregular container scheduling or virtual machine migration can disrupt the stability of the attack chain or the reachability of the AS and increase the difficulty of cognition and attack on the target object in a relatively simple and inexpensive way.

7.2   The Attack Surface of DHR

In summary, either typical or atypical DHRA can be regarded as a one-dimensional or multi-dimensional mobile AS system of integration defense deployed in an uncertainty manner unpredictable to an attacker, whether it is a typical configuration of DHR or atypical structure. For example, it can randomly sample, dynamically call, or randomly combine elements from the heterogeneous virtual machine with multiple FEs using the multi-virtualization technology or from the component pool consisting of nonhomogeneous software/hardware elements of multiple FEs to select or generate the executors or diverse defense scenarios required by the current service sets. Given the diversity or pluralism in the operating environment of each executor in the space-time dimension, the resources available for the attacker are uncertain in the same dimension, which is macroscopically showed in the irregularly moving attack surface. Figure 7.4 illustrates the different time points t_1 and t_2, where the dynamic changes in the elements in the HES (service scenarios) cause the apparent AS of the target system to move in an uncertain mode.

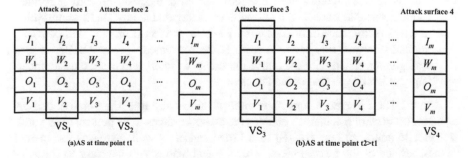

Fig. 7.4 Illustrations of AS moved by DHRA

The mobile AS effect of DHRA can be concluded as follows:

(1) In the DHR architecture, the system generally presents an attack surface that is irregularly changing and unpredictable. At some time point or period, an attack surface and resources that the attacker has "locked" may be "offline or unaccessible," or a load balancer (such as a scheduler) may direct the probe/attack packet to another attack surface. Therefore, it is difficult for an attacker to detect and launch the pre-planned attack by locking the attack surface.

(2) Curb the potential damage caused by 0 day vulnerabilities or unknown backdoors. The DHR iterative convergent environment makes it impossible to sustain the attack "from within and without" against specific executors (service scenarios) in the current service set, because once the MR-based NFCM detects an executor OV abnormality, its host system may be forced offline by a certain strategy and get cleaned or restored or reconstructed. Even if there is no abnormality, the system will perform periodic or non-periodic offline cleaning or even reorganization and reconstruction operations on the executors according to a certain preventive policy. In other words, even if an attacker can build an attack chain based on some "0day or Nday" backdoors within the DHRA, it is not worth of use due to the difficulty to obtain the cross-executor (service scenarios) cooperative attack effect.

(3) Unlike the defense methods that focus only on the dynamics or randomness of a particular aspect of the system or a component (e.g., instruction randomization, address/port dynamization, etc.), DHR utilizes the iterative convergent control mechanism based on strategic scheduling and MDR under the MR conditions, so that diversified technology products or open COTS-level software/hardware components present a robust control effect dynamically, which is diverse, random, and integrated via the structural effect. This makes it hard for attackers to find resources to use in a coordinated way, whether they are hardware/software code design flaws within the structure or deliberately implanted hardware/software codes.

(4) Because of the MR-based negative feedback mechanism, DHRS always makes full use of strategic scheduling, and MDR means to maximize the role of generalized dynamization, so that the frequency of OV abnormalities can be controlled within a given threshold range. Correspondingly, the AS of the DHRA will move convergently with the activated NFC mechanism until it reaches some metastable state. The result of actions targeted at any "fault perturbation or attack disturbance or outside commands" in the feedback loop will converge gradually toward this state. Therefore, the movement law of DHRA is also strongly correlated with the above disturbance factors and has a classic uncertainty effect on the side of the attacker.

Since DHRA in nature constructs multiple attack surfaces, the iterative convergent robust control mechanism enables the attack surfaces to be presented as parallel mobile attack surface (MAS) in a broad sense, as well increases the overall attack surface of the target object, which goes obviously in contrary to the AST "shrinking AS for security," therefore, the classic AST targeted at a given attack surface is not suitable for the security analysis of such an iterative convergent parallel MAS. This assertion is backed by three reasons: (1) The classic MAS itself breaks the assumption of the AS evaluation theory, namely, "AS remains unchanged." (2) The attacker cannot control the attack packet (to uploaded virus Trojans or interactive information) and is directed to the specified executor, thus failing to meeting the assumption that the target attack surface is always reachable to the attacker. (3) DHR's iterative convergent MAS requires the attacker to be able to simultaneously lock multiple non-similar MASs (see Chap. 2 for the relevant content).

7.3 Functionality and Effectiveness

We mainly have the following perceptions of the capabilities and endogenous effects of DHR GRC.

7.3.1 Creating a Cognition Dilemma for the Target Object

In the I[P] O model, the existence of HR executors (service scenarios) of FE means that the expression of the service function P can be implemented by executors of a fixed structure or algorithm or by definable software/hardware means of the executor itself, such as reorganization, reconstruction, and virtualization. Of course, it can also be implemented by a complex combination (including iterative convergent combination) of executors. For example, when there are three REs in HES of the service function P (represented by A, B, and C, respectively), the executor set of P (represented by P_A) can be constructed in multiple ways, such as $P_A = A$, $P_A = B$, $P_A = C$, $P_A = A \cup B$, $P_A = A \cup C$, $P_A = B \cup C$, $P_A = A \cup B \cup C$, etc. In other words, as long as the normal expression of the service function P is guaranteed, the actual mapping relationship (structural representation) of the executor's implementation structure (or algorithm) will no longer have a deterministic relationship in the theoretical sense. This uncertain structural representation leading directly to the increase of information entropy inside the architecture can be a dilemma or form a defense mist for the attacker to recognize the target object's running environment or defense scenario, which increases the difficulty of system fingerprint and vulnerability detection, backdoor spotting or locating, uploading virus Trojans, hidden communication, and other attack operations, thus posing a serious challenge on accurate delivery of attack packets (data or executable codes) targeted at backdoors and the like. In any attack operation that causes multimodal OV inconsistency, DHR's strategic scheduling and MDR negative feedback mechanism will not stop changing or transforming the response of current service or executor reconfiguration scenarios until such inconsistency disappears in this attack scenario or its occurring frequency is limited below a certain threshold. This is the endogenous security effect that comes up from within the system. In theory, any "differential-mode attack" or non-cooperative attacks will become invalid. "Even if the attack is successful, it is impossible to keep on." The planability of attack actions and the usability of attack results will become impossible to be assessed.

7.3.2 DFI to Present Uncertainty

The DHR structure and its operating mechanism can make the hidden vulnerabilities, backdoors, or malicious functions (such as virus Trojans) in a component present or exhibit a considerable degree of uncertainty without changing the service functionality of the component. The backdoors and other DFs parasitic on the executors (service scenarios) will, on the one hand, change (but generally will not disappear forever) with the reconstruction, reorganization, virtualization, or C&R operations of the host environment and, on the other hand, show an apparently uncertain state both inward and outward macroscopically when the host is subject to strategic scheduling, random selection, and combined presentation, making it difficult for an attacker to achieve the imagined effect of the attack mission based on the information obtained in advance or on-site, the drafted attack strategy, and the technical means adopted.

The DFIs (including backdoors, etc.) hidden in the relevant executors (service scenarios), if no longer static, definite, and sustainable, their exploitability for the attacker will inevitably slump or even become completely invalid. The challenge lies in that in the DHR structure, it is almost an impossible mission to achieve synchronous coordinated attack and escape to heterogeneous dynamic multi-target under non-cooperative conditions of executors or service scenarios by employing these uncertain DF intersections under non-cooperative conditions across the executors.

7.3.3 Making It Difficult to Exploit the Loopholes of the Target Object

The MR-based strategic scheduling and MDR negative feedback mechanism greatly reduce the visibility, reachability, and stability of potential DFs in the executor, even if there is an unknown backdoor:

(1) DHR makes it difficult for an attacker to analyze and control loopholes by changing the input motive and getting the output response mode. DHRA requires the physical or virtual hardware/software modules in the executor to be heterogeneous (at least diverse), which theoretically greatly reduces the possibility for different modules to have the same DF (except for social

engineering factors). In this scenario, if an attacker attempts to detect and analyze a loophole by changing the input motive and checking the abnormal output response, it is difficult for any personalized performance of the multimode OV to be recognized and normally presented by the MR mechanism since the executor DFs are mostly or completely different. This means that any attack behavior, including scanning and detection, is not mechanically effective unless it can get fully consistent or mostly identical OVs. In other words, personalized or differentiated design defects and backdoors or virus Trojans are hard to be accurately perceived by the outside world. It is also impossible to establish a reliable information exchange mechanism on the attack surface.

(2) The MR mechanism significantly increases the difficulty for an attacker to implement a cooperative attack by exploiting the vulnerabilities of the target system. On the one hand, DHR generates the current HES from the component element pool through strategic scheduling, which causes the executor (service scenario) in the service set to obtain various degrees of uncertainty in the space-time dimension, so that an attacker finds it hard to "track and lock" the specific vulnerability or backdoor target in the executor through the attack surface, making it extremely difficult to accurately upload a virus Trojan through a loophole. On the other hand, the strategic scheduling and MDR mechanisms of the executor (service scenario) make it difficult to perform cross-executor "malicious collaboration." In other words, even if there are personalized or differentiated vulnerabilities from one executor to another, it is difficult to make the relevant executor cooperatively generate the exact same OV under the generalized dynamic and non-cooperative conditions. Therefore, any attempt to exploit backdoor or virus Trojan to export or tamper system sensitive information, or attempt to use attack surface to issue error instructions can be easily detected and blocked by the MR mechanism constructed by the DHR. In theory, if the DFs in each executor or service scenario are completely independent, DHRA and its operating mechanism will be able to be naturally immune from any attack based on or using these DFs ("downtime" attacks or attack-caused failure excluded). Namely, DHR has a natural information security role, which is somehow like a nuclear launch button that only works when all the authorized officials press it at the same time.

7.3.4 Increasing the Uncertainty for an Attack Chain

Due to the DHRA's strategic scheduling and MDR negative feedback mechanism, the target system exhibits a strong uncertainty in constructed scenarios or defense behaviors. Once an attack is perceived by the MR segment or the current constructed application scenario is replaced by a random robustness control process, it is difficult to get identical output response from the same attack sequence or operation package. For example, the target system's loopholes and their characteristics that an attacker sees in a static structure will be uncertain in the DHR environment, which will seriously affect the authenticity of the information obtained in the loophole detection and back-linking phase and will also make attack codes like the virus Trojan uploading process unreliable, where any hidden or latent attack will lose its supportable and reliable environment, and the historical scenario of successful attack is unlikely to be reproduced in the space-time dimension. Obviously, a target object with too many "service scenario and defensive behavioral uncertainties" or an attack chain that cannot be used stably will be catastrophic for the attacker.

7.3.5   Increasing the Difficulty for MR Escape

The DHRA may, by nature, forestall any cooperative attack through trial and error. This is because the MR segment forces any non-cooperative attack "tampering" OV to be mechanically ineffective. However, to achieve successful and even sustainable escape, it is difficult to obtain the desired OV of the dynamic multi-target coordinated attack without using trial and error. In addition, the MR-based strategic scheduling and MDR negative feedback mechanism have a stronger response to OV abnormalities caused by trial-and-error attacks, such as:

(1) If any operation or attack behavior against the dark functions (DF) of the executors or service scenarios fails to generate a coordinated OV within a given MR window time, the "abnormal" OV executor operating scenario will be discovered and replaced or cleaned based on strategies. If necessary, even all executors in the current service set can be replaced as an entirety, unless the attack is not relevant to the OV (such as causing a downtime or tunnel-through).

(2) The executor set (service scenario) or target operational scenarios that provide external service functionality are extracted from the entire heterogeneous resource pool. If the extraction is performed according to the given design strategy or user-defined rules, it is quite difficult to meet the escape conditions of the current executor set's homologous combined loopholes.

(3) Even if an attacker successfully escapes in an attack with or without the arbitrator's awareness, the DHR operation mechanism based on the backward verification or the control loop disturbance mechanism based on the external command will change the current executor set or operation scenarios in an iterative approach so as to make sure that there are few chances for the attacker to reproduce the OV-identical escape scenario based on the escape experience obtained at time t in subsequent attacks due to the time-spatial change attributes of the DHR operating environment or the executor set. In other words, even if an attack-and-escape event happens, this is not an ensured event. This is the one of the core differences between DHR and DRS.

7.3.6   Independent Security Gain

The organic DHRA security mechanism is mainly embodied in three aspects. First, it can transform any uncertain attack against an individual executor (service scenario) into a system-perceivable attack event. Second, it can transform a perceivable attack event into a reliability event with redundancy processing, where the attack-and-escape probability will be kept below the expected threshold by the strategic scheduling and MDR negative feedback mechanism. Furthermore, its surface defense effect is reflected in the fact that the OVMR mechanism can completely suppress known or unknown non-cooperative attacks, and its point defense effect is reflected in the fact that once the MV segment finds an OV abnormality, it will trigger a generalized dynamic negative feedback control mechanism with an iterative convergent functionality. It can be seen that the effectiveness of DHRA defense is determined only by its organic security mechanism and HR resources rather than by understanding and obtaining the prior knowledge or behavioral characteristics of the attacker nor is it based on the effectiveness of any additional security means. In fact, DHR is usually not clear about the details of the attack actions blocked in principle. It has a typical feature: "I know it's so but I don't know why." Therefore, in a strict sense, its security gain belongs to neither passive nor active defense. Also, it is neither static nor dynamic defense. It may be appropriate to describe it as "combined point-and-surface" integrated defense.

It should be specially pointed out that, because the effectiveness of DHRA depends to a large extent on the nonintersection of DFs among the executors or service scenarios, any means that can enhance the cross-executor heterogeneity or diversity can nonlinearly improve the effectiveness and reliability of the target object, which is determined by DHR's inherent structure and operational mechanism, such as various methods and means of dynamization, diversification, and randomization, which, of course, include the application of intrusion detection, prevention and isolation, encryption authentication, virus and Trojan hunting, backdoor closure, trusted computing, and other conventional security technologies or measures. There is no doubt that the smart application of these classic defense technologies in the executors or the service scenarios will help achieve twice the result with half the effort.

7.3.7   Strong Correlation Between the Vulnerability Value and the Environment

In theory, as long as the loopholes of different executors (service scenarios) in DHRA are strictly heterogeneous, they will hardly be utilized cooperatively under double-blind conditions, which is determined by the MR mechanism. Similarly, even if there exist vulnerabilities of the same lineage or origin in the DHR environment, the attacker often needs to study and develop different utilization methods to overcome the cooperative challenge caused by environmental differences. Although different methods of exploiting vulnerabilities may be successful in the corresponding execution, on the whole, as long as the "cooperative attack effect of dynamic multiple targets under uncooperative conditions" can not be achieved, there will be no substantial harm to the DHR target system (except for the vulnerabilities independent of environmental factors). In other words, the DHR architecture can, to a large extent, tolerate DFs, such as backdoors, in an executor (service scenario). Even if an attack-and-escape event happens to the DFs with small probability, it will be of no sustainable use value due to the existence of strategic scheduling and MDR mechanism. For example, an attack that needs to be achieved by exploiting loopholes requiring a specific operating system or supporting environment will fail due to a change in the constructed scenario; a deliberate attack entering the system via application software loopholes and achieved by authority enhancement of the operating system will also be difficult due to environmental factors; application layer attacks implemented by side-channel effects such as executor memory and cache will no longer be effective as defense behavior changes; and backdoors and virus Trojans customized to the personalized environment will also lose the desired functionality as the reachability of AS and exploitable resources change.

7.3.8 Making It Difficult to Create a Multi-target Attack Sequence

Available loopholes in different parts or components are inseparable from appropriate environmental conditions and "tailor-made" attack procedures and specialized tools. The discovery and utilization of loopholes also require the introduction of an input motive sequence (called attack sequence) that conforms to grammatical and semantic and even pragmatic rules through the external channel of the AS. The reason for emphasizing this premise is that the functionality of the target surface attack channel is not specifically tailored to the attacker, that is, the construction of any attack sequence must be within the range allowed by the normal input rules of the system and the receiving state; otherwise it will be automatically filtered or discarded by the inspection mechanism for service request compliance. To achieve the escape effect in DHR architecture, the attacker must make the same error occur to the majority of executors at the same time or similar moments. The loopholes of these executors are often different, with different available scenarios (or processes), tools, and internal and external triggering mechanisms. How can an attacker simultaneously inject these attack sequences into the input channel with compliance limitations and accurately distribute them without exposing the attack trace due to reachability errors? Under non-cooperative conditions, such precise coordinated action is not an easy task for any attacker.

7.3.9   Measurable Generalized Dynamization

The negative feedback control mechanism of the DHR architecture not only enables the dynamic, random, and diverse defense elements to present a systematic effect but makes the generalized dynamization process generally converge iteratively to an operational scenario that can effectively respond to the impact of current attacks or random failures. Any abnormality perception obtained by multimode ruling will trigger the strategic scheduling and MDRFC mechanism, which will subject the abnormal executor to C&R or R&R or resort to generalized dynamization means (i.e., standby executor replacement) and negative feedback operation, trying to eliminate the abnormality perceived by the arbiter or lower the frequency of occurrence of abnormal state within a given time window, thus almost invalidating any attempt to achieve a cooperative attack through trial and error, which is an inherent attribute of the DHR architecture. Therefore, for a given DHR system, based on the reliability verification theory and methodology, we can, under the white-box testing, "inject" the test samples in the executor (operational scenario) and arrange the matching input sequence (IS) and then transform the IS and the test samples to observe the state of the multimodal arbiter and indirectly measure the security level of the target object via the occurrence frequency of statistical abnormal rulings within a given test period. Obviously, unlike most security defense technologies, DHRA's anti-attack performance is designed to be calibrated and testable and can be measured by the probability value like the reliability parameters (see Sect. 9.5). It should be emphasized that DHR can fully perceive and shield multimodal OV abnormalities caused by non-cooperative attacks or DMF or cooperative attack mismatches in mechanisms, but the processing procedure for the recovery or avoidance of the abnormality has to suffer from the cost of reducing the availability of the executors or operational scenarios in a short period of time. In extreme conditions, there may appear poor services or the suspension of service providing. Although the cost is high, it can obviously increase the difficulty for attackers to launch attacks from the same source or from other sources. Thus, it can reduce the perceivable attack-and-escape probability (with the exception of offline attacks and tunnel-through).

7.3.10   Weakening the Impact of Homologous Backdoors

Today, the ubiquitously applied "open and open-source technology and industrial development model" has made the spread of homologous backdoors or malicious codes a reality rather than an assumption. Many facts show that the information product ecology based on open-source software/hardware codes has been under serious security threats. Regrettably, problems such as state explosions make it difficult to effectively implement formal proofs of DFs checking backdoors, which, in turn, encourages the vicious trend of using open-source community technology innovation and industrial development models to threaten cyberspace security. What makes the IT sector extremely frustrated is that there is not yet a solution with predictable security effects. Fortunately, in theory, even if there is a DF such as homologous backdoors between REs, due to the uncertain change of the host environment and the irregular AS movement of the target system, in respect of exploiting the backdoors, etc., unless the REs can be precisely coordinated and the premise of attack reachability can always be guaranteed, it is difficult to achieve the "coordinated attack effect of dynamic multi-objectives under the non-cooperative conditions." In other words, the scheduling policy of the executors in the DHR architecture or the R&R or MR policy of an executor itself (service scenario) will significantly weaken the coordinated utilization value of the homologous backdoors based on the environment dependency. It is easy to conclude that the heterogeneity design requirements for DHR structure and its operating mechanism are much more lenient than that for DRS.

From the engineering practice already carried out, the cost of generalized dynamic implementation based on feedback control (FC) is indeed much lower than the heterogeneous design of DRS.

7.4   Reflections on the Issues Concerned

DHR enables the systematical presentation of defensive elements, such as dynamicity, diversity, and randomness, based on the MR-encouraged strategic scheduling of and the negative feedback control mechanism of MDR, maximizing the convergence effect of these organic or integration defense factors. It is easier to obtain nonlinear security gains or exponential-enhanced defense effects and to accurately measure the attack resistance and reliability by directly injecting test samples under the white-box conditions. Theoretically, even if the DHR's RE carries viruses, it is difficult for any form of attack to compromise the stability robustness and quality robustness of the target system as long as the host components are not able to show majority or identical consistent errors in the space-time dimension.

7.4.1   Addressing Uncertain Threats with Endogenous Mechanisms

In engineering practice, it is often difficult to achieve problem zeroing in the realization of complex systems due to design flaws or loopholes, "backdoors," and "trapped doors." Therefore, the target objects of cyberspace always have to face uncertain threats via backdoors of unknown loopholes and other DFs from the design, tool, manufacturing, supply, and maintenance chains. The DHR endogenous architectural mechanism can guarantee that as long as the heterogeneity (diversity) of backdoors or virus Trojans (such as Windows-specific vulnerabilities relative to other operating systems) in each executor is large enough, it will not cause attack-and-escape problems under the MR mechanism. In other words, any problem that is "correctable and fault-tolerant" with the DHR architecture endogenous mechanism will not affect the functionality, performance, and security of the target system. Namely, DHRA has the characteristics of "tolerating design and implementation defects" and intrusion-tolerant attributes based on known or unknown threats without changing or removing backdoors and virus Trojans. That is to say, DHRA itself can produce "nonlinear destruction and suppression" from the endogenous integration defense mechanism for certain or uncertain threats beyond the given service functionality. The MR-based GRC structure and mechanisms result in the following three nonlinearly enhanced defense functions:

(1) It is easy to disintegrate the cooperative attack under non-cooperative conditions. On the one hand, the attacker must exclude the interference of the system's generalized dynamic mechanism and accurately lock the target to ensure the reachability of the attack and carefully construct the cooperative attack under non-cooperative conditions to achieve the successful escape of this operation. On the other hand, the completion of the final attack mission depends heavily on the escape success rate of each segment and each stage of the attack chain. In other words, the defender will put the attacker in a dilemma where "one careless move loses the whole game."

(2) Since the application of algorithms such as trial and error or exclusion is based on the premise of unchanged background circumstances or conditions, the adaptive NFCM makes any perceptible trial and error, or exclusion attacks lose the implementation prerequisites as the current construction scenario changes, unless the attacker is able to attack with precisely collaboration in non-cooperative conditions (escape after a single attack) or to perform a derogatory attack on all executors of the target system.

(3) Introducing conventional defenses can result in "ultra-nonlinear defense gains." For example, add some conventional security measures such as intrusion isolation, testing, and prevention to some or all of the REs, or use measures such as firewalls, honeypots, smart sandboxes, virus and Trojan hunting, and backdoor closure, or apply active and passive defense technologies such as dynamization and virtualized migration and encryption verification, where the EF effects are directed to enhanced heterogeneity between the executors (service scenarios), thus causing greater uncertainty in the cooperative attack under non-cooperative conditions, which ultimately leads to an exponential decrease in the attack success or escape probability.

7.4.2 Reliability and Credibility Guaranteed
         by the Structural Gain

DHRA has an endogenous trait of high reliability. In the field of reliability, DRS provides a construction-level reliability gain proven by practical applications. The introduction of strategic scheduling and the RE-based MDR negative feedback mechanism based on DRS not only further improves the reliability but also significantly enhances the robustness of the system to curb uncertain disturbances. At the same time, it is also capable of sustainable confrontation against artificial attacks based on known or unknown backdoors in cyberspace. In general, there are always GRC requirements for high-reliability applications. The traditional approach is to increase the credibility of the system by adding external or additional security layers on the highly reliable structure. Although the inception of DRS has not taken into account the scenarios of addressing artificial attacks, it is inherently, though not persistently, intrusion-tolerant. The DHR structure, which combines generalized dynamics (including strategic scheduling, executor reconstruction, recombination, rebuilding, virtualization, and randomization) and negative feedback mechanisms, obtains time-independent intrusion-tolerant capabilities while enhancing the reliability of DRS, thus significantly increasing the target system's ability to suppress uncertain disturbances, including general unknown threats.

7.4.3 New Security-Trustable Methods and Approaches

In theory, the DHR system can achieve safe and credible goals not only in the component layer but also in the structural layer. In practice, due to differences in algorithm structure and resource configuration, different elements in the RE set (service scenarios) required to meet the same service functionality will inevitably differ in performance, effectiveness, and maturity. If these differences can be utilized consciously in the scheduling strategy when the current DHR service set is built, it is possible to obtain economic and technical complementarity that is not available in a non-DHR architecture. For example, some heterogeneous parts or components are designed with safety and credibility factors taken into consideration, and some may not consider security-controllable requirements; some may have higher performance and relatively lower maturity, while some exhibit moderate performance but relatively high maturity; some have strong comparative advantages in terms of advancement and economy, but the credibility cannot be guaranteed. In the DHR architecture, high-performance but insufficiently security-controllable components can be provided with regular service provision through strategic scheduling, while the parts or components with weaker performance or insufficient maturity but higher security and credibility are used as the "concomitant surveyor" of the former for real-time, quasi-real-time, or concomitant observation of the service compliance provided by the former, triggering the system cleaning, initialization or restart, and reconfiguration mechanisms in time, and if necessary, the latter can be used for service functionality or performance derating replacement. Of course, it is also possible to adopt a combined or mixed environment configuration. For instance, the motherboard can be x86, the operating system can be Linux or Kirin OS, and the database can be a Jinshan product, etc. or dynamically configured defense scenario resources if possible. In short, the DHR can use the architectural technology to achieve the "security-controllable and credible" goal at the system level when the credibility of the component supply chain cannot be ensured; and, on the other hand, even if the component level can meet the requirements for production and supply safety, the DHR architecture can still be applied to achieve higher levels of robust control effect in terms of high reliability, high availability, and high credibility. Without loss of generality, DHR answers the representative question raised in the *US National Defense Authorization Act for FY 2017* from the theoretical and technical level: "How to ensure the credibility of components from the global market, commercial grade, and non-trusted source?" Under the premise of safeguarding the national sovereignty and information security in cyberspace, the launch of "white-box measurability and user's customized security" inside DHR has a very positive constructive role and significance in eliminating technical exchange and policy-oriented trade barriers in IT and related fields, as well as in constantly advancing the economic and technological globalization strategy.

7.4.4 Creating a New Demand in a Diversified Market

The cyberspace market usually follows the "first-becomes-dominant and winner-takes-all" rule of the zero-sum game. The promotion and application of DHR as the general robust control (GRC) structure needs support from a market-oriented "isomeric" diversified supply mechanism and an industrial ecological environment. This is tantamount to creating a new demand on the supply side, transforming the traditional "homogeneous jungle competition" into "isomeric and diversified competition," providing broader innovative space for the emerging development of subsequent market participants, technology followers, or transcenders. At the same time, it can greatly promote the evolution of the information system from the "black-box" to the "white-box" transparent development model. In fact, the higher the transparency, the more thorough the standardization, and the more diversified the supply is; the DHR architecture based on the commercial environment and COTS-level hardware/software components will enjoy a broader space for technology and industry innovation when fighting against known or unknown threats, improving system robustness, and reducing full-life cycle product cost of use, which will profoundly change the existing cyberspace game rules (including military, commercial, legal, and technical rules.). In particular, the GRC structure and mechanism enable IT (or related) software/hardware products to have stability robustness and quality robustness with quantifiable design and verifiable measurement against uncertain disturbances, including unknown security threats, and removes the "non-actuarial" commercial barrier for financial capital from the insurance sector to enter the cyber security field. At the same time, it has a subversive impact on the "seller's market" offensive strategy based on "backdoor engineering and concealed loopholes" in the cyberspace, a monopoly attempt of technology or industry pioneers. No one can possess the capability of enjoying absolute freedom in cyberspace any longer.

7.4.5   The Problem of Super Escape and Information Leaking

If an attacker has enough attack resources and strong attack capabilities, masters the high-risk vulnerabilities in most REs, and gets access to the supreme control authority (such as the OS system-state or security-state authorization) and can accurately complete the upload or injection of virus Trojans, the attacker is not subject to the constraints of the inter-executor heterogeneity theoretically at this point and can achieve any attack function that satisfies the computability requirements, i.e., with the ability to set any DF that is completely equivalent or even identical in the majority of executors. Even so, to achieve attack and escape, it is necessary to overcome the cooperative attack problem in a non-cooperative environment. In theory, the external input channel can be employed to send the internally and externally agreed-upon motivation sequences that meet the input rules so as to achieve the synchronization between the executors. The synchronized operation is called super-escape (SE). However, the challenge hard to overcome is that an attacker does not know when and what uncertain actions the FC loops motivated by external control parameters will produce, as well as how many executors will be affected thereby, for instance, operations such as triggering replacement, migration, initialization or even recombination and reconstruction of the executor's structural scenarios, which can render the attacker's carefully constructed attack scenarios (especially a memory-resident-based attack scenario) valueless. In other words, it is difficult to overcome the uncertain disturbance effects of the DHR even if the attacker has SE capabilities. In fact, one of the biggest challenges encountered in DHR engineering practice is how to complete the compliance comparison operation of multimodal executor-related OVs within the smallest ruling window, because there are too many factors that will affect the time at which the multimodal executor OVs reach the comparator. For example, the implementation algorithmic differences in OS process scheduling, output/input buffers, and protocol stacks can result in considerable impact. Therefore, in the absence of strict time synchronization means, it is not easy to achieve stable escape even under the SE conditions. However, the SE problem does give us an important hint: loopholes that seem to be completely different in form (in theory, heterogeneity can be assumed to be infinity), such as loopholes in a nonhomologous operating system, may still implement the arbitrary DF through the possibly acquired super authority in a CPU program re-orientation or other on-site programmable intelligent processing environment. From this perspective, it is possible to achieve attack and escape even if the backdoors across executors are completely different. Even though they may be small probability events, in some application protection fields with high-security and high-sensitivity requirements, it is necessary to set higher bars preventing attackers from getting super privileged in the engineering practice. This can obviously increase the difficulty for the attacker to obtain super-privileges or sensitive information by limiting the on-site program or data modification (such as using the trusted-area function in the SGX or arm CPU in the x86 CPU or separating physical space or limiting the protected program or data read-write permissions).

Due to the mechanisms of DHR, when all the redundant executors or service scenarios in the current service set have the identical data environment, if an attacker has the capability of launching side-channel attacks or can avoid the inner and outer communication of the multimode ruling service set, the data confidentiality in the current operating environment cannot be guaranteed. For example, if there exist malicious codes in the OS, the separation of the container cannot be expected; if the attacker can use acoustics, photoelectric, and magnetic heating for side-channel attacks or establishes a stealthy information leaking channel to bypass the ruling process, sensitive information can be transmitted. In such a case, the information confidentiality in the DHR cannot be guaranteed.

7.5 Uncertainty: An Influencing Factor

In the information field, information entropy can be used to measure the degree of uncertainty of the system. The higher the uncertainty of a system, the larger the information entropy will be. On the contrary, the higher the information system order is, the lower the information entropy will be. For defenders, the more uncertain the apparency of the DHR architecture, the more difficult and costly it will be for the attackers or for the decrease of information entropy. There are two kinds of uncertainty-influencing factors in the DHR architecture, one is generated by the structure itself, which we call the endogenous factor of the DHR architecture, and the other is imported or attached to the structure from outside, which we call the introduced factor. The combined effect of organic and introduced factors can significantly increase the information entropy of the defense scenario.

7.5.1 DHR Endogenous Factors

DHR endogenous factors can be understood as structural factors that produce uncertain effects (i.e., increase information entropy) and also act as the root cause of DHR's ability to obtain endogenous security gains independent from conventional defense means. The system security gains brought about by the endogenous factors of DHR can be interpreted from the following aspects:

(1) The diversity of reconfigurable or software-definable executors under the FE conditions can cause the uncertainty of the apparent relationships between the external functionality of the target object (the hardware/software system that needs security and reliable functions) and its inherent structure or algorithm, where an attacker cannot derive the nature and structure of P in the I[P]O model by input motivation and output response relationships. Based on the FE relationship, P itself can be any element in the HES E ($E = \{E_1, E_2,\ldots, E_n\}$ (see Fig. 7.2), or it can be one or a combination of the unions or intersections of these heterogeneous elements. Obviously, by adding new elements or changing the implementation algorithms and structures of existing elements or by changing the selection and calling strategies of the elements in the executor set, we can affect the uncertainty of the apparent structure of the target object. As a result, an attacker will struggle to build a stable attack chain due to the inability to control the operating environment of the defense object.

(2) The ideal goal of the DHR architecture is to defuse the attacker's resources, means, or methods that threaten the target object. Thus, the DHR architecture or application policies (including static redundancy, dynamic redundancy, hardware redundancy, software redundancy, spatial redundancy, temporal redundancy, and other possible combinations thereof) and iterative changes in RE internal structural scenarios and resource allocation relationships can affect the certainty of the attack effect to varying degrees.

(3) Dynamics and randomness are the origin of uncertainty. In the DHR architecture, there are a variety of influencing factors, such as scheduling timing and strategy of executor structural scenarios, reconstruction timing and algorithm selection of running scenarios, task migration and container provisioning in virtualized environment, R&R algorithm inside the programmable arbiter, and timing of addition and subtraction of elements in the FE component set. All these aim at expanding the uncertainty (diversity) of the apparent structure of the target object as much as possible in the space-time dimension. Just like nucleic acids and DNA in biology, although there are only four types of nucleobases (adenine A, thymine T, cytosine C, and guanine G), they can get almost infinite DNA sequences in spatial dimension.

(4) The OV-based MR mechanism makes it difficult for DFs hidden in REs to function independently. In the DHR environment, DFs distributed in the executors usually belong to functions beyond the given functionality (i.e., those that the designer does not want or does not know) and are strongly correlated to the structural scenarios and operating mechanisms configured by the executor. In theory, the effects of completely heterogeneous DFs in the DHR executor are shielded by the MR mechanism. In practice, since absolute heterogeneity cannot be guaranteed (so far, the problem of theoretical and engineering implementation of heterogeneity design has not been fundamentally solved), the existence of inter-executor dark function intersection (DFI) is not excluded. However, there is no substantial harm to the target system, unless these DFIs can cause a

consistent multimodal OV presentation in the space-time dimension. For example, in the phase of the attack chain scanning, even if the probing operation of querying the operating system's version number is responded to by all the executors in the current service set, the heterogeneity of the OS version necessarily makes the given OVs different, and the scan, query, and output result will naturally be rejected or mistitled by the MR mechanism. It can be seen that MR's requirement for spatiotemporal consistency makes it necessary for an attacker to overcome the bottleneck of cooperative utilization under uncertain environment and non-cooperative conditions even if the same DF exists. In particular, as the semantic complexity and information abundance of the OV extends, the difficulty of cooperative attack will rise exponentially.

(5) The MR mechanism nonlinearly amplifies the difficulty of cooperative attack under the non-cooperating conditions. Whether it is dynamicity, diversity, or randomness or information abundance of the OV or selection and combination strategies of executor structural scenarios in DHR, as long as it is a factor that increases the defense behavioral uncertainty of the target environment, it will turn the attack action based on the executor DF to an uncertain event or a small probability event in front of the MR mechanism under the non-cooperative conditions. As we all know, "without perfect synchronization and negotiation mechanisms, precise coordination between multiple individuals is almost impossible, and its success rate is inversely proportional to the complexity of the collaborative mission." Therefore, it is difficult for non-cooperative domain-crossing and coordinated attacks to achieve certainty effects in the DHR architecture.

(6) The NFCM of DHR causes the target object to persistently maintain an uncertainty effect. This is because any non-compliant performance of multimodal OV will be discovered by multimode ruling and will trigger the strategic scheduling and MDR mechanism to change the current structural scenario of the target object. The negative feedback mechanism always attempts to eliminate the abnormal state of the multimode ruling segment or make the frequency of the abnormality lower than a given threshold. This is a self-adaptive process of dynamic iterative convergence. When the attack scenario remains unchanged, the convergence speed of the closed-loop control segment is strongly correlated not only to the diverse quantity of configured redundant resources and the heterogeneity of the structural scenarios but also to the FC algorithm and ruling strategy. The metastable state effect of the negative feedback mechanism will select a structural scenario meeting the current security requirements (a certain combination of REs) for the target object, but this is by no means a steady state. Since the feedback controller will be reactivated at any time by new output abnormalities (including sporadic failures caused by physical and design flaws in hardware/software), it can also be triggered by a consciously added randomness strategy in external control parameters. The latter will increase the implementation difficulty of tunnel-through or the reliability of "invisible transmission."

7.5.2   DHR-Introduced Factors

DHR-introduced factors refer to factors that are not inherent to the system and need to be added externally and can be nonlinearly amplified by the structural effect of "heterogeneous multi-target coordination in non-cooperative conditions." The following factors have a positive impact on system security gain or uncertainty:

(1) Such operations as instructions, addresses, data, and ports can be randomized, dynamized, and diversified in the RE.
(2) A diversified compilation and translation mechanism introduced to the RE software version will result in N versions in the redundant form or polymorphic modules. If these versions are activated according to a certain strategy or the modules are randomly called or imported into a monitoring layer to execute multiple variants according to strict sequence and check whether their conducts are different [3], the apparent uncertainty of the DHR system will increase.
(3) Some methods can be used to change the program functionality to eliminate security vulnerabilities [3], including input correction, function removal, function replacement, dynamic configuration by using dynamic interfaces, and the like.
(4) Symbiotic mechanisms [3], conventional active and passive security measures added into the RE, such as intrusion prevention, intrusion examination, intrusion isolation, or loophole scanning and virus and Trojan hunting, will significantly increase the uncertainty during the attack plan implementation process.
(5) A more diversified voting algorithm can be introduced in the MV segment, or measures such as "System Fingerprint" can be added in the output agent (OA) segment. For example, the adoption of the fixed response delay helps to make the apparent output characteristics of the target object lack regularity so that "tunnel-through" side-channel attacks will become invalid.

7.5.3   DHR-Combined Factors

We can naturally conclude from the above analysis that if the endogenous and introduced factors of DHR are used in a combined way, the uncertainty (information entropy) of the operating environment of the DHR system will undoubtedly increase. In this case, even if there is a DF such as a "homologous" backdoor at the part or component level, the attacks through the same backdoor are difficult to result in the same attack effect as in different service scenarios (unless the backdoor is environment-independent). It is because of the uncertainty of the execution environment; the strategic scheduling and combination relationship of the executor are difficult to predict in each service scenario, and since the backdoor availability is usually strongly correlated with the application scenario. When these effects appear as OV content differences in the RE set, the attack will be perceived by the MR segment and trigger strategic scheduling and MDR NFCM (negative feedback control mechanism); the suspected problematic executor in the current service set may be replaced and put offline or cleaned and restarted or subject to in-depth recovery by invoking the related back-end functions or assigned to other structural-diversified scenarios in a software-definable manner. In particular, the DHR architecture requires its executors to have reconfigurable or software-definable features, so that there are more structural-diversified scenarios for feedback mechanisms to make dynamic selections when the external functionality (or performance) remains unchanged. To a considerable extent, this helps to lower the requirements for heterogeneity within the executor or across executors, including the maturity requirements, and therefore is of great significance in engineering application.

7.5.4 Challenges to a Forced Breakthrough

In an encryption mechanism, once the encryption algorithm is certain, the computational complexity of the decryption is usually created by the randomness of the key. By analogy, DHR features similar encryption algorithm uncertainty, key uncertainty, scene-reproducing difficulty, and limited IS structure. Attackers trying to break through the defense barrier by brute force will face great challenges.

(1) The FE encryption algorithm is uncertain. Generalized dynamization technology makes "encryption algorithms" not only diversified but also dynamic and random.
(2) The equivalent key is uncertain. The uncertainty values in the target system, such as current memory, CPU occupancy, number of active processes or threads, network input/output traffic, access frequency, etc., can be used as the selection parameters for the strategy scheduling algorithm and the R&R Scheme.
(3) The scene is difficult to reproduce, and the attack effect is difficult to evaluate. Once the MR segment spots an abnormality, it will shield the abnormal output and meanwhile trigger the NFCM based on strategic scheduling and MDR to automatically replace, re-initialize, or reconstruct the suspected problematic executor in the current executor set until the abnormality is eliminated or its occurrence frequency drops below a given threshold. At this point, the attacker's trial and error or exclusion or other attack means are unlikely to work.
(4) The IS structure is limited. In fact, the structural conditions of IS are often limited. For example, we can't exhaust the domain values of all messages in a certain period of time or subjectively expect to repeat certain messages. There are usually strict causal or temporal correlations between the messages, and those not compliant with logic and temporal correlations may be automatically filtered out by the grammatical and semantic checks of the network or service system.

As a result, the inherent uncertainty of DHR architecture is mostly not a problem of computational complexity, and forced breakthrough is not applicable either in theory or in practice.

7.6   Analogical Analysis Based on the Coding Theory

Shannon's information theory focuses studies on data compression and information transmission rates in communications. The source coding theorem makes it possible to effectively compress information. The channel coding theorem converts the question from transmission of communication codes into probability theory by introducing redundancy and achieves reliable information transmission on uncertain noise channels. It can be seen from the foregoing that the DHR architecture can suppress model perturbation caused by physical factors and control uncertain disturbances caused by unknown attacks. This is very similar to the Shannon channel characteristics based on the coding theory and methodology. The latter can theoretically prove that, as long as the appropriate coding method and acceptable redundancy are selected, the random white noise existing in the channel and the artificial additive interference introduced by artificial factors can be dealt with simultaneously. Therefore, the author attempts to use Shannon's coding theory and methodology to make an analogical analysis of the DHR characteristics (without strict equivalence proving).

If we regard the harm or impact caused by various cyberspace attacks on the information system as the interference noise on communication and transmission channels, the DHR systemic effect can be analogized to some channel error-correcting code, and the target object under attack can be equivalent to a scenario where the communication channel suffers additive noise, while the process of determining whether the defense target can provide the correct service function is similar to the process of whether the error-correcting code can accurately reproduce the input information. Therefore, analogical analysis of the basic properties of the DHR architecture by means of information and the coding theory can provide useful reference for the engineering design and technical realization of DHR architecture.

7.6.1   Coding Theory and Turbo Codes

Since Shannon established the subject of Information and Coding Theory in 1948, the channel coding technique necessary for reliable communication over unreliable channels has raised broad-based attention and in-depth research. Shannon first proposed the communication system model in his long paper entitled *A Mathematical Theory of Communication* published in 1949 [4], pointing out that the central problem of system design was how to transmit information effectively and reliably in interference noise and stating that this goal could be achieved by coding.

To find out the ultimate transmission capability of the channel, Shannon introduces the concept of channel capacity, namely, the maximum amount of average information that can be transmitted over the channel. In order to improve the validity and reliability of information transmission, Shannon proposed the source coding and channel coding theorems. The lossy source coding theorem gives the rate-distortion function of the discrete memoryless stationary source as $R(D)$. When the information rate $R>R(D)$, as long as the source sequence L is sufficiently long, there must be a code with a decoding distortion less than or equal to $D + \varepsilon$, where ε is an arbitrarily small positive number. According to Shannon's channel coding theorem, for a discrete memoryless stationary channel whose channel memory capacity is C, when the information transmission rate $R<C$, as long as the random code length n is sufficiently long, there will always be a code that can make the average mis-decoding probability arbitrarily small. Source coding requires a reduction in average code length and a change in code word structure in order to improve the effectiveness of information transmission, while channel coding pays the cost of increased coding redundancy in order to improve reliability.

It has been a hot issue in theoretical and technological research of channel coding to find and construct a channel code that can approximate the channel capacity limit and its practical and effective decoding algorithm. So far many excellent coding schemes have been created in this regard, including Turbo codes, LDPC codes, and Polar codes. Modern channel coding usually follows a quasi-random redundancy coding method, which, in combination with iterative soft decoding, achieves or approximates the maximum likelihood decoding performance. In general, the distance from the Shannon limit is used to measure the performance of the coding method.

At the International Conference on Communications (ICC) held in 1993, C. Berrou, A. Glavieux, et al. proposed a new coding method: Turbo code, also called the parallel concatenated convolutional code (PCCC). It cleverly combines the component (recursive systematic convolutional, RSC) code with the pseudo-random interleaver, constructs the pseudo-random long code with short code by parallel cascading, and adopts the soft decision mechanism in multiple iterative decoding to approximate the maximum likelihood decoding. The experimental results show that when the BPSK modulation method is used in the additive Gaussian white noise channel, if the iterative decoding is performed 18 times on the Turbo code with a code rate of 1/2 and a random interleaver length of 65,536, the error code rate will be BER $\leq 10^{-5}$ when SNR $E_b/N_0 \geq 0.7$dB, an outstanding performance only 0.7 dB from the Shannon limit.

Observe the principle of Turbo code encoding and decoding from the perspective of redundant fault tolerance and error correction, as shown in Fig. 7.5; when Turbo code is encoded, the signal is divided into three channels, one is the original signal C^s, one is the component (RSC) code encoding signal C_1^p, and one is the encoding signal C_2^p encoded by the interleaver (intra-line and interline interleaving) and the component code encoder. Before the three-way signals are multiplexed, the parity can be deleted according to the transmission rate requirement, i.e., during parallel/serial conversion, alternately selecting two channels for code outputs on the parity bit will enhance the code rate to 1/2 from 1/3.

As shown in Fig. 7.6, when the Turbo code is compiled, the multiplexed signals are converted through parallel/serial conversion at the receiving end into a two-channel check signal and one-channel system bit information, the received system bit information y_k^s and the coding check information y_k^{1p} are sent into Decoder 1 first, and the external information $L_{21}^e(U_k)$ deinterleaved by Decoder 2 is simultaneously

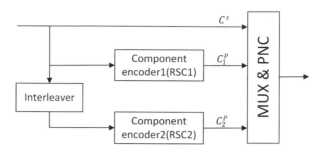

Fig. 7.5 Turbo code encoder

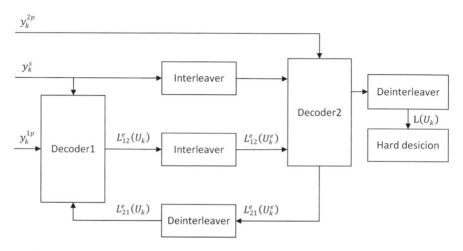

Fig. 7.6 Turbo code decoder

sent to Decoder 1 as a priori signal (the external information of Decoder 2 upon the first decoding is 0). The output information comprises two parts. One is the external information related to the code word, and the other is the decoding output corresponding to the system bit. The three input signals of Decoder 2 are respectively the external information interleaving signal of the Decoder 1 $L_{12}^e(U_k)$, the original post-interleaved check signal y_k^{2p} and the received system bit information interleaving signal. The output signal of decoder 2 subtracts the external signal $L_{12}^e(U_k^e)$ and system bit signal provided by decoder 1 to obtain the external signal $L_{21}^e(U_k)$ of decoder 2. The deinterleaved system bit information of Decoder 2 is used as a log likelihood ratio decoding for output result. After several iterations, the external information of Decoders 1 and 2 tends to stabilize, the asymptotic value of likelihood ratio approximates the maximum likelihood code of the entire code, and the system output will be subject to the soft decision mechanism throughout the iterative process, except for the final output step, where hard decision will be imposed on this likelihood ratio to produce the best evaluation sequence U_k of each bit in the information sequence C_k.

The decision principle of the MAP algorithm is described below. The MAP algorithm, namely, the maximum posteriori probability algorithm, aims to estimate the maximum likelihood bit. Generally, a decoding algorithm adopts a soft or hard decision mechanism. The hard decision decoder has only binary quantized input, and the implementation method is simple, while the soft decision improves performance using an octal quantized sequence or a simulated sequence sent by the demodulator. The Turbo code uses a decoding algorithm of soft input and soft output, where the soft output contains the decision value of the system bit information and the credibility of making such a decision, represented by $L(u)$ and $L^e(U_k)$,

$$L(U_k) = \ln \frac{P(U_k = +1 / y_1^N)}{P(U_k = -1 / y_1^N)}$$

$$L^e(U_k) = \ln \frac{P(U_k = +1)}{P(U_k = -1)}$$

where y_1^N is the decoder receiver sequence $\{y_1 \cdots y_N\}$.

After n times of iteration:

$$L_1^n(U_k) = L_c y_k^s + \left[L_{21}^e(U_k) \right]^{n-1} + \left[L_{12}^e(U_k) \right]^n$$

$$L_2^n(U_k^e) = L_c y_k^s + \left[L_{12}^e(U_k^e) \right]^{n-1} + \left[L_{21}^e(U_k^e) \right]^n$$

The first item on the right side of the equation is the output value of the system bit channel; the second item is the priori information provided by the previous iteration, where $L_{12}^e(U_k^e)$ is provided by Decoder 1 and $L_{21}^e(U_k^e)$ provided by Decoder 2; the third item is the external information required for subsequent iterations, and $L_{21}^e(U_k)$ and $L_{12}^e(U_k)$ are deinterleaving signals.

7.6.2 Analogic Analysis Based on Turbo Encoding

From the perspective of randomness, Shannon's channel coding theorem transforms the communication coding problem into an issue of the probability theory and achieves reliable transmission on uncertain noise channels. As described in this chapter, the DHR's multimodal voting system can transform the definite "one-way transparent attack from within and without" against an individual executor in the service set into an attack event with identifiable uncertain cooperative attack effects and can re-transform the event into a random event with a controllable probability. Therefore, an attack event targeted at the DHRA system can be analogized to the random noise introduced in the information transmission process, so that the DHRA system analytically satisfies Shannon's randomness codec conditions.

The DHR structure system mainly consists of core mechanisms such as input distribution, RE, strategic scheduling, MDR, and MR. It uses the multi-heterogeneity of the executors in the space-time dimension to break the static, similarity, and certainty patterns in the technological structure of the information system, so that it can, to some extent, tolerate external attacks based on known and unknown vulnerabilities and penetration attacks of unknown virus Trojans and can strive for the goal of integrated processing against both external and internal penetration attacks with improved reliability and attack resistance. The Turbo coding process distributes the input data to encoders of different structures, which is very similar to the input distribution, strategic scheduling, and RE mechanism of the DHR structure system; there are various kinds of basic units in Turbo coding, including interleavers, component encoders, and combined deciders, and they can be united to dynamically construct various heterogeneous encoding and decoding structures, which can be analogized to the FC-based dynamic reconstruction mechanism of the DHR structure system; the decoding segment of Turbo coding includes the iterative superposition mechanism, the soft and hard decision mechanism, and so on, and the credibility is configured for each decision result, which can be analogized to the multimodal strategic adjudication mechanism in the DHRS system. Figure 7.7 shows the analogy between Turbo coding and the DHR architecture. In the following, we will conduct a simulation analysis of the similar coding heterogeneity, coding redundancy, encoding OV, and decoding and ruling and encoding/decoding dynamics of the DHRS system in the additive white Gaussian Noise (AWGN) channel model.

7.6.2.1   Coding Heterogeneity

In the DHR architecture, it is necessary to meet the FE requirements for each executor yet with structural independence according to the principle of design heterogeneity, so that multimodal arbitration can tolerate non-cooperative attacks and reduce

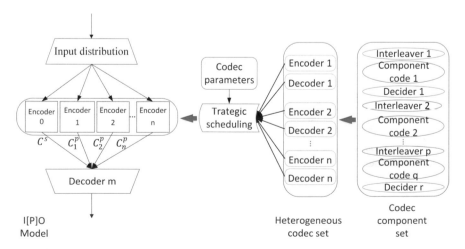

Fig. 7.7   Codec analogy of Turbo coding and DHRA

the cooperative attack effect. Namely, the greater the heterogeneity between the executors, the stronger the DHRA is against non-cooperative attacks. The same requirement is also found in the Turbo encoding system: the smaller the correlation between the original and code signals, the stronger the error correction capability of the code.

The encoded output signal of the Turbo code is converted into three channels of heterogeneous signals by the original signal through heterogeneity processing, which are, respectively, the original signal C^s, the component encoder encoding signal C_1^p, and the encoding signal C_2^p encoded by the interleaver (in-line and interline interleaving) and the component code encoder. Such signals are multiplexed and transmitted to prevent the uncertainty noise from simultaneously disturbing the same position of the information, so that the decoding can be based on the credibility of the information, whereas:

$$L^e\left(U_k\right) = \ln \frac{P\left(U_k = +1\right)}{P\left(U_k = -1\right)}$$

In addition, heterogeneous interleavers and heterogeneous component encoders can be combined within the coding system. The interleavers are divided into the packet interleaver, the packet spiral interleaver, the random interleaver, and the like. Packet interleaving refers to code element interleaving performed via reading in information sequences in rows and reading them out in columns. The interleaving function is expressed as follows:

$$\pi\left(i\right) = \left[\left(i-1\right)\bmod n\right] \times m + \left[\left(i-1\right)/n\right] + 1, \quad i \in C$$

The method of packet interleaving is simple, and the interleaving effect on short sequences is better, while the information remains correlated to some extent after such interleaving. In random interleaving, the information sequences are read out from the interleaver in a random manner. An index array is introduced to store N random numbers to correspond to N random addresses, where each random replacement address $\pi(i)$ needs to be compared with the previous s values and shall meet the following conditions:

$$\pi\left(i\right) - \pi\left(i-j\right) \geq s, \quad j = 1, 2, \ldots, s$$

The longer an information sequence, the more evenly distributed the random number, and the less correlated between the information bits. As shown in the simulation result in Fig. 7.8, compared to the null code and null interleave and the packet interleaver, the information sequences processed by the pseudo-random interleaver are more random, and thus the error correction effect is also better. Table 7.1 lists the other parameter statuses corresponding to different interleavers.

The following conclusions can be drawn from the simulation results: In the DHR architecture, the higher the executor heterogeneity (the lower the correlation), the stronger the error correction (attack resistance) capability of the system; when the

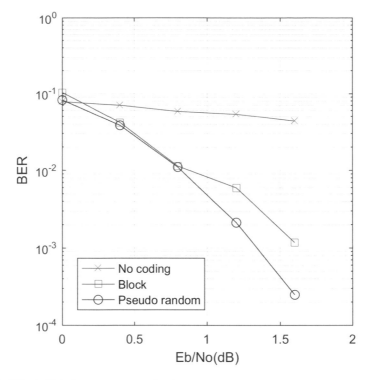

Fig. 7.8 Effect of interleaver on error code rate

Table 7.1 Other parameter statuses corresponding to the interleaver

Impact factor	Decoding algorithm	Code rate	Component code	Interleave length	Number of iterations
Interleaver	Log-MAP	1/3	(7, 5)	1024	3

heterogeneity is improved to a certain extent, the enhancement in the anti-attack performance of the system is not remarkable. Since the cost is high in research and development of highly heterogeneous executors, the executor heterogeneity is only supposed to reach a moderate rather than a high level in case of excessive system R&D cost.

7.6.2.2   Coding Redundancy

In a typical DHR architecture, redundancy requires multiple FE REs at the same time, each executor can perform its service functions independently or collectively, and a combined presentation can lead to the attacker's cognitive ambiguity with regard to the structure and properties of the target object. In the Turbo coding system, the combined coding of multi-channel redundant signals is also used, wherein the coding signal of each channel is not a simple conversion of the original signal but the parity information generated by the interleaving and the encoder, which prevents the uncertainty noise from simultaneously disturbing specific information bits of the original data and recovery information bits of the interleaved and encoded data. For a Turbo code with a code rate of 1/3, the code receiving signal can be expressed as follows:

$$y_k = \left(y_k^s, \ y_k^{1p}, \ y_k^{2p} \right)$$

Although increasing the coding redundancy can increase the reliability of the ruling, it will be followed by the reduced transmission performance. Therefore, we should take a comprehensive consideration. When the code rate is 1/3, the decoding outperforms the code rate of 1/2. As shown in Fig. 7.9, raising the redundancy by reducing the code rate can improve the performance of the Turbo code, yet at the same time, the transmission efficiency will be reduced. Moreover, the lower the code rate, the less obvious the performance improvement. So we must weigh the gains and losses between transmission efficiency and transmission quality when selecting the code rate. Other parameter statuses corresponding to the code rate are shown in Table 7.2.

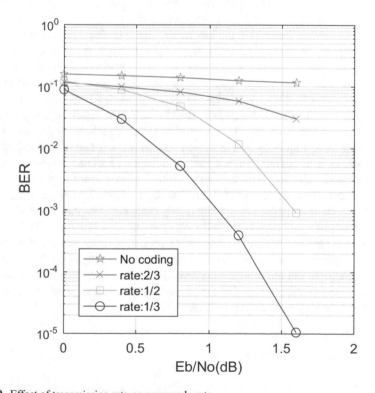

Fig. 7.9 Effect of transmission rate on error code rate

Table 7.2 Other parameter statuses corresponding to the code rate

Impact factor	Interleave pattern	Decoding algorithm	Component code	Interleave length	Number of iterations
Code rate	Pseudo-randomness	Log-MAP	(13, 15)	1024	3

The following conclusions can be drawn from the simulation results: In DHRA, the system's attack resistance can be effectively improved by increasing the redundancy. When the executor has a margin of 2–3 degrees, the system has a relatively high attack resistance. A higher-margin executor does not obviously enhance the system's attack resistance performance. In order to reduce the system cost, a different redundancy can be selected based on the importance of the modules, with the core modules endowed with a relatively high redundancy and the peripheral modules with a lower or even a zero redundancy.

7.6.2.3   Coding OV

Mechanically, DHR MR can exponentially increase the difficulty for an attacker to launch a coordinated and consistent attack by expanding the OV space, so that the attacker cannot implement precise coordination under the non-cooperative conditions, not to mention attack and escape. In the Turbo coding system, the interleave length is typically used to characterize the OV. As shown in the simulation results in Fig. 7.10, the increase in interleaver length can significantly improve the performance of the Turbo code, reduce the error code rate of decoding, and improve the transmission quality. However, the interleaver introduces a time delay, and the longer the interleave length is, the greater the time delay will get. In this case, when determining the interleave length, we must consider both the transmission quality and the time-delay requirement. For those services that tolerate relatively large time delay (such as data traffic), we can choose a larger interleave length to ensure a lower error code rate; for those that do not tolerate large time delay (such as voice traffic), we can choose a short interleave length to ensure a low time delay. Other parameter statuses corresponding to the interleave length are shown in Table 7.3.

The following conclusions can be drawn from the simulation results: In DHRA, the increase in the OV length can effectively improve the system's attack resistance; as the OV length increases, the output time delay and the ruling cost will be increased to some extent. In the engineering implementation, with affordable technical and economic costs, an appropriate OV length should be selected to make it harder for an attacker to launch a cooperative attack.

7.6.2.4   Decoding and Ruling

In the DHR architecture, a consistency or majority decision-making strategy is typically applied to the multi-channel executor OV. Due to the heterogeneity between the executors, it is difficult for any attack to simultaneously break each HE and

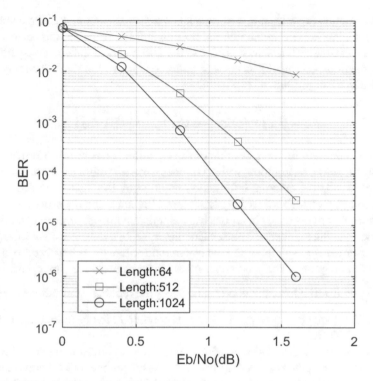

Fig. 7.10 Effect of interleave length on error code rate

Table 7.3 Other parameter statuses corresponding to the interleave length

Impact factor	Interleave pattern	Decoding algorithm	Code rate	Component code	Number of iterations
Interleave length	Pseudo-randomness	Log-MAP	1/3	(13, 15)	5

generate the same OV content, thus making the effect of the system DF attenuate exponentially. When a Turbo code is decoded, the two component code decoders (corresponding to the two component encoders) work, respectively, by adopting a soft output decoding algorithm (such as the MAP algorithm) and obtaining the performance approximating the Shannon Limit.

The Turbo code sends the external information $L_{21}^{e}(U_{k})$ deinterleaved by Decoder 2 to Decoder 1 as a priori signal, and through multiple iterative feedbacks, the external information upon each judgment is subject to credibility superposition, thereby constantly alleviating the noise-led distortion. Corresponding to the DHRA system, there is a historical confidence parameter for different executors, and the parameter of the corresponding executor will change after each MR. This change records the historical performance of the executor. Once the multimodal OV is completely inconsistent, MR arbiter will call this parameter into strategic ruling so as to select or form a more credible OV from the multimodal vectors. Therefore, the defense function is valid unless an attack can affect all REs and cause completely consistent error content at the relevant multimodal OV bits (i.e., the occurrence of attack and escape). This is very similar to the Turbo code approach constantly alleviating the noise-led distortion through multiple iterative feedback decoding.

In the MAP algorithm, after n iterations

$$L_1^n(U_k) = L_c y_k^s + \left[L_{21}^e(U_k)\right]^{n-1} + \left[L_{12}^e(U_k)\right]^n$$

The second item on the right side of the equation is the priori information provided by the previous iteration, which provides a trusted reference for this decoding; the third item is the external information required by the next-stage decoder. The high credibility of such a decoding can tolerate the noise-led distortion.

The number of decoding iterations has a significant impact on error detection and correction. As the number of iterations increases, $L^e(U_k)$, the external information of the encoder tends to stabilize, and the asymptotic value of likelihood ratio approximates the maximum likelihood code. After being iterated five times, BER $< 10^{-5}$ when $\dfrac{E_b}{N_0} = 1.6\text{dB}$. Other parameter statuses of the number of iterations are shown in Table 7.4. Other parameter statuses corresponding to the decoding algorithm are shown in Table 7.5.

As shown in the simulation results in Fig. 7.11, a certain number of iterations will largely correct the noise-led distortion, but as the number of iterations increases, the information outside the system tends to stabilize, and the output of the system will remain almost unchanged. Increasing the number of iterations at this time will further increase the time delay, making it unable to meet the user requirements of a low-time-delay system.

As shown in the simulation results in Fig. 7.12, selecting a high-complexity decoding algorithm generally helps to improve performance greatly but also increases the time delay and the amount of calculation.

The following conclusions can be drawn from the simulation results: In DHRA, increasing the number of iterations in the ruling and selecting a more complicated ruling algorithm can effectively improve the attack resistance of the system. Since

Table 7.4 Other parameter statuses of the number of iterations

Impact factor	Interleave pattern	Decoding algorithm	Code rate	Component code	Interleave length
Number of iterations	Pseudo-randomness	Log-MAP	1/3	(13, 15)	1024

Table 7.5 Other parameter statuses corresponding to the decoding algorithm

Impact factor	Interleave pattern	Code rate	Component code	Interleave length	Number of iterations
Decoding algorithm	Pseudo-randomness	1/3	(13, 15)	1024	3

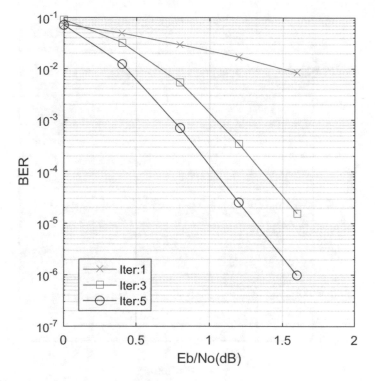

Fig. 7.11   Effect of the number of iterations on the error code rate

increasing the number of iterations can reduce the redundancy requirement for the executor, the implementation cost of the system can be reduced by appropriately increasing the number of iterations. As the number of iterations and the complexity of the ruling algorithm increase, the time delay and the calculation amount will also go up. Therefore, it is necessary to select the appropriate number of iterations and ruling complexity to meet the system's cost-effective requirement.

7.6.2.5   Codec Dynamics

In the DHR architecture, dynamics can be fully introduced in the codec process by means of strategic scheduling, dynamic executor reconstruction, executor online operation and offline cleaning, alternation of the ruling algorithm, adjustment of executors, and ruling parameter configuration. Turbo codes can adapt to all kinds of channels upon adjustments to the codec algorithms and parameters. As for the dynamics of codec, the code executors are spatially represented in serial, parallel, serial-parallel, and other forms and temporally expressed as static, dynamic, pseudo-random, etc. In terms of strategy, historical performance, structural performance excellence, and randomization initiatives can be introduced, while in respect of the generation mode, reorganizable, reconstructable, and reconfigurable mechanisms can be applied.

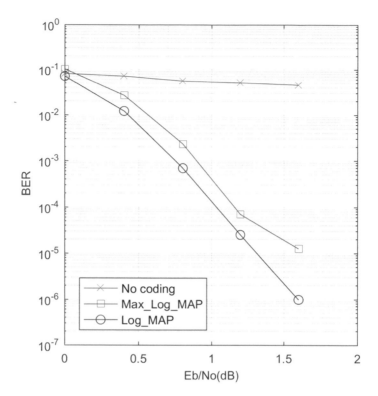

Fig. 7.12 Effect of the decoding algorithm on error code rate

The DHR architecture supports active and passive strategic scheduling and possesses nonspecific and specific threat perception capabilities. With the executors and MR subject to dynamic adjustment, it has an uncertainty effect on the attacker, which greatly improves the system's anti-attack capability. The simulation results in Fig. 7.13 show that in the coding system, corresponding measures can be taken for different channel environments based on the variations of external information. For example, different interleavers and component encoders can be dynamically combined according to historical information, while the decoding algorithm, the number of decoding iterations, and the coding interleave length can be dynamically changed. Different noise channels can be dynamically adjusted to achieve the most cost-effective interference countermeasure or error correction effect.

It can be concluded from the simulation results that in the DHR architecture, if the system parameters are dynamically adjusted, it is possible to make the MR inconsistency go below a certain threshold in a statistical sense. This threshold should be one that takes into full account the overhead cost and anti-attack performance requirements of the system, and the MR-based strategic scheduling and MDR NFCM should be the process of dynamically fitting this goal.

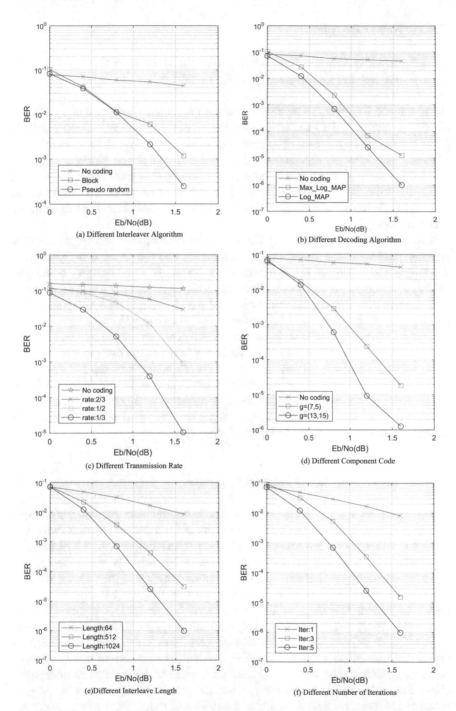

Fig. 7.13 Effect of dynamic parameters on error code rate. (**a**) Different interleaver algorithm. (**b**) Different decoding algorithm. (**c**) Different transmission rate. (**d**) Different component code. (**e**) Different interleave length. (**f**) Different number of iterations

7.6.3   Some Insights

7.6.3.1   Randomness and Redundancy Serving as the Core Elements for Solving Cyberspace Security Problems

If cyberspace facilities are regarded as a communication system and cyberspace attacks as additive noise on transmission channels, the cyberspace security defense architecture may be regarded as an error correction coding system for channel transmission. With reference to Shannon's theory of information and coding, if a random redundancy coding mechanism is introduced in the defense architecture and a ruling method similar to maximum likelihood decoding is used, we can evaluate the correlation between the success probability of a given attack and the randomness and redundancy.

Since the DHRA model is very similar to the Turbo coding model, it also satisfies the random codec conditions in Shannon's noisy-channel coding theorem. Therefore, the coding is implemented by introducing a redundantly configured FERE, the decoding is realized by the MR mechanism, and the interleaved coding and iterative decoding are realized by the FC-based strategic scheduling and MDR mechanisms, so that the issue of defending known or uncertain attacks can be transformed into a matter of reliable transmission at a given error code rate. The simulation results show that the sameness or similarity between the two models can serve as a basis for qualitative or quantitative reference analysis of the DHRA-related properties and the relations between the elements. As a conjecture, is there a cyberspace security defense structure that is analogous to LDPC or Polar codes? Furthermore, is there any other security defense architecture that is equivalent to the modern channel codec theory and methodology?

7.6.3.2   Uncertainty Effect Brought by DHRA

In physics or any other seemingly unpredictable natural phenomenon, the randomness may appear to be just noise or may originate from deeper complex dynamics. For DHRA that adopts the MR-based strategic scheduling and the MDR negative feedback mechanism, any situation that causes multimodal OV inconsistency or a dynamic scheduling action will trigger the transformation of the defense scenario until such inconsistency disappears or its occurring frequency is limited below a certain threshold. Therefore, as far as an attacker is concerned, whether it is scanning and sniffing, uploading a virus Trojan, or stealing, destroying, or deleting sensitive information, once any attack action triggers the defense mechanism, the target object will immediately change the defense scenario in the current service environment, so DHRA can be viewed as a random system or device with an uncertainty effect. Similarly, any injected white-box test sample can hardly achieve the purpose of attack and escape unless it can avoid disturbing or triggering the defense mechanism. Obviously, this uncertainty effect can significantly enhance the cooperative attack difficulty in attack chain construction, target scanning, channel verification, attack packet uploading, access to sensitive information, and permission elevation, leaving few chances for the accumulation or recurrence of any staged attack effect or experience or knowledge in the time dimension, thus allowing DHRA to be tested and quantified for evaluation directly by white-box injection testing means.

7.6.3.3 Flexibility and Self-restoring Capability of DHRA

By introducing dynamic and randomized elements into the dissimilar redundancy mechanism, DHRA transforms the latter's static and diverse processing space into dynamic and diverse randomized processing space through nonspecific/specific sensing, dynamic reconstruction, and negative feedback mechanisms of strategic scheduling, which has fundamentally addressed the flaws of DRS in terms of its "unsustainable intrusion-tolerant attributes." Just as channel coding makes it possible to achieve reliable transmission on channels with noise or even jamming, the DHR structure system not only has powerful fault-tolerant and error correction capabilities similar to error correction coding but also has nonspecific/specific threat perception, flexibility, and self-recovery capabilities, which enable the DHR structure to conduct autonomous judgment, automatic response, and self-restoration against known and unknown attacks.

7.6.3.4 Insufficiency in Analogical Analysis Using the Turbo Code Model

The Turbo code model is identical or similar to the DHR in architecture and functioning mechanism, especially in the metastable scenario of DHR. However, intuition tells us that cooperative attacks through the backdoor of the target object are like the additive interference on transmission channels. Different error correction coding or configuration parameters shall apply to different interference scenarios, while the MR-based strategic scheduling and the MDR negative feedback mechanism are intended to achieve this in principle. It is not difficult to imagine that this is a dynamic game process that self-adaptively varies with the MR threat perception (OV inconsistency) until the occurring frequency, scale, or proportion of multimodal OV inconsistency drops below a certain threshold in a given time, where the defense scenario in which the system is located can finally match the current attack scenario, i.e., meeting the requirements at a certain security level. In the above analogical analysis, we also introduce parameters, such as dynamically changing the number of decoding iterations, the complexity of the decoding algorithm, and the interleave length, and give the simulation results of adjusting the relevant parameters. However, if redundancy and heterogeneity supporting real-time changes can cause advanced-form changes in error correction coding, what impact will it have on the ability to withstand jamming, especially syndrome and spot jamming? This deserves our special attention in the follow-up research.

7.7   DHR-Related Effects

7.7.1   *Ability to Perceive Unidentified Threats*

For the DHRA operation mode under the FE conditions, as long as the RE input can be normalized, namely, the input is limited to the required range by a certain algorithm, the output can also be subject to multimode ruling with the normalized or standardized result. In theory, all abnormalities occurring on the OVs of the DHR multimode ruling interface can be identified, making it engineeringly feasible to perceive and locate an abnormal executor in real time. This ability of perceiving unidentified threats based on relativity judgment is brought about by the HR endogenous mechanism and can, if used appropriately, conduct syndrome monitoring of the security of "black-box" devices with standardized input/output protocols or procedures (such as routers, domain name servers, etc.). Unlike the application scenarios of trusted computing, both the monitoring and monitored devices require only the FE relationship under the standard protocol, and it is unnecessary to accurately know the implementation details and expected behaviors or states of the monitored devices or to conduct any in-house "intervention," such as setting a "probe" or "hook" or the like.

7.7.2   *Distributed Environmental Effect*

In the distributed network environment, the network elements (i.e., domain name servers, file storage servers, routers/switches, content servers, and various general-purpose or dedicated servers within a data center or cloud platform) are usually deployed in the distributed or redundant mode to enhance service robustness, load balancing, or data security. These network element devices are usually provided by multiple vendors. All the configured hardware/software components are not single-sourced, but they are often used to provide the same service functionality and even share the same data resources. In respect of the DHR principle, this deployment and service delivery method has an intrinsic attribute of FE, HR, and dynamic resources calling. From the perspective of the abstract DHR model, these network elements can be viewed as REs within a given service set. To provide highly credible and reliable robustness services, we simply need to add the service request proxy and output arbitration components. Take the highly credible and reliable service of domain name resolution as an example: The user request is directed to the request proxy server, and the server issues the same request to multiple domain name servers in the same domain according to a certain strategy based on the resource distribution of the domain name resolution service network. The domain name response information of these servers is directed to the output arbitration component, and the information meeting the security level requirements upon strategic adjudication is returned to the requesting user; otherwise the process will be repeated until the service requirements are met. Obviously, as long as the transmission information security is guaranteed from the user requests to the proxy server (e.g., using a VPN tunnel or taking encryption measures, etc.), the domain name resolution service will be highly credible, and this has nothing to do with whether there is a known/unknown security threat of this kind or that in the intranet domain name server or whether there is any hijacking or counterfeiting conduct on some transmission path.

7.7.3 Integrated Effect

(1) As DHR's attack difficulty increases exponentially, it is easy to achieve the super-nonlinear defense effects. This conclusion is easily drawn from the perspective of either qualitative/quantitative analysis or the "white-box" evaluation.
(2) The failure rate of the DHR system also decreases exponentially. According to the reliability model analysis, when the same degree of redundancy n ($\geq$3), the reliability of DHR is increased by several orders of magnitude compared to that of the typical DRS.
(3) From the visual analysis of the general model, the defense cost within the DHR architecture is at most proportional to the number n of REs, and the cost increase is at most linear. The strategic area defense mode, however, has only limited impact on the total cost of the target system due to the deployment of pass and core area defense. However, in respect of the integrated life-cycle cost performance ratio, a device with intensive functionality evidently outperforms the traditionally deployed additional security devices or technical means where security devices are separately deployed from the target system, and, due to its natural immunity to unknown threats, such as 0 day (or Nday) attacks, it can significantly reduce the work of security maintenance, which is not only costly but hardly done in real time.
(4) As DHR can form REs using COTS software/hardware components and can even directly use open or open-source cheap products, the development and after-sales costs of DHR's heterogeneous parts or components are easy to be offset due to the market price advantage of scale when compared to specially designed security systems or parts and components.
(5) While the traditional practice lies in attaching or piling up dedicated security defense facilities to protect network service systems, the DHR system integrates high-credibility service function provision, high-reliability guarantee, and high-availability support, so its application features better cost performance ratio or cost-effectiveness.

7.7.4 Architecture-Determined Safety

Today, globalization is irreversible. Actually, no country or business can completely control the entire ecology of its products or industries or expect any "black technology" to ensure the credibility of design chains, manufacturing chains, supply chains, tool chains, and service chains of all software/hardware parts, components, modules, and systems. In addition, as the new format of open-source mass innovation is widely accepted, "mutual convergence" has become an irresistible trend. Therefore, cyberspace security is one of the top international political issues and one of the biggest concerns in the cyber economy. However, even if the integrity of the IT or related industry supply chains can be ensured, it is hardly to guarantee the security of cyberspace. From the first chapter of this book, we know that the design flaws or loopholes of information systems cannot be completely solved by self-controllable supply chains and the backdoor in a non-credible supply chain cannot be completely avoided by self-controllable policy. We need to look for a new model whose safety is determined by its architecture. Just as the geometrically stable triangle in Euclidean space, the DHR architecture is a sharp tool that can solve the problem of non-credible hardware/software components at the structural level. In other words, even if the hardware/software components are "virus-bearing," as long as they are built on the DHR architecture or form a DHR environment, it can effectively suppress security threats based on the known or unknown backdoors of the target object in both theory and practice. Since the control loop noumenon is part of the DHR architecture, it can inherently obtain the endogenous security gains. Its concise and special functionality is transparent and independent compared to the REs within the structure and usually not attack-reachable. Its designed functional safety generally allows the use of formal proof tools to ensure null backdoors and minimal loopholes or trusted computing or dynamic defense techniques to enhance robustness and can be applied by way of a standard embedded module. Like commercial cipher machines, you can ensure their credibility if you assign customizable privacy algorithms, definable data or sensitive area access control authorization to feedback control (FC), strategic adjudication, or trusted storage areas.

7.7.5   Changing the Attack and Defense Game Rules in Cyberspace

"Vulnerabilities are inevitable, backdoors cannot be eliminated, yet technologies are still unable to detect all the backdoors. This is the only fact that threatens today's cyberspace safety." However, the launch of DHR architecture is a step to "change the rules of the attack and defense game" based on software/hardware code defects of the target system and can reverse the current scenario of "easy to attack and difficult to defend":

(1) The completely heterogeneous backdoors within the architecture become useless and have nothing to do with the known or unknown nature.
(2) The completely heterogeneous virus Trojans within the architecture become structurally ineffective and have nothing to do with the known or unknown nature.

(3) Even if there are diversified backdoors within the structure, they will not endanger the security of the system if not used in a coordinated and consistent manner.

(4) Even if there are diversified Trojans within the architecture, they will not be practically useful if they fail to achieve a stable attack and escape.

(5) The uncertainty effect of the architecture can block or disrupt any attempt to make a cooperative attack through trial and error or exclusion.

(6) The effectiveness of defense functionality inside the architecture is not correlated to the attacker's a priori knowledge, behavior, and characteristics. If conventional security defense means are integrated, its defense effectiveness can increase exponentially.

(7) "Even if the attack and escape happen, it is impossible to hold on," as the attack experience and outcome are neither reproducible nor inheritable.

(8) The effectiveness of the defense functionality is weakly correlated to the real-time security maintenance and the technical qualification of the related personnel.

(9) The generalized robust architecture makes the security level of the hardware/software system configurable and measurable against cyber attacks.

(10) Endogenous security will become a landmark of the next-generation IT products.

(11) The gray or black industrial chain and trading market based on loopholes, backdoors, viruses and Trojans, etc. will be severely squeezed, and the cyberspace environment will be purified.

Obviously, the traditional attack theory and methodology based on software/hardware code vulnerability will be toppled, and the theory and methodology of cyberspace attack and defense will be rewritten. The "easy to attack but hard to defend" asymmetry will be reversed, while the safety, openness, advancement, and credibility of information products will be redirected to a new stage of sustainable development featuring the "unity of opposites" rather than the "at daggers drawn" status quo.

7.7.6 Creating a Loose Ecological Environment

The rapidly developing information technology provides a fertile ground for software/hardware diversity, and it is not difficult to find a functionally equivalent or similar alternative for almost every component. Abundant diversity also brings about all kinds of backdoors, which certainly is a disaster for precise defense based on the acquisition of feature information. However, it is a rare development opportunity for DHRA-based defense, as the rich diversity provides a good heterogeneity basis for the engineering application of DHR defense. As far as the web is concerned, 1550 heterogeneous web service executors can be easily implemented through the five-layer software stack heterogeneity (see Chap. 12 for details).

Furthermore, the hardware platform that carries the software stack can also be heterogeneous, with different processor platforms adopted to produce more types of web executors. As is quoted from *A Tale of Two Cities* by Charles Dickens, "it was the best of times, it was the worst of times."

7.7.6.1   "Isomeric and Diversified" Ecology

The DHR architecture model often requires the support of a commodity market or a format of standardized and diversified parts or components of FE, while open-source communities, cross-platform computing, customizable computing, heterogeneous computing, hardware/software definability, functional virtualization, etc. can naturally support such development requirements. The former creates new demands on the supply side. The market for isomeric products is no longer filled with exclusive zero-sum competition, as DHR injects strong openness, complementarity, and diversification momentum into the market and doubles the market capacity for similar but nonhomologous products.

7.7.6.2   New Ways to Accelerate Product Maturity

The DHR architecture can be used to improve the functional performance of the under-mature heterogeneous parts or components. In general, there is definitely a difference between FE parts or components in terms of technical maturity, which is especially true for the latecomers to the market or the products in the early stages of use. After the log analysis function is added to the DHR architecture, the design defects and performance weaknesses of new products can be quickly discovered online or quasi-online by comparing and identifying the operations of the executors. By doing so, we can expect at least three benefits:

(1) Users no longer need to worry about the diversified parts or components in the market that need to be perfected for higher maturity.
(2) The emergence of "isomeric" demand can greatly lower the threshold for late-comers to enter the isomeric market.
(3) Tolerating design defects to a certain extent can significantly reduce the verification workload in the design phase, shorten the experiment and trial time and cost of the product, and speed up the new product time to market.

7.7.6.3   Self-controllable Complementary Form

The DHR architecture requires the construction of a service environment with diversified physical or virtual hardware/software but does not require the components themselves to be "virus-free" or "absolutely credible." This allows us to adopt a working model of hybrid configurations and real-time (or syndrome) monitoring in a global industrial ecosystem, using some commercial products with good functionality, performance, and maturity yet without guaranteed supply chain credibility and credible products with a high degree of self-control yet not sufficiently advanced or mature. In the multimodal judgment process, the introduction of weight parameters and the addition of refined, intelligent ruling strategies, such as performance priority, credibility priority, new function priority, and historical performance priority, will give full play to the self-control and credibility in the two technical aspects of the architecture and components, ensuring the complementary advantages between products and avoiding security defects or systemic risks. This aims to reduce the life-cycle utilization and maintenance cost of the target system by mixing or assorting the high and low advanced and security approaches.

7.7.6.4 Creating an Integrated Operating Environment

The cyberspace information systems not only have to prevent against conventional security issues but also have to address nonconventional security threats. This is particularly true in industrial control, embedded applications, special application fields, and sensitive industries.

DHR's GRC architecture can tolerate design defects to a large extent in mechanism (it is theoretically proved that DHR boasts of improvement in reliability by orders of magnitude over the reliability of typical DRS) as long as these defects do not simultaneously produce a common-state fault between the HEs. In other words, whether it is a random physical failure or design vulnerability or a malicious function implanted during engineering implementation or operation, unless the HEs can simultaneously produce identical OVs (to realize controllable escape at the operating step level) and achieve the "controllable escape at all operating steps," otherwise it will be subject to "intrusive tolerance" or "strategic cleaning." Therefore, the DHRA can cope with conventional and nonconventional security threats at the same time. This GRC architecture can be used to create a highly reliable, highly available, and highly robust information service system or control device that combines flexibility and credibility.

7.7.7 Restricted Application

As a robust control architecture for innovation in the information domain or cyberspace, DHR is universally applicable. Its advantage of integration is demonstrated in application scenarios requiring high reliability, high availability, and high credibility, so it is even indispensable. However, DHR is by no means an "all-mighty technology that works everywhere," and it may be restricted in the following application scenarios, including but not limited to:

7.7.7.1   Micro-synchronous Low-Time-Delay Operating Environment

Unlike the Isomeric Redundant (IR) mode, due to the structural or algorithmic heterogeneity requirements for the DHRA hardware/software implementation, MR's
judgment of OV strategy is a complex operation. It is difficult to achieve accurate
multimodal OV synchronization in time (e.g., at the microsecond, nanosecond, or
sub-nanosecond level) in the engineering implementation process or to avoid the
time-delay effect introduced in the judgment process. For applications with strict
time-delay and micro-synchronous requirements, this indeed constitutes a considerable engineering implementation challenge. In fact, even the IR system cannot
dodge the insertion time-delay and high-precision synchronization of the majority
voting mechanism.

7.7.7.2   Time-Delay-Constrained Scenarios That Cannot Be Corrected

DHRA is unlikely to achieve the multimode ruling of multiple executor OVs in real
time under severe response conditions upon a definite input incentive. Therefore,
the methodology for the DHR architecture is often designed to pre-select the output
of an executor's operating results according to some weighting strategies or historical performance, to ensure the real-time requirements of the application system as
much as possible, and to compare with the subsequently arriving OV within an
acceptable time range to determine whether to make corrections or modifications to
the preorder output results. DHR applications will be constrained, or special cost
has to be paid wherever there is a strict time-delay requirement and no corrections
or modifications are allowed at all.

7.7.7.3   Lack of a Normalizable Input/Output Interface

The DHR architecture requires the input/output interface of function executors or
service scenarios to be a normalized or standardized one. At this interface, there is
a high probability for the multimodal OVs of the FERE under a given input sequence
(IS) incentive to be consistent. In other words, this interface enables the judgment
of the equivalence between executors by a consistency test method for a given function or performance. In fact, for complex systems, it is often impossible to give a
complete test set. As device manufacturers often follow different exception handling algorithms when it goes beyond specific protocols or standards, so the functionality or performance equivalence of the executors is only valid within the range
covered by the test set and may be negative in the case of a small probability.
Therefore, an important prerequisite for the application of DHR architecture is an
interface that can be normalized (not necessarily with open standards), and its
equivalence testing can be implemented, coupled with the related resources.
Suppose that an executor with the encryption function only loads encrypted information on a certain field of the OV, the DHR architecture will not be applicable if it
cannot perform multimode ruling after shielding with mask.

7.7.7.4   Lack of Heterogeneous Hardware/Software Resources

The HR configuration of hardware/software processing resources is a prerequisite for the applicability of DHR architecture and generally requires the executors to be independent in physical and logical space. At present, COTS products with a broad-based market, such as underlying and supporting hardware/software and middle-ware, embedded systems, and IP cores, are increasingly diversified. In particular, with the development of the open-source community format, the homogenous diversified market threshold has been greatly reduced. However, there still lacks a market to supply a variety of personalized third-party applications or service software, especially in the business environment where only binary-version files are released without providing the source codes. This largely constrains the application of DHR architecture in spite of the finite-degree differentiation processing means, such as binary decompilation and diversified compilation.

7.7.7.5   "Blackout" in Software Update

Generally speaking, the software version upgrading of the FE executors in the system may be out of sync (a situation often unavoidable when using commercial products), which may affect the MR of DHR, except the version updating that does not change the content of the normalized input/output interface (such as patching, performance improvement, algorithm optimization, etc.). For upgraded versions involving normalized interface changes, a function can be introduced in the DHR system to set the output weights of the relevant executors based on the software updates. When the versions are different, higher output weights can be assigned to the functions expanded in the higher version (because the majority of the functions of the new and old versions are the same). When there is any inconsistency in the multimodal voting (MV) segment, the weights should be introduced before further ruling. Of course, this strategy will more or less reduce the target system's ability to deal with security risks during the version upgrading process.

7.7.7.6   Cost-Sensitive Area

Consumer electronic products, such as portable terminals, handheld terminals, wearable devices, and personal desktop terminals, are generally sensitive to purchase prices, upgrade cost, or power supply capacity. The application of DHR components will be constrained in these areas unless they become miniaturized, embedded, diversified, integrated, and low-power-consuming (it is expected that 3Q can mass-produce the SIP-packaged mimetic MCU devices in 2019). Of course, software or programmable DHRA products based on the multicore operating environment are excluded. Nevertheless, in an age of electronic information, especially when the design and manufacturing complexity of software/hardware products are no longer the main factor of their market pricing, DHR applications will no longer be plagued by factors such as cost price or volume and power consumption due to the development of technologies, including virtualization, heterogeneous mass-core processors, user-customizable computing, software-definable functions, CPU + FPGA, software-defined hardware (SDH), etc. However, it may bring about new problems to the maintainability of software/hardware versions.

7.7.7.7 Concerns Regarding the Highly Robust Software Architecture

In theory, the DHR architecture is essential for enhancing the reliability and credibility of software products in application. It is indeed difficult to completely eliminate software vulnerabilities in the current technical context, while the open-source community's mass innovation model may give rise to problems such as a trapdoor (inadvertent malicious code), and the state explosion problem makes the formal proof not always feasible in engineering. In principle, DHR is not bothered by these seemingly thorny troubles. The biggest challenge at present is that DHRA software is running in a quite inefficient way, because one interface operation may require multiple FE heterogeneous modules to be strategically invoked and multiple output results to be arbitrated, making the execution overhead unaffordable in a unilateral processing space and shared resource environment. Fortunately, with the spread of new computing architectures such as multicore, mass-core, and user-definable computing, we can use parallel processing technology to solve the problem of simultaneous calculation across multiple heterogeneous modules, and the output ruling overhead issue can be less complex in the arbitration phase through preprocessing of the output results in the FE heterogeneous module (such as encoding). The rest is the inherent problem of the HR system, namely, the complexity of design and maintenance and the cost increase (in fact, except the version upgrade maintenance, DHRA software can greatly reduce the life-cycle cost of security maintenance for the manufacturers and users alike, because there is no urgent patching and upgrading requirements when 0 day/Nday vulnerabilities do not exist). The author believes that if DHR can be applied at scale, these will not be our concerns. In particular, given today's severe cyberspace security challenges, we need to innovate the traditional product design model that only pursues functional performance without considering cyberspace safety, restore the basic commodity economic law where "product providers shall be held accountable for product design defects (including security defects)," and establish a new design and use philosophy—"safety is a service function as well as product quality."

7.7.7.8 Issue of Ruling

The heterogeneity in the executor environment can cause executable codes from the same source program version to produce different output responses. For example, the TCP initial sequence number in the IP stack of the operation system is random; when options in the IP packets or undefined extension contents use different stacks, there exists uncertainty to some extent; the output data packets use the encrypted algorithm and are strongly related with the host environment parameters. In theory, no transparent ruling "unrelated with the syntax or semantics" is allowed between the multimodal output vectors containing uncertain contents. At this time, some innovative imagination is needed to solve the problem. When conditions are allowable, we need to adjust functional arrangement between the output agent and the executors, adopt a unified stack version, use the mask code shielding technology, implement the ruling of detection vectors generated by the symbiotic system, compare parameters related with the environment, and use the output results of the intrusion detection. What we need to pay attention to is that there are many things which are needed to be compared. This can result in considerable ruling delay. At this time, such functions like output vector checking or hash value may need to be added so as to decrease the time overhead of the ruling.

References

1. Franklin, G.F.: Feedback Control of Dynamic System. Electronic Industry Press, Beijing (2004)
2. Kaminogawa, S., Tiecong, L., Zhongpu, S.: The Body and Immune Mechanisms. Science Press, Beijing (2003)
3. Jajodia, S., Ghosh, A.K., Swarup, V., et al.: Moving Target Defense: Creating Asymmetric Uncertainty for Cyber Threats (translated by Yang Lin). Beijing: National Defense Industry Press (2014)
4. Chung, J., Owen, H., Clark, R.: SDX architectures: a qualitative analysis. IEEE Southeast Con. 1–8 (2016)

Part II

Chapter 8
Original Meaning and Vision of Mimic Defense

DHR, as an innovative and simplified system architecture, can provide general robust control which is not seen in a conventional mode. Not relying on any a priori knowledge and additional security support, it can normalize all uncertain disturbances caused by random faults or attacks based on vulnerabilities or backdoors as classic reliability problems and handle them in the same approach. With its inherent "uncertainty" effect, the DHR framework has an intrinsic defensive function—"invisibility." Such invisibility is related not only to the redundancy, heterogeneity, and multimode output vector grammar and semantic abundance of the executors under functional equivalence conditions but also to the multimode decision algorithm set; the executor scheduling policy; the cleaning, restoration, and reorganization and restructuring mechanism; as well as the application of traditional security technology. So now the question is: what driving mechanism and gaming strategy should be adopted to get the desired endogenous security defense effect? The earth biosphere that has been naturally evolving for hundreds of millions of years usually provides a perspective of solving the problem and enlightening us on our creative thinking.

8.1 Mimic Disguise and Mimic Defense

8.1.1 Biological Mimicry

In 1998, Professor Mark Norman from the University of Melbourne discovered a mimic octopus, scientifically named as *Amphioctopus marginatus*, in the waters of Indonesia's Sulawesi Island, which is a natural master of mimicry. As Fig. 8.1a shows, when meeting a shark, the octopus will fold its tentacles to form an oval and swim slowly at the ocean bottom like a flatfish, because sharks dare not risk attacking a flatfish as the latter can eject a kind of venom to poison the former. When

© Springer Nature Switzerland AG 2020
J. Wu, *Cyberspace Mimic Defense*, Wireless Networks,
https://doi.org/10.1007/978-3-030-29844-9_8

(a)   (b)

(c)   (d)

Fig. 8.1 Mimicry in the biological universe. (**a**) *Amphioctopus marginatus*. (**b**) The disguise of a walking stick. (**c**) The disguise of a stealth frog. (**d**) A plaid hawkfish hiding in a red coral thicket

swimming in open water, it will unfold its tentacles evenly, looking like a spinous and poisonous lionfish in order to frighten its enemies away. Another trick played by a mimic octopus is that it will put six tentacles into a cave and then stretch out the rest two tentacles, mimicking a poisonous *Laticauda colubrina*. In addition, it can also mimic a reef or a hunting fish. By color-changing techniques, its mimicry can be flawless. Research shows that a mimic octopus can not only actively change its body color and texture but also mimic the shapes and behaviors of other creature, for instance, it can hide perfectly in gravel seafloor and coral reefs, for it can at least mimic 15 kinds of marine organisms. Mimicking similarly structured references, it creates cognitive difficulties or mistakes for the predators to acquire the survival advantage and security assurance.

In the long process of evolution and variation, many creatures have earned the name of "master of mimicry," such as walking stick shown in Fig. 8.1b, stealth frog in Fig. 8.1c, and plaid hawkfish in Fig. 8.1d. They can "hide themselves" by taking advantage of their own physical structure and biological characteristics, which is like military stealth. Such biological mimicry has been constantly evolving under the natural law of "the survival of the fittest" through a lengthy process that is almost as long as the life history. In biology, we call this "mimicry," a phenomenon commonly seen in the biosphere, which helps to effectively improve the defensive or attack ability of creatures. Biological mimicry can be categorized into Müllerian mimicry, Batesian mimicry, Wasmannian mimicry, and Poultonian mimicry. In Müllerian mimicry, the simulator and simulated target are both poisonous and inedible. Mimicry may help them survive better in the feeding period, which is a good thing for both species. In Batesian mimicry, the simulated target is poisonous and inedible, while the simulator is nonpoisonous and edible [1]. Wasmannian mimicry, in a broad sense, refers to the insects mimicking their living environment. For instance, *Kallima inachus* mimic leaves, and walking sticks mimic twigs. In a narrow sense, Wasmannian mimicry refers to parasitic insects mimicking their hosts. Poultonian mimicry refers to poisonous insects mimicking nonpoisonous creatures; it is a kind of aggressive mimicry to hide themselves and confuse their prey [2].

Endogeny and mimicry are the two cores of biomimesis. Endogeny means that biological mimicry relies on the physical appearance or functional behavior of a creature, rather than on other external tools or devices. Mimicry means that a living creature develops the appearances, colors, textures, or behavioral characteristics that are similar to the objects, enemies, hosts, or other creatures in its surroundings to confuse other creatures and survive better.

Biological mimicry includes static mimicry and dynamic mimicry. Static mimicry only takes effect in a particular environment and scenario. For instance, a *Kallima inachus* can hide itself very well in a broad-leaved forest, while it is exceptionally obvious in a coniferous forest. Static mimicry goes without any perception, cognition, decision, or execution. In contrast, dynamic mimicry does not have a fixed representation state; it takes on a random form with a simulated target dynamically chosen according to the environmental characteristics. The simulated target and behavior also vary with environments. In dynamic mimicry, creatures adapt better to the environment, such as a mimic octopus. They need to perceive the environmental changes, gather the characteristic information, and then determine the mimic behaviors. As an advanced form of mimicry, it includes a complete process of perception, cognition, decision, and execution.

8.1.2 Mimic Disguise

From the point of defense, we may refer to the endogenous mimicry of a living being as mimic disguise (MD). In MD, a creature can hide its appearance according to the surrounding environment, including concealing its inherent functions and nature. In essence, it aims to increase the apparent uncertainty to reduce the effect of an attack so as to create an advantage of self-security and survival.

The features of MD can be summed up as the following five aspects:

(1) MD is transparent to the entity or meta-function of a creature. In other words, a creature will not change its basic function due to diversified forms of mimicry. Its meta-function is unchangeable.
(2) MD is an integral part of the meta-function. The effect of mimicry does not depend on any affiliated devices or tools. It is endogenous.

(3) MD is directional. It refers to the mimicries of a particular target or environment.
(4) MD is identical with or similar to the mimicry of colors, textures, appearances, and behaviors of the simulated targets.
(5) MD is limited in that the targets or categories of mimicry are limited. It is related to the basic structure and surrounding environment of the disguiser.

As a natural association, to increase the apparent diversity of the software and hardware systems (e.g., computers, control units, networks, platforms, and modules) to survive and defend against the external attacks, it seems that it can draw upon the concept of biological MD, for instance, it simulates the defense effect of mimicry by using the honeypot net and other additional defense technologies based on environment cognition. We might as well call it narrow mimic defense (NMD). In fact, the effect of NMD is often restricted by cognitive ability of the disguiser and whether its disguise is true to life, and it is strongly related to the mode of deployment among the protected targets. Of course, the random and dynamic security technologies, such as encrypted authentication and moving target defense (MTD), can effectively defend against non-internal-external collaborative attacks without relying on threat cognition, but they technically do not belong to the mimic defense scope. The points need to be stressed are:

(1) MD helps conceal the exterior functions, performance, characteristics, and behaviors of an entity instinctively. However, the objects and service functions of most information processing systems on computers and in network spaces are not allowed to or cannot be hidden. For example, web servers, routers and switches, file storages, and data centers cannot be disguised or hidden. Instead, their functions and usage should be crystal clear to the users. Any additional defense measures should not affect or change the users' original operation habit.
(2) The operational environment of a computer, such as the execution process, resource use, and service status, is constantly changing, while the potential attack threats and security risks are unpredictable to the defenders. Under the circumstances of specific scenario concealment or ambient environmental perception, MD can barely defend against uncertain threats based on unknown vulnerabilities without any a priori knowledge.

Therefore, to simulate the mimicry function through the attached NMD technology without affecting the transparency of the functions and performance of the defense target, and to conceal the system architecture, operating mechanism, implementation methods, abnormal performance, and the possible known or unknown vulnerabilities, backdoors, or Trojans, it is necessary to address two issues: one is to perceive the uncertain threats and the other to generate appropriate defense scenarios, both of which, however, cannot be solved simultaneously by the additional NMD technology. This is because MD is unable and impossible to be done relying totally on external means; just like the magic mimic octopus, its endogenous, strongly related ambient perception and scenario-iteration mechanism are the key factors.

8.1.3 Two Basic Security Problems and Two Severe Challenges

Today, cyber security has been drawing wide attention from all walks of life, and all kinds of innovative cyber security technologies are emerging from time to time, but, on the contrary, the number of vulnerabilities and backdoors is also on the rise. In accordance with the *China Internet Cyber Security Report* [3] and *Review of China Internet Cyber Security Trends* [4] released by China's National Computer Network Emergency Response Team/Coordination Center (CNCERT/CC), the average annual increase rate of security vulnerabilities collected by China National Vulnerability Database (CNVD) has been 21.6% since 2013. Compared with 2016, the number of vulnerabilities collected in 2017 increased by 47.4% to 15,955, making a new historical record. The vulnerabilities in the information system are key elements inducing cybersecurity events. In fact, cyber security risks are much more serious than these open data. Those discovered ones are just a tip of the iceberg. Most vulnerabilities are unknown or hidden or created wittingly or unwittingly. So it is impossible to provide any data or assessment with any statistical meaning.

As mentioned in Chap. 1, vulnerabilities are design flaws that can be exploited maliciously by an attacker. In theory, we may find out and eliminate all the bugs in software and hardware design in line with a given standard by applying formal correctness proof techniques (FCPT)—or completely eradicate all vulnerabilities. Yet in engineering practice, it is almost impossible to build a complicated bug-free system. First, the correct perception of a given standard (particularly a clear and rigorous assumption) may change along with the technological development. Codes with no flaws in the past could become a vulnerability in the future. Second, it is difficult to develop a formal scientific inspection standard for a complicated system, as the standard itself is so intricate that it will be hard to ensure the correctness and completeness of its design. Third, formal verification tools for bugs or errors (including specific vulnerabilities, such as buffer overflow) are constrained by the code quantity, code complexity, and state explosion of the verified targets, so its integrity, completeness, and ergodicity are inevitably compromised. Fourth, whether a design flaw can become an exploitable vulnerability is related to the resource and experience owned by an attacker, as well as to its operating environment and resource configuration. Fifth, the development of science and technology can be divided into different phases. As a result, people's perception is often constrained within a certain time period. So we cannot and are unable to recognize those vulnerabilities which will be utilized in the future. Therefore, vulnerabilities are essentially an unavoidable problem to today's science and technology world, while backdoors are "malignant tumors" in the digital economic ecosystem under the division of labor among different countries and within an industry in the formation and development of a global value chain. Backdoors are more harmful to technologically underdeveloped countries and those relying on the international market. What's worse, there has been no other technical means and engineering measures to completely detect and eliminate vulnerabilities and backdoors other than emergent patching, fixing, and experience-based detection, let alone giving scientific diagnosis of "zero vulnerability" to the given software/hardware systems. All these reasons have ultimately led to insufficient security quality control in the design, production, supply, maintenance, and use of software and hardware and the inability to control or manage the cyber security problem from the industry chain sources. Consequently, the cyber security ecosystem has fallen into a vicious cycle.

Theoretically, the inevitability of vulnerabilities and backdoors makes it impossible to have "virus-free" information system and control components and parts. It brings two severe challenges to our informatization drive. First, the autonomous and controllable strategy cannot provide the necessary mature, stable, and high-performance components and parts on a timely, all-directional, and continuous basis over a long period, which are basic and supporting software/hardware products needed in informatization drive. Moreover, the autonomous controllable strategy is not able to fundamentally eliminate vulnerabilities and control backdoors. Second, to completely solve the problem of "security and credibility" of the design chain, tool chain, manufacturing chain, supply chain, and service chain, it not only requires major sci-tech breakthroughs but also involves a series of tough challenges in multilateral trade, ideology, national interest, and cyber economy. In fact, cyber security is caused, to a great extent, by the information technology and industry development mode. If people in the past were only compelled to overlook the security problem in a long period due to the fact that the hardware processing capacity had been a "rare high-end resource," then, as Moore's Law has been taking effect for more than half a century, it is totally unreasonable for the industry today to still exclude the security and credibility design of software and hardware from the functional and performance considerations. The author believes that the intention of "autonomous and controllable strategy" is to break through the technical and market monopoly in the pursuit of a constructive way of realizing competitive economic and technological development. The original purpose is to address the security problem of the product supply chain. Therefore, it cannot completely solve the problem of information or cyber security and should not become an excuse for trade protectionism and "de-globalization." Cyber security calls for a change of the information technology development conception and innovation of design philosophy. In an era of digital economy, we are expecting a "game-changing" technological reform and industry mode innovation.

With that purpose, it is necessary to take stock of the "interwoven and interdependent" industry and technology conditions, change the conventional security protection design philosophy in due course, and employ the innovative general robust control architecture and endogenous security mechanism to address the concern that the credibility of the software and hardware components and parts cannot ensure the security of the supply chain. In other words, by drawing upon the system engineering theories and methodologies, we must explore a new way to offset or reduce the impact of the components' dark features with the structural layer technology and carry out research on the innovative method to contain or undermine general uncertain interruptions (including uncertain security threats) via the structure technology and endogenous effect.

8.1.4 An Entry Point: The Vulnerability of an Attack Chain

So far, the research on cyber security defense has covered several topics, including static defense, dynamic defense, passive defense, active defense, and combined defense, and achieved notable technological progress. However, cyber security still has not seen a fundamental change due to the aforementioned two basic security issues and two severe challenges. Hence, it is an urgent mission to develop a new type of revolutionary, inclusive, open, and integrated defense technology system through technical innovations in information system or control device structure. The aim is to eliminate the attack threats based on the software and hardware code, to significantly reduce the "negative energy" injected to the cyber space in the age of globalization with an open industry chain, to greatly offset the effectiveness of cyber attacks based on "pre-embedded backdoors" and "hidden vulnerabilities", to substantially raise the threshold and cost of cyber attacks, and to completely change the status quo where some "computer geniuses," a couple of hackers, several cyber criminal organizations, or even government bodies can challenge the basic cyber and real-world order by taking advantage of some unknown vulnerabilities, backdoors, and other technical means.

In Chap. 2, we discussed that a cyber attack can usually be divided into several steps, including system scanning detection, feature recognition, correlation, coordination, vulnerability mining and attack, attack effect evaluation reporting, and information acquisition and propagating. It is known in the industry as an attack chain, as shown in Fig. 8.2.

Each link of the attack chain is often closely related to the implementation structure (algorithm), resource configuration, and operational mechanism. For two complete heterogeneous executors, even if their functions are identical or equivalent, their attack chains could be totally different. Therefore, in each step, an attacker must depend on a static, certain, and similar operating environment of its target in order to realize an effective and stable attack. For instance, they may use cyber scanning and digital fingerprint tools to obtain the target's IP address, platform type, version information, protocol port, and resource occupation and even to discover the vulnerabilities of the target system and flaws of the firewall rules. So, we could naturally conjure up a scenario: given that the target's cyber service functions and performance are secured, we may introduce a mimic disguise function or stealth based on an endogenous security mechanism to cause a cognition difficulty for the attacker, or mislead the attacker to continuously follow the target's deceptive defense actions, or disable it from perceiving and evaluating the results of an attack, disrupt its attack plan, and destabilize the attack chain. The dynamic heterogeneous redundancy (DHR) architecture has such a role and function, and its endogenous uncertainty effect can effectively hide the vulnerabilities of defensive actions and security, making it more difficult to form and maintain an attack chain.

Fig. 8.2 A typical attack chain

8.1.5   Build the Mimic Defense

If the nature is a real world built by the Creator, then the cyber world can be considered as a virtual world jointly created by experts in the computer and communication fields. Like the natural world where all kinds of uncertain threats exist, the cyber world also sees many known or unknown threats. When re-examining the methods used to handle uncertain threats in the natural world and real or potential threats in the cyber space, we may find some inspirational phenomena that will inspire us to address some tricky problems.

Active and passive defense actions, such as vulnerability mining, feature extraction, intrusion detection, intrusion prevention, and intrusion toleration, have long been trapped in a basic thinking pattern: only by obtaining the accurate a priori knowledge of the attacking methods and features can we realize effective defense, which means that we should get the feature information or even conduct big data analysis to defend against threats based on feature matching or model recognition technologies. Even so, people can only prevent the known or certain attacks to some extent. As to the massive unknown security risks or threats in the cyber space, there are no other effective solutions except strict but inhuman encryption and authentication. As the cyber attack and defense gaming rage on, the defenders see a quickly inflating attack feature database (e.g., vulnerabilities, backdoors, viruses, Trojans, blacklists, and white lists), with rocketing defensive resource cost and solution complexity at the same time. Moreover, the management and updating of the feature database not only relate to the defense effectiveness but also bring a huge burden for security maintenance. At the same time, new attack methods and modes are emerging, with almost no effective solution to the attack modes not included in the attack feature database. Neither typical active and passive defense nor various dynamic defensive measures have changed the status quo characterized by passive defense. Ex post remedy and threat warning are still the major technical strategies adopted. The passive defense mode is very similar to the specific immune system of vertebrates (see Chap. 9). As a specific immune antibody can only clear specific antigens, its diversity is highly related to the type of infected antigens. Functionally speaking, it is a kind of point defense. Unfortunately, there is no other effective extensive defense mechanism other than point defense measures based on perception and cognition. There are no other effective defensive means against unknown security threats except encrypted authentication with a less friendly user interface and relevant dynamic defense measures. More seriously, even additional security facilities (vulnerability scanning, virus scanning and removal, encrypted authentication, packets, and firewalls) themselves cannot escape the impact of vulnerabilities, backdoors, and Trojans, often leaving people helpless and overwhelmed.

As we know, there are many creatures in the natural environment with different forms, functions, and roles, including known or unknown harmful biological antigens. However, the reason why specific immune activities are infrequent in healthy organisms is that most invading antigens are resisted by the non-specific selective cleanup mechanism inherent in vertebrates. It is an ability of "identification of friend or foe" that "kills any antigens" without hurting the organism. Biologists call this congenital self-defense system "non-specific immunity" (see Chap. 9). It is an extensive defensive function of living organisms, providing a basis for non-specific immunity for constantly improving itself.

In recent years, new types of defensive systems have been emerging, with a range of theories and technologies represented by Moving Target Defense (MTD), Trusted Computing, and customized trusted cyberspace. The core idea is to fully utilize dynamic, random and diversified methods and non-public mechanisms to fix the systematic defects of current target systems (i.e., staticity, certainty, and similarity, which are discussed in Chaps. 3, 4 and 5). To create cognitive difficulties or taking unpredictable defensive actions against an attacker, you can basically build an uncertain target scenario in the cyberspace to acquire a broad defensive capacity that does not rely on the acquisition of a priori knowledge or behavior features of the attacker or significantly increase the difficulty and cost of an attack by narrowing the scope of attack.

The arduous development of cyberspace defense technologies shows that to genuinely change the current predicament of confronting massive security threats via extensive defense technology, the target system must be equipped with an inherent defensive mechanism similar to the non-specific and specific immunity. Only with such an integrated defensive system that combines both point and extensive defense actions can we effectively address unknown security risks or cyber threats. It is a truth that has been proven time and again by the biological evolution history.

In exploring the solutions to the abovementioned problems, the biological mimic disguise system can provide important implications. First, the endogenous mechanism of mimic disguise is to create cognitive difficulties for an attacker through the prior environmental (rather than threat) perception and iterative disguise of forms. As a kind of active defense, it can effectively change the inaction of passive defense before launching an attack. Second, the mechanism based on mimic disguise can integrate both the intrinsic and security defense functions. Having changed the practice that a security defense function should be realized through conventional resource configuration or deployment methods, such as addition, plug-in, parasitism, or symbiosis, NMD—under the intensive structural effect—can lower the operational cost of the target's cyber security defense and information security assurance functions throughout its entire life cycle. NMD, therefore, is an optimal choice to draw on the MD mechanism and its endogenous structural effect to defend against or prevent certain and uncertain cyberspace threats and risks. Nonetheless, the first move is to find an appropriate physical or logical architecture (algorithm) in order to produce the desired "endogenous security" effect. NMD, as mentioned in Sect. 8.1.2, still hinges on prior knowledge, with a necessity for pre-threat perception of the environment to determine the mimic change. A purpose of such a change is to hide the target's exterior form, including its intrinsic functions, as much as possible. However, the target's service functions often can't be hidden, while the "one-way transparent and colluded" attacks launched by a hacker are usually not intended for apparent destruction but to obtain, tamper, manipulate, and even "lock" the target's sensitive information (e.g., ransomware). It indicates that cyber attacks (rather than the countermeasures or deterrence) are somehow hidden or concealed, with unknown or imperceptible results. Therefore, NMD needs either to depend on the a priori knowledge of threat perception and attack feature extraction or blindly increase the target's resource cost to implement inefficient dynamic or random changes (by reinforcing the randomization and dynamization of commands, addresses, data, and ports as in MTD).

It has been discussed in Chap. 7 that by utilizing DHR (dynamic heterogeneous redundancy) under the condition of functional equivalence, policy and schedule based on multimode ruling, and multi-dimensional dynamic reconfiguration negative feedback mechanism, we may perceive or discover the differential mode failures caused by random internal unknown bugs of the target system or non-coordinated attacks based on dark features. Without knowing when and where the failure or attack will take place, we can count on the endogenous defensive function of DHR to normalize the "one-way transparent and colluded" uncertain risks into predictable and controllable events at the system level. As this general robust control structure can take advantage of the MD mechanism and integrate all kinds of existing security technologies to create a defensive environment with nonlinear complexity or uncertain visibility, we may build in the cyberspace a general integrated mimic defense system under the condition of functional equivalence that does not rely on the prior knowledge and behavior features of an attacker. We call it cyberspace mimic defense (CMD). The term MD referred to in the following chapters generally means CMD, unless otherwise stated.

Conceptually, CMD is based on the cognitive fact that uncertain attack is the biggest threat to a network, and its defense is grounded in one axiom and two basic principles. First, with the logic expression of TRA in the functionally equivalent heterogeneous redundant scenario, non-coordinated uncertain attacks can be measured, perceived, and converted into reliability problems equivalent to DM failures and be handled collectively. Second, CMD will have an endogenous uncertain effect by drawing on the biological MD system and importing multiple ruling and scheduling policy so that an attacker may find it difficult to clearly recognize the target as he becomes "blind." Third, the DHR architecture can throw an attacker into the predicament of "coordinated attacks on multiple dynamic targets under non-coordinated conditions," thanks to the policy and schedule based on multi-mode ruling and the multi-dimensional dynamic reconfiguration negative feedback iterative control mechanism. The three factors combined will form a mimic defense system based on the DHR architecture. Its anti-attacking performance and reliability are quantifiably programmable and verifiably measurable and can support the general robust application scenarios featuring high reliability, high usability, and high credibility. Figuratively, if we compare a DHR architecture to a six-degree-of-freedom "magic cube," then the MD system is the "player" of the cube, while the attacker can only be an "audience" of this fabulous magic cube show.

Obviously, the introduction of an MD system to the DHR architecture is the core of building a general robust control function (including security defense) in the target. CMD has brought about a new defense mode characterized by "multi-dimensional heterogeneous perception → multimode strategic ruling → defense scenario rebuilding → iterative convergent problem avoidance" as an extension of the classic passive defense mode of "detection and perception → judgment and decision → removal of problem" and of the active defense mode represented by MTD. Unlike the accurate threat perception whose effectiveness is regarded as the precondition for traditional defense, CMD's effectiveness does not rely on acquisition of a priori knowledge or behavioral features, nor does it have to implement a "blind action" mechanism to the commands, addresses, ports, and data as MTD does. The service scenario of a DHR architecture only "migrates dynamically in the form of iterative convergence according to the feedback control loop driven by the decision status or execute mandatory changes through the external control commands amid temporal, frequency, and strategic uncertainties." The goal is to increase "the difficulty of coordinated attacks on multiple dynamic targets under non-cooperative conditions" via diversified defense tactics. Unlike the use of dynamic, diversified, and random defense elements, the general robust control mechanism of CMD can not only instruct an "abnormal or suspiciously problematic" executor to change its defensive environment but also assess and determine in real time the effectiveness of migration or switching in the current defensive scenario. Moreover, it can autonomously decide whether it is necessary to alter iteratively the running environment and turn the attack escape rate quantifiably programmable and verifiably measurable at the structural level. As the DHR control loop is weakly related to the functional complexity of an executor, we can add a restructuring function or traditional security technologies, such as error detection and recovery, intrusion detection and prevention, and encrypted authentication in the executor, so as to increase the dissimilarity between the executors and then greatly improve the ability of the target system to perceive unknown threats or malfunctions. Unlike the traditional deployment mode made possible by separating the highly reliable or available service provision from the highly credible service guarantee, CMD can enable the target system to provide highly reliable, available, and credible robust services. To sum up, it is well expected that, exactly as the stealth technology is introduced into the military area, the adoption of CMD in the cyberspace will pose a fundamental challenge toward the current attack theories and methods based on vulnerabilities and backdoors, thoroughly revolutionize the traditional defense philosophy, reverse the current trend of easy attack and hard defense, offset the "inter-generation gap" in the attack-defense gaming technology, and rebalance the cyberspace security landscape.

8.1.6   Original Meaning of Mimic Defense

According to the laws of the universe, "change is absolute and constancy is relative," and "right and wrong, known and unknown are all relative." Only the "uncertain defense for attackers can be used to guard against the uncertain threats for defenders," and the "TRA can be used to convert uncertainties into certain and controllable events." Therefore, we must fundamentally transform the cyberspace defense philosophy. Rather than pursuing a vulnerability-free, backdoor-free, bug-free operating scenario, or defensive environment against various cyberspace security threats, we should adopt a dynamic control strategy of iterative convergence to develop the policy and schedule based on multimode ruling and to build a multi-dimensional, dynamic reconfiguration and self-adaptive feedback mechanism. Based on a simplified structure, we must integrate all the traditional defense theories and techniques, including static, dynamic, passive, and active defense modes, while combining the basic dynamic, diversified, and random security elements for an inherent effect. Through these efforts, we will be able to normalize the defense against uncertain threats into a target which can be handled by reliability theories and technologies. In this way, we can create a system which is manageable and controllable to the defenders but unpredictable to the attackers. Under invisible defense, an attacker may find it "hard to detect vulnerabilities, to penetrate the system, to actuate attacks, to exploit the attack outcome, and to sustain an attack." Meanwhile, we can make use of the well-established reliability verification theories and techniques such as white-box instrumentation verification in injection testing, to carry out measurable tests and verification of the security level of the target system.

   In addition, we can introduce the MD mechanism based on the DHR architecture to enhance the subtlety of the attack-defense game. Drive the iterative convergent dynamic mechanism to obtain the uncertainty effect of the target's defensive scenario by using the negative feedback control strategy based on multimode ruling; provide the point-surface-integrated robust control function for general uncertain disturbances, such as certain or uncertain threats and random failures; use the non-compliant state of multimode ruling to trigger the policy and schedule and multi-dimensional dynamic reconfiguration, and normalize the problems that are complicated and unable to be solved in real time into simple ones to avoid being handled, so as to economically and effectively utilize the limited software and hardware resources to enlarge the entropy space in the defensive scenario. What's more, the systematic effect based on MD can bring the rate of superposition state error rate included in the multimode output vector ruling state below the given threshold value. It can make full use of the endogenous security function of the mimic structure to largely reduce the scale of security facility deployment and updating frequency and dramatically lowering the threshold and cost of use and maintenance of the information system and control device in their full life cycles.

The dissimilar redundant architecture (DRA) not only brings high reliability and high availability to the target system but increases the design complexity, equipment cost, energy loss, and maintenance burden. Likewise, the creation of an MD architecture and mechanism is also costly. For instance, while being able to generate an uncertainty impasse for the attackers, the "replacement and migration" of heterogeneous executors or their restructuring and reorganizing functions will add to the complication of the system design, engineering implementation, application deployment, upgrading, and maintenance. Given the globalized and open supply chain, in particular, many technical challenges may arise, including the testing and evaluation of the heterogeneous executors' function and performance equivalence, the rapid clearing and restoration of the executors, the migration and synchronization of the defense scenarios, restructuring and reorganizing, normalization of the multimode vector output, and the complexity and timeliness of strategic ruling. Although these costs are affordable for public (shared) service platforms or high-value control facilities, such as data centers, cloud computing/service platform, and information infrastructure (since there has been no other technology with better cost performance than MD), the cost of heterogeneous redundant resource configuration or deployment is indeed a problem which needs to be taken into serious consideration in engineering practice. However, if MD only plays a role in defending the key part of the target system, the cost increase will actually not be a major barrier to the system application and promotion. Especially when the software and hardware components or modules of the defense architecture are mainly COTS-level products, we can genuinely benefit from the cost-effectiveness brought about by Moore's Law, as well as by the IT industry that pursues a standardized, massive, and diversified market. Even so, we still have to carry out an in-depth analysis and research on how to identify the problematic scenario, choose the fortified links, and take into consideration multiple other factors (e.g., design complexity, ease of maintenance, system volume, energy consumption, and comprehensive cost). Try to avoid dogmatism and formalism in solving application problems concerning in specific industries and areas, but creative thinking is often indispensable.

8.2   Mimic Computing and Endogenous Security

8.2.1   The Plight of HPC Power Consumption

As an important branch of computer science and engineering, high-performance computing (HPC) mainly refers to the research and development of high-performance computer technology in terms of system architecture, parallel algorithm, and software development [5]. HPC can be applied in a number of fields, such as nuclear weapon research, nuclear material storage and simulation, oil exploration, biological information technology, medical care and new medicine research, computational chemistry, meteorology, weather and disaster forecast, industrial process improvement, and environmental protection. It has become an important tool to push forward sci-tech innovation and social progress.

The latest global HPC TOP500 list was released during the International Supercomputing Conference 2017 (ISC 2017) held in Frankfurt, Germany, on June 19, 2017. China's Sunway TaihuLight, a supercomputer developed by China National Parallel Computer Engineering Technology Research Center and deployed at the National Supercomputing Center in Wuxi, maintained the crowned champion again for 3 consecutive years [6]. It consists of 40 cabinets and 160 supernodes. In each node, there are 256 computational nodes, and each computational node is equipped with a 1.45 GHz, 260-core Sunway 26010 processor. The total peak performance of the system can reach 125.4359PFLOPS, and the Linpack measured performance 93.0146PFLOPS, 2.75 times that of Tianhe-2—the world's second fastest computing system [7]. The computational capability of HPC has reached the 100P level.

In June 2018, the Oak Ridge National Laboratory (ORNL) under the United States Department of Energy (USDOE) announced the successful development of HPC "Summit." With a peak performance of floating-point operations hitting 200,000 trillion, Summit, the IBM-manufactured HPC, was 60% faster than Sunway TaihuLight, replacing the former champion that held the first place for 5 years in a row as.

However, an unavoidable fact about HPC is that it is highly power-consuming, with high operational cost and a series of other problems, including thermal dissipation, reliability, maintainability, and usability. As the operational speed of HPC increases from level E to level Z, the problem of power consumption has become one of the crucial barriers to its further development. Though with multilayered multidirectional consumption reduction techniques, the operational power consumption of P-level HPCs can still reach as high as 17 MW.

Among today's top HPCs, Tianhe-2 is the largest power consumer, with its overall power consumption reaching 17.81 MW (excluding air conditioners and other supporting facilities); Piz Daint of Switzerland is the least power-consuming HPC, with the overall power consumption of 2.272 MW [8]. Professor Haus Femny from the US Lawrence Berkeley National Laboratory believes that the development of supercomputation technology is still inadequate to break through the E-level barrier.

The 1000P-level supercomputation, which was thought to be achieved in 2018, now stands little chance to be realized in 2020 or even 2022. The foremost challenge lies in the difficulty to control the power consumption. The current IC technique has not seen any substantial improvement compared to its predecessor. It is estimated that, based on the current computer architecture and realization technique, the power consumption of E-level supercomputers in the future may reach hundreds of MWs at an engineering cost of hundreds of millions of dollars. Besides that, the economic efficiency would also be an important issue. For instance, the annual electric bill of Tianhe-2 can reach 100 million yuan, and even rise to 150 million yuan if it runs at full speed. Therefore, it has become one of the most noteworthy problem to lower the power consumption of unit computing performance and improve the computational efficiency in the development of HPC.

8.2.2 Original Purpose of Mimic Calculation

The practice of computer application shows that under a fixed hardware architecture, the improvement of software algorithm can help raise the processing speed and computational efficiency to some degree. Meanwhile, common sense suggests that there are usually several different solutions to one problem. In different stages of a solution, there are multiple realization algorithms to choose from. With varied characteristics and attributes, each solution has starkly different requirements on the computational environment and cost, delivering hugely distinctive computational performance and operational efficiency. If we can choose the proper realization solution or algorithm at the right occasion and time, we may get close to the optimal value of computational efficiency under the constraint conditions.

Supported by current technique and acceptable cost conditions, software and hardware systems can also be intelligently generated through well-designed executive variations or computational plans with equivalent functions and different efficiencies. In actual operations, the system may choose a proper plan at the proper occasion and time to automatically generate a proper computing or executive environment in line with the application needs and achieve a higher energy efficiency ratio via the dynamic variable structure collaborative computing based on active cognition. A range of technologies under development can support this idea, including user-customized computing, reconfigurable computing, CPU + FPGA fusion architecture, software-defined hardware (SDH), domain-specific architecture collaborative computing (DSAs) or joint inter-domain management.

Under the condition of functional equivalence, the multi-dimensional dynamic reconfigurable functional architecture which contains an array of software and hardware variations or basic modules is called "mimic structure calculation" (MSC) or mimic calculation in short. It can choose or generate an appropriate computing environment with equivalent functions, in line with the cognitive information of the current operational status. In MSC, we can utilize a variety of hardware and software variations with equivalent functions and different calculating efficiencies to solve calculable questions. In the time-space dimension, we may choose the proper executive variation or module based on the intelligent decision made according to the energy target and environmental cognition and utilize traditional calculation based on instruction stream, pipelining processing mode based on data stream and control flow, and even neural network acceleration or domain-specific software and hardware co-computing mode to raise the system computational efficiency (compared to purely physical or software methods) to the greatest extent. This is the original intention of mimic calculation.

What needs to emphasize is that, different from the user programmable or customizable heterogeneous accelerated computing, such as the combinations of CPU + GPU, CPU + FPGA, GPU + OTP, or CPU + GPU + FPGA, the MSC is adaptive variable structure computing which is all about the fitting of best energy efficiency. Under the condition of equivalent function, the design and deployment of software and hardware algorithm and calculating mode should follow the principle of highest energy efficiency ratio.

8.2.3   Vision of Mimic Calculation

MSC does not attempt to build any high-efficiency computing system independently, nor does it exclude the energy conservation, consumption reduction, and efficiency increase realized through the progress of device processing and introduction of material technology or resource management. Moreover, it does not reject the benefit of software algorithm optimization and computing efficiency improvement. Abiding by the axiom that "application determines the structure and structure determines the energy efficiency," MSC intends to optimize the system operational energy efficiency through the computation structure, executive environment, or calculating mode formed by the multi-dimensional dynamic reconfiguration system based on cognitive decision.

Taking advantage of the multi-dimensional dynamic reconfigurative functional architecture, we may choose or generate relatively ideal calculating environment and mode to solve application problems, including software-defined, reconfigurable, and reorganizable functional components, such as computation and control, storage and cache, exchange and interconnection, as well as input and output; we can also co-manage and cooperatively schedule the tasks or operations among various hardware executors (or environments) and the related diversified software variations at different time, under different resource conditions, for different service quality requirements, and with different processing loads and operational efficiencies so as to achieve the trans-environmental dynamic migration of such tasks or operations, including the combined use of instruction stream and control stream.

Yet the structural change or reorganization of a hardware system is subject to the constraint of device technology, design complexity, cost of structural changes, and price factors, so any random or arbitrary change is impossible. Biological mimicry could be a good idea to describe structural dynamic changes to a limited degree and on a limited scale: on the one hand, it requires the system to have a "similar" structure in order that "the structure matches the applications"; on the other, it also calls for an ability in the system to take identical "actions" so as to realize the notion that "the structure changes dynamically." The similarity, sameness, and limitation of mimicry best depict the essence that a functional architecture based on multi-dimensional dynamic reconfiguration (or mimic transformation) can achieve high-efficiency calculation in specific domains. In addition, the multi-variation executive environment with equivalent functions boasts an advantage of natural redundancy and innate high reliability and security.

The multi-dimensional dynamic variable structure collaborative computing under the condition of functional equivalence is a proven way of implementing the "structural adaptive application." It has gone through three major stages of evolution: (1) variable structure or heterogeneous computing (CPU + XPU + FPGA); (2) variable mode computing, i.e., dynamic switching or migration between multiple given computing modes; and (3) mimic environment computing, i.e., a meta-structure pool intelligently generating or changing the necessary computing environment according to a group of given set of structural schemes, as shown in Fig. 8.3. MSC normally contains N meta-structures consisting of computing components with different computing performance and equivalent functions, an intelligent scheduler which can choose a proper computing component from the meta-structures by implementing the operational environment parameter strategy,

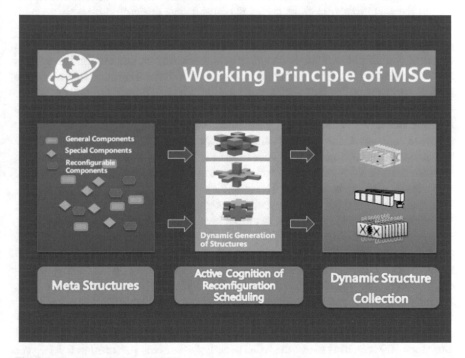

Fig. 8.3 Working principle of MSC

and a real-time reconfigurable computing structure (a physical computing environment generated dynamically according to the templates) in terms of the structure granularity, structure layer, and time dimension. Obviously, MSC can realize both high performance and efficiency through the dynamic variable structure and software and hardware collaborative computing under the condition of functional equivalence.

What's worth noting is that, as the effect of Moore's Law and Dennard Scaling is easing and even stagnant, the core performance increase of uni-processor has been reduced to about 3% annually. Therefore, the development of application-specific IC/architecture and special programming language is of great significance for improving the computing performance and efficiency in specific domains and accelerating system development. Moreover, the outbreak of "Spectre" and "Meltdown" vulnerabilities based on the computer system structure design flaws in 2018 fully demonstrated that the security problem of system structures has not been taken into serious consideration for decades by researchers and designers.

Figure 8.4a–c shows a typical mimic reconfigurable structure, including the processor mimic reconfiguration, memory-level mimic reconfiguration, and connected network-level mimic reconfiguration. To address the distinctive processing needs of different applications, the basic processing operators (possibly application specific) will become processing components with variable granularity, quantity, and structure through mimic reconfiguration; the basic memory unit will become memory modules with variable capacity, structure, address, and access patterns which can best fit the specific application storage needs and even structured memory components with processing functions (e.g., stacks, arrays, and pointer lists); the basic network unit will become the connection components with variable interconnection topology, variable interconnection agreement, variable interconnection bandwidth, and processable transmission content through mimic reconfiguration to address the need of specific applications for interconnection. The key points to realize MSC are shown in Fig. 8.5. Counting on limited processing resource, memory resource, and interconnection resource, we can coordinate the processing scenarios for specific software and hardware through mimic reconfiguration. While improving the resource utilization ratio, we can reduce the total amount of resources needed and increase the efficiency of system resource management and collaborative operation.

In 2008, the "Research and Development of New Concept High-efficiency Computer Architecture and System" was included in the State High-Tech Development Plan (863 program) as one of the major science and technology projects co-implemented by a joint group consisting of members from NDSC, Fudan University, Shanghai Jiao Tong University, Tongji University, and No.32 Research Institute of China Electronic Technology Group Corporation (CETC). The development of prototype was started in 2010. In May 2013, as commissioned by the State Ministry of Science and Technology, the Linpack testing team of Tianhe-2 selected 3 typical applications (i.e., output/input intensive Web service, computation-intensive N-body problem, and storage-intensive image processing problem) with more than 500 scenarios to test the MSC prototype. The best performing IBM server in that year was used as a reference. The test result showed that the efficiency of

Fig. 8.4 Typical mimic reconfiguration. (**a**) Processer-level mimic reconfiguration. (**b**) Memory-level mimic reconfiguration. (**c**) Connected network-level mimic reconfiguration

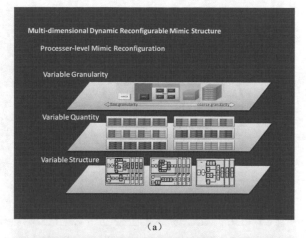

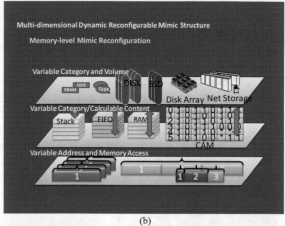

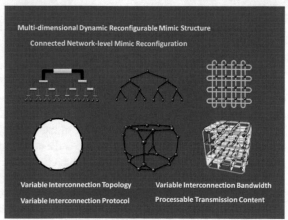

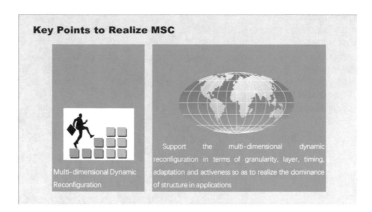

Fig. 8.5 MSC based on multi-dimensional dynamic reconfiguration

MSC could be 13.6–315 times higher and that, in given computing tasks, the processing efficiency could be greatly improved through dynamic generation or intelligent transformation of computation structures (mode and environment) and executors in spite of different processed resources, service quality, running time, and execution phases—serving as a vigorous illustration of the axiom "structure determines efficiency" in engineering practice. The mimic calculation structure, as an innovation application-specific computing structure that integrates instruction stream with control stream and combines parallel data processing with parallel pipeline processing, has opened a new trail of computer system research and development. In the short term, the application-specific computational structure and software-hardware collaborative design will be the new way out of the predicament of Moore's Law and Dennard Scaling.

8.2.4   Variable Structure Calculation and Endogenous Security

We know that the dynamic variable structure mimic calculation based on active cognition can dynamically form a calculating environment that matches with different conditions or parameters of tasks, time periods, workloads, efficiency requirements, and resource utilization. Though its original purpose is to improve the processing efficiency of the system, it can cause irregular changes to the operating environment and disrupt the staticity, predictability, and similarity of the target's operating environment on which the attack chain relies. As a result, the attack chain will lose the stable condition or environment it needs. Under such reinforced defense, the complexity and cost of an attack will be consequently increased.

From the perspective of security defense, the mimic calculation system is a typical device of "time-varying computation structure" and is both dynamic and random. In the eyes of an attacker, the mimic calculation system is actively transforming or rapidly switching between diversified and dynamic computing environments on an irregular basis. Its computing environment is hard to observe, operational mechanism hard to predict, and software and hardware code vulnerabilities hard to be discovered. Consequently, it will become more difficult to establish, utilize, and maintain an attack chain based on the target's dark features. So we can easily infer that all active variable structure calculation environments, such as heterogeneous computing, variable mode computing, and mimic computing, should be visually uncertain and dynamically defensive to different degrees.

Fortunately, the negative impact caused by computer system structure design flaws in recent years, such as Spectre and Meltdown, has finally made the industry see the importance of system structure security, as a result of suffering from the consequence of its indifference to the design of a secure system structure and paying too much attention to system performance. We have neither established a measurable security index nor realized the joint optimization of system performance, efficiency, and security. Hence, the development of secure computer system structures has become highly urgent. As early as in 2012, the author put forward the idea of endogenous security that can be realized by variable structure mimic computing system architecture and is far better than the traditional defense technology. In 2016, the so-called mimic defense verification system passed the test and evaluation organized by the Ministry of Science and Technology (MOST). The result showed that the "verification system completely meets the theoretical function and performance expectation," proving the feasibility of the highly creative forward-looking idea presented 4 years earlier.

8.3 Vision of Mimic Defense

Without any overarching security goals or intention of ubiquitous application, CMD simply tries to address the unpredictable threats of vulnerability-based dark features (not including those for lack of diversification or heterogeneous redundancy). It cannot solve the problems related to "protocol vulnerabilities and backdoors," "anti-spyware," "brute-force attacks or cryptographic algorithm interpretation," and user compliance auditing. CMD only looks to develop a universally applicable robust control structure and mechanism, which can provide highly reliable, usable, and credible robust service and control function for the software and hardware components, equipment, systems, or control devices and also form a defensive mechanism with its endogenous structure that does not depend on (but not rejecting) conventional security tools and can naturally accept current or future security technology to acquire "super nonlinear" effect of active defense. Through a security-measurable system structure, it expects to enable the target to be endogenously uncertain or invisible, so as to greatly increase the difficulty in establishing and exploiting the attack chain, significantly lower the reliability of the attack chain, and immensely reduce the usability of "0day" vulnerabilities or hidden backdoors. CMD aims to crack the universal problem of unassured credibility of software and hardware component supply chain through a reform of the system-building technology and to break through the predicament of a general lack of robust control functions among the software and hardware products; therefore, like its reliability, the credibility of an information system can be designed, scaled, verified, and measured.

Mimic defense expects to create a theoretical system that can integrate static defense with dynamic defense, as well as passive defense with active defense. In engineering practice, it intends to generalize the problem of uncertain threats into the technical structure of typical reliability problems. Specifically, the goals of mimic defense include:

(1) Use the TRA logic expression to establish an IFF (identification of friend or foe) mechanism that does not rely on the prior knowledge or behavioral features of an attacker.
(2) Use the inherent invisibility of a dynamic heterogeneous redundancy structure to change the staticity, similarity, and certainty of the defense environment.
(3) Create the "difficulty in coordinated attacks on dynamic diversified targets under non-cooperative conditions" through the given vector space strategy decision mechanism to achieve the surface defense function.
(4) Use the policy and schedule based on output vector ruling and the multi-dimensional dynamic reconfigurable adaptive feedback mechanism to provide point defense function.
(5) Create a scenario of uncertainty through point-surface integrated active defense to prevent or deny any trial-and-error attacks targeted on the dark features of a system's executor.
(6) Use the simple general robust control structure and mechanism to endow the target system with triple traits of high reliability, high credibility, and high availability.

8.3.1 Reversing the Easy-to-Attack and Hard-to-Defend Status

Mimic defense should be a game-changing technology with the following advantages:

(1) It enables the target defense scenario to benefit from the effect of equivalent uncertainty or invisibility:

    (a) Under the condition of functional equivalence, it can change the uncertainty of the diversified structure or algorithm to decouple the target system structure as much as possible from its external functional characteristics, so that an attacker may find it hard to discover or exploit the potential vulnerabilities and backdoors through the mapping relations between the target system's I [P] O functions and its structure or algorithms. It also makes it difficult to detect or predict the defense actions and environments.

    (b) Without relying on the prior knowledge of an attacker or his behavioral features, it can endow the target system with a reliable ability of IFF and an endogenous ability to provide point-surface defense to tolerate unknown attacks based on the vulnerabilities and backdoors and disable an attacker to perceive or evaluate the result of attack under the system's endogenous invisibility effect.

(c) With the attack surface being hard to perceive, it can automatically control the differential mode failures or non-collaborative attacks through a general robust control structure that integrates service provision with security defense to naturally address the impact of nonpermanent common mode failures or non-paralyzing collaborative attacks.

(d) It can take advantage of the invisibility function of a DHR structure at the component, equipment, node, system, platform, and even network level and achieve an exponential increase in the defensive effect with the cascading deployment of DHR.

(e) It accepts the results of current or future information and security technological progress at the structure and mechanism level and notably enhances the inherent uncertainty effect.

(2) It can reverse the pattern of "easy-to-attack and hard-to-defend" in the cyberspace:

(a) In an interdependent eco-environment and under the globalized open-source innovative mode, it can address the conundrum of building a safe and credible system based on COTS software and hardware products through the innovation of system architecture and mechanism and reduce or neutralize security threats at the component level by independent controllable system building technology.

(b) Given the vulnerability of an attack chain and the complexity of the attacking steps, it increases the difficulty of attacks nonlinearly in major stages of an attack, greatly reducing the constructability of attack chains as well as the reliability and effectiveness in exploiting the dark features.

(c) It can force an attacker to face the "difficulty in coordinated attacks on multiple dynamic targets under non-cooperative conditions" and decrease exponentially the reliability of attacks and the usability of attack results under the iterative effect of an endogenous security mechanism.

(d) It can enhance the dynamism, diversity, and randomness of the basic defensive elements through the architecture technology to eliminate the inheritability of the attack experience, to turn the result of an attack into an uncertain event, and to decouple the security of the structural layer from that of the component layer.

(e) It contain or squeezes the living space of the gray or underground industry chains for vulnerabilities, backdoors, and viruses, greatly offsetting the "negative impact" of individual hackers, illegal organizations, and even government bodies in the age of cyber economy and vastly increasing the cost of an attack.

(f) It fundamentally rocks the theories and methods of attacks based on the target software and hardware code flaws and overturn the asymmetric attacking strategy of "hiding vulnerabilities and reserving backdoors" implemented by technical pioneers or market leaders in the cyberspace to achieve the rebalance of cyber security.

8.3.2   A Universal Structure and Mechanism

It is expected that the "privacy, completeness, and effectiveness" of an information system or control device no longer depend on the autonomous controllability, security, and credibility of individual components, devices, and modules nor depend on the design, production, operation, and management of their individual software and hardware. In other words, the target system shall be able to adapt to the open ecological environment where the software and hardware components cannot "ensure the credibility" of their industrial chain. Given the "beach" with no security guarantee, or an "ecosystem with no thorough enclosure," it is advisable to design and build manageable, controllable, and credible information systems or control devices, fundamentally eliminating opposition and contradiction between "security and openness," "advancement and credibility," as well as "independent controllability and globalization" in today's cyberspace.

What's more, we should create a general robust control structure and mechanism that combines "service provision, assured reliability, security, and credibility," which can adapt to the defensive needs in application-specific or most general-purpose systems. To this end, we can incorporate or integrate the reliability, security, and credibility into the hardware and software through the structural endogenous mechanism during the design period, without demanding that the software and hardware be free from viruses. Without relying on (but not rejecting) conventional security means, the target system can suppress the threats against its dark features solely through its endogenous security effect. This structure can naturally integrate the advantages of current defensive technologies and security tools to form a syncretic defense system with both congenital and acquired immunities. It can not only render accurate point defense against threats with clear features but also provide surface defense against uncertain threats in unknown forms.

8.3.3   Separation of Robust Control and Service Functions

The robust control and service function are expected to be separated in the target system, for, if the two are separable, the control structure will be able to use third-party COTS products or custom-made service components, benefiting from exploiting the global technologies and supporting capabilities in the open source environment to lower the cost of design, engineering implementation, use, and maintenance. On the other hand, the separation attribute enables the system to effectively control the impact of problematic codes in third-party products whose credibility cannot be guaranteed, specifically to block contamination of or penetration into the robust control components and the host environment. In addition, the upgrading and revision of the system service function should not influence or have less influence on the functions of the robust control structure, except when the standards and protocol on the defense interface have changed.

8.3.4   Unknown Threat Perception

Today, threat perception or behavior prediction based on a priori knowledge has developed into diversified intrusion detection and trusted computing technology, including intelligent honeypots and honeynets, trusted behavior and characteristic state auditing, and even new ways of active threat detection and feature extraction, such as deep learning (DPL) based on historical big data. However, most attacks based on unknown vulnerabilities and backdoors (e.g., 0day and Nday vulnerabilities) often mingle with normal service processes or utilize the normal operation and interconnection relations to launch attacks (e.g., DDOS attack). So in high-speed links or an environment of massive information, it is quite a challenge to analyze or discover malicious acts in the real time and keep the false alarm rate and missed detection rate at a low level (although the analytic tools based on cloud or big data have already become powerful enough). The core issue is still the difficulty in distinguishing legal operations from illegal ones "under the mechanism of shared resource in a single processing space." Especially when the control of CPU, operating system, database, virtual management, and other layered supporting environments is lost, an attacker may build an accurate collaborative relation which is "one-way transparent." By uploading attack packets, the attacker can constantly improve the intelligent attacking ability and even put such an attacking mechanism under self-directed learning. Meanwhile, in these asymmetrical games, the threat detection device should keep monitoring the system in an uncoordinated way without affecting the normal service function and performance of the target system. In short, the additional security technology based on the traditional architecture or operating mechanism is correlated to the environmental factors. It can neither prove its own effectiveness independently nor have it proven by other evidence.

However, in line with TRA and its logical expression mechanism, a heterogeneous redundancy architecture under the condition of functional equivalence can detect unknown threats or identify "friend or foe" under certain credibility conditions, without relying on the traditional detection method and a priori knowledge of the attackers. In other words, as the heterogeneous redundancy mechanism itself does not distinguish the known threats from the unknown ones, or pays much attention to the specific attack features and behavior information, it only executes multimode ruling based on the semantics and grammar in the given output vector space: "total consistent output" is regarded as zero threat; "mostly consistent output" is regarded as avoidable threats; and "no consistent output" is regarded as severe threats. In fact, the first two scenarios could a little likely be misjudged, because when heterogeneity and redundancy are not high enough, the judgment of "total consistent output" and "mostly consistent output" could possibly be the effect of common mode failures or coordinated attacks. As for the third scenario, there needs to be more reference information and more complicated ruling strategies to make the judgment.

Therefore, we can easily assume that if the TRA-based heterogeneous redundancy expression mechanism can be embedded into a threat detection device, we will likely design a device that no longer relies on the knowledge of behavioral rationality. In fact, the rationality of behaviors is often hard to be defined. For a sophisticated system, it could even be difficult to prove the correctness of its formalization due to state explosion. Hence, when the detection device and target system are completely separated (even under serial or parallel connection), it will be hard for the current threat detection mechanism based on the rationality expectation or state analysis to grasp the relations between the false alarm rate and missed detection rate.

8.3.5   A Diversified Eco-environment

The extension of mimic defense depends to a great extent on the maturity of the diversified market of software and hardware components, particularly that of standard COTS products, such as user customizable CPU (e.g., open structure RISC-V), software-defined hardware (SDH), mimic calculation, diversified OS/DB, FPGA that supports data stream processing, database, cross-platform apps, standard virtual software, diversified function libraries, middleware or embedded systems, software and hardware compliers, gene mapping analytical tools, and various integrated functional designing platforms. Nonetheless, it often requires large-scale application or utilization before we can optimize a product or technology. To achieve market maturity, we have to overcome the challenge of current business rule of "winner takes all" and "zero-sum game." Without increased market demand, it will be extremely difficult to change the current business environment under the mode of "crowd innovation for win-win results." That is to say, mimic defense needs diversified software and hardware components to create new market demands for diversified products with equivalent functions. In particular, the permission to use the virus-carrying executors makes it possible to use the products which are less mature and underperformed but more or less able to satisfy the need of heterogeneity. Meanwhile, as system manufacturers are technically and financially unable to supply diversified components independently, it helps to fuel the maturity of products through the mixed use of standard third-party products and to discover underlying component design flaws more easily via the multimode ruling mechanism. Moreover, the appropriate use of policy decision algorithms can avoid the risk caused by less mature executors on the target system's service function and performance. A diversified product eco-environment created by the diversified rigid demand of technology is beneficial to the innovative development of products and technologies with equivalent functions.

8.3.6 Achievement of Multi-dimensional Goals

We expect that CMD can technically support the mode of "overlapping develop-ment" to naturally utilize and inherit current technologies; support the "genetically modified" development mode to gradually advance the upgrading of current models of equipment; support the mode of "incremental growth" to ensure an nonlinear increase of the defensive effect along with the increase of application locations, lay-ers, and scale; support the mode of "expansion into blue ocean" to play an indis-pensable role in the emerging markets and to penetrate deep into the traditional markets; and limit the cost within the scope of user affordability, and we make sure that under the condition of equal reliability, usability, and credibility, the system's life-cycle cost-effectiveness ratio should be higher than that of separate deployment of the target system and additive security defense devices.

The development of technology and product should be put in a globalized open environment while relying to the least extent on the closed links or security mea-sures in the process of technological development and product manufacturing. The protecting effect of the target system should mainly depend on its own structure or endogenous effect.

Obviously, the effect, scalability, and scope of application of CMD rely to a large degree on the abundance in the market of heterogeneous redundancy software and hardware components, modules, and middleware, as well as the progress of hetero-geneous design theories, methods, and tools. Therefore, CMD welcomes a diversi-fied open-market development mode based on standards and regulations. For instance, nonhomologous operating systems, databases, CPU suites or modules, applications, tools, embedded middleware, or IP cores can all find their own com-mercial positions in the emerging mimic defense market.

As Fig. 8.6 shows, we expect the mimic calculation and defense technology to become one of the major impetuses of the development of a new generation of information technology.

It is noteworthy that, with the emergence of the open-source communities and open-structure products in recent years, the development of open-source software and hardware is also expanding. Therefore, it has become especially necessary to study a series of problems related to the availability of homologous-source code flaws under the dynamic heterogeneous redundancy mechanism and multi-dimensional dynamic reconfiguration and strategy decision mechanism, including the research on automatic identification of homologous-source elements and code gene mapping analytical tools. As we have discussed in Chap. 1, how the code flaws become exploitable vulnerabilities is inextricably related to the ability and technical conditions of an attacker in a given environment. The characters and uti-lization of homologous-source code flaws in different environments are often dif-ferent as well. It is expected that the CMD environment based on a DHR structure can turn these differences into one of the major challenges which needs to be addressed in "coordinated attacks on dynamic diversified targets under the non-coordinated conditions." As to the dark features within the target system which do not rely on the environmental factors, they are usually independent events in a

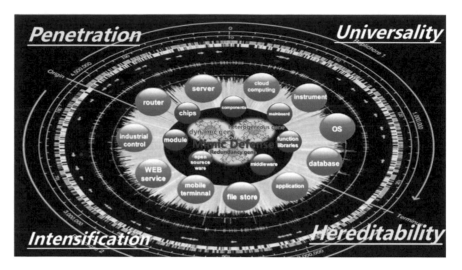

Fig. 8.6 Multi-dimensional industrial goals of Mimic Technology

mimic defense structure. As long as the executors are different enough from each other, the dark features will be difficult to be utilized effectively. Yet there are still two exceptions: (1) heterogeneous executors have the same logic flaws or malicious codes which can be reflected as continuous coordinated attack evasions and (2) utilize side channel attacks (e.g., going through channels) without changing the content of multimode output vectors to send sensitive data in the executors in a "hidden" way. For mimic defense, all these problems require in-depth research and need to be prevented.

8.3.7   Reduce the Complexity of Security Maintenance

Making remedies is the mode adopted by mainstream defensive technologies in the cyberspace nowadays. The effectiveness of such a defensive mode hinges to a large degree on the accuracy of security management and timeliness of daily maintenance; it is also closely related to the skill and security awareness of the network administrators. Typical technical and intelligent remedial operations include appropriate adjustment of the firewall rules, installation of vulnerability patches, updating of software versions, regular/irregular change of the keys/passwords, real-time analysis of the running logs, timely updating of vulnerability libraries/virus databases, and updating of antivirus software, among others. It will take a huge amount of manpower, materials, and financial resources to perfectly complete these tasks. Sometimes, the mistakes and oversights of the operating staff may even lead to disastrous consequences. In 2013, more than 20 million user check-in records of hotels which had established partnerships with Cnwisdom, China's largest digital hotel room service provider, were leaked online due to security vulnerabilities. In 2014, JPMorgan was hacked in spite of a cyber security team consisting of over 1000 people and a security budget of US$250 million. In 2017, Equifax, one of the largest credit reporting agencies in the United States, had the data of 143 million consumers leaked due to a vulnerability in its website. Now it faces a compensation of US$450 billion from a class action. Therefore, we expect that mimic defense can significantly transform the current security maintenance and management mode at the technical level. The key lies in the autoimmunity of the target system from unknown threats of vulnerabilities, backdoors, and Trojans. In this way, the intensity of daily security maintenance could become only weakly related to the management mode itself.

References

1. Yuchang, S.: Mimicry of butterflies. Bull. Biol. **42**(7), 14–15 (2007)
2. Xiao, Z., Shiwei, F., Dong, R., et al.: History of insect mimicry. J. Environ. Entomol. **31**(4), 365–373 (2009)
3. CERT. China internet cyber security report. http://www.cert.org.cn/publish/main/upload/File/2017annual(1).pdf (2 Aug 2018)
4. CERT. Review of China internet cyber security trends. http://www.cert.org.cn/publish/main/upload/File/situation.pdf (25 Apr 2018)
5. Genguo, L., Yadong, G., Xin, L.: Brief discussion on the position and application of high performance computing. Comput. Appl. Softw. **9**, 3–4 (2006)
6. IT168 Steadily ranks among world top 4 supercomputers, and Sugon "liquid-cooling HPC" under spotlight. http://server.it168.com/a2017/0621/3134/000003134948.shtml (21 June 2017)
7. HPC Top100, China HPC TOP100 in 2016 released. http://www.hpc100.cn/news/21/ (29 Oct 2016)
8. Top 500. https://www.top500.org/list/2017/06/ (21 June 2017)

Chapter 9
The Principle of Cyberspace Mimic Defense

9.1 Overview

As pointed out in the previous chapters, cyberspace mimic defense (CMD) is technically based on an innovative general robust control structure (GRCS) and a deceptive mimic camouflage or invisibility/stealth mechanism that can produce a physical or logical scenario similar to the uncertainty principle of quantum mechanics, and this structure and mechanism can be employed to provide the intensified "trinity" functionality: application service provision, reliability guarantee, and secure and trusted defense in the target object. Its GRCS and endogenous security mechanism can provide Integrated Point-Surface (IPS) suppression without relying on prior knowledge of the attacker and conventional active/passive protection and control measures against uncertain threats based on unknown vulnerabilities or virus Trojans within the target system. Physical or virtual software/hardware bodies or virtual physical combined bodies designed based on this control structure and camouflage mechanism, such as IP cores, chips, middleware/embedded hardware and software, modules, components, devices, systems, platforms or networks, as well as all kinds of virtualization processing scenarios, can effectively control general uncertain disturbances (GUD), including known unknown security risks or unknown unknown security threats. However, CMD aims to solve the deal with uncertain threats from target-unknown vulnerabilities and cannot and will never completely solve all security-related problems in cyberspace, nor does it attempt to independently support the entire protection and control system for cyber security or hinder the inheritance or adoption of the relevant technological achievements and emerging technologies in future. As the world faces a conundrum of unsecured credibility of supply chains at the component layer, CMD only seeks to solve it in the uncertain operational scenarios taking shape on the basis of an innovative GRCS and a mimic camouflage mechanism (MCM),at the software/hardware structural level, and against the background of fully open ecology characterized by national division of labor, industrial division of labor, or even intro-product division of labor; also, it

© Springer Nature Switzerland AG 2020
J. Wu, *Cyberspace Mimic Defense*, Wireless Networks,
https://doi.org/10.1007/978-3-030-29844-9_9

tends to change the current "cyber game rules" by turning over the cyber-attack theories and methods based on software/hardware code design vulnerabilities; promote major innovations in cyber defense theories, game rival models, and technical means and methods; reverse the strategic pattern of seriously imbalanced attack-defense postures; reshape the new cyber security order in the information age; and develop a new generation of information technology and industry featuring endogenous security functionality. In this way, we can eradicate the Achilles heel by addressing the "overflow" of product vulnerabilities and other dark features polluting the cyber ecosystem as a result of uncontrollable security or quality factors in the design or manufacturing process of hardware/software products.

9.1.1   Core Ideology

Given the "zero trust" global, open industry or product ecology and independence of (not excluding) conventional security protection and control measures [1], such as intrusion detection, intrusion prevention, and intrusion tolerance, and with TRA's logic expression as the means of measuring and perceiving multimode output vectors (MOVs), the author proposes to establish a new and "simple" defense system based on endogenous security effects on the GRCS by introducing the MD-based mechanisms of ruling, policy and schedule (PAS), negative feedback control (NFC), and multi-dimensional dynamic reconfiguration (MDR).To this end, we need to achieve two basic goals: the first goal is to increase the "difficulty of a cooperative attack on dynamic and diverse targets in non-cooperative conditions" to achieve the "uncertainty" effect on the defense scenarios or behaviors, creating a model of IPS defense without relying on prior knowledge and behavioral characteristics of the attacker; the second goal is to construct a GRC system/mechanism that integrates reliable service provision and trusted service guarantee to normalize the control or management of GUD in the architecture and make the target system's reliability, availability, and anti-aggression subject to quantitative design and experimental measurement. Its basic principles and methods are universally applicable in scenarios featuring high reliability, high availability, and high credibility.

We know that the DHR architecture is able to manage and control GUD and effectively suppress uncertain disturbances including known unknown security risks or unknown unknown security threats. Through the introduction of the MD into the DHR processes, such as multi-modal ruling (MMR), PAS, NFC, and MDR, the fundamental system of CMD can be established. The closed-loop feedback control mechanism based on mimic strategy not only endows systematic defense with gains in defensive elements and technical means, including dynamization, diversification, and randomization but also quantifies the unknown threat perception function and the endogenous uncertainty mechanism in the design and verifies the measurement through the injection test method of classical reliability theories. Compared with the DRS architecture, the DHR feedback control process has a policy distribution and scheduling mechanism, which ensures the uncertainty of the apparent construction of the target object under the condition of functional equivalence, making it exponentially difficult for an attacker to detect, perceive, or predict the defense behaviors. As it is reconfigurable, reorganizable, reconstructable, redefinable, and virtualized according to the mimic ruling strategy, the iteratively convergent MDR mechanism [2] creates more targeted structural scenario changes in the defense scenarios and the executors themselves relative to the attacker's behaviors, as its inherent uncertainty effect poses insurmountable challenges (non-decreasing initial entropy) to the inheritability and reproducibility of the attack experience, preventing cyber attacks from generating the planned or predicted effects; the MD-based integrated control mechanism functions in both the non-specific surface defense, independent from threat attribute perception and the point defense based on specific identification, and can also effectively block the cooperative and consistent escape in the space-time dimension through trial-and-error attacks via in-house heterogeneous executors.

The organic combination of the DHR and the MCM not only has a subversive impact on the attack theories and methods based on the vulnerabilities of software/hardware codes but also enables the information system or control facilities to simultaneously obtain endogenous security featuring "quantitative design and verifiable measurement" and GRC performance.

9.1.2 Eradicating the Root Cause for Cyber Security Problems

In a modern society with a highly developed commodity economy, product quality issues and consequent user losses have long been borne by the providers/manufacturers or insurance institutions. This has become a basic norm and a common sense in the modern commodity society. Unfortunately, cyberspace is an exception, as a seemingly ridiculous unwritten practice or unspoken rule in the industry today goes like this: "A product provider is not held liable in an economic, legal, or even moral sense to any security flaws existing in the design, manufacture, and maintenance of IT (including information security) and relevant technical products, as well as for their possible consequences." It is only an obligation rather than a legally binding liability to help users find or dodge product security problems or reduce their losses. For example, Microsoft has never taken any legal responsibility for user losses caused by design flaws or security loopholes in its Windows series, nor has Intel ever assumed any commercial duty for the hardware vulnerabilities of its CPU products. The weird reason for such violation of the basic laws of the modern commodity economy, as pointed out in Chap. 1 of this book, lies in that humans cannot eradicate software/hardware design flaws at the current technological level, nor can they prevent the implantation of deliberate codes in software/hardware components. What's more, we simply cannot afford a thorough investigation to screen out the vulnerabilities and other dark features of a complex system. Therefore, the issue of product robustness control, which shall be originally borne by the manufacturer, is helplessly subject to an "overflow and diffusion" and thus becomes the most dangerous uncertainty threat to cyberspace. Therefore, it seems to follow a logical train to accept that the equipment provider cannot guarantee a zero-security-defect product design, nor can it ensure the security and credibility throughout the manufacturing, procurement, or entrusted processing of the equipment. Although it seems to be biased to attribute the root cause of all security threats in cyberspace to the robust control of hardware/software products, it is undeniable that due to the distorted outlook of informatization and security development, the growth of robust control technology for suppressing GUD has not drawn sufficient attention across the industry and even the scientific research community, which is among the fundamentals of seriously imbalanced security postures in cyberspace. Therefore, both scientific and industrial communities must change the security development outlook against the clock to face the crisis where cyberspace is being overwhelmed. It is necessary to find a theory and methodology to overcome the GRC functionality vacancy in hardware/software products; otherwise, we can never expect cyber security landscape to be fundamentally improved. In other words, to overcome the target object's hardware/software vulnerabilities, we must first solve the root cause of the GRC problem across hardware/software products; to offset or weaken the biggest uncertainty threat in cyberspace, we have to construct a defense system with endogenous security effects. Therefore, rather than introducing the latest cyber security defense theories and methods, this book can be more accurately described as a publicity campaign recommending the GRC principle and endogenous security to readers.

It is not difficult for readers to find that the DHR introduced in Chap. 7 is a typical GRCS. As GRCS has the function of effective control over the target object's general uncertain disturbances, including known unknown security risks and unknown unknown security threats, given the known action mechanism of the attack chain and the defined possible impact range of the attack, the system can perceive GUD (including unknown unknown threats) through the multimode ruling mechanism and detect or locate the "suspected problematic executors or defense scenarios" through the relativity ruling or the convergeable iterative comparison mechanism. The PAS and the MDR-NFM (negative feedback mechanism) endow the service function or performance of the target system with stability as well as quality robustness. In other words, the hardware/software systems based on the DHR architecture are born with GRC attributes to suppress man-made attacks such as vulnerabilities or virus Trojans and the impact of conventional model perturbations. Even so, the attack-defense game in cyberspace is the highest-level confrontation of intelligence, where a strategic or tactical advantage always stems from a better strategy or means. "There is neither an impregnable defense nor an invictus attack," and mimic defense is no exception.

9.1.3   Biological Immunity and Endogenous Security

Biology tells us that the history of biological evolution is also the development process of the immune system. The non-specific immunity and specific immunity mechanisms of vertebrates and the hinge correlation between them are miracles of life, with countless mysteries to be explored. Among them, the innate non-specific immunity acquired from genetic attributes is the most magical for its "non-specific clearance" response to the absolute majority of invading pathogenic microorganisms. Scientific research reveals that pathogenic microorganisms in the nature are constantly mutating (at a speed counted even by minutes or hours). Then, what factors act to ensure that nonspecific immunity depends solely on biological genetic data to keep the organism selectively and non-specifically wiping out all kinds of ever-mutating invading pathogenic microorganisms from the real world? What circumstances and conditions are required to activate the specific immunity mechanism, and by what means? Genetic information is relatively stable, but does it need to be updated throughout the life cycle of the organism, and when and how if so? How good is the memory effect of specific immunity, and how does it affect the genetic information of nonspecific immunity? The enlightenment generated herefrom is that whether we can design an endogenous defense ability similar to vertebrates' immune mechanism in cyber hardware/software devices or systems, so as to "generate a cleaning function without specific selection targeted at unknown attacks on the target object's various vulnerabilities," as well as to trigger a point defense function similar to the specific immunity mechanism at the right time. The author believes that this defense function originating from the structure of the target object itself may be appropriately described as the endogenous safety and security (ESAS).

9.1.3.1 Non-specific Immunity

Nonspecific Immunity (NI) in the biological sense, also known as inherent immunity or innate immunity, is a born genetic trait that vertebrates (or humans) have acquired during the age-old evolution, while specific immunity (SI) derived from NI needs to be acquired through repeated antigen stimulation. The inherent immunity built in the relevant species in the natural world can respond to a variety of invading pathogenic microorganisms with surface defense and also play an important role in SI activation and antibody formation. Its traits are as follows:

(1) An extensive scope of action. The organism is not specifically selective in the clearance of invading antigenic substances.
(2) Fast response. An antigenic substance is immediately rejected and removed by the organism upon contact.
(3) Relative stability. It is neither affected by nor increased or decreased with the strength or frequency of the invading antigenic substances.
(4) Inheritance. An organism is born with NI that can be passed on to its offsprings.
(5) The SI of vertebrates is based on NI, which means that SI and NI shall not be studied separately. From an individual's point of view, when an antigenic substance invades an organism, NI will react first, and then SI will be generated.

The "identification friend or foe (IFF)" mechanism of NI features "non-specific selection in the clearance of invading antigenic substances" for a wide variety of changeable microorganisms, but is this amazing mechanism the result of long-term evolution, or the outcome of the "biological singularity" similar to that of the big bang theory, or the creation of the "first push of God's hand?" Scientific research has not reached a clear conclusion yet. However, the author makes a "cross-border" guess that, logically, this IFF mechanism constrained by some conditions is by no means a full-sample spatial comparison. Instead, it relies on the main gene fragments of the organism itself as the basis for sampling comparison, and these fragments are often also major feature fragments of the invading antigen. It is highly probable for this fingerprint-comparison-like mechanism to detect the absolute majority of invading antigens and achieve "clearance without specific selection." Unfortunately, there are too many similarities and even identicals in cyberspace, making it difficult to extract "feature fragments." Even if we attempt to attach an encrypted authentication measure as the "unique fingerprint identifier," as long as there are unknown attacks based on unknown vulnerabilities, the "trustworthiness of trusted roots" will be inevitably questioned. Therefore, in a strict sense, cyberspace cannot shoot the surface defense problem by directly employing the IFF mechanism of spinal cord organisms. God has not given cyberspace the slightest care in this regard.

9.1.3.2   Specific Immunity

Specific immunity (SI), also known as acquired immunity or adaptive immunity, is usually targeted at certain pathogens alone. It requires acquired infection (healing or asymptomatic infection) or artificial vaccination (bacteria, vaccine, immunoglobulin, etc.), so that the organism can obtain antibodies against infection. It is usually formed through repeated stimulation by microorganisms or other antigens (such as immunoglobulins and immune lymphocytes) and can specifically respond to certain antigens. Its traits are as follows:

(1) Specificity. The second response of the organism is targeted only at the antigens reentering the body rather than other antigens entering the body for the first time.
(2) Immunological memory, as the immune system memorizes the information of the initial antigen stimulation. For example, while a portion of lymphocytes becomes effector cells fighting against intruders, another portion will differentiate into quiescent memory cells, which will produce antibodies when they encounter the same antigen that reenters the body.
(3) Individual traits. SI is an individual protection function established through repeated stimulation by antigens on the basis of NI after the birth of an organism, which differs from NI in terms of quality and quantity, i.e., immunity differ between individuals.
(4) Positive reactions and negative reactions. General cases where a specific antibody is produced and the immunity functions are called positive reactions. Some cases where no more targeted antibodies are produced upon restimulation by antigens are called negative reactions, also known as immunologic tolerance.

(5) A variety of cells are involved, as the response to antigen stimulation is mainly made by T cells and B cells, while some other cells, such as macrophages and granulocytes, are also involved in the process of immunization.

What puzzles us is why the empirical information obtained in SI cannot be expressed at the genetic layer, and why the biological offspring needs to rebuild or reconstruct acquired immunity. There is a view that had all the SI information acquired over the generations be recorded on DNA, then the DNA—the combination of only four base elements (ATCG) might not provide sufficient capacity to carry the data. To this end, it may be a last resort to inherit the SI-learning mechanism rather than all immunological memory information. In fact, the point defense function with the specific immunity has a rich library in cyberspace, evident in all kinds of active defense, passive defense, static defense, and dynamic defense; the only thing missing is the surface defense function with the NI mechanism.

9.1.3.3   Non-prior-Knowledge-Reliant Defense

We know that NI is innate genetic-based trait that provides non-specific selective clearance of various invading pathogenic microorganisms. When this concept is introduced into the field of cyber defense, as long as the target object can mechanically produce a "clearance function without specific selectivity" targeted at unknown attacks based on vulnerabilities, we call it non-prior-knowledge-reliant defense (NKD).Obviously, this is different from conventional defense reliant on threat trait awareness and cognition, since it is not necessary to understand or accurately grasp the specific features of the attack action. Its defense effectiveness should be based on the TRA or consensus mechanism, by sensing potential threats via the strategic ruling of functionally equivalent MOVs and endowing it with a "non-specific selective clearance" effect on random DMFs or non-cooperative attacks. In addition, there should also be mechanisms serving as the first defense line to trigger specificity or point defense.

9.1.3.4   Endogenous Security

Explained literally, being endogenous refers to an effect or function generated inside rather than acquired from external factors. Endogenous security (ES) is a conventional or unconventional security attribute acquired by employing the internal factors of a system, including but not limited to its architecture, mechanisms, scenarios, and laws. For example, vertebrates' NI-/SI-learning mechanisms are covered by the ES scope.

Generally, passive defense devices in cyberspace do not have a tight coupling relationship with the function/performance or even the control structure of the defense target object. In most cases, they are connected in parallel or in series to the target object's service loop as external or additional devices. Also, the issue of cyber security threats based on unknown vulnerabilities is also rarely considered in the target object's functionality design. More often than not, no NI-like surface defense function is outlined or deployed. Moreover, there can even be a lack of interfaces for the attached defense devices can be ignored, making it hard to trigger their functionality and mechanisms (similar to how NI triggers SI).

However, the DRS mentioned in Chap. 6 is an exception. In order to address the physical failures of components and the uncertain faults caused by software/hardware design defects, the DRS adopts the heterogeneous redundancy system (HRS) architecture and multimode voting (MV) mechanism under the FE conditions. In theory, as long as the majority of heterogeneous executors do not fail simultaneously, the system will not lose the given functionality and performance. This is the reliability or robustness based on the structural design of the system, which the author calls "endogenous reliability" (consistent with the statement that reliability is "designed").

Similarly, in view of the intention of CMD, in the design process of the target object's software/hardware architecture and operation mechanism, there should be a focus on the ecology of globalized industries/products, the technological development trend across open-source communities, the status quo where vulnerabilities cannot be eradicated, the fact that the credibility of hardware/software component supply chains cannot be guaranteed, and the system security problems in the presence of known or unknown cyber threats, with independent, reliable, IPS defensive effects achieved without relying on prior knowledge and additional defense facilities. This security attribute based on the structural design of a hardware/software system is the "endogenous security" the author expects. The concept of "DiS" (Designed-in Security, proposed by the Obama administration in the *Trustworthy Cyberspace: Strategic Plan for the Federal Cybersecurity Research and Development Program*) should also belong to the same category.

It should be emphasized that how to use endogenous factors in a cost-effective way to achieve endogenous security functions is worthy of further research. For example, the complexity of electromagnetic wave transmission paths in open space often results in disparate channel parameters in different receiving positions, which can be called "channel fingerprints." If the communication parties can properly use a parameter to implement channel scrambling, pilot redefinition, etc., it will be difficult for an eavesdropper to receive a de-modulable signal even several wavelengths beyond. Besides, most of data service centers and cloud service platforms adopt a virtual machine of pooled resources or a virtual container service mechanism, while the general servers, dedicated processors, network switching devices, file storage devices, basic tool software, and the environment support software in the resource pool are often from multiple vendors and HR-configured. In addition, the scheduling policy of virtual resources is strongly correlated to indicators such as the service type, traffic flow, service quality assurance, economic utilization of resources, etc., which means the target objects are inherently heterogeneous, redundant, and dynamic. The conscious or systematic use of these intrinsic features (such as the DHR architecture) significantly enhances the ability to respond to vulnerabilities, virus Trojans, and other security threats of the target object.

In short, whether it is to provide services with high credibility, high availability, and high reliability based on the integrated design of the ES control structure or to enhance the defense capabilities of the target system using endogenous structure mechanisms, it can be called "ES technology." But DHR or the mimic structure has the great significance. What needs to be pointed out is that the endogenous security can be quantitatively designed and verified like reliability and availability.

9.1.4 Non-specific Surface Defense

From the descriptions in Sect. 9.1.2, we know that NI of vertebrates has a non-specific selective trait in the clearance of invading antigenic substances. Meanwhile, it is relatively stable, neither affected by nor increased or decreased with the strength or frequency of, the invading antigenic substances. The first trait emphasizes that any invasive antigenic substance, regardless of its specificity, will be "indiscriminately killed" (which can also be considered as a surface defense function), and the second trait is that this immune function is affected neither by the type of invading antigenic substance nor by the frequency of intrusion. However, the lack of the necessary GRC function across today's hardware/software products in cyberspace has left the entire security environment completely out of control from the source. Conventional passive defense based on threat perception and trait extraction is similar to SI in nature, both belonging to point defense (PD) featuring "medication and injection," and is thus impossible to effectively deal with vulnerability attacks that are endless, diversified, ubiquitous, and cannot be eliminated. Similarly, the collecting or updating rate of the attack features, virus Trojans and other library data cannot keep up with the emerging rate of the vulnerabilities and virus Trojans due to the locality and hysteresis of threat perception. The core of non-specific defense (ND) is to invent a GRCS and an endogenous security mechanism (ESM), so that the effectiveness of the defense function can be irrelevant to the type/feature information abundance of the vulnerabilities and virus Trojans, as well as to the way in which they are detected and exploited. It aims to enable the hardware/software system itself to acquire a non-prior-knowledge-reliant surface defense function.

9.1.5 Integrated Defense

In a conventional defense system, a key difference between active defense and passive defense lies in whether it is an active move of "forward-out" attack detection or a passive perception of the attack features. The former includes isolation monitoring technologies such as honeypots, honeynets and sandboxes, and cloud security based on big data drives, etc., while the latter includes firewalls, security gatekeepers, process guardians, and Trojan screening. However, the common ground for the two is also remarkable, as both need to be able to accurately perceive security threats and precisely extract attack trait information from the target entity or its virtual environment. The mechanism can still be classified into the category of point-defensive SI. As mentioned above, vertebrates have NI functions in addition to SI functions, and they are organically integrated and cannot be completely segregated. When an antigenic substance invades, the organism first implement surface defense using NI. Once an antigen "slips through the net," it will stimulate the production of an SI antibody accordingly or use the SI antibody acquired in the previous stage to implement point defense. The immunization process of the organism perfectly fits in with the integrated defense mechanism, which not only reflects the layout of IPS hierarchical defense but also highlights the beauty of precise defense via active-passive integration, where we have to sincerely praise the greatness wonder of the Creator.

CMD is aimed at developing such defense systems and mechanisms with an integrated effect.

9.1.6   GRC and the Mimic Structure

According to the descriptions in Chap. 6, Robust Control (RC) methods are suitable for applications prioritizing stability or reliability, and the perturbation range of known or uncertain factors of the dynamic traits in the process can be estimated. It is especially applicable to the process control applications where the target object plays a relatively sensitive role and has a wide range of uncertainties and a narrow margin of stability. From the perspective of engineering implementation, a feedback control system (FCS) is designed to find a controller or algorithm based on a given control object model to ensure the stability of the FCS while achieving the desired performance and meanwhile to endow the target system with robustness against the uncertainty of the model and disturbances [3].

Narrow robust control minimizes the sensitivity of the controller or algorithm to model uncertainty (including external disturbances, parameter disturbances, etc.) to maintain the original performance of the system, while GRC refers to all control algorithms using a certain controller to deal with systems that contain uncertain factors. The TRA-based DRS can normalize uncertain disturbances into probability-controllable reliability events to a certain extent, thus belonging to the scope of GRC. However, constrained by its non-closed loop control structure and inherent defects of certainty and staticity, DRS is unable to effectively deal with "co-cheating" or "trial-and-error" or TT attacks, and cannot maintain stability robustness in suppressing man-made disturbances. In other words, if human disturbance exceeds the given range, the system will lose stability.

A mimic structure combines DHR's ruling, scheduling, FC, and MDR processes with the MD mechanism, filling the gap of stability robustness of conventional robust control in the field of cyber defense applications. Because the mimic disguise strategy-based feedback control loop (FCL) will use the PAS and MDRM driven by the relative ruling mechanism to self-adaptively select "the appropriate running scenario" in a progressive or iteratively convergent manner to cope with the current threat, seeking "a quick dodge" of a problematic scenario instead of overemphasizing "quick return-to-zero" of the scenario. Therefore, once any trial-and-error or attempted attack behavior is perceived in the ruling process, the mimic structure will transform the current operating environment so that the trial-and-error attack will lose the premise of an unchanged target scenario.

You can easily see that a mimic structure can keep stability, asymptotic adjustment, and convergent dynamicity(the feature of non-decreasing initial entropy) unchanged under some specific uncertainty conditions, while its multimode ruling mechanism can perceive GUD, including unknown unknown threats, its mimic ruling mechanism with an iterative algorithm can identify or locate "executors or scenario defense with abnormal output vectors," and its PAS and MDR-NFM can endow the in-house services or performance of the target structure with stability robustness and quality robustness.

9.1.7 Goals and Expectations

9.1.7.1 Development Goals

CMD attempts to create a "vertical and nonintegrated, closed-loop and open" systemic structure, which can not only support the service functionality of information networks/systems (components, software or hardware), but also, with the help of the endogenous security mechanism, effectively curb or control the certainty/uncertainty threats arising from the vulnerabilities, Trojans, and other deficiencies introduced in the design, manufacture, maintenance, and other related aspects of the hardware/software components in the mimic interface (MI), while significantly improving the resilience, vigorousness, and robustness of the target system's service functions and greatly diluting the strong correlation between security maintenance management and defense effectiveness, as shown in Fig. 9.1.

To meet the development trend of technology, industry, or product in the globalization era, we aim to create an emerging non-enclosed, autonomous, controllable, and sustainable ecology in the win-win cooperation environment, so as to better fulfill the open-source crowd innovation business and technology development model. The expected goals are as follows:

1. Goal of intensification

The "vertical and nonintegrated, closed-loop and open" structure will be the core of mimic defense applications (MDAs), with a vision of natural incorporation of existing or coming security protection technologies, which allows the components or building blocks within the MI (which can be software/hardware facilities or devices at the levels of network element (NE), platform, system, part, component,

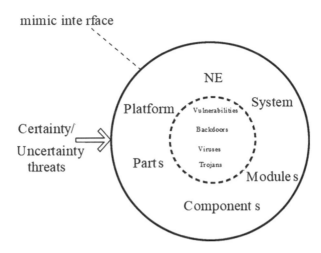

Fig. 9.1 Mimic interface

and module) to adopt COTS-level products or open-source methods, with an expectation to achieve the goal of "five-in-one" intensification:

(1) Integration of service functions and security functions.
(2) Integration of NI and SI.
(3) Integration of anti-external attacks and anti-internal penetrant attacks.
(4) Providing quantifiable design and measurable reliability, availability, and trustability function.
(5) Integration enables the improvement of self-immunity and reduced complexity of security maintenance.

2. Closed-loop, open, and superimposable

A closed loop means that CMD itself has a control structure with system robustness and a FC-based operating mechanism, and its nonlinear security gain is not acquired through a patchwork or stack of multiple technologies, methods, or facilities; being open means not to impose excessively stringent security and credibility requirements on the design and development, procurement, or integrated manufacturing of components or reconfigurable executor elements within the structure; being superimposable means that CMD can naturally carry on or accept the existing or coming achievements in information/security technology and can increase the attack difficulty exponentially through iterative applications at different levels or granularities or operational steps.

3. Structure-level autonomous control

Engineering implementation refers to the problem-solving approach through software/hardware system construction techniques targeted at technical or engineering problems that cannot or should not be solved at the subsystem, module, part, component, and device layers. The structure-level autonomous control means that technological innovation in RCS enables the target system to acquire a GUD suppression effect from its ESM, which is a difficult or impossible mission for conventional security defense measures, and the system is weakly correlated or irrelevant to the credibility of the hardware/software supply chain or technology chain within the structure. Obviously, the mimic structure with a GRC attribute has an endogenous mechanical suppression effect on uncertain threats within the target system. In addition, the mimic structure and working mechanism is relatively independent from the algorithm construction and service functionality of the executor, irrelevant to the level of the executor or defense scenario granularity (module, component, system, platform, or network), and not correlated to the physical or logical implementation of the executor or defense scenario.

4. Transparent management and control process

The management and control process herein specifically refers to the physical or virtual feedback control loop hardware/software settings to improve the reliability and anti-aggression ability of the system (or its parts or hardware/software components).Transparency means that the controlling loop physical or virtual body is invisible not only to the target object's service functionality but also to the reconfigurable executors of the target object's redundant configuration. It, in the DHR environment, refers to the addition of management and control levels or processes, including but not limited to input distribution and PAS control coordinating multi-channel or multicomponent work, MDR of the executor and MMR, which, in principle, should not affect the original functions and services of the target system. In other words, by adding a feedback control device (FCD) that is irrelevant or weakly correlated to the defense object's service functionality and can formally prove that there are no malicious vulnerabilities (while loopholes are allowed), the hardware/software resources in the target object covered by mimic brackets (MBs), which call for "autonomous, controllable, secure, and trusted" functionality, can be implemented in a FE manner through "manageable, controllable, and credible MBs. As shown in Fig. 9.2a, this control device should be a theoretically open and functionally transparent intermediate. In implementation, it does not have artificially designed vulnerabilities (except for loopholes) or Trojans(in less complicated conditions, it is not difficult to be available both in theory and technology) nor will it become an explicit attack target given its attack-unreachable attribute; in engineering, it relies mainly on the ES effects generated from the GRCS, as well as one-way information transmission, independent processing space, and the dynamic random operating mechanism (including the mimicry of the management and control device itself) to ensure the maximum defense effect (as shown in Fig. 9.2b), but less dependent on confidentiality measures or enclosed production methods [4]. The attacks in the social engineering sense or side channel attacks based on acoustic, photoelectric, magnetic, and heating effects or applications based on strict security requirements do not fall into this category.

5. Hierarchical iterative effect

The aim is to cover or affect the various stages of the attack chain with the innovative GRCS and MCM, including: during the vulnerability scanning stage, affect the integrity and authenticity of the attacker's detection of the target object's defense scenario information and disturb the reachability or credibility of the information

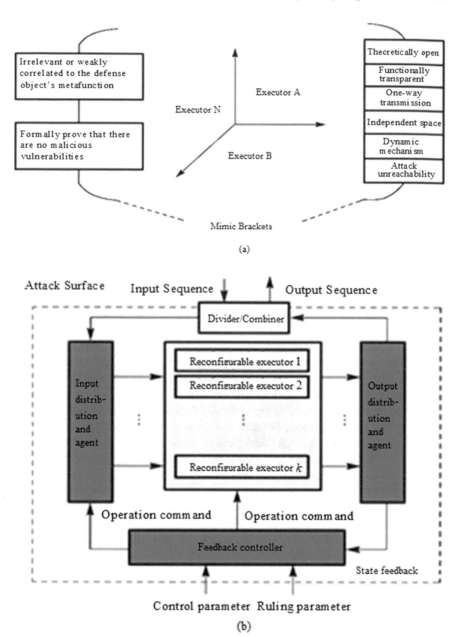

Fig. 9.2 Principle of the MB. (**a**) Mimic brackets. (**b**) MB with ES effects

passed back by built-in scanning tools such as the probe; disturb or block malicious code insertion via exploitable vulnerabilities during the stage of attack injection or attack package uploading, making it difficult to reliably build an attack chain/environment; disrupt the reachability of the AS during the vulnerability exploitation stage, or severely distort the vulnerability information mined offline in a mimic environment; when the attacker has reached his goal in the ongoing attack period, he then expands the outcome by attempting to implant backdoors for next attack, or upgrade the Trojan, or transmit back the sensitive information secretly. The above results should reflect the iterative effect based on the structural mechanism and should be able to, through the hierarchical software/hardware deployment of heterogeneous redundant resources under the FE conditions, make the mimic construction environment particularly sensitive to the information and communication frequency of internal-external and cross-hierarchical interactions of unknown attack actions, with an ESM-based nonlinear enhancement effect on any technical measures contributing to the uncertainty effect of system (whether dynamicity, randomness, and diversification or the incorporation of traditional security means).

9.1.7.2 Technical Expectations

We hope that a hardware/software system adopting the MD principle can not only afford the fusion guard (integrated defense) function able to conceal or deceive external attackers and internal penetrators but also provide service functions with high availability, high reliability, and high credibility in an integrated manner. Its fundamental principles and methods should be universally applicable to information systems or the related control devices.

(1) Able to implement the non-prior-knowledge-based surface defense (NKSD) and the precise perception-based point defense (PD) integratedly through the normalized GRCS and MDM (mimic disguise mechanism).

(2) The endogenous uncertainty effect can independently (but not excluding other technical means) suppress attacks based on the vulnerabilities of hardware/software executors (defense scenarios) within the MB (but with the exception of halt and side-channel attacks based on acoustic, optical, electric, magnetic, and heating effects).

(3) Introduce traditional security means into the reconfigurable or software-definable executors or defense scenarios to exponentially increase the difficulty of attacking the target object.

(4) Able to transfer the attack complexity to the multi-dimensional target space from the one-dimensional target space to a dynamic software-reconfigurable/definable scenario from a single static certain scenario and to cooperative attacks on plural heterogeneous targets under the non-cooperative conditions from the breakthrough of individual goals, escalating the attack difficulty nonlinearly across the above three stages. As for non-cooperative/DM or mismatched attacks, the attack defense effects should reach 100%, and the

common mode escape probability can be quantitatively designed, and defense effects can be verified and measured.

(5) The uncertainty effect formed within the MB can effectively block potential cooperative attacks achieved through the output vector (OV) "trial and error" method. "Even if an attack succeeds, it will be difficult to be maintained steadily."

(6) The non-traditional security threats within the MB can be normalized to traditional reliability issues, which can be consolidated for handling. The effectiveness of the defense is irrelevant or weakly correlated to the security maintenance and management model, remarkably reducing the maintenance cost.

(7) The MMR (multimode ruling) and output agent (OA), IDA, FC, and other modules should have user-definable policy control functions.

(8) Allow users to independently select custom or non-customized executor hardware/software that meets the MI requirements of the target product.

(9) The rigid thresholds for the target system can be lowered in respect of the completeness and dissimilarity in the design stage of the parts and components as well as of the security management in the manufacturing process. While the credibility of a hardware/software supply chain cannot be ensured in the context of globalization, we expect the bottleneck of security and credibility in the "era of network economy and technology" to be addressed through technology at the structural level.

(10) It can technically support "superimposed R&D" based on inheritance and innovation and commercially support "incremental deployment" from point to surface. As for the defense effect, it supports the "economic coverage" model by way of key nodes, sensitive paths, and hierarchical deployment and can achieve "super-nonlinear" systemic defense results as the application scale expands and the application level deepens.

9.1.8 Potential Application Targets

In theory, there are the following preconditions for implementing CMD: The rigid use demand with the intention to prevent against unknown attacks based on unknown vulnerabilities and virus Trojans, and the willingness to pay the necessary price against uncertain threats, the possibilities and conditions for diversified or pluralized FE software and hardware components, the technical conditions to implement the policy-modal ruling and the known properties and processes, as well as the predictable variation range of GUD inside the target object. That is, under the I 【P】 O model that meets the "input-processing-output" pattern and the RC constraints, there is an MI that is equivalent to a given function (performance) and can pass conformance testing based on a standard (or normalizable) interface and consistence testing of the interaction process; a virtualized running environment for plural or diversified software/hardware executors or mirrors that meets the MI requirements; the necessary conditions for implementing MMR, PAS, and MDR based on the MI; and no design flaws or malicious features existing in the standards, procedures, and protocols on which the MI relies. Obviously, the use of CMD is constrained in application scenarios frequently updating services or upgrading software/hardware versions to affect the MI interface interactive information, or application fields lacking a standardized or normalizable interface, or the scenarios where services are obtained through the downloaded "black box" software, or the application scenarios over-sensitive to the device's volume, energy consumption, and cost. Currently, the application objects envisaged include but are not limited to the following:

① Network information communication infrastructure: various high-robustness-oriented network routers, switches, transmission systems and wired/wireless access systems, network management (center) facilities, content delivery network (CDN), Web access services, domain name systems (DNS), software-defined network (SDN), file system, data center, gateways, etc.
② Highly standardized professional/dedicated network service platforms with robustness requirements, such as office networks, firewalls/gatekeepers, mail servers, file servers, dedicated network routing/switching devices, Internet of Things (IoT), Internet of Vehicles (IoV), etc.
③ File management and data storage or disaster recovery backup system, as well as encryption and authentication system
④ Various highly robust and specialized mass information service systems, cloud service (computing/storage) environments, IDCs, or CDNs
⑤ Highly robust hardware/software core equipments in the field of industrial control(such as distributed control system, field bus control system, PLC, CNC, RTU, and edge-gateway), intelligent nodes, embedded/middleware components, network adapters, sensors, switching and interconnected components, series/parallel-connected security devices, embedded systems, etc.
⑥ A new generation of highly reliable, available, and credible robustness service and control systems able to address both traditional and non-traditional security threats
⑦ Relevant software products that are highly reliable and secure, such as operating systems, databases, tool software, middleware software, virtualization software, embedded software, etc.
⑧ Threat perception and security monitoring and exploration systems or devices based on the mimic principle

It should be emphasized that although the mimic structure principle has universal significance in the management and control of vulnerabilities and virus Trojans in the fields of information control, information communication and information physics (and can also be used for reference in other application fields), in view of engineering implementation or cost-effective aspects, it may be more cost-effective for the principle to be used in hardware/software systems with "a fortification at the pass for key-site defense" or the protection of sensitive processes. In addition, the mimic defense structure cannot and will not replace the role of information security technologies, especially information encryption, identity authentication, and password encryption. However, the use of mimic-structured encryption devices can remarkably reduce or simplify the overweighted engineering management burden of design, manufacturing, maintenance, and use.

9.2   Cyberspace Mimic Defense

The CMD principle can be memorized in reference to a set of numbers "8122," specified as follows:

Focused on ONE premise, uncertain threats based on unknown vulnerabilities in the target software and hardware

Based on ONE axiom, the true relatively axiom (TRA), which can conditionally perceive uncertain threats

Find ONE mechanism, any self-adaptive mechanism whose "increase of information entropy" can stably guard against uncertain threats

Inventing ONE structure, a DHR architecture featuring GRC performance

Introducing ONE mechanism, the MDM, which can form a mimic defense fog

Forming ONE effect, an uncertainty effect which makes trial attacks lose the required preconditions

Acquiring ONE function variety, ES function with relying on a priori knowledge or additional defense technologies

Producing ONE nonlinear defense gain, the introduction of any kind of security technology can exponentially enhance the defense effect within the structure

Solving TWO categories of problems through normalization, to handle traditional and non-traditional security issues in an integrated manner

The mimic defense theory (MDT) is premised on four basic security issues and two regularity cognitions regarding cyberspace. The four issues at this stage are: inevitable software/hardware code design vulnerabilities; ineluctable backdoors in cyber ecology; insufficient technological capability to screen out all vulnerabilities and backdoors; and quantifiable management and control of the security and quality issues of hardware/software products from the origin. The two regularity cognitions are (1) in the absence of prior knowledge and behavioral characteristics of the attacker; a non-redundant or isomorphic-redundant structure can't in real time perceive and roughly locate uncertain threats based on unknown backdoors within the structure; (2) the heterogeneous redundant structure and the feedback control mechanism based on MMR are indispensable defensive elements for dealing with uncertain threats.

The foundation of the MDT includes an axiom and two mature theories and the related methods. The axiom is the TRA (also known as the consensus mechanism). The two theories are the heterogeneous redundancy reliability (HRR) theory and the closed-loop feedback robust control (CFRC) theory, including the channel coding/decoding method and the reliability verification test method.

When analyzing and studying the existing security defense technologies and methods, the author finds that an uncertain threat will eventually achieve the attack purpose as long as it manages to constantly reduce the information entropy of the defender (i.e., realizing the process of entropy reduction), which is expressed in the technical system as the lack of stability robustness across the related defense functions. For example, the DRS can in real time perceive and roughly locate uncertain threats and shield abnormal information output therefrom. However, due to the staticity, certainty, and similarity of its logical structure and majority-rule algorithm, when dealing with blind or trial-and-error attacks based on unknown vulnerabilities, its intrusion tolerance function will gradually abate with the entropy reduction process if not repaired in real time.

A DHR architecture "grafts" on the "rootstock" of DRS the "screw" of the self-adaptive FCS and the PAS mechanism based on multi-dimensional iterative ruling and gives redundant executors diversified functionality, reorganizability, reconfigurability, and software-definability, so that a "plant with affinity" takes shape and exhibits functional/performance stability robustness and quality robustness against the GUD within the structure.

It is easy for you to see that if the MDM is introduced into the FCL of the DHR architecture, the uncertainty effect based on the diversified scenario set can be generated within the structure. Appropriately used, this effect can endow all kinds of service functions on the DHR architecture with ES attributes, making it possible to solve the problem of traditional and non-traditional security threats within the structure in a normalized way. In other words, the MDM makes the DHR architecture-based target environment present an uncertain defense scenario/behavior to the attacker, and the iterative convergence function based on the ruling strategy, the ruling state, and the feedback control function can utilize the self-adaptive selection mechanism of the diversified defense scenarios (targeted at the increase of entropy) to suppress cyber attacks based on hardware/software vulnerabilities and virus Trojans in the MI while keeping the physical or logical perturbation impacts of the structural model within the expected range.

Obviously, the effectiveness of MD neither relies on the acquisition of prior knowledge and behavioral traits of the attacker, nor does it need to ensure the security and credibility of the hardware/software component supply chain or technological chain within the MI, and it is even not premised on any additional security technology or real-time maintenance effect. However, the mechanism can prove that if the relevant defensive elements or mature security technologies are comprehensively used, the level of anti-aggression and reliability in the MI can be exponentially elevated (namely, the information entropy of the defender is improved).This kind of structure and mechanism can normalize the GUD problems of the target object, including known unknown security risks or unknown unknown security threats, into classic reliability problems for integrated processing, making it possible to quantify the design of security defense expectations (including the full lifecycle cost-effectiveness of the relevant devices) within the MI backed by reliability, RC, and the channel coding theory. It is not difficult to discover that the differential form or transient stability of the DHR architecture is very similar to the DRS, except that the ruling strategy is not limited to majority-rule/majority voting. However, on the whole, the DHR architecture is like the integral form of various DRS scenarios or the convergent dynamic iterative process based on the re-equalization of the ruling state (i.e., the increase of initial entropy process).Therefore, MD belongs to neither conventional static defense nor classical dynamic defense, nor can it be classified into active or passive defense. It is only an application of the innovative GRC framework and mechanism applied in cyberspace defense. However, in view of the mechanism, the DHR-based mimic structure is substantially characterized by static/dynamic defense and passive/active defense. This is the magic feature of the mimic structure "non-decreasing initial entropy." The author believes that the term "positive defense"(PD) is more appropriate.

9.2.1   Underlying Theories and Basic Principles

The main theoretical basis of CMD is an axiom and two underlying theories and related methods. The main technical traits include the MCM-based DHR architecture and five endogenous effects derived from the structural characteristics.

1. Main theoretical basis

   TRA: "Everyone has one shortcoming or another, but it is most improbable that they make the same mistake simultaneously in the same place when performing the same task independently." With its logical expression, we can perceive and measure uncertain disturbances, including human disturbances such as the target object's unknown vulnerabilities, backdoors, and Trojans. This mechanism can be used to normalize the known unknown or unknown unknown threats reflected on the MI to the events that can be expressed by probabilities.

   The two underlying theories and the related methods are: the HRR theory, the CFRC theory, and the channel coding/decoding method and the reliability verification test method.

2. Main technical traits

   (1) Other functions designed within the DHR framework. In principle, functional equivalent heterogeneous executors (FEHEs) or defense scenarios within the structure should have reconfigurable or software-definable capabilities subject to command assignment.

   (2) The interior of the DHR architecture complies with the "de-cooperative" deployment standards and the one-way contact operating mechanism.

   (3) The MDM is introduced into MMR, PAS, NFC, MDR, and the related IA/OA processes of the DHR, forming the mimic ruling based on policy iteration and the robust control feedback loop based on backward verification.

   (4) The DHR security (including reliability, anti-attack feature, and availability) has the "uncertainty" structural effect subject to quantifiable design,

thus supporting the realization of highly reliable, available, and credible robustness services and IPS defense functions within the structure.

(5) The DHR architecture can achieve an exponential defensive gain by accepting existing or coming security technologies.

3. Endogenous effects derived from the structure

(1) Non-specific defense effects, which originate from the policy iterative ruling function implemented on the MOVs within the structure by the MB. Regardless of the causes of GUD occurring to the hardware/hardware components in the MI, as long as the MOV policy iterative ruling state is affected, its multiple policy iterative ruling modes (apart from choosing multiple algorithms, it can be combined with historical-behavior-based weight ruling, trustability-based ruling, and ambiguous ruling) may trigger the robust control mechanism(RCM) based on backward verification, leading to the progressive iteration of the algorithm or structure under the EF conditions in the current operating environment.

(2) The apparent uncertainty effect, which is a unique structural effect produced by the integration of the MDM and the DHR architecture. Any attempt by an attacker, such as a trial and error or a blind attack, unless the MOV consistency on the MI is not triggered (e.g., TT based on response latency modulation will not affect the content information of an output vector); otherwise any non-cooperative attack or DMF will be shielded by the MI, as trial-and-error attacks require an unchanged environment, and the MI does not support that constraint. In addition, even in the event of a perceptible attack and escape or non-permanent CMF, the RCM based on backward verification will drive the current problematic scenario to converge or iteratively "migrate" to prevent the escape and CMF states from getting temporal stability.

(3) Automatic evasion of problematic scenarios. Once an executor's defense scenario in the MB is broken by some attack action or affected by some model perturbation, the robust control feedback mechanism (RCFM) based on backward verification can progressively find a suitable option from the diversified defense scenarios to replace the current "suspected problematic" application scenario executor or reconfigure the problematic executor or even change the entire operating environment within the MB. This method of convergent problem dodging rather than problem zeroing significantly improves scenario defense resource utilization.

(4) An exponential security gain. The anti-attack and reliability indicators in the MB are strongly correlated to the operating mechanism of the system structure. Any of the following methods has a potential to yield an exponential security gain: Increase the number of executors within the structure; expand the inter-executor dissimilarity; introduce any traditional or non-traditional security technology into the executors; enrich the semantic abundance of the output vectors; or extend the scope of ruling algorithms.

(5) The unreachability effect of the MB vulnerabilities. The MB adopts a strict
one-way contact mechanism, making it difficult to exploit the loopholes in
the processes of IDA, Output Agent and Ruling (OAR), and FC from the
AS. As shown in Fig. 9.3, the MB components within the dashed line and the
associated reconfigurable executors are strictly unidirectional, and the
bracket function itself can be sheltered by the structural effect.

What needs to be pointed out is that the reconfigurable entity in DHR has
diversified operation scenarios or multiple defense scenarios with various kinds of
transformations in the given FE conditions. The feedback controller can instruct
them to change into the expected operations or defense scenarios. Therefore,
when the number of executors is k, there are generally m kinds of utilizable operations or defense scenarios in normal conditions. If the current DHR operation
needs n kinds of redundant scenario configurations, C_m^n kinds of combinations
can be selected in ideal conditions. However, the number of k and m is under the
economic and complexity control. The engineering implementation often needs
compromise.

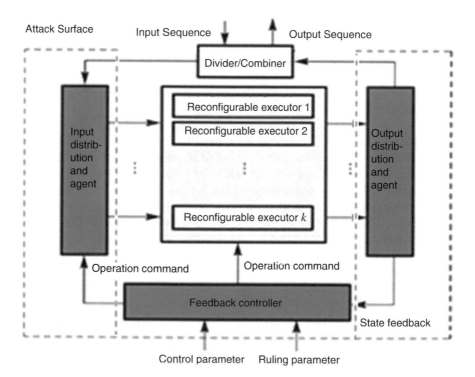

Fig. 9.3 MB's one-way contact mechanism

9.2.1.1   FE Common Sense and the TRA

"All roads lead to Rome" is a well-recognized philosophy. The corresponding common sense is that "there are often multiple implementation algorithms or structures under given functional and performance conditions." That is, an information system or hardware/software facility with a given service functionality/performance can usually be implemented with a variety of equivalent heterogeneous executors or algorithms.

According to the logical expression (or say the common sense mechanism) of the TRA, where "everyone has one shortcoming or another, but it is most improbable that they make the same mistake simultaneously in the same place when performing the same task independently," we can replace individuals (without any cooperative ties) of a group with one shortcoming or another by the abovementioned FEHEs and set up a majority-rule voting mechanism of MOVs regarding input responses (corresponding to the result of a given task) under the HR configuration. Then, the target system constructed according to the logical TRA expression may, through the HR and consensus mechanisms, transform its internal non-cooperative certain or uncertain threats and disturbances into probabilistic events where majority or consistency errors occur to MOVs in the space-time dimension. In this case, the relative comparison mechanism is used as a means of perceiving security situation (although, theoretically, there is a small probability of misjudging the superposition state), so that it becomes possible to provide within the MI a defense not relying on the prior knowledge and behavioral traits of the attacker. In view of the TRA and the relevant inferences (see details in Chap. 6), the trustworthiness of the ruling results can be affected by a change in the number of executor configurations, or an increase in the dissimilarity between executors, or the modified complexity of OV semantics or grammar, or the adjusted ruling strategy.

It should be emphasized that unlike the dissimilar redundancy (DR) majority-rule voting mechanism. The state of completely inconsistent MOVs on the MI is not the worst case. Since mimic ruling is an iterative processing mechanism that not only iterates according to a given ruling strategy but also makes iterative decisions based on the backward verification information (e.g., historical performance, trustworthiness status, priority, the relevant weights, and other parameters of each executor derived from the operation log), including directly specifying the executor currently with the highest trustworthiness to continue to provide service functions outward. In fact, the most serious case is precisely the "blind zone" of the majority-rule voting algorithms. At this point, most of the executors may simultaneously produce the output content expected by the attacker, namely, entering the attack and escape state, although, in general, the probability for such an occurrence is sufficiently small in a "non-cooperative DHR" environment.

9.2.1.2   DHR Architecture

As described in Chap. 6, DRS has the "surface defense (SD) and intrusion-tolerance properties strongly correlated to duration," but in respect of the mechanism, due to the vulnerability arising from its static and certain in-house defense scenarios, it will eventually get completely disintegrated by repeated trial-and-error attacks based on prior knowledge (e.g., a TT attack adopting a completely consistent voting algorithm).A DHR architecture can use the backward verification feedback iterative control mechanism and the external mandate disturbance command based on the ruling state and log information, so that any trial-and-error attack will lose the basic condition of the unchanged background, thus acquiring the attributes of "time-irrelevant PS defense functionality and intrusion tolerance." At the same time, since the DHR mechanism must follow an executor structuring approach that is reconfigurable, reorganizable, or software-definable, it can obtain reliability indicators higher than that of the DRS when the redundant resources of the executor are identical (theoretically there should be an exponential comparative advantage; see the subsequent analysis in this chapter).

Using the DHR as the MD infrastructure, we expect this integrated structural technology to provide the given service functions of the target object with guaranteed reliability and availability, as well as to make the attack and escape targeted at unknown vulnerabilities and other dark features an uncertain event within the relevant space or hierarchy of the target system under the FE premise. The application of the MDM on the DHR architecture is intended to create an almost insurmountable challenge for the attacker—"the difficulty of cooperative and consistent attacks on dynamic and plural targets under the non-cooperative conditions." Because once the attack action is perceived by mimic ruling, the iterative convergence mechanism of the defense scenario within the MI will definitely damage the stability or reliability of the attack chain. In particular, for an attack task that requires multiple passes through the MI, unless each time a cooperative and consistent escape is achieved in the mimic environment, otherwise any cooperation error will be perceived by the MI and trigger PAS and the MDR-NFM to change the current defense scenario. In other words, in the "dis-cooperation" DHR space, in order to achieve a precise, cooperative, and consistent attack on the MI and obtain identical OVs, the trial-and-error method or the use of homologous dark features are the two indispensable basic means. The necessary premise of trial and error is that the P algorithm of the I [P] O model is constant during the implementation stage of a trial-and-error attack, such as the DRS. However, for the mimic structure, the P algorithm (structure) of the I [P] O model may lead to a change in the defense scenario under the conditions of the unchanged functionality and the non-perceived attacker in response to the result of the trial-and-error attack, which will then not only fail to get a trial-and-error reaction but also lead to the missing of the prerequisite for continued trial and error. In fact, the redundancy features of the DRS can also shield f trial-and-error results under the condition of $N \geq 2f + 1$, but the number N of available redundant executors or defense scenarios will be decremented by the success frequency of trial-and-error attacks as the problem gets zeroed until the precondition of $N \geq 2f + 1$ is no longer satisfied. The DHR architecture has no such concern, because the MMR-based PAS and MDRM will progressively remove or dodge the impact of trial-and-error attacks. For example, the "suspected problematic executors" or even the current defense scenarios can be replaced, cleaned, or reconfigured via progressive convergence or iteration. The dynamic process will continue from the mechanism until the replaced or reconfigured defense scenarios can effectively disintegrate the current attack action, regardless of whether these scenarios have had a "problematic experience" or whether "the problem itself" has been zeroed. In other words, if Executor A (a defense scenario) is invalid only in the attack scenario x and Executor B only invalid to the attack scenario y, the defense invoked to Executor B (a defense scenario)in the attack scenario x will definitely be valid, and the scenario y invoked to Executor A should generate the same defense effect. We call this dynamic iterative invoking a "problem avoidance" processing mode (PAPM).

It is true that the DHR architecture can theoretically perceive the "attacks and escapes that affect the ruling state" but cannot perceive "those that do not affect the ruling state." The pluralistic strategy iterative ruling and backward verification FCM as well as the loop disturbance command based on the external mandate strategy (see Sect. 9.2.3) help to effectively reduce the probability or duration of the two types of attack and escape, while the defense effect of the DHR is uncertain for side channel attacks based on the acoustic, optical, electric, magnetic, heating, and other physical radiation mechanisms of the executors.

9.2.1.3 Security Effects Brought About by Endogenous Mechanisms

Using the MMR-based PAS and MDRM under the condition of "dis-cooperation," the problem of how to reduce the difficult and costly engineering technology on the AS of complex systems can be transformed to how to reduce the hardware/hard component AS of the MB with a relatively controllable complexity and how to "de-correlate" between reconfigurable executors; to use the PAPS, it is only necessary to replace or clean the problematic executor or to reconfigure the current operating environment or carry out the related processing until the statistical frequency of the non-compliant state found by the arbiter falls below the expected threshold. At this point, the operating environment in the MB usually avoids only the impact of the current attack scenario rather than completely wipe out the problematic factors from the executor (in fact, many sporadic or uncertain disturbances in complex systems are often not zeroed); combine or integrate the existing security defense technologies, such as encrypted authentication, firewall filtering, vulnerability scanning, virus screening, Trojan removal, and other measures for intrusion detection, prevention, isolation, and elimination, as well as introduce dynamic, diverse, and random defensive elements to enhance the dissimilarity between executors in a FE manner for exponential security gain or defense effect. The endogenous mechanism of the DHR architecture reduces the difficulty of autonomously controllable engineering implementation of complex systems (networks, platforms, systems, parts or modules, components, etc.) to an individual process/factor level from the whole-industry-chain environment level, to the simple level from the complex functional level, and to the decoupled "single-line/one-way contact" processing level from the strong correlation processing level (discussed in Sect. 10.2.2).Based on this mechanism and backed by a few key components, a secure and credible control structure is formed on the executors, where software/hardware components in the MI can be autonomously controllable, secure, and credible at the structural level even when their trustworthiness cannot be ensured. To be precise, the introduction of DHR into the target object helps to find a lower-dimensional solution to security issues of complex systems within the structure.

9.2.2 Mimic Defense System

It can be seen from the foregoing text that the CMD system aims at uncertain threats based on unknown vulnerabilities or backdoors and is based on "a fortification at the pass for key-site defense," with a purpose to safeguard the robustness of the meta-service functions (including control functions) of the target object within the MI (possibly networks, platforms, systems, parts, components, hardware/software modules, etc.) centered on the system service functionality in the MB and its implementation structure of diversified and redundant configurations or the introduction of uncertainty mapping relationship between algorithms. It takes an approach of enhancing the availability of defense resources within the MI through the PAS and MDR-NFM of HEs under the given resource conditions and uses MOV mimic ruling of the operating environment within the MB and the RCM based on backward verification as a means to block and shield non-cooperative attacks (or DMFs). With a focus on the reachability and cooperative exploitability of hidden or invisible parasitic vulnerabilities in the MB, the CMD system obtains the stability robustness and quality robustness for the target object's functionality and performance through the endogenous uncertainty mechanism of the structure.

It is important to emphasize that a target system needs to lay out the MB and set out the types and quantities based on security and cost-effectiveness requirements as well as on the probability of engineering implementation. The MB can be set either in an independent or distributed form or in a hierarchical or nested form. In terms of mechanism, the latter has a stronger anti-aggression capability. Similarly, for a network or platform system with multiple nodes, the more mimic-structured nodes it has, the safer it will be. In general, a system with innate dynamic, diverse, and random attributes (a cloud platform/data center/distributed service network, etc.) can achieve the cost-effective mimic defense if it incrementally deploys mimic control components.

The CMD system described in this section consists of the conception, rules, and models [5], as shown in Fig. 9.4.

(1) Conception refers to a collection of structured concepts. For example, with service providers, service consumers and service access points in the same OSI systemic structure. The CMD conception includes FE reconfigurable or software-definable executors, MDI, MB, and the related core mechanisms.

(2) Rules tell us how to use the conception. An example of a rule is that a service consumer in the OSI architecture must use the services provided by the underlying layers through service access points. The CMD rules include: the target object's

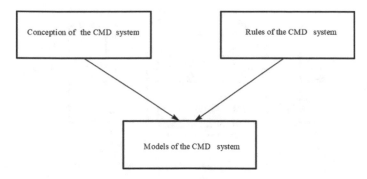

Fig. 9.4 CMD system

external service requests must be distributed to executors according to the set policy; the MOVs must be subject to ruling and output in accordance with the given policy; and the MDR and scheduling of the executors must also be operated on the basis of the feedback control policy (FCP), etc.

(3) The models describe how to use the above concepts and rules to guide the design of the CMD system. An example is the OSI reference model, which can be used to guide the design of an information system. The CMD model is the mimic I 【P】 O model [6].

CMD is suitable for the occasions where there are dual requirements for the reliability and availability of the system or device, namely, conventional security and non-conventional cyber security, and the following conditions must be met: having a functional input/output relationship or satisfying the I 【P】 O model; satisfying the RC constraints for the known dynamic processes and mechanisms as well as the predictable impact range of uncertain disturbances; providing important information resources related to service functions and performance in a dominant and robust manner; not being over-sensitive to initial investment or product price; and having standard or normalizable functional interfaces and protocol specifications that can pass compliance and conformance testing and judgment, as well as technical conditions that can be processed in plural or diverse ways. It should be emphasized that when the above conditions are met, the CMD system is suitable for any application scenario where a given target algorithm coexists with other equivalent algorithms. Even for disciplines such as scientific computing or specific applications in the field of industrial control, different algorithms that are functionally equivalent and compliant with the accuracy standard may lead to different calculation results (e.g., corrected to X decimal places), or the threshold of the control amount is a range, and policy decisions can also be made by means of "precision mask" or "range judgment." Figure 9.5 is an abstract CMD architecture (CMDA) model.

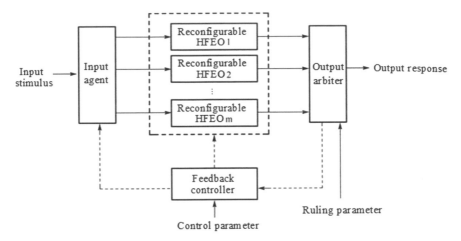

Fig. 9.5 Abstract CMDA model

9.2.2.1   The Main Concepts and Core Mechanisms of CMD

1. FEO

Function equal objects (FEOs) of the CMD can be networks, platforms, systems, parts, modules, components, or devices or facilities of a different level/granularity, or pure software/hardware/hybrid hardware and software implementation objects, or a virtualized implementation object. Therefore, unlike the DHR that stresses on the reconfigurable or software-definable traits of executors, CMD pays more attention to the diversity of executors under the FE conditions and the existence of COTS.

In most of the circumstances, a given function (regardless of the complexity) often has multiple hardware/software implementation structures or algorithms, which is a natural conclusion of the "structure-determined functionality" axiom and also a logical expression of the proverb that "all roads lead to Rome." In general, whether the heterogeneous executors (HEs) or operational scenarios with a given function (or performance) are equivalent can be determined by standard or normalizable interface testing. However, the ideal dissimilarity requirement is that the given functional intersection is equal to the maximum functional intersection, that is, there is no other functional intersection between the HEs or operational scenarios except for the intersection of the given functions. This means that in addition to normal functionality, any dark features that exist in these environments (such as undesired associated functions or side effects in the design, including unknown vulnerabilities, virus Trojans, etc.) will not produce identical output response due to the Attack Surface Input Stimulus (ASIS). In other words, as long as the output response fails to satisfy the relativity judgment conditions, it means that the dark features (including random physical failures, etc.) in the operational scenarios may have been triggered, which is also an ideal function expected by the CMD structure.

Occam's razor is the problem-solving principle that essentially states that "simpler solutions are more likely to be correct than complex ones." Launched by William of Ockham, a Franciscan friar and a great logician of the thirteenth century, the principle emphasizes a philosophy that "Entities are not to be multiplied without necessity." Kolmogorov complexity, proposed by famous former Soviet mathematician Kolmogorov, gives a formulaic expression of the principle, that is, the Kolmogorov complexity $C(s)$ or $K(s)$ of a string s is the length of the shortest description of the string. In other words, the Kolmogorov complexity of a string s is the length of the shortest computer/Turing machine program that can output and only output this string. In theory, we can use that complexity to define the redundant interface of the system, where a system is considered as a pure one without redundant dark features when its system complexity is equal to the Kolmogorov complexity. Suppose there is an ideal system without any redundant function, with all its I 【P】 O in the set defined by the system application target, there will be neither vulnerabilities and other dark features nor security problems concerned. However, in reality, we cannot design an ideal complex information system without any redundant function. Therefore, the security of a complex system is always an unavoidable dilemma in engineering practice. To some extent, the attack-defense game in cyberspace is built around the exploitation and suppression of these dark features. For example, the firewall sets out rules to filter out the operational content factors beyond the application target, the intrusion detection/defense system monitors abnormal behaviors (redundant functions) beyond the target system's application functionality to detect security problems, and the trusted computing system needs to check the expected results at the agreed stage/node to eliminate the undesired state disturbances and to screen the virus Trojan by monitoring "excessive" code functions in the target system to remove or restrict them. However, limited by the cognition of human beings and the periodicity of scientific and technological progress, we encounter insurmountable obstacles in our attempts to fully constrain or limit the redundant functionality of information systems technically and economically. More unfortunately, while adding defense functions, we are likely to unconsciously introduce undesired redundant functions, which may become new exploitable attack resources for an attacker. In this technology era, the market price of a hardware/software product seems to be strongly correlated to the scale of application while weakly correlated to its functional complexity. "Using a sledgehammer on a gnat" is no longer an anachronism. People are more willing to develop target products in an integrated way using COTS-class hardware/software components that are eco-friendly, function-powerful, standard-matched, performance-reliable, and inexpensive, making it impossible to do a better job in eliminating redundant/dark features in the foreseeable future. Therefore, MD must be built on systemic and institutional arrangements for the possible redundant/dark feature intersections of HEs beyond the given functional intersections.

CMD's FEOs can exist in either a logical or a physical sense and can also be expressed by a combination of the two. Namely, they can take a variety of forms, static or dynamic, reconfigurable or hardwired, software-definable or fix-configured, physical or virtual, software or hardware, firmware or middleware, system or component, platform or network, etc. Hybrid models are also allowed. Besides, CMD does not care whether the function executors are presented in a "white box" or "black box," because its mechanism does not exclude the possibility of existing security flaws and embedded third-party hardware/software products for both autonomously designed components without a guaranteed credibility. Therefore, a well-developed plural or diverse FEO ecology is a key element to ensure the effectiveness of CMD. In addition, attractive implementation means and approaches including user-customizable computing, software-definable hardware (SDH), mimic computing, CPU + FPGA, reconfiguration, reorganization, environment rebuildability, and functional virtualization (e.g., virtual machines/containers) can be used to enrich defense scenarios.

2. Mimic Interface and Mimic Brackets

In Fig. 9.4, the boundary of the area covered by the input agent and the output arbiter is called the mimic interface (MI). It generally consists of several sets of service (operational) functions that are well-defined and protocol-critical, with conformance or consistence testing implemented through standardized protocols or normalizable specifications, and able to judge the equivalence of multiple HEs or operational scenarios (regardless of the complexity) in terms of a given service (operational) function or even performance. Namely, the equivalence of a given function between the executors can be studied and judged through consistence testing based on the MI input/output relationship, while conformance testing serves to judge the functional identity to some extent, including abnormality processing. The integrity, validity, and security of the functions defined on the MI are prerequisites for the establishment of mimic defense, which does not cover the functions not explicitly defined on the MI (such as deep-embedded backdoors and other dark features) and those with known/unknown defects (such as design loopholes in the protocol/procedure), etc., though they may have a derivative defense effect).In other words, if an attack action fails to make the MOV on the MI non-compliant, the iterative ruling and FCM on the MI will not respond. An attack not causing an abnormal reaction of the interface's MOV will only be regarded by the target system as a tolerable intrusion process or an unperceived threat. For those attacks based on complex standby (such as APT) patterns, even if they are not recognized in the mimic ruling process, the FCL needs to mandatorily change the current defense scenario in the MI by adding external control parameters or random-order incentive mechanisms so as to disintegrate the stability of the attack chain. Therefore, an appropriately installed, partitioned, or identified MI helps to enable the early perception of unknown threats, which is critical for engineering implementation. Like the encrypted authentication function generally set to rely on a pass "guarded by one to stop ten thousand from getting through," the setting of the MI is based on similar if not identical requirements for "a fortification at the pass for key-site defense." Of course, it is also necessary to simultaneously consider the reasonable trade-off between the demand, possibility, and implementation cost.

The MB is generally used for the engineering representation of the MI. It is usually composed of the IDA, the feedback controller and the Output Agent and Arbiter

(OAA). It is the defense boundary of the HE set or operational scenarios that may contain uncertain disturbance factors such as vulnerabilities or virus Trojans. According to the definition of I 【P】 O, the IDA component represented by the left bracket strictly follows the single-line contact and PAS mechanism, responsible for distributing the external input stimulus information to the executors or operational scenarios or providing other related functions in the bracket. The OAA represented by the right bracket follows the same single-line contact mechanism to collect the MOVs of the current executors in the bracket or some algorithmic expression based on algorithm processing (such as the checksum or hash value of the OV) to make policy decisions and determine the output response. When the arbiter detects an unexpected state, the log data will trigger a BVM (backward verification mechanism)-based feedback controller and its response policy to determine whether the current operating environment should be subject to clearance and recovery (C&R), PAS, MDR, etc. or subject to imported background technologies and tools for analysis and verification and C&R. To ensure the effectiveness of the defense interface, the MI and the executor set elements should be as independent as possible in the physical (or logical) space, with the MB functions transparent to any executor or operational scenarios in principle. These functions include the input policy distribution, the output policy ruling, the inserted agent function, the executor PAS, the executor reconfiguration and reorganization strategy, and the abnormal C&R mechanism. It should be emphasized that since the MB in theory is a "non-perceived transparent middleman" not resolving the semantics and grammar of input stimulus sequences and OVs on the AS, the attacker cannot guarantee the accessibility of the attack against MB. However, in engineering implementation, CMD requires that apart from adopting the single-direction connection mechanism and trying to eradicate vulnerabilities, every effort must be made to ensure that there are no vulnerabilities (except for loopholes) or virus Trojans left maliciously. A simple and easy way is to simplify the functional complexity of the MB and reduce its AS, taking measures such as limiting or controlling the authority of online software/hardware code modification to meet the requirement for formalized correctness certification. Since each executable version of the MI is relatively stable, any measure that enhances the integrity of the currently running program version (e.g., using wired logic or firmware to the maximum extent in the implementation) is beneficial. Conditionally, there can be enhanced processing measures such as trusted computing and distributed ruling, while mimic processing can also be imposed on the MB themselves.

From the RC perspective, the MI based on the one-way contact and CLFM should also obtain the corresponding security gains from the structural effects of MD.

It should be noted that the MD scope does not cover security issues outside the MB, including but not limited to security threats not triggered by factors such as the vulnerabilities within the MI: phishing; bundling malicious functions in APP software; pushing virus Trojans in cross-platform interpretive execution files; pushing veneniferous codes through user download channels, inherent flaws in network interconnection protocols, and imperfect customized service functions; and using legitimate commands to form a malicious attack process or obtaining access and operation authorization by deciphering encrypted codes, stealing passwords, and other social engineering methods, because attackers can obtain the access privilege from the "front door" without relying on the vulnerabilities or backdoors in the mimic border [7]. In other words, the mimic defense cannot deal with the front door attacks independently. However, in principle, the MI-introduced GRCS will indirectly bring a security gain to the mitigation and solution of these problems. For example, DDOS, the most unscrupulous conduct with an established underground industrial chain in cyberspace, has to exploit the vulnerabilities and virus Trojans within the "target objects" in principle to produce massive "broilers." The GRC brings daunting challenges to the utilization of these attack resources unless poisonous software and hardware are maliciously transmitted through marketing.

3. *Core mechanisms concerned*

   (1) Input assignment and output agent

As for the left and right boundary functions of the MB, each input stimulus on the MI needs to be imported into the corresponding executor, while the OVs of the relevant executor into the output arbiter, both according to the scheduling policy. This means that both need to have multiple switching units so as to flexibly choose the corresponding operating objects according to the scheduling policy (e.g., randomly selecting k operational targets from the set M). Not only that, as HEs are required to be theoretically independent, and there are often differences in response time and process in an application environment with a dialog mechanism or a state transition mechanism (e.g., the initial value state of TCP sequence numbers in different executors in the IP protocol), or there are differentiations such as alternatives, extensions, and even computing precision, these differences can be maintained only inside the MB, while presentations outside the MB have to be in the same appearance. In addition, the output response time (ORT) of each executor is also subject to a certain degree of dispersion, and the OV can be very long (such as a long IP packet or a Web response packet). In order to simplify the subsequent ruling complexity, the MOVs may need to be preprocessed (e.g., setting a buffer queue, conducting mask processing, using regular expressions for option extraction, or calculating the checksum, CAPTCHA, hash value of an IP packet, etc.). Thus, the bridging function relevant to these processes must also be included in the OA functionality.

(2) Mimic ruling

   In the DRS, the MV object's configuration is generally certain, and the voting content is usually relatively simple, preferring the law of majority-rule or conformance voting with a low implementation complexity, and usually adopting a hardwired logic. Mimic ruling (MR) is different from the DRS voting mechanism that supports error detection functions such as inconsistent detection. The former can only judge between majority and minority, or identical and different, on the basis of the TRA. In theory, there is always a superposition-state probability of error, though in historical observations, it is agreed in the preconditions that the probability of "being relatively correct" is much higher than that of "being relative incorrect." Therefore, MR needs to not only face the engineering implementation challenge in the MOVs but also enrich the ruling information and strategy to reduce the relative error probability. Besides, in the attack-defense game environment, it should hide the defensive behavior of the target object as much as possible, including the attacker-invisible ruling algorithm and the fingerprint information on system output operations. To this end, MR adopts an iterative ruling approach based on a definable policy set and the backward verification information (BVI).

(a) *Definable Policy Set.* It consists of three aspects. First, an identical ruling algorithm can define multiple engineering implementation approaches. According to the different arrival time of the MOVs, typical majority-rule voting algorithms can define different implementation methods. For example, in the case of N-redundancy voting, as long as an MOV meeting the number threshold for the voting algorithm arrives, the voting operation can be performed to determine the output then carry out strict subsequent N-redundancy ruling in the backward verification. If the application scenario permits, the strategy of "first come, first output" can be applied, with strict N-redundancy ruling performed in the backward verification to determine whether it is necessary to update the preceding output response; of course, we can also perform the voting operation and determine the output response when N OVs are in place, which is a classical practice. Second, define the method of multi-dimensional information-assisted ruling. For example, when the MOVs are completely inconsistent, the ruling operation needs to make an output response decision using the assistant ruling parameters, including the backwardly verified historical information, the consistent performance of the executor combination, the dynamic weight information, and so on. Third, define the method of joint ruling. For example, while majority ruling is implemented, parameters of the executor concerned should also be judged accordingly, including, but not limited to its credibility level, historical performance record, version maturity, and priority level, paving way for joint ruling in conjunction with the majority ruling results.

(b) *Backward Verification Information.* It refers to the judgment using the MOV information recorded and saved in the log after output ruling (OR) is completed (in event of massive OV information, the associated data such as checksum, CAPTCHA, hash value, etc. can be retained)according to the given ruling algorithm or refers to the multi-dimensional, multi-algorithmic judgment based on the relevant historical parameters, dynamic weights, credibility, etc., including the adoption of more complex combinatorial or iterative judgment. The results of the judgment will affect the backward verification library data and may activate the RC loop to change the current operating environment. It is not difficult for readers to find that getting BVI at the time t may not only change the defense scenario in the MI but also affect the ruling output at the time $t + x$.

In short, MR is divided into two levels. One is output ruling (OR) to select the OVs satisfying the information requirements of the ruling algorithm and the assistant decision-making from the MOVs of the current HE. The other is backward verification (BV). After the output response at the time t is achieved, the multi-dimensional information and multiple comparison algorithms are used to verify or judge the OR result at the time t. In the event of non-compliance, including undesired differences between the MOVs, or incompliant OR/BV results, it may be necessary to initiate C&R, replacement and migration, or reconfiguration or reorganization of the problematic executor according to a given RC algorithm, or even to update the entire operating environment within the MI. It should be pointed out that the problematic executor is treated as a "priority" because it is likely to encounter a random failure or a deliberate attack, but we cannot rule out the possibility of misjudgments, so a BVM is needed. However, if the problem is still not circumvented when the problematic executor is processed, it indicates that there may be attack and escapes in the operating scenario within the MB, and the RC algorithm will select executors free of explicit problems and deal with them in the problematic scenario. Whether the result is negative or positive, this mechanism will disintegrate the attack and escape state based on the MR blind zone. If an executor itself is endowed with diagnostic and maintenance functions, such as log records, on-site snapshots, and reconfiguration and reconstruction history of hardware/software components, or equipped with proactive or negative defense measures, such as vulnerability scanning, data collection, virus Trojan screening, sandbox isolation, cloud defense, etc., then using the mimic ruling mechanism (MRM) to trigger related supporting functions or back-office facilities is helpful to improve the pertinence and effectiveness of fault screening and security troubleshooting of the executor or components concerned and is especially vital for the identification of 0 day vulnerabilities and unknown virus Trojans.

(3) Multi-dimensional dynamic reconfiguration (MDR) and policy and schedule

The MDR (Multi-dimensional dynamic reconfiguration) objects include all physical or virtual resources of reconfigurable or software-definable executors within the MB. New FE executors can be generated by extracting component elements from the heterogeneous resource pool according to the pre-set reconfiguration/reorganization plan, while some components can be replaced in the existing executor, or the current runtime environment can be reconfigured via increased or decreased component resources in the existing executor, or new algorithms can be loaded to the programmable, definable components to change the operating environment of the executor itself, or the back-office tasks of the executor can be added or reduced, change their working scenarios, and so on. Its functional significance lies in two aspects. One is to change the dissimilarity of the operating environment in the MB to destroy the cooperation of the attack and the inheritability of the staged results. The other is to transform the apparent traits of the hardware/software vulnerabilities in the current service set or to invalidate the reachability of the AS by means of reconfiguration, reorganization, redefinition, etc. In fact, just as patching can also cause new vulnerabilities (backdoors), reconfiguration and reorganization have not achieved the goal of problem zeroing in a strict sense, but it circumvents the temporary intrusion attack events by changing the current defense scenarios of the executor, which essentially belongs to problem avoidance. Therefore, in the CMD system, as long as an executor output anomaly is not an "unrecoverable" problem, it should not be "suspended" for processing like the DRS. Instead, guided by the principle of "assign the appropriate person to complete the appropriate task," we can use the combination of executors unsuitable for Scenario A to address the attacks in Scenario B or X and vice versa. Such an operational strategy follows a hypothesis consistent with that of a CMD system that allows the executors to work under the "virus-bearing" conditions. The scope of PAS not only reconfigures and reorganizes the executor elements themselves but also realizes the heterogeneous change of the defense scenarios in the MB by cleaning [8], replacing, or migrating the current service set's executor elements so as to achieve the purpose of problem avoidance.

It is important to emphasize that the trigger of both the MDR and PAS is based on the MR-based abnormality perception mechanism (MPM), including whether any BV operation needs to be added upon a change in the defense scenario. In order to deal with a "latent" standby attack in the MB (which tries to avoid affecting the OVs during the incubation period) or even a cooperative attack already prepared for an escape, it also needs to use the internal dynamic or random parameters of the target system and the external control commands or parameters generated by the preprocessing policy, periodically or irregularly triggering the feedback control loop (FCL) to implement preventive switching of the current defense scenario so that the operating environment can better resist latent, standby, and cooperative attacks (see Sect. 10.4.6 Benchmark MD functionality test).

(4) Cleaning and recovery and state synchronization

Cleaning and recovery (C&R) is used to handle HEs with abnormal output vectors (AOVs) at three levels: about to restart the problematic executor, reinstall or rebuild the executor operating environment, and put the restarted or reinstalled executor in the standby state as fast as possible. In general, HEs within the MI should be periodically or irregularly subject to different levels of pre-cleaning or initialization, or reconfiguration and reorganization, to prevent the attack code from persisting in the memory or implementing complex attacks based on state transitions or maintaining the attack-and-escape state. In particular, once the executor output is found to be abnormal or not functioning properly, it should be removed from the available queue timely for mandatory cleaning [8]or reconfiguration. Different executors are usually designed with a variety of abnormality recovery levels to provide flexible options in line with the context.

Reconfiguration and reorganization are not only effective means to move, change, or dodge the current environmental vulnerabilities of the executors but also important measures to change or enhance the inter-executor dissimilarity for enhanced cooperative attack difficulty. Among them, FPGA-based, SDH, mimic computing, user-definable computing, virtualization technology, and mimic compilation all feature a relatively high cost performance, easier for the engineering implementation.

After C&R, an executor needs to re-synchronize with the online executors in terms of state or scenario to maintain synchronization with the MRM, which is called state or scene synchronization (SSS).It can usually refer to the mature abnormality handling and recovery theory and mechanism of the DRS in the field of reliability. However, synchronous processing in different application environments will be challenged at varying difficulty levels, with many tough problems in engineering implementation. It should be emphasized that in the DRS, there is generally a mutual-trust relationship between the executors (unless in an abnormal state) and the recovery operation can be simplified through mutual learning (e.g., using an environment copy method).Since MD allows HEs to bear viruses, it is necessary to isolate the transmission routes among them or block cooperative operation of all forms. In principle, HEs are required to operate independently and eliminate the impact of "hidden channels" or side channels as much as possible. Of course, while forcing an attacker to face the challenge of "cooperative and consistent attacks on dynamic and plural targets under the non-cooperative conditions," it will bring engineering and technical challenges to the rapid recovery and re-synchronization of the executors.

(5) Feedback control

The functional relationship between the feedback controller and the related components is the arbiter sends the ruling state information (including the BVI) to the feedback controller, which then forms operational instructions for IA/IO and reconfigurable executors according to the algorithms and parameters given by the control channel or the control policy generated from the self-learning mechanism, wherein the input distribution instructions to the IA are used to direct the external input information to a specified HE (which can affect the reachability of the AS), so that the HE elements can be dynamically selected to form the current service set of the target object. The operational instructions sent to reconfigurable or definable hardware/software executors are used to determine the reconfiguration object and the relevant policy for reconfiguration. It is not difficult to see that the above functional components constitute a closed loop, but they need to operate in the negative feedback (NF) model. That is, once the feedback controller finds that the arbiter has any non-compliant state output, then, according to the FCP, the IA/OA can be instructed to "replace" the executors whose OVs are not compliant in the current service set or "migrate" the service to the executors in the standby state. If the arbiter has not been restored to the previous state after the "replacement or migration," the foregoing process will continue. Similarly, the feedback controller can also instruct the executors with non-compliant OVs to perform cleaning, initialization, reconfiguration, or redefinition, etc., until the arbiter state returns to the normal level or falls below a given threshold. We call this process "PAS and MDR-NFM based on arbiter output." It should be emphasized that, in theory, once the negative feedback loop (NFL) enters the steady-state process, the defense scenario in the service set is similar to the "unsustainable intrusion tolerance" of the DRS and may be exploited by an attacker incubating or hidden in the target object (e.g., a side channel attack). Therefore, it is necessary to force the feedback controller to impose preventive changes on the defense scenarios in the current service set through the external command channel.

The negative feedback mechanism (NFM) is advantaged in its ability to timely evaluate the effectiveness of a defense scenario organized through dynamic, diverse, random, or traditional security measures and to automatically select an appropriate defensive scenario in the current attack scenario, avoiding unnecessary system overhead for defense behaviors such as the MTD, which "constantly migrates or transforms addresses, ports, commands, and data over time." However, if the attacker's ability can lead to frequent activation of the NFM, even if the attack and escape cannot be realized, it may cause bumping service performance in the target system due to the constantly changing defense scenario. In this case, it is necessary to introduce a control law function or an intelligent processing policy (including a machine learning and inference mechanism) in the feedback process to cope with such DDOS attacks on the CMD control process. Fortunately, the service functions are separated from the robust control functions in the CMD system, namely, the MB's control function is separated from the executor's service function. Even if an attacker can take control of the feedback loop, the intended purpose is hard to be achieved if the service provider executors are not cooperatively captured. According to the one-way contact mechanism of the MI and the AS theory, since the loopholes in the FC process itself are attack-unreachable, an attack based on design vulnerabilities will be a mission impossible for an attacker without tremendous imagination and creativity, of course, excluding a social engineering approach or a backdoor left by the designer or even the maintainer. In fact, FCL can also be subject to functional splitting for the standardization of MR, PAS, MDR, and other functional components, which will then be designed, manufactured, and delivered by multiple vendors in a market-oriented manner. End users can selectively purchase and configure personalized policies or algorithms in a customized way to dodge problems such as "how to ensure the credibility of trusted root vendors" in trusted computing. Of course, if conditions are allowed, the feedback control loop can embedded in the firmware, or users can adopt the customization as an optional solution.

(6) De-cooperation

An attack based on the target object's vulnerabilities can be considered a "cooperative action." We can start with analyzing and identifying the target object's structure, environment, operating mechanism, and hardware/software components; find as much as possible the exploitable defects; analyze the defense vulnerability; and study the exploitation methods, including the required mechanism for synchronization or cooperation. Similarly, in malicious code setting, we should also study whether a code can be stealthily embedded in the specific environment of the target object and how can it remain unidentified even at work. The more complex and sophisticated the attack chain and the more processes or paths to go through during the period, the more demanding the conditions for the attack will be, and the harder it is to ensure the reliability and stability of the attack. In other words, the attack chain is actually highly fragile and strictly reliant on the staticity, certainty, and similarity of the target object's running environment and attack paths. In fact, the defender only needs to appropriately add some synchronization mechanisms controlled by random parameters or establish necessary physical isolation domains in the processing space, sensitive paths, or corresponding processes (such as SGX, trusted-area or offline modification of FPGA and PROM, etc.) to disintegrate or reduce to varying degrees the impact of attacks based on the vulnerabilities. For example, even if the same dark feature exists across the HREs within a mimic structure, it is extremely challenging to achieve a cooperative attack under the non-cooperative conditions. Therefore, the core mission of dis-cooperation is to prevent a penetrator from exploiting a possible synchronization mechanism for a CM or homomorphic attack in a spatiotemporal dimension. In addition to the common input stimulus conditions on the AS, the possible communication paths or interaction functions as well as contents between the HEs should be removed as much as possible (e.g., invisible communication links or side channels, uniform timing or time service, and mutual handshake agreements or synchronization mechanisms, etc.), making it difficult for the "isolated and segregated" dark features in the HEs to form a cooperative attack and escape in the MI. However, in the theoretical sense, the input stimulus sequence in I 【P】 O is "synchronously reachable" for each and every P_i, which deserves special attention and careful handling in dis-cooperation.

(7) MB's one-way communication mechanisms

The MB's communication mechanisms are mainly those (1) between the IA/OA and the HEs; (2) between the HEs and the arbiter; (3) between the feedback controller and the executors; (4) between the arbiter and the feedback controller; and (5) between the feedback controller and the IA/OA. The reason for adopting one-way communication mechanisms is that an MB component itself can obtain the defense gain brought by the mimic structure. Even if there are design defects or loopholes, they should be transparent or invisible when observed from outside of the MB, namely, with an attribute of attack unreachability.

It must be emphasized that, in all MB aspects, we should avoid as much as possible the two-way conversation mechanism (with the executors) and the function of receiving uploaded executable files. Especially when the input and output agents are located in front of the MB, arrangements must be made to mechanically and completely block all possible attack code injections from the AS in mechanism.

4. *Examples of the MI settings*

As for the MI settings, in addition to the given constraints and application scenarios, the following principles must be considered: ① MD has special effects on known unknown security risks or unknown unknown security threats in the MI, and we should give full play to its strength. ② Analyze the core security requirements of the target object, select pass to build a fortification, and determine the key-site of defense. ③ It is cost-effective with a low implementation complexity.

(1) Defense Interface (DI) based on core data. If the storage or reading of core data and directory management is the protection focus of a security system, it is appropriate to set the access and operation interface for core data as the MI. The introduction of a FE-DHRS ensures that, under the same input stimulus, when accessing directories and data or forming routing tables and forwarding flow tables, FEHEs can make real-time, escort, or follow-up judgments on the conformance of the related operations. Typical applications include various file storage servers, routing switches, domain name systems, data centers, SDNs, and the like.

(2) DI based on critical operations. In the field of industrial control, control systems often have to send various operational instructions and control parameters to the executing components. The legitimacy as well as the parameter domains of the instructions and data formats are verifiable, but the combined form of instruction/data streams often fails to support the completeness verification due to a state explosion, which provides an opportunity for compliant yet abnormal combined operations using normal instructions or data to deliver a deliberate attack. After the introduction of the DHR architecture, the question whether multiple FEHEs have sent correct instructions and data to the same stimulus signal is transformed into an issue of conformance ruling between the MOVs. Similarly, given an input sequence, the question whether multiple FEHEs have produced the same instruction stream and data stream can also be translated into an issue of conformance between the output response sequences of the executors. The premise is that a deliberate attack in a combined form will find it hard to cause the HEs to produce output response sequences featuring majority conformance.

(3) DI based on core functions. Suppose the goal is to endow robustness or flexibility protection in a given function, as long as the function to be protected can be described in a standardized or normalized manner, and this description can be used to test whether the diversified hardware/software executors satisfy the given equivalent conditions, the CMD architecture will be mechanically eligible for achieving the goal. For example, multiple FE heterogeneous CPUs (HCPUs) can be selected through conformance testing of a given set of arithmetic computing functions. These HCPUs can be used to build a system or device with a DHR architecture, which, in respect of a given arithmetic computing functions, is insensitive to design flaws of the HCPUs within the structure or the vulnerability-based attacks.

(4) DI shielding unknown vulnerabilities in the interface. If there is a standardized sensitive application to be based on a complex hardware/software system with high reliability, availability, and credibility, it is appropriate to adopt a DHR architecture at the system level. If it is impossible to ensure "virus-free" conditions, the system architecture technology will be the only choice to shield the impact of attacks based on known or unknown vulnerabilities on upper-layer applications, for instance, distributing sensitive applications in the HR environments whose credibility cannot be guaranteed and using the MB to implement MD in the given interfaces of sensitive applications.

(5) DI for deliberately designed backdoors. Unlike the vulnerabilities and backdoors passively introduced to the target object without notice via the IP cores, middleware, shared modules, open-source software, entrusted processing, and other pathways, the backdoors are deliberately implanted by the software/hardware code designer. The former can incorporate these elements without a guaranteed credibility into the scope of the MB in a FE-HR manner, while the latter can be effectively protected by the MB if the implantation is intended by only a few designers of executor elements in the MB. However, if the backdoor implantation is a deliberate action of the MB designer, you can hardly expect

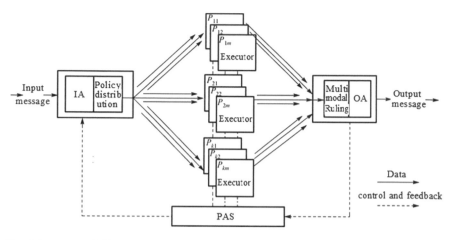

Fig. 9.6 Mimic I 【P】 O model

that it will be well prevented. Of course, the MB itself can also be made mimic and made sure that the designer is not the same person/organization. If there are standardized COTS-level hardware/software components for the MB, users can purchase such components to avoid the risk (unless the attacker has strong social engineering capabilities and resources).

It should be pointed out that the above methods are, in principle, also applicable to the applications guarding against virus Trojans in the MI.

9.2.2.2 CMD Model

The CMD model is an equivalent I 【P】 O model. Assume that in a defense target (regardless of its size or granularity) that satisfies the I 【P】 O model, P can be a complex network element, a hardware/software processing system, or a part, module, component, etc. and the technology and economy allow the application of the CMD architecture, then the model can be expressed as: I 【$P_1,P_2,...,P_n$】 O, wherein the left bracket (IA) connected to Input I is given an input distribution assignment function (the pattern can be one-to-one, one-to-many, broadcast, etc., and the mode may be either static or dynamic or random) and the right bracket connected to Output O is given a MMR (or policy ruling/weight selection) function and other AO functions, including the normalization or preprocessing of the HE output information. In the MB, P_n is an HE functionally equivalent to P, as shown in Fig. 9.6.

Compared with the complex processing functions of P_n, IA and OA have mainly simple functions such as multiplexing-demultiplexing, normalization, MMV, and selective output; their design defects (including hidden backdoors) can usually be detected by formalized correctness proof method and tools. Since the IA/OA generally does not parse the semantics of the input and output information (this may involve some grammatical content such as the message format) and has no AS-reachable internal or external addresses or ports, its functions are completely transparent for the attackers or HREs outside the structure. Besides, with the strict one-way contact mechanism, attack reachability is a rare occurrence even if there are design flaws or vulnerabilities. In addition, IA/OA typically requires functional and physical independence from P_n to avoid as much as possible cooperative attacks (including those in the social engineering sense). Similar to the engineering implementation risks that may exist in an encrypted device, if a backdoor with a particular trigger is implanted in an IA/OA, the MD effect will be uncertain. Of course, in addition to the one-way contact mechanism, FCL can also add some PAS and control functions to enhance its own immunity, such as the reconfigurable (reorganizable) and multi-redundancy deployment of functional components, the acquisition of dynamic operation schemes using policy selection and randomized parameters, etc., and the design of specialized private components or trusted roots in a trusted custom and trusted computing approach. It should be emphasized that once IA/OA functions get impaired, such as stopping dynamic scheduling or no longer conducting policy ruling, the object of defense will lose the uncertain attribute. Therefore, it is necessary to attach a "watchdog" to the IA/OA. In special cases, the security of the MB per se can also be enhanced through trusted computing or the mimic structure.

In engineering implementation, the instructive functions of the HEs, such as scheduling, reconfiguration, reorganization, and C&R, as well as the "watchdog" function, are usually included in the MB functionality.

9.2.3 Basic Features and Core Processes

1. Basic features

There are four basic features in the MI: (1) the threat perception mechanism based on the TRA; (2) the GRCS based on the DHR format; (3) policy ruling, PAS, and MDR-NFM based on the mimic disguise philosophy; and (4) the endogenous structural effect of MD which can be verified and where its anti-aggression and reliability should be subject to quantitative design and measurable verification. The specific presentations include:

(1) The mimic structure adopts the iterative ruling model, which is different from the traditional majority-rule voting method. The threat perception space migrates to a multi-dimensional space from a single space, and the cognitive domain of threat perception moves to heterogeneous redundant targets from individual objects. The threat perception model ranges from the matching based on feature library information to the iterative ruling based on plural policy, while the threat perception error rate can be adjusted via the number of individuals engaged in the relativity ruling, the dissimilarity between individuals, the information abundance of output vectors, and the number of ruling policies.

Generally, the superposition state attribute of the relative ruling result needs to be confirmed through the backward operation. For example, in a three-redundant structure, the arbitrator will have a 2:1 state, and there may exist two judging possibilities. ① In a large probability, it can be regarded as DM failure or independent attack event;② In a small probability, a common mode failure or cooperative escape may have happened. Therefore, it is necessary to use multiple-aided arbitrating algorithms to verify the current state of the arbitrator accordingly. If the verification result fails, the feedback control loop must be activated to change the current operation environment to dodge the common mode failure or disrupt the escape state through the iterative mechanism of defense scenarios. If the verification result is correct, the decision can be made about whether the current scenario needs to be changed according to the given control state. In summary, on the one hand, we have to prevent the occurrence of common mode failure in the superposition state or cooperative escape. On the other hand, it is necessary to avoid the system overhead caused by frequent switchovers.

(2) The GRCS in the DHR form is based on the mature heterogeneous redundancy reliability (HRR) theory and practice and takes the DRS scenarios as a transient steady-state (TSS) manifestation of dynamized and diversified defense scenarios. Since the MR has a more diversified ruling policy and multi-dimensional BVI than the simple majority ruling, it features a higher reliability and anti-attack performance, as well as design calibration and verifiable measurement. In principle, these TSS defense scenarios deployed on the time axis are the differential forms of the DHR architecture. The DHR at this time has identical or similar advantages and disadvantages as the DRS does. However, the DHR has the ability to adaptively select the appropriate defensive scenarios, thus avoiding the genetic defects of the DRS, which struggles to maintain anti-attack performance in fixed defense scenarios. As an innovative GRCS, the DHR can mechanically and effectively suppress the impact of GUD factors in the MI, including unknown DM and non-consistent cooperative attacks.

(3) The endogenous structural effect of CMD can be verified, and the reliability and anti-aggression of the structure is subject to quantitative design and measurable verification. This is because the CMDA can normalize GUD to a reliability issue, and it in essence has an RC function with a CFM. Therefore, most of theories and methods for reliability testing and verification are suitable for the security and reliability testing and evaluation of mimic structural objects, including classical injection or destructive testing principles and methods applied under the "white box" conditions.

(4) The DHR architecture, PAS, and MDR-NFCM based on MR can produce a variety of endogenous effects:

  ① Uncertainty effect. Since any abnormal situation found in the MR process will trigger the feedback controller to generate operational instructions for the PAS of the defense scenarios or the reconfiguration/reorganization of the executors within the service set. Any information or experience acquired in the current scenario will lose the reusable value under this ICM.

② "Fuzzy pattern recognition" effect. Due to the relativity ruling in the FEHR scenarios, as long as the multimodal executor's output vectors exhibit an abnormality in space and time, the system does not care whether it is a known or unknown threat, nor does it care whether it is a cyber attack event or a random fault, not to mention whether the attack originates from external or comes from internal penetration attacks (much like the big data analysis scenario where "I know but I don't know why").Therefore, all of these problems can be normalized or say avoided by a defense scenario replacement that can automatically converge under abnormal conditions. However, using this perceptual effect, it is also possible to develop a "know-why" analysis function. This is of engineering implementation significance for detecting 0 day vulnerabilities and virus Trojans.

③ Integrated Point-Surface (IPS) defense effects. Since MD adopts relativity ruling, which is weakly correlated or even uncorrelated with the specific attack behaviors and features and only strongly correlated with the spatio-temporal consistency of output vectors of the executors or operative environments, it has the surface defense capability similar to NI. With the MR-based PAS and MDR-NFM, for any DMF or any explicit attack against an executor in the current service set, as long as it involves changes in the OV semantics or contents, it can be detected in real time by the MR process, and the current problem scenario will be replaced or circumvented by the PAS and MDR-NFM, thus providing a point defense (PD) function not relying on the vulnerability library and the attack signature database. It is easy to observe that the abovementioned IPS defense functions are derived from the operating mechanism of the same technology architecture.

④ The effect of "difficulty of cooperative attack under the non-cooperative conditions." In view of the basic definition, all mimic defense elements in the MB should be as heterogeneous as possible, and there should be no synchronization or cooperation between them, except that the functionality and performance on the MI conforms to the consistency requirements. Ideally, there are no other functional intersections between the HEs except for the given ones. Although this is an unrealistic requirement in engineering, the "dis-cooperation" of the mimic environment requires that any attack attempting to exploit the dark feature intersection for escape has to break the "cooperation barrier under the non-cooperative conditions." The more complex and diverse the attack environment, the more system resources the attack chain depends on, where any mismatch in the attack action, once perceived by the MR-based FC process or triggered by a random external control command, will inevitably end up with all labor lost. Therefore, the effectiveness of CMD essentially depends on "the difficulty of cooperative and consistent attacks on dynamic and plural targets under non-cooperative conditions," while the technical structures, operational mechanisms, and all possible measures of CMD are focused on the enhancement of cooperative attack difficulty. In other words, instead of thinking that CMD is aimed at destroying the stability of the attack chain, it is more appropriate to say that the target is

MD infrastructure

FEHE set

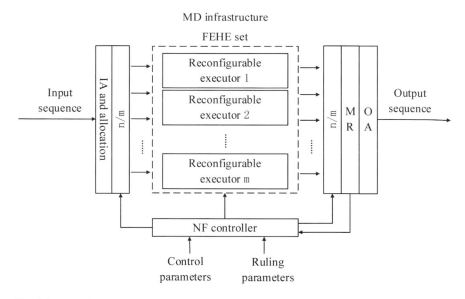

Fig. 9.7 MD infrastructure

to significantly increase the difficulty of coordinated attacks through position defense.

⑤ Robustness service effect. The CMDA can normalize the anti-aggression issue into a reliability issue, so that the two can be consolidated and handled in parallel, enabling the target system to provide highly reliable, available, and trusted service/control functions in an integrated manner, and its mechanism for segregated closed-loop control and service provision can incorporate or integrate traditional and non-traditional security technologies unimpeded for exponential defensive gain.

2. Core process

The mimic defense infrastructure based on the I 【P】 O model is shown in Fig. 9.7. The software and hardware resources in the MB at least have two kinds of implementation functions: one is the service function provided by the reconfigurable entity (including diversified or form-variable operational scenarios), and the other is the security function cooperatively provided by the utilizable resources in the MB. The latter is the endogenous security function without relying on a priori knowledge, and its core mechanism originates from the uncertainty effects in the mimic structure. If it is combined with conventional security technologies, the defense capability can grow exponentially.

The basic process of CMD is shown in Fig. 9.8. It should be pointed out that due to the differences of mimic guise policies in the feedback control loop in engineering implementation; the differences of implementation (algorithms) methods in the input agents/allocation, output agents/allocation, and the arbitration and feedback

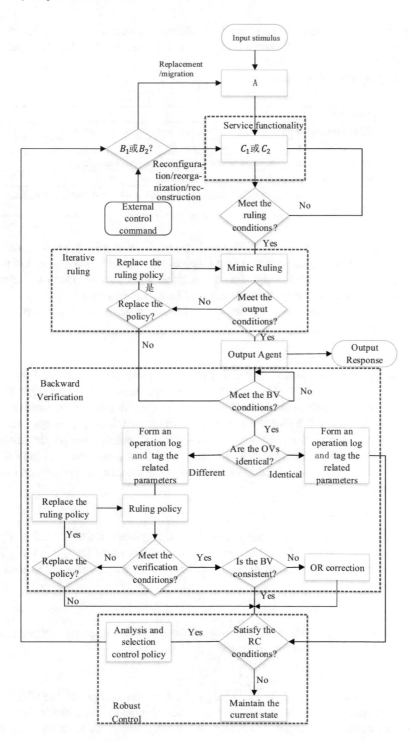

Fig. 9.8 Basic process of MD

control; as well as the differences among different service scenarios in the recon-figurable executors, the following figure only serves to describe the general procedure.

Note 1:Symbol description

Operation	Meaning
A	Send the input sequence based on the input assignment and distribution policy agent
B1	Abnormality replacement/migration executor
B2	Abnormal executor reconfiguration/reorganization/reconstruction
C1	Corresponding executor processing input sequence
C2	Perform the related commands such as initialization, reconfiguration, reorganization, reconstruction, etc.

Note 2: The system generates operating instructions with a random nature according to some uncertain parameters (such as the current number of active pro-cesses, CPU/memory occupancy, etc.) and a given algorithm, mandatorily activat-ing the loops in a stable state through external control channels.

The above chart consists of five top-down functional areas: service, iterative rul-ing, OA, BV, and RC.

(a) Service

This area has four main functions: According to the instructions and policies sent from the RC area, it (1)completes the assignment and distribution of the input stim-ulus source to the HEs in the MB; (2)implements the functions of the instructions on the migration and C&R of the executors or the reconfiguration, reorganization, and reconstruction of the operating environment; (3)forcibly changes the current running environment in response to an external control command; and (4) deter-mines whether the number of the OVs received meets the threshold stipulated in the ruling policy.

(b) Iterative ruling

The iterative ruling area is mainly made according to a given set of ruling algo-rithms and the related information obtained in the BV. Once there is an eligible output vector, it will be sent to the OA process or otherwise transferred to the BV functional area.

(c) OA

This area is responsible for completing the OR(operation response)-related oper-ations and proceeding to the next step.

(d) BV

There are four main functions: (1) judge whether the given number of OVs have been received; (2) judge whether these OVs are identical and complete the state transition based on the result; (3) impose BV on non-compliant OVs based on the policy library; and (4) make policy decisions based on the BV and iterative ruling results to determine whether to send a corrective notice and enter the RC process.

(e) RC

There are at least two main functions: (1) determine whether to activate the FCL according to the control law and other policies; (2) once the loop is activated, make a comprehensive analysis of the log information, the related parameters and the resource well-being, and change the operating environment in the MB accordingly following the policy of fastest problem avoidance. Herein there may exist two kinds of conditions: the current attack state can last long enough to guarantee that the iterative effects of the defense scenarios get confirmed; otherwise, the attack state cannot last long enough to ensure whether the replaced defense scenarios can be effective, which is only the action of defense scenario change. What is of optimal significance is that the defense can utilize the confirmed information related to the former and establish the visual map of relationship between the attack scenarios and the defense scenarios through self-learning mechanisms so as to remarkably reduce the average times of scenario iteration.

It should be emphasized that the "external control command" branch process in the upper left corner of the flowchart is specifically designed to disintegrate the stability of the N-model escape state. For example, once an attacker has reached the point where the MOV comparison proves consistent and in an escape state, it is unperceivable for MR, and the FCL will not be activated. The escape state at this time can be maintained mechanically. Therefore, it is necessary to introduce a random strategic disturbance mechanism irrelevant to the ruling state into the FCL of the DHR architecture, namely, forcibly changing the current operating environment according to the external control command, including preventive C&R or restoration, reconfiguration, and reorganization of the replaced executors. Such strategic disturbances should be unperceivable to normal service functions on the MI but very fatal to the attack and escape. We will see the indispensable role of this mechanism in the 10.4.6 Benchmark Functional Test for Mimic Defense.

9.2.4 Connotation and Extension Technologies

9.2.4.1 Connotation Technologies

The logical connotation of MD lies in the generation of the endogenous uncertainty effect based on the DHR architecture and the biological MDM within the MI, which delivers an exponential defense gain. Its unique endogenous security (ES) function can effectively deal with non-cooperative known or unknown security threats in the MI in the absence of prior knowledge of the attacker and can also manage and control the uncertainty faults stemming from random physical failures or design defects in real time so as to enhance or improve the structural reliability and availability. In essence, mimic defense belongs to the category of automatic control and RC, but the DHR architecture employing the MDM can better normalize traditional and non-traditional security issues into reliability issues for processing. Since CMD adopts the GRC technology, the target object can strictly limit the impact of its unknown vulnerabilities within its controllable or traceable range so as to avoid the "overflow effect" polluting the cyber security environment. In theory, the mimic structure can fully prevent unconventional and conventional security threats of the different-mode nature and constrain the attack-and-escape and common mode failures of the common mode nature within the given threshold value. Therefore, the author herein makes a bold prediction that a new generation of hardware/software devices will have an iconic function of ES effects based on the systemic structure and subject to "quantitative design."

9.2.4.2 Extension Technologies

Although the RC functionality of the CMD is separated from the target object's service functionality, its defense effect largely depends on the general dissimilarity between the executors and the cooperation under the non-cooperative conditions in the current service set. Therefore, any extension method that increases the dissimilarity and cooperation difficulty can significantly improve the target object's anti-attack performance and reliability indicators. At least four types of extension methods can be taken into account:

1. Executor incremental extension

It is to introduce defensive elements such as dynamicity, diversity, or randomness on the executors to increase the cooperative attack difficulty in the multi-dimensional space, including but not limited to the related methods mentioned in the mobile target defense, as well as traditional security measures such as additional intrusion detection, intrusion prevention, and intrusion isolation.

2. Executor reconfiguration extension

An executor itself selects a variable architecture technology system featuring redefinable software/hardware, reconfigurable resources, and a reorganizable processing environment/operation mechanism, such as software-definable functions, user-defined computation, hardware/software-cooperative mimic structure computation (MSC), CPU + FPGA, SDH, etc.

3. Executor virtualization/cyberization extension

The vogue for today's information and communication networks is the DHR architecture with an outstanding feature of "separating control from service." Therefore, as long as the HEs in the structure meet the FE service conditions, they shall be acceptable regardless of their form, granularity, composition, and technical maturity. That is, a HE can be a virtual server or data storage at the data center, or a virtual component or environment in a pooled resource form such as a virtual container of a cloud service or cloud computing platform, or an independent network element device or a functional platform, such as a domain name server in the same autonomous domain on the Internet or several mirror nodes of a distributed file storage system. This means that the CMDA can not only be implemented in a more cost-effective technical approach but also achieve "super-value" robustness and security by inexpensive and incremental MB software/hardware components deployed in a network or cloud service environment.

4. Executor detection technology extension

Although CMD does not depend on the ability of individual executors to accurately detect unknown faults and uncertain threats, it does not exclude the corresponding detection functionality in the executors. Conversely, if it is sufficiently economical, an executor is expected to have a rich capability in fault detection or threat identification, including various functions of error correction or fault tolerance or abnormal situation avoidance. The combined use of these functions can make the PAS and MDR operations more oriented or targeted, the dynamic convergence of closed-loop control faster, and the cooperative attack more complex, so as to greatly alleviate the target system's response to bumpy service performance including DDOS attacks.

In short, any technique or method that can prevent or disrupt the attacker's cooperative attack on the target object, including internal snooping, penetration, internal-external collusion, and other factors damaging information security, can be used as the extension technology for CMD, such as introducing a defender-controllable uncertainty (dynamic programs, data, addresses, ports, protocols, networks, etc., or even system fingerprints) in a static (fixed) relationship; minimizing the interactive information between the associated parties; seeking data file storage security through an intentional fragmented and decentralized storage and multi-path transmission; conducting various practices increasing the difficulty of perceptions and attacks by external attacks or internal penetrators; or increasing the difficulty of reproducing the attack results by reducing their exploitation sustainability. Naturally, it also includes techniques and methods involved in the current cyber security defense landscape, such as firewalls, honeypots, honeynets, and trusted computing. These security measures attached to the CMDA equivalently increase the intrinsic dissimilarity of the CMD environment and the difficulty of cooperative attacks under the non-cooperative conditions in the multi-dimensional space, thus nonlinearly enabling the target object to disturb and destroy cooperative attacks by way of the structural effect. In other words, it reduces the probability of attack and escape nonlinearly.

9.2.5 Summary and Induction

(1) In general sense, the core objective of CMD is that in view of four challenges in cyberspace, namely, "the design flaws are inevitable," "the backdoors cannot be eliminated," "the vulnerabilities cannot be thoroughly detected," and "the

product security and quality can't be controlled," and in the context of globalization where the credibility of a hardware/software supply chain and the opensource innovative technology and industrial ecology cannot be guaranteed to reliably resist man-made attacks (i.e., backdoor and Trojan attacks) in the MI by employing the GRCS-DHR and its endogenous security effects without relying on the prior knowledge and behavioral traits of the attacker and to apply quantitative design and verifiable measurement to the defense effect. This will disrupt the attack theories and methods based on software/hardware code vulnerabilities, changing the rules of cyber games.

(2) Technical pathways of CMD: Enhancing the endogenous uncertainty effect of the DHR by means of the mimic disguise strategy and the reconfigurable or form-variable mechanisms of the executors or defense scenarios and transforming the issue of unknown security threats targeted at individual executors in the MB to the issue of DMF/CMF and further to the issue of stability robustness and quality robustness that can be uniformly processed by the relevant control theories and methods, so that the target object can eventually acquire integrated, highly reliable, available, and credible service functionality.

(3) Basic theories and methods of CMD: The TRA, HRR theory, and CFRC theory and related methods.

(4) Basic operational model of CMD: Apply the mimic disguise strategy to DHRs MMR, OA, ID, FC, and the executor's MDR process, so that the defense environment will attain endogenous security attributes such as uncertainty or "increase-of-entropy" and display an external defense feature for an attacker, which can't be locked or controlled and difficult for cooperation. At the same time, it can provide IPS defense irrelevant to the attack behavior and feature information and can also effectively sabotage the effort to achieve cooperative attacks through trial and error or exclusion methods. It emphasizes the effect of introducing traditional security technology based on the nonlinear enhancement of structural effects. We can give full play to the fusion effects of the DHR structure and introduce other security technologies so as to strengthen the defense fog in a nonlinear mode

(5) Provision of the universal GRC function. In the context of globalization where the credibility of a hardware/software supply chain cannot be ensured, the GUD, including man-made attacks such as those based on the vulnerabilities of the target object, can be suppressed or controlled by the theories and methods of system engineering. As an innovative enabling technology, it can find a new way for IT/ICT/CPS systems to access the ES functions via the law of the structure-determined nature.

(6) Improvement of autoimmunity and reduction of the cost of security maintenance and management. The effectiveness of traditional defense technology relies heavily on costly, time-consuming, and laborious manual guarantee (management) methods. The root cause lies in the defense model of "mending the fold after a sheep is lost." The mimic structure is naturally immune to the vulnerabilities and Trojans inside the structure, and its defensive effectiveness is not or weakly correlated to the artificial intervention factors, so it can greatly reduce

the full life-cycle comprehensive use cost, including the security maintenance, and guarantee. However, it is beneficial to utilize the conventional technologies to have a routine (monthly and quarterly) checkup and diagnosis of the software and hardware resources in the MB.

9.2.6   Discussions of the Related Issues

9.2.6.1   CMD Level Based on the Attack Effect

In comparison with the confidentiality, integrity, and validity of information security [9], CMD defines three mimic defense levels (MDLs) that don't rely on traditional security measures based on the attack effect.

1. All shielding

It refers to a scenario where the functions, services, or information protected in a given mimic defense interface are not explicitly affected in any sense when attacked by an external intrusion or an "insider penetrator" and the attacker cannot give any valuable assessment of the attack effectiveness from outside the MI. It is like an entry to an "information black hole" called All Shielding (ASD), which is the highest level of CMD.

2. Unsustainable

It refers to a scenario where the functions or information protected in a given CMDinterface, when attacked from inside or outside, may be subject to "post-correction" or self-recovery with an uncertain frequency as well as duration. As for the attacker, even if a successful attack and escape can hardly maintain the attack effect, or give any meaningful paving for the subsequent attack operations, called Unsustainable (USB). At this point, the defender's responsive action can be perceived by the attacker from outside the MI.

3. Hard to reproduce

If a given CMDinterface is attacked from inside or outside, the protected functions or information may appear to be out of control at the time t, but it is difficult to reproduce an identical scenario if the same attack is repeated. In other words, as far as the attacker is concerned, the successful attack scenario or experience achieving a breakthrough is not inheritable and lacks programmable tasks and certain strike effects in the space-time dimension, thus called hard to reproduce (HTR). In this case, the "privacy, integrity, and validity" protection functionality in the MI may have a "temporary leakage" with an uncertain frequency, and the defender's responsive action can be perceived by the attacker from outside the MI.

The above three defense levels emphasize the degree of uncertainty caused by the attack process. ASD prevents the attacker from obtaining valid information of the defender across all stages of the attack chain; USB makes the attack chain lose indispensable stability; HTR prevents historical experience based on detection or attack accumulation from being used as prior knowledge in planning attack tasks. However, existing security theories and test methods can hardly quantify the degree of uncertainty imposed by the mimic defense level on the attack process.

9.2.6.2 Measurement Based on Reliability Theories and Test Methods

As pointed out by the reliability design theory, "the reliability of a product is above all designed." Some special techniques are used in the system design process to "design" the reliability into the system to meet its reliability requirements. Similarly, the anti-attack of mimic defense is also "designed" to obtain ES defensive performance by applying RC technologies including the DHR architecture and the MR-based PAS and MDR-NFM. Typically, reliability is calibratable so as to assess or verify the reliability level of a product in a quantitative manner during the design, production, test verification, and use. Similarly, CMD also has to give quantifiable anti-aggressive indicators, which can be tested and verified. There are mainly the following tests:

(1) Single Event Defense (SED). No matter what form of independent security events (both traditional and non-traditional) happen in the MI, except for permanent failures, the CMD system should reach the ASD level, especially effectively blocking the ORT-based modulated TT attacks. This means that the ASD of the target system can be judged by injecting an independent test case into an executor by an injection method similar to reliability verification without losing the service functionality (performance) of the executor. Among them, the same test case can be implanted into any executor, but not into more than two (including two) executors. In theory, once the test case is activated, it should be able to be perceived or detected by the MR process and can be removed or circumvented (but not disappeared) from the current defense environment by an NFCM. This process should be unperceivable from outside the MI. Therefore, in the open-loop state, the activation operation is performed on the test case N times from outside of the MI, and the MR process should be able to identify an abnormal condition not exceeding m ($\leq$ N) times and not perceived from outside the interface. When the FL is closed, the number of abnormal conditions spotted in the MR process in the same test scenario is $\leq$i(i < <N) and can be completely shielded by the MI.

(2) Multiple Independent-event Defense (MID). If the loop is open, N ($\leq$the number of the service set executors) test cases that are functionally independent and non-cooperative are respectively injected into the corresponding executors without affecting the normal functionality (performance) of the host executors. Each test case should be able to be activated normally, with the OV abnormalities perceived in the ruling process yet opaque to outside of the MI. Once the CL is closed, no matter how the test cases are activated from outside of the MI, the number of abnormalities found in the MR process should be less than or equal to N and cannot be perceived from outside.

(3) Common Mode Event Defense (CMED). A homologous test case is injected into more than half of the executors to test the probability of attack and escape in the case of a closed loop, which is expected to be as low as possible in terms of engineering.

In principle, the test cases injected in the MB are generally permissible as long as they are non-cooperative and differential mode, but there are also some limitations. For instance, test cases in which the CMD may not work are not included, such as the halt paralyzing and side-channel attack cases. For another example, the injected test cases should satisfy the precondition that do not destroy the CMD non-cooperation. Therefore, if there are multiple DM test cases that can be operated cooperatively in the MB or can obtain "super privilege" over the majority of the executors, such test cases are also not permitted. In addition to the above restrictions, it is also stipulated that direct injection test method is banned in the processes, such as input distribution, policy ruling, feedback control, and output agent (but AS injection is permitted). On the one hand, since these processes have relatively simple functions (thus it is possible to avoid the problem of state explosion in formal proof), and even if design flaws or loopholes are unavoidable in engineering implementation, backdoor functions can generally be identified by way of formal detection. On the other hand, the MB control process is often insufficient to form an AS, or the AS is too small to be exploited. For example, when the relevant software/hardware code does not support online modification, the attacks by exploiting vulnerabilities to upload Trojans or by injecting instruction codes cannot take effect. Of course, if a test case can be used to indirectly obtain the super privilege of the host executor, the DM properties between the test cases are no longer guaranteed in the design sense.

9.2.6.3 Security Situation Monitoring of the Target System

According to the principle of CMD, as long as non-compliance happens in the ruling of MOVs, it indicates that a tolerable security problem occurs to the executors in the current service set, which may be a failure caused by a physical fault or design defect or an abnormality caused by a known or unknown attack. All the subsequent FC operations, such as C&R, reconfiguration, reorganization, and PAS, attempt to make the arbiter's output state no longer abnormal or to make the OV non-compliance frequency below a certain threshold in a given period. Therefore, the nature, frequency, and distribution law of these abnormalities can be analyzed against statistical units of hour, day, week, month, or year, and the security situation of the target object can be gauged and measured. Although MMR can only perceive the performance of non-cooperative attacks or sporadic faults on the OVs, the DHR-NFM can indirectly reduce the probability of attack escape. In other words, according to the TRA, as long as MMR does not find any case of OV non-compliance, it indicates that the target object is in a relatively secure state, with a minimum probability of negating this conclusion. Unless the attacker can achieve constant attack and escape without leaving traces within a given observation time, this factor can be excluded from the security situation assessment. Therefore, the statistical monitoring of OV non-compliance serves to indicate the target object's security situation to a certain extent. The author believes that the indicator to measure the effectiveness of CMD should be "the difficulty of cooperative and consistent attacks on dynamic and plural targets under non-cooperative conditions," while what method shall be used in a given time and how to directly measure this indicator remain to be further studied.

9.2.6.4   Contrast Verification

The security of MD objects can also be verified through contrast tests: via the "rotten persimmon" test software with a known problem already loaded, we observe the vulnerability scan or the vulnerability-based test (attack) case reaction under two scenarios of the target object, respectively, the mimic mode and the non-mimic mode, or observe the difference between the reactions of the executors within the MB and on the MI, where the perceived difference is accepted as the standard for security ranking. Of course, it is also possible to segment the security levels of security of the relevant mimic hierarchy according to the type of test software (e.g., the application software, basic supporting software, tool software or library function software, etc.). However, there are still many issues to be studied on how to construct diversified test specifications and how to scientifically set up test cases.

In addition, other principles can also be used in the CMD level classification, such as the difficulty of non-cooperation, the HR margin, the richness of reconfigurable options, the available defense scenario combinations of the service set executors, the number or hierarchical number of the MI and MB, the size of the policy ruling library and the ruling window time, the abundance of dynamic PAS, and the convergence time of the FC process.

9.2.6.5   Information Security Effect

Since the MR objects are the OVs of all executors of the current service set, if an attacker attempts to acquire or tamper with any sensitive information in the target object (where the TT based on the output response latency modulation is another case), it must be able to control all or most of the executors to read or modify sensitive information in a cooperative manner; otherwise the covert actions implemented independently will be spotted and removed by the MI. For example, in a mimic-structured router, the creation, updating, and deletion of entries in the routing table shall be determined by the consistent output result of most or all of the heterogeneous routing controllers; in a mimic-structured SDN controller, the storage management of the flow table shall also be determined by the consistent output result of most or all of the heterogeneous controllers; in a mimic-structured file storage system, the reading and modification of file directories or contents also requires the consistent operational result of most or all heterogeneous controllers. This effect of CMD has nothing to do with information encryption technology, but effective only for information security problems caused by the vulnerabilities or Trojans based on the target object, which do not belong to acoustic, optical, electrical, magnetic, or thermal side channel attacks. As a host technology with guaranteed security in mechanism, even if there are software/hardware code design flaws or malicious coding behaviors in mimic-structured encryption devices or information security products, they would have little chance to affect the security of the encryption algorithms or the control mechanisms. It enables the relevant equipment manufacturers to confidently reply to the users' doubts about the inherent security of the product in respect of the technical structure rather than merely in terms of component or production process supervision.

Unlike the blockchain technology, which only restricts the modification or addition/deletion of data blocks through the consensus mechanism, the mimic defense technology can not only restrict the addition/deletion of protected information in the MI but also constrain its read/write operations, protecting the privacy of data blocks without seeking a dedicated encryption process as the blockchain does. Therefore, CMD features a considerable degree of information security effect.

9.2.6.6 Mimic Defense and Mimic Computation

The essence of CMD is to create a certain degree of uncertainty for cooperative attacks based on vulnerabilities through the PAS or reconfiguration of the FEHR executors, with a purpose to significantly reduce the reliability of a cooperative attack. When the required defense level is not demanding, the MB only needs to follow the FE principle and call independent HEs upon a given algorithm or randomness policy to provide outbound services, and no complicated MR function is needed.

This simplified CMD is essentially MSC (see Fig. 9.9). Although both have a common FE-DHR requirement, they have different purposes or uses. The former creates impediments to cooperative attacks by decoupling the service functionality from the implementation structure, while the latter seeks high-performance indicators in the task-resolving process by strongly correlating the processing power to the computing structure. The difference between the two is also outstanding, as the former pursues the randomness changes in the service environment, and the latter expects the computational environment migration or reconfiguration/reconstruction function with the best performance fitting.

The degenerated form of MSC is heterogeneous calculation. Both require HR computing resources, MSC strives for the high performance of hardware/software co-processing by employing different heterogeneous computing resources under the FE conditions, while heterogeneous calculation is only focused on how to use heterogeneous processing resources to achieve the expected acceleration performance.

MSC: Dynamic heterogeneous calculation under FE

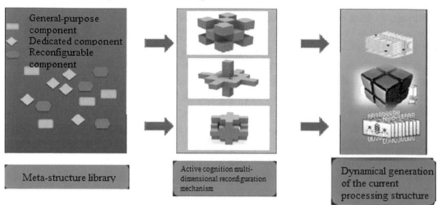

Fig. 9.9 Illustration of the MSC principle

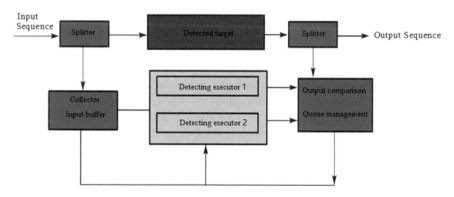

Fig. 9.10 An unknown threat detection device

9.2.6.7   Unknown Threat Detection Devices

Using the mimic structure, we can design unknown threat detection devices, as shown in Fig. 9.10.

Assume that the executor in the MB is composed of two parts: the detecting executor and the detected target. The input splitter is responsible for introducing the IS into the collector, the input buffer, and the detecting executor, while the output splitter will introduce the detected target's OV into the output comparator, and all the output of the detecting executor will also be connected to the output comparator. In this mode, the detected target is in the "black box" working state. Since the detecting executor is completely equivalent in respect of the given functionality and is in the escort or tracking mode, the output comparator herein only provides the OV comparison between the executors and the buffer queue management functions, as well as the command functions for collectors and detecting executors. With the stimulus of the same IS, if the output sequence of the detected target is consistent with that of the detecting executor, it is considered "normal," otherwise "abnormal." If there is only one detecting executor in this case, the detected target is not identified as abnormal unless the OVs of the two detecting executors are identical and inconsistent with the detected target. In this mode of operation, if it is determined that a detecting executor has higher trustworthiness and there are more than two detecting executors, then the entire security situation throughout the operation period of the detected target can be perceived online in real time or quasi-real time, while the relevant data collection and analytical tools can even perceive highly secretive attacks or clarify the nature of unknown threats. This FE-based and feature-library-independent ability of threat perception is unique to mimic models. Security detection devices designed according to this principle can be used in security checks on some dedicated "black box" devices subject to standardized protocols or uniform procedures, such as routers, domain name servers, industrial controllers, and other sensitive devices.

9.2.6.8 "Halt-Restart" Bumps

Assume that an attacker has sufficient resources, knows and masters the relevant vulnerabilities of all (or most) FEOs or all defense scenarios in the mimic defense interface (MDI), and exploits these vulnerabilities to reduce the target executors into an "endless loop" or an equivalent "halt problem." During a given time interval, an attacker can send all the "halt sequences" so that the executors "halt" successively, and in the MDI the restart or cleaning or reconfiguration is outrun by the transmission of attack sequences, the executors of the target system will enter the bumpy state of "halt-restart," and the service will flop ineffectiveness, even unable to provide any valuable service. This is equivalent to a successful cooperative attack to the mimic defense target in the C&R time window of the target object (effect-equivalent to the DDOS attack). However, this assumption will be rare in practice, since the possession of sufficiently abundant and targeted attack resources means that the cost of the attacker will be greatly elevated, even requiring various resources and means in the social engineering sense. In this case, even powerful hacker associations, government-backed bodies or state-level institutions have to premeditate their actions to play the game, not to mention individual hackers.

9.2.6.9 Standby Cooperative Attacks and External Command Disturbances

Assume that an attacker masters the high-risk vulnerabilities of the majority of the executors in the MB and has the ability to exploit these vulnerabilities to obtain the supreme control, achieve the purpose of cooperative attacks, or acquire identical sensitive information and content or has sufficient backdoor resources to achieve the intended purpose directly through the "super-user privileges" of the system. In order to achieve mimic escape, the Trojans in each executor can take out the required sensitive information in advance as a standby, and then the attacker sends a unified request command through the input channel of the shared interface along with the normal protocol message, and the Trojans in each executor take a "unified action" to send sensitive information in the standby state to the MI for the purpose of attack and escape. Although this assumption seems to hold true in theory, it is still very difficult to achieve stable and cooperative escape. First, only by mastering and effectively exploiting the vulnerabilities in the majority of the executors of the target object can an attacker guarantee that most of the defense scenarios on display are in full control, no matter what kind of PAS or reconfiguration/reconstruction of the executors in the current service set or the algorithm/structure of the executors themselves. Second, it has the ability to exploit different executors and different vulnerabilities to achieve the same goal. Third, it must fully grasp the multimode vector selection and algorithm policy of the arbiter and the ruling window time, which is the precondition for the "cooperative attack" and escape. Obviously, it is basically impossible to meet the above three requirements simultaneously. Finally, since the feedback loop has a mechanism triggered by a compulsory external command, where the system replaces some executor or even all the executors in the current service set and performs pre-cleaning operations periodically or irregularly according to some strategy, this is a huge disaster for memory-injection attacks, because "even if an escape succeeds, it will not be sustained steadily."

9.2.6.10   Superimposable and Iterative

In general, CMD can be implemented independently at multiple levels, hierarchies, and granularities or iteratively used in a defined target system. If CMD is laid out in a nonintegrated vertical or in-depth iterative manner, the complexity of a cooperative attack will ascend exponentially and bring non-linear defense gains to the defender.

9.2.6.11   Granularity of the Target Object

In theory, the complexity of HR elements in the MB is not rigidly defined. For example, the configured HR elements may be multiple FE domain name recursive servers or authoritative domain name servers on the Internet, or FE physical servers in different cloud platforms or FE virtual servers in the same cloud platform, or disk arrays provided by different manufacturers in the file storage system, or FE heterogeneous processors in the SDN controller, or FE heterogeneous hardware/software modules, middleware, and smart devices/components. In short, any FE (perhaps with additional performance requirements) physical or virtual hardware/software, regardless of its size, nature, and geographical distribution, can be used as an element in the MB.

9.2.6.12 Natural Scenarios of DHR

"Natural scenarios of DHR" are common in an information system. For example, there may be multiple physical or virtual FEHR processing resources at a cloud platform or anIDC, which are always subject to dynamic scheduling as the service load changes; In a distributed cloud storage system, there are often FE storage devices provided by different manufacturers, and it is impossible to completely resort to static allocation due to resource sharing or random failures, etc.; In user-customizable computing or reconfigurable processing or MSC environments, there are usually FE hardware/software modules or algorithm schemes with different performance and efficiency, which can be dynamically called by the resolving target, and so on. There are different degrees of dynamicity, diversity, and randomness in the abovementioned scenarios, while the scheduling mechanism has both external and internal elements, which will seriously weaken or hinder the cooperative attack actions. If the MCS is consciously used to increase the uncertainty confronting collaborative attacks under the non-cooperative conditions, or the simplified MB functionality is incrementally deployed on the basis of these resources (e.g., only retaining IA/OA and MR functions without considering PAS and MDR), even if the cooperative attack difficulty is not leveled up to that in a classic DHR architecture, it will significantly increase the cost and toll of the attacker.

9.2.6.13 About the Side Channel Attack

By definition, all the executors within the MB have the same functional privileges and data access authority. Assume that an executor is also able to control the loudspeaker pronunciation frequency or the cooling fan speed, the signal transmission rate of a given connecting cable, the LED indicator light on/off, and the CPU operating frequency and power consumption, etc., as long as the attacker can control any executor within the MI, it is possible to bypass the MOV arbiter to send the sensitive information in the MB out of the MI (even if the information transmission rate is very low in this way) by way of "modulating" the sound/ultrasonic waves given out by a loudspeaker or fan, or "modulating" the luminance frequency and intensity of the LED indicator light, or "modulating" the electromagnetic radiation on the cable by controlling the rate and interval at which information is transmitted, or "modulating" the radiant heat of the CPU surface, etc. Conversely, the preset backdoor functionality can also be triggered in the network isolation state via the acoustic, optical, or electrical sensors hidden in the system. In theory, the mimic defense mechanism has no certainty effect on such attacks based on vulnerabilities and physical effects. An attack mode mentioned in Sect. 6.5.3 is the "modulation" of the OV response latency for TT. Although it belongs to direct traversal rather than MIbypass, it can still be classified as a side channel attack based on a shared mechanism. There is no considerable technical bottleneck for solving such side channel attack problems.

9.3   Structural Representation and Mimic Scenarios

9.3.1   Uncertain Characterization of the Structure

The basic means of CMD is to introduce the MDM through the DHR architecture to trap the attacker in a dilemma of cognition and cooperative action, resisting known or unknown threats from inside and outside, through an unobtrusive or deceptive mimic method. In the process of MD, it is necessary for the defender to study how to create a complex and variable structural environment, to use plausible deceptive measures or specious false phenomena for confusing and misleading the opponents, or to strategically change the defense scenarios so as to make it more difficult to understand or cognize the defensive behaviors to create or maintain defense fogs. For the attacker, it is necessary to study how to accurately identify the defects or the embedded hardware/software backdoors of the defense scenario from the complex and variable target environment, build a channel with guaranteed reachability for internal and external information interaction to maintain the attack chain stability, and achieve a series of challenging tasks such as cooperative attacks under the non-cooperative conditions.

As a generalization of the given axiom "that the structure determines the functionality, performance, and efficiency," a function often has diverse implementation methods. In other words, there are often diversified execution structures or implementation algorithms for a given function. If a redundantly configured information service device is constructed on the basis of this principle, there is no longer a one-to-one correspondence between its external functionality and internal structure. Or the system of one I 【P】 O model can make the functional P have multiple corresponding algorithms in the determined input/output relationship, as is shown in Fig. 9.11.

Backed by this feature, these executors can be subject to dynamic reconfiguration and reorganization or random scheduling, so that the service functions provided by the target device and their apparent structures have uncertain attributes, which impose challenges to the attempt to find and exploit the vulnerabilities through the processing structures or algorithms for input/output relationship detection and inversion. Structural representation can be divided into random structural representation and complex structural representation according to different HE scheduling policies.

1. Random Structural Representation

If the structural representation of the functional device of a heterogeneous redundancy architecture only randomly and dynamically selects between multiple (e.g., the number of N) HEs in a time dimension, it is called "random structural representation," the apparent structural complexity of which is simply correlated with the number of HEs (including the diversified or form-variable scenarios) linearly. Due to the lack of a more changeable scheduling mechanism such as combined display and policy display, the apparent structural complexity and uncertainty are not satisfactory.

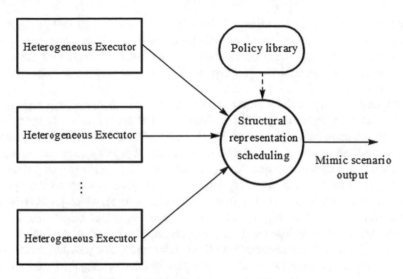

Fig. 9.11 Illustration of structural representation

2. Complex Structural Representation

If a dynamic redundancy control mechanism based on PAS is introduced into a heterogeneous redundancy architecture device, the executors engaged in service provision are therefore subject to dynamic, diverse, and random changes in terms of quantity, type, time, space, and even their own structure or operating environment. When any two or more executors are stimulated in parallel, the target object's structural representation scenario can be either an independent representation of these executors' structures or a representation of the union or intersection of them, where the complexity will be exponential. The representation scenarios of different complexities in a spatial structure, coupled with the random expansion in the temporal dimension, can make the DHR environment maximize the invisibility effects on the vulnerabilities or backdoors which are parasitic therein.

3. FC-Based Structural Representation

There is a common defect in the above two structural representation methods: the PAS of all executors can be said to be an active conduct with a "blind action" nature, where the engineering implementation cost is high and the operating effect cannot be quantitatively evaluated. Unlike the above two methods, the uncertainty of the DHR structural representation depends on the PAS of MMR and the MDR-NFM. The PAS or the MDR mechanism will be started only when the current structural representation is threatened (where MMR spots an abnormal state), changing the element configuration in the current service set of the target object or reconstructing the operating environment of the "problematic executor" so that the structural representation of the system will change accordingly. If the current threat cannot be avoided upon the change (where the MR process can still perceive the abnormal state), the above operation will be iterated to a point where the system automatically picks out the structural representation able to cope with the threat. This process has progressively convergent RC attributes and uncertainty environmental characteristics.

Obviously, the dynamicity requirements (change rate) for structural representation in CMD are completely different from that for the frequency hopping anti-jamming communication system or MTD. Only upon an OV non-compliance or when stimulated by some scheduling policy, the target object's structural representation will have dynamized and diversified changes with convergence properties, i.e., only after the "disturbances" (which can also be processed according to some policies), there will be a possible dynamic response process, which tends to be stabilized when a certain security threshold is reached. As vulnerabilities, backdoors, and Trojans are largely dependent on the operating environment, the fundamental purpose of introducing dynamicity into CMD is to undermine possible cooperation among the heterogeneous executors by changing or transforming the structural representation scenarios within the MB, with emphasis laid on efficient action rather than inefficient blind operation.

It should be noted that if economically affordable, it is vital to increase the abundance of the structural representation and the intensity of iterative convergence algorithms to reduce the difficulty of CMD engineering implementation.

9.3.2   Mimic Scenario Creation

Each apparent structural form of a mimic structure information system or device is called a mimic scenario (MS). Regardless of its type, each MS has equivalent and perceptible meta-service capabilities, including dark features or side effects such as unknown vulnerabilities beyond the given features.

The number and complexity of the defender's MS determine the scope and effect of the concealable camouflage, while finite MFs can exhibit more complex and multi-dimensional defense scenarios (structural representation) through mimic display. Typical creation approaches are as follows:

1. Dynamized/randomized virtualization

In this approach, as required by the functional intersection of the greatest common divisor, a plurality of virtualized MS and functions is planned or designed in advance and applied through a dynamic and randomized operation scheduling mechanism. The disadvantage lies in the considerable performance loss, which may also affect the flexible expression or extension of the upper-layer functions (such as many practices in MTD). In addition, the security guarantee on the software and hardware carrier that implements virtualized scenarios will become a new sensitive target.

2. Reconfigurable/reorganizable/software-definable meta-structures

In this approach, as required by the functional intersection of the greatest common divisor, a plurality of structural schemes for MS is planned or designed in advance and through multiple reconfigurable, reorganizable, or software-definable meta-structures (such as all kinds of software-defined everything, SDx [10, 11]), field-programmable gate array (FPGA) or user-customizable calculation (applying structures such as CPU + FPGA, etc.) or software-defined hardware (SDH), and mimic display processes, real-time or non-real-time generating or presenting MFs that meet the service functionality and performance requirements. The disadvantage is that the more MS are designed, the higher the development cost will be, and the development method is relatively enclosed.

3. FE COTS-Level Element Utilization

In this approach, as required by the functional intersection of the greatest common divisor, a plurality of source-different software/hardware entities with service functions or performance intersections is designed (or picked up) in advance (especially the more standardized middleware, embedded systems, or open-source products, etc.), which can also be cloud services, virtual/physical heterogeneous servers on data center platforms, storage components, or heterogeneous net element devices on the network. By adding control processes such as IA/OA institutions, MMR, and PAS, we can dynamically call or request or display the desired MFs or stack the MFs composed of some physical or virtual bodies. The disadvantage is that usually commercialized elements can only be used in a black box. Modifications and upgrades of various software/hardware versions may involve adaptive design or modification of the MB. For instance, undefined or user-defined fields used in some communication or interconnection protocols may affect the implementation complexity of mimic ruling (MR). In general, the implementation of the MR function is not theoretically affected as long as the upgraded content of each executor software/hardware version does not involve grammatical or semantic changes on the MI.

4. Mixed Creation

In practical application, the above three creation approaches can be mixed as needed.

9.3.3 Typical Mimic Scenarios

1. Mystified scenarios

In a mystified scenario, an attacker does not find in its knowledge base any information matched to the target object's defense scenarios, which makes the attacker confused or overwhelmed. Such fields are often difficult to be constructed, because human activities are highly dependent on the inheritance of knowledge or experience, and anything beyond experience will be difficult to imagine and construct. In theory, mimic structure scenarios are quite mysterious to the attackers. First, the DHR architecture can turn uncertain threats into events with controllable probabilities, contrary to cognitive common sense. Second, even if the trial and error method is permitted under the white box conditions, it is difficult to stably construct cooperative and consistent attacks on dynamic and plural targets under the non-cooperative conditions as a result of the strong uncertainty effect. Third, the NFM makes the number, nature, and availability of the detected vulnerabilities uncertain. It is impossible to establish a reliable communication or reachability relationship even if a Trojan can be implanted. Fourth, the detection cognition or attack experience obtained at the time t may not be inheritable or reproducible. In short, the uncertainty effect of the mimic structure can lead to the mystification of the MS.

2. Specious scenarios

In a specious scenario, the defense scenario an attacker sees has multiple matched information in the knowledge base involved, yet is unable to form a complete and clear understanding, ending up with an-elephant-and-blind-men experience. This scenario construction is not only the inherent function of the typical mimic structure but can also be done using atypical CMD approaches as long as there are FEHEs.

3. Concealable camouflage scenario

In a concealable camouflage scenario, an attacker will neither find the defense features or laws of the target object even when exhausting his/her own experience and knowledge in the field of vision nor establish any functional contact with a pre-implanted hardware/software code. The highest state of CMD is to completely shield the defender's defense actions. Whether it is a scanning probe or a heuristic attack, whether it is a random fault or an internal disturbance caused by man-made factors, attackers cannot get any exploitable feedback information from the stimulus-response operations of the mimic structure, which is like falling into the "information black hole."

4. Sliced or fragmented scenario

A sliced or fragmented scenario displays a target in a non-panoramic and sliced (segmented or fragmented) manner to an attacker from different angles, times, spaces, and paths, who thus finds it difficult to cognize the overall architecture or running laws of the target. Such scenario engineering is easier to implement.

5. Overfitting scenario

In data analysis and prediction, it is often the case that "a hypothesis can outfit others on the training data, but does not fit well on data sets other than the training data." This kind of scenario is likely to mislead the attacker's prediction and evaluation based on machine learning [12]. The key point is that the implementation process should consciously expose the majority of the target features and add feature noises, which can be accomplished through the properly designed PAS and MDR mechanism.

6. Shielding and disguising-abnormal scenario

An attacker will perform detection and attack effect evaluation [13, 14] both before and after (or even during) the attack in most cases. One of the original intentions of the mimic structure defense scenario transformation is to make the attacker hesitate or terminate the subsequent attack actions due to the difficulty in obtaining the real attack effect. For external exploration or reconnaissance activities [15, 16], the information or conditions detected by the attacker cannot properly drive the automaton-based analysis mechanism due to the unperceivable transformation (switching) of the mimic structure defense scenarios, thus unable to develop a deeper and more detailed understanding of the target object. In particular, by shielding or disguising the operational abnormalities resulting from external probes (and possibly heuristic attacks) or internal errors or from the accumulated probing data contributing to mismatches, too many errors or uncertain representations are introduced into the attacker's reconnaissance and attack effect evaluation.

9.4 Mimic Display

Mimic display is an intelligent scheduling function based on the DHR principle, a process of combining HE sets in the dimensions of time, space, and policy. In essence, mimic display is a structural representation method based on mimic features. Dynamicity, diversity, randomization, and the NFM are its basic means for making complex MFs. The mimic display quality (i.e., the strangeness and ambiguity of the structural representation) has an important influence on ensuring the mimic defense effect.

9.4.1 Typical Modes of Mimic Display

In the mimic display mode, the FEHEs under the same input stimulus generally have multiple OVs temporarily dispersed to some extent, and with a large probability of the same information content. The output content is subject to mimic ruling or a consistency review according to a given algorithm, based on which the corresponding OVs are selected according to a given policy set as the response output of the MI. Several typical modes are as follows:

1. Front-end/back-end display mode

The mimic structure service functionality is independently provided by the front-end HEs, while there are multiple FEOs at the back-end, synchronously or asynchronously simulating or escorting the task processing of the front-end executors, checking and reviewing in real time or in a tracking manner whether there are any abnormalities in the result (data) or state (flow) processed by the front-end executors, and determining whether to correct the historical output of the front-end executors, ready to replace the executors that currently provides the service functionality under the scheduling and control of the mimic display process. FEHEs, whether abnormally suspended or randomly put offline, will require recovery or preventive operations such as cleaning and restarting, as well as reconfiguration and reorganization when necessary. The structural representation complexity in the front-end/back-end display mode is linearly correlated with the type and number of the MFs.

2. Combined display mode

In a specific service scenario, the HEs in the MB perform specific service functionalities concurrently or sequentially according to some preset-combined policy, such as polling and random combination. Since the HEs of redundant configurations are equivalent in a given function, they will theoretically always display the same service functionality (perhaps different in performance) to the outside no matter they are motivated in whatever combinations. However, the defender has therefore acquired the ability to dynamically and randomly combine and display complex MFs, making it difficult for attackers to distinguish the backdoors on the implementation structure of the service functionality in the MI merely by functional perception or system fingerprints, or to build a reliable attack chain in the non-static, non-certain, and non-persistent environment resulting from the PAS and MDR-NFM based on MR. The structural representation complexity in the combined display mode is nonlinearly correlated with the number of MS.

3. Ruling output mode

The MS or HE components synchronously receive service requests from outside of the MI and perform MMV or policy ruling on the generated multichannel OVs through the display control components, such as the IA/OA, etc. Any OV that meets the voting/ruling rules can be used as an alternative vector for the OA, which will then select the output from the candidate vectors according to some policy (that can seriously interfere with smart attacks based on system fingerprint/performance analysis). As for the "suspected problematic" executors or operational scenarios, the back-end processing mechanisms can be used in time, including but not limited to vulnerability scanning, virus Trojan screening, fault analysis and positioning, C&R, and reconfiguration and reorganization.

In the strict sense, the ruling and judgment process will usually impose different degrees of impact on the real-time service response. As a compromise in engineering implementation, the output ruling (OR) target on the MI can also be a coarse-grained option, such as the output content of a thread, process or task-level preset checkpoint, or the completeness of the routing information reviewed by way of the neighbor broadcast mechanism in a routing protocol.

Sometimes the OVs may contain a large amount of information (such as the packets in an IP protocol), and the ruling and processing time overhead may affect the service performance of the target object. In this case, a preprocessing mechanism prior to the ruling session can be added. For instance, the functions added to the output end of the executor or the input end of the arbiter, such as a cyclic check code requesting the packeting information or the hash value of the routing table entry/domain name information, the time complexity of the ruling process can be reduced to a mere numerical comparison of the cyclic check code or hash value from the OV domain-wide comparison. From another perspective, the separation of the ruling from the output agent function can help to strengthen the difficulty of vulnerability detection and utilization in the MB.

From the perspective of engineering realization, the establishment of the ruling site will inevitably be confronted with a "blind zone." On the one hand, attacks based on complex state transitions always try to avoid triggering the MI. On the other hand, from the technical and economic point of view, it is not appropriate to make a bit-level domain-wide comparison of the OVs or to set too many MIs to affect the cost performance of the target system. Therefore, in the MS, special measures should be taken to tackle the blind zone problem. For example, regularly or irregularly cleaning or initializing online/offline executors will make an intruder unable to maneuver complex attack mechanisms and can clear abnormalities once they are detected or implement algorithms and changes in the execution environment. It is worth noting that both the amount of the judgment content and the timing of the judgment will affect the probability of attack and escape, calling for compromised or balanced design considerations.

4. Policy display mode

A policy-based scheduling mechanism can apply in the FC process. The optional scheduling policies include weighting, polling, randomness, performance priority, maturity priority, high version and new function priority, balanced load priority or historical performance priority, and so on. These scheduling policies, when used dynamically or randomly, can increase the apparent complexity of the mimic display scenarios in multiple dimensions.

On a standardized or normalizable MI, the introduction of manageable IA/OA and PAS mechanisms, as well as the mimic ruling mechanism of MOV under the non-cooperative conditions, the mimic display form is even more strange and ambiguous, making it more easily to conceal the internal traits of the defender (e.g., disrupting the detection attacks that attempt to derive system response functions through the input/output relationship), or to shield abnormal reactions triggered by an external stimulus (such as a significant increase in response time or the output of incorrect results), or to block attempts to detect software version data through undefined fields in the protocol packets.

It should be noted that the implementation of mimic display based on the adjustment policies will inevitably introduce a corresponding scheduling cost. In engineering practice, since special processing capacity or system overhead needs to be added to display, transform, identify, and hide an abnormal state in an MF, these scheduling operations should not be implemented too frequently and should not be too complicated. In particular, it should be noted that dark features on the feedback scheduling process should be minimized in the design to reduce the attack reachability and the AS. In addition, it is beneficial to set up functional components such as "watchdogs" [17] to monitor possible faults in the scheduling process.

5. OA mode

When choosing COTS or non-proprietorially designed hardware/software products as the HEs, the OVs may be consistent at the grammatical and semantic level but different in respect of the option, undefined domain, padding field, response latency, packed message amount, and even encrypted content. In addition, operations such as dynamic online/offline processing and pre-cleaning, or forced reset, or reconfiguration or reorganization, may result in a synchronized state, process, and serial number between the executors or in their external presentation. All these will bring much trouble to the MR-based policy adjustment. The adoption of the OA mode helps to simplify the engineering complexity of the mimic display.

In addition, the agent process can effectively shield the outside world from detecting and investigating the "system fingerprints" and block the conducts of indirectly peeping at the internal features of the system or implementing TT through the service response time or protocol packet details; or, when the encryption of IPSec, HTTPS, and SSL, [18] etc., causes completely inconsistent output/input results of the executors in the MB, the encryption process will be moved backward/forwarded to the OA/IA process. At this point, efforts to reduce the OA/IA AS, and attack reachability become essential.

Since there are OV judgment and selection mechanisms, we have to decide the setting of the checkout point and the content to be reviewed. Excessive checkpoints not only increase the cost of HR resources but also add to the judgment overhead and affect the normal service response performance. In particular, the ORTs of the executors are usually discrete (there should be measures of dispersion requirements in engineering implementation), and it is likely that there are different grammars with the same semantics, which will introduce additional system processing latency in the judgment and selection processes. The mimic display granularity generally has two divisions, namely, the process level and the task level. The former is more sensitive to GUD in the MB, but the overhead is too high, just opposite to the latter. For example, in a mimic router, the semantic level consistency of the HE routing table entries is the major target judged to ensure the reliability and credibility of data forwarding routing. Those attacks or actions that have not yet affected the correctness of the routing table can be considered as a process of intrusion tolerance. For the file storage system, the review point can be set in the operating process of modification/addition/deletion of the disk file index directory (metadata file). Any operation to modify/add/delete the file directory must pass the MOV compliance examination. The GUD without any explicit reaction to the disk file directory operation can be ignored. In engineering implementation, there are still many topics to be studied in depth regarding the MRM and output selection policy, as well as the fault/intrusion tolerance mechanisms.

9.4.2   Considerations of the MB Credibility

Based on the MR's policy adjustment and MDRM, the MB's reliability and credibility is always one of the core issues of CMD. In addition to reference to high-reliability redundant control and robust control practices, the MB (software/hardware) entities

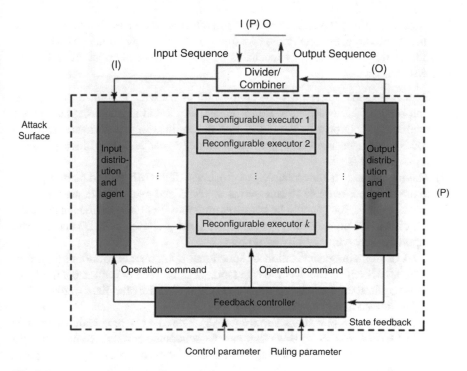

Fig. 9.12 Pipeline-structured MB

should, if possible, be kept independent from any executor (structural or algorithmic) entities. The FC algorithms and policies do not require strict confidentiality, but at least some of their parameters should come from the random information inside the system or the executors (such as the current network traffic, the memory/processor resource occupancy, the number of active processes, and the frequency of interruptions, etc.).The ability to obtain a panoramic view of systemic resources or operational postures at all processes of the MB must be limited by a "single-line contact or one-way transmission mechanism," strictly following the principle of "worse-is-better control information," decomposing centralized control points (e.g., using the distributed ruling algorithm), removing redundant information, blurring the correspondence between physical and logical resources, and using watchdogs to monitor the activities of the MD process (e.g., starting the alarm mechanism upon the loss of dynamicity or the display of some fixed change laws).In addition, the Input Distribution and Agent(IDA), the reconfigurable executors and the Output Agent and Ruling (OAR) processes should be handled through the "one-way pipeline" in the strict sense, with a core purpose to prevent the executor functionality in the MB from being "bypassed." In other words, only when the MB (including the reconfigurable executors) obtains the structure-based ES effect, the exploitability of the vulnerabilities in the FC process (in the blue module) can be significantly reduced, and the CMD system can mechanically avoid security-related short boards, as shown in Fig. 9.12.

At the same time, the dark features of the relevant processes should be reduced as much as possible, especially the backdoors and other "spy" functions where the conditions permit. The security mechanisms that the MB can adopt are at least as follows:

(1) Reducing the AS. Minimize the complexity of the MB, minimize the application scale of hardware/software resources, and avoid any unnecessary smart/intelligent functions. In particular, online modification of software/hardware codes should be used in a prudent manner to mechanically block potential injected attacks.

(2) Making the attacks unreachable. All processes of the MB should be transparent to any HE and invisible to an external attacker, and even if there are unknown vulnerabilities, they should be hardly exploitable for attackers, thus making the AS of the MB unreachable. For example, some practices of MTD can be used (upon the zero-vulnerability assumption).

(3) SDH or domain specialization of functional components. Configuration-file-driven SDH or domain-specific accessories can be used to implement the functions required for the MB, making the configuration file pluralized or diversified at free choice of end users.

(4) Importing the mobile attack surface (MAS). The MB per se is realized by multiple FEHEs, and these executors can be scheduled through some uncertain internal parameters in the device, making it difficult for an attacker to cooperatively utilize the vulnerabilities therein.

(5) Watchdogs. The MB should have preset checkpoint information for active output so that the specially configured watchdog device(s) can detect possible anomalies in time.

(6) Trusted computing. For the MB, the functions and states of the relevant processes are generally predictable, and trusted computing techniques can be used to monitor whether their operation is as expected.

(7) If necessary, mature HRR technology can be adopted.

9.5   Anti-attack and Reliability Analysis

9.5.1   Overview

The CMDA [19] has the ability to detect non-specific threats, resist attack events, and tolerate reliability faults without relying on prior knowledge. It is based on the premise of non-exclusion of conventional intrusion detection/prevention and other security measures, based on the DHR architecture under the FE conditions, and centered on the MR-based PAS and MDR-NFM. Introducing the uncertainty relationship between the target object's service functionality and its external structural representation as a means, the CMDA employs the endogenous uncertainty effect to form a "two-way defense fog," which causes a cognitive dilemma for the attacker and significantly reduce the probabilities for a successful attack under the non-cooperative conditions. It is an information system infrastructure featuring GRC that integrates "service provision, reliability guarantee, and security protection." In particular, CMD can deal with the problem of model perturbation reliability caused by physical factors and the problem of uncertain threats caused by unknown attacks via a normalized approach through channel coding/decoding theories and methods, just like what happens to the white noise and additive interference in the Shannon channel [20].

In this section, the General Stochastic Petri Net (GSPN) is used as a basis for CMDA's anti-attack model, as well as for the reliability models of three typical information systems: non-redundant architecture (NRA), dissimilar redundant architecture (DRA), and CMDA. The isomorphism between the GSPN model's reachability graph and the continuous-time Markov chain (CTMC) [21] is then employed for quantitative analysis of the steady-state probabilities of the systems (the analytics identical to that in Sect. 6.6.2). Finally, MATLAB-based simulation is used to further verify the nonlinear gain of the CMDA in terms of anti-attack and reliability. As for the evaluation indicators, three parameters are proposed to comprehensively characterize the anti-attack of the target architecture, respectively, the steady-state availability probability (SSAP), the steady-state escape probability (SSEP), and the steady-state non-specific awareness probability (SSNSAP), while the reliability of the target architecture is characterized by the SSAP, the reliability R(t), i.e., at the normal working probability in the time interval of (0, t]. The mean time to failure (MTTF) and the degradation probability are also included in the features. The models and methods presented in this section can be used to analyze the robustness, reliability, availability, and anti-attack of the architecture of an information system. Relevant analysis conclusions can help guide the designing of a robustness information system highly reliable, available, and credible.

9.5.2 Anti-attack and Reliability Models

1. Anti-attack considerations

We believe that the anti-attack and reliability of a mimic-structured system can be normalized in design, while its behaviors and results can be expected through model solution. It should be able to achieve systematic goals with designable system security performance, grade calibratability, and effect verifiability, where the behavioral states can be monitored, the behavioral outcomes can be measured, and the abnormal behaviors can be controlled.

As shown in Fig. 9.13, CMDA introduces the mimic disguise mechanism (MDM) on the basis of the DHR to make the attack-defense game more proactive and applies games of strategy to drive the MDRM in search of the uncertainty effect regarding the target object's defense deployment or behaviors. Making use of the DHR mechanism, CMDA provides surface defense (SD) for certain or uncertain threats,

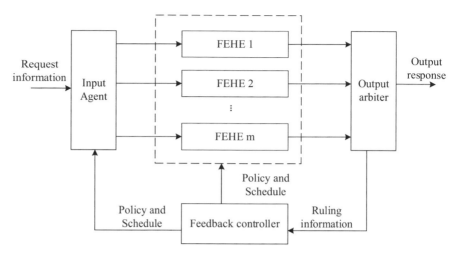

Fig. 9.13 CMDA that adopts the MR-based PAS and MDR-NFM

triggers the "point clearance" mechanism based on the dissimilarity statistics of multimode ruling (MMR), and achieves the goal of minimizing the MOV non-consistency expression probability based on the systematic effect of CMD. It should be emphasized that the CMDA first converts the uncertainty attacks based on the target object's executor vulnerabilities and Trojans, etc., into events at the system level, the attack effect of which can be expressed by probabilities, and then normalizes certain or uncertain "DM and CM attacks" and random DFMs to reliability issues. Furthermore, it should be able to greatly enrich the CMD scenarios against non-cooperative attacks through the MR-based policy adjustment and MDR-NFM, so that SSAP, SSEP, and SSNSAP can be controlled at a reasonable level. The correlation between the above transformation steps is: the more certain the attack effect on an individual HE, the lower the probability of a successful system-level cooperative attack will be, coupled with a nonlinear connection between the two; the richer the spatial-temporal connotation of an OV, the more oriented MMR, the lower the probability of a cooperative escape will be, also coupled with a nonlinear connection between the two. To simplify the analysis, in the modeling analysis, we simplify the various MR methods that are subject to dynamic selection, such as priority, historical weight, scenario trustability, etc., and only choose a "majority-rule" voting method. Besides, we normalize all kinds of MDR means, such as the restart, reconstruction, recombination, and reconfiguration of an executor, into a means of executor "recovery" of the executor. At the same time, traditional "point" defense measures such as intrusion detection, firewalls, and other specific fault detection and encrypted authentication are not added into the target structures. Under this assumption, the threshold of a CMDA's anti-attack can be produced. In practical application systems, the above methods or means are often added or incorporated, which, in the equivalent sense, adds to the dissimilarity of the mimic system, thus enabling CMDA's anti-attack exponentially. In other words, the actual anti-attack performance of the CMDA should be much higher than the threshold of the hypothetic analysis.

For a n-redundancy/m scenarios (usually m ≥ n) CMDA, the probability of a system-level permanent fault/failure is extremely low due to the functionality of the MR-based PAS and MDR-NFM, and the system will be required to enable the relevant re-ruling policy to produce an output with a reduced confidence level (i.e., a downgrade fault occurs) only when the MR ends up with completely non-compliant OVs of the three executors. An escape failure refers to the failure where an attacker controls the majority of the executors or operational scenarios in the system and causes their OVs to be consistent and erroneous after the ruling.

In order to simplify the analysis and verify the anti-attack of the architecture itself, the surface defense methods for specific fault detection and encrypted authentication such as intrusion detection and firewalls are not added into the executors of the abovementioned system, while the rates of fault detection, false alarm, and missed alarm of the various fault perception and defense mechanisms are not considered. The system hardware and software devices are not separated, and their mutual influence is not taken into consideration. The specific values of the defense time cost for different structures and defense scenarios are detailed in Part 3 of this section.

In this section, we also assume that, whereas the TRA is established, the threat issues based on the target object's vulnerabilities can be normalized to reliability issues by the defense system's structural effect, and also assumes the executor's attack success and recovery time to be subject to the exponential distribution, so that analysis of the structure's anti-attack can be done with the GSPN model. As far as CMDA is concerned, the uncertainty effect caused by the MR-based PAS and MDR-NFM makes it difficult for an attacker to attack and escape by means of trial and error or exclusion, that is, the attack experience is hardly inheritable, and the attack scenario that is successful at the time t cannot be reproduced at the time t + x, so the actual anti-attack performance of the CMDA tends to outgo the threshold generated from the hypothetic analysis.

2. Anti-attack modeling

Anti-attack is the ability of a system to constantly provide effective services and recover all services within a specified time period in the event of an invisible failure upon an external attack. The assumptions about the anti-attack model are as follows:

Assumption 1 Assume that the target system is a tri-redundancy DHR system (referred to as a CMD system) not including specific perception and defense measures such as intrusion detection and firewalls in its executors. The mimic ruling mechanism helps to identify faulty executors caused by general/halt attacks (since their OVs are different from that of the majority of executors) and can perform recovery operations on them, including dynamic reconstruction and the like; also, with a specific probability, the system can periodically or irregularly recover potential errors or attack and escape that the mimic ruling mechanism fails to detect, while the success/recovery time of the attacks to the executors is subject to an exponential distribution.

The anti-attack GSPN model for each system is defined as follows:

Definition 9.1 The GSPN model for system attack failures

$$\text{GSPN}_C = \left(S_C, T_C, F_C, K_C, W_C, M_{C0}, \Lambda_C \right)$$

where, the executor state of the tri-redundancy mimic system can determine the library set $SC = \{P_{C1}, P_{C2}, \ldots, P_{C24}\}$; $T_C = \{T_{C1}, T_{C2}, \ldots, T_{C42}\}$; FC is the arc set of the model. WC is the set of arc weights, with each arc weighing 1. $K_C = \{1,1,\ldots,1\}$ defines the capacity of each element in the S_C. $M_{C0} = \{1,1,1,0,\ldots,0\}$ defines the initial state of the model. The time-relevant average transition rate set $\Lambda_C = \{\lambda_{C1}, \lambda_{C2}, \ldots, \lambda_{C15}\}$ is determined according to the action implementation characteristics.

The probability indicators used to evaluate the system's anti-attack include availability probabilities, escape probabilities, non-specific awareness probabilities, dormancy probabilities, and degradation probabilities. The definitions of the probabilities are as follows:

Definition 9.2 Availability probabilities (AP): The probabilities of the system in the state of normal service. In a tri-redundancy CMD system, AP refers to the probabilities that all executors of the system are in a vulnerability dormant state, or a single executor is in a fault state.

3. Reliability modeling

Reliability refers to the ability of a component, device, or system to perform the specified function(s) under the specified conditions within a predetermined period of time. The assumptions regarding a reliability model are as follows:

Assumption 2 An NRS does not have a fault recovery mechanism and will degrade when a single executor fails; the time of a random fault occurring to an executor is subject to an exponential distribution.

Assumption 3 Through the multimode ruling mechanism, a DRS can find out the random faulty executors and suppress their output, recover the faulty executors (a single executor in a tri-redundancy case), and will degrade when the majority of the executors fail; the time of a random fault occurring to an executor is subject to an exponential distribution.

Assumption 4 Through the mimic ruling mechanism, a CMD system can find out the random faulty executors or operational scenarios and recover them in a normalized mode; the time of a random fault occurring to an executor is subject to an exponential distribution.

The reliability GSPN model for the abovementioned systems is defined as follows:

Definition 9.3 The GSPN model for NRS reliability failures

$$\text{GSPN}'_{N} = \left(S'_{N}, T'_{N}, F'_{N}, K'_{N}, W'_{N}, M'_{N0}, \Lambda'_{N} \right)$$

where, $S'_{N} = \{P_{N1}, P_{N2}\}$; $T'_{N} = \{T_{N1}\}$; F'_{N} is the arc set of the model. W'_{N} is the set of arc weights, with each arc weighing 1. $K'_{N} = \{1,1\}$ defines the capacity of each element in the S_{N}. $M'_{N0} = \{1,0\}$ defines the initial state of the model. $\Lambda'_{N} = \{\lambda_{N1}\}$ defines the average implementation rate set associated with the time transition.

Definition 9.4 The GSPN model for DRS reliability failures

$$\text{GSPN}'_{D} = \left(S'_{D}, T'_{D}, F'_{D}, K'_{D}, W'_{D}, M'_{D0}, \Lambda'_{D} \right)$$

where, $S'_{D} = \{P_{D1}, P_{D2}, \ldots, P_{D7}\}$; $T'D = \{T_{D1}, T_{D2}, \ldots, T_{D9}\}$. F'_{D} is the arc set of the model. W'_{D} is the set of arc weights, with each arc weighing 1. $K'_{D} = \{1,1,\ldots,1\}$ defines the capacity of each element in the S_{D}. $M'_{D0} = \{1,1,1,0,\ldots,0\}$ defines the initial state of the model. $\Lambda'_{D} = \{\lambda_{D1}, \lambda_{D2}, \ldots, \lambda_{D6}\}$ defines the average implementation rate set associated with the time transition.

Definition 9.5 The GSPN model for CMDA reliability failures

$$\text{GSPN}'_{C} = \left(S'_{C}, T'_{C}, F'_{C}, K'_{C}, W'_{C}, M'_{C0}, \Lambda'_{C} \right)$$

where, $S'_{C} = \{P_{C1}, P_{C2}, \ldots, P_{C10}\}$; $T'C = \{T_{C1}, T_{C2}, \ldots, T_{C16}\}$; T'_{C} is the arc set of the model. $W'C$ is the set of arc weights, with each arc weighing 1. $K'_{C} = \{1,1,\ldots,1\}$ defines the capacity of each element in the S_{C}. $M'_{C0} = \{1,1,1,0,\ldots,0\}$ defines the initial state of the model. $\Lambda'_{C} = \{\lambda_{C1}, \lambda_{C2}, \ldots, \lambda_{C10}\}$ defines the average implementation rate set associated with the time transition.

The indicators used to evaluate system reliability are reliability, mean time to first failure (MTTFF), AP, and DP, where the definitions of AP and degradation probabilities (DP) are the same to definitions 6.3 and 6.7, while reliability and MTTFF are respectively defined as follows:

Definition 9.6 Reliability R(t): The probability that a system will work properly in the time interval (0,t].

Definition 9.7 MTTFF: Mean time to first failure of a system.

9.5.3   Anti-attack Analysis

1. Analysis of the CMD System Against General Attacks

By adjusting the failure rate λ, the OV dissimilarity parameter σ, and the recovery rate μ, the CMD system can use a variety of OR-based negative feedback

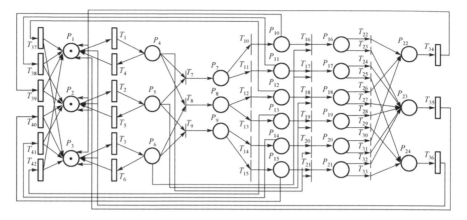

Fig. 9.14 GSPN model for the CMD's general attack failures

scheduling policies (NFSP). This section analyzes the CMD system performance in resisting general attacks, respectively, upon the stochastic scheduling policy (SSP), the rapid recover scheduling policy (RRSP), the flaw channel scheduling policy (FCSP), and the rapid recover and flaw channel scheduling policy (RRFCSP).

Based on the definition of the $GSPN_C$, the GSPN model for the CMD system with a tri-redundancy and the executor number equivalent to the operational scenario number (n = m) under the general attack conditions is shown in Fig. 9.14. There are 24 states and 42 transitions. P_1, P_2, and P_3 each contain a token to indicate that the executor is in the dormancy state, that is, all executors have vulnerabilities not yet exploited by an attacker; P_4, P_5, and P_6 each contain a token to indicate a post-attack executor failure; P_7, P_8, and P_9 each contain a token to indicate two simultaneously failed relevant executors with abnormal output vectors; P_{10}, P_{12}, and P_{14} each contain a token to indicate two faulty executors with consistent abnormal output vectors; P_{11}, P_{13}, and P_{15} each contain a token to indicate two faulty executors with non-compliant abnormal output vectors; P_{16}, P_{17}, P_{18}, P_{19}, P_{20}, and P_{21} contain a token each to indicate three simultaneously faulty executors with abnormal output vectors; P_{22} contains a token to indicate three faulty executors with consistent abnormal output vectors; P_{23} contains token to indicate three faulty executors with abnormal output vectors, with two of them having consistent output vectors; and P_{24} contains a token to indicate three faulty executors with abnormal and totally non-compliant output vectors.

T_1, T_2, and T_3 indicate that the executor is faulty due to the attack; T_4, T_5, and T_6 indicate that the executor is restored to the dormancy state at the rate μ_1 due to the changed attack/defense scenario; T_7, T_8, and T_9 indicate that the two executors simultaneously fall in the faulty state; T_{10}, T_{12}, and T_{14} indicate that the abnormal output vectors of the two faulty executors enter the consistent state with the selection probability σ; T_{11}, T_{13}, and T_{15} indicate that the abnormal output vectors of the two faulty executors enter the inconsistent state with the selection probability

$(1-\sigma)$; T_{16}, T_{17}, T_{18}, T_{19}, T_{20}, and T_{21} indicate the simultaneous fault of three executors; T_{22}, T_{26}, and T_{30} indicate that the abnormal output vectors of the third executor enter the consistent state with the same abnormal output vectors of the first two executors with the selection probability σ; T_{23}, T_{27}, and T_{31} indicate that the abnormal output vectors of the third executor enter the inconsistent state with the same abnormal output vectors of the first two executors with the selection probability $(1-\sigma)$; T_{24}, T_{28}, and T_{32} indicate that the abnormal output vectors of the third executor enter the consistent state with the abnormal output vectors of either of the first two executors with the selection probability 2σ; and T_{25}, T_{29}, and T_{33} indicate that the abnormal output vectors of the third executor enter a state inconsistent with the abnormal output vectors of either of the first two executors with the selection probability $(1-2\sigma)$. T_{34} indicates that all executors are restored to the dormancy state at the rate μ_3 via the periodic or random recovery mechanism of the system (when the abnormal output vectors of the three executors are fully consistent); T_{35} indicates that all executors are restored to the dormancy state at the rate μ_2 via the negative feedback and recovery mechanism (when there are three completely inconsistent and two abnormal output vectors); T_{36} indicates that all executors are restored to the dormancy state at the rate μ_4 via the negative feedback and recovery mechanism (when the abnormal output vectors of the three executors are completely inconsistent); T_{38}, T_{40}, and T_{42} indicate that the two relevant executors are restored to the dormancy state at the rate μ_4 via the negative feedback and recovery mechanism (when the abnormal output vectors of the three executors are completely inconsistent); and T_{37}, T_{39}, and T_{41} indicate that the two relevant executors are restored to the dormancy state at the rate μ_2 via the negative feedback and recovery mechanism (when the abnormal output vectors of two executors are consistent).

As for the recovery of faulty executors, there are a variety of negative feedback scheduling policy (NFSP) options. In the following analysis, we adopt SSP, namely, the anti-attack of a CMD system upon the stochastic selection of backup executors restored to the dormancy state via the reconstruction mechanism or the like.

The analyzed CMD system is a tri-redundancy DHR system, assuming that the attack type is a general attack, and the security gain resulting from the introduction of specific awareness and defense means (such as intrusion detection and firewalls) can be replaced by the endogenous cognitive module of the mimic architecture, while a single executor has fault recovery capabilities.

The CTMC model for general attack failures of the CMD system is shown in Fig. 9.15. To unify the preconditions, we assume that the success and recovery time of an executor-oriented attack are subject to the exponential distribution, then for a CMD system, the actual situation is compliant with the assumption, since the mimic structure can first turn vulnerability-based attacks into a probability event and then keep the probability below a minimum expected threshold. The steady states of the CTMC model are shown in Table 9.1.

The meanings of the parameters of the CTMC model are shown in Table 9.2. $\lambda_1 \sim \lambda_3$, respectively, represent the average transfer rate of a fault caused by a general attack on each executor, σ represents the uncertainty of consistent output vectors of two executors after being attacked, and $\mu_1 \sim \mu_4$, respectively, represent the transfer rate of the recovery of a faulty executor in different attack and defense scenarios.

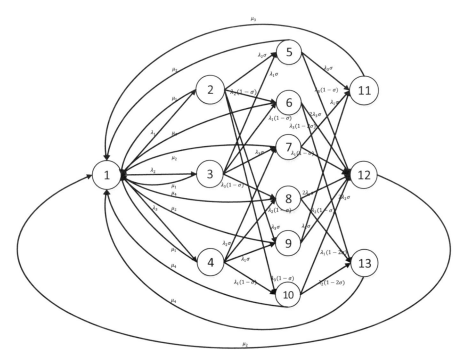

Fig. 9.15 CTMC model for general attack failures of a CMD system

Table 9.1 Parameters of the CTMC model for the CMD's general attack failures

State serial No.	Meaning
1	Each executor is in normal operation and in the dormancy state of vulnerabilities and backdoors
2	Post-attack failure of Executor 1
3	Post-attack failure of Executor 2
4	Post-attack failure of Executor 3
5	Failure of Executors 1 and 2 with consistent abnormal output vectors
6	Failure of Executors 1 and 2 with inconsistent abnormal output vectors
7	Failure of Executors 2 and 3 with consistent abnormal output vectors
8	Simultaneous failure of Executors 2 and 3 with inconsistent abnormal output vectors
9	Failure of Executors 1 and 3 with consistent abnormal output vectors
10	Failure of Executors 1 and 3 with inconsistent abnormal output vectors
11	Failure of the three executors with consistent abnormal output vectors
12	Failure of the three executors with consistent abnormal output vectors between two of them
13	Failure of the three executors with totally inconsistent abnormal output vectors

Table 9.2 Parameters of the CTMC model for the CMD's general attack failures

Parameter	Value	Meaning
$\lambda_1(h)$	λ	Assume that the average abnormal output vector time of Executor 1 upon an attack is 10 hours/10 minutes/10 seconds
$\lambda_2(h)$	λ	Assume that the average abnormal output vector time of Executor 2 upon an attack is 10 hours/10 minutes/10 seconds
$\lambda_3(h)$	λ	Assume that the average abnormal output vector time of Executor 3 upon an attack is 10 hours/10 minutes/10 seconds
σ	1.0×10^{-4}	The uncertainty of consistent output vectors between the two CMF executors attacked
$\mu_1(h)$	3.6×10^6	One executor's OVs are abnormal (perceived non-escaping), when the attack scenario changes, assuming that the average time for the executor to restore to the dormancy state is 0. 001 seconds
$\mu_2(h)$	30	Assume that when two executors' OVs are abnormal and consistent (perceived escaping), the average time for an executor to recover to the dormancy state is 2 minutes
$\mu_3(h)$	2	Assume that when three executors' OVs are abnormal and consistent (non-perceived escaping), the average time for an executor to recover to the dormancy state is 30 minutes
$\mu_4(h)$	60	Assume that when three executors' OVs are inconsistent (perceived degrading), the average time for an executor to recover to the dormancy state is 1 minute

Note: $\lambda = 0.1, 6, 360$

μ_1 represents one output vector abnormality caused by a general attack. When the attack scenario changes, the average time for the executor to recover to the dormancy state is assumed to be 0.001 seconds (perceived non-escaping). μ_3 represents that when three executors' OVs are abnormal and consistent (non-perceived escaping), the average time for an executor to recover to the dormancy state is 30 minutes. μ_4 represents that when three executors' OVs are abnormal and inconsistent (perceived degrading), the average time for an executor to recover to the dormancy state is 1 minute. μ_2 represents that when two executors' abnormal OVs are identical, yet different from the third executor's abnormal OV, the average time for an executor to recover to the dormancy state is 2 minutes (perceived escaping).

σ represents the proportion of consistent output vectors of two faulty executors after being attacked. NVD-based literature [22] analyzes the vulnerabilities of 11 operating systems in 18 years and finds that the operating systems from the same family (such as Windows 2003 and Windows 2008) share a relatively large number of CM vulnerabilities, which is a rare occurrence (which in many cases is 0) for operating systems from different families (such as between BSD and Windows). For this reason, for two executors with sufficient heterogeneity, in event of a general attack, the proportion of consistent abnormal output vectors σ can be set out as a reasonable small value of 1.0×10^4. In the development of specific products, the engineering management methods similar to those used in the development of flight control systems can be employed to ensure that the actual value of this parameter is much smaller than 1.0×10^4.

Based on the CTMC model of the CMD system, the state transition equation can be obtained:

$$
\begin{bmatrix}
\dot{P}_1(t)\\
\dot{P}_2(t)\\
\dot{P}_3(t)\\
\dot{P}_4(t)\\
\dot{P}_5(t)\\
\dot{P}_6(t)\\
\dot{P}_7(t)\\
\dot{P}_8(t)\\
\dot{P}_9(t)\\
\dot{P}_{10}(t)\\
\dot{P}_{11}(t)\\
\dot{P}_{12}(t)\\
\dot{P}_{13}(t)
\end{bmatrix}
=
\begin{bmatrix}
-\lambda_1-\lambda_2-\lambda_3 & \mu_1 & \mu_1 & \mu_1 & \mu_2 & \mu_4 & \mu_2 & \mu_4 & \mu_2 & \mu_4 & \mu_3 & \mu_2 & \mu_4\\
\lambda_1\beta & -\mu_1-\lambda_2-\lambda_3 & 0 & 0 & 0 & 0 & 0 & 0 & 0 & 0 & 0 & 0 & 0\\
\lambda_2\beta & 0 & -\mu_1-\lambda_3-\lambda_1 & 0 & 0 & 0 & 0 & 0 & 0 & 0 & 0 & 0 & 0\\
\lambda_3\beta & 0 & 0 & -\mu_1-\lambda_2-\lambda_1 & 0 & 0 & 0 & 0 & 0 & 0 & 0 & 0 & 0\\
0 & \lambda_2\sigma & \lambda_1\sigma & 0 & -\lambda_3-\mu_2 & 0 & 0 & 0 & 0 & 0 & 0 & 0 & 0\\
0 & \lambda_2(1-\sigma) & \lambda_1(1-\sigma) & 0 & 0 & -\lambda_3-\mu_4 & 0 & 0 & 0 & 0 & 0 & 0 & 0\\
0 & \lambda_3\sigma & 0 & \lambda_1\sigma & 0 & 0 & -\lambda_2-\mu_2 & 0 & 0 & 0 & 0 & 0 & 0\\
0 & \lambda_3(1-\sigma) & 0 & \lambda_1(1-\sigma) & 0 & 0 & 0 & -\lambda_2-\mu_4 & 0 & 0 & 0 & 0 & 0\\
0 & 0 & \lambda_3\sigma & \lambda_2\sigma & 0 & 0 & 0 & 0 & -\lambda_1-\mu_2 & 0 & 0 & 0 & 0\\
0 & 0 & \lambda_3(1-\sigma) & \lambda_2(1-\sigma) & 0 & 0 & 0 & 0 & 0 & -\lambda_1-\mu_4 & 0 & 0 & 0\\
0 & 0 & 0 & 0 & 0 & 2\lambda_3\sigma & 0 & 2\lambda_2\sigma & 0 & 2\lambda_1\sigma & -\mu_3 & 0 & 0\\
0 & 0 & 0 & 0 & 2\lambda_3\sigma & 0 & 2\lambda_2\sigma & 0 & 2\lambda_1\sigma & 0 & 0 & -\mu_2 & 0\\
0 & 0 & 0 & 0 & \lambda_3(1-2\sigma) & \lambda_3(1-2\sigma) & \lambda_2(1-2\sigma) & \lambda_2(1-2\sigma) & \lambda_1(1-2\sigma) & \lambda_1(1-2\sigma) & 0 & 0 & -\mu_4
\end{bmatrix}
\begin{bmatrix}
P_1(t)\\
P_2(t)\\
P_3(t)\\
P_4(t)\\
P_5(t)\\
P_6(t)\\
P_7(t)\\
P_8(t)\\
P_9(t)\\
P_{10}(t)\\
P_{11}(t)\\
P_{12}(t)\\
P_{13}(t)
\end{bmatrix}
\tag{9.1}
$$

Where M_{C0} is known, the stability probability of each state can be figured out:

$$
\begin{cases}
P_{M_0} = \dfrac{G_0}{G} \\[2mm]
P_{M_1} = \dfrac{G_1}{G} \\[2mm]
P_{M_2} = \dfrac{G_1}{G} \\[2mm]
P_{M_3} = \dfrac{G_1}{G} \\[2mm]
P_{M_4} = \dfrac{G_2}{G} \\[2mm]
P_{M_5} = \dfrac{G_3}{G}
\end{cases}
\qquad (9.2)
$$

$$
\begin{cases}
P_{M_6} = \dfrac{G_2}{G} \\[2mm]
P_{M_7} = \dfrac{G_3}{G} \\[2mm]
P_{M_8} = \dfrac{G_2}{G} \\[2mm]
P_{M_9} = \dfrac{G_3}{G} \\[2mm]
P_{M_{10}} = \dfrac{G_4}{G} \\[2mm]
P_{M_{11}} = \dfrac{G_5}{G} \\[2mm]
P_{M_{12}} = \dfrac{G_6}{G}
\end{cases}
$$

where,

$$G_0 = (2\lambda\mu_3\mu_2^{\ 2}\mu_4^{\ 2} + \mu_3\mu_4\mu_2^{\ 2}\lambda^2 + 2\mu_3\mu_4\mu_2^{\ 2}\lambda^2 + \mu_1\mu_3\mu_4\mu_2^{\ 2}\lambda + 2\mu_3\lambda^2\mu_2\mu_4^{\ 2}$$
$$+ \mu_1\mu_3\mu_2\mu_4\lambda + 2\mu_2\mu_3\mu_4\lambda^3 + \mu_1\mu_2\mu_3\mu_4\lambda^2)$$

$$G_1 = \left(\lambda\mu_3\mu_2^{\ 2}\mu_4^{\ 2} + \mu_3\mu_4\mu_2^{\ 2}\lambda^2 + \mu_3\lambda^2\mu_2\mu_4^{\ 2} + \mu_2\mu_3\mu_4\lambda^3\right)$$

$$G_2 = 2\left(\sigma\mu_3\lambda^2\mu_2\mu_4^{\ 2} + \sigma\mu_2\mu_3\mu_4\lambda^3\right)$$

$$G_3 = 2\left(\mu_2\mu_3\mu_4\lambda^3 + \mu_3\mu_4\mu_2^{\ 2}\lambda^2 - \sigma\mu_3\mu_4\mu_2^{\ 2}\lambda^2 - \sigma\mu_2\mu_3\mu_4\lambda^3\right)$$

$$G_4 = 6\left(\sigma^2\mu_4^{\ 2}\mu_2\lambda^3 + \sigma^2\mu_4\mu_2\lambda^4\right)$$

$$G_5 = 6(\sigma\mu_3\mu_4^{\ 2}\lambda^3 + 3\sigma\mu_3\mu_4\lambda^4 + 2\sigma\mu_2\mu_3\mu_4\lambda^3 - \sigma^2\mu_3\mu_4^{\ 2}\lambda^3$$
$$- 3\sigma^2\mu_3\mu_4\lambda^4 - 2\sigma^2\mu_2\mu_3\mu_4\lambda^3)$$

$$G_6 = 6(2\sigma^2\mu_3\mu_2^{\ 2}\lambda^3 + 2\sigma^2\mu_3\mu_2\lambda^4 - 3\sigma\mu_3\mu_2^{\ 2}\lambda^3 - 3\sigma\mu_3\mu_2\lambda^4$$
$$+ \mu_3\mu_2^{\ 2}\lambda^3 + \mu_3\mu_2\lambda^4)$$

$$G = G_0 + 3G_1 + 3G_2 + 3G_3 + G_4 + G_5 + G_6$$

In the GSPN model, according to the implementation rules of time transitions and the no-latency feature of immediate transitions, Table 9.3 gives the reachable set of states of the CMD system, as well as the steady-state probabilities of each state identifier according to the relevant parameters of the transition process, with Pu, Pk, and Ps, respectively, representing the steady-state probabilities of the weak attack, medium attack, and strong attack scenarios.

According to the definition of the system state probabilities, each steady-state probability of the CMD system can be calculated:

(1) Steady-state AP (SSAP)

$$AP = P(M_0) + P(M_1) + P(M_2) + P(M_3)$$
$$= \begin{cases} 9.999999 \times 10^{-1}, & \text{Weak attack scenarios} \\ 9.999989 \times 10^{-1}, & \text{Medium attack scenarios} \\ 9.964137 \times 10^{-1}, & \text{Strong attack scenarios} \end{cases}$$

(2) Steady-state EP (SSEP)

$$EP = P(M_4) + P(M_6) + P(M_8) + P(M_{10}) + P(M_{11})$$
$$= \begin{cases} 5.574345 \times 10^{-14}, & \text{Weak attack scenarios} \\ 2.364045 \times 10^{-10}, & \text{Medium attack scenarios} \\ 1.947108 \times 10^{-6}, & \text{Strong attack scenarios} \end{cases}$$

Table 9.3 Reachable state set of CTMC model for the CMD's general attack failure

Steady-state probabilities

Mark	Steady-state probability P_u	Steady-state probability P_k	Steady-state probability P_s
M_0	9.999999×10^{-1}	9.999939×10^{-1}	9.961149×10^{-1}
M_1	2.777777×10^{-8}	1.666651×10^{-6}	9.959158×10^{-5}
M_2	2.777777×10^{-8}	1.666651×10^{-6}	9.959158×10^{-5}
M_3	2.777777×10^{-8}	1.666651×10^{-6}	9.959158×10^{-5}
M_4	1.845834×10^{-14}	5.555504×10^{-11}	1.838614×10^{-8}
M_5	9.242927×10^{-11}	3.029972×10^{-7}	1.707113×10^{-4}
M_6	1.845834×10^{-14}	5.555504×10^{-11}	1.838614×10^{-8}
M_7	9.242927×10^{-11}	3.029972×10^{-7}	1.707113×10^{-4}
M_8	1.845834×10^{-14}	5.555504×10^{-11}	1.838614×10^{-8}
M_9	9.242927×10^{-11}	3.029972×10^{-7}	1.707113×10^{-4}
M_{10}	4.681853×10^{-18}	5.002546×10^{-14}	9.928482×10^{-10}
M_{11}	3.629402×10^{-16}	6.968935×10^{-11}	1.890956×10^{-6}
M_{12}	4.620544×10^{-13}	9.088097×10^{-8}	3.072189×10^{-3}

Tangible states

	P_1	P_2	P_3	P_4	P_5	P_6	P_{10}	P_{11}	P_{12}	P_{13}	P_{14}	P_{15}	P_{22}	P_{23}	P_{24}
M_0	1	1	1	0	0	0	0	0	0	0	0	0	0	0	0
M_1	0	1	1	1	0	0	0	0	0	0	0	0	0	0	0
M_2	1	0	1	0	1	0	0	0	0	0	0	0	0	0	0
M_3	1	1	0	0	0	1	0	0	0	0	0	0	0	0	0
M_4	0	0	1	0	0	0	1	0	0	0	0	0	0	0	0
M_5	0	0	1	0	0	0	0	1	0	0	0	0	0	0	0
M_6	1	0	0	0	0	0	0	0	1	0	0	0	0	0	0
M_7	1	0	0	0	0	0	0	0	0	1	0	0	0	0	0
M_8	0	1	0	0	0	0	0	0	0	0	1	0	0	0	0
M_9	0	1	0	0	0	0	0	0	0	0	0	1	0	0	0
M_{10}	0	0	0	0	0	0	0	0	0	0	0	0	1	0	0
M_{11}	0	0	0	0	0	0	0	0	0	0	0	0	0	1	0
M_{12}	0	0	0	0	0	0	0	0	0	0	0	0	0	0	1

(3) Steady-state NSAP (SSNSAP)

$$\text{NSAP} = 1 - P(M_0) - P(M_{10})$$
$$= \begin{cases} 8.361113 \times 10^{-8}, Weak\ attack\ scenarios \\ 6.000062 \times 10^{-6}, Medium\ attack\ scenarios \\ 3.885044 \times 10^{-3}, Strong\ attack\ scenarios \end{cases}$$

It can be seen that even when the CMD system encounters strong attacks, its SSAP remains highly close to 1, and its SSEP is very low. The low SSNSAP does not mean that the system's perceptual ability is poor. Instead, it represents a short

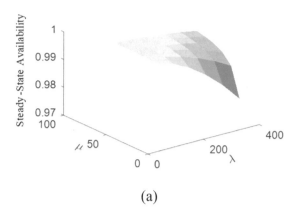

(a)

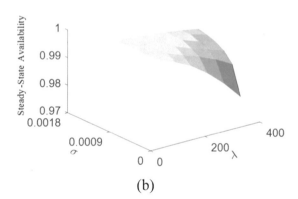

(b)

Fig. 9.16 Varying surfaces of the SSA of a mimic system. (**a**) Steady-state availability (SSA) varies with the attack disturbance rate (ADR)λ and the dynamic reconfiguration rate(DRR)μ. (**b**) SSA varies with the ADRλ and the dissimilarity σ

period of time for the system to be in a state of cognition upon a general attack, since the faulty executor can be quickly recovered once it is perceived.

The relationship between the CMD's state probabilities and the parameters σ, μ, and λ are shown in Figs. 9.16, 9.17, and 9.18. According to the simulation results, similar to the DRS system, SSEP is much impacted by σ, the uncertainty of two consistent OVs emerging upon the attacked CMD system. The larger the dissimilarity between the executors, the lower the SSEP will be. In respect of the CMD system, when the executor recovery rate μ increases, the system's SSEP and the SSNSAP will decrease. When the executor failure rate λ increases, the system's SSAP will decrease, while its SSEP and SSNSAP will increase.

Fig. 9.17 Varying curved surfaces of the SSNSAP of a mimic system. (**a**) SSNSAP varies with the ADRλ and the DRRμ. (**b**) SSNSAP varies with the ADRλ and the dissimilarityσ

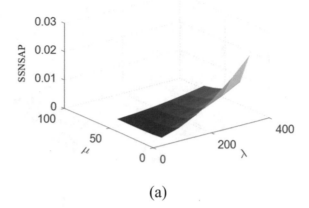

(a)

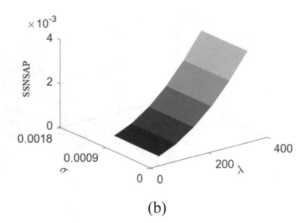

(b)

The CMD system has a high SSAP/SSNSAP and a low SSEP, as well as an agile, accurate, and persistent anti-attack capability. The state probabilities of the system are greatly affected by the parameters σ, μ, and λ. The larger the dissimilarity σ between the executors, the higher the SSAP and SSNSAP, and the lower the SSEP will be; the faster the executor's fault recovery rate μ is, the higher the SSAP, and the lower the SSEP and SSNSAP will be; the lower the failure rate of the executor, the higher the SSAP and the lower the SSEP will be. Therefore, in order to improve the anti-attack of a CMD system adopting the SSP, we should try to choose the executors with larger dissimilarities, higher fault recovery rates, fewer vulnerabilities, and more defensive measures.

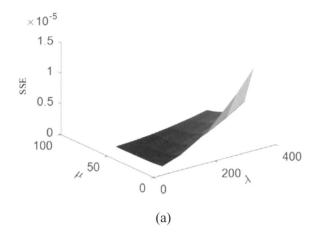

(a)

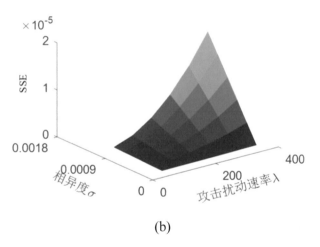

(b)

Fig. 9.18 Varying curved surfaces of the SSE of a mimic system. (**a**) SSE varies with the ADRλ and the DRRμ. (**b**) SSE varies with the ADRλ and the dissimilarityσ

Robustness can be divided into stability robustness and performance robustness according to different definitions. The former refers to the degree to which a control system maintains asymptotic stability when its model parameters vary greatly or when its structure changes. The latter refers to the ability of a system's quality indicators to remain within a certain allowable range under the disturbances of uncertain factors. As shown in Figs. 9.16, 9.17, and 9.18, when the parameters σ, μ, and λ vary, the state probabilities of the CMD system can remain asymptotically stable, and finally reach a steady state, so that the system does well in stability robustness.

Similarly, for man-made disturbances, including those based on vulnerabilities, namely, "general" uncertain disturbances (GUD), the "TRA" transforms the problem of uncertain component failures in the CMD system into the robustness problem of probability attributes at the system level in FE and HR relativity ruling scenarios. In view of the AP, EP, and NSAP of the CMD system at the steady state, it can be known that when the executor's dissimilarity σ, fault recovery rate μ, and fault rate λ are maintained within a relatively extensive and reasonable range, the anti-aggression indicators of the CMD system can be maintained within a quality permitted scope, so that the system will have excellent performance robustness.

2. Rapid Recover Scheduling Policy (RRSP)

The simulation conclusions in the previous section indicate that in the CMD system the executor's recovery rate μ has an impact on the AP, EP, and NSAP. Since the hot backup executor used for rapid recovery is a core resource of the CMD system, it must be used efficiently, that is, the RRSP should be used to quickly recover the executor in the most serious faulty situation (where the output of each executor is inconsistent), such as the shift to container-based executor rapid scheduling (ms/sec) from virtual machine-based executor slow scheduling (min/sec).

The parameters of the CTMC model applying the RRSP are as shown in Table 9.4. The parameters other than μ_4 are the same as those of the CMD system applying the SSP, that is, when the output of each executor is inconsistent; the assumed average time for restoring the three executors to the dormancy state is changed to 10 seconds from 1 minute (Rapid Recovery).

Table 9.4 Parameters of the CTMC model for the CMD system introducing the PPSP

Parameter	Value	Meaning
$\lambda_1(h)$	λ	Assume that the average abnormal output vector time of Executor 1 upon an attack is 10 hours/10 minutes/10 seconds
$\lambda_2(h)$	λ	Assume that the average abnormal output vector time of Executor 2 upon an attack is 10 hours/10 minutes/10 seconds
$\lambda_3(h)$	λ	Assume that the average abnormal output vector time of Executor 3 upon an attack is 10 hours/10 minutes/10 seconds
σ	1.0×10^{-4}	The uncertainty of consistent output vectors between the two CMF executors attacked
$\mu_1(h)$	3.6×10^6	One executor's OVs are abnormal (perceived non-escaping), when the attack/defense scenario changes, assume that the average time for the executor to recover to the dormancy state is 0. 001 seconds
$\mu_2(h)$	30	Assume that when two faulty executors' OVs are abnormal and consistent (perceived escaping), the average time for the executor to recover to the dormancy state is 2 minutes
$\mu_3(h)$	2	Assume that when three faulty executors' OVs are abnormal and consistent (non-perceived escaping), the average time for the executor to recover to the dormancy state is 30 minutes
$\mu_4(h)$	360	Assume that when three executors' OVs are inconsistent (perceived degrading), the average time for an executor to recover to the dormancy state is 10 seconds

Note: λ=0.1, 6, 360

Table 9.5 Steady-state probabilities of the states in the CTMC model for the CMD system adopting the RRSP

Mark	Steady-state probability P_u	Steady-state probability P_k	Steady-state probability P_s
M_0	9.999999×10^{-1}	9.999948×10^{-1}	9.990996×10^{-1}
M_1	2.777777×10^{-8}	1.666652×10^{-6}	9.988999×10^{-5}
M_2	2.777777×10^{-8}	1.666652×10^{-6}	9.988999×10^{-5}
M_3	2.777777×10^{-8}	1.666652×10^{-6}	9.988999×10^{-5}
M_4	1.845739×10^{-14}	5.555508×10^{-11}	1.844123×10^{-8}
M_5	1.542627×10^{-11}	5.463888×10^{-8}	9.988000×10^{-5}
M_6	1.845739×10^{-14}	5.555508×10^{-11}	1.844123×10^{-8}
M_7	1.542627×10^{-11}	5.463888×10^{-8}	9.988000×10^{-5}
M_8	1.845739×10^{-14}	5.555508×10^{-11}	1.844123×10^{-8}
M_9	1.542627×10^{-11}	5.463888×10^{-8}	9.988000×10^{-5}
M_{10}	4.136379×10^{-18}	4.999123×10^{-14}	9.958232×10^{-10}
M_{11}	2.096094×10^{-16}	3.988639×10^{-11}	1.382954×10^{-6}
M_{12}	1.285317×10^{-14}	2.731398×10^{-9}	2.995801×10^{-4}

Table 9.5 shows the steady-state probabilities of the states in the CMD system adopting the RRSP, as well as the steady-state probabilities of each state mark according to the relevant parameters of the transition process, with Pu, Pk, and Ps, respectively, representing the steady-state probabilities of the weak attack, medium attack, and strong attack scenarios.

According to the definition of the system state probabilities, each steady-state probability of the CMD system can be calculated:

(1) SSAP

$$AP = P(M_0) + P(M_1) + P(M_2) + P(M_3)$$
$$= \begin{cases} 9.999999 \times 10^{-1}, & \textit{Weak attack scenarios} \\ 9.999989 \times 10^{-1}, & \textit{Medium attack scenarios} \\ 9.993993 \times 10^{-1}, & \textit{Strong attack scenarios} \end{cases}$$

(2) SSEP

$$EP = P(M_4) + P(M_6) + P(M_8) + P(M_{10}) + P(M_{11})$$
$$= \begin{cases} 5.559030 \times 10^{-14}, & \textit{Weak attack scenarios} \\ 2.066016 \times 10^{-10}, & \textit{Medium attack scenarios} \\ 1.439273 \times 10^{-6}, & \textit{Strong attack scenarios} \end{cases}$$

(3) SSNSAP

$$\text{NSAP} = 1 - P(M_0) - P(M_{10})$$
$$= \begin{cases} 8.337967 \times 10^{-8}, & \textit{Weak attack scenarios} \\ 5.166812 \times 10^{-6}, & \textit{Medium attack scenarios} \\ 9.003283 \times 10^{-3}, & \textit{Strong attack scenarios} \end{cases}$$

Upon the adoption of the RRSP, the system's steady-state unavailability probability (SSUP) is reduced by a magnitude. When the CMD system is deployed in practice, the adoption of the RRSP can significantly reduce the SSUP of the system, making the system more available.

3. Flaw Channel Scheduling Policy (FCSP)

The uncertainty σ of consistent output vectors between two executors of the CMD system attacked has a great impact on the SSEP and SSNSAP. The higher dissimilarity between the executors, the lower the SSEP will be. Therefore, we consider increasing the dissimilarity between the executors by deliberately introducing an executor with more defects so as to reduce the SSEP of the system, which is called the flaw channel scheduling policy (FCSP).

The parameters of the CTMC model applying the FCSP are as shown in Table 9.6. The parameters other than σ are the same as those of the CMD system applying the

Table 9.6 Parameters of the CTMC model for the CMD system introducing the FCSP

Parameter	Value	Meaning
$\lambda_1(h)$	360	Assume that the average abnormal output vector time of Executor 1 upon an attack is 10 seconds
$\lambda_2(h)$	λ	Assume that the average abnormal output vector time of Executor 2 upon an attack is 10 hours/10 minutes/10 seconds
$\lambda_3(h)$	λ	Assume that the average abnormal output vector time of Executor 3 upon an attack is 10 hours/10 minutes/10 seconds
σ	1.0×10^{-5}	The uncertainty of consistent output vectors between the two CMF executors attacked
$\mu_1(h)$	3.6×10^6	One executor's OVs are abnormal (perceived non-escaping), when the attack scenario changes, assuming that the average time for the executor to recover to the dormancy state is 0. 001 seconds
$\mu_2(h)$	30	Assume that when two executors' OVs are abnormal and consistent (perceived escaping), the average time for an executor to recover to the dormancy state is 2 minutes
$\mu_3(h)$	2	Assume that when three executors' OVs are abnormal and consistent (non-perceived escaping), the average time for an executor to recover to the dormancy state is 30 minutes
$\mu_4(h)$	360	Assume that when three executors' OVs are inconsistent (perceived degrading), the average time for an executor to recover to the dormancy state is 1 minute

Note: $\lambda=0.1, 6, 360$

Table 9.7 Steady-state probabilities of the states in the CMD system introducing the FCSP

Mark	Steady-state probability P_u	Steady-state probability P_k	Steady-state probability P_s
M_0	9.998992×10^{-1}	9.998563×10^{-1}	9.961158×10^{-1}
M_1	9.998992×10^{-5}	9.998530×10^{-5}	9.959166×10^{-5}
M_2	2.777220×10^{-8}	1.666258×10^{-6}	9.959166×10^{-5}
M_3	2.777220×10^{-8}	1.666258×10^{-6}	9.959166×10^{-5}
M_4	6.643782×10^{-12}	3.332682×10^{-10}	1.838615×10^{-9}
M_5	3.327252×10^{-7}	1.817807×10^{-5}	1.707269×10^{-4}
M_6	1.630235×10^{-16}	5.127190×10^{-13}	1.838615×10^{-9}
M_7	1.322474×10^{-11}	4.760689×10^{-8}	1.707269×10^{-4}
M_8	6.643782×10^{-11}	3.332682×10^{-10}	1.838615×10^{-9}
M_9	3.327252×10^{-7}	1.817807×10^{-5}	1.707269×10^{-4}
M_{10}	3.965647×10^{-15}	2.570533×10^{-14}	9.928482×10^{-10}
M_{11}	9.405052×10^{-14}	2.963091×10^{-10}	1.890956×10^{-6}
M_{12}	1.188409×10^{-9}	3.921177×10^{-6}	3.072189×10^{-3}

SSP, that is, the uncertaintyσ of consistent output vectors between the two CMF executors attacked changes to 1.0×10^{-5} from 1.0×10^{-4} (where the dissimilarity between the defected executor and other executors gets bigger).

Table 9.7 shows the steady-state probabilities of the states in the CMD system adopting the FCSP, as well as the steady-state probabilities of each state mark according to the relevant parameters of the transition process, with Pu, Pk, and Ps, respectively, representing the steady-state probabilities of the weak attack, medium attack, and strong attack scenarios.

According to the definition of the system state probabilities, each steady-state probability of the CMD system can be calculated:

(1) SSAP

$$AP = P(M_0) + P(M_1) + P(M_2) + P(M_3)$$
$$= \begin{cases} 9.999993 \times 10^{-1}, & \textit{Weak attack scenarios} \\ 9.999596 \times 10^{-1}, & \textit{Medium attack scenarios} \\ 9.964146 \times 10^{-1}, & \textit{Strong attack scenarios} \end{cases}$$

(2) SSEP

$$EP = P(M_4) + P(M_6) + P(M_8) + P(M_{10}) + P(M_{11})$$
$$= \begin{cases} 1.338574 \times 10^{-11}, & \textit{Weak attack scenarios} \\ 9.633839 \times 10^{-10}, & \textit{Medium attack scenarios} \\ 1.946386 \times 10^{-7}, & \textit{Strong attack scenarios} \end{cases}$$

(3) SSNSAP

$$NSAP = 1 - P(M_0) - P(M_{10})$$
$$= \begin{cases} 1.007121 \times 10^{-4}, & \textit{Weak attack scenarios} \\ 1.436437 \times 10^{-4}, & \textit{Medium attack scenarios} \\ 3.884172 \times 10^{-3}, & \textit{Strong attack scenarios} \end{cases}$$

As for the CMD system, its SSEP is reduced by 1~3 magnitudes upon the adoption of the FCSP compared to the SSP. When the CMD system is deployed in practice, the adoption of the FCSP can significantly reduce the SSEP, making the system more resilient to attacks.

4. Rapid Recover and Flaw Channel Scheduling Policy (RRFCSP)

We resort to the rapid recover and flaw channel scheduling policy (RRFCSP) to comprehensively improve the state probabilities of the system. The parameters of the CTMC model is as shown in Table 9.8, where σ and μ_4 are reduced simultaneously, with the rest of the parameters identical to that of the CMD system applying the SSP.

Table 9.9 shows the steady-state probabilities of the states in the CMD system adopting the RRFCSP, as well as the steady-state probabilities of each state identifier according to the relevant parameters of the transition process, with Pu, Pk, and

Table 9.8 Parameters of the CTMC model for the CMD system introducing the RRFCSP

Parameter	Value	Meaning
$\lambda_1(h)$	360	Assume that the average abnormal output vector time of Executor 1 upon an attack is 10 seconds
$\lambda_2(h)$	λ	Assume that the average abnormal output vector time of Executor 2 upon an attack is 10 hours/10 minutes/10 seconds
$\lambda_3(h)$	λ	Assume that the average abnormal output vector time of Executor 3 upon an attack is 10 hours/10 minutes/10 seconds
σ	1.0×10^{-5}	The uncertainty of consistent output vectors between the two CMF executors attacked
$\mu_1(h)$	3.6×10^6	One executor's OVs are abnormal (perceived non-escaping), when the attack scenario changes, assuming that the average time for the executor to recover to the dormancy state is 0. 001 seconds
$\mu_2(h)$	30	Assume that when two executors' OVs are abnormal and consistent (perceived escaping), the average time for an executor to recover to the dormancy state is 2 minutes
$\mu_3(h)$	2	Assume that when three executors' OVs are abnormal and consistent (non-perceived escaping), the average time for an executor to recover to the dormancy state is 30 minutes
$\mu_4(h)$	360	Assume that when three executors' OVs are inconsistent (perceived degrading), the average time for an executor to recover to the dormancy state is 1 minute

Note: λ=0.1, 6, 360

Table 9.9 Steady-state probabilities of the states in the CMD system introducing the RRFCSP

Mark	Steady-state probability P_u	Steady-state probability P_k	Steady-state probability P_s
M_0	9.998998×10^{-1}	9.998899×10^{-1}	9.991009×10^{-1}
M_1	9.998998×10^{-5}	9.998866×10^{-5}	9.989011×10^{-5}
M_2	2.777222×10^{-8}	1.666314×10^{-6}	9.989011×10^{-5}
M_3	2.777222×10^{-8}	1.666314×10^{-6}	9.989011×10^{-5}
M_4	6.643783×10^{-12}	3.332794×10^{-10}	1.844125×10^{-9}
M_5	5.553123×10^{-8}	3.278123×10^{-6}	9.988911×10^{-5}
M_6	1.629566×10^{-16}	5.127365×10^{-13}	1.844125×10^{-9}
M_7	7.714439×10^{-12}	2.777162×10^{-8}	9.988911×10^{-5}
M_8	6.643783×10^{-12}	3.332794×10^{-10}	1.844125×10^{-9}
M_9	5.553123×10^{-8}	3.278123×10^{-6}	9.988911×10^{-5}
M_{10}	3.979703×10^{-15}	2.571908×10^{-14}	9.956616×10^{-12}
M_{11}	5.577243×10^{-14}	1.723537×10^{-10}	1.383080×10^{-6}
M_{12}	3.856437×10^{-11}	1.370396×10^{-7}	2.996613×10^{-4}

Ps, respectively, representing the steady-state probabilities of the weak attack, medium attack, and strong attack scenarios.

According to the definition of the system state probabilities, each steady-state probability of the CMD system can be calculated:

(1) SSAP

$$AP = P(M_0) + P(M_1) + P(M_2) + P(M_3)$$
$$= \begin{cases} 9.999998 \times 10^{-1}, & \textit{Weak attack scenarios} \\ 9.999932 \times 10^{-1}, & \textit{Medium attack scenarios} \\ 9.994005 \times 10^{-1}, & \textit{Strong attack scenarios} \end{cases}$$

(2) SSEP

$$EP = P(M_4) + P(M_6) + P(M_8) + P(M_{10}) + P(M_{11})$$
$$= \begin{cases} 1.334748 \times 10^{-11}, & \textit{Weak attack scenarios} \\ 8.394510 \times 10^{-10}, & \textit{Medium attack scenarios} \\ 1.438503 \times 10^{-7}, & \textit{Strong attack scenarios} \end{cases}$$

(3) SSNSAP

$$NSAP = 1 - P(M_0) - P(M_{10})$$
$$= \begin{cases} 1.001566 \times 10^{-4}, & \textit{Weak attack scenarios} \\ 1.100432 \times 10^{-4}, & \textit{Medium attack scenarios} \\ 8.991428 \times 10^{-4}, & \textit{Strong attack scenarios} \end{cases}$$

The CMD displays excellent stability robustness and performance robustness upon adoption of control laws such as the SSP, RRSP, FCSP, and RRFCSP. When the CMD system is deployed in practice, the adoption of the RRFCSP can significantly increase the SSAP and SSNSAP and reduce the SSEP, making the system more resistant to attacks.

9.5.3.1 Analysis of CMD's Resistance Against the General DM/CM Attacks

A fault is the cause of an error. When a fault causes an error, it is active; otherwise it is dormant. A failure occurs when an error reaches the system service interface and changes the service. An information system is in the failure state when it fails to perform the functions required by the specification. A system does not always fail in the same way, while the way the system fails is called the failure mode.

In an information system with high reliability requirements, the system is often designed into a redundant structure to improve its reliability and security. A common cause fault (CCF) is a dependent fault event because there are two or more redundant components that fail simultaneously due to some common cause. It is called a common mode failure (CMF) when the fault modes are the same or a differential mode fault (DMF) when the fault modes are different. Similarly, it is called a common mode error (CME) when the modes of the error caused by the fault are the same or a differential mode error (DME) when the modes of the error caused by the fault are different. An attack that causes a CME is called a common mode attack (CMA), while an attack that causes a DME is called a differential mode attack (DMA). CMFs/CMEs seriously affect the anti-attack capability and reliability of the redundant system and become the main cause of failures of redundant systems.

In the simulation of the previous section, we analyzed the ability of the CMD system to resist general attacks at a relatively short executor recovery latency in the case of integrated DMAs and CMAs. In this section, we will analyze the ability of the CMD system to resist general attacks at a relatively long executor recovery latency in the case of independent DMAs and CMAs.

The DMAs stated in this section mainly fall in two categories, namely, time-coordinated DMAs and single DMAs. Since the executor recovery latency is assumed to be relatively long, before the C&R of the executor, there will be a case where multiple executors are successfully attacked one by one, but with erroneous and inconsistent output vectors, this is called a time-coordinated DMA. A DMA that causes only one executor to output vector errors is only an exception of the time-coordinated DMA, and it is called a single DMA.

For CMAs, the common mode degrees can be distinguished according to the proportion of executors generating CMEs or the proportion of executors with common vulnerabilities. An executor's common mode proportion is inversely proportional to its heterogeneity. As described in Chap. 6, a DRS system must have $f \leq (n-1)/2$ number of executors to be in the error state; otherwise it will fail. However, as far as the

CMD system is concerned, it can completely defend against $1 \leq f \leq n$ DMAs, allow relatively low escape probabilities of $f \leq (n - 1)/2$ CMAs, and render $n - 2 \geq f > (n - 1)/2$ CMAs unable to succeed continuously, with the escape probabilities under control when the recovery speed or the PAS satisfies certain conditions.

Assume that the CMD system has K executors or running scenarios, and the current executor service set has a redundancy of 3. In the case of a DMA, it is assumed that all the executors or running scenarios of the system have DM vulnerabilities/backdoors, and there is no CM vulnerability/backdoor in any executor or running scenario. In the case of a CMA, it is assumed that CM vulnerabilities/backdoors exist in 3 of the K executors or running scenarios of the system, while DM vulnerabilities/backdoors exist in the rest of the executors or running scenarios.

Each and every of the hypothesized vulnerabilities/backdoors mentioned above has a "kill-with-one-thrust" effect. Any attacker aiming at controlling the majority of the executors and modifying their output vectors has to grasp a large number of high-risk zero-day vulnerabilities unknown to the defender, which in fact is an extremely small probability event. Even if we assume that the attacker can go so far as to wage time-coordinated DMAs and single DMAs that are hard to be defended by conventional defense means, he or she will end up with the mimic defense system not even outputting one result upon request (i.e., the mimic defense system can withstand 100% of DMAs) after spending highly on exploiting a series of high-risk zero-day vulnerabilities and backdoors. As for CMAs with higher attack difficulty, the attacker needs to utilize the high-risk zero-day vulnerabilities shared by multiple executors. The structural effect of the strategic ruling, the PAS, and MDR-NFM based on the mimic disguise idea makes the CMAs on the CMD system unable to sustain their success and brings the escape probabilities under control.

A CMD system can adopt a variety of multimode ruling algorithms, such as basic majority voting, plurality voting, maximum likelihood voting (MLV), weighted voting, mask voting, majority voting based on historical information, quick voting, watched voting, Byzantine failures, AI-based advanced judgments, etc. For the convenience of simulation and calculation, it is assumed that the basic majority voting strategy and the stochastic scheduling policy (SSP) are adopted.

The parameters of the CTMC model for the CMD system against DM/CM general attacks are shown in Table 9.10.

The attack intensity scenario is represented by parameter $\lambda(h)$. According to the statistics of the pilot operation of the CMD products on the current network, on average, each device will encounter several attacks per month resulting from inconsistent arbiters. To simulate the considerable attack intensity, we assume that it takes an average of 10 minutes to successfully attack a single executor.

σ represents the probability of a consistent faulty output vector for two CMF executors attacked. By setting different parameters, it can represent different CMA types, wherein σ_1 represents the selection probability for a CMF to occur to two (and only two) faulty executors; σ_2 represents the selection probability for a CMF to occur to the three faulty executors; and σ_3 represents the selection probability for a CMF to occur to the third faulty executor and either of the other two faulty executors.

Table 9.10 Parameters of the CTMC model for the CMD system against DM/CM general attacks

Parameter	Value	Meaning
$\lambda_1(h)$	6	Assume that the average time of abnormal output vector of Executor 1 upon an attack is 10 hours/10 minutes/10 seconds
$\lambda_2(h)$	6	Assume that the average time of abnormal output vector of Executor 2 upon an attack is 10 hours/10 minutes/10 seconds
$\lambda_3(h)$	6	Assume that the average time abnormal output vector of Executor 3 upon an attack is 10 hours/10 minutes/10 seconds
σ_1	DM: 0 CM: 1/15, 3/28, 1/5, 1/2	The selection probability for a CMF to occur to two (and only two) faulty executors When CM, $\sigma_1 = \dfrac{C_N^2}{C_K^2}$ (CM proportion = N/K, where K = 10, 8, 6, 4, N = 3)
σ_2	DM: 0 CM:1/8, 1/6, 1/4, 1/2	The selection probability for a CMF to occur to the three faulty executors When CM, $\sigma_2 = \dfrac{C_{N-2}^1}{C_{K-2}^1}$ (CM proportion = N/K, where K = 10, 8, 6, 4, N = 3)
σ_3	DM:0 CM: 1/8, 1/5, 3/8, 1	The selection probability for a CMF to occur to the third faulty executor and either of the other two faulty executors When CM, $\sigma_3 = \dfrac{\left(C_K^2 - C_N^2 - C_{K-N}^2\right)}{\left(C_K^2 - C_N^2\right)} * \dfrac{C_{N-1}^1}{C_{K-2}^1}$ (CM proportion = N/K, where K = 10, 8, 6, 4, N = 3)
$\mu_1(h)$	60, 6, 3	One executor's OVs are abnormal (perceived non-escaping), when the attack scenario changes, the average time for the executor to recover to the dormancy state is 60/600/1200 seconds
$\mu_2(h)$	30, 3, 1.5	When two executors' OVs are abnormal and consistent (perceived escaping), the avewwrage time for the executors to recover to the dormancy state is 120/1200/2400 seconds
$\mu_3(h)$	20, 2, 1	When three executors' OVs are abnormal and consistent (non-perceived escaping), the average time for the executors to recover to the dormancy state is 180/1800/3600 seconds
$\mu_4(h)$	60, 6, 3	When three executors' OVs are inconsistent (perceived degrading), the average time for the executors to recover to the dormancy state is 60/600/1200 seconds

$\mu_1{\sim}\mu_4$ represent the average recovery time for a faulty executor in different attack and defense scenarios, with their respective meanings identical to that in the previous section. The recovery scenarios are divided into three categories by the average time for a single executor to recover to the dormancy state: fast (60 seconds), medium (600 seconds), and slow (1200 seconds).

In order to describe the probabilities that the output vectors of all executors of the current service set are in an inconsistent state, we add the steady-state not response probabilities (SSNRP), defined as the steady-state probabilities for the system, to find the output vectors of all executors of the current service set in an inconsistent state through the ruling.

Table 9.11 Steady-state probabilities for the CMD system against DM/CM general attacks

Steady-state probabilities	DM	CM			
		CM proportion 3/10	CM proportion 3/8	CM proportion 3/6	CM proportion 3/4
Fast recovery scenario					
SSAP	0.9615385	0.9586382	0.9568882	0.9529035	0.9403049
SSEP	0	0.005979384	0.009558527	0.01764315	0.04259866
SSNSAP	0.2307692	0.2329297	0.2341477	0.2367243	0.2430546
SSNRP	0.0384615	0.03538246	0.03355322	0.02945338	0.01709645
Medium recovery scenario					
SSAP	0.5	0.4693611	0.4528302	0.4195804	0.3428571
SSEP	0	0.1199478	0.1832884	0.3076923	0.5714286
SSNSAP	0.7500000	0.7574967	0.7574124	0.74825175	0.6571429
SSNRP	0.5	0.4106910	0.3638814	0.2727273	0.08571429
Slow recovery scenario					
SSAP	0.3142857	0.2847774	0.2697968	0.2414399	0.1839465
SSEP	0	0.1836382	0.2747022	0.4424934	0.7491639
SSNSAP	0.8571429	0.8581291	0.8521374	0.8270413	0.6755853
SSNRP	0.6857143	0.5315844	0.4555011	0.3160667	0.06688963

Table 9.11 shows the steady-state probabilities for the CMD system against DM/CM general attacks, and data analysis leads to the following conclusions:

- The escape probability for all DMAs is 0, that is, the CMD system can defend against 100% of the DMAs.
- Even if the executor has a certain proportion of CM faults/errors or vulnerabilities/backdoors, the escape probability for CMAs remains low when the CM proportion is less than 50%; when the CM proportion is greater than or equal to 50%, a CMA cannot escape on an ongoing basis.
- The recovery speed of an executor has a great influence on the anti-attack of the CMD system. A high recovery speed helps to effectively reduce the escape probabilities and is much better economically than increasing the number and heterogeneity of executors.
- If the CMD system adopts a confidence-based or weighted scheduling policy, when the number of high-confidence executors is greater than or equal to 2, the scheduling policy will be able to ensure higher probabilities for the high-confidence executors to be in the current service set, while the system's escape probability approximately is equal to the escape probability of the DMA, namely, close to zero.

All in all, the CMD system can defend against 100% of the DMAs and prevent the CMAs from continuing to take effect. Its structural effect can greatly reduce the strict requirements for the heterogeneity of the executors and keep the escape probabilities of CMAs at a relatively low level by comprehensively adopting a faster executor recovery speed and a confidence-based or weighted scheduling policy, contributing to the overall anti-attack capability of the CMD system. Since each of the above parameters is set to the bottom threshold in the worst case, far outperformed by the corresponding parameter value in actual cases, the actual anti-attack performance of the CMD system should be much higher than the quantitative design value.

9.5.3.2 Anti-special Attack Analysis of the CMD System

A special attack makes the time for the executor to remain in the fault state, and the number of faulty executors increases greatly (i.e., the attack effect can be cumulatively cooperative in the space-time dimension), and the abnormal output vector heterogeneity of the executors decreases greatly (for a halt attack, the functionally equivalent heterogeneity length of the abnormal output vectors is 1 bit; for a standby cooperative attack, the functionally equivalent heterogeneity length of the abnormal output vectors is 0 bit, where a successful attack will not cause any changes in the output vectors of the executor). The above two reasons lead to the greatly reduced steady-state AP and the greatly increased steady-state EP of the system. The background we need to understand is that special attacks are based on a thorough understanding of the real-time status of online and offline executors. They have mastered the "killer" or "super" vulnerabilities for most executors, as well as the effective use of the related attack chains. On this basis, it will be able to overcome the bottleneck of "cooperative attack to diverse dynamic targets under the non-cooperative conditions," while addressing all uncertainty factors. Obviously, this "super attack" ability can only exist in an "ideological/theoretical game."

The following is an analysis of the performance of the NRS, DRS, and CMD systems in defending against special attacks (halt attacks and standby cooperative attacks) on the assumption that the attacker has the above capabilities.

The GSPN model for special attack faults of a CMD system is shown in Fig. 9.19. There are 10 states and 16 transitions. P_1, P_2, and P_3 contain a token each to indicate that the executor is in a faulty dormancy state; P_4, P_5, and P_6 contain a token each to, respectively, indicate a special attack fault occurring to the executor; P_7, P_8, and P_9 contain a token each to indicate a special attack fault simultaneously occurring to two executors, while P_{10} contains a token to indicate a special attack fault simultaneously occurring to three executors. T_1, T_2, and T_3 indicate a special attack fault occurring to the executor; T_4, T_5, and T_6 indicate that two executors simultaneously enter the special attack fault state; T_7, T_8, and T_9 indicate that three executors simultaneously enter the special attack fault state; T_{10} indicates that three executors recover to the fault dormancy state; T_{11}, T_{12}, and T_{13} indicate that two executors recover to the fault dormancy state, while T_{14}, T_{15}, and T_{16} indicate one executor recovers to the fault dormancy state.

The CTMC model for the CMD system's two types of special attack is shown in Fig. 9.20.

1. Downtime attacks

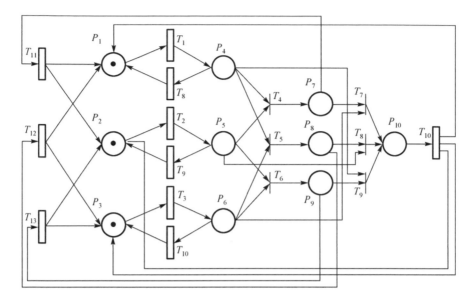

Fig. 9.19 GSPN model for the CMD system's special attack faults

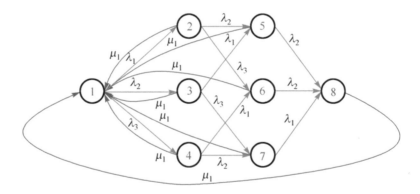

Fig. 9.20 CTMC model for the CMD system's special attack faults

The mimic structure increases the available defense scenarios and improves the system's ability to recover from downtime attacks by dynamically combining or reconstructing the same amount of resources. The steady states of the CTMC model are shown in Table 9.12. The meanings of the parameters of the model are shown in Table 9.13.

Table 9.12 Steady states of the CTMC model for the CMD system's halt attack faults

State serial No.	Meaning
1	Each executor is in normal operation and in the dormancy state
2	One executor is downtime after an attack
3	Two executors are downtime after an attack
4	Three executors are downtime after an attack
5	Executor 1 and Executor 2 are downtime after an attack
6	Executor 1 and Executor 3 are downtime after an attack
7	Executor 2 and Executor 3 are downtime after an attack
8	Post-attack halt of three executors

Table 9.13 Parameters of the CTMC model for the CMD's halt attack faults

Parameters	Value	Meaning
λ_1	6/360	Assume that the average halt time of Executor 1 when attacked is 10 minutes/10 seconds
λ_2	6/360	Assume that the average halt time of Executor 2 when attacked is 10 minutes/10 seconds
λ_3	6/360	Assume that the average halt time of Executor 3 when attacked is 10 minutes/10 seconds
μ_1	60	Assume that the average self-recovery or reconstruction time of the executor is 1 minute

Based on the CTMC model of the CMD system, the state transition equation can be obtained:

$$
\begin{bmatrix}
\dot{P}_1(t) \\
\dot{P}_2(t) \\
\dot{P}_3(t) \\
\dot{P}_4(t) \\
\dot{P}_5(t) \\
\dot{P}_6(t) \\
\dot{P}_7(t) \\
\dot{P}_8(t)
\end{bmatrix}
=
\begin{bmatrix}
-\lambda_1-\lambda_2-\lambda_3 & \mu_1 & \mu_1 & \mu_1 & \mu_1 & \mu_1 & \mu_1 & \mu_1 \\
\lambda_1 & -\mu_1-\lambda_2-\lambda_3 & 0 & 0 & 0 & 0 & 0 & 0 \\
\lambda_2 & 0 & -\mu_1-\lambda_1-\lambda_3 & 0 & 0 & 0 & 0 & 0 \\
\lambda_3 & 0 & 0 & -\mu_1-\lambda_1-\lambda_2 & 0 & 0 & 0 & 0 \\
0 & \lambda_2 & \lambda_1 & 0 & -\mu_1-\lambda_3 & 0 & 0 & 0 \\
0 & \lambda_3 & 0 & \lambda_1 & 0 & -\mu_1-\lambda_1 & 0 & 0 \\
0 & 0 & \lambda_3 & \lambda_2 & 0 & 0 & -\mu_1-\lambda_2 & 0 \\
0 & 0 & 0 & 0 & \lambda_3 & \lambda_1 & \lambda_2 & -\mu_1
\end{bmatrix}
\begin{bmatrix}
P_1(t) \\
P_2(t) \\
P_3(t) \\
P_4(t) \\
P_5(t) \\
P_6(t) \\
P_7(t) \\
P_8(t)
\end{bmatrix}
\tag{9.3}
$$

Table 9.14 gives the steady-state probabilities of states and gives the steady-state probabilities of each state identifier according to the relevant parameters of the transition process.

According to the definition of the system state probabilities, each steady-state probability of the CMD system can be calculated:

(1) SSAP

$$AP = P(M_0) + P(M_1) + P(M_2) + P(M_3)$$
$$= \begin{cases} 9.615383 \times 10^{-1}, & \text{Medium attack scenarios} \\ 1.255061 \times 10^{-1}, & \text{Strong attack scenarios} \end{cases}$$

(2) SSEP EP $= 0$
(3) SSNSAP NSAP $= 1$

When a CMD system guards against halt attacks, the SSNSAP is 1, and the SSEP is 0, while the SSAP is proportional to the recovery speed of the executor and inversely proportional to the attack intensity and the attacker's ability. When the recovery speed outraces the halt rate, the CMD system will have better defense performance against halt attacks.

2. Standby cooperative attacks

The steady states, parameters, and reachable set of the CTMC model for the CMD system are shown in Tables 9.15, 9.16, and 9.17.

Based on the CTMC model of the CMD system, the state transition equation can be obtained:

Table 9.14 Steady-state probabilities of states of the CTMC model for the CMD's halt attack faults

Identifier	M_0	M_1	M_2	M_3	M_4	M_5	M_6	M_7
Steady-state probability P_k	7.692307×10^{-1}	6.410256×10^{-2}	6.410256×10^{-2}	6.410256×10^{-2}	1.165501×10^{-2}	1.165501×10^{-2}	1.165501×10^{-2}	3.496503×10^{-3}
Steady-state probability P_s	5.263158×10^{-2}	2.429150×10^{-2}	2.429150×10^{-2}	2.429150×10^{-2}	4.164257×10^{-2}	4.164257×10^{-2}	4.164257×10^{-2}	7.495662×10^{-1}

Table 9.15 Steady states of the CTMC model for the CMD's standby cooperative attack faults

State serial No.	Meaning
1	Each executor is in normal operation and in the dormancy state
2	Executor 1 is attacked, where a standby cooperative fault occurs without abnormal output vectors
3	Executor 2 is attacked, where a standby cooperative fault occurs without abnormal output vectors
4	Executor 3 is attacked, where a standby cooperative fault occurs without abnormal output vectors
5	Executors 1 & 2 are attacked, where a standby cooperative fault occurs to both without abnormal output vectors
6	Executors 1 & 3 are attacked, where a standby cooperative fault occurs to both without abnormal output vectors
7	Executors 2 & 3 are attacked, where a standby cooperative fault occurs to both without abnormal output vectors
8	The three executors are all attacked, where a standby cooperative fault occurs to them all without abnormal output vectors

Table 9.16 Parameters of the CTMC model for the CMD's standby cooperative attack faults

Parameters	Value	Meaning
λ_1	6/360	Assume that the average standby cooperative fault time of Executor 1 when attacked is 10 minutes/10 seconds
λ_2	6/360	Assume that the average standby cooperative fault time of Executor 2 when attacked is 10 minutes/10 seconds
λ_3	6/360	Assume that the average standby cooperative fault time of Executor 3 when attacked is 10 minutes/10 seconds
μ_1	2	Assume that the average recovery cycle of an executor is 30 minutes

Table 9.17 Reachable state set of the CTMC model for the CMD's standby cooperative attack faults

标识Identifier	M_0	M_1	M_2	M_3	M_4	M_5	M_6	M_7
Steady-state probability P_k	9.999999×10^{-2}	4.285714×10^{-2}	4.285714×10^{-2}	4.285714×10^{-2}	6.428571×10^{-2}	6.428571×10^{-2}	6.428571×10^{-2}	5.785714×10^{-1}
Steady-state probability P_s	1.848429×10^{-3}	9.216543×10^{-4}	9.216543×10^{-4}	9.216543×10^{-4}	1.833125×10^{-3}	1.833125×10^{-3}	1.833125×10^{-3}	9.898872×10^{-1}

$$
\begin{bmatrix}
\dot{P}_1(t) \\
\dot{P}_2(t) \\
\dot{P}_3(t) \\
\dot{P}_4(t) \\
\dot{P}_5(t) \\
\dot{P}_6(t) \\
\dot{P}_7(t) \\
\dot{P}_8(t)
\end{bmatrix}
=
\begin{bmatrix}
-\lambda_1-\lambda_2-\lambda_3 & \mu_1 & \mu_1 & \mu_1 & \mu_1 & \mu_1 & \mu_1 & \mu_1 \\
\lambda_1 & -\mu_1-\lambda_2-\lambda_3 & 0 & 0 & 0 & 0 & 0 & 0 \\
\lambda_2 & 0 & -\mu_1-\lambda_1-\lambda_3 & 0 & 0 & 0 & 0 & 0 \\
\lambda_3 & 0 & 0 & -\mu_1-\lambda_1-\lambda_2 & 0 & 0 & 0 & 0 \\
0 & \lambda_2 & \lambda_1 & 0 & -\mu_1-\lambda_3 & 0 & 0 & 0 \\
0 & \lambda_3 & 0 & \lambda_1 & 0 & -\mu_1-\lambda_1 & 0 & 0 \\
0 & 0 & \lambda_3 & \lambda_2 & 0 & 0 & -\mu_1-\lambda_2 & 0 \\
0 & 0 & 0 & 0 & \lambda_3 & \lambda_1 & \lambda_2 & -\mu_1
\end{bmatrix}
\begin{bmatrix}
P_1(t) \\
P_2(t) \\
P_3(t) \\
P_4(t) \\
P_5(t) \\
P_6(t) \\
P_7(t) \\
P_8(t)
\end{bmatrix}
\tag{9.4}
$$

According to the definition of the system state probabilities, each steady-state probability of the CMD system can be calculated:

(1) SSAP

$$
\begin{aligned}
AP &= P\left(M_0\right) + P\left(M_1\right) + P\left(M_2\right) + P\left(M_3\right) \\
&= \begin{cases}
2.285886 \times 10^{-1}, & \text{Medium attack scenarios} \\
4.619800 \times 10^{-2}, & \text{Strong attack scenarios}
\end{cases}
\end{aligned}
$$

(2) SSEP

$$
\begin{aligned}
EP &= P\left(M_4\right) + P\left(M_5\right) + P\left(M_6\right) + P\left(M_7\right) \\
&= \begin{cases}
7.714114 \times 10^{-1}, & \text{Medium attack scenarios} \\
9.953802 \times 10^{-2}, & \text{Strong attack scenarios}
\end{cases}
\end{aligned}
$$

(3) SSNSAP NSAP = 0.

Under the premise that the attacker's resources and capabilities are not constrained, the CMD system has a relatively low defense capability against a standby cooperative attack. For instance, in a strong attack scenario, its SSNSAP is 0, SSEP is larger than 0.99, and SSAP is approximately 0.05. Obviously, for the CMD system and the attacker alike, there is neither "absolutely credible" defense capabilities nor "unconstrained" attack resources.

9.5.3.3   Summary of the Anti-attack Analysis

The anti-general attack abilities of NRS, DRS, and CMD systems are as shown in Table 9.18. Taking a strong attack scenario as an example, we can find that, compared to an NRS/DRS system, a CMD system has an infinite gain in terms of the SSAP, while in terms of the SSEP, the CMD system has a gain of 5–6 magnitudes over the NRS system and 2–3 magnitudes over the DRS system. In respect of the SSNSAP, the CMD system has an infinite gain over the NRS system and a similar gain to the DRS system.

Upon a general attack, an NRS system is basically not resistant if the attack is constant; the DRS has a higher SSNSAP and a lower SSEP, but its SSAP is 0, that is, the DRS has an anti-attack capability that is relatively agile and accurate yet not persistent. A CMD system has an extremely high SSAP, an extremely low SSEP, and an extremely high SSNSAP, namely, an anti-attack capability agile, accurate, and persistent.

As shown in Table 9.19, for special attacks, an NRS is unable to resist halt attacks and standby cooperative attacks. A DRS can detect all the halt attacks but cannot escape, but its SSAP is 0, that is, the DRS system is unable to persistently resist halt attacks and standby cooperative attacks. A CMD system can detect all the halt attacks but cannot escape, but its SSAP is correlated with the attacker's capability

Table 9.18 Steady-state probabilities of the three systems' resistance to general attacks

Steady-state probabilities	NRS system	DRS system	CMD system			
			SSP	RRSP	FCSP	RRSP & FCSP
Weak attack scenarios						
Steady-state backdoor dormancy probability (SSBDP), $P(M_0)$	0	0	9.999999×10^{-1}	9.999999×10^{-1}	9.998992×10^{-1}	9.998998×10^{-1}
SSNSAP	0	9.999999×10^{-1}	8.361113×10^{-8}	8.337967×10^{-8}	1.007121×10^{-4}	1.001566×10^{-4}
SSAP	0	0	9.999999×10^{-1}	9.999999×10^{-1}	9.999993×10^{-1}	9.999998×10^{-1}
SSEP	5.00000×10^{-1}	2.999700×10^{-4}	5.574345×10^{-14}	5.559030×10^{-14}	1.338574×10^{-11}	1.334748×10^{-11}
Medium attack scenarios						
SSBDP, $P(M_0)$	0	0	9.999939×10^{-1}	9.999948×10^{-1}	9.998563×10^{-1}	9.998899×10^{-1}
SSNSAP	0	9.999999×10^{-1}	6.000062×10^{-6}	5.166812×10^{-6}	1.436437×10^{-4}	1.100432×10^{-4}
SSAP	0	0	9.999989×10^{-1}	9.999998×10^{-1}	9.999596×10^{-1}	9.999932×10^{-1}
SSEP	5.00000×10^{-1}	2.999700×10^{-4}	2.364045×10^{-10}	2.066016×10^{-10}	9.633839×10^{-10}	8.394510×10^{-10}

(continued)

Table 9.18 (continued)

Steady-state probabilities	NRS system	DRS system	CMD system			
			SSP	RRSP	FCSP	RRSP & FCSP
Strong attack scenario						
SSBDP, $P(M_0)$	0	0	9.961149×10^{-1}	9.990996×10^{-1}	9.961158×10^{-1}	9.991009×10^{-1}
SSNSAP	0	9.999999×10^{-1}	3.885044×10^{-3}	9.003283×10^{-4}	3.884172×10^{-3}	8.991428×10^{-4}
SSAP	0	0	9.964137×10^{-1}	9.993993×10^{-1}	9.964146×10^{-1}	9.994005×10^{-1}
SSEP	5.00000×10^{-1}	2.999700×10^{-4}	1.947108×10^{-6}	1.439273×10^{-6}	1.946386×10^{-7}	1.438503×10^{-7}

Table 9.19 Steady-state probabilities of the three systems' resistance to special attacks

Downtime attacks

Steady-state probabilities	NRS systems		DRS systems		CMD systems	
	Medium attacks	Strong attacks	Medium attacks	Strong attacks	Medium attacks	Strong attacks
SSBDP, $P(M_0)$	0	0	0	0	7.692307×10^{-1}	5.263158×10^{-2}
SSNSAP	0	0	1	1	1	1
SSAP	0	0	0	0	9.615383×10^{-1}	1.255061×10^{-1}
SSEP	1	1	0	0	0	0

Standby cooperative attacks

Steady-state probabilities	NRS		DRS		CMD	
	Medium attacks	Strong attacks	Medium attacks	Medium attacks	Strong attacks	Medium attacks
SSBDP, $P(M_0)$	0	0	0	0	9.999999×10^{-2}	1.848429×10^{-3}
SSNSAP	0	0	0	0	0	0
SSAP	0	0	0	0	2.285886×10^{-1}	4.619800×10^{-2}
SSEP	1	1	1	1	7.714114×10^{-1}	9.953802×10^{-1}

and the executor's recovery speed: The SSAP is relatively high when the executor's halt rate is outraced by its recovery speed. While the CMD system is unable to resist standby cooperative attacks theoretically, due to its uncertainty effect resulting from the PAS and MDR-NFM based on MR and controlled by external parameters, an attacker may not afford the cost required for a standby cooperative attack. Therefore, the mimic defense system has an unparalleled anti-cooperative-attack ability over the NDR/DRS system.

The adoption of the RRFCSP can significantly increase the SSAP and SSNSAP and reduce the SSEP of a CMD system, whose steady-state probabilities are sensitive to the parameters σ, μ, and λ. When formulating the mimic scheduling policy, you should try to choose non-cognate executors with larger dissimilarities in the implementation algorithm and construction and higher quality and fault recovery rates.

In view of the fact that CMD systems can effectively defend against various non-cooperative attacks on the target object's vulnerabilities and Trojans, etc., it can be predicted that the future attacks will be forced to turn their focus to the extremely difficult standby cooperative attacks or the costly social engineering approach aiming at the control processes of CMD systems.

All in all, in respect of anti-attack, the CMD system has excellent stability robustness and quality robustness: Compared to the NRS/DRS system, the CMD system has an infinite gain in terms of the SSAP, while in terms of the SSEP, the CMD system has a degradation of 5–6 magnitudes over the NRS system and a degradation of 2–3 magnitudes over the DRS system. In respect of the SSNSAP, the CMD system has an infinite gain over the NRS system and a similar gain to the DRS system. It is evident that the highly reliable, available, and credible CMD system has an exponential or super-nonlinear anti-attack gain.

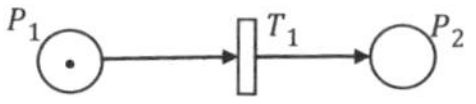

Fig. 9.21 GSPN model for the NRS system's reliability faults

$$\bigcirc\!\!{\scriptstyle 1} \xrightarrow{\ \lambda_1\ } \bigcirc\!\!{\scriptstyle 2}$$

Fig. 9.22 CTMC model for the NRS system's reliability faults

9.5.4   Reliability Analysis

1. Reliability analysis of the NRS system

   Based on the definition of GSPN'$_N$, the GSPN model for an NRS system (of a single-redundancy static structure) upon a reliability fault is shown in Fig. 9.21. There are two states: P_1 contains a token indicating that the system is in the fault dormancy state, with all the executors in normal operation without any reliability fault, while P_2 contains a token indicating that the system is in the degrading state. T_1 indicates that the system transits to the fault state from the fault dormancy state (Fig. 9.22).

   The CTMC model for the NRS's reliability faults isW shown in Table 9.27. Assume that the time of reliability faults occurring to the executors are subject to the exponential distribution, the steady states of the CTMC model are shown in Table 9.20.

   The meanings of the parameters of the CTMC model are shown in Table 9.21. The reachable state set is given in Table 9.22, together with the steady-state probability of each state identifier based on the parameters relevant to the transition process. Table 9.23 gives the degradation probability of the NRS system.

   As for the NRS system, the SSAP = $P(M_0) = 0$, the reliability $R(t) = e^{-\lambda_1(t)}$, the MTTF=λ_1^{-1}, namely, 1×10^3 hours.

   The NRS system's SSAP is 0, MTTF is 1×10^3 hours, and system DP is 1×10^{-3}. The reliability is the same for an NRS system and an executor.

2. Reliability analysis of the DRS system

   Based on the definition of GSPN'$_D$, the GSPN model [23, 24] for a DRS (of a tri-redundancy static HRS) upon a reliability fault is shown in Fig. 9.23. There are seven states and nine transitions: P_1, P_2, and P_3 contain a token each to indicate that the executor is in the faulty dormancy state; P_4, P_5, and P_6 contain a token each to, respectively, indicate a random reliability fault occurring to the executor; and P_7 contains a token to indicate a random reliability fault simultaneously occurring to two or more executors. T_1, T_2, and T_3 indicate a random reliability fault occurring to the executor; T_7, T_8, and T_9 indicate the executor recovered to the fault dormancy state, while T_4, T_5, and T_6 indicate two or more executors simultaneously entering the random reliability fault state.

Table 9.20 Steady states of the CTMC model for the NRS system's reliability faults

State serial No.	Meaning
1	One executor is in normal operation, and the system is in a fault dormancy state
2	One executor encounters a reliability fault, and the system is in a fault state

Table 9.21 Parameters of the CTMC model for the NRS system's reliability faults

Parameters	Value	Meaning
$\lambda_1(h)$	1×10^{-3}	Assume that the average time for a random fault generated from an executor is 1000 hours

Table 9.22 Reachable state set of the CTMC model for the NRS system's reliability faults

Mark	Steady-state probability P_f	P_1	P_2	State
M_0	0.000000	1	0	Tangible
M_1	1.000000	0	1	Tangible

Table 9.23 DP of the NRS system

Name of the system	Executor	NRS
Degradation probability (DP)	1×10^{-3}	1×10^{-3}

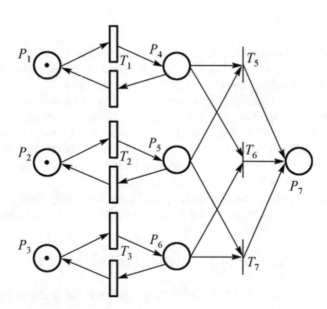

Fig. 9.23 GSPN model for the DRS system's reliability faults

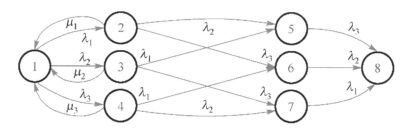

Fig. 9.24 CTMC model for the DRS system's reliability faults

Table 9.24 Steady states of the CTMC model for the DRS system's reliability faults

State serial No.	Meaning
1	Each executor is in normal operation and in a fault dormancy state
2	Executor 1 is faulty
3	Executor 2 is faulty
4	Executor 3 is faulty
5	Executors 1 & 2 are faulty
6	Executors 1 & 3 are faulty
7	Executors 2 & 3 are faulty
8	Three executors are simultaneously faulty

The DRS system analyzed is a tri-redundancy static HRS system [23–25], and it is assumed that no fault monitoring means is added to the executors, while each single executor is endowed with the fault recovery capability.

The CTMC model for the DRS system's reliability faults is shown in Fig. 9.24. In the event of a reliability fault, the system and the executors may be degraded. Assume that the time of reliability faults occurring to the executors is subject to the exponential distribution; the steady states and parameters of the CTMC model are shown in Tables 9.24 and 9.25, respectively, with $\lambda_1 \sim \lambda_3$ representing the average time of a reliability fault occurring to an executor.

Table 9.26 gives the reachable state set and gives the steady-state probability of each state identifier according to the parameters relevant to the transition process. Table 9.27 gives the DP of the NRS system.

$$\text{SSAP} = P(M_0) + P(M_1) + P(M_2) + P(M_3) = 0.$$

Based on the CTMC model for the DRS system's reliability faults, its reliability can be calculated as follows:

Table 9.25 Parameters of the CTMC model for the DRS system's reliability faults

Parameters	Value	Meaning
$\lambda_1(h)$	1×10^{-3}	Assume that the average time for a random fault generated from Executor 1 is 1000 hours
$\lambda_2(h)$	1×10^{-3}	Assume that the average time for a random fault generated from Executor 2 is 1000 hours
$\lambda_3(h)$	1×10^{-3}	Assume that the average time for a random fault generated from Executor 3 is 1000 hours
$\mu_1(h)$	60	Assume that the average time for Executor 1 to be self-recovered or restarted for recovery is 1 minute
$\mu_2(h)$	60	Assume that the average time for Executor 2 to be self-recovered or restarted for recovery is 1 minute
$\mu_3(h)$	60	Assume that the average time for Executor 3 to be self-recovered or restarted for recovery is 1 minute

Table 9.26 Reachable state set of the CTMC model for the DRS system's reliability faults

Mark	Steady-state probability P_f	P_1	P_2	P_3	P_4	P_5	P_6	P_7	State
M_0	0.000000	1	1	1	0	0	0	0	Tangible
M_1	0.000000	0	0	0	1	0	0	0	Tangible
M_2	0.000000	0	0	0	0	1	0	0	Tangible
M_3	0.000000	0	0	0	0	0	1	0	Tangible
M_4	0.000000	0	0	0	1	1	0	0	Tangible
M_5	0.000000	0	0	0	1	0	1	0	Tangible
M_6	0.000000	0	0	0	0	1	1	0	Tangible
M_7	1.000000	0	0	0	1	1	1	1	Tangible

Table 9.27 DP of the DRS system

Name of system	Executor	DRS system (requiring two executors to be in normal operation
DP	1.0×10^{-3}	3.0×10^{-6}

$$R(t) = 1 - P_{M_4}(t) - P_{M_5}(t) - P_{M_6}(t) - P_{M_7}(t)$$

$$= \frac{\left(\mu_1 + 5\lambda_1 + \left(\mu_1^2 + 10\mu_1\lambda_1 + \lambda_1^2\right)^{\frac{1}{2}}\right)}{2\left(\mu_1^2 + 10\mu_1\lambda_1 + \lambda_1^2\right)^{\frac{1}{2}}} e^{-\left(\frac{\mu_1 + 5\lambda_1 - \left(\mu_1^2 + 10\mu_1\lambda_1 + \lambda_1^2\right)^{\frac{1}{2}}}{2}\right)t}$$

$$- \frac{\left(\mu_1 + 5\lambda_1 - \left(\mu_1^2 + 10\mu_1\lambda_1 + \lambda_1^2\right)^{\frac{1}{2}}\right)}{2\left(\mu_1^2 + 10\mu_1\lambda_1 + \lambda_1^2\right)^{\frac{1}{2}}} e^{-\left(\frac{\mu_1 + 5\lambda_1 + \left(\mu_1^2 + 10\mu_1\lambda_1 + \lambda_1^2\right)^{\frac{1}{2}}}{2}\right)t} \quad (9.5)$$

While its MTTF can be calculated as follows:

$$\text{MTTF} = \int_0^\infty R(t)\,dt$$

$$= \frac{1}{2\left(\mu_1^2 + 10\mu_1\lambda_1 + \lambda_1^2\right)^{\frac{1}{2}}}\left(\frac{\mu_1 + 5\lambda_1 + \left(\mu_1^2 + 10\mu_1\lambda_1 + \lambda_1^2\right)^{\frac{1}{2}}}{\mu_1 + 5\lambda_1 - \left(\mu_1^2 + 10\mu_1\lambda_1 + \lambda_1^2\right)^{\frac{1}{2}}}\right.$$

$$\left. - \frac{\mu_1 + 5\lambda_1 - \left(\mu_1^2 + 10\mu_1\lambda_1 + \lambda_1^2\right)^{\frac{1}{2}}}{\mu_1 + 5\lambda_1 + \left(\mu_1^2 + 10\mu_1\lambda_1 + \lambda_1^2\right)^{\frac{1}{2}}}\right) \tag{9.6}$$

$$= 1.000083 \times 10^7\,\text{h}$$

As for a tri-redundancy DRS system, its SSAP is 0, MTTF is 1.000083×10^7 hours, and system DP is 3.0×10^{-6}. A DRS system outperforms an NRS system by a three-magnitude lower DP and a four-magnitude higher MTTF, indicating that the DRS system is more reliable.

3. Reliability analysis of the CMD system

Based on the definition of GSPN'$_C$, the GSPN model for a CMD system (of a tri-redundancy DHR system) upon a reliability fault is shown in Fig. 9.25. There are 10 states and 16 transitions: P_1, P_2, and P_3 contain a token each to indicate that the

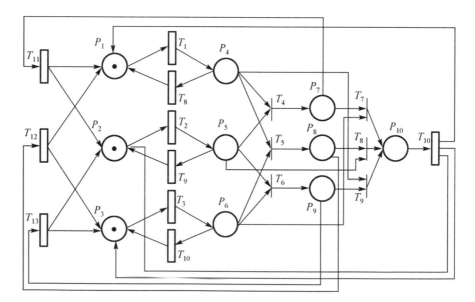

Fig. 9.25 GSPN model for the CMD's reliability faults

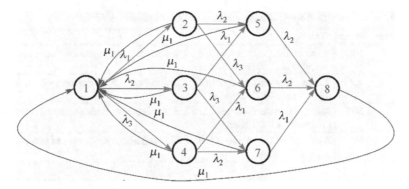

Fig. 9.26 CTMC model for the CMD system's reliability faults

Table 9.28 Steady states of the CTMC model for the CMD's reliability faults

State serial No.	Meaning
1	Each executor is in normal operation and in a fault dormancy state
2	Executor 1 is faulty
3	Executor 2 is faulty
4	Executor 3 is faulty

executor is in the faulty dormancy state; P_4, P_5, and P_6 contain a token each to, respectively, indicate a random reliability fault occurring to the executor; P_7, P_8, and P_9 contain a token to indicate a random reliability fault simultaneously occurring to two executors; and P_{10} contains a token to indicate a random reliability fault simultaneously occurring to three executors. T_1, T_2, and T_3 indicate a random reliability fault occurring to the executor; T_4, T_5, and T_6 indicate two executors simultaneously entering the random reliability fault state; T_7, T_8, and T_9 indicate three executors simultaneously entering the random reliability fault state; T_{10} indicates three executors recovered to a fault dormancy state. T_{11}, T_{12}, and T_{13} indicate three executors recovered to the fault dormancy state, while T_{14}, T_{15}, and T_{16} indicate one executor recovered to the fault dormancy state.

The CMD system analyzed is a tri-redundancy static HRS system, and it is assumed that no fault monitoring means is added to the executors, while each executor is endowed with the fault recovery capability.

The CTMC model for the CMD system's reliability faults is shown in Fig. 9.26. Assume that the success and recovery time of an executor-oriented attack are subject to the exponential distribution; the steady states and parameters of the CTMC model are shown in Tables 9.28 and 9.29, respectively.

Table 9.30 gives a reachable state set of the CTMC model for the CMD's reliability faults, and Table 9.31 gives the degrading probability of the CMD system.

Table 9.29 Parameters of the CTMC model for the CMD's reliability faults

Parameters	Value	Meaning
$\lambda_1(h)$	10^{-3}	Assume that the average time for a random fault generated from one executor is 1000 hours
$\lambda_2(h)$	10^{-3}	Assume that the average time for a simultaneous fault generated from two executors is 1000 hours
$\lambda_3(h)$	10^{-3}	Assume that the average time for a simultaneous fault generated from three executors is 1000 hours
$\mu_1(h)$	60	Assume that the average time for the executor to be recovered to a dormancy state is 1 minute

Table 9.30 Reachable state set of the CTMC model for the CMD's reliability faults

Mark	Steady-state probability P_f	P_1	P_2	P_3	P_4	P_5	P_6	P_7	P_8	State
M_0	9.999500×10^{-1}	1	0	0	0	0	0	0	0	Tangible
M_1	1.666528×10^{-5}	0	1	0	0	0	0	0	0	Tangible
M_2	1.666528×10^{-5}	0	0	1	0	0	0	0	0	Tangible
M_3	1.666528×10^{-5}	0	0	0	1	0	0	0	0	Tangible
M_4	5.555000×10^{-10}	0	0	0	0	1	0	0	0	Tangible
M_5	5.555000×10^{-10}	0	0	0	0	0	1	0	0	Tangible
M_6	5.555000×10^{-10}	0	0	0	0	0	0	1	0	Tangible
M_7	2.777363×10^{-14}	0	0	0	0	0	0	0	1	Tangible

Table 9.31 Degrading probability of the CMD system

Name of system	Executor	CMD system (requiring two executors to be in normal operation)
DP	1.0×10^{-3}	3.0×10^{-6}

$$\text{SSAP} = P(M_0) + P(M_1) + P(M_2) + P(M_3) = 9.999999 \times 10^{-1}.$$

Based on the CTMC model for the CMD's reliability faults, its reliability can be calculated as follows:

$$
\begin{aligned}
R(t) &= 1 - P_{M_4}(t) - P_{M_5}(t) - P_{M_6}(t) - P_{M_7}(t) \\
&= \frac{\left(6\lambda_1 e^{-t(\mu_1 + 2\lambda_1)}\right)}{\mu_1 + 2\lambda_1} - \frac{6\lambda_1^2}{(\mu_1 + 2\lambda_1)(\mu_1 + 3\lambda_1)} - \frac{\left(6\lambda_1 e^{-t(\mu_1 + 3\lambda_1)}\right)}{\mu_1 + 3\lambda_1} + 1
\end{aligned}
\tag{9.7}
$$

While its MTTF can be calculated as follows:

$$\text{MTTF} = \int_0^{\infty} R(t)\,dt \to \infty \tag{9.8}$$

Table 9.32 Reliability of the NRS/DRS/CMD systems

Probabilities	NRS system	DRS system	CMD system
SSAP	0	0	9.999999×10^{-1}
MTTF	1.0×10^3	1.000083×10^7	∞
FP	1.0×10^{-3}	3.0×10^{-6}	3.0×10^{-6}

As for a tri-redundancy CMD system, its SSAP is 9.999999×10^{-1}, MTTF tends to infinity, and system DP is 3.0×10^{-6}. For a tri-redundancy CMD system faced with random reliability faults, its SSAP approximates 1, and its MTTF tends to infinity, far outperforming the NRS/DRS systems, while its DP is similar to that of a DRS and outperforms an NRS by a three-magnitude lower level. The above indicators show that the reliability ceiling of a CMD system is constrained only by the maximum lifespan of the hardware, namely, the system has an extremely high reliability.

4. Summary of the reliability analysis

As shown in Table 9.32, a tri-redundancy CMD system is outstanding in reliability, as its SSAP approximates 1, while that of the NRD/DRS is 0; its MTTF far outperforms that of the other two systems, while its DP is similar to that of a DRS system and outperforms an NRS system by a three-magnitude lower level. It is evident that the highly reliable, available, and credible CMD system has an exponential or super-nonlinear reliability gain.

9.5.5   Conclusion

In this section, the GSPN is used as a basis for the anti-attack models, as well as for the reliability models, of three typical information systems: NRS, DRS, and CMD. The isomorphism between the GSPN model's reachable set and the CTMC is then employed for quantitative analysis and simulation of the steady-state probabilities of the systems. As for the evaluation indicators, the three parameters, SSAP, the SSEP, and the SSNSAP, are proposed to comprehensively characterize the anti-aggression of the target architecture, while the reliability of the target architecture is characterized by the SSAP, the reliability R(t), the MTTF, and the DP. According to the simulation and analysis, the CMD system has excellent stability robustness and quality robustness. The models and methods presented in this section can be used to analyze the robustness, reliability, availability, and anti-attack of the architecture of an information system. Relevant analysis conclusions can help guide the designing of a robustness information system highly reliable, available, and credible.

In addition, in the anti-attack and reliability analysis, to refine the analysis for a better guidance on the engineering practice, the tools such as the colored GSPN and the DSPN can be used to refine the attack type/effect analysis. Besides, anti-attack analysis is performed on the integrated system that incorporates specific threat detection and defense methods, such as intrusion detection and firewalls. Finally, the interactions between the hardware and software are taken into account, and the parameters such as the rates of fault detection, false alarm, and missed alarm are introduced, contributing to the comprehensive analysis of resilience and reliability.

9.6 Differences Between CMD and HIT (Heterogeneous Intrusion Tolerance)

From the foregoing chapters, we know that the fault-tolerant DRS lacks stability robustness against the GUD, but compared to a classic intrusion-tolerant structure that combines intrusion detection IDS and HR, it has considerable advantages regarding its intrusion tolerance performance. The CMDA growing out of the DRS is far superior to the classic intrusion tolerance structure in respect of anti-attack and reliability. Although they all have basic defensive elements such as heterogeneity, diversity, and redundancy, as well as a multimodal consensus voting mechanism, the "structure-determined nature" makes the graphite incomparable to diamond in terms of hardness. Even so, it is still necessary to distinguish between HR-based classic intrusion tolerance and CMD from multiple perspectives, so that the readers can fully understand the revolutionary significance of CMD in comparison to conventional intrusion tolerance. We use the typical intrusion tolerance systems listed in Sect. 3.4.2 of this book as a reference to showcase the major differences or disparities between the two as objectively as possible.

9.6.1 Major Differences

① *Different fault models:* unlike the security target of heterogeneous intrusion tolerance (HIT) and the system fault model, the CMD emphasizes the ability to manage and control or suppress the GUD occurring inside the mimic structure for whatever reason. That is, it should be able to defend against external attacks against internal vulnerabilities of the system and DF-based internal-external cooperative attacks, as well as to tolerate various known/unknown DMFs or non-cooperative attacks of an unlimited number and to relieve or interfere with unknown CMFs or cooperative attacks. Moreover, quantitative analysis of the two regarding their anti-attack and reliability based on the GSPN model leads to conclusions differing at the.

② *Different tolerance mechanisms and policies:* For CMD, intrusion tolerance or fault tolerance is a function stemming from the endogenous uncertainty effect of the system structure, while the MR-based PAS and MDR-NFM is the root cause of this effect. Unlike HIT that applies a monotonous consensus judgment mechanism, mimic ruling is an iterative process adopting plural voting algorithms. The non-consistent voting state generally does not affect the output response function but will stimulate the BVM of the current structure scenario changes and make the comparison difference between the MOVs gradually converge within the desired threshold range.

③ *Different system architectures:* HIT uses a combined structure of intrusion detection and heterogeneous redundancy, i.e., IDS + HRS; CMD adopts the mimic disguise strategy-based DHR architecture, where any additional facilities of detection, monitoring, compliance checks, or protective measures are nonessential configurations.

④ *Different Hypotheses:* HIT assumes that the "bad guys" in the HR components as the "last line of defense" belong to the minority, while CMD assumes that all the hardware/software resources in the MB can be in a "virus-bearing" state. The effectiveness of the HIT-IDS relies heavily on the prior knowledge and behavioral traits of the attacker, but such information is not an indispensable prerequisite for CMD.

⑤ *Different voting significance:* The mimic ruling result is a relative concept with a superposition state attribute. As long as the MOV comparison suggests any non-compliance, it indicates a possible attack and escape (although a small probability event). So, there should be not only the iterative ruling supported by multiple algorithms but also the FCL-based iterative BVM. The HIT voting state is considered correct and does not require any verification operations as long as the majority identical conditions are met; HIT relies on the majority-rule voting algorithm (including the Byzantine failures), while mimic ruling requires a set of voting algorithms that can be dynamically iterated. So, in addition to the majority-rule voting algorithm, it can meanwhile include a variety of algorithms such as maximum likelihood voting (MLV), weighted voting, mask voting, large number voting based on historical information, and Byzantine failures.

⑥ *Different control method:* CMD adopts the ruling state-driven GRC mechanism, which has the property (or the "increase-of-entropy" property) of progressively selecting appropriate defense scenarios in the diversified scenario set, "making it difficult to maintain the attack and escape steadily even if the attack succeeds," while HIT does not have the corresponding functionality due to the "entropy-decreasing process."

⑦ *Different abnormality handling:* When an executor's output fault is spotted, HIT will immediately suspend the executor, that is, if the condition that $N \geq 2f + 1$ is satisfied, the faulty executor removal or suspension strategy is adopted. Since CMD adopts the relative ruling method, an OV inconsistency is not a proof that the relevant executor is the faulty one. Instead, the problem evasion strategy is adopted to put the suspected problematic executor or any current executor offline for cleaning or reconstruction to get it into a standby state, unless the executor has an unrecoverable problem or failure. Therefore, in the case of a given redundancy N, a state in which every two MOVs are inconsistent is not the worst situation. Conversely, a completely consistent ruling result may contain a small probability event of a potential attack and escape. So, it is necessary to mandatorily alter the mechanism for the current operating environment by adding external control commands.

⑧ *Different verification methods:* CMD's ability to resist known or unknown DM/CM attacks is subject to quantifiable design and can be confirmed generally by the "white box" benchmark function testing. HIT is unable to resist unknown cooperative attacks, and even if it manages to cope with a few unknown DM attacks, its ability is limited by the precondition of $N \geq 2f + 1$.

⑨ *Different effectiveness guarantee method:* The effectiveness of HIT depends on the abundance of knowledge possessed by IDS and is strongly correlated with the real-time updates of the information libraries concerned. The effectiveness of CMD is mechanically irrelevant to whether it has prior knowledge and behavioral traits of the attacker. Therefore, the requirements for the two in terms of anti-aggression performance maintenance and guarantee are completely different.

9.6.2 Prerequisites and Functional Differences

① HIT assumes that there are only a few exploitable vulnerabilities in the HREs; CMD assumes that all executors in the MB may have heterogeneous vulnerabilities or other dark features, regardless of the number.

② HIT assumes that f, the number of abnormalities occurring to N HREs, cannot exceed $N \geq 2f + 1$; otherwise the service response will be stopped; CMD agrees that even if the MOVs in the current service set are completely inconsistent, as long as the iterative ruling standards in the ruling algorithm set is satisfied, an output with reduced trustworthiness can still be produced.

③ HIT assumes that an output is normal as long as it satisfies the majority voting condition; CMD agrees that unless the completely consistent voting condition is met, even if an output vector is generated, the arbiter will trigger the FCL-BVM, change the operating environment or defense scenarios in the MB in a dynamic and convergent manner, and operate iteratively until the arbiter's non-compliance state disappears. In particular, even if there is a "standby cooperative escape" completely non-perceived by mimic ruling, the loop mandate disturbance mechanism based on external policies will make it difficult to be exploited steadily.

④ HIT assumes that IDS's own vulnerabilities may be exploited by attackers, so dynamic reconfiguration is a necessary function; CMD allows vulnerable components of the MB, such as the IA/OA, the arbiter, and the feedback controller, but they are not exploitable or attackable for external attackers, because the MB itself can obtain a systemic defense gain derived from the DHR structural effect.

⑤ HIT stipulates that IDS needs to have the relevant configuration information including compliance check parameters, and the completeness of configuration parameters will directly affect the intrusion tolerance performance. In theory, the CMD effect does not depend on the IDS process and does not need the support of relevant configuration parameters. Its anti-attack performance is only strongly correlated with the heterogeneity, redundancy, OV abundance, and ruling algorithm diversity between the executors.

⑥ HIT agrees that IDS does not perform any intervention on the COTS-level HREs of the back-end, while CMD requires the executors themselves to be endowed with reconfigurable functions (i.e., diversified scenario requirements), at least for the COTS-level executors' internal resources, such as configurations where alternatives can be made in an execution environment running multiple operating systems, databases, and virtualization software or application software variants.

⑦ HIT assumes that the system is in a normal state if not any abnormality is detected, and no additional operations are required in this case; CMD believes that even if no abnormal state is spotted, there is a small probability of attack and escape, which requires random or periodic activation of the FCL to change the current operating environment or defense scenarios so as to prevent against or disintegrate standby escape events.

⑧ HIT needs IDS functions, including but not limited to compliance checks and Byzantine failures, while CMD needs to perceive uncertainty perturbations through relative ruling functions. Therefore, there may be "blind spots" in both HIT and CMD. However, CMD has a BVM that disintegrates the stable escape state, while the classical HIT does not.

⑨ In addition to FEHR executors, CMD can achieve the integrated goal of "high reliability, high credibility, and high availability" of the object through one architecture alone, namely, the DHR architecture. In contrast, a classic HIT system is much more cumbersome and redundant and far less in functionality and cost effectiveness.

9.6.3   Summary

In principle, based on heterogeneous fault-tolerance technology, HIT still falls in the range of passive defense, because it needs an intrusion detection process to obtain the prior knowledge necessary for fault-tolerant functionality. Of course, if IDS + DRS is used as a combined intrusion tolerance structure, even if there is an intrusion attack that the IDS fails to detect, the DRS still has the ability to resist unknown attacks. This conclusion can be evidenced by quantitative analysis in Sect. 6.7. However, even such a combined structure fails to deal with more than DM attacks or any CM attack, and it is impossible to avoid TT or attack and escape situations that may result from trial and error attacks. In other words, as long as the staticity and certainty of the HRE and the current operating environment remain unchanged, the stability robustness of the target object cannot be established. Once an attack and escape is achieved, the state can be maintained or reappeared for a long time. The quantitative analysis conclusion in Sect. 9.5 of this chapter also suggests that the DHR-based CMD still has excellent stability robustness and quality robustness to GUD without relying on prior knowledge. It should be pointed out that as far as the full life-cycle comprehensive cost performance is concerned, the CMDA also has the functional and performance advantages that the classical HIT structure does not have.

References

1. Stallings, W.: Network and internetwork security: principles and practice. Prentice Hall, Englewood Cliffs (1995)
2. Malik, J., Arbeláez, P., Carreira, J., et al.: The three R's of computer vision: recognition, reconstruction and reorganization. Pattern Recogn. Lett. **72**, 4–14 (2016)
3. Zhou, K.M., Doyle, J.C.: Essentials of robust control. Pearson, New York (1998)
4. Ceccato, M., Nguyen, C.D., Appelt. D., et al.: SOFIA: an automated security oracle for black-box testing of SQL-injection vulnerabilities. In: Proceedings of the 31st IEEE/ACM International Conference on Automated Software Engineering, 2016, pp.167–177
5. Pras, A.: Network management architectures. University of Twente, Enschede (1995)
6. Littlepage, G.E., Schmidt, G.W., Whisler, E.W., et al.: An input-process-output analysis of influence and performance in problem-solving groups. J. Pers. Soc. Psychol. **69**(5), 877 (1995)
7. Newman, R.C.: Cybercrime, identity theft, and fraud: practicing safe internet-network security threats and vulnerabilities. In: Proceedings of the 3rd Annual Conference on Information Security Curriculum Development, 2006, pp. 68–78
8. Zhang, S., Wang, Y., He, Q., et al.: Backup-resource based fault recovery approach in SDN data plane. In: Network Operations and Management Symposium (APNOMS), 2016, pp.1–6.
9. Chang, E.S., Ho, C.B.: Organizational factors to the effectiveness of implementing information security management. Ind. Manag Data Syst. **106**(3), 345–361 (2006)
10. Chung, J., Owen, H., Clark, R.: SDX architectures: a qualitative analysisIEEE Southeast Conference, 2016, pp. 1–8
11. Ventre, P.L., Jakovljevic, B., Schmitz, D., et al.: GEANT SDX-SDN based open eXchange point. In: NetSoft Conference and Workshops (NetSoft), 2016, pp.345–346
12. Michalski, R.S., Carbonell, J.G., Mitchell, T.M.: Machine learning: an artificial intelligence approach. Springer, Berlin (2013)
13. Pipyros, K., Mitrou, L., Gritzalis, D., et al.: A cyber attack evaluation methodology. In: Proceedings of the 13th European Conference on Cyber Warfare and Security, 2014, pp. 264–270
14. Shi, Z., Zhao, G., Liu, J.: The effect evaluation of the network attack based on the fuzzy comprehensive evaluation method.In: The 3rd International Conference on Systems and Informatics (ICSAI), 2016, pp.367–372
15. Tuvell, G., Jiang, C., Bhardwaj, S.: Off-line mms malware scanning system and method: U.S.Patent Application 12/029,451. [2008-2-11]
16. Orebaugh, A.D., Ramirez, G.: Ethereal packet sniffing. Syngress Publishing, Rockland (2004)
17. Alfieri, R.A., Hasslen, R.J.: Watchdog monitoring for unit status reporting: U.S.Patent 7,496,788. [2009-2 24]
18. Barylski, M., Krawczyk, H.: Multidimensional approach to quality analysis of IPSec and HTTPS applicationsThe Third IEEE International Conference on Secure Software Integration and Reliability Improvement, 2009, pp.425–430
19. Jiangxing, W.: Meaning and vision of mimic computing and mimic security defense [J]. Telecommun. Sci. **30**(7), 1–7 (2014)
20. Shannon, C.E., Weaver, W.: The mathematical theory of communication. University of Illinois Press, Urbana (1949)
21. Chuang, L., Yuanzhuo, W., Yang, Y., et al.: Research on network trustworthiness analysis methods based on stochastic petri net [J]. Acta. Electron Sin. **34**(2), 322–332 (2006)

22. Garcia, M., Bessani, A., Gashi, I., et al.: Analysis of OS diversity for intrusion tolerance[J]. Softw. Pract. Exp. 1–36 (2012)
23. Yeh, Y.C.: Triple-triple redundant 777 primary flight computer[C]. IEEE Aerosp. Appl. Conf. Proc. **1**, 293–307 (1996)
24. Xudong, Q., Zongji, C.: Reliability analysis of the dissimilar redundancy flight control computers based on petri net [J]. Control Decis. **20**(10), 1173–1176 (2005)
25. Lin, C., Marinescu, D.C.: Stochastic high-level petri nets and applications [J]. IEEE Trans. Comput. **37**(7), 815–825 (1988)

Chapter 10
Engineering and Implementation of Mimic Defense

In Chap. 9, we described the basic principles of mimic defense. In this chapter, we will focus on the basic conditions and constraints for mimic defense engineering and implementation, the key implementation mechanisms, major issues, and possible approaches and solutions. We will also briefly explore how to test and evaluate the mimic defense.

10.1 Basic Conditions and Constraints

10.1.1 Basic Conditions

The application of the CMD architecture requires the following basic conditions.

(1) Inelastic demand. The target has a rigid demand to suppress or control broad uncertain interferences, like security threats based on vulnerabilities and backdoors, and lays emphasis on cost-performance value during the entire life cycle of the system, including the cost for security maintenance and management.
(2) Application environments meet the requirements of I[P]O mimic interface model.
(3) Presence of a standardized or normalizable functional interface. A test sequence could be sent through the input channel through such an interface, and a response sequence could be obtained through the output channel of the interface. In accordance with the given interface function or performance and relevant test criteria and procedures, you can recognize the conformance of the functions or features of the heterogeneous executors (HEs) or operational scenarios or judge the consistency and other abnormal processing functions. Such kind of interfaces shall meet the basic setting requirements of the mimic interface (MI).

© Springer Nature Switzerland AG 2020
J. Wu, *Cyberspace Mimic Defense*, Wireless Networks,
https://doi.org/10.1007/978-3-030-29844-9_10

(4) Mimic interface can be closed off. Aside from the prescribed input/output interface, there are no other explicit interactive channels between the heterogeneous executors or operational scenarios in the mimic brackets (MB) and the outside world.

(5) Meeting the basic requirements of heterogeneity. Unlike the classic DRS, DHR does not require its software/hardware components to be strictly heterogeneous, but it does require physical or logical heterogeneity (or multiple functionally equivalent scenarios to be used) for the components of sensitive nodes, key redundancy paths, or core redundancy module, including heterogeneity of the vulnerabilities and backdoors on these components (at least it is difficult to form identical attack chains or to exploit them).

(6) Tolerating insertion delay. A certain level of difference in performance is allowed among diverse executors or operational scenarios. The related input/output scheduling function and the mimic ruling policy would inevitably lead to insertion delays. The value of the delay depends on many factors such as whether it uses input/output agent, whether the output results can be corrected, the location of mimic interface, abundance of output vectors that need connected analysis, as well as affordability of the algorithms, policies, and systems being used.

(7) Acceptable related conditions. The application system is not too sensitive to the cost of securing the environment, energy supply and cost of initial investment.

(8) There are necessary software/hardware resources and industrial and market conditions to support diverse application needs.

10.1.2   Constraints

(1) Security of interface functions. When applied to fields that have standard requirements or agreements for protocols, rules, or interfaces, the mimic defense relies heavily on the security of the standard or rule designs. More generally, if any normalizable logical or physical interface has security flaws or hidden malicious functions (not including vulnerabilities and backdoors during the realization process of a protocol or rule) in the function design, then the effect of mimic defense cannot be guaranteed.

(2) Relative independence. Usually, there are no unified or cooperative relationship between the functionally equivalent heterogeneous redundancy executors or operational scenarios apart from the service functions and performance requirements, as well as relevant standards, protocols, procedures, and interfaces. In other words, mimic defense has independence screening on the project execution process and strict "dis-cooperation" requirements (in order to eliminate the possibility of colluded cheating in social engineering [1]).

(3) Genetic screening. If COTS-level products are used as heterogeneous elements, we suggest that the "software gene genealogy" analytics be used to help study and determine the "extent of homology" and "hereditary relationship," so as to identify and locate the same or most similar source code segments, evaluate

their potential security risks, and develop prevention measures (such as increasing the difficulty of building the vulnerability-based attack chain).

(4) Modification of identical genes. If independently designed heterogeneous elements share the same sub-modules or identical "genes" or identical software/ hardware source codes (for instance, from the same open-source or shared community), they, in principle, need to be diversified (e.g., by using N-version or diverse compiler, binary translator, or other tools) or to be recompiled using a specialized mimic compiler. In addition, the application environment, allocation mechanism, and incentives of the homologous sub-modules in heterogeneous elements must be differentiated in order to lower the probability of mimic escape.

(5) Priorities. Besides providing the highly reliable, available, and credible robust services, DHR shall be primarily applied to the fields insensitive to cost and environment factors because it needs the configuration of heterogeneous redundant resources and guarantee conditions.

10.2 Main Realization Mechanisms

The guidelines on achieving mimic defense are as follows: *set* the functions of MI (or network units or platforms or software/hardware components or modules) on the DHR architecture. Introduce the mimicry strategy to diversify, randomize, and dynamize the mapping relationship between the running environment and the functions of the MI, puzzling attackers as to identifying the vulnerabilities, backdoors, and defense scenarios in the MB. Try to block or disguise MI abnormalities – regardless of the causes – to lower the detectability and predictability of defense behaviors. Destroy or undermine the certainty and stability of attack chains based on unknown vulnerabilities or backdoors to significantly increase the difficulty of coordinated attacks on dynamic and diverse targets under the condition of noncooperation. Contain and control security threats such as vulnerabilities and backdoors in the MB while at the same time strengthening or ensuring the robust stability and robust performance of target system functions.

The technical roadmap of mimic defense is as follows: use the DHR architecture with endogenous security as the basis for achieving intensive functions. Take the uncertainty in the mapping relationship between the apparent structure and the functions as the goal. Create a fog by using the mimic disguise mechanism to conceal or block the general uncertain disturbance inside the structure. Adopt a one-way or unidirectional connection mechanism of the sensitive pathway to distribute control functions, simplify control relations, and minimize control information. Take the fragmented, distributed, and dynamic transmission and storage of important information as the method. Adopt the mimic ruling-based PAS and the negative feedback control loop (NFCL) of multi-dimensional dynamic reconfiguration (MDR) as the core. Regard the minimization of the probability of escape of "coordinated attacks on dynamic and diverse targets under the non-cooperative conditions" as the purpose.

10.2.1   Structural Effect and Functional Convergence Mechanism

As previously discussed, the target object based on DHR architecture has an integrated attribute of high reliability in function provision and high credibility in service guarantee. This is because (1) its basic structure derives from the DRS architecture of the high reliability technology system, which has a solid foundation in both theory and engineering practice, and (2) the quantifiable endogenous security effect is achieved through MR-based PAS and MDR-negative feedback control mechanism (NFCM), which gives the elements in the executor set the convergent dynamic and random deployment policy, and through the diversified restructuring or reconfiguration policy of the elements. In addition, the I[P]O model also makes it possible that the failure rate of attacks would grow exponentially based on the frequencies of input/output through the MI. During the attack, if any maneuver that goes through the mimic brackets is detected through mimic ruling or influenced by the "irregular disturbance" in the feedback control phase, then the entire attack could be jeopardized, unless the attackers have the amazing capability of coordinated escape. Otherwise the more dependent an attack is on complicated coordination, the more difficult it is to accomplish. In the same vein, any security-bolstering measures (such as traditional security defense measures) in the heterogeneous executors or operational scenarios in the mimic brackets would augment dissimilarity or diversity due to the DHR architecture's inherent nonlinear effect. Therefore, not only is mimic defense inherently convergent with traditional security technologies, but it can also exponentially improve the effect of defense.

It should be emphasized that, although we expect that the heterogeneous executors – regardless of their nature or granularity – can be reconstructed and reorganized, or software-defined, it is not mandatory that the executors must have the DHR structure. In general, as long as the structure meets the functional or performance requirements of the MI, it does not matter whether the heterogeneous executors are white-box- or black-box-structured. That said, if the DHR structure can be adopted in the design of heterogeneous executors, then the target can achieve mimic defense with iterative effects. If the layered mimicry structure can be applied to complex systems such as networks or platforms, then we can expect to see some "super nonlinear" defense gains.

10.2.2   One-Way or Unidirectional Connection Mechanism

In order to improve the security of mimic bracket components (including the functions such as input agent, output ruling, output agent, PAS, and multi-dimensional dynamic reconfiguration control), it is necessary to eliminate or limit the use of two-way interactive mechanism between mimic brackets and executors or operational scenarios. Instead, the assembly line mode should be used in key control areas to divide the control functions into segments, creating an "absolutely credible" one-way or one-dimensional connection mechanism that does not rely on control segments, so as to control the potential influence of unknown threats and contain the possible area of dissemination. Another common function of control function segmentation and unidirectional control chain mechanism is to impede the reachability of the attack channel, making it difficult to build or maintain the necessary information transmission or Trojan uploading in coordinated/cooperative attacks. Eventually, even if there are vulnerabilities in the relevant components in the mimic brackets, it would be difficult for the attackers to take advantage of them. Many requirements are necessary to ensure the security of mimic brackets. For instance, to ensure the security of the mimic brackets, the isolation or interception is necessary, which requires that the mimic ruling and negative feedback control be "invisible" to all executors or that the executors or operational scenarios be "unknown to each other."

10.2.3   Policy and Schedule Mechanism

Policy and schedule (PAS) is an important part of the DHR's negative feedback control mechanism. Its basic function is to instruct the input agent to distribute external request sequences to designated executors or operational scenarios and form a service set of current executors in the mimic brackets. By changing the designation, it is possible to choose or change the executors or operational scenarios in the service set, thereby replacing executors or operational scenarios, putting them offline, or migrating their services. One of the purposes of introducing this mechanism is to make the mimicry presentation trickier and craftier and to make it easier to hide the features of the scene within the mimic brackets (for instance, to confuse or disrupt any attempts to export system response functions based on the input/output relations). The second purpose of this mechanism is to make the scheduling of executors more targeted and to increase the efficiency of the executor resources by combining the historical performance of executors or operational scenarios and scenario analysis of related abnormal events. For example, Executor A has a relatively high probability of abnormality in problem scenario k (because the reliability of attack chains is strongly related to the specific defense scenario), while it shows zero or low probability of abnormal performance in other scenarios, we should try to avoid using Executor A in problematic scenario k during the scheduling. Obviously, PAS can take advantage of executor resources as much as possible by using avoidance instead of a closed-loop approach to problems. This is especially important when resources are limited and abnormal problems cannot be eliminated in time. The more sufficient the information on the problem scenario and the clearer the analysis on data characteristics, the more accurate the avoidance, the shorter the autonomous iterative response time, and the more efficient the utilization of executor resources.

10.2.4 Mimic Ruling Mechanism

The multimode ruling mechanism serves as the cognitive function of DHR architecture's general uncertain disturbance. Through consistency ruling on multimode output vector with preset syntax and grammar in a standardized or normalizable mimic interface, it is possible to effectively sense any non-cooperative attacks or random failure reflected onto the mimic interface. The arbiter transmits the relevant status information to the feedback controller, which forms instructions for input/output agents and reconfigurable executors based on preset policies or machine learning results. Importing a policy similar to mimic disguise in the multimodal arbiter can not only avoid the superposition state of majority ruling (refer to Sect. 9.2.3) but also acquire more abundant functions. For instance, in order to increase the level of defense, we can use the iterative ruling methods such as consistency comparison, weight judgment, and policy perimeters, or we can dynamically and randomly change the number and target of executors involved in the ruling. Even if the ruling results show that the output vectors are all different from each other, it is still possible to use time and space iteration to come to a decision, based on weighed values such as executor credibility and historical performance. If we find same syntax but different grammar in the multimode output vectors, we can carry out a domain-based ruling. When there are undefined domains in output vectors or when the threshold is uncertain or when the threshold deviation is allowed, it is possible to use the mask ruling. We can also preprocess output vectors to facilitate the ruling (e.g., by comparing only the check code of data fields in IP packets or the hash value based on vector contents). In different application scenarios, we can choose synchronous or asynchronous ruling or concentrated or distributed ruling. In addition, adding up the abnormal outputs in the multimodal arbiter, recording and analyzing the corresponding problem scenario can help to determine the security status of the software/hardware resources of the target system, as well as to measure its effect against the non-cooperative attacks. It should be stressed that mimic ruling can significantly increase the difficulty of side-channel attacks that take advantage of the arbiter sharing mechanism to tunnel through.

10.2.5 Negative Feedback Control Mechanism

Unlike MTD, mimic defense integrates different defense elements – including dynamicity, diversity, and randomness – into a unified negative feedback control architecture and uses the system engineering approach and systematic architecture effect to achieve high cost-efficient reliability and resilience. The feedback control loop consists of mimic ruling/output agent, feedback control, input agent, and reconfigurable executors. An abnormal status output from the arbiter would trigger changes in the level of heterogeneity or diversity of the defense scenario in the current service set within the mimic brackets, a process that involves two stages. The first stage involves altering the abnormal executor elements in the service set, reconfiguring the software/hardware components or algorithms of the abnormal executor, and forcing the abnormal executor to change its status (such as entering the cleaning or rebooting status) or to reconfigure the executor elements in the service set. The second stage involves first observing whether the abnormal status has been eliminated or whether the abnormal frequency is below the threshold. Repeat the first stage until it meets the requirements. Otherwise the feedback loop would enter a temporarily stable status. The defense environment of the target object is simultaneously dynamic, diverse, and random, and its operation process is the adaptive iteration convergence approach. Therefore, the effect of mimic defense can be designed, planned, tested, and measured.

10.2.6 Input Allocation and Adaptation Mechanism

Since the heterogeneous executor elements in the MB service set are formed dynamically in accordance with the PAS instructions, the input incentive channel outside the MI can only be flexibly connected to the corresponding heterogeneous executor through an allocation process. Generally speaking, it would be an ideal configuration to have both connection and exchange functions. Transparency requires adaptation of the differences between executors in terms of communication sequence numbers (such as TCP sequence numbers), custom fields, timestamps, and reserved fields, it also includes introducing active or passive defense measures or mechanisms, such as adding "internal fingerprints" information, normative verification of input sequence syntax and semantics, compliance control of input frequency, black/white list filtering, and verification.

10.2.7 Output Agency and Normalization Mechanism

In theory, when the selected functionally equivalent heterogeneous executors are under the same input incentive norms, it is highly probable that a majority or all of the multimode output vectors are the same. But due to differences in algorithms, supporting environments and platforms, the output responses would be different. For example, the message domain of the protocol may contain certain self-defined elements (such as communication sequence number, priority, version information). There may also be differences within the preset range of accuracy (such as approximation error in scientific computing). There may be problems when processing reserved or expansion bytes that are not subject to strict regulations (such as whether the unified filling code should be used). Especially in a mixture of protocols, there may be version differences when heterogeneous executors use the reserved or expansion fields.

Preprocessing algorithms are often used to reduce the ruling delay. For example, hash processing may be used on output vectors (could be more than 1000 bytes) so as to compare hash values of the limited length, instead of comparing each byte. To avoid influencing the multimode voting of output vector, some encryption or disturbance function may need to be transferred from outside the executors to behind the mimic ruling process. Engineeringly, it may be necessary to add the output vector normalization and output agent function before the mimic ruling in order to ensure the mimic brackets can not only block all outward differences of defense scenarios but also allow certain differences between the executors (especially when the COTS-level software/hardware is used). In other words, the mimic brackets need to be transparent to the inside heterogeneous executors, but they also need to "hide" from the external environment.

10.2.8   Sharding/Fragmentation Mechanism

Besides bolstering resilience by using the heterogeneous redundant resources, the introduction of the sharding and fragmentation mechanism, as appropriate, may augment the results. For example, when performing mimicry reform on file system or data centers, we can transmit information through multiple paths in a fragmented way, use the heterogeneous nonhomologous devices to achieve shard storage or shard encrypted storage, or use one-way connection mechanism for the logical expression of file directory and the separate management of physical mapping. The endogenous mechanism of the architecture guarantees the security of private or sensitive information, making it difficult for the attackers to obtain the complete sensitive information.

10.2.9   Randomization/Dynamization/Diversity Mechanism

Mimic defense takes full advantage of multimode ruling-based policy scheduling and multi-dimensional dynamic reconfiguration negative feedback mechanism, and as a result, it creates defense scenarios that are random, dynamic, and diverse, characterized by adaptive iteration convergence processes and designable methods. It significantly increases the "difficulty of coordinated attacks on dynamic, diverse targets under the non-cooperative conditions," making it difficult for the attackers or penetrators to establish a continuous and dependable attack chain. While in the case of a single executor, the originally certain vulnerabilities and backdoors as well as the effect of vulnerability-based attacks will no longer be certain and sustainable in its attack effect in the mimic environment created by negative feedback mechanism. This is done in several ways: randomly substitute the heterogeneous executors in the current service set within the mimic brackets or change the structural (algorithmic) scenario of an executor; clean or initialize at varying level the abnormal executors;

trigger the traditional security mechanism to clean up viruses and Trojans, or initiate the functions to automatically block backdoors and fix vulnerabilities; or use certain policies to reconfigure and reorganize relevant executors under the condition of functional equivalence. Such changes could significantly lower the escape probability of coordinated attacks on dynamic and diverse targets under the dis-cooperative conditions.

In fact, in the service provision intensive areas (such as cloud service platform, data centers), heterogeneity, redundancy, dynamism, and functional equivalence already exist to varying degrees in the allocation and scheduling of resources. As long as the mimic defense approach is properly used to design an appropriate resource control architecture and scheduling policy, it could significantly improve the defense of target system without setting the mimic ruling. Therefore, if the goal is simply to cause uncertainty to the results of the attack, then the mimic ruling mechanism is not a must. But it is indispensable if the goal is to make attack escape probability controllable.

10.2.10   Virtualization Mechanism

As Fig. 10.1 shows, by using virtualization mechanism, in particular, the emerging container technology, we could set the mimic field in an economic and flexible way to support the needs of diverse defense environments.

Similarly, virtualization mechanism can support the realization of a dynamized and randomized execution environment in an economic way. For instance, the virtual container can support the physical (logical) segregation and division of heterogeneous functional modules for easy management and control of data and service functions. At the same time, the container technology can be used to allocate the dynamic resources and increase the flexibility of deployment and the utilization of resources. In terms of lowering the cost of mimic field realization and the cost of management, the virtualization mechanism effectively reduces the number of physical modules, hides the internal complexity, enables automatic and centralized management, simplifies recovery and synchronization, and facilitates the seamless transfer of services across multiple platforms. In particular, the virtualization mechanism can facilitate a simplified version of mimic defense in networks, cloud platforms, or systems via incremental deployment of control components inside the mimic brackets. For example, one can use private or public cloud or use the heterogeneous redundancy processing resources provided by different public clouds, load the diversified service software with multiple variants, form functionally equivalent virtual heterogeneous executors, and distribute external service requests to different virtual executors through the incremental deployment of controllable and trustworthy mimic bracket components (can be in virtual format) while at the same time choosing output vectors that meet the ruling requirements as service request response. As the allocation of heterogeneous and pooled resources on the cloud and its dynamic calling and allocation mechanism

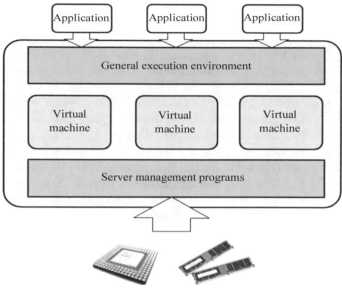

Computing and storing resources

Fig. 10.1 Illustration of implementation of the virtualization mechanism

are influenced by user online requests and system scheduling algorithms, it is difficult for attackers or penetrators to identify the changing patterns of cross-platform mimic service fields or to take advantage of vulnerabilities in a consistent way. This method is universally applicable in the network-based distributed service modes.

10.2.11   Iteration and Superposition Mechanism

The iteration and superposition mechanism can exponentially increase the complexity of the defense environment of the target as well as the uncertainty of defense behavior. After mimicry reformation of the platform and mimicry-based redundancy deployment and policy and schedule of the functionally equivalent heterogeneous executors, the structural manifestation of the processing platform could create nonlinear cognitive difficulty for the attackers. Similarly, the heterogeneous combination of the supporting environment, such as operating system, database, function library, and assistant software, can alter the dependency of attack chains [2, 3]. Further mimicry reformation of the executors' components, modules, units, and devices can significantly increase the apparent uncertainty of the device or system. Either vulnerability-based known or unknown threats or Trojan-based certainty attacks can hardly achieve consistent attack outcomes under the non-cooperative conditions in a multi-dimensional heterogeneous dynamic space due to the apparent uncertainty of defense scene and extreme instability of attack chains. If other protective measures are added to the mix (such as encryption or authentication, instruction set/address/port randomization, and dynamization [4, 5]), more significant benefits can be achieved through the combinations.

10.2.12 Software Fault Tolerance Mechanism

The idea of software fault tolerance design is to reduce the possibility of the same design errors occurring simultaneously in program codes by using the heterogeneous redundancy. There are possibly design flaws in a software code that can be used for malicious purpose. Such flaws, however, can be tolerated under the diversified and redundant configurations. Furthermore, if malicious codes are embedded in a software module, they can also be detected through consistency ruling in the multimodal conditions. Since the effectiveness of fault tolerance mechanism is predicated on the guarantee of least relevance between functionally equivalent heterogeneous redundancy modules and versions, what kind of method is used to achieve module heterogeneity has become the focus of study in the current fault tolerance design. The traditional dissimilar redundancy software design is carried out by "back-to-back" teams which, under the requirements of the same function and performance specifications, use a strict heterogeneity background management model to design their separate modules or versions, and then the system integration group will complete the redundancy configuration of heterogeneous software modules or versions in a tolerance framework. Currently there are other methods such as automated generation of logic maps, mutation, genetic algorithms, virtualization, software reuse library, and N-version programming, all of which significantly reduce the level of design complexity of redundant modules or versions under the fault tolerance model, but it is still hard to guarantee the performance and efficacy.

It is encouraging that with the development and improvement of the design environment, including mimic structure program design language, compiler, and debugger, it is no longer unimaginable to run a classic mimic architecture software on the generalized, multi-core, many-core, heterogeneously accelerated, and field-specific software/hardware coordinated computing platforms. The multi-core homogeneous and heterogeneous many-core parallel processing approach even enables the software-defined hardware acceleration technology to effectively support the parallel processing and multimode ruling of heterogeneous redundant software modules. This can greatly ease the problem of low performance of mimic architecture software in the single-CPU environment. If the non-classic mimic defense mechanism is used to design high security level software system, the efficiency of software execution on the whole is not affected as there is no need for multimode ruling, but its attack resistance level is far superior to the software developed using traditional design approaches.

10.2.13   Dissimilarity Mechanism

One of the prerequisites for the establishment of the "true relatively axiom (TRA)" is that everyone has independent judgment and ability to act and any fraudulent acts such as joint fraud, bribery election, and canvassing will undermine the premise of correctness of the axiom. Therefore, as a design concept and management method, the dissimilarity mechanism is adopted to avoid common or homomorphic failures caused by design flaws. For example, you can engage design teams with different training backgrounds and technical experiences to simultaneously develop multiple functional modules by using different development environments and tools. Such modules can independently fulfill the system's predefined requirements and form a dissimilar redundancy architecture. With the multimode voting method, one can detect, isolate, and locate random faults within the architecture and shield the impact of faults on system service functions and performance as much as possible. In theory, both hardware and software can use the dissimilarity mechanism to overcome common or homomorphic faults caused by human or nonsubjective factors to improve system reliability and security. The dissimilarity mechanism not only suppresses attacks based on known or unknown vulnerabilities or malicious code of the target object but also effectively overcomes common defects caused by development tools and environments. Of course, no one can always afford the cost of engineering implementation in the classical dissimilar redundancy design method.

Although we know that, in theory, there is neither absolute sameness nor absolute difference, so far humans are regrettably unable to accurately test or evaluate the degree of dissimilarity between redundant bodies, theoretically and technically, nor do we even know what kind of dissimilarity criteria should be met under a given condition of reliability and attack resistance. Fortunately, the product technology and commodity market are bound to be diversified along with the economic and social development (as is the case with the nature). Differentiated competition will inevitably lead to the formation of a diversified ecology. However, while using the COTS-level products to construct a diversified defensive scenario for mimic defense, we should not only pay attention to avoiding the same hardware and software components or algorithms in executors of the same service set but also focus on "discooperating" the executors in the mimic brackets, so as to make it difficult for the attackers to form or maintain a consistent misrepresentation (i.e., attack and escape) in the space-time dimension even if they can use the homologous vulnerabilities and backdoors.

In fact, when the number of heterogeneous redundant executors and the strategy of the mimic ruling are determined, we can only rely on the strategic scheduling, reconfigurable mechanism, and adaptive feedback to achieve differentiated changes of defensive scenarios in the mimic brackets; thus the abundance of spatial and temporal diversity and iterative convergence algorithms are critical to the effectiveness of mimic defenses. In a sense, it is the heterogeneity or diversified scenarios of dynamic heterogeneous redundant space that enable mimic defense to eliminate the impact of differential-mode faults or "lone-wolf attack" and suppress the damage of common-mode faults or increase the difficulty of collaborative attacks.

10.2.14   Reconfiguration Mechanism

The DHR architecture definition emphasizes that the executor must have reconfigurable or software-definable functional attributes. The fact whether the elements in the mimic brackets have reconfigurable, reorganizable, or software-definable functions is also directly related to the dissimilarity (diversity) implementation means of the target system and the abundance of the defense scene resources. If the system has only three fixed-structure heterogeneous executors, the changes in defense scenarios can only be controlled through software resource configuration. Conversely, if the executors have software-definable hardware structure, combined with multivariant, multi-version, or even mimic architecture software configurations, then the degree of diversification of the defense scenarios is highly expectable. At present, the technology supporting the reconfigurable reorganization mechanism is developing rapidly, from programmable array logic (PAL) to field-programmable gate array (FPGA), eFPGA, from CPU + FPGA acceleration structure to SKL + FPGA integrated package structure [6], from reconfigurable computing [7] to mimic computing, from rigid computing architecture to software-defined hardware (SDH) architecture, and from general-purpose computation of pure instruction streams to domain-specific software and hardware collaborative computing by converging instruction streams and control stream. In particular, mimic computing under the functional equivalence conditions is particularly suitable for the basic architecture for reconfigurable executors. Of course, the reconfigurable, reorganizable, and modular software architecture is also an important choice.

10.2.15   Executor's Cleaning and Recovery Mechanism

The cleaning and recovery mechanism of executors is very effective for eliminating the occurrence of sporadic hardware and software malfunctions, memory-resident (or injection) Trojans, and standby cooperative attacks. It can be used as a way to handle abnormalities, as an active prevention measure, or as a way to change the defense scenario. Different levels of cleaning and recovery, and even reorganization and reconstruction, can produce a variety of defensive scenarios. Of course, the operation time should be as short as possible and the operation effect as high as possible, in order to reduce the probability of "downtime-bumping" attacks. The cleaning and recovery operations can generally be divided into four categories:

(1) Active cleaning or reconfiguration. Since the external control parameters of the feedback controller include random information of the target system, such as

the number of currently active processes/threads, CPU and memory occupancy, and network port traffic, the resulting PAS instruction would non-periodically instruct the executors to perform different levels of initialization or the reconfiguration, reorganization, and reconstruction of the operating environment. This mechanism is quite a challenge for standby attacks or for maintaining an escape state.

(2) Event driven. When the mimic arbiter detects inconsistencies in the output vectors of an executor, it will activate the feedback controller, which will decide whether to issue a cleaning and recovery instruction to the executor in question according to a given evaluation strategy. It will also decide what level of cleaning and recovery needs to be performed, including whether to initiate reconfiguration, reorganization, and reconstruction of a particular form.

(3) Cleaning based on historical information. When an executor exhibits a fairly high frequency of abnormalities in a given statistical period or has never had an abnormal situation (because we cannot rule out the possibility of a dormant attacker), it is necessary to reduce the chance of this particular executor to go online, increase the depth of cleaning, and carry out reconfiguration and reorganization when conditions permit.

(4) Cleaning based on security situation and system security index. If the arbiter frequently exhibits inconsistent status output, or if the iteration convergence of the feedback control loop is too slow or unable to converge, especially if the multimode output vector is completely inconsistent, then it indicates that the system has a low security index or is in a high-risk status, which requires increasing the scheduling frequency of executors and frequency of active cleaning, so as to reduce the online exposure of the executor and enhance the depth and breadth of software and hardware reconfiguration.

One tough issue in executor cleaning and recovery is how to synchronize as fast as possible to the standby or "always on" status. In particular, the service functions or session functions involving complex status transition mechanisms are likely to be strongly associated with historical status, which requires innovative methods and mechanisms. In addition, operations such as synchronous recovery of the working database or reconstruction of the environment are also subject to the requirements of "dis-cooperation." Traditional neighbor data and environment copying methods may lead to the proliferation of potential security problems. However, the rapid cleaning and recovery of executors is also important to increase the difficulty of the "downtime-bumping" attacks. However, there may not be a unified solution for different fields of application, but rather, each case should be treated differently based on the specifics.

It should be noted that the cleaning mechanism is only effective for memory injection attacks or standby attacks, but not for backdoors or malicious codes inherent in hardware and software. Fortunately, either vulnerability or backdoor attacks usually demand the upload of a Trojan or Trojans to the memory before achieving the expected purpose. In a mimic environment with redundant configurations, it is a good choice to implement memory cleaning, configuration register formatting, or even executor reboot, not only because it is simple and effective but also because there is no need to worry about system service interruption.

10.2.16   Diversified Compilation Mechanism

The multi-level and diversified compilation method [8] can help achieve the code obfuscation, equivalent transformation, and randomization of relevant layers at three levels – source code, intermediate code, and object code – as well as stack-related diversification techniques, thereby generating multiple target code variants that are transformed from the same source code, as shown in Fig. 10.2. This method can not only increase the difficulty to reversely analyze a single target code and to take advantage of vulnerability generalization but also provide an effective development tool for dissimilarity design of single-source or self-developed software.

1. Diversified compilation defense mechanism

The compiler-based diversified technology strategy generates many different variants through a variety of methods at different levels, such as parallelism recognition, library program selection, memory allocation policy, and data structure diversity algorithms, thus posing a huge challenge to the coordinated attacks based on mimicry targets.

In the mimic environment, the attacker must first of all have the ability to crack these multivariants and, secondly, have the ability to use different vulnerabilities of the variants to implant Trojans (often times, an attack method applicable to one variant may not be applicable to another). Finally, in the mimic environment, it is often difficult to achieve a coordinated attack of multiple targets or to achieve consistent results. In addition, the diversified compiler method also creates obstacles for attackers to crack the software through decompilation. Attackers need two

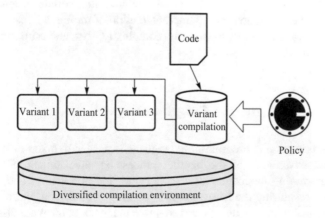

Fig. 10.2 Diversified compilation mechanism

important pieces of information to break the software, namely, the software version and the corresponding patch information. In a diverse software environment, each instance of the software is special, with distinct binary files, which increases the difficulty of cracking the software.

2. Defense effectiveness of diversified compilation

The Address Space Layout Randomization (ASLR) technology based on diversified compilation tools has been widely used. At present, it has been effectively applied in dealing with stack address overflow attacks, decompilation, and cracking of software. In addition, the diversified software versions generated by diversified compilation tools can significantly increase the complexity of attacks based on analysis of target version software vulnerabilities. "One version for one device" or "one version for one instance" can effectively block the spread of Trojans as well as its working mechanism, curbing the occurrence of massive cyberspace attacks. It is important to emphasize that if the source program itself has design flaws or malicious functions, then it is impossible for any diversified compiler to change it. However, if these problematic codes still need to rely on the operating environment conditions to function, they may not be reliably utilized in the mimic system.

Since the compiler can automatically diversify the code when translating the high-level source code into low-level machine code across platforms, it is therefore possible to create multiple variants for one program with the same external function but different internal codes. There are two ways. One is to create an operating environment for multivariant programs, with a monitoring layer that executes multiple variants in strict order and checks whether the results or behaviors are different in order to identify attacks. The other way is to use the large-scale software polymorphism, with each user having their own polymorphic version, so that it is difficult for an attacker to know the software version code of others from the version they have obtained.

In short, any mechanism that can increase the dynamic heterogeneous redundant space and the difficulty of multi-target coordinated attack under non-cooperative conditions can be considered in the implementation of mimic defense engineering, but it is necessary to carefully balance the complexity, cost, and economic and technical effectiveness.

10.2.17   Mimic Structure Programming

We all know that parallel programming language is needed to take full advantage of the technical advantages of the parallel processing environment. Similarly, the mimic programming language can also consider the application and efficacy of mimic structures during the hardware and software code design phase. Of course, heterogeneous redundancy design and adaptive feedback control are also essential. The premise of mimic structure software design is multi-core or many-core processing resources at least. Ideally, it is desirable to have distributed processing resources with multiple processors or heterogeneous redundancy and even use the processing resources on the cloud.

10.3 Major Challenges to Engineering Implementation

10.3.1 Best Match of Function Intersection

In order to prevent possible escape events in heterogeneous redundant systems, it is desirable that the functional intersection of a given executor is equal to the maximum functional intersection, that is to say, there should not be any other functional intersections. However, it is almost unattainable in engineering practice. For example, an ×86 CPU can solve a Fourier transform problem, and an ARM CPU or Loongson CPU can solve the same problem. However, these CPUs also have the ability to handle many other tasks (including dark features that are not desired by the target object environment). In other words, in engineering practice, it is generally not possible to guarantee that a given functional intersection equals the maximum functional intersection.

When the target object itself is a complex system, it becomes extremely challenging to eliminate the influence of the intersection of redundant functions between the executors in a mimic defense system with a given function intersection. Obviously, the requirement that a given functional intersection exactly matches the maximum functional intersection should not and cannot be a criterion for engineering practice. The second best practice is that the implementation algorithms for a given functional intersection should be as different as possible, with interface or threshold limits. For example, $2 \times 3 = 6$, and it can be calculated manually or by using abacus, slide rule, or the multiplication table. But for $12 \times 12 = 144$, we cannot get the result directly from the multiplication table. Therefore, the scope of calculation on a multiplication table can be used as the constraint of the abovementioned equivalent multiplication function. That is to say, when it cannot be guaranteed that the given function intersections are an exact match, then we must firstly strive to achieve maximum dissimilarity of the algorithm or the physical platform (e.g., using CPU, FPGA, ASIC, respectively, for the host computing components), pay special attention to improved designs on the basis of open source or same source, and consciously increase dissimilarity or heterogeneous components. But it is not easy to achieve this in an economical way. Secondly, the definition of the intersection function should be as clear as possible. The symbiotic mechanism or additional fingerprint information can be used to increase the semantic or information abundance of the output vector, if necessary. Moreover, even if a homologous malicious code exists in the executor, as long as the heterogeneous execution environment has different execution authorities or storage operation authorities, it would be impossible for the malicious code to obtain "super privilege" in the majority of the executors for cooperative escape. For example, using the trusted computing function SGX of the ×86 CPU, or the Trusted-area function of the ARM CPU, or the FPGA function limited to the offline modification mode, we can heterogeneously embed an executable code, which controls key passes, in the protected region of various executors, invalidating any attempts to obtain "super privilege" by modifying or injecting the protected area program code. Similarly, some core algorithms that are frequently used but do not change frequently may be implemented through different equivalent methods for different executors; for some, you may take the software implementation method based on instruction stream, and for others, you may choose the hardware implementation method based on control flow or a combination of the two. In short, setting a variety of "super privileges" in addition to traditional system privileges can effectively increase the difficulty of cross-domain coordinated attacks.

10.3.2   Complexity of Multimode Ruling

Generally speaking, the basic conditions for mimic defense can be found in any functional or operational interface with standardization or normalization. However, different application scenarios and different performance requirements and security standards all have a significant impact on the complexity of the implementation of multimode ruling. Since multimode ruling, policy and schedule, or input/output agent are always concatenated between the input and output ends of the mimic interface (MI), and the output vectors of the diverse redundant executors in the interface have different response time, calculation accuracy range, or allowed version difference (syntax, options, default values, extended domain padding), among others, there is not a uniform method to overcome these differences from an engineering point of view, nor is it possible to have one single method. Therefore, we must use different solutions according to specific application scenarios. In particular, if there is a large grammatical expression difference between the multimode output vectors, or if there is a lot of content to be ruled on (e.g., the object is a very long IP packet), then the processing complexity of the arbiter will be greatly increased. For example, we need to add a ruling buffer queue to overcome inconsistencies in the arrival time of the output vector; use regular expressions to mask grammatical or even semantic differences; use some sophisticated iterative ruling algorithms to "mask the defensive behavior" or improve resource utilization; introduce some preprocessing algorithms (such as the checksum or hash calculation of the output vector) to reduce the ruling time; or add extra agent functions due to the presence of encryption options, random serial number, and multi-address expression. In general, synchronous or centralized rulings are less difficult than asynchronous or distributed rulings, but their robustness of the defense (the arbiter would become the focus of new offense and defense) and the impact on the performance of the target system services are also significantly less than the latter. Since the asynchronous ruling can adopt the method of "first come, first output" and allows for additional decision and correction, it has less impact on service performance of the target system (mainly insertion delay) and wider choices of algorithms. It should be emphasized that the output vector, as defined, can choose environmental parameters, resource usage parameters, or process features that are indirectly affected by applications, so as to avoid dealing with semantics analysis conundrum, for instance, using the test status vectors of virtual honeypot as the ruling subjects.

In short, even though the diversification of ruling methods can help confuse or disrupt side-channel attacks such as tunneling through, the setting of mimic interface and the realization of mimic ruling are relevant not only to achieving protection purposes but also to the performance expense and cost. There are many difficult engineering implementation problems that need to be solved creatively.

10.3.3 Service Turbulence

Suppose that all the heterogeneous redundant entities in the mimic brackets have a vulnerability or backdoor of a simple trigger mechanism (such as the "China Chopper" attack) and that the attacker has fully grasped the vulnerability/backdoor and can implement a continuous attack under non-critical conditions, then the mimic defense mode would repeatedly migrate, switch, or conduct offline cleaning because abnormalities frequently occur to the executor resources, leading to the violent bumping in external service performance provided in the mimic brackets. In fact, this constitutes a coherent attack on service functions and performance (a differential-mode attack initiated continuously by most executors with determined effects), with similar effects to the denial-of-service (DOS) attack. Although this is only a "thought game" or a theoretical assumption, we cannot rule out the possibility that an attacker with strong technical resources may have such practical ability, and thus mimic defense needs to focus on the prevention of such problems.

In engineering practice, on the one hand, the shutdown state of the executors should be resolved by cleaning and rebooting as soon as possible (may only be effective for memory-resident Trojans), and on the other hand, the restart or fault recovery time of the executors should be reduced as much as possible to ensure that the system has enough available executable resources or defense scenarios. Special attention should also be paid to the use of mechanisms concerning the mandatory changes in the operating environment, such as software and hardware reconfiguration, reorganization, and reconstruction. It is also necessary to use intelligent analysis of operational log to find the behavioral characteristics of the corresponding "downtime attacks" in a timely fashion while combining different defense measures to block the reachability of subsequent similar attacks (such as adding compliance checks and rule-based interception functions in the input agent) when appropriate. Mechanically, as the vulnerabilities in the feedback control loop are difficult to be use, they will not easily pose security threats, but problems such as bumps or downtime caused by backdoors are difficult to address. This is why we always emphasize that mimic brackets may allow for vulnerabilities but not backdoors. This is reasonable because the functions of the mimic brackets are relatively simple and straightforward, and the elimination of the back-door function through compliance check generally would not encounter the dilemma of state explosion. As far as vulnerabilities are concerned, due to cognitive and technological limitations of our time, we do not expect to completely get rid of them for the time being.

It should be pointed out that as a relatively independent functional component, the mimic brackets may develop into a diversified and open COTS-level product for end users to choose freely. Therefore, there is no need to face the philosophical conundrum of "how to prove credibility of root of trust." However, by using technical and human means including social engineering methods (such as the installation of spies), it is not impossible to have such a "superpower" backdoor attack. "Technology can never replace management," and this issue is beyond the scope of this book.

10.3.4   The Use of Open Elements

The mimic defense device allows for open or even open source to a considerable extent due to the design, development, or allocation of heterogeneous redundant executors. Therefore, it is necessary to ensure special independence and dis-cooperation requirements in the preparation process and identify and avoid using the same component or the same space. If necessary, the systematic or distributed technical measures and heterogeneous environments, such as different occasions, different time periods, and different incentives, can be used to eliminate the coordinated escape problem that may be caused by using the same component. The problem at the moment is how to discover the homologous elements in heterogeneous execution. It is expected that the software fingerprint analysis technology under vigorous development will be able to find homologous problems through big data search and deep learning techniques so that targeted prevention measures can be taken.

It is worth noting that if COTS-level components are used in the heterogeneous redundant space of hierarchical deployment or configuration, the unknown hardware and software vulnerabilities (not ruling out the possibility of backdoors either) will inevitably exist in the basic hardware and software components, such as processors, operating systems, and databases. Homologous vulnerabilities are usually also found in heterogeneous executors composed of these components (e.g., there is a high statistical probability of such cases being found in Windows, Linux, ×86, ARM, or similar products). However, the way these vulnerabilities are used often differ with the host executor environment, and as a result, the effect and scope of impact of attack may also differ depending on defense scenarios. When these differences are large enough to influence the consistency of multimode ruling of target object output vectors, then it would be very difficult to achieve an attack escape. In other words, the uncertainty effect of mimic defense – formed on the basis of ruling-based PAS, multi-dimensional reconfiguration, and adaptive feedback control mechanism – can greatly improve the difficulty of coordinated attacks under the non-cooperative conditions, so that even if homologous vulnerabilities and backdoors exist in the target object executors, it is difficult for Trojans to take advantage of such vulnerabilities/backdoors to dependably and effectively launch an attack (unless it is independent of environmental factors). This is also the reason why mimic structure can reduce the heterogeneity requirements between executors (in engineering, it is possible to compensate with the abundance of diversity). In fact, even under normal circumstances, compatibility is to be considered when compiling homologous codes into a different operating environment or using them across platforms (this is usually strongly related to environmental factors such as the operating system or CPU). Therefore, the heterogeneous environment has a strong impact on the execution of homologous codes. In addition, coordination is always strongly associated with harshness. Even the high-performance fault-tolerant computers with various cooperative mechanisms will face lots of insurmountable engineering obstacles to guarantee the precise synchronization of completely isomorphic computational components at the beat level or instruction level.

10.3.5   Execution Efficiency of Mimic Software

In theory, software with mimic architecture is promising in terms of both reliability and security, but its execution efficiency in a single-core processing environment cannot be depended upon; and even in a multi-core or many-core environment, it is difficult to achieve the desired performance requirements unless the parallel design and compilation tools are used. Therefore, the mimic architecture software must first use the parallel design and compilation tools to ensure that the heterogeneous modules of the redundant configuration can run in parallel in a multi-core environment. Secondly, the heterogeneous modules must also engage in preprocessing for multimode output vector ruling (such as forming checksums, hash values, feature vectors, etc.), in order to reduce the processing complexity of the ruling module. Furthermore, the ruling-based PAS feedback control must also take into consideration the issues of module replacement, synchronization, and recovery. These will not become challenging issues for the domain-specific hardware and software co-computing environment that is being developed. It should be pointed out that the mimic structure-based software can significantly reduce design flaws and the security impact that vulnerabilities or traps have on the products due to the use of third-party components. Meanwhile, its unique robustness – high reliability, high availability, and high credibility – is unmatched by current software technologies. In other words, software producers finally have a new way to ensure the generalized robustness of products through structural techniques.

10.3.6   Diversification of Application Programs

As with the dissimilar redundancy architecture, the diversification of application program often leads to unbearable technical and cost pressures, and it also poses serious challenges to the promotion and application of mimic defense, to which there are roughly seven possible solutions:

(1) The preferred method is to diversify compilation. This is to introduce at the source code, intermediate code, and object code levels the diversified parameters, such as prioritizing efficiency, parallelism or dynamism, and redundancy module scheduling policy. Although the diversified version cannot overcome its own logical defects (vulnerabilities), it creates great obstacles to the exploitation of vulnerabilities, even if the exploitation and timing of even the same vulnerabilities may differ in different heterogeneous redundant environments, and these vulnerabilities are difficult to be used as resources for collaborative attacks under the multimode ruling mechanism. However, if the supporting operating environment can be fully heterogeneous, the diversely complied applications can play an important role in blocking attacks – including the attacks launched by running the vulnerabilities and backdoors dependable on environmental factors or running the "dark features" in the operating environment attempting to take advantage of the application fragility – unless there are dark features, like backdoors, in the source application that are not dependent on environmental functions. This is very important for determining the location of problematic resources in the mimic brackets.

(2) Adopt an economically affordable multi-team "back-to-back" independent development model only in the design of key aspects of the application or the design of sensitive paths or shared modules; use different functional equivalent modules developed by different teams to compile differentiated, heterogeneous, multivariant implementation version (*N* version).

(3) Introduce a dynamized and randomized scheduling mechanism based on redundant functional modules in the application and ensure that at least one scheduling parameter must be taken from the dynamic information of the current execution environment (such as processor, memory usage, or number of active processes). In other words, maximize the differences of execution scenarios in the application. In this way, even if the executors run the same version of application, they invoke different heterogeneous equivalent modules in their running since the parameters taken from the heterogeneous execution environment may not be consistent. This results in a significant difference among executors in terms of the operating status or processing flow or dynamic structure, which can often be used to reduce the probability of occurrence of co-state events and increase the difficulty of attack escape.

(4) More generally, mimic architecture can be adopted in the design of supporting software such as operating systems and databases (especially modules or function libraries related to system calls or management), as well as in the design of middleware and embedded components. It is difficult to predict the

dynamic, diverse, and random performance of even the same source program version in different execution environments.

(5) For some COTS-level executable code programs, the binary code translation and transplant techniques can be used for multivariant transformation. Programs with source codes can also generate multivariants via the diverse compiler, but the effect of such measures is uncertain against functional backdoors that are independent of the supporting environment (such as high-risk vulnerabilities and Trojan viruses).

(6) Use FPGA of the CPU + FPGA heterogeneous computing architecture of executors to set up the TPM function similar to trusted computing, so that different executors can independently check the behavior characteristics and status information on the relevant links of the same application, and submit the inspection results as part of threat perception to the multimode ruling.

(7) If we do not consider the special purpose hardware and software codes that are deliberately implanted by the designer (usually can be used maliciously), then we can also use instruction randomization or address space randomization for SQL injection attacks or cross-site scripting (XSS) attacks using the application design flaws.

(7) Similar to (6), the heterogeneous design of mimic systems can be greatly simplified with the help of SGX technology (see Chap. 4). If some of the core modules of the application meet the setting requirements of the mimic interface (MI), we can use the classic dissimilar redundancy to design these functionally equivalent, heterogeneous redundant core modules, which can be respectively compiled with other parts of the application to form partially heterogeneous multivariant application software versions. Such versions are then loaded into each isomorphic executor and the corresponding SGX area, thus forming operating environments with different defense scenarios. The SGX function determines that programs and data stored inside cannot be accessed by any program (including operating systems) other than itself. So theoretically it is impossible for any program to modify the code and data in SGX unless it has such needs or there are hardware vulnerabilities or malicious code in the CPU – the guardian of the security perimeter. However, the mimic architecture does not care if the SGX region of an executor is deliberately modified, unless the attacker has the ability to modify across platforms the heterogeneous program code or data in the executor SGX within the MI under the non-cooperative conditions. The important engineering value of this approach is that the SGX-like features can greatly simplify the dissimilarity design of operating scenarios in mimic brackets, including allowing for the same operating system, CPU environment, auxiliary support software, and even most of the same program codes, which can significantly reduce the implementation cost of the mimic architecture and lower the application threshold.

(8) It is also possible to effectively enrich the diverse scenarios by increasing the labels that are only available at the ruling phase and by adding fingerprint verification information (including related encryption methods) in the executors.

10.3.7   Mimic Defense Interface Configuration

Like the commonsense concept of defense theory and means, mimic defense must also follow the principle of "guarding the pass and defending the strategic point," but there are no golden rules as to how to choose the right place to guard. The usual practice is first to understand which resources or functions of the target system need to be prioritized and then to know whether the target is under threats from vulnerabilities/backdoors. After making sure that mimic structure could achieve the desired effect, we can then consider how to economically set up the mimic interface. In principle, the mimic defense interface is generally set up in a standardized or normalizable operation interface. In cases where it is recognized through the functional conformity and consistency test and the degree of centralized control is high, the mimic brackets are the functional entities. In general, the size or granularity of elements contained in mimic brackets is strongly related to security objectives, economic indicators, and implementation complexity. For complex systems, mimic defense may require multiple mimic defense interfaces for different security standards. For heterogeneous components (which may also be "black-box" components) that are functionally equivalent but not trusted or whose supply chain security cannot be guaranteed, we can also achieve controllable, secure, and reliable purposes by assembling according to the DHR architecture. The following five application examples illustrate the setting of the mimic interface.

10.3.7.1   Route Forwarding Based on Mimic Defense

At present, the protocol processing code of the backbone router is generally up to tens of millions of lines, which inevitably contain design defects or backdoors, as well as various dark features, but the routing table generated by the standard protocol for packet forwarding may be probably normal [9]. In the mimic defense architecture, it is necessary to set up diverse protocol processing components and a system routing table component that has independent processing space, supports packet forwarding, and is updated by multimode ruling. When the routing update request is received, each protocol processing component outputs an updated routing entry, and the system routing table component determines whether to update the routing forwarding entry according to the multimode ruling result, thereby effectively addressing the non-cooperative router hijack attacks that are based on vulnerabilities, backdoors, or Trojans of the router protocol executors.

10.3.7.2   Mimic Defense-Based Web Access Server

Design flaws are inevitable due to the large quantity of software codes of web services of various versions. In the mimic defense architecture, it is necessary to set up at least a diversified web service executor and a physically separate file server that operates according to the multimode ruling mode. Generally speaking, the normal response of the diverse executor to the same web service request is that identical operation requests are sent to file servers. Multimode ruling on these operation requests (including content) can effectively address non-cooperative attacks [10] based on web executor vulnerabilities, backdoors, and Trojans.

10.3.7.3   File Storage System Based on Mimic Defense

The CMD-based file storage system consists of a metadata server (MDS), an object-based storage device (OSD), an arbiter, and a client. The MDS is used to manage metadata in the entire network storage system, including file control block inodes, directory trees, and so on. The OSD is used to store fragments of file data. Files generated by clients are stored, respectively, in the MDS and the OSD. Specifically, the file management information is stored in MDS, and the file data is strategically divided into segments and stored (or encrypted and stored) on different OSDs. The MDS in the system should consist of at least three functionally equivalent heterogeneous servers. The arbiter forwards requests from the client to at least three MDSs, which work independently and return the processing results to the arbiter. The arbiter performs a mimic ruling on the processing results of each metadata server and selects a consensus result and returns it to the client. The arbiter can also trigger a post-processing mechanism when it detects an abnormal metadata server and its anomalous behavior.

10.3.7.4   Mimic Defense-Based Domain Name Resolution

Assume that there are m domain name servers provided by the same carrier in an autonomous domain network and that the domain name servers are provided by different vendors. Assume also that any domain name server can provide the same domain name resolution function. In theory, there may be different vulnerabilities, backdoors, or Trojans in these domain name servers, so their domain name resolution services may be attacked. A mimic domain name service system can equivalently be formed by deploying mimic brackets incrementally and through networked application of these virus-bearing domain name servers. When a user initiates a domain name resolution request, the request will be first directed to the dynamic assignment unit in the mimic brackets, and then it will be sent to k ($2 < k < m$) number of servers that are randomly selected from the m domain name servers using the dynamic assignment algorithm. The results from the k servers are sent back for mimic ruling. The ruling results that meet the given policy requirements will be used as the user domain name resolution response. The consensus result can also be stored in the trusted domain name resolution record database in the mimic brackets. Obviously, such a defense architecture makes it almost impossible to attack an arbitrary domain name resolution service node.

10.3.7.5   Mimic Defense-Based Gun Control System

In the gun control system, there is always a communication and control interface between the control unit and the execution unit to transmit the firing data or operational commands to the artillery actuator or to obtain relevant information from distributed sensors. It is usually appropriate to set the mimic defense interface based on this interface, because the control part is generally composed of complex hardware and software and uses a large number of mature COTS-level components, and there are functionally equivalent diversified conditions. The control part can adopt the DHR architecture and therefore obtain reliability and safety gains. The output of multiple heterogeneous executors can be aggregated into independent multimode ruling stages and then be connected to the communication and control interface. Comparing the output information can help detect and block random malfunctions (incidental bounces) in executors or errors caused by vulnerabilities/backdoors or Trojans.

In each of the five examples, there are mimic interfaces that can be normalized or standardized and can be tested and measured, all of which require the setting of gate components with multimode ruling mechanisms, and have the conditions for diverse execution components at the open-source level, COTS-level, or redundant deployment. Their purpose is to protect against the impact of software and hardware vulnerabilities/backdoors, Trojans, and physical or logical disturbances in the mimic brackets. However, the diversification conditions are not always met in engineering practice, and the inserted multimode ruling will inevitably affect the service response performance (assuming the processing conditions remain unchanged). Therefore, the choice and setting of the mimic interface are related to whether a mimic defense architecture can be effectively constructed, as well as to the cost and performance loss.

10.3.8   Version Update

Software or hardware version updates or upgrades are a problem for diversified executors within the mimic interface, especially COTS-based or open-source versions. In general, as long as the update and upgrade of the relevant hardware and software do not change the semantic and grammatical expression of the given mimic interface, they will not affect the function or performance of the mimic ruling and will usually be left alone, for instance, hardware upgrades or software patches or additions and improvements that do not affect the functionality of the mimic interface. Any update or upgrade involving the mimic interface, such as changes in information content, output vector format, and difference in update time of heterogeneous executor versions (especially when multiple vendors are involved), is related to real-time modification of multimode ruling algorithms or backward compatibility. Fortunately, most version upgrades are only in forms of additions or patches, and the base or standard parts are relatively stable. Even if there are some local changes, the multimode ruling can use mask method to compare only the invariant parts or perform similarity ruling, or give the new version more weight, or compare the thresholds of range or interval. It should be noted that the software and hardware versions in professional fields such as industrial control, weapon systems, command and control, and energy transportation are relatively stable, without frequent upgrades and updates, so mimic defense technologies should be primarily applied in these fields. When compared with systems that use traditional means and approaches, the facilities of a mimic structure can not only improve the reliability and availability of the target objects, but there is no need for frequent "patchwork" maintenance on security vulnerabilities in software and hardware versions, thereby lowering the risks of human error and requirements on the staff's security skills, significantly improving the cost efficiency of the system over the entire life cycle.

10.3.9   Loading of Non-cross-Platform Application

Loading non-cross-platform applications in a target system based on the mimic defense architecture generally requires a source program or an intermediate code source program. The diverse heterogeneous redundancy version is automatically generated by a specialized compilation tool based on the user-defined isomerization parameter set and then loaded onto the relevant executors by a special installation tool. There are two problems with this approach in terms of using COTS-grade software products as components. Firstly, there is a conflict with the mainstream non-cross-platform software version protection and the product sales models; in particular, it cannot adapt to the business model where products are released in binary code executable files. Secondly, diversified compilation methods cannot be expected to fully eliminate software function design defects, nor is it possible to change the deliberate code function. For software version upgrades with only executable code, the manufacturers would generally release new versions that support the mainstream operating systems or CPUs and could be directly loaded as long as they are compatible with the executor environment. If the environment is not suitable, it needs to be adapted by using a complex binary translation tool. Therefore, it is advisable to select the mainstream COTS-grade hardware and software components to avoid the problem of version upgrade caused by the unstable maintenance chain of less-popular products.

In general, the CMD model is inconvenient for occasions where non-cross-platform applications are frequently loaded (such as on desktops and mobile terminals). Here the non-cross-platform application loading problem is emphasized because it is different from the basic supporting software and tools, the degree of standardization is low or not standardized at all, and there are usually no nonhomologous but functionally identical, diverse third-party software versions. Even if the conditions for using diverse compiler tools are met, it still needs to deal with possible compatibility issues between the operating environment and the new program version. It should be emphasized that diversified compilation can only generate multivariant versions rather than diversified versions without homologous components. In this case, even if the mimic structure related layers are isomerized, it cannot be effectively defended. Multivariant application software attacks that do not rely on environmental factors such as backdoors, malicious code, can thus become one of the weaknesses of mimic defense. However, in this case, if an attack escape event occurs, the application software itself is almost the only suspect. A possible workaround is that, when using a single-source non-cross-platform application, you select a CPU with trusted computing capabilities (e.g., ×86 with SGX functions, ARM with Trusted-area functions) in each heterogeneous executor, or use the CPU + FPGA computing architecture that provides trusted computing functions or deploy the traditional invasion detection functions to monitor the operational behavior and state of the application software at any given stage or in specific environment. If conditions permit, it may be better to have different executors monitor the status or information of the application software in different stages or phases, so as to maximize the challenge that the mimic architecture poses to coordinated consistent attacks under the non-cooperative conditions.

10.3.10   Re-synchronization and Environment Reconstruction

In a mimic environment, the heterogeneous executors or service scenarios may be taken offline for cleaning due to abnormal outputs, or may be reset for preventive purposes out of policy and schedule reasons, or forced to restart or reconfigure. In summary, it is important to re-synchronize the standby state with the online executors or service scenarios as soon as possible after the cleaning, reset, or reorganization and reconstruction, because it concerns the effectiveness of the heterogeneous redundancy mechanism and the availability of defense resources and service scenarios. Different application scenarios may require completely different re-synchronization strategies and methods as well as different time costs and operational complexities, which are directly related to the robustness of the mimic defense system. In particular, the Zero Trust Architecture, proposed by Google, emphasizes dynamic and multiple-step authentication based on user identity and managed devices. It also boasts a scoring mechanism based on real-time fingerprints and user historical behaviors and has functions such as precise and layered resource access control. Therefore, the copy recovery algorithm based on the normal executor field environment, which is commonly used by the traditional fault-tolerant mechanism, cannot be used. This is because the mimic defense mechanism emphasizes "dis-cooperation," while the methods such as environmental replication can lead to failure of the "dis-cooperation" efforts. In addition, if the environment reconstruction process involves the repair or migration of a large number of data files, the timing of processing will also become a difficult issue. Therefore, how to quickly achieve re-synchronization or environment reconstruction is a key problem that needs to be addressed in the mimic defense engineering.

10.3.11   Simplifying Complexity of Heterogeneous Redundancy Realization

Just like the difficulties encountered at the engineering implementation of the dissimilar redundancy structure (DRS) based on the basic principles and initial definitions, the mimic defense will inevitably encounter similar challenges from technologies and market rules if put to practice strictly in accordance with the theoretical definition. In this section, we explore possible ways and methods to simplify the implementation complexity while reaching certain quantifiable security level of mimic defense (still being able to deal with the differential-mode attacks based on target's vulnerabilities, backdoors, or Trojans).

10.3.11.1   Commercial Obstacles to Heterogeneous Redundancy

1. Burden of diversified application software

Unlike the supporting software, the application software usually directly faces the end user, so it inevitably requires frequent maintenance, version upgrades, and function expansions throughout its life cycle. Therefore, it is difficult to imagine that the system provider or service integrator would gladly use the functionally equivalent, diversified application solutions from multiple parties, because it is difficult for them to coordinate multiple software providers to synchronously modify, update, and upgrade the nonstandard and diverse application software. But this is one of the ideal conditions for mimic defense. For example, Huawei cannot use the protocol processing and control software from its competitors – such as Cisco, Juniper in the United States, or ZTE (even if the other party is willing) – as the functional equivalent executors in its mimic routers, nor can it afford the burden of managing multiple versions of application software of third parties.

2. Provision of cross-platform heterogeneous software versions

Except for Java-type script files that are not related to the operating environment, the executable codes compiled and generated by the traditional source program vary in different environments such as operating systems, CPUs, and databases. For example, they differ in operating system library functions, application programming interfaces (APIs), and even custom macros. Such differences usually demand some extra modifications and debugging if the source programs are to run on a cross-platform heterogeneous environment. This will undoubtedly increase the burden of version management and maintenance of software products and therefore make it less appealing to manufacturers to issue heterogeneous versions unless there are sufficient commercial incentives. In other words, as the law of the market dictates, even when there is a need for diversification, the cross-platform software versions usually tend to support only a few mainstream operating systems or processing environments.

3. Composability of heterogeneous scenarios

The issue of composability of heterogeneous defense scenarios must be taken seriously. In theory, the more the mimic defense scenarios and the greater the dissimilarity, the stronger it will be in defending against attacks. However, the real-world information ecosystem does not allow arbitrary combination of various CPUs, OSs, databases, application software, and related supporting software, due to difficulties such as the unsolvable software legacy issues, tough problem of forward and backward compatibility, and technical barriers manually setup. Therefore, the abundance and dissimilarity of the defense scenarios can hardly be achieved by arbitrarily combining the basic hardware and software component resources. The engineering implementation of mimic structure must try to find the optimal solution under the constraints.

4. Single-source software/hardware components

In theory, if single-source hardware and software components, units, or subsystems cannot be avoided in a diversified scenario of mimic defense configuration, it may affect the anti-attack performance of the mimic architecture because of the hidden design flaws or malicious codes, for even if there is a problematic homogeneous component in the "dis-cooperation" space of dynamic heterogeneous redundancy, it is difficult to form or maintain a stable attack escape without coordinating mechanisms or available synchronization conditions. A possible exception is that the same problematic component exists in most or all of the defense scenarios in the current service set and can satisfy the attack reachability condition at the same time.

In practice, a single-source situation is often inevitable. In particular, because of the pursuit of higher, faster, and better user experience in the Internet era, and because the pioneers of product technology always try to extend the high-profit window of first mover and strive for exclusive competition or other means − which is no longer a commercial secret anymore − it will inevitably lead to a personalized and specialized monopolistic pattern of market supply, which will further aggravate the impact of single-source issues. In addition, users will naturally feel reluctant to use multiple versions of hardware and software in order to reduce maintenance complexity and cost. Therefore, the promotion of mimic systems must address the issues of providing, using, and maintaining integrated software and hardware versions.

10.3.11.2   Locking the Robustness of the Service

The fundamental goal of mimic defense is to ensure the robustness of a given service, that is, to meet the functional and performance requirements of reliability, credibility, and availability. Therefore, the service functions to be protected by the mimic brackets should be clear, as is the direction. For example, the mimic router should protect the route-forwarding table from malicious tampering. The mimic file system should protect the metadata used for directory management from unauthorized reading or modification. The mimic domain name service system must ensure that the domain name resolution function is reliable and credible. The mimic web server must ensure the correctness of the access and response services. We know that the traditional system requires complex processing links and related supporting systems to achieve the above objectives, and any mishap at any link may affect the function or performance of a given service. However, the mimic system is different in that its heterogeneous redundant processing environment has not only fault tolerance but also intrusion tolerance functions, unless the processing environment can produce completely consistent output errors under non-cooperating conditions. In other words, as long as the mimic system can lock the robustness of the service, non-cooperative mishaps of any cause within the mimic boundary would not be an issue.

10.3.11.3   Achieving Layered Heterogeneous Redundancy

It is required that in the mimic system, each executor must provide the same service function, and its performance must meet the indicator specifications, while the implementation algorithm or structure should be as different as possible. In general, there are at least three layers of heterogeneous redundancy. The first layer requires that all functions between the executors are heterogeneously redundant, the second one requires that the service functions under protection are heterogeneously redundant, and the third requires that only the core functions or data of the services under protection are heterogeneously redundant. Obviously, the three levels of heterogeneous redundancy vary widely in cost and security levels. Based on the definition of the principle of mimic defense, we do not want too many problematic homologous components in either the executor itself or its component entity or virtual body, even though there is no "absolute reliability and credibility" constraints in them. The first layer is ideal, but it is difficult to avoid problematic homologous components in the other two layers. Our goal is to use the "dis-cooperative" DHR environment to make it difficult for attackers to achieve mostly consistent or completely consistent coordinated attacks based on problematic components. To this end, it is crucial to deal with the following two problems: (1) the heterogeneous redundant program code for a given service cannot be easily read or modified by unauthorized operations (including super privileged operations), and (2) it cannot be easily bypassed or short-circuited by unauthorized operations.

10.3.11.4   SGX and the Protection of Heterogeneous Redundant Code and Data

The mechanism for time-sharing processing resources is a classic mechanism of most information systems. The operating system or management program provides support for the application layer software to economically utilize or share the computing, storage, communication, and other software and hardware resources of the system, and as a result, it generally has "super control" over the target environment. Because of this, traditional attack methods always try to gain control over the target operating system in order to acquire, tamper, and destroy the user program code or sensitive information or even control the overall behavior of the target object. In addition, due to the time-sharing mechanism, users often lack effective means to control key operating codes and sensitive data for the full period of time. But the SGX architecture based on trusted computing proposed by Intel in 2013 (see Chap. 4) makes sure that only the users have the right to invoke their program codes and data stored in the SGX. Other programs, including the operating system, do not have access to this area. In theory, its security perimeter is only related to the CPU, unless if there is an unsafe or malicious code in the SGX. Therefore, the use of such functions as SGX can greatly simplify the implementation complexity of the mimic system. We only need to add heterogeneity to the algorithm through which the service functions (or critical path, core code, sensitive data) protected in the mimic brackets are realized and install it with the relevant data in the SGX area corresponding to each executor. Due to the existence of the mimic ruling, except for the given service function or relatively correct operation result, it is difficult for the dark features in each execution environment to create sustainable attack escape by reading or tampering with program codes and data within the SGX (although there have been publicized instances where SGX were hacked, here in the mimic environment, we require that the attacks can lead to coordinated and consistent compromise). Of course, the prerequisite is to ensure that the program code in the SGX to which each executor belongs is not "bypassed" or "equivalently replaced."

10.3.11.5   Avoiding the "Absolutely Trustworthy" Trap of SGX

In fact, at the outset, the trusted computing technology has to face the question of whether the trusted computing root is credible. SGX is no exception, because SGX is also composed of CPU privileged instructions. In theory, unintentional negligence or intentional behavior by CPU manufacturers (including relevant tool providers) can cause security problems. In addition, it is also technically challenging for users to ensure that the program code loaded into SGX has no vulnerabilities or traps (backdoors or even Trojans that come along with the middleware, reusable modules, or open-source code). The issue may be especially serious when function complexity and code size in SGX reach a certain level. However, in the mimic environment, the functionally equivalent heterogeneous program code and the completely independent key are stored in the SGX associated with each executor. According to the mimicry principle, as long as the code and data stored in the SGX have no homogeneous security issues, even the exploitable vulnerabilities in the "God's eye" or memory management unit (MMU) of the x86 CPU will find it difficult to affect the generalized robustness of service function of the target system. This is determined by the endogenous security effect of the mimic architecture, which is based on relative correctness or consensus mechanism.

In summary, by using the SGX architecture or similar technologies and through effective anti-bypass or "replacement" measures, we can simplify the implementation complexity of the mimic system and reduce its dependence on the diversity of the ecological environment and on heterogeneity of the supply chain of the hardware and software components. However, this also means that there are a large number of homologous or even identical hardware and software codes (including problematic codes) in the system, which theoretically would lead to risk of decline in security level, and it is therefore necessary to carefully weigh the pros and cons, gains, and losses.

10.4 Testing and Evaluation of Mimic Defense

10.4.1 Analysis of Mimic Defense Effects

There are roughly three independent levels in traditional information security: secure management, secure protocols, and secure implementation. Each level requires different defense mechanisms and methods, and naturally they have corresponding attack mechanisms and methods. For example, TCP semi-connected DDoS attacks are for protocol-level vulnerabilities. SQL attacks target the implementation-level vulnerabilities, while weak password attacks target the management-level vulnerabilities. Mimic defense covers all three levels, with different effectiveness or certainties of the defense.

10.4.1.1 Definitive Defense Effect Within the Interface

Ideally, the mimic architecture theoretically has a definitive defense effect on known or unknown threats based on backdoors or vulnerabilities in the mimic interface. When dissimilarity, redundancy, dynamism, and randomness can be guaranteed, it is a controllable small probability event that attack escape can be attempted under the condition of output vector ruling-based PAS and multi-dimensional dynamic reconfigurable negative feedback mechanism. The inherent uncertainty effect makes the frequency and repeatability of the same escape event uncertain; therefore, any attack effect is not useful in the planning sense. If a password-protected system is cracked, then everything is exposed. But for mimic defense system, the apparent uncertainty attribute makes the dark features appear irregular, whether explicit or implicit, which can greatly increase the difficulty of collaborative utilization under the non-cooperative conditions. It cannot be emphasized enough that CMD can only effectively suppress or control the generalized uncertain disturbance problem of heterogeneous executors in the mimic interface.

10.4.1.2   Uncertain Defense Effect on or Outside the Interface

If there are defects, backdoors, or trapdoors in the design of protocols, procedures, interfaces, functions, algorithms, and even applications in or outside the mimic defense interface of a device, then the same issue may show up in all defense scenarios or implementation structures within the mimic brackets (that is to say, uniform defects may undermine the independence requirements of the design of the dissimilarity between the executors). If attackers can successfully exploit such defects or backdoors (such as TCP/IP protocol vulnerabilities or hidden backdoors in encryption algorithms), then the CMD mode should not have any deterministic effect at the mechanism level. In that case, it can only be compensated according to the principle of human society, "legal loopholes can only be amended afterwards."

It should be noted that some of the problems on the mimetic interface are often translated into problems within the interface. For example, in software programming, issues such as whether the upper and lower interfaces of the array structure are checked for limitation, or whether the stack is checked for maximum capacity, are a matter of compliance. However, in engineering practice, experienced programmers and inexperienced code writers may have completely different approaches, and the errors caused by such differences may be detected and blocked by the mimic defense mechanism. For another example, the protocol does not strictly prohibit the request for querying the target object version information, and attackers can generally obtain the information through scanning. However, for the mimic system, the version information of the executors is definitely different due to the heterogeneity; therefore, the mimic ruling will reject such scanning requests. Generally speaking, as long as cooperation from executors inside the mimic system is involved (such as elevation of privilege), any exploitation of software and hardware vulnerabilities outside the mimic interface could be blocked by the mimic mechanism unless it can ensure coordinated escape. In other words, mimic defense may be effective against security issues on or outside the mimic interface, but the effect is uncertain.

10.4.1.3   Uncertain Defense Effect Against Front Door Problems

Intentional or unintentional, the effect of the mimic defense against security problem introduced by the user from the front door is uncertain. For example, there may be vulnerabilities and backdoors in the executable file or script file downloaded by users and accepted by most heterogeneous executors, or the file itself could be a virus or Trojan. There may also be cases where the malicious function is pushed through the "front door" in a "user-approved" cross-platform script file. Such situations should all be regarded as security issues outside the mimic interface. In theory, there is the possibility of escaping the mimic interface, but the vulnerability/backdoor problem is generally strongly associated with the operating scene and operating mechanism. Escape effect will also be uncertain due to the difference in the "dis-cooperative" DHR execution environment and the difficulty in achieving collaborative effect of escape. It should be emphasized that the "front door" function provided by the mimic device manufacturer for remote service support should not be listed here.

10.4.1.4 Uncertain Social Engineering Effects

(1) Difficult to determine the effect of "ferry attack." For example, insiders may violate the security regulations (or in an act of espionage) by using a USB flash drive or network ports to "ferry" the attack code from external networks into the intranet system. The heterogeneity of mimic defense makes the environment of the executors diverse or versatile, the redundancy makes the available quantity of executors uncertain, and the dynamic and randomness enhance the heterogeneity and the uncertainty of the number of redundant bodies both in space and time. Therefore, it is very challenging to import the appropriate attack code into the appropriate executor at the right time. In other words, even if the attack code can be successfully imported into an individual executor, or the same code can be imported into different executors, such "ferry attacks" will be mechanically ineffective as long as the "coordinated attack on dynamic diverse targets under non-cooperative conditions" cannot be achieved. By consciously utilizing this intrinsic effect, it is possible to guarantee security by requiring several different keys to open the same lock at the same time.

(2) Invalid infiltration or propagating attacks. Due to the isolation requirement of "dis-cooperation" and the heterogeneous operating environment, there is no direct communication link or identical execution environment between executors. Therefore, the wormlike replication and propagation mechanism are blocked or disabled. Even if a sophisticated attacker can exploit side-channel or hidden-channel transmission techniques like CPU caches, the replication and propagation would still become invalid because of heterogeneity.

(3) Weak spot against coordinated attacks. The DHR architecture emphasizes that design flaws available to attackers are spatially independent and cannot be temporally exploited by the same attack or used continuously. This means that if there is an unknown vulnerability with strong correlation or a backdoor that can achieve the effect of coordinated attacks, then a common state fault may occur in a single attack or a group of attacks. For example, if most of the executors are implanted with "shutdown backdoors," or if the attacker has taken hold of high-risk vulnerabilities of most executors and can obtain "super privilege," then ergodic persistent attacks can also cause "reboot" of executors within the mimic brackets, thereby causing the situation that the system service function to be effectively blocked or downgraded. In other words, how to prevent opponents from using social engineering methods to learn about the vulnerabilities of most executors or to prevent opponents from prepositioning lethal or "China Chopper" attacks in most executors becomes a new security issue. Even if the attacker lacks the ability to paralyze all the executors, the

service ability of the target object will be degraded due to the continuous cleaning of the executors, and the result could be equivalent to denial-of-service attacks of varying degrees.

The author believes that the biggest security threat in the mimic system is the kind of continuous paralysis or attacks that disrupt service functions of executors, especially attacks from "auto-shutdown or inside-job function" hidden in a single-source application software. More generally speaking, if there is an environment-independent vulnerability or backdoor in a single-source application software, or if the software itself is a malware with Trojan function, and the attack target is limited to its own service range, then the mimic defense is invalid in theory, but it can indicate that the problem can only come from the application itself and is not related to environmental factors. In short, developments such as the diversification of market supply, open source of technology products, turning black-box products into white-box products, and the standardization of network services are not only problems for the network economy but also constitute the diverse ecological environment that need to be created or cultivated for the mimicry technology industry and the cyber security industry.

10.4.2   Reference Perimeter of Mimic Defense Effects

The DRS architecture can be seen as a DHR architecture in which the heterogeneous executors are statically configured or in a temporarily stable state. There are already mature mathematical models and reliability analysis methods under the "nonhomologous" conditions, such as Markov processes and generalized stochastic Petri net. The DHR architecture is essentially a DRS architecture with additional PAS and multi-dimensional dynamic reconfiguration negative feedback mechanism. Under the same circumstances, when dealing with external attacks or internal infiltration based on known or unknown vulnerabilities, backdoors, and Trojans, the DHR architecture is, in theory, more effective than static DRS, though its best performance is no better than the case where random physical faults occur to all heterogeneous executors in the corresponding time. The considerations here are based on the assumption that the executors under attack can be recovered after malfunction, including active cleaning or random reboot, or reconfiguration, reorganization, reconstruction, of executors in the offline state and that the recovery time is negligible (in practice, it must be considered that the attack interval is likely to be shorter than the recovery time).

It should be emphasized that in the DRS architectural failure analysis, it is assumed that the executions are completely different and there is no common-mode fault (in engineering, since dissimilarity is difficult to quantify, this assumption is only for theoretical purposes), and then there can be only two cases that may cause an architecture reliability to fail: one is when most executors (more than half) have multiple differential-mode faults at the same time, and the other is when the number of faulty executors is greater than $(N - 1)/2$. Unlike DRS, DHR does not emphasize that the executors be completely different, so the common-mode failures are possible. However, the complete inconsistency of multimode output vectors does not mean that the architecture's security defense and reliability have failed, because policy ruling mechanism can enable parameters, such as confidence, weight values, and scene counters, for backward verification decisions. Conversely, when multimode output vectors show mostly consistent or identical errors, the attack escape can be detected in DRS but not in DHR, even though the probabilities are low or extremely low in both cases. Therefore, the purpose of introducing the mimicry strategy based on the negative feedback mechanism, such as dynamicity, diversity, and randomness, and of introducing the command disruption loop mechanism in the DHR architecture is to reduce the probability of common-mode faults or attack escape through the uncertain operating scenarios within the structure.

10.4.2.1   Ideal Effects of Mimic Defense

In theory, the heterogeneous executors within the mimic interface can achieve the "ideal effect" as long as there is no dark feature intersection outside the given functional intersection and that they can always be recovered after the attack. In this case, the possibility of mimic ruling escape has nothing to do with the complexity of the dark feature and is only related to the consistent expression of the output vector in time and space formed by the dark feature between the heterogeneous executors. The DHR architecture significantly enhances the dynamicity, heterogeneity, and randomness attributes through PAS and multi-dimensional dynamic reconfiguration negative feedback mechanism based on the output vector ruling, focusing on reducing the escape probability of most or consistent output vectors due to dark features or limiting the possibility of escape achieved through dark features carrying sensitive information of the target object through a normal output response sequence (such as tunnel through).

10.4.2.2   Reference Range of Defense Effect

As the DHR architecture increases uncertainty in the DRS by introducing dynamization, randomization, and diversification as well as adaptive feedback control mechanism, its effectiveness against network attacks should be significantly higher than the DRS, but its effectiveness cannot go beyond the limitations of hardware reliability of dissimilar redundancy structure under the same redundancy condition. If the lower interface of defense effect is based on the probability that a scenario is compromised by coordinated attacks, the upper interface should be the probability that all scenarios can be successively compromised, but this probability is likely to be lower than the probability of simultaneous permanent physical failure in systems with same redundancy, which is why it is appropriate to use the latter as the upper interface.

1. Reference lower interface of defense effect

The dissimilarity redundant system (which can be seen as a differential scenario of the DHR architecture) has the minimum protection capability with service function protection as the goal. If we consider both the vulnerabilities induced through design flaws and the implanted backdoors as vulnerabilities, then the system under this scenario is secure, as long as attacks based on the vulnerabilities in the mimic brackets cannot make the majority of the heterogeneous executors exhibiting completely or mostly identical faults or assuming that different vulnerabilities need to be exploited in different ways and that there is no possibility of coordinated attacks. With reference to the DRS reliability evaluation model, we can assign the assumed probability that all executor output vectors show completely or mostly identical errors to the failure rate calculation formula under the condition of a given redundancy and non-incentivized feedback loop. The value we get from the formula is the lower reference interface of mimic defense effect when dynamic heterogeneity is disregarded. This reference value may be quite conservative in practice because the probability of error consistency in the output vector of executors is strongly related to the dynamic, diverse, and random mechanism of the negative feedback control, that is to say, the consistency error itself can be described with high-order probability function (related to dissimilarity of the executors, abundance of semantics and information of the output vectors, quantity of executor redundancy, ruling policy, and so on).

In fact, in the dynamic heterogeneous and dis-cooperative operating environment, it is unimaginable to achieve consistent escape in the mimic interface without trial and error. Besides, the negative feedback control mechanism based on mimic ruling is always trying to replace, clean, recover, or even reconfigure and reorganize abnormal executors in order to converge or iteratively "eliminate" the inconsistencies of output vectors, disabling the trial-and-error application for lack of the prerequisite that the scenario remains unchanged. Different from DRS, mimic defense can effectively resist this kind of trial-and-error-based collaborative attacks. In addition, for the attackers, because of too many uncertain factors, even if an attack escape succeeds in one mimic scene, the successful experience is often not replicable in other mimic scenarios. Therefore, once the attack task is so complex that it needs to overcome the mimic defense system composed of multiple heterogeneous scenarios, the probability of successful escape can only get slimmer.

2. Reference upper interface of defense effect

Although the DHR system is a DRS system with a dynamic feedback structure, it still has a dissimilar redundancy attribute. Therefore, no matter how uncertain the system structure representation – caused by dynamized, diversified, and randomized scheduling policy and reconstruction strategy – is, its protection effect cannot exceed the limitation of physical reliability of DRS system under the same redundancy condition, with the limitation of all executors experiencing irreversible physical failure at the same time. Therefore, it is suitable to use the reliability limit of the dissimilar redundancy as the reference upper interface of the mimic defense effect.

10.4.3   Factors to Be Considered in Mimic Defense Verification and Test

Many problem scenarios of a mimic system are isomorphic or similar to failure or failure scenarios of a fault-tolerant system. In particular, in the mimic structure, the non-cooperative differential-mode attack scenarios (based on personalized vulnerability backdoors or virus Trojans) can be normalized to system-level differential-mode faults, the coordinated common-mode attack problem can be normalized to the problem of homomorphism or common-mode faults in the reliability field, and the unknown security threats based on components can be normalized into reliability processing problems with controllable failure probability at the system level. In summary, the   true relatively axiom (TRA) can translate known or unknown security threats against individual members into system-level reliability issues, thus stopping the traditional and nontraditional security threats in an integrated manner in a mimetic architecture with reliability attributes. There is no doubt that reliability theory and related test verification methods can be used as the basic theory and methods for CMD to test and measure its anti-attack and reliability. The author believes that the test method of the future mimetic system may be diversified, but the basic test verification process should follow the following steps.

(1) The mimic structure system is first and foremost a service system with a given functional performance, it needs to meet the relevant equipment standards and test specifications, and its service function performance needs to pass the conformity and consistency test and verification.
(2) Based on the mimicry principle, perform routine security test and injection verification according to the test specification and related test set of target object's mimicry function performance. For the injection verification in the test set, the functions and effects of the test cases should be controllable and verifiable to the host, and the test set must include at least individualized test cases and homologous variant test cases.
(3) Select the injection test case in the test set to verify the completeness of the mimic mechanism of the target object. For example, inject individualized test cases into different executors while ensuring that there are no identical or homologous test cases between executors, regardless of the configuration or combination of test cases; the effectiveness of the mimic defense mechanism should not be affected.
(4) Select the homologous variant test cases in the test set, inject them into more than half of the heterogeneous executors in the system, observe, and count the attack escape probabilities under various combination tests (similar to the common-mode rejection ratio test in the differential amplifier).
(5) Besides verifying the completeness of the mimicry mechanism, it is also necessary to test the performance of the mimic defense, which includes the convergence speed of the mimic ruling-based PAS and multi-dimensional dynamic reconfiguration negative feedback mechanism, the service response delay

caused by the input and output agents and the ruling, the perceptibility on the mimic interface of the defensive behavior in the mimic brackets, the preparation time for the abnormal executors to clean and recover to the available state, and the speed at which the executors reconfigure and finish synchronized tracking and enter the standby state.

10.4.3.1 Background of Testing

In theory, the environment inside the mimic brackets allows for viruses, but will not cause substantial reduction or complete removal of the inherent vulnerabilities/ backdoors (traps) (under certain combination conditions they might reemerge). Different from the traditional security review approaches, the mimic defense effect emphasizes the lockability of the backdoor/vulnerability in the mimic brackets, the reliability of the attack chain based on the backdoor/vulnerability of the defense target, the detectability of the defense behavior on the mimic interface, the continuous impact on system function and performance, and the protection of the integrity of sensitive information. The core is to increase the detectability of vulnerability/ backdoor and the difficulty of exploitation and reduce the threat from Trojans in the mimic brackets while ensuring the service function (or performance) of the target object. The effect of mimic defense has a strong correlation with the system reliability index. As the inconsistency of multimode output vectors caused by external attacks is almost identical to the phenomenon caused by random differential-mode fault of heterogeneous executors, they are handled in similar ways.

The fault injection method is usually adopted in the reliability test and verification. It is a process through which a fault is manually and deliberately generated according to a given fault model and applied to a specific target system in order to accelerate the occurrence of error and failure of the system. In the meantime, it collects information about system's response to the injected faults and analyzes the recovered information to give out results. The mimic structure can normalize the differential-mode attack into a differential-mode fault, the unknown threat into a random fault, and the cooperative attack into a common-mode fault. Therefore, the basic principles and methods such as fault injection and related simulation tests are also applicable to the test and verification of anti-attack function and performance of the mimic system. To this end, when conditions permit, you should try to inject the test case into the executors in the mimic brackets, confirm whether the reaction of executors is consistent with the expected result, and then observe the appearance in the mimic interface under the same conditions. The injection test method under the "white-box" condition can be used as the basic test method of mimic defense equipment. But scientific and technical issues remain as to the rationality of test case, appropriateness of injection position, and the accurate quantification of mimic defense results.

It should be emphasized that, in a purely academic sense, the traditional security protection measures should be excluded when testing the "endogenous security gains of mimic defense." However, it can also be proved both in theory and in practice that convergence of mimic defense and traditional defense measures can often significantly enhance the dissimilarity in the mimic brackets, thereby exponentially improving the defense capability of the target.

10.4.3.2   Principles of Testing

As the endogenous security of the mimic system is based on a trusted security mechanism under the reliability architecture, it needs to be compared with the control system under the same functional and performance requirements (e.g., comparison between mimic routers and traditional routers). It should be reviewed in terms of efficiency, cost-effectiveness, safety, availability, and maintenance according to the same reliability and availability indicators as required for the traditional counterparts. In particular, the combined cost of the target system throughout its life cycle should be considered. In addition, when reviewing the endogenous security mechanisms, it is often necessary to eliminate the effects from traditional security measures or methods such as firewalls, encryption authentication, black and white lists, and sandbox honeypots. However, in practice, it is difficult to strip off these security technologies that are already being used in combination with each other. We can draw the conclusion that the injection test should probably be the main method for mimic system review.

1. Robustness of target system

Fundamentally speaking, mimic defense protects the service function and related performance of the target system (rather than the security of a certain hardware and software execution component). This means that mimic defense aims to solve the problem of robustness of service provision and tries to respond to security threats based on vulnerabilities and backdoors of the target while, at the same time, improving reliability in the traditional sense, including enabling the defender to manage or respond to the threat of known threats or uncertainties, making the target system more flexible or resilient. In other words, mimic defense converts the effect of executor's vulnerability-based non-coordinated attacks into the issue of effectiveness of attacks at the target object level and normalizes it into a random failure caused by physical failure and by hardware and software design defects.

2. Disruption of attack chain

In theory, the endogenous security effects constructed by mimicry defense can cover all stages of the attack chain. The dynamics, diversity, and randomness established in the adaptive feedback architecture make the detected information (obtained during the vulnerability scanning phase) lose its authenticity and even become uncertain; the same mechanism will affect the effectiveness of any traditional methods and tools when exploiting the vulnerabilities. In the attack implantation phase, due to the uncertainty of the presentation and locking of the apparent vulnerability, the accurate implantation or injection of the attack code becomes a new challenge that is difficult to overcome. At the attack breakthrough stage, due to the existence of the output vector mimic ruling mechanism under the condition of "dis-cooperation," any successful attack must accomplish accurate coordination in a dynamic redundant space under non-cooperating conditions in order to complete a mimic escape. In the attack maintenance phase, the mimic escape can be achieved only if each attack is successful during the entire life cycle of the attack mission. However, the PAS based on mimic ruling and the multi-dimensional dynamic reconfiguration negative feedback mechanism means the attack escape is surely a highly uncertain event.

3. New mechanism of integrated defense

Assuming that the nontraditional security risks of sensitive paths or related links are not fully controllable or fail to meet the requirements of high credibility, the introduction of mimicry mechanisms allows us to use a unified architecture to integrate defense technologies to fully impact the various stages of the attack chain. It combines the non-specific endogenous defense mechanism with specific active and passive security measures and combines surface defense mechanism with point defense means, to effectively avoid the known unknown security risks and the unknown unknown security threats and nonlinearly increase the cost for the attackers. The one-way or unidirectional communication mechanism transforms centralized processing into linear or treelike decentralized processing and splits or constrains the information acquisition authority of a single node, so that no node can obtain the overall view or global data. The dynamization and randomization mechanism based on adaptive feedback control in the DHR architecture make the mimic field (including the backdoor or Trojan virus attached to it) as dissimilar as possible, which can significantly increase the complexity of the spatial and temporal presentation of the mimic fields. By adopting the output vector mimic ruling mechanism under the condition of "dis-cooperation," the deliberate attacks from inside and outside must overcome a number of obstacles of mimic escape, such as feasibility and redundancy, dissimilarity, semantics and information abundance of output vectors, and the diversity of ruling algorithms. The attacks must also face the challenge of accurate coordinated attacks and successive escape between domains under the non-cooperative conditions in the dynamic heterogeneous space (across physical and logical domains). All of these new mechanisms have a disruptive impact on traditional means of attack.

4. Cleaning and preventive treatment

Attacks based on vulnerabilities and backdoors can be normalized to reliability and availability issues in target systems with the DHR architecture. In mimic brackets, if output vectors can be obtained from the mimic interface and if they could be determined to be the behavior of an individual or a small number of executors, then they can be regarded as unstable working components (in the case of small probability, such an approach may not be very appropriate) and then subject to forced halt, roll back and repetition, initialization at corresponding level or offline cleaning check, or even reconfiguration and reorganization. More generally, an executor needs to be pre-cleaned or initialized at the corresponding level or carry out environment reconstruction as long as it is offline due to dynamic random scheduling. It also includes a background processing mechanism such as scanning and clearing of triggering Trojans to avoid the influence of potential vulnerabilities/backdoors or uploaded Trojans in the current scenario or to block latent (standby) attacks based on state transition mechanisms. This means the attackers have to carefully plan each step of the collaborative attack; otherwise any mishap may be "cleaned up" once it is discovered by the mimic ruling. At the same time, it is also necessary to effectively avoid the destructive effect of the pre-cleaning mechanism on the stability of the attack chain. Obviously, these are very challenging issues for attackers.

10.4.3.3 Major Testing Indicators

1. Apparent complexity of attack chain

For the attackers, the unknown nature and complexity of the mimic presentation (structural representation) of the target environment pose a series of nonlinear difficulties in terms of detecting and exploiting vulnerabilities, setting up and triggering backdoors, and the coordination of attack behaviors. One of the quantifiable indicators in mimic defense is to objectively evaluate the complexity of the attack chain.

Open the defense mechanism of all the executors in the mimic brackets by following the "white-box instrumentation" approach of the injection test case and artificially construct an attack chain according to the service function, performance, and incentive rules of the given external channel (attack surface) and system so that it can exhibit a prescribed attack effect according to certain expected probability, without altering the service function of the system under test and without destroying the defense mechanism. An example of such an attack effect would be an event that can cause an impact on the information security of the target system. Obviously, the apparent complexity of this attack chain is the black-box expression of all the conditions, assumptions, and software and hardware code revision and all the resources necessary for the attack chain to be effective. It is not hard to imagine that an attack that could cause the same functional problem in all heterogeneous executors, respectively, or simultaneously (such as the shutdown-dead loop problem) seems to be less complex than the apparent complexity of an attack that causes identical errors in a given output vector space simultaneously. In fact, however, even if only the effect of the former attack is to be achieved, it requires that each executor must have all the resources, methods, and means to enable this function, which is a serious challenge for any attacker. It is even more difficult to achieve continuous or high probability mimic escape from the target object.

2. Difficulty of coordinated attacks

In theory, to break through the mimic defense, an attacker must be able to cause the output vector of most heterogeneous executors to exhibit identical errors under the condition of uncertainty and based on the defined attack path and conditions. This requires the attacker to not only accurately understand the vulnerabilities and backdoors of multiple diverse targets but also be able to lock or control the vulnerabilities/backdoors inside the target object in real time and to steer these loopholes to achieve coordinated mimic escape. The dynamic and random nature of mimic defense creates difficulties for vulnerability scanning or locking. Heterogeneity differentiates the vulnerabilities and the ways they are to be exploited. Redundancy creates time-space independence or isolation between heterogeneous executors of the target object. The mimic ruling-based adaptive feedback control mechanism forces the attacker to face the challenges of dynamic space, heterogeneous multi-target, multimode, and coordinated escape. These are extremely demanding conditions. Under the influence of so many uncertainties, the success rate of attacks is very low with the exception of functional failure such as shutdown or dead loop. However, if an attacker has multiple resources in the mimic brackets and can properly exploit the attack surface reachability, then mimic escape is not completely impossible. Please take care not to let all the executors go online simultaneously in engineering practice.

3. Difficulty in apparent vulnerability detection

Design flaws are inevitable for complex systems. System vulnerabilities can be categorized as shown in Fig. 10.3 based on the presentation of system flaws in the mimic or non-mimic situations.

(1) Inherent vulnerabilities: vulnerabilities detectable in mimic structure systems when in non-mimic mode or in all components inside the mimic brackets.
(2) Apparent vulnerabilities: vulnerabilities scanned and detected in the same way as inherent vulnerabilities from outside the mimic brackets or from attack surface when the system is in the mimic mode.
(3) Exploitable vulnerabilities: apparent vulnerabilities that can be exploited and meet the following conditions: the apparent vulnerabilities can be presented

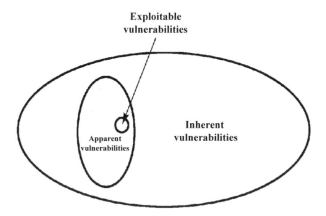

Fig. 10.3 Categories of vulnerabilities

stably in a predictable pattern; the exploitable nature of the apparent vulnerabilities matches the corresponding vulnerabilities of the inherent vulnerability set.

For the attackers, in the mimic defense mechanism, the external representations of vulnerabilities in the mimic brackets – including the presentable quantity, the display frequency of the same vulnerability, and the nature of exploitation – are subject to the influence of a number of mechanisms, such as dynamism, randomness, virtualization, and multi-dimensional dynamic reconfiguration and pre-cleaning. As a result, the apparent nature of the vulnerabilities would inevitably go through uncertain changes.

According to the traditional vulnerability classification method, the general vulnerabilities, low security level vulnerabilities, and high-risk security vulnerabilities in the non-mimic mode may no longer show the original form or nature in the mimic environment, with the high-risk vulnerabilities possibly changing to low security level ones or even completely becoming unavailable. Surely, new security vulnerabilities may also arise (due to dynamic factors such as reconfiguration, reorganization, virtualization, etc.). Therefore, in the white-box test, not only the number of test cases (vulnerabilities) should be confirmed, but also the nature of the vulnerabilities should be checked as much as possible.

After the mimic mechanism is activated, if an apparent vulnerability is detected (it may not be reproducible within a given time), we should try to distinguish whether its nature matches the prescribed test case and to determine the characteristics of the new vulnerability, its spatial and temporal stability, and its security level. The mimic mechanism should be able to change or influence the manifestation of the vulnerabilities and reduce the level of security threat, but its scientific measurement remains to be further studied.

It should be noted that since a rigorous scientific model has not been established, it cannot be rigorously proven whether the mimic ruling-based PAS and negative feedback control mechanism of multi-dimensional dynamic reconfiguration are linear transformations, whether they would lead to new security loopholes, especially when the environment inside the mimic brackets contains viruses. What kind of new vulnerabilities may appear after the mimic changes is currently unpredictable. However, what is certain is that even if a vulnerability of new nature is generated under the negative feedback control mechanism, it should equally have an attribute of uncertainty and will be difficult to be used in a coordinated way.

4. Difficulty of vulnerability exploitation

According to GB/T 30279-2013 "Information Security Technology – Vulnerability Classification Guide," vulnerability classification is mainly based on three factors: access vector, exploitation complexity, and degree of impact.

Access Vector (AV): Attackers use security vulnerabilities to influence the path premise of the target system, which is divided into three modes: remote, adjacent, and local.

Exploiting Difficulty (ED): The difficulty of using security vulnerabilities to affect the target system can be divided into two categories: simple and complex.

Damage Potential (DP): The degree of damage on a target system caused by exploiting security vulnerabilities. The damage potential can be divided into four levels, "complete," "partial," "minor," and "none," based on the degree of impact on any of the three aspects, namely, confidentiality, integrity, and availability.

The traditional vulnerability classification guide is useful for vulnerability rating to a certain degree, but the division is relatively rough. For the mimic structure system, due to the uncertainty of the apparent vulnerabilities and the environmental factors of vulnerability exploitation, the complexity of exploitation and degree of impact are also different.

Complexity of exploitation (or difficulty of exploitation) mainly includes the following factors:

(1) Complexity of attack chain (CAC): the complexity of a directed graph consisting of the process of attacks by exploiting apparent vulnerabilities. It is determined by the implementation cost of each chain.
(2) Re-exploitability of attack: probability that apparent vulnerabilities can be successfully exploited again. According to the probability based on test results, re-exploitability can be divided into three levels: not re-exploitable, occasional, and inevitable. Apparent complexity of attack chain is determined by the number of medium states and the cost of each attack chain.

5. Mimic escape

In theory, the mimic structure system expects that the potential vulnerabilities/backdoors of the heterogeneous executors are strictly dissimilar and independent; therefore there is no vulnerability-based mimic escape in the DHR environment. In practice, it is impossible to ensure absolute dissimilarity between the executors, so there may be dark feature intersections between the executors. As the dynamism, randomness, and re-configurability of the operating environment are based on policy ruling, policy and schedule, and adaptive feedback mechanisms, even the same malicious codes (or dark features such as Trojans) existing between the heterogeneous executors in the mimic brackets cannot guarantee a successful escape from the mimic boundary. The "white-box instrumentation" method of test case injection – a reliability verification technique – may be used to quantify the escape effect to a certain level.

Assume that an executable code with a corresponding test function is injected into a heterogeneous executor in a given mimic structure system. In the open-loop scenario, we can get the attack success rate under the non-mimic operating mechanism, which is measurable, i.e., F_o = the number of successful exploitation/number of tests. In the closed-loop scenario, we can also get the attack success rate under the mimic mechanism, F_c = number of successful uses/number of tests. Thus, the effectiveness of the mimic defense is $F_e = F_o/F_c$, and the larger the value, the better. Of course, a more comprehensive approach is to build a test set for a given application system, then traverse all test cases using the abovementioned method, and use a suitable statistical algorithm to derive F_e under a certain degree of confidence. In the same way, we can inject the multi-version execution code generated by the test case source code in the test set into the corresponding position of all the heterogeneous executors, and after the mimic mechanism is activated, we can obtain the escape success rate by following the same method and steps, E_r = number of successful escapes/number of tests. Obviously the smaller the value, the higher the tolerance of mimic environment for the same function source code (similar to the common-mode rejection ratio test of the differential-mode amplifier).

Of course, even if the execution code of the same test function is injected, the result will have different spatial and temporal performance due to different environmental conditions such as the implantation location, the operation scheduling policy, parameter assignment, and the operation mechanism in the heterogeneous executor. It is still difficult to evaluate the completeness of the test set.

6. Security of the mimic brackets

Similar to the impact of multimode voting on system reliability in the dissimilar redundancy structure, the mimic brackets is also very crucial to the mimic reliability for they employ the PAS and MDR-NFCM of mimic ruling. In theory, the mimic brackets should be completely transparent to the content, syntax, and semantics of the attack channel in the attack surface, as well as the function and performance of the defense target. That is to say, the brackets do not parse any content in the attack channel at the semantic level; therefore, they do not have the attack reachability, unless there are already malicious codes inside the brackets.

However, it is impossible to be completely transparent in engineering. For example, in order to overcome the time difference in the output vector response of the heterogeneous executors, an output buffer queue needs to be added to the mimic ruling link; in order to solve the problem that the accuracy of equivalent algorithms may be different, a mask comparison function needs to be added; in order to increase the dynamics and randomness, it is necessary to have input exchange and assignment functions; in order to solve the situation with same semantics but slightly different grammar, it is necessary to increase the regular expression matching function; in order to support the strategic ruling, timestamp, single TCP serial number, and IPSec encryption, the proxy function is added; in order to reduce the delay of the decision insertion, some preprocessing (such as checksum or hash calculation) of the multimode output vector is required to reduce the complexity of the ruling. All of these considerations may increase the complexity of the engineering implementation of the mimic brackets, and design flaws or errors are inevitable. Potential safety issues are particularly prominent when the powerful, inexpensive, and high-performance COTS components are used. Therefore, there are several principles to follow in the mimic brackets design, such as avoid using complex general-purpose hardware and software; eliminate mixed usage of program and data space; use hardware processing as much as possible; in terms of function and performance of the mimic brackets, "sufficiency" should be the principle; implement such functions in private ways such as trusted customization or trusted self-reconfiguration or trusted computing; minimize the attack surface as much as possible, and if necessary, consider mimicry of the bracket itself; increase the transparency of the mimic brackets to their internal heterogeneous executors (to prevent the conditions for coordinated attacks that may be caused by the session function) as much as possible by introducing the one-way or one-directional communication mechanism. In short, specialized research on security testing (including the impact from the social engineering perspective) is needed, and it may be necessary to draw on relevant theories, experiences, and methods in cryptographic engineering.

7. Endogenous security gain

Unlike the traditional defense mechanisms such as intrusion detection, intrusion prevention, vulnerability mining and utilization, Trojan elimination, and encryption authentication, the mimic defense mechanism does not involve the accurate perception or timely elimination of vulnerabilities or Trojans in the executor resources inside the mimic brackets. The endogenous security gain mainly manifests itself as an iterative effect by increasing the difficulty of coordinated attacks in a multi-dimensional heterogeneous space, meaning that any non-coordinated steps of an attack that involve entering and exiting the mimic interface in a given attack task may be blocked by the mimic mechanism. This is mainly reflected in the following four aspects:

(1) It changes the nature of presentation of the target system vulnerabilities and backdoors. The PAS mechanism of the defense scenario in the mimic brackets will disrupt the lockability of the backdoors/vulnerabilities and the accessibility (or reachability) of the attack chain, increase the difficulty of exploiting the apparent vulnerabilities/backdoors, and destroy the temporal validity and functional reliability of attacks based on vulnerabilities and backdoors.
(2) It prevents attacks from spreading. Trojans or worms hidden inside the executors cannot spread the infiltration attacks due to the independence requirement of executors inside mimic brackets, that is to say, even if every executor in the system contains a Trojan virus, the "dis-cooperative" design can isolate the transmission of virus-bearing functions, so that the Trojan cannot be consistently expressed in the spatial-temporal dimension (unless in the event of a shutdown), and in theory, it would not seriously affect the system service function.
(3) It has the unknown threat detection and warning function. The policy ruling function of the mimic interface can detect or perceive unknown non-coordinated attack events targeted at heterogeneous executors. If necessary, it can trigger corresponding background support mechanisms such as log recording, on-site snapshots, tracing, and search. The proper use of data collection and analysis can efficiently discover 0day vulnerabilities and backdoors or new attack methods and even achieve the goal of traceability. This is important for increasing the awareness and early warning capabilities against unknown threats.

10.4.3.4   Considerations of Test Methods

The security test of the mimic structure system mainly follows the reliability verification theory and methods and adopts the "white-box instrumentation" comparison test method of the injection test case. The test cases in the test set are preferably designed for the vulnerability of a given target object or by simulating the attack path that an attacker may utilize as a prerequisite.

There are several basic considerations when establishing a mimic system security evaluation model:

(1) Collect the device information of weak points in the target, including information about software and hardware vulnerabilities inside the mimic interface and information about open service and product information. Obtain open information about the target's network topology structure and relevant components or parts.
(2) Generate an atomic model for each atomic attack behavior that constitutes a combined attack (see Chap. 2), and strictly define the conditions required for the transition to occur.
(3) Define the initial state of a network attack, and start from the initial state to use three operations – sequence, concurrency, and selection – to describe the process of the combined attack based on the relationship between the protocol information and the attack behavior on the mimic interface.
(4) Study the insertion mode and insertion position of test cases to achieve the purpose of reducing test complexity and explaining problems by using model-related properties or introducing hierarchical ideas, using composite transitions to represent an atomic attack behavior or combined attack.
(5) Verify if the target is capable of defending against non-cooperative or cooperative attacks.

Based on these considerations, we propose a conceptual framework of vulnerability analysis of mimic structure system, which consists of three core modules, namely, network parameter abstraction, model construction and verification, and vulnerability quantitative analysis and evaluation.

(1) Network parameter abstraction module. Abstract parameters of components of the target's resource pool, heterogeneous executor set, PAS, and other information, including potential vulnerability information, service information, and connection relationships. Such information mainly comes from vulnerability scanning tools (such as X-Scan and Nessus) or vulnerability mining techniques, big data analysis based on machine learning, or information related to network topology and target defect analysis. At the same time, build a vulnerability knowledge base according to the collected system vulnerability information, preparing the data for building a target evaluation model.
(2) Model construction and verification module. To begin with, preprocess the data of the obtained target parameters, and then use the formatted data as the input parameters for the model and use the modelling algorithm to generate the evaluation model, before finally verifying the model for correctness. After the target evaluation model is established, we can use the Petri net reachability graph and reachability tree to verify the validity of the model.
(3) Vulnerability quantitative analysis and evaluation module. Interact with the security targets of the network administrators based on the foundation mentioned

above, and select different methods to analyze the existing vulnerability in the target object according to the administrator's security needs. The approach of analyzing the best attack path can be employed to predict the optimal attack path in the target object and verify the security of the system.

In the meantime, we propose three reasonable assumptions based on the model.

Assumption 1: As a smart agent, an attacker is very familiar with the vulnerabilities in the target object and has the ability to exploit these vulnerabilities to perform attacks, which means, as long as the attack conditions are met, the attacker can successfully attack and always prefers the fastest path.

Assumption 2: In an attack event launched by the attacker, any changes in attack behavior can only be performed once in the attack path. The attacker will not launch another attack for security elements that they have already obtained.

Assumption 3: Attackers have clear targets and launch attacks to achieve their goals. The attacks are always target-oriented rather than done randomly.

Attack path usually refers to the process from the mimic interface of the target to the operational node defined by the relevant heterogeneous executors in the mimic brackets. When an attack is launched, there may be multiple attack paths based on the mimic interface (possibly unguarded) in the target object, but each attack path may be used by attackers in a different way due to the differences in difficulty of exploiting vulnerabilities and the dynamic defense scenarios. Therefore, it is necessary to analyze and compare the attack paths to find the path of the greatest security risk, namely, the path that can measure the security performance of the system.

In this respect, most of the researches at home and abroad try to find the best attack path by analyzing the probability of success. In practice, however, due to the diversity of systems, servers, and websites, it is not easy to assign an attack success probability in these methods. If the probability is not set reasonably, the result of analysis will deviate greatly. Therefore, from the perspective of attack-defense cost, in reference to GB/T 30279-2013 "Information Security Technology: Vulnerability Classification Guide," we use access path, exploitation complexity, and impact level as the basic indicators.

In order to make the evaluation results more accurate, the quantitative analysis is used to compare and analyze the test results, and the evaluation criteria for testing are proposed according to GB/T 30279-2013 "Information Security Technology – Vulnerability Classification Guide" and the Common Vulnerability Scoring System (CVSS). The evaluation is based on a number of evaluation factors including the difficulty of vulnerability discovery, the difficulty of exploiting the vulnerability, the degree of vulnerability and its constituent factors, and the factors are quantified by scoring and then compared. The level of each of the influencing factors relies on subjective estimation in the actual test, so there are certain errors, which can be corrected by referring to the existing vulnerability assessment criteria.

10.4.3.5 Qualitative Analysis of Defense Effectiveness

Most mainstream cyber-attacks exploit the vulnerabilities of specific systems (such as vulnerabilities, backdoors), the use of which requires certain conditions, and the attacker needs to have the relevant "knowledge" (e.g., knowledge of the operating system or software version, memory address information of a relevant target, etc.) before they can effectively exploit the vulnerabilities. A complete attack task consists of several steps, and each step needs to use the conditions created by the previous step, obtain the relevant "benefits," and create conditions for subsequent attacks, thereby forming a stable attack chain.

Therefore, as long as we can destroy the conditions for cyber-attacks, and prevent the attackers from obtaining the "knowledge" necessary for the attack (or make the obtained knowledge invalid during the attack), then the attack can be effectively prevented, despite the fact that the vulnerability may not be truly removed. One important difference between the mimic defense mechanism and the traditional defense mechanism is that the traditional mechanism focuses on removing vulnerabilities and blocking the attack paths (such as the network channel), while the mimic defense mechanism focuses on destroying the preconditions on which the attack depends (without limiting to the network channel) and the accurate recognition of the target object. The uncertain defense scenario created through the DHR architecture and the mimicry mechanism makes it difficult for the attacker to stably construct the coordinated and consistent scenario required for the mimic escape. Therefore, even if the target bears viruses, the robustness and credibility of the target's meta-service can still be maintained.

However, different types of attacks exploit different vulnerabilities, so their prerequisites and the "knowledge" that attackers need to master are also different. Therefore, we can sort out and classify the current mainstream attacks; summarize and categorize the preconditions that the attacks rely on, the "knowledge" that the attacker needs to obtain beforehand, the potential benefits, and so forth; then analyze the various mechanisms of mimic defense; and identify effective strategies and methods against different types of attacks.

10.4.4 Reflections on Quasi-stealth Evaluation

Stealth technology (ST), also known as invisibility technology or low observable technology (LOT), uses a variety of technical means to change the detectability information characteristics of their own targets and to minimize the probability of the detection by the opponent's detection system. Stealth technology is an application and extension of traditional camouflage technology.

The typical application of stealth technology in the field of aircraft is the stealth aircraft, which uses various technologies to weaken the radar echo, infrared radiation, and other characteristic information. At present, there are mainly three ways to create a stealth aircraft. The first is to reduce the radar reflective surface of the aircraft by technical means, like designing a reasonable exterior of an aircraft, using absorbing materials, active cancellation, and passive cancellation. The second is to reduce infrared radiation mainly through measures such as heat insulation and cooling down on aircraft parts that are prone to generate infrared radiation. The third is to use the camouflage colors to reduce the visibility to the naked eyes.

Some of the rough indicators when evaluating stealth capabilities include the size of radar cross-section (RCS), the intensity of infrared radiation, and the visible distance to the naked eye. Some finer indicators include the cross section of an aircraft in different positions and against different wavelengths of radar echoes, the spectral and amplitude-frequency characteristics of infrared radiation, the visible distance of different visibilities and illumination angles, and so forth.

Mimic defense is also a type of stealth technology. It hides its own design flaws or vulnerabilities by changing the defense scenarios inside the mimic brackets and creating ambiguity in the reachability of vulnerabilities/backdoors, thereby posing obstacles to the attackers' cognition and behavior by creating an endogenous uncertainty effect.

Correspondingly, the evaluation of mimicry capability is also the evaluation of the visibility of known (or given) vulnerabilities inside the mimic brackets (including vulnerability characteristics, nature, and changes in stability of presentation) and the evaluation of visibility of the same vulnerability from outside the mimic brackets (including vulnerability characteristics, nature, and changes in stability of presentation). The difference between the two should be one of the main indicators of mimicry capability. Similarly, the tester evaluates the exploitability of backdoors in the mimic brackets (including backdoor characteristics and accessibility verification, reliability of Trojan upload, effectiveness of Trojan control, etc.), as well as the exploitability of the same backdoor from outside the mimic brackets (including backdoor characteristics and accessibility, Trojan upload, and changes in the stability of the Trojan control). The difference between the two should also be used as one of the main indicators of mimicry capability. By analogy, you can get the evaluation data of the target's mimic defense capability by injecting test cases of different functions or test cases of the same function in each executor and collecting and analyzing the output vectors inside and outside the mimic brackets.

In summary, the testing and evaluation of mimic ability can be modeled against a control or contrast model in the invisibility test. However, factors such as the construction of the test case and the functional position of the inserted object will affect the quality of the final evaluation. Further studies are needed to scientifically test and measure the mimic defense capabilities.

10.4.5 Mimic Ruling-Based Measurable Review

According to the principle of mimic defense, mimic ruling can sense the non-coordinated attacks from the non-compliance of the multimode output vectors (including differential-mode fault), but cannot sense the escapes caused by coordinated consistent attacks. Because of the uncertainty effect, the mimic defense mechanism can block any trial-and-error type of coordinated attacks. In other words, it is generally true that trial-and-error coordinated attacks can be discovered by mimic ruling. Thus, we can measure the defense performance of the mimic system through white-box instrumentation by injecting test cases and testing the status of the arbiter. When the multimode output vectors are inconsistent (usually not perceived by the outside due to fault tolerance mechanism) or completely inconsistent, the adaptive feedback control mechanism based on PAS and MDR would always try to replace the abnormal executor in the current service set, or clean and restore, or reconfigure and reorganize the executors, in the hope that output inconsistency would disappear in the mimic brackets or that the frequency of occurrence of inconsistencies in a given observation period can be controlled within a certain threshold. Obviously, when the attack scenario stays the same, the response time to inconsistency from its occurrence to disappearance is an important indicator of defense capability of a mimic system.

The use of this indicator can be further divided. Firstly, it indicates which current defense environment of the executors can be breached by this attack, whether there is a minor or major inconsistency in the multimode output vector. Secondly, this indicator can help determine which strategy or defense scenario can cope with this attack, because the output arbiter can sense the effect of replacement of the defense scenario no matter whether an abnormal executor is replaced, cleaned, and recovered or an executor is reconfigured and reorganized. Thirdly, the indicator can measure how many iterations the defensive scenario have gone through before the convergence of output arbiter status is achieved. Conversely, when the feedback loop is disconnected, it can identify whether the output vector inconsistency caused by the attack has spatiotemporal stability and its changing trend. It is also possible to use the white-box instrumentation method in the open-loop state to change the attack scenario and monitor the status and output vector of the output arbiters. If it can lead to an attack escape and maintain it, then observe how long it takes for the escape phenomenon to disappear after closing the loop. This index can measure the anti-escape capability of the mimic defense system.

Thus, based on a given test set (which can be open to the tested object), we can categorize and add up the inconsistencies perceived in the mimic ruling phase on a timeline. For example, we can count in a given observation period the frequency with one output inconsistency; frequency with multiple output inconsistencies; frequency that is completely inconsistent; or even the frequency of a particular abnormal output of an executor. These figures could represent the current security status of the defense target, just like the MTBF in the reliability field, or symbol error rate or bit error rate (SER or BER) in the field of communication transmission. Of course, the configuration condition of the test set and the completeness of its configuration directly affect the scientific quality of the evaluation, although the nature of the problem at this time is far beyond the scope of the existing attack theories, methods, and influence. This is what we mean by saying that mimic defense mechanism "changes the rules of cyberspace."

10.4.6 Mimic Defense Benchmark Function Experiment

The test based on the expected effect of the mimic defense theory is called benchmark function experiment, also called definition-based effect test of mimic defense. There are usually two types of tests for mimic defense, one is the traditional security test, which mainly tests the vulnerability display and attack reachability, and the other is the test case injection experiment, also called "white-box instrumentation" experiment, which requires to accurately identify whether the tested target has the basic mimic defense function and whether it has the capabilities to deal with attack escape according to examples for product test standards, without relying on the experimenter's experience and skills. This section will focus on the white-box instrumentation experiment.

1. Prerequisite and arrangement

On the basis of the threat analysis and security design of the given mimic structure of a tested object, with the input path, syntax and semantics, and rules and methods in the mimic brackets as the normalized attack surface, the experiment examines if the design of security, reliability, and availability of the tested object can meet the alleged quantifiable indicators in the "white-box testing" by injecting some given test cases or test case sets through the attack path on the mimic bracket surface. It should be pointed out that the mimic structure usually contains two big types of functions and performance. The white-box injection approach can only be used to check the endogenous security function and relevant performance of the mimic structure, without causing unrecoverable damage to other functions or performance of it. The test of security, reliability, and availability outside the mimic structure is not included in the experiment.

Experiment arrangement:

(1) One agreed test case consists of two parts, namely, the test interface embedded in the target source program and the test code injected through the attack-reachable path.

(2) Any code functions of test interfaces embedded in the tested target executor or operational scenario are the same, including the functions of receiving instruction code through the attack surface path and compliant method or of updating the test content.

(3) The test codes injected through the test interface are memory-resident executable codes and should not lead to unrecoverable damage to the tested target function.

(4) The test codes shall have the ability to influence the output vector content of the resident executor or operational scenario, or to control its output ability.

(5) The test cases whose every two test code functions are different are called "differential-mode test cases," and those whose every two test code functions are the same are called "common-mode test cases."

(6) All the executors or operational scenario F can, in principle, have test interfaces, but at the same time, the number f of resident test codes shall satisfy $n \leq f \leq F/2$, and n is the redundancy of the current service scenario in the mimic brackets.

(7) During the experiment stage, open the system management interface to observe the entire experiment process.

Structure of Test Cases

Usually, the hardware and software supporting environment of the test target's heterogeneous executor or operational scenarios has only the executable target codes or even physical devices, so the test case is not only difficult to be constructed but often cannot be injected. Even if the test case can be implanted, its function verification is rather challenging. Therefore, it's a possible choice to construct the test case interface and design the invoking function at the application source code level. The test interface should be designed to have a "backdoor function" that is activated by input channels and compliance method of the attack surface in the mimic brackets and can receive the uploaded test code through the attack surface. It can perform online control of injection test code through applications. Of course, under certain conditions, the test interface and related functions may also be set up at other levels (e.g., operating system level). It should be noted that in addition to the test interface code, the uploaded target test code should be the executable code resident in the memory to flexibly define or change the test function and enhance the completeness of injection test.

It should be explained that there are two goals for setting up an experiment interface at the application level and for injecting test codes through the attack surface. One is that the experiment result is credible. Any attacks based on vulnerabilities and backdoors at any layer cannot directly or indirectly threaten the security goal of the system until they can precisely control the expression of output vectors of the target executors or operational scenarios (the ultimate goal the attacks can reach), ensuring that the application software invoke the test interface function as per the agreed requirements and activate the injected test code. The other is to observe the resource configuration of the executors or operational scenarios through the system management interface. By changing the injection test code, you can verify the CUP type, OS version, and other information of the operating environment and specify the target execution environment for the test code.

2. Differential-mode test case injection experiment

Prerequisite: Assume that the test target function satisfies $I[p_1, p_2, p_3, \ldots, p_n]O$, among which all functions p_i are the same, i.e., $p_1 = p_2 = \ldots = p_n$, but the p_i implementation algorithms are all different, that is, $p_{c1} \neq p_{c2} \neq \ldots \neq p_{ci}$. If there is a test case e_1 that could make, p_1 generate an output vector outside the normal response sequence s_1. By analogy, test cases $e_2\ e_3\ e_i$ can lead $p_2\ p_3\ p_i$ to generate output vector $s_2\ s_3\ s_i$ and $s_1 \neq s_2 \neq s_3 \neq s_i$; then, according to the definition of mimic defense effect, if test cases $e_1\ e_2\ e_3\ e_i$ which do not have a cooperative relationship with each other are injected into executors $p_1\ p_2\ p_3\ p_i$ or defense scenarios, respectively, through the attack surface, then no s_i should appear on the mimic interface of $I[P]O$. The mimic defense function of the tested target should be able to clearly indicate that, except for the shutdown-paralysis event that cannot recover automatically, the suppression effect of the differential-mode test cases can be 100% achieved as long as the "lone-wolf attack" differential-mode test case does not have a cooperative relationship,

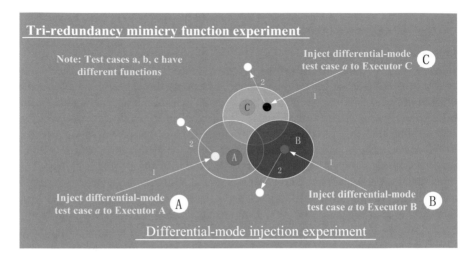

Fig. 10.4 Injection experiment of differential-mode test cases

and the test case can generate an action perceivable by the arbiter, which can be cleared or removed before the activation of another test case.

Figure 10.4 illustrates an experimental scenario with a tri-redundancy mimicry function. All the differential-mode (with different functions) test cases that fall within the three executor regions A, B, and C should be designed as scenarios that could be perceived by the mimic ruling phase. According to the feedback control strategy and the iterative convergence backward verification mechanisms of mimic structure, another expected result of the experiment is that the problematic executor or operational scenario is either replaced, reconfigured, or reconstructed or cleaned, recovered, or restarted by latching onto the executor or operational scenario with injected test cases. The related experimental scenarios and operation processes should be observed.

3. Experiment of the time-cooperative differential-mode test cases

Experiment goal: based on injection experiment of the differential-mode test cases, distribute f differential-mode test cases evenly among m defense scenarios of n executors in the mimic structure ($m \geq n$), the ratio of DM test cases f to m is $F = f/m * 100\%$, subject to $34\% \leq F \leq 50\%$. The experimenter sends the activation instruction to f test cases consecutively through the attack surface to observe the functionality and performance of the target object, focusing on (a) possible differential-mode escape in the experiment process; (b) degradation of functionality and performance of the target object in the process; and (c) how long it takes for the problem avoidance mechanism to keep relevant defense scenarios from being re-invoked.

Experiment plan: ① test the recovery time of the sampled executor defense scenarios to determine the consecutive activation strategy of f test cases; ② decide on the deployment strategy (e.g., even distribution) of f test cases in m scenarios, given the ratio F is 34%, 40%, or 50%; ③ measure the functionality and performance of the target object in the experiment process to verify the design indicators of reliability and availability, with emphasis laid on the measurement of the time of the same group of time-cooperative differential-mode test cases from "attack reachability to unreachability," which tests the validity of avoidance mechanism of the target object in problematic scenarios; ④ for other methods and operations, please refer to the injection experiment of differential-mode test cases. Please note that the values of F shall be higher than the failure tolerance limit of the classic DRS, $f \leq (n - 1)/2$. For instance, if $n = m = 3$, DRS allows $f \leq 1$, and the corresponding ratio $F \leq 33.3\%$, while the experimental values of the time-cooperative differential-mode test cases are $F = 34\%$, 40%, and 50%, which correspond to the worst condition of the DRS at $n = 3, 5$ and infinite redundancy.

4. $n - 1$ mode test case injection experiment

The experiment aims to check that the mimic structure should still have the ability to recover from the escape state in case a $n - 1$ common-mode perceivable escape (n is the redundancy of the current service set) occurs in the mimic interface and prove that the mimic structure still have the expected probability even a common-mode escape event occurs. The $n - 1$ common-mode escape is a small or tiny probability event in quantifiable design of security; however, failure to get rid of common-mode escapes in the engineering implementation of the object will make the attack experience replicable but no longer be a probability event whenever a first common-mode escape successful happens. The $n - 1$ common-mode escape experiment not only verifies the correctness of the mimic function but also test the duration of the recovery process; the latter can quantitatively evaluate the performance of resisting common-mode escapes. Also, the whole experiment process and effect shall be observable.

The injection experiment of $n - 1$ common-mode test cases stipulates that, when the number of executors or defense scenarios in the current service set of the mimic structure is n, suppose a test case t_i can be injected into $n - 1$ executors or defense scenarios and can be activated through the attack surface in a way to generate $n - 1$ same output vector s_i, then, as per the mimic defense definition, mimic escape may probably occur in the I[P]O mimic interface, but the arbiter can perceive the inconsistency between multimode output vectors. In other words, the $n - 1$ common-mode escape is a kind of perceivable escape. As a result, the feedback control phase will decide how to change the defense scenario in the mimic brackets by asymptotic or iterative convergence method according to the pre-designed backward verification strategy (the relevant auxiliary ruling strategy is needed to tell the difference between differential-mode scenarios and $n - 1$ common-mode scenarios), degenerate it into the differential-mode experiment form as shown in Fig. 10.4, and finally remove or clear it.

Figure 10.5 illustrates the experiment of an anti-n-1 common-mode escape in a tri-redundancy mimic structure. It shows that common-mode test cases a, b, and c are injected, respectively, into the intersections of executors (defense scenarios) A and B, A and C, and B and C. It also shows the path (shown in dotted lines of

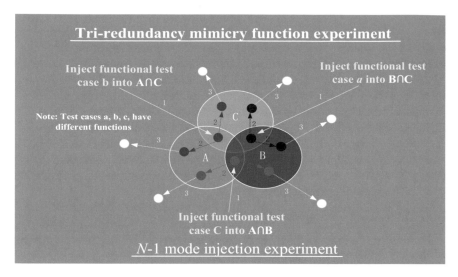

Fig. 10.5 $n - 1$-mode test case injection experiment

different colors) from which these test cases were removed or cleared from the executors or operational scenarios. Assume that the $n - 1$ mode common-mode test case can be injected into the functional intersection of any two executors or operational scenarios and be enabled to generate the same output vector by inputting channel compliant activation messages through the attack surface; then, according to the definition of mimic defense, the arbiter can perceive the inconsistency in the multimode output vector but cannot directly identify the problematic executor or operational scenario and needs to distinguish and eliminate the effect of the injected test case through the backward verification mechanism. To do this, the first step is to clear, restart, or replace the executors or operational scenarios with inconsistent output vectors. If the status of the arbiter remains unchanged, choose one object from among the executors or operational scenarios with the same output vectors according to a certain strategy to repeat the previous step. Secondly, observe the status of the arbiter. If the status reverses, it would degenerate into the differential-mode status shown in Fig. 10.4. In this case, repeat the above step on executors not updated in the service set, and the injected test case will be cleared or removed from the current service set. Therefore, according to the mimic defense definition, even if the $n - 1$ common-mode attacks succeed, their escape states are not stably robust.

It should be pointed out that, when the $n - 1$ mode test cases can be activated simultaneously, they may be recognized as differential-mode attacks due to the inherent mechanism of the mimic structure and they cannot ensure steady display of the $n - 1$ mode escape statue, so they shall be repeatedly injected and activated until the expected $n - 1$ mode escape status appears in the experiment. The whole process of the experimentation should be observable.

It is clear that the mimic structure system has a feature that "an escape cannot be steadily sustained even if it is successful." Although the system cannot replace all the functions of the traditional information security measures, it has the resilience that the latter does not. Especially when dealing with attackers who attempt to obtain sensitive information or sabotage the integrity of information through differential-mode attacks, the mimic defense may be even more advantageous than general encryption measures. As mimic defense is not a computable problem, it will not fall into the dilemma "where once it is breached by brute force, the whole defense system will collapse."

5. n-mode test case injection experiment

As its definition dictates, the mimic structure supports quantifiable design and features a tiny probability of n-mode escape, which, unlike the $n - 1$ mode escape, is mechanically unperceivable to the mimic structure. Therefore, to ensure the potential occurrence of the n-mode escape, it is necessary to employ a certain external or internal strategy to disturb the feedback control loop, change the current operating environment inside the mimic brackets into the $n - 1$ mode perceivable form, and enter the relevant disengagement or recovery process, ensuring that the n-mode escape, if happens, is still a probability event. The function needs to be checked through the white-box test of injected test cases and measured and verified for the nominal time it takes to finish the disengagement according to the preset loop perturbation strategy.

As shown in Fig. 10.6, assume that the injected n-mode test case can produce consistent output vectors within the functional intersection of the three executors $A \cap B \cap C$, and then it should not be perceived theoretically in the mimic ruling phase. However, according to the definition of mimic defense, even if the arbiter does not find any output vector anomaly, the executors or defense scenarios in the current service set may experience a forced and non-deterministic replacement or cleaning and restart operation due to an external control instruction. This means that

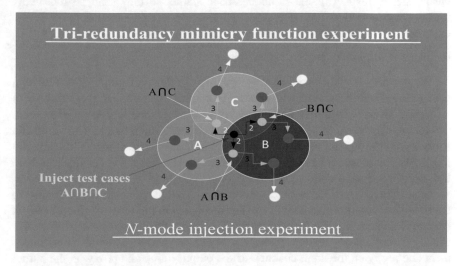

Fig. 10.6 n-mode test case injection experiment

under the n-mode injected test conditions, the escape state is surely unstable, and when the executors or defense scenarios themselves have the cleaning, restart, reconfiguration, or reconstruction functions, the n-mode unperceivable escape should automatically turn into the $n - 1$ mode perceivable escape status as the recovery progresses and eventually degenerate into a differential-mode context and be removed unperceivably. That is, the n-mode test cases injected through the attack surface into the service set are removed since the host executor or operational scenario is strategically cleaned and restarted, or reconfigured and reconstructed, by an external instruction. If the backward verification policy prescribes that executor C goes through a routine cleaning and then rejoins the current service set (may also directly reconfigure or replace executor C) and then if the output vectors of C (or its substitute) are still inconsistent with those of A and B, the feedback loop will prioritize cleaning or replacement of the executor with the longest running duration (such as A).

Thus, when the arbiter finds that the AC output vectors are different from those of B, it indicates that an escape phenomenon (i.e., the scenario in Fig. 10.5) has occurred in the scenario and the above steps are performed on B until the arbiter no longer perceives the inconsistency. Tracing the migration trajectory shown by the dotted line in the above figure, the n-mode test case scene will regress into the $n - 1$ mode test scenario and finally to the differential-mode test scenario until all test cases are removed. We can see that in a mimic defense environment, even if the attacker has the ability to implement a n-mode or cross-domain coordinated attack and achieve a temporary escape, he will not be mechanically capable of maintaining a stable escape, and therefore the robustness of attack effect remains an insurmountable challenge. It should be pointed out that, when the n-mode test cases can be activated simultaneously, they may be recognized as differential-mode or $n - 1$ mode attacks due to the inherent mechanism of the mimic structure and they cannot ensure steady display of the n-mode escape statue, so they shall be repeatedly injected and activated until the expected n-mode escape status appears in the experiment.

6. Injection test in the feedback control loop

The mimicry feedback control loop consists of three parts: input allocation and proxy, output ruling and proxy, and feedback control. According to the definition of mimic defense, the input/output channel between the feedback loop and the executors and mimic brackets allow for "one-way communication mechanism" only and for one prerequisite that the feedback control loop may have vulnerabilities (really inevitable) but must not have malicious codes (this can be satisfied engineeringly in low complex cases) (as shown in Fig. 10.7).

According to the strict unidirectional communication mechanism, the Trojans in the executor should not be able to use the loopholes in the feedback loop to inject the attack code or achieve tunnel through. Similarly, the mimic bracket function is usually "transparent" to external attackers, which means that either the input assignment and proxy on the input channel, or the output ruling and proxy on the output channel, or the feedback control for internal policy and schedule should all be "invisible."

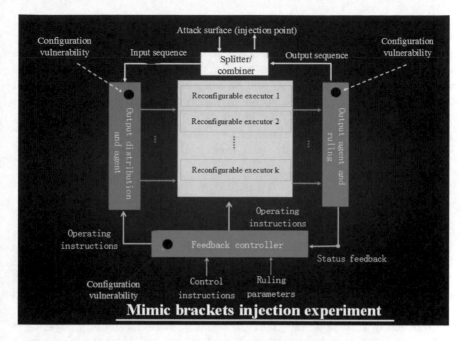

Fig. 10.7 Unidirectional mechanism verification in mimic brackets

In theory, the mimic brackets neither resolve the input incentive sequence content nor care about the syntax and semantics of the multimode output vector, so they are unreachable to threat targets. However, in engineering implementation, you have to care about the isolated import and load balance of the input excitation sequence and have to import the agent system. Oftentimes, the mimic ruling cannot avoid the uncertainty of some optional value ranges and communication serial numbers of multimode output vectors of the heterogeneous redundant executors. This, together with the difference in computation precision, requires that the mimic brackets must be opaque in syntax and semantics. There will be an increasing probability of the existence of unknown vulnerabilities as the function of bracket parts gets more complex or their intelligent processing capability gets stronger. However, mimic brackets can be considered to be up to the basic security requirements if the design can ensure that their parts will not be exploited even if they contain viruses. The white-box testing aims to check if it is possible to set vulnerability test interfaces on some parts, if the test codes can be uploaded through the vulnerability interfaces, and if mimic escape can be achieved via the uploaded test codes. Sometimes, this method may not work due to the technological form of the object. For example, the vulnerability test interfaces cannot be set, and the test codes cannot be uploaded and executed when passive optical splitters or wiring logic devices are used as input agent parts or the encryption mechanism is introduced into them. If so, it can be considered that the input agent parts are born with an attribute that their vulnerabilities are unavailable. Similarly, failure to upload the test codes through the mimic brackets-specified path via the input agent parts and heterogeneous executors to the output agent parts or the arbiter for execution signifies that the output agent parts or the arbiter also has an attribute that their vulnerabilities are unavailable. Obviously, this attribute is not inherent in these parts; it derives mainly from the mimic structural effect. It should be noted that event if the test cases can be injected with test vulnerabilities of relevant mimic bracket parts from outside the mimic interface, the brackets may still be considered to meet the assumption that the vulnerabilities are unavailable on condition that no mimic escape can be achieved. In short, the security technologies are often required in engineering practice to guarantee that the mimic brackets shall not become the "short plate" of defense for the set security goal even if they contain design flaws and vulnerabilities.

7. Performance measurement

In the abovementioned tests, the time from activation of the test case to disabling them should be measured in order to evaluate whether the converging speed, actuating performance, or scenario avoidance precision of the feedback control loop have met the requirements of the system design.

No matter what types the common-mode test cases are, the mimic defense shall always be able to get out of the escape state. Maybe different design schemes lead to different time of disengagement and to different cost of implementation. However, the function of disengagement from the common-mode escape is indispensable because "failure to maintain a successful attack escape" is the high-availability goal of mimic defense. Similarly, you can test the reliability design of the target object with the differential-mode test cases. The author declares that it is feasible to set confidentiality inspection function in the injected test cases in the mimic structure on condition that there are no side-channel attacks launched by taking advantage of the physical (acoustic, optical, electrical, magnetic and thermal) effects or the resources shared with other non-mimic structure tasks

10.4.7   Attackers' Perspective

From the perspectives of the attackers, they might consider the following ways to deal with the mimic defense system.

10.4.7.1   Mining or Setting up Vulnerabilities/Backdoors in the Mimic Interface

As social order is to the flaws in laws and rules, so is mimic defense to the vulnerabilities in attack surface channels, methods or rules, it is ineffective theoretically. One of the effective ways for attackers to break through the mimic defense is to develop abilities to target areas not usually covered by protocol security checks, including the ability to mine and exploit design flaws of mimic interface protocols or procedures; the ability to embed deep backdoors in highly transparent, standardized algorithms or interfaces; and in particular, the ability to mine and exploit potential specific or combined attack resources under the multi-protocol, multi-standard, and multimode convergence conditions. Obviously, this is a challenging job for attackers.

10.4.7.2 Creating a Homologous Ecosystem with the Development Tools and the Open-Source Community Model

For a mimic defense system, the greater the dissimilarity between executors, the better. In theory, if all the dark features are never intersected, there will be no attack escapes. But homology may erode the effect of diversity, reduce the significance of heterogeneity, and increase the possibility of attack escape. Possible attack options include importing malicious codes through development tools and an open-source community, or implanting backdoors using reusable hardware and software products, or hosting special functions in the COTS products by using the integrated service model.

10.4.7.3 Black-Box Operations Using "Irreplaceable" Advantage

Mimic defense needs the guarantee of a heterogeneous redundant environment under the functional equivalence conditions. If it is a single-source black-box product, then it is hard to ensure the heterogeneous redundancy condition, or at least the defense effect will be greatly lowered. In this sense, in order to ensure the healthy and orderly development of the cyberspace, we must resolutely oppose any form of technological and market monopoly, and create a market ecosystem conducive to the development of diversified technologies and products by introducing standardized and white-boxing technological development approach, which shall be advanced, open and secure. This should become the shared responsibility of the international community to solve issue of cyberspace security, or even that of network economy.

10.4.7.4 Developing Attack Codes that Are Not Dependent on the Environment

Mimic defense makes it difficult for attackers to use vulnerabilities, backdoors and Trojans in a coordinated way by creating iterative changes of service environment or running scenarios under the functional equivalent conditions. Therefore, attackers can influence the effect of mimic defense by hiding bad code in cross-platform script files, embedding malicious code that is not related to the execution environment in a single-source application, or designing special mechanisms in diverse software and hardware with homologous codes to facilitate coordinated attacks under the non-cooperative conditions.

10.4.7.5   Coordinated Operation Under Non-cooperative Conditions Using Input Sequence

Mimic environment emphasizes "dis-cooperation" and is always trying to avoid communication, consultation or synchronization mechanisms between heterogeneous executors. However, it cannot prevent attackers from achieving coordinated operation through normal channels of the attack surface, input sequence and information packet. For instance, the dark features in relevant executors can carry out relatively accurate cross-range coordinated operations (attacks) according to predetermined and compliant arrangement of input sequence or content. Of course, this is done on condition that the attackers can understand completely the dark features and dispatching procedures in the diverse executors.

10.4.7.6   Trying to Bypass the Mimic Interface

Using the side-channel effect or other means to bypass the mimic interface is an effective way to break through the mimic defense. In particular, if the dark features in the executors can access sensitive information in the executors, then the mimic interface can't perceive their stealthy transmission of information through physical forms (side channels) like sound waves, light waves, electromagnetic waves or thermal radiation. Conversely, attackers can also to receive or inject the attack code through the side channels, which usually takes longer. For instance, use APT attack mechanism.

10.4.7.7   Attacking the Mimic Control Aspect

Undoubtedly, the core aspects of the mimic structure, such as input/output agent, mimic ruling, and feedback control, will become the new target of the attackers. Since design flaws are actually inevitable in these aspects, and they may also contain the implanted malicious code, there are no absolutely credible exceptions. As these functions are relatively simple and strictly unidirectional (as shown in Fig. 10.7), they will be difficult to be exploited from the attack surface even if they have high-risk vulnerabilities. However, if the purpose is to destroy or paralyze the mimic control mechanism, attackers may achieve some effect like denial of service by, for example, letting the feedback control operation act frequently in order to keep the executors constantly in a policy and schedule state, so that the executors do not have time to provide normal or degraded service functions (service bump); or attackers may disable the arbiter, disrupt its normal function, or invalidate the input/output agent.

It should be emphasized that the processing complexity and working state of these aspects are relatively simple and the state transition path can be estimated. In general, the majority of backdoors can be discovered through relatively complete formal inspection methods. In other words, the thorough investigation of malicious codes (except the thorough check of vulnerabilities) here is usually not impeded by the state explosion. Apparently, attackers will face insurmountable challenges if the software and hardware codes of this aspect are open sourced and standardized, with added private functions such as user-definable and customizable functions, and are put together with executors to constitute a diversified ecological environment in product form, or simply designed as a hardened "black-box" part.

10.4.7.8 DDoS Brute Force Attacks

Like other non-network defense technologies, mimic defense is not effective against DDoS attacks that block the communication links (In reality, this is still a universal conundrum for defenses against this mode of attacks in network environments or shared transmission resource conditions). However, it is also difficult for "smart DDoS attacks," which consume processing resources of a system, to achieve the expected results because the attackers cannot accurately sense the resource overhead of the mimic-structured target object or the reconfiguration of the redundant service scenarios.

10.4.7.9 Social Engineering-Based Attacks

In theory, if a fatal Trojan or backdoor is embedded in the diversified component supply chain of a mimic structure product, it can cause "permanent shutdown" once it is activated, for its running will exhaust the executor resources in the mimic interface, and may have a catastrophic impact on the mimic defense. Mimic defense is theoretically ineffective on paralyzing, disruptive backdoors or Trojans that are launched or activated by side-channel attacks because they cannot touch the mimic interface at all. However, it is unlikely to achieve the goal if the attacker relies only on cyber technologies without strong resources and capabilities of social engineering.

10.4.7.10 Directly Cracking Access Command or Password

From the mechanic point of view, mimic defense has no effect on attackers entering from the "front door or main entrance" (those who have cracked the authentication code or acquired the authority of operation. However, as for the traditional security functions such as identity authentication, compliance check, or password application based on mimic architecture, as well as password algorithm, blockchain, zero trust security architecture, there is no need to worry about the risk of them being bypassed or short-circuited by dark features such as backdoors in the host software/hardware or environment. This is also a fundamental role of security of the mimic structure.

References

1. May, G.A.: Social Engineering in the Philippines: The Aims, Execution, and Impact of American Colonial Policy. Greenwood Press, New York (1980)
2. Dawkins, J., Hale, J.: A systematic approach to multi-stage network attack analysis. In: Proceedings of The Second IEEE International Information Assurance Workshop, pp. 48–56 (2004)
3. Ye, N., Zhang, Y., Borror, C.M.: Robustness of the Markov-chain model for cyber-attack detection. IEEE Trans. Reliab. **53**(1), 116–123 (2004)
4. Wang, S., Zhang, L., Tang, C.: A new dynamic address solution for moving target defense. In: Proceedings of Information Technology, Networking, Electronic and Automation Control Conference, pp. 1149–1152 (2016)
5. Evans, N., Thompson, M.: Multiple operating system rotation environment moving target defense. U.S. Patent 9,294,504, 22 Mar 2016
6. Intel. Skylake server processor, external design specification addendum for SKL + FPGA. www.intel.com/design/literature.htm (22 Mar 2016)
7. Wei, S., Liu, L., Yin, S.: Reconfigurable Computing. Science Press, Beijing (2014)
8. Hataba, M., Elkhouly, R., El-Mahdy, A.: Diversified remote code execution using dynamic obfuscation of conditional branches. In: 2015 IEEE 35th International Conference on Distributed Computing Systems Workshops (ICDCSW), pp. 120–127 (2015)
9. Hailong, M., Yiming, J., Bing, B., et al.: Router mimic defense capability test and analysis. Acad. J. Inf. Secur. **2**(1), 43–53 (2017)
10. Zheng, Z., Bolin, M., Jiangxing, W.: Test and analysis of Web server mimic defense principle verification system. Acad. J. Inf. Secur. **2**(1), 13–28 (2017)

Chapter 11
Foundation and Cost of Mimic Defense

11.1 Foundation for Mimic Defense Realization

As pointed out in the previous chapter, the mimic defense is essentially an endogenous security effect formed by the combination of the generalized robust control DHR architecture and the mimic disguise mechanism. It is especially suitable for highly reliable, available, and credible applications that require stability robustness and quality robustness and enjoy better universality in areas such as IT, ICT, CPS, and intelligent control. The related technologies can be used for low-complexity applications such as various software and hardware components, modules, middleware, and IP cores and can also be used for various information processing in control equipment, devices, systems, platforms, facilities, network elements, and even in network environments. Its endogenous security functions should be one of the iconic features of a new generation of information systems. However, the target systems based on the DHR architecture require dedicated or COTS IT products, advanced development tools, and diverse product ecosystems to reduce implementation complexity and cost. The following technological advances or trends have laid an important foundation for the engineering application of mimic defense.

11.1.1 Era of Weak Correlation of Complexity to Cost

In the field of microelectronics, the prediction by Moore's Law has been going on for more than half a century [1], and circuit integration has achieved rapid development, as shown in Fig. 11.1. This allows chips to accommodate more and more transistors at relatively low costs and can carry more and more complex logic functionality. The improved chip capability in turn drives large-scale development of software; as a result, its pricing reflects the sales volume rather than the complexity of design and fabrication, a "wield" phenomenon in the market. This is because, in

© Springer Nature Switzerland AG 2020
J. Wu, *Cyberspace Mimic Defense*, Wireless Networks,
https://doi.org/10.1007/978-3-030-29844-9_11

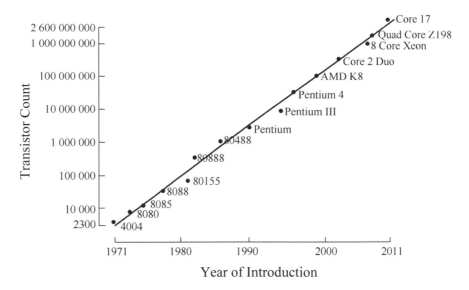

Fig. 11.1 Moore's law

the IT industry, the rules of the game are different from those of the machinery industry. On the other hand, the CPU architecture is evolving from the single-core to the multicore, many-core, multimode hybrid architecture. For example, the 4004 processor introduced by Intel in 1971 carried only 2300 transistors, with low single-core performance, but its latest generation of i7-6700K processor uses the Skylake architecture [2], with up to 1.9 billion transistors, quad core, eight threads, and high-end performance. At present, diversified COTS software and hardware products, such as CPUs and operating systems, are quite common. The dynamic heterogeneous redundancy environment, as required by mimic defense, coincides with the development trend of software and hardware.

11.1.2   High Efficiency Computing and Heterogeneous Computing

The processing devices with different architectures have different energy efficiencies, and the computing systems with different architectures vary largely with loads. For example, a large-scale streaming processor is more suitable for computing-intensive applications than a general-purpose processor, so the "Tianhe-1" high-performance computer used the heterogeneous mode of CPU + GPU [3] to cope with the LINPACK test and proved a huge success in the field of performance applications. There is also the "Sunway TaihuLight," which adopted the multicore instruction stream parallel technology architecture and headed the top 500 high-performance computers in the world for 2 consecutive years. In June 2018, a new generation of supercomputer "Summit" (the apex) of Oak Ridge National Laboratory in the United States became the new king in the field. Despite this, such heavy electricity consumers are lingering headaches for users, just as a saying goes, "you can afford to buy a horse, but with no money left for a saddle." In fact, 10 years ago, the National High Technology Research and Development Program (863 Program) deployed a major task of "New Concept High-Efficiency Computer Architecture and System Research and Development" in Shanghai. The mission was undertaken by the National Digital Switching System Engineering Technology Research Center (NDSC). In 2010, the Center proposed a "multi-dimensional dynamic function architecture based on active cognition, PRCA," which enables the same task to acquire the expected performance through software and hardware cooperative computing done by real-time combination of modules or algorithms of different energy efficiency ratios in different time periods, different loads, different resources, and different running scenarios. As its working principle is very similar to the biomimic phenomenon, it is called Mimic Structure Calculation (MSC) and thus becomes the third energy-saving means through the computer system architecture, the other two being through technology and operational state control. In September 2013, China successfully developed the world's first MSC prototype, which perfectly integrates high-performance and high-efficient computation through the variant structure (heterogeneous) MSC based on active cognition under the functional equivalence conditions. Tests show that for the three classic CPU-bound, storage-intensive, and input-output-intensive computing cases, its energy efficiency is 10–300 times higher than that of the mainstream IBM server in the same period. At the same time, its "multidimensional dynamic function structure based on active cognition under the functional equivalence conditions" has naturally become one of the basic techniques of mimic defense. The difference between the two is only the application scenario. The former is to achieve high-performance computing and the latter to deal with the security threats based on the "dark features" of the target object's vulnerabilities and backdoors. At present, CPU + FPGA-based "heterogeneous computing," "user-customizable computing," domain-specific software and hardware collaborative computing (DSAs), and software-defined hardware (SDH) are emerging. For example, Intel's newly introduced Xeron + FPGA architecture based on QPI (QuickPath Interconnect) can provide the function of data parallel and pipeline parallel acceleration processing. From the defense perspective, new functions such as customizable computing environments and dynamic heterogeneous processing have greatly improved the staticity, certainty, and similarity of the information system or control device, which is then not as vulnerable as before. In particular, when the target system uses noninstructional flow control methods such as FPGA and SDH and the parts or components with reconfigurable algorithm or online reconfiguration functions, it is almost possible to significantly increase the difficulty and cost of attacks without increasing much of the defense cost or the compromising too much of the system performance. It is expected that new parts or components such as user-definable heterogeneous computing and domain-specific hardware and software cooperative computing can be used as the basic components of the mimic structural system.

11.1.3   Diversified Ecological Environment

In the mimic interface, there is a need for diversified or pluralistic functional equivalent executors or a variety of heterogeneous defense scenarios, so that it can reduce the attack escape probability and increase the difficulty of invasion and penetration of attack through the policy and strategy based on mimic ruling and the adaptive feedback control mechanism of multi-dimensional dynamic reconstruction. Usually, there are multiple COTS products available on the market for an information system or control device, which range from the top-level application software through the middle layer supporting software (such as operating system, database, virtualization software, etc.) to the underlying hardware storage, processing equipment, and switching interconnection network, and such products are well compatible with each other. For example, currently available operating systems are Windows, Linux, UNIX, etc. The distribution version of Linux includes Debian, Ubuntu, Red Hat, Kirin, etc.; PC processors like Intel, AMD, ARM, IBM, Loongson, Sunway, Phytium, and other manufacturers; mobile phone processors like Intel, Qualcomm, Samsung, Huawei, MediaTek, etc.; routers like Huawei, Cisco, Juniper, H3C, etc.; and the inter-domain routing protocol which has multiple standardized routing control protocols (such as RIP) and the open shortest path first (OSPF), etc. Standardized and diversified COTS hardware and software products provide the basic support for the heterogeneous redundant environment required for mimic defense.

From the perspective of new business formats, the open-source community has become a highly sought-after new model for technological innovation and product development, and the influential ones are open-source RISC-V and Linux-China. Since open-source codes are mainly developed by programmers scattered around the world, the programmer's cultural environment, thinking habits, development tools, and technical methods make the hardware and software modules result in different implementation versions for the same function. This provides a diversified foundation for the creation of heterogeneous resource pools required for mimic defense and injects vitality into the market.

From the point view of tool technology advancement, the diverse compilation tools can lead to multiple variations of the source program. With the modern compilation or interpretation tools, a source program can be compiled into executable files with multiple kinds of object codes (cross-platform instruction sets) or be generated into a variety of executable files with the same target instruction set but different compilation results by changing the compilation parameters (especially those related to parallel recognition and automatic optimization). One of the effective ways to obtain the heterogeneous functional equivalents is by generating a plurality of different executable files from the same source code through the compiler, thereby realizing the multi-variation of the software. Similarly, this method can also be used in the field of programmable or software-defined hardware. It must be emphasized that it is impossible to eliminate design flaws or dark features in the source program, by using either a variety of compilers or using different compilers or interpreters. For example, if an application does not filter or review the user's input, then you won't be able to solve the SQL injection problem even if you create a SQL query in a different language or use a different compiler or interpreter. However, the diversified compilation method can indeed change the attack surface of the target execution file [4], increasing the difficulty of the injection attack.

From the perspective of engineering implementation, the dissimilarity design of the executor under the functional equivalent conditions is realized through engineering management, and its classic example is "Dissimilar Redundancy Structure (DRS)," which requires the guarantee that, in engineering implementation, the design flaws in each functionally equivalent executors are not coincident. At present, the dissimilarity design of the DRS system is mainly realized through strict engineering management. In other words, multiple working groups composed of people with different educational backgrounds and work experience work on multiple functional equivalent devices independently in different development environments, according to different technical routes, and by following a common functional specification. For example, the flight control components based on the DRS architecture have better dissimilarity or difference between the heterogeneous redundant bodies [5], so that the failure rate of B-777 flight control system can be kept lower than 10^{10} and that of the F-16 flight control system can be less than 10^8.

11.1.4   Standardization and Open Architecture

Standardization and open architecture have become the consensus of the technical and industrial circles, which is conducive to the division of labor and collaboration of producers and promotes the rapid development of productivity. At the same time, standardization and open architecture also provide important technical support for mimic defense.

Open system architecture (OSA) [6] makes the application system portable and tailorable, ensuring interoperability between components or modules, expanding the range of hardware and software products for users to choose, and avoiding the risk of monopoly or "single-source, black box use" in the market. Also, OSA can promote the development and improvement of new technologies and new functions. Therefore, OSA can provide diversified, tailorable, and scalable hardware and software components or modules for mimic defense.

Standardization is the basis for openness. In order to ensure the performance of interconnection and interoperability, some standard specifications must be formulated. The ISO subcommittee on information processing systems has developed a series of open system interconnection standards that cover almost every important area of information processing. The world's major computer manufacturers and users support the OSI standard, the environment built on OSI is called OSIE, and the OSA is also the technical foundation for implementing OSIE. The standards have standardized the interfaces and protocols that can be used as optional executable components in the mimic interface, shielded the differences in their internal implementation, facilitated the integration of executor components from different vendors (especially nonhomologous products) into the DHR architecture, and guarantee no difference in service functions or performance. Standardization greatly reduces the workload of fusing multiple heterogeneous executors within a mimic structure, while also allowing teams or trusted third parties to independently develop different (nonhomologous) software and hardware executors without the need for close communication and collaboration.

The open architecture and standard technologies and interfaces have assisted the formation of specialized, diversified market segments, which provide sustainable and progressive conditions for functionally equivalent COTS hardware and software parts or components for DHR architecture systems. For example, in the field of network technology, the emergence of technologies and standards such as SDN and network function virtualization (NFV) has enabled the traditional market such as routing exchange and domain name service to realize the "black box to white box" open development, which helps to form a diversified market pattern and provided a sustainable ecological environment for mimic defense.

11.1.5 Virtualization Technology

In computer science, virtual technology [7] is a way to combine or partition existing computer resources (CPU, memory, disk space, etc.) so that these resources behave as one or more operating environments, providing better flexible access than the original resource allocation. Virtualization refers to the act of creating a virtual (rather than actual) version of something, turning it into a kind of physically irrelevant resources that can be uniformly managed in the logic sense, can be flexibly deployed, and can be convenient for users to use. Virtual resources are free from the limitation by deployment method, geographic location, or physical configuration of existing resources (but at the expense of processing performance and efficacy). Virtual technology has an important engineering significance for the realization of mimic defense, especially for the improvement of its economy.

For example, the representative operating system virtualization technology, Docker [8], uses the C/S architecture Docker daemon as a server to accept requests from clients and process these requests (create, run, and distribute containers) to implement the functions similar to virtual machines (VMs), for the purpose of employing the resources more efficiently and flexibly. Virtualized environments often require coordination of multiple technologies, such as the virtualization of servers and operating systems, storage virtualization, and system management, resource management, and software delivery. Virtual technology is essentially based on a virtual space, where it can create virtual resources, build virtual structures, form virtual processing environments, and implement virtual functions.

Admittedly, virtualization technology brings flexibility at the expense of processing performance and efficiency overhead. The problem is that if local overhead can bring about or improve global benefits, the rationality of virtualization can be supported at the application layer. For example, given the diversity of network services and the uncertainty of load distribution, the traditional static configuration of resources has a low cost-effectiveness ratio in that either the processing resources allocated cannot satisfy the user experience or the unbalanced busy states of different servers seriously affecting the effective use of the resources. Virtualization technology can transform the static allocation into dynamic allocation, thus increasing the comprehensive utilization efficiency of the processing resources and greatly improving the user experience. Although this approach adds to the overhead of virtualization itself, it is quite appealing because it can increase the cost-effectiveness ratio of the system as a whole.

Virtualization technology is of great significance for the economic realization of mimic defense. Firstly, it can economically provide the heterogeneous executor resources. Suppose there are three multicore processors, ×86, ARM, and POWER; three kinds of OSs, Windows 10, Linux, and Android; and three types of virtualization software, VMware, Virtual Box, and MiWorkspace, and they can theoretically be combined into dozens or even hundreds of virtual operational scenarios. Secondly, the virtualization technology better solves the problems of dynamic scheduling such as virtual machine switching, service migration, and synchronous recovery. Thirdly, highly shared network service environments (such as data centers, cloud platforms, etc.), as well as their inherent dynamics, diversity, and uncertainty, can be economically utilized by the mimic structure and can achieve generalized robustness with less effort. It should be emphasized that the mimic defense effect will be better if the mimic interface intentionally selects the cross-physical domain or entity domain executor that is composed of virtual machines, because this will greatly lower the requirements for virtual control protection.

11.1.6 Reconfiguration and Reorganization

In recent years, with the rapid development of network communication technology and multimedia technology, the demand for computing resources is becoming more and more complex. In particular, various computing-intensive multimedia applications and data-intensive big data applications are increasingly demanding computing environments, promoting the development and application of reconfigurable technology. At the same time, the development of reconfiguration technology provides an indispensable technical support for the diversification of DHR architecture defense scenarios.

Reconfigurable technology [9] combines the universality of software and the efficiency of hardware. It is widely used in information technology, manufacturing, intelligent services, and other fields to enable limited resources by flexibly changing system processing structures or algorithms, so as to adapt to more application needs to achieve efficiency, reduce costs, and shorten development cycles. In the field of information technology, reconfigurability mainly includes hardware reconfigurability and software reconfigurability, while hardware reconfigurability can be specifically divided into FPGA-based, eXtreme processing platform (XPP), mimic computing platform, software-defined hardware (SDH), hardware and software cooperative-computing platforms, and reusability for chip design. Software reconfigurability can be divided into reconfigurable technologies based on the separation of module proxy and implementation, reconfigurable dynamic framework, and control plan program. Based on the above reorganizable or reconfigurable technology, different element sets can be constructed in the mimic system resource pool, and a variety of different heterogeneous executors or defense scenarios can be constructed, so that the limited resources can provide richer structural representation for the target object. The reconfiguration process of the system or the executors can also be regarded as the process of dynamic, diverse, and random change of the mimic defense scenarios. The equivalent function cannot only expand the dissimilarity between the implementations in the mimic brackets but also provide a means of flexible implementation for the problem avoidance mechanism. Therefore, both theoretically and technically, the reconfigurable or software-definable variant structure (algorithm) technologies have important engineering practical significance for reducing the escape probability of mimic bracket attacks. However, the reconfigurable control aspect may also become a new target for attackers.

11.1.7   Distributed and Cloud Computing Service

A typical application of distributed technology [10] is a grid-based computer processing technology, which studies how to solve a problem that requires a very large computing power to be broken down into many small parts (this method is not always feasible in practice), which then can be assigned to many computers for processing, and finally which combines these calculation results to get the final processing results. Distributed technology can not only solve bottleneck problems with limited capacity of a single computer but also improve system reliability, availability, and scalability. The mimic defense emphasizes the importance of heterogeneity, de-coordination, and distributed processing of the operating environment and tries to avoid the situation of "uniform resource management," "panoramic resource view," and "centralized processing environment." Dynamic segmentation, fragmentation, and decentralization of key nodes should be generally followed without affecting or least affecting system functionality and performance. The classic distributed theory, methods, and techniques can all be used as the important basis for the implementation of mimic defense.

With the widespread application and deep development of distributed technologies, the concept of cloud computing services has emerged. Cloud computing [11] is an on-demand, pay-per-use model which, through a configurable computing resource sharing pool (including network, server, storage, application software, services, etc.), enables users to use network and computing resources easily and on-demand or even specify or configure a virtual container environment. These resources can be quickly provisioned with little administrative effort or little interaction with service providers, as shown in Fig. 11.2. Foreign Amazon EC2, Google

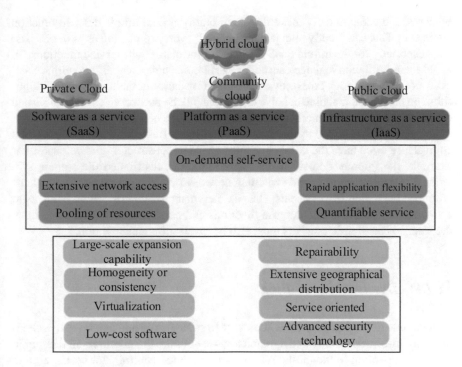

Fig. 11.2 Cloud computing

App Engine, and domestic Alibaba Cloud have achieved great results in commercial applications. The application scope of various "cloud computing" is expanding day by day, with its influence immeasurable.

The scale effect of cloud computing makes the "cloud collection" provision of services more economical than traditionally decentralized management and terminal usage and is likely to bring about major changes in the way smart terminal services are provided. Similarly, cloud systems are not sensitive to platform infrastructure costs, including security investments. Therefore, the DHR architecture can technically support cloud computing, cloud service stability robustness, and quality robustness requirements. Due to the "cloud-based" implementation of the service function, in theory, the terminal device only needs to complete functions such as network access and input/output interface management, thereby greatly simplifying the security protection design of the terminal or the desktop system.

Because cloud services can provide flexible, convenient, cheap, efficient, and scalable virtualized processing resources and functional implementation environments and have endogenous dynamics, randomness, and diversity, these pooled resources and virtualized heterogeneous executors can be formed into highly reliable, highly credible, and highly available mimic cloud service system with generalized robust control functions, which, by incrementally deploying specialized mimic components (or a virtualized component in the cloud), can effectively protect the security threats caused by virus-bearing soft components or dark features in the brackets, and economically make the cloud platform with mimic defense function come true. This can greatly alleviate customers' "worry of placing eggs in one basket," especially for the mimic SaaS that orients various certain service functions. To avoid the information leakage caused by "virtual machine escape," the mimic SaaS shall not only consider cross-physical domain or entity domain distribution algorithm on the virtual machine but also ensure that all the user services are established on the mimic structure. In other words, there is not the non-mimic SaaS.

In general, distributed systems or information infrastructures with functional equivalence often have the basic elements needed to construct mimic defenses. For example, the Internet domain name service systems, distributed file storage systems, data centers, routing and switching networks, distributed processing systems with load balancing requirements, etc. can incrementally deploy mimic brackets to economically obtain mimic defense functions and performance. At the same time, the safety of the mimic brackets must also be given sufficient attention.

11.1.8   Dynamic Scheduling

In the mimic defense process, it is necessary to strategically and dynamically schedule or multi-dimensionally and dynamically reconfigure the executors in the mimic interface according to the mimic ruling state or related control parameters, aiming to change the current target object defense scenario through iterative convergence to eliminate incompliance of multimodal output vectors of executors in the mimic interface. The efforts to achieve a coordinated attack by means of trial and error have significantly reduced the stability and reliability of the attack chain. Unlike the dynamic process of MTD that looks too blind or not so scientific, the dynamic scheduling generally refers to the progressive or autonomous updating of the scheduling strategy based on the gradually obtained information in the case of unpredictable disturbances in the scheduling environment and tasks. Compared with static scheduling, dynamic scheduling can generate more targeted decision-making schemes for actual situations. In particular, policy scheduling with functions such as feedback control and dynamic convergence can improve system performance and resource utilization. Dynamic scheduling has many classic research cases [12]. For example, Ramash studied the dynamic scheduling simulation method, Szelde studied the knowledge-based response scheduling method, and Suresh studied the single-part workshop dynamic scheduling problem. In recent years, dynamic scheduling theory and applied research have made great progress, including optimization methods, simulation methods, and heuristic methods, especially the newly developed artificial intelligence (AI) scheduling, covering various methods such as expert systems and artificial neural networks, intelligent search algorithm and multi-agent. Although most of the above methods are used to optimize the performance and efficiency of the system, they should also be applied to the dynamic scheduling requirements of mimic defense if their objective function is defined as the ability of the system to resist generalized uncertain disturbances or the security and credibility of the service functions.

11.1.9   Feedback Control

The mimic defense has a typical negative feedback mechanism. In the mimic ruling process, the feedback control loop, triggering policy and scheduling, and multi-dimensional dynamic reconfiguration mechanism that can be defined by the objective function form a closed-loop control system characterized by adaptive dynamic convergence and generate relevant operating instructions according to the incompliance of output vectors and the given ruling parameters, in order to eliminate the output vector inconsistency or to control it within a given threshold range by iterative change of the mimic field. With the control theory and method established by American scientist Wiener in 1948, it is possible to quantitatively analyze the relationship between the nature of mimic defense and various elements and to qualitatively and quantitatively evaluate the quality of defense, attack behavior, security situation, adjudication strategy, convergence speed, etc. Evaluate and establish continuous improvement mechanisms for tracking, control, feedback, decision-making, adjustment, etc. and continuously enhance the optimization process of mimic defense scenarios. It should be pointed out that the change of only the mimic ruling algorithm or feedback control function (control law) can alter the defense scenario and action even if other conditions remain unchanged. The attackers will find it an insurmountably difficult obstacle if the system designer or user intentionally exploits the feature.

11.1.10   Quasi-Trusted Computing

Chapter 4 of this book briefly introduces the main points of SGX technology. Following Intel, ARM also released a similar feature (trusted area) product. Although the security boundary of SGX still includes CPUs whose trustworthy cannot be guaranteed, users are still doubting if "the root of trust" (Trusted Platform Module – TPM) of vendors is trusted, just like trusted computing. However, you don't have to worry about it when using the SGX technology in the mimic environment. As the endogenous effect of the mimic architecture is determined by the redundant configuration of the heterogeneous resources and the relative decision mechanism, the output vector ruling mechanism can detect "any mishaps of the cooperative attack" on the mimic interface and can timely update the content of the SGX of the problematic executor on condition that the CPUs distributed on different executors cannot coordinately modify the core code of the heterogeneous configuration in the respective SGX storage areas. Similarly, the CPU + FPGA structure can also achieve similar functions. If the target system design specification does not assign the CPU with the authority to modify the FPGA bit file online or in real time, and rigidly stipulates that certain core functions of the heterogeneous executor can only be completed independently by the FPGA module, the security perimeter function and credibility of the CPU can be ignored completely, and whether the FPGA itself is credible or not will not be a prominent problem in the mimic environment, because the FPGA bit files in each executor are characterized by functional equivalence and diverse algorithms, and the "dis-cooperation" mechanism can ensure the independence of the executors. It is therefore difficult to achieve mimic escapes by modifying the physically isolated FPGA code file online for lack of special capabilities of cooperative interference under the noncooperative conditions. Such methods are of great significance for improving the technical economy of the mimic defense system and accelerating the application of scale. It should be pointed out how to find the strategic pass (usually the same as or similar to the encryption authentication setting scenario), which core function codes or sensitive information should be stored in the SGX area, which functions must be executed by the FPGA or ASIC, and which isolation method is adopted between the executors in the service set may eventually affect the quality of the closed defense of the mimic brackets and the possibility of being "bypassed" by attackers. Therefore, in-depth threat analysis and careful technical planning are required to achieve the desired results.

11.1.11 Robust Control

Robust control is an important theoretical and methodological basis for mimic defense. According to the concept of cybernetics, various faults or disturbances in the target system will the model uncertain. In other words, the uncertainty of the model is ubiquitous in the control system. In applications where stability and reliability are primary objectives, the robust control method typically ensures that the uncertainty of the target object can still meet the quality control requirements even if the model exhibits a certain range of parameter perturbation when the dynamic features of the fault or disturbance process are known and the range of uncertainty is predictable. Based on the adaptive feedback control structure, the commonly used design methods are INA method, simultaneous stabilization, integrity controller design, robust PID control, robust pole placement, and robust observer. It is not difficult for you to find that if the non- anthropogenic perturbation concept is extended to the generalized uncertainty perturbation category including anthropogenic perturbation, the goal of robust control is no longer limited to high-reliability and high-availability applications. The "trinity" of high reliability and credibility will be the latest application height. In a sense, the theories, methods, and techniques related to robust control can even be directly applied to the design and engineering implementation of mimic structures.

11.1.12 New Developments of System Structure Technologies

The core of mimic defense is to use the innovative robust control architecture and the synergistic mechanism of diverse reconfigurable executors to obtain endogenous uncertainty effects, achieve the generalized uncertain disturbances including known unknown risks and unknown unknown threats in the mimic framework, and attain the goal of quantifiable control of security. In May 2018, four development trends were predicted at the International Symposium on Computer Architecture (ISCA):

(1) The full stack design and software and hardware synergy in the domain-specific architecture will be a promising direction out of the Moore's Law dilemma. This is also one of the implementation technologies for the future needs of diversified reconfigurable executors within the mimic domain.
(2) The neural network accelerator will be moving from the research design phase to the landing application phase.
(3) In addition to software and systems in the security field, sufficient attention should be paid to the architectural level. This makes it possible for us to embed the core ideas of mimic defense into the security architecture in the future.
(4) The RISC-V open instruction set architecture makes the agile development possible and will hopefully become the foundation of open-high-reliability source hardware in the future. Like the diversified compiler, the technology can be used to quickly generate a variety of RISC-V versions from a hardware source code or even one version for one executor, and the tape-out cost is affordable. The technology, coupled with software diversity compilation or even mimic compilation technology, will enable us to collaboratively design the mimic structural scenarios with diversified and higher performance.

11.2 Analysis of Traditional Technology Compatibility

We know that mimic defense mainly targets the uncertain threats based on dark features like unknown vulnerabilities and backdoors, that its basis is the endogenous security function formed by generalized robust control structure DHR and mimic disguise mechanism, and that it functions to enhance the anti-attack performance of the system, the robustness and resilience of the services with the architectural technology. Therefore, mimic defense is well compatible with the traditional security technologies of the additive forms mainly in the following aspects.

11.2.1 Naturally Accepting Traditional Security Technologies

Traditional security technologies are often weakly related or unrelated to the target object's functionality and are mostly external or add-on technologies that often do not involve the underlying architecture and functionality of the target system, for example, firewalls, cryptographic authentication, intrusion detection, threat perception, etc. They all intervene by means of outer cladding or series and parallel connection, as shown in Fig. 11.3. The "uncertain effect" or endogenous security function of mimic defense is not based on traditional security technologies, but if combined with additional security or protective measures, the mimic defense effect can be enhanced nonlinearly through dynamic scheduling and reconfigurable mechanism based on policy and schedule. In other words, the DHR architecture has the nature of naturally accepting additional security technologies. For example, the MTD instruction, address, data, and port randomization measures are not as good as expected in the traditional architecture. However, in the CMD architecture, the technical methods to break the Windows randomization defense may not have the same effect in the Linux environment. At least, the methods vary with difficulties, and the unidentical attack method is expected by the mimic structure to create "difficulty for coordinated attack under noncooperative conditions." Without losing their generality, most traditional security technologies can greatly enrich the diversified defense scenarios of mimic systems. In other words, the introduction of traditional security technologies in the mimic structure may significantly reduce the difficulty of engineering implementation and greatly improve the technical and economic indicators of mimic defense.

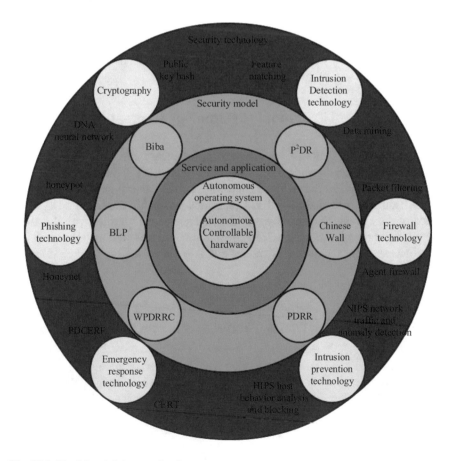

Fig. 11.3 Traditional defense technology system

(1) The traditional security technology mostly acts on the target object in an additional form, and DHR is an architecture technology whose service provision and feedback control function are separated. It is allowed to add any security technology to the executors of the mimic brackets as long as you can guarantee that the output vector is "transparent" to the mimic ruling.

(2) At present, there are many kinds of mainstream active and passive defense technologies, and most of them are functionally invisible to the system. If they are consciously dispersed in the heterogeneous executors in the mimic brackets, their technical effect is equivalent to enhancing the generalized dissimilarity or dynamicity between the executors, thus improving the overall anti-attack escape ability of the mimic structure.

(3) The traditional defense technology has been improved in terms of precise perception and precise defense. It can activate the emergency response mechanism (especially with better precision response to known attacks) before detecting an abnormality in the mimic interface, or it is used to precisely check, isolate, or clean the abnormalities detected by mimic ruling.

(4) The mimic defense has both non-specific immune surface defense function and specific immune point defense function, while most of the traditional defense techniques belong to the specific immune point defense technologies, and the combination of the two can nonlinearly enhance the effect of integrated point-surface defense.

It should be pointed out that mimic defense is much effective against the "crooked attacks," but it usually does not have a sure effect on the "front door attacks" made through the paths, rules, and methods on the attack surface or in the mimic interface, for example, attacks made in the capacity of a lawful identity obtained by way of stealing authority, password guessing and code crack, or penetration attacks launched by exploiting the flaws in communication protocols and interoperability specifications or even the deep-hidden backdoors. These attacks shall be dealt with by using the traditional security technologies.

11.2.2 Naturally Carrying Forward the Hardware Technological Advances

The performance of mimic defense technology at the hardware design level is simply to add or modify existing modules, for example, adding functionally equivalent heterogeneous redundancy modules and mimic bracket controllers; or transforming many-core, multicore processors into functionally equivalent heterogeneous processors or programs or data storage devices with redundant configuration settings; or introducing the "one-way contact mechanism" into the traditional control chain; or importing the "space fragmentation" mode in the data file storage; or using the reconfigurable and reorganizable technology or even software-defined hardware (SDH) to increase the abundance of mimic field; or using the user-definable computing components such as CPU + FPGA in executors. An open, standard, and diverse hardware ecosystem contributes to the universal application of mimic technology.

11.2.3   Strong Correlation to Software Technological Development

The main changes reflected on software design are adding functionally equivalent redundancy modules and corresponding dynamic selection (calling) and feedback control to the critical or sensitive security paths; or transforming centralized processing to decentralized processing; or converting static environment calculations to variable environment calculations; or consciously distributing control functions of sensitive control links, simplifying control relationships and minimizing control information; or using multipath, fragmented, sharded transmission and storage, etc. (much less difficult than massive parallel software design and operational control). Of course, mimic software is best built on multicore, many-core, or multiprocessor operating environments to avoid introducing too much latency in service response. In addition, the diversified compilation, software gene mapping analysis, virtualization, automatic program generation, redundant design language, binary compilation, cross-platform technology, and even mimic construction programming tools will greatly enrich the engineering implementation of mimic defense and reduce the threshold for technical applications.

11.2.4   Depending on the Open and Plural Ecological Environment

Informatization has entered the era of open architecture, especially the newly architectured high-performance server + high-speed network interface + virtualized universal platform technology is fully infiltrating the traditional special equipment field, and functional virtualization has become an unstoppable development trend. For example, closed network routers are being replaced by open SDN architecture and network function virtualization (NFV), the control plane function can be implemented on the basis of general server environments; the pool-based resource management and virtual container technology breaks the "chimney"-style ecosystem, facilitating the management and use of heterogeneous redundant resources. The open- and cross-platform architecture makes the hardware and software components portable and tailorable and ensures interoperability, thus providing diversified heterogeneous components and modules for mimic defense, especially laying technical and market basis for addressing the diversification of application software.

11.3   Cost of Mimic Defense Implementation

The endogenous security effect of mimic defense can unify high reliability, high availability, and high credibility into one architecture, so that the target system has attained the endogenous generalized uncertain perturbation ability to suppress random failure or human attack. However, mimic structure and related control mechanism will inevitably introduce certain technical costs and implementation overhead while bringing endogenous security effects. Below we will analyze the cost of its implementation from nine aspects.

11.3.1   Cost of Dynamicity

The mimic field presentation needs to strategically schedule the executors or defense scenarios in the current service set and dynamically reorganize and config-ure the corresponding executors in multiple dimensions, in order to expand or maintain dissimilarity (or diversity) within the service environment as much as possible, so that the mimic ruling mechanism can perceive the multimode output vector inconsistency, the expected uncertainty attribute appears in the mimic brackets, and the attacker cannot accurately perceive the current defensive sce-nario of the target object, thus increasing the difficulty of attacks under the same non-cooperating conditions. Since the implementation of functions such as dis-play, transformation, migration, identification, or anomality concealment of a mimic field requires additional and special processing resources, these functional operations may not be implemented as frequently and densely as the anti-jamming frequency hopping communications nor may it be like the mobile target defense the moving target defense (MTD) to quantify the dynamicity of the calibration which is difficult to perform. Therefore, it is necessary to introduce an adaptive feedback control mechanism based on mimic ruling in the mimic defense, so that the dynamic change of the executor or defense scenario has an iterative conver-gence property. At the same time, for additional processing resources, the dark features in this process should be reduced as much as possible, the attack surface should be narrowed, and the attack reachability can be reduced. However, in some application areas, scenario switching and synchronous recovery are likely to bring unbearable processing overhead, requiring innovative technical methods and pro-cessing strategies.

In addition, in order to monitor the failures that may occur in the scheduling process, it is also necessary to set an auxiliary function module such as watchdog or root of trust, which also increases the complexity and cost of the system.

11.3.2   Cost of Heterogeneity

At present, although there are many ways to obtain diversified heterogeneous func-tional equivalents, both theoretical breakthroughs and quantifiable engineering and design methods are required to ensure the dissimilarity and even the degree of dissimilarity. Some components (especially the special ones) still rely on strict engineering management methods to achieve their dissimilarity design. That is to say, multiple functionally equivalent components are developed independently by multiple working groups composed of people with different educational backgrounds and work experience, in different development environments, according to different technical routes, and following a common functional specification. This design method is relatively expensive and may not be sensitive to the market price of the product (which may be the reason for the large volume), but it will definitely increase the burden on maintenance and upgrading. Therefore, in the design of the mimic defense system, it is impossible to perform heterogeneous redundancy processing on all parts of the target system, and only the diversified design and heterogeneous redundancy configuration can be implemented for system weak points, sensitive hotspots, and critical paths, like the pass or strategic fortress. However, this will bring another problem, that is, how to prove that the selected hotspot or critical path or defensive pass is impossible for an attacker to bypass. These are not easy items for a complex system.

In addition, functionally equivalent heterogeneous executors differ more or less in performance (although they may be within the scope of the system), such as latency in request or response. Even the executors, e.g., IP stack of different OSs, that conform to standard functions may differ in implementation. This is because the current states of the system output buffer are different, they are not operated in the strict request-response I[P]O mode, and their "sticky" function may combine multiple requests from the same source address into one response packet to reduce the frequency of response operations. These differences will bring about troubles of varying degrees to ruling operation for multimode output vectors.

In particular, the DHR architecture does not require the same rigorous dissimilarity design of DRS (ensuring that both hardware and software are nonhomologous). First, DHR emphasizes to avoid the simultaneous occurrence of majority or complete consistency errors (of a common mode property) of multimode output vectors, while DRS is more concerned with avoiding the occurrence of multiple faults at the same time (or even differential mode faults). Unlike DRS, DHR does not consider the complete inconsistency of multimode output vectors to be the "worst case" because the latter can also perform iterative decisions using more complex backward verification strategies such as weight values. Secondly, DHR is based on the policy and schedule of mimic ruling and multi-dimensional dynamic reconstruction adaptive feedback control mechanism, so that the current defense scenario will vary or alter iteratively and convergently with the ruling state, and the DRS scenario is usually fixed or just undergoing the "zero-sum" changes. In the case of DRS architecture, sufficient trial-and-error attacks can theoretically destroy the constraints of $N \geq 2F + 1$ or easily achieve the attack escape (such as tunnel through), and the obtained attack experience can be used to break through the defense line of the object with "one blow." The hole penetrates the defensive bottom line of the target object. In the DHR environment, the preconditions of the trial-and-error attack no longer exist due to the role and influence of the adaptive feedback mechanism, so it is difficult to achieve or maintain a sustainable attack escape through trial and error or exclusion measures. This is why the mimic ruling-based policy and schedule and the multi-dimensional dynamic reconfiguration negative feedback mechanism can reduce the DHR system dissimilarity design requirements.

11.3.3   Cost of Redundancy

In general, it is necessary to configure heterogeneous redundant physical or virtual executors in the mimic brackets, and the cost or processing overhead caused thereby has a linear relationship with the redundancy. In theory, if the redundancy of heterogeneous executors in the mimic interface gets greater, the equipment development, production cost, operating cost, and power consumption may multiply accordingly. Therefore, the use of mimic brackets is like "defense of pass and fortress." For example, the control plane of the router is more sensitive to the attack based on the backdoor of the vulnerability than the data forwarding plane, so you can use the strategy of only strengthening the control plane. Fortunately, the control plane is important, but its implementation cost is usually only a small port of the system price (e.g., in the router at the core of the network, the most expensive component is the high-speed interface board, each of which costs several times or tens of times the COTS control components). Even though the control plane of the mimic architecture has increased several times the cost, it is still a small proportion of the total system cost (the proportion of different types of products may be different), and the cost increment part generally does not exceed 25% of the system cost (see Chap. 14 for related analysis), so it has little effect on the system price, but the improvement of safety can significantly enhance the market competitiveness of the product.

At the same time, we also notice that there are a large number of redundant resources in some centralized or networked information service facilities or control devices, and it is not necessary to deliberately add extra heterogeneous redundant executors to achieve mimic defense. For example, the data center IDC is equipped with a large number of heterogeneous redundant computing and storage resources. The network often includes data links with heterogeneous redundant backups, storage devices, or network element devices provided by different vendors. The SDN network is equipped with dynamically backup master-slave controllers and heterogeneous redundant data forwarders, and the cloud computing service environment has a large number of pooled resources of various types. By rationally planning and utilizing these heterogeneous redundant resources according to the mimic architecture, it is very economical to provide robust services with high-reliability and high-security levels. In other words, even in these environments, it is only necessary to incrementally deploy some physical or virtual devices related to input distribution, feedback control, and output decisions on the resource management layer to achieve a high-security level of mimic effect without special attention to issues such as source, configuration, and dynamic management of heterogeneous redundant executors.

11.3.4   Cost of Cleanup and Reconfiguration

Reconstruction, recombination, reconfiguration, or cleaning (including varying degrees of initialization) operations are fundamental functions that mimic defenses ensuring the diversity and availability of their scenario resources. When the multimode output vectors are found incompliant in the mimic ruling process, its negative feedback control mechanism can have multiple operational strategies, such as replacing or migrating the output incompliant executors or defense scenarios, or cleaning or initializing the suspected problematic executors, or reconstructing, reorganizing, or reconfiguring the exceptional executors. These operations are designed to interfere with or disrupt the stability of the attack chain by changing the defense scenario, so that the attack results are not inheritable, and the attack experience is difficult to replicate. However, cleaning requires additional resources and time overhead and even involves complex processing issues such as checking and recovering the data file legality. In particular, deep cleaning also includes the application of security measures such as vulnerability and Trojan scanning, virus-killing, and door sealing. In addition, reconstruction not only involves the availability of resources, cross-platform compatibility, or adaptability of software but also is related to the reconstruction module or the composability of software and hardware components and may bring security issues to reconstruction control aspect.

11.3.5   Cost of Virtualization

In recent years, virtualization technology has made great progress, which has played an important supporting role in the economic realization of mimic defense. These technologies include the virtualization such as data centers, cloud computing and cloud services, the migration of tasks, jobs and processes, and newly developed virtual container technologies; popularization of distributed network environments; software-based dynamic technologies; cross-platform scripting explanatory execution technologies; and diverse or multi-version compilers and other technologies. However, in addition to the inherent loss of processing efficiency in virtualization, there are many challenging technical issues to be addressed, such as security of virtual underpinning infrastructure, security and recovery technologies in virtualized environments, and creation of heterogeneous virtual machine and mimic applications.

11.3.6   Cost of Synchronization

When some executors in the mimic brackets produce sporadic anomalous outputs due to generalized uncertain disturbances, in order to fully utilize or use these executors in subsequent work, they need to be rolled back or cleaned. Therefore, in normal operation, the system needs to retain the necessary state and data in a phased manner according to a certain mechanism, so that the rollback and cleaning operations based on the state machine principle can synchronize the operations of other normal executors as soon as possible and shorten the system derating operation time. This requires an additional synchronization recovery mechanism that not only preserves the operating state and data (such as scenario snapshot) at a certain point in time but also sends the status and data to the executor that implements the roll-back or cleaning operation. It can enter the standby state consistent with the pace of other normal executors as soon as possible. Related synchronization operations and data environment recovery strategies will inevitably introduce a certain (and possibly costly) resource overhead, so reasonable planning and careful design are required.

In general, mimic field switching is accompanied by movement of stored data. If the amount of data migration between defense scenarios is too large, it not only consumes system performance and efficacy but also degrades the user service experience. Therefore, controlling the amount of data migrated is a question that needs to be carefully discussed in the implementation of mimic defense. In addition, in the mimic system, in order to prevent the cooperative cheating between multiple executors or to maliciously spread the virus Trojan, the design requirements of "dis-cooperation" are emphasized to avoid the session mechanism and data interaction between the executors. This will bring challenges to the cleaning and recovery operations of the executor in an abnormal situation, although the degree of challenge varies in applications.

It should be pointed out that the synchronization problem does not occur only in the process of executor replacement, scenario migration, cleaning, and recovery in the case of multimode output vector anomalies. For example, when there is a TCP/IP protocol stack in the executor, since the initial TCP sequence number of each executor is generally determined randomly, the output arbiter not only associates the selected IP packet sequence number with the peer device but also the input dis-tributor makes peer IP confirmation number associated with each executable. In particular, applications that are strongly associated with the "long session state transfer mechanism" (such as network routers) may require additional proxy opera-tions that match the TCP sequence number. Therefore, the heterogeneous redun-dancy architecture and its operating mechanism are strongly related to the synchronous operation problem.

11.3.7 Cost of Ruling

Since the heterogeneous executors in the mimic brackets may have differences in the output vector response delay, length, content, and even fields (especially when using COTS components), even semantically identical grammatical differences may exist. The mimic brackets are connected in series on both sides of the mimic interface. Therefore, the processing complexity, time overhead, and self-security (anti-attack) of the mimic decision may become new challenges. The advantages and disadvantages of different ruling methods need to be weighed, for example:

11.3.7.1 Synchronous Judgment

It can start with the output vector response that has the latest reachable output response of the executor or the arrival of the output vector that satisfy the minimum number request (e.g., ≥ 2). When there are many contents to be judged and the output time is uneven, the content of the output vector that arrives successively can be segmented and judged to minimize the delay caused by the judgment itself. This approach requires that the semantics and syntax of the multimode output vector be clear or that the content or range of values involved in the study is clear. Since the delay of the multimode output vector reaching the ruling aspect is different, it may be necessary to set a buffer queue and a control mechanism.

11.3.7.2 Agreed Output

Based on historical statistics or the degree of autonomous credibility, the output vector with the best performance or reliability is selected as the agreed output. The output vector of other executors is only used as the "confirmation" or backward verification reference system of the agreed output vector. If the conclusion of the judgment does not meet the policy requirements, it is necessary to start the operation of negating the previous output (e.g., sending the correction result, etc.) and suitable to applications that "allow correction" within a given time window.

11.3.7.3 First Come, First Output

In fact, the response of each executor to the service request is discrete in time and has certain random factors, so it is difficult to say that an executor is the fastest in response to all services. Therefore, it may be reasonable to use the "first come, first output" strategy in the output decision. In the same way, the advancement of software technology, especially the advancement of virtualization technology, cross-platform scripting technology, and middleware and embedded technology, has led to a reduction in the software integration innovation threshold and an enrichment of technical resources.

11.3.7.4 Regular Judgment

For multimodal output vectors that are substantially identical in syntax semantics, a regular expression algorithm can be used to judge only the fields, domains, or loads that are of particular interest. The information of the output vector arrived first is used as a template, and the output vector arrived subsequently is regularly matched, and the output vector required for compliance is selected as the arbiter output. This method is suitable for multimode output vector cases where there are undefined expression fields such as undefined items, extended fields, timestamps, priorities, communications serial numbers, and custom fields. However, this also provides an opportunity to tunnel through non-judgment fields.

11.3.7.5   Mask Decision

Similar to regular judgment, when there are uncertain content options in the multi-mode output vector protocol message or message format (typically like calculation accuracy error, TCP serial number, etc.), or only concerned with the state of some defined domains, you can use mask first. The code template masks or extracts the corresponding fields from the output vector and then uses a suitable algorithm for decision processing. The problem with its existence is the same as above.

11.3.7.6   Normalized Pretreatment

When the output vector is long (e.g., an IP long packet), or if the semantics have the same syntax, or there are non-fully defined fields such as undefined, custom, extended fields, etc., the output vector can be normalized (such as data stream-based preprocessing). For example, a mask template is used to extract the fields of interest in the output vector for subsequent processing; the algorithm for checking and summing the hash values is used to convert the cropped output vector into a determined value to reduce the decision delay (its attached effect is able to mask the injection information from the attack surface); set the output vector buffer queue to sense the multimode output vector arrival; and set the template and weight parameter library to handle complex decisions. In short, normalized preprocessing can simplify the implementation of the ruling process, but in the distributed environment, it will increase the processing overhead of the executors, and in the centralized processing mode, it will increase the technical cost of the arbiter accessibility function.

11.3.8   Cost of Input/Output Agency

Setting the input/output agent links on both sides of the mimic brackets usually helps in the implementation of the mimic function. For example, to assist the feedback control, select or activate the executor in the standby state, select the specified executor output vector to participate in the mimic ruling, realize some normalized conversion or synchronization operation inside and outside the mimic interface, provide an input request or an output response buffer, and offer function to provide a supporting environment for various possible preprocessing operations. However, the introduced insertion delay and technical complexity will be inevitable, and its own security flaws may also become a new target for attackers. It should be emphasized that setting the input agent can increase the difficulty of attack accessibility and effectively suppress the attack input of known features or behavior rules.

11.3.9 Cost of One-Way Connection

One-way connection is usually on a hidden front, and it is a secure contact mechanism adopted to prevent the possible unpredicted events, like being captured, from causing damage to the organizational system. The one-way connection mechanism is that a node has only one at upper level and one at lower level and does not associate with any other node. In case any node has an accident, it is only necessary to transfer its superior and subordinate members to ensure the security of the organization system. The one-way connection mechanism of the mimic brackets needs to ensure that there is no cooperative cheating between multiple executors and there should be no two-way conversation or communication mechanism between the mimic brackets and the executors. However, since the one-way connection needs to introduce the distribution of control functions, the simplification of control relationships, and the minimization of control information, some service functions that can be processed or controlled centrally in the same space need to be configured with additional processing resources, storage space, or increasing the length of the control chain, etc.

The one-way connection mechanism uses an engineering approach to ensure that the credibility of the system does not depend on the absolute loyalty of any aspect, which can greatly simplify the security design and verification of each aspect. However, the application of this serial working mechanism for single-line contact will inevitably increase the complexity of system reliability design, and it needs careful consideration.

In summary, the cost of defense of passes or strategic points is usually increased linearly in the mimic brackets. Fortunately, although the COTS products such as hardware processing resources and environmental support software play a significant control role in the target system, their proportion of the total cost of the target system is not high. Even if the material resources of the heterogeneous executors in the mimic interface increase by three to five times, their impact on the total cost may still be affordable (see Chap. 14 for some case studies). If we consider the indispensable role of the target system in the entire security defense system, and the comprehensive use cost in the target system life cycle, the cost-performance ratio of the mimic structure system is almost unparalleled because mimic defense cannot only effectively control (whether from "surface or point") the risks or threats caused by dark features such as known or unknown backdoors, Trojans, etc. but can, without depending on any additional security technologies, significantly improve the robustness and flexibility of the information system service function with only the endogenous structural effect. This endogenous effect of high-reliability, high-credibility, and high-availability robustness is unique to mimic defense technology, and it cannot be made by just stacking the existing technologies, equipment, and financial and human resources.

11.4 Scientific and Technological Issues to Be Studied and Solved

Frankly speaking, the advancement of microelectronics technology makes the design and manufacturing cost of technical systems only closely related to the scale of market applications but weakly related to complexity or sometimes seemingly less relevant. The benefit is that we can widely adopt in engineering implementation. The practice of "breaking a fly upon a wheel" avoids the troubles caused by specialized design. The rich, mature, powerful, reusable, and standardized COTS products are inexpensive, with complete ecological environments such as supply chain, service chain, and tool chain, laying a solid foundation for innovations of technology and product such as derivatives and integration. In the same way, the advancement of software technology, especially the advancement of virtualization technology, cross-platform scripting technology, and middleware and embedded technology, has led to a reduction in software integration innovation threshold and an enrichment of technical resources. The open-source community and other crowd funding models have not only accelerated the development and maturity of software and hardware but promoted the development of diverse ecological environments and technology markets.

However, the rapid advancement of contemporary information technology has laid a solid foundation for the engineering realization of cyberspace mimic defense. However, there are still many challenging theoretical and engineering issues that need to be resolved.

11.4.1 Scientific Issues Needing Urgent Study in the CMD Field

The scientific issues to be explored in the CMD field mainly include:

(1) What scientific understanding is lacking in cyberspace mimic defense?
(2) What research is needed to achieve these scientific understandings?
(3) What are the major theoretical and technical challenges facing the realization of cyberspace mimic defense?
(4) What is the development roadmap for cyberspace mimic defense?
(5) How to construct the attack theory and model in the mimic defense environment?
(6) How to scientifically set the mimic defense granularity and security level and make them measurable?
(7) How to give the simplest design of calibratable metrics according to the mimic defense security level?
(8) How to give or establish the mathematical principle of mimic defense?
(9) How to quantify the heterogeneity of the defensive scenarios in the mimic brackets?
(10) Test standards and verification methods for mimic defense systems.

11.4.2   Engineering and Technical Issues Needing Urgent Solution in the CMD Field

11.4.2.1   Dissimilarity Design and Screening Theory

In theory, the functionally equivalent heterogeneous executors should be absolutely dissimilar, that is, the following three requirements can be met to ensure that the mimic defense has an idealized defense effect (Fig. 11.4):

(1) The maximum functional intersection between a given functional intersection and a heterogeneous redundant executor is perfectly matched.
(2) The implementation algorithms for a given functional intersection are completely different.
(3) The implementation algorithm for a given functional intersection is completely independent in the physical space.

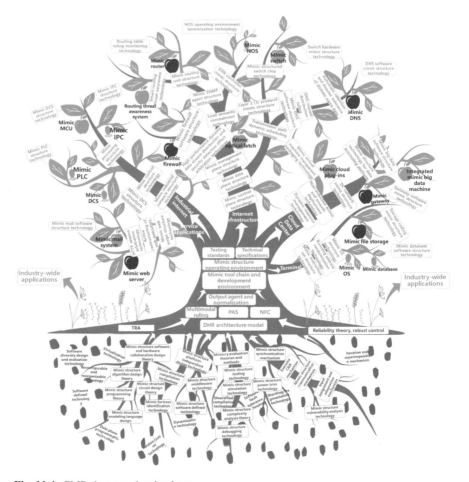

Fig. 11.4 CMD theory and technology

However, in practice, these requirements can neither be fully satisfied nor require such strict dissimilarity, which can be confirmed by the engineering practice and application history of the dissimilar redundancy architecture in the aerospace field. However, the following theoretical and engineering issues are also unavoidable:

(1) What kind of models and methods can be used to design a heterogeneous or diversified heterogeneous executor that only has intersections in a given feature set, while other functions (including dark features) do not intersect?
(2) For diversified executors (such as COTS products) that satisfy the functional equivalence requirements of the mimic interface, what methods can be used to identify or measure whether there are other intersection functions (possible dark features) between them?
(3) Under the condition of given input excitation, what degree of quantifiable dissimilarity is required between heterogeneous executors to ensure that the multimode ruling escape probability meets the expected requirements?
(4) What kind of safety requirements can be met if each executor implements the dissimilarity design only in the sensitive part, critical path, or core link of the input stimulus?
(5) Theoretically, "dynamic heterogeneous redundant space, coordinated consistent attack under noncooperative conditions" does not allow trial and error, so how to mathematically prove that "refusal of trial and error" can "block collaborative attacks" and so on?

11.4.2.2   Pluralistic and Diversified Engineering Issues

In the process of implementing mimic defense technology, there will also be some new pluralistic and diversified problems, including:

(1) The intersection of equivalent functions between diversified executors can be determined by conformity and conformance testing of standardized or normalizable interface functions, but how can we judge whether there are other dark features intersections?
(2) Is there a certain intersection function based on the source between the soft and hard variants of homologous diversity? If so, how can we determine that there is an origin-based intersection function between a given heterogeneous hardware and software component?
(3) In the noncooperative independent space and the same restricted excitation source, in addition to the equivalent intersection function between the given heterogeneous executors, if there are other intersection functions, how can these additional intersection functions simultaneously activate and produce the same consistent output vector?
(4) How to evaluate the availability of dark features such as inherent backdoors and Trojans after the hierarchical and dynamic combination of diverse hardware and software components?

11.4.2.3   Assessing the Security Impact of the "Homologous" Component Vulnerability on the DHR Architecture

Today, when the open-source innovation model is ubiquitous, the same genetic defects are likely to exist in functionally equivalent diverse or pluralistic hardware and software components. How to find out whether there are "other homologous" problem codes in these components, whether they have the same available conditions, what kind of impacts may be on the corresponding executors, and how much side effects on the generalized robustness of the DHR architecture, and what are the possible methods and measures, and so on?

If effective analytical methods and techniques are formed, we can reliably and economically avoid or weaken the impact of the "same homologous" problem code.

11.4.2.4   How to Establish a System Design Reference Model

Different from the static architecture of dissimilar redundancy, the mimic defense architecture is dynamic, diverse, and random and has feedback control for two purposes. One is to minimize the noncooperative conditions, and each executor can produce most identical or completely identical error output vectors through the same input stimulus. The other is to obtain the maximum degree of endogenous security effects with as few heterogeneous redundant resources as possible. At the same time, the existence of the mimic ruling-based policy and schedule and multi-dimensional dynamic reconstruction negative feedback mechanism deprives the attacker of the chance of trial and error in the "noncooperative blind coordination" scenario. Both intuitive and theoretical analysis show that the higher the degree of dissimilarity between the executors, the stronger the dynamics of the defense scenario in the mimic brackets. The more heterogeneous scenarios that can be utilized, the richer the semantics of the output vector, and the more complex the mimic ruling strategy, the lower the escape probability based on the output vector. However, in order to obtain engineering guiding principles and evaluation criteria, under the given economic and technical conditions, how to use the above factors to achieve the desired uncertain disturbance suppression ability within the DHR framework, there is no precise design method up to now.

11.4.2.5   How to Prevent Standby Attacks

In practice, we have noticed that the heterogeneity between heterogeneous executors in the mimic defense (including dissimilar redundancy) architecture is not so harsh in engineering implementation. Firstly, the diversified heterogeneous environment itself is very different, such as differences in the instruction system, addressing mode, address space, cache scheduling, working frequency, bus control, supporting chipset, and other hardware environment. Also, the operating systems and related support software have been charged with more differences in algorithms and functions. Secondly, although the basic functions of non-single-source applications meet the requirements of the relevant standards, the implementation architecture and the methods or algorithms used may be different. Moreover, even if the same source code passes through different versions of the compiler and combines the executable files generated by the host operating system, the difference between them is actually quite significant. For example, it is hard to tell the degree of commonality among the parallel and nonparallel environments, isomorphism and heterogeneous redundancy design, stack and storage management, library functions, file systems, drivers, and even power management, which, together with different settings of optimization and parallelism identification parameters, will affect the structure and operational efficiency of the executable file generated by the compiler. The differences included or not included in the above examples will not only cause temporal dispersion of the expected multimode output vector, but difficult to get full or most of the same error output vector at the same time, and meet the mimic escape probability of the attack mission planning requirements, besides a common algorithm (equivalent function) or method with consistency errors (not excluding the deliberate setting). However, as far as the current level of cognition is concerned, if an attacker can obtain the "super privilege" (such as OS privilege) of most executors or has the ability to initiate a standby collaborative attack, it has a mechanism to possibly crack the mimic defense. Therefore, preventing attackers from achieving cross-domain coordinated attacks through "super privilege" has become issues that need to be solved first theoretically and engineeringly in mimic defense. Such issues include how to prevent the "super privilege" of most executors from being stolen, how to establish a function or information protection zone that is not related to "super privilege," how to disrupt the stability under the "standby synchronization state" of the attack chain, how to use the multi-dimensional heterogeneous dynamic space to disintegrate the preconditions of the standby attack, and how to forcefully change the current operating environment through the external channel command of feedback control.

11.4.2.6   Mimic Ruling

The mimic ruling is the key aspect of mimic defense. By comparing the consistency of multimode output vectors with a given ruling strategy, you can timely discover the abnormality or abnormal performance of the executor and prevent further destruction of the target system from possible deliberate attacks. However, mimic ruling does not deal with such operations as greater than, less than, equal, or unequal; in other words, the engineering implementation usually cannot be "unrelated to the output vector syntax or semantics," an ideal goal, because multiple heterogeneous executors cannot guarantee from time to time that the output vectors are always the same in syntax or even semantics even if they are functionally equivalent. For example, the TCP/IP protocol stack has designed a "sticky packet processing" algorithm to improve processing performance, that is, multiple requests from the same source are responded with a "super long" packet instead of a strict request-response mode. For example, due to the uncertainty of the starting sequence number of each TCP, the nonmandatory requirements of IP packet options, the difference in calculation accuracy of different algorithms, and the opacity introduced by encryption operations, etc., the complexity of the mimic ruling will increase. In addition, multimode output vectors are surely discrete in delay, which may more or less obstruct the implementation of mimic ruling. If the ruling strategy also includes conditional judgment parameters such as weight value, historical performance, and version confidence, the engineering complexity of the mimic ruling will be challenging. Therefore, although the setting of the mimic interface is theoretically transparent, it is difficult to be completely independent of the protocol specification and semantic grammar in engineering practice, so it is still necessary to thoroughly analyze the problems related to the realization of the mimic ruling. In a general sense, the introduction of an output proxy mechanism may help to simplify the engineering difficulty of the decision itself, but it also causes new problems such as insertion delay, additional cost, and reliability of the output agent component.

11.4.2.7   Protection of the Mimic Control

Due to different application scenarios, the implementation complexity of the input/output agent, mimic ruling, and negative feedback control of the mimic architecture will be very different. In addition to the technical challenges that mimic rulings may face, the implementation of feedback control will encounter many thorny issues. For example, it is not easy to replace, migrate, clean, repair, resynchronize, reconstruct, reorganize, and reconfigure abnormal executors, but it directly decides on or affects the iterative convergence speed of the feedback control, and the quality of reconstruction and reorganization will involve the dissimilarity of executors in the current service set. Considering the economic compromise of failure rate, escape rate and uncertainty, we need to make differentiated arrangements or deployments according to confidence, reliability, compatibility and diversity when configuring service set elements. In addition, the complexity and importance of the mimic control will undoubtedly become a new target for attackers. Of course, there is also an "ultimate question" like whether "the root of trust is trustworthy" in trusted computing. The answer to the above question is preliminarily answered in Sect. 9.4.2. The key is to use the one-way connection or unidirectional processing mechanism to link the mimic brackets function to the entire feedback loop including the service function P so as to it can better obtain endogenous security attributes based on mimic structure. In addition, the following measures can be considered: first, open the design of control components to diversify the supply due to its feature of separation of control from service; second, add the private user-defined functions and hardware reconfigurability; third, use the uncertain parameters inside the target object to achieve the variable structure processing; fourth, the control component itself uses trusted computing or re-mimic technology; and fifth, implement the control function completely by non-instruction stream hardening technology.

11.4.2.8   Mimic Structural Design Technology

It mainly includes basic theory and technology such as mimic structure modeling language, mimic structure programming theory, mimic structure circuit design theory, mimic structure algorithm design theory, and mimic structure software and hardware collaborative design theory.

(1) Mimic structure modeling language

Similar to UML, it is a modeling language used to describe mimic structure models, providing modeling and visualization support for mimic structure development. The modeling language for mimic structure describes the requirements, methods, and development process of architecture, configuration, interfaces, data flows, and mimic structure implementation. The modeling language should support the abstract mimic structure, the description methods for specific systems, hardware and software components, and the graphical language for modeling and software system development.

(2) Mimic structural programming theory

Similar to the parallel programming theory based on parallel distributed computing architecture, it mainly researches how to design the mimic structure program based on the mimic defense architecture and how to transform the existing non-mimic software programs and run them in the mimic structural system, including message transfer mechanism, instruction set level, code level, component level, module level, and unit level programming.

(3) Mimic structure circuit design theory

Similar to the dissimilar redundancy design theory in the flight control system, it mainly studies how to design the mimic structure at the hardware circuit level, including circuit level, component/device level, etc., such as digital circuit, industrial controller, CPU, and so on. In addition, the research should also include how to design the hardware program for the mimic structure based on its logic device, such as FPGA, Verilog, and other hardware programs with mimic structure processing characteristics based on ASIC, FPGA, and other architectures.

(4) Mimic structure algorithm design theory

Similar to the parallel programming theory based on parallel distributed computing architecture, the mimic structural programming theory mainly researches how to design the CMD-based mimic algorithm and data structure and how to transform the existing non-mimic algorithm into the mimic structure system, for example, description and expressions of the same structured data with different data structures, such as arrays and FIFOs; then it uses different algorithms to achieve the same functions, such as using different routing calculation methods and so on.

(5) Mimic structure software and hardware collaborative design theory

Similar to the parallel computer architecture and programming theory, it mainly studies how to design the mimic structure synergy through the bottom-up hardware and software level to maximize the endogenous security effects and efficiency of the mimic structure, such as threats, security, functionality, and performance requirements of the specific system. It also studies which ones are suitable for mimicking at the software level, which are suitable for mimicking at the hardware level, and which require hardware and software cooperative mimicry.

11.4.2.9   Mimic Construction Implementation Technology

It mainly includes mimic fortress (or pass) recognition technology, mimic structure compilation technology, mimic structure platform-independent design technology, mimic structure software definition technology, mimic structure middleware technology, and mimic structure power control technology.

(1) Mimic fortress (or pass) identification technology

The technology explains how to seek sites for mimic protection, such as a given system model, actual system, or software code. A typical technical implementation vision is to input a piece of code or protocol into the mimic system of pass identification, and the system will automatically identify the vulnerabilities through the software code and give out mimic recommendations.

(2) Mimic structure compilation technology

Similar to the diversified compilation technology, the technology deals with how to automatically construct an executor that meets the requirements of heterogeneity (diversification) under certain conditions of heterogeneity or safety performance and perform mimic assembly to generate the mimic structure software. For example, input a piece of source code or two progress codes into the mimic structure compilation system, and the system will automatically convert to different software versions that meet certain heterogeneity (diversity) and quantity requirements and give out the mimic structure model of the software code.

(3) Mimic structural platform-independent design technology

Similar to the Java virtual machine, the technology focuses on how to design a mimic runtime environment executor to ensure that the mimic program can seamlessly migrate between different underlying architectures, non-mimic and non-mimic operating environments, such as mimic software running on a non-mimic structural operating system, or running on a processor with a different architecture such as X68 or ARM.

(4) Mimic structure software definition technology

Similar to the software-defined network, the technology implements the definition, reconstruction, executor switching, mimic structure configuration, and data flow in the mimic field. The software-defined mimic structure technology uses the out-of-band control technology, which accepts the input information of the mimic feedback control strategy and performs the dynamic adjustment of the defense strategy and the field.

(5) Mimic structure middleware technology

Similar to the software middleware technology and related to the mimic platform-independent design technology, it carries out the abstract encapsulation, resource management and mimic assembly, functional programming, executor cleaning, etc. of the underlying heterogeneous resources in large-scale hardware and software systems such as cloud computing environment to make the mimic structure transparent to system management, users, and applications.

11.4.3 Defense Effect Test and Evaluation

For information systems or control devices, the measurability of functional performance (including functional indicators and safety indicators) is a prerequisite for its promotion and application. For example, the performance test of the firewall can refer to the benchmark defined in RFC 3511 [13]; the performance of the software-defined controller can be based on the RFC benchmark of the SDN control plane [14]. Mimic defense is a revolutionary technical architecture. Prior to this, there was no technical system based on DHR architecture, so there are no references for its measurement. Therefore, how to confirm that the system under test is a mimic architecture, including how to test and measure the defense effect of the mimic system, is the most realistic and urgent issue. In the system of mimic architecture, the mimic ruling-based policy and schedule and the multi-dimensional dynamic reconfiguration adaptive feedback mechanism make the attack surface uncertain, the reachability of the attack packet to the target cannot be guaranteed, and the reproducibility of the attack effect cannot be ensured. So the traditional theory based on vulnerability detection, discovery and utilization is no longer applicable to the mimic system for the existing security detection theories and methods do not work for it, and new theories and methods and indicator systems need to be created. Fortunately, as the mimic defense effect is derived from the generalized robust control architecture and operation mechanism of DHR, we can establish reference indicators and functional test sets for the anti-attack and reliability of the mimic system with mature robustness, reliability assessment theory, and reference models and simulation analysis tools, plus the verification metric of the "white box" injection test case method (see Sect. 10.4.6). However, the completeness of the test set, the appropriateness of the test case injection, and the scientificity of the index system also have problems that need further research and discussion. Please pay attention to the following issues:

(1) Mimic structure debugging technology

Similar to the parallel program debugging technology, the mimic structure debugging technology is a technology that debugs the consistency of the state, data, and results of the heterogeneous code in the parallel execution process in the mimic structure. For example, how to sample the internal state of a heterogeneous executor, check on rotation, and track by single-step different executor states for a mimic structure software with multiple executors running in parallel.

(2) Theory of mimic structural complexity analysis

Similar to the algorithm complexity analysis theory, it explores the ways to analyze the time complexity and space complexity of a mimic structural hardware and software system, including space complexity and hardware implementation complexity. A typical technical vision is to produce the optimal design of the mimic structure based on system performance and security requirements.

(3) Mimic structural vulnerability analysis technology

Similar to the vulnerability analysis theory and technology of hardware and software, it explores the fragility of the mimic structure itself, the size of the attack surface, etc., including the collaborative attack problem for mimic structure and the vulnerability introduced by mimic structure in engineering implementation. In general, mimic structural strength can be used to describe the fragility and attack surface size of the mimic structure itself.

(4) Simulation of mimic structure

Similar to the flight control simulation technology, it explores how to use computer simulation technology to comprehensively analyze and evaluate the function logic, operation performance, and safety performance of the mimic structure before engineering development of the mimic structure system, so as to lay a foundation for the subsequent project realization. In addition, the mimic structure simulation technology can also simulate possible anomalies.

(5) Mimic structure testing technology

Similar to the hardware and software testing technology, it mainly tests the functional performance, reliability, availability, and security of the mimic structure. The functional performance and availability of the mimic structure mainly depends on the service functions it carries. The reliability test can be carried out referring to the reliability theory. The security test mainly refers to the "white box testing" methods, including white box injection test method, white box test case construction technology, white box code insertion technology, and collaborative attack test.

(6) Theory and method of mimicry assessment

Similar to the concept of redundancy, it measures the dissimilarity between two hardware and software with same functions but different methods of implementation and studies how to measure the mimicry of a mimic structure and the metrics and how to design different hardware and software executors when the mimicry is certain. Usually, the higher the mimicry, the higher the security of the system.

11.4.4 Comprehensive Use of Defense Capability

Usually, there are many "basic defense elements" such as dynamicity, diversity, randomness, relativity ruling, and feedback control in the mimic architecture, including traditional threat perception, behavior analysis, feature extraction, and other active and passive defense measures (such as intrusion detection and intrusion prevention), as well as generalized robust control methods for dealing with uncertain disturbances. How to allocate these resources economically, deploy these capabilities reasonably and effectively, and form an integrated defense effect require comprehensive consideration and arrangement. For example, if you shorten the iterative convergence time of the mimic control loop and reduce the cleaning recovery time of the problem executor as a single measure to counter the network attacks targeting functional paralysis and performance degradation, the defense cost and effect will certainly not be as good as adding a "problem message" filter to the input distribution; the latter will be more effective and economical. Any hope that policy and schedule or dynamic reconfiguration can properly play its role, or how to reorganize and reconfigure more diversified scenarios with the least amount of heterogeneous resources, or how to reduce the scene iterations to "precisely avoid problematic scenarios," will not come true without the support of artificial intelligence analysis and decision-making based on running logs and historical scenario data.

It should be emphasized that the mimic defense provides the technical means of security for the system to avoid the case that the "fortress is often broken from the inside" and implement a security model of "multiple keys for a lock" at the manageable level, like "the nuclear button" that require the simultaneous action of multiple weapon operators, and the maintenance engineers can operate only under mutually constrained conditions.

11.4.5 Issues Needing Continuous Attention

The research contents that need continuous attention include scientific argumentation and concept abstraction of mimic structure, operation mechanism and their effectiveness; formal description method of vulnerability exploitable mechanism in mimic environment, proving the effect of mimic system on attackers who exploit the vulnerabilities and backdoors across domains; the suppression of the "homologous" dark features of the target system by the mimic ruling-based policy and schedule and the multi-dimensional dynamic reconstruction adaptive negative feedback mechanism; and the scientific description of the uncertainty in the mimic architecture.

11.4.6 Emphasizing the Natural and Inspired Solutions

There are many biological systems in nature that are far more robust, flexible, and efficient than human cyberspace systems. In the defense against bacterial and viral invasion, the division of labor and synergy between non-specific immunity and specific immunity has important enlightening effects on improving the integrated surface-point control mechanism of the mimic defense. To do this, we have to figure out the surface defense mechanism of non-specific immunity and the effect of its change of genetic diversity and to explore the working principle between the point defense and natural evolution of the specific immunity. For example, distributed processing, pathogen recognition, multi-layer protection, decentralized control, diversity, signal characterization, and "IFF" or "accidental injury" prevention of non-specific selection and removal have shown many inspiring mechanisms that can become new ideas for solving cyberspace security issues.

References

1. Schaller, R.R.: Moore's law: past, present and future. IEEE Spectr. **34**(6), 52–59 (1997)
2. Rotem, E., Engineer, S.P.: Intel architecture, code name skylake deep dive: a new architecture to manage power performance and energy efficiency. Intel Developer Forum. pp. 1–43 (2015)
3. Zhu, X., Meng, X., Yan, X., et al.: "Tianhe No.1" massively parallel application testing. In: National Conference on High Performance Computing, pp. 265–269 (2011)
4. Jajodia, S., Ghosh, A.K., Swarup, V., et al.: Moving Target Defense: Creating Asymmetric Uncertainty for Cyber Threats. Springer, Berlin (2011)
5. Yeh, Y.C.B.: Triple-triple redundant 777 primary flight computer. In: Proceedings of Aerospace Applications Conference (AAC' 96), pp. 293–307 (1996)
6. Heinecke, H., Schnelle, K.P., Fennel, H., et al.: Automotive open system architecture-an industry-wide initiative to manage the complexity of emerging automotive e/e-architectures. In: Convergence, pp. 325–332 (2004)
7. Wang, L., Tao, J., Kunze, M., et al.: Scientific cloud computing: early definition and experience. In: The 10th IEEE International Conference on High Performance Computing and Communications, pp. 825–830 (2008)
8. Docker, M.D.: Lightweight Linux containers for consistent development and deployment. Linux J. **2014**(239), 2 (2014)
9. Duan, R., Fan, X., Gao, D., et al.: Reconfigurable computing technology and its development trend. Compu. Appl. Res. **21**(8), 14–17 (2004)
10. Braun, T.D., Siegel, H.J., Beck, N., et al.: A comparison of eleven static heuristics for mapping a class of independent tasks onto heterogeneous distributed computing systems. J.Parallel Distrib. Comput. **61**(6), 810–837 (2001)
11. Armbrust, M., Fox, A., Griffith, R., et al.: A view of cloud computing. Commun. ACM. **53**(4), 50–58 (2010)
12. Ouelhadj, D., Petrovic, S.: A survey of dynamic scheduling in manufacturing systems. J. Sched. **12**(4), 417–431 (2009)
13. Hickman, B.: RFC 3511-Benchmarking methodology for firewall performance. http://www. ietf.org/rfc/rfc3511.txt (22 Mar 2016)
14. Vengainathan, B., Basil, A., Tassinari, M., et al.: Benchmarking methodology for SDN controller performance. https://tools.ietf.org/html/draft-bhuvan-bmwg-sdn-controller-benchmark-meth-00.html (22 Mar 2016)

Chapter 12
Examples of Mimic Defense Application

12.1 Mimic Router Verification System

12.1.1 Threat Design

As the hub node of the Internet facility, the router decides on the packets forwarding path through network calculation to achieve end-to-end data transmission. It is a basic element of cyberspace; covers the core layer, convergence layer, and access layer of the entire network; and connects multiple heterogeneous networks. Therefore, its service reliability and credibility are crucial for cyberspace security. However, the current router security is not optimistic at all. Snowden revealed that "the US National Security Agency is monitoring Chinese networks and hosts through Cisco routers" [1]. The National Computer Network Emergency Response Technical Team/Coordination Center of China (known as CNCERT or CNCERT/CC) analyzed the routers of mainstream vendors, like Cisco, Linksys, NETGEAR, Tenda, and D-Link, confirming the existence of preset backdoors in their products [2]. The investigation shows severe security problems with the routers.

Due to their location and route forwarding function, routers are ideal entry points for attackers to launch attacks. Once an attacker controls a router, he will cause incalculable damage to cyberspace security. If he attacks the host, only the host will be damaged; if he attacks the router, he will jeopardize the entire network connected with the router. As is known to all, routers are network devices through which attackers can easily get users' private data, monitor users' online behavior, obtain their account and password, tamper critical user data, push, and spread fake information or Trojans, or divert network data flows, paralyze the interaction of information on the network, or even derail the entire target network.

In theory, the threat to router security comes from two aspects. The first is the vulnerabilities that are unavoidable in the system design and realization; the second is the trapdoors brought in unintentionally by open codes, and the backdoors deliberately planted. The uncertainty related to vulnerabilities or backdoors threatens the router security the most.

© Springer Nature Switzerland AG 2020
J. Wu, *Cyberspace Mimic Defense*, Wireless Networks,
https://doi.org/10.1007/978-3-030-29844-9_12

The difficulty to defend routers—a closed special system—is threefold. Firstly, lack of auxiliary security means. Unlike the general system, routers do not and will not have additional protection measures like firewalls, so most of the routers are not or unable to guard against malicious attacks; Secondly, more exploitable vulnerabilities. Huge amount of codes (sometimes even up to tens of million lines) in its design and implementation periods have left numerous potential vulnerabilities that may be hard to be discovered in the closed system; Thirdly, deeply concealed backdoors. The fact that routers are designed, manufactured, and applied in non-open environments makes it easier to implant backdoors or Trojans deeply in the normal functions.

In view of these problems, this chapter presents a case of defending routers in a mimic manner, where its endogenous security mechanism is mobilized to resist uncertainty threats caused by unknown vulnerabilities, backdoors, or virus Trojans. This approach deals with security issues brought about by untrusted virus-bearing components from the architectural level of the object target, dramatically changing the traditional passive, patch-making defense model of "unearthing vulnerabilities, blocking backdoors, analyzing characteristics, screening viruses and eradicating Trojans."

12.1.2   Designing Idea

A router can be functionally divided into three planes: the data forwarding plane, the routing control plane, and the management plane. The data plane involves checking the inflow data packets and forwarding them according to the check results. The control plane carries out route computing by running different router protocols (such as RIP, OSPE, BGP) [3–5] and sends the generated router table entry to the data plane. The functions on the control plane are configured by the management plane applying rules such as CLI, SNMP, or Web. The logic function model of router system is shown in Fig. 12.1.

The control plane receives router advertisements from neighboring nodes, calculates routes, generates routing entries, and sends them to the neighboring nodes. The configuration management plane receives the management configuration request from the network administrator, configures and manages the router, and outputs the result; the forwarding plane receives the data packet input through the interface from the interface unit and then checks the local forwarding table. If the address information carried by the data packet matches that in the table, the forwarding plane will immediately follow the instructions to output the data packets through the specified interface.

Each plane of the router faces different security threats for their varied functions. The control plane is subject to the risk of tampering the routing table for hijack attacks and release of fake routing information. The configuration management plane faces the risk of man-in-the-middle attack or router hijack and release of fake routing information. The configuration management plane faces the threat of being

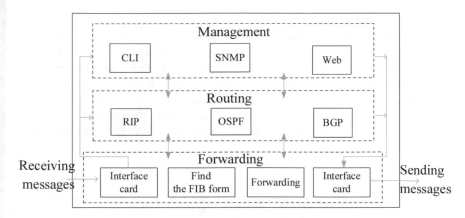

Fig. 12.1 The logic function model of router system

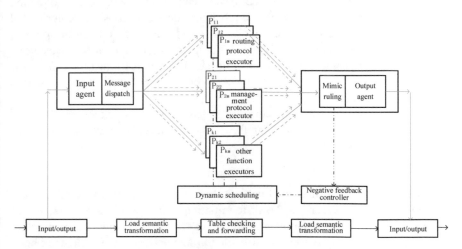

Fig. 12.2 DHR-based router mimic defense model

deprived of control authority for administrator operation or theft of configuration information. The data forwarding plane faces the threat of the preset backdoor (lurker) in the data packet being triggered in the system to steal sensitive user data or directly paralyze the system paralysis.

To deal with the security threats facing each plane, the mimic interface is set up to clarify their protection targets. The mimic interface for the data plane is set up in the checking and forwarding aspect; the mimic interface for the control plane is set up in the entry modification process; and the mimic interface for the management plane is set up in the administrator's information configuration process.

Figure 12.2 is a model of DHR-based router architecture for defense (DHR2), which is designed to deal with the protection targets of the mimic interfaces by combining the router function structure and DHR architecture of mimic defense. This DHR-based model introduces multiple entities or virtual executors on different routing planes, for example, multiple entity or virtual management executors for the management plane, multiple entity or virtual route computing executors for the control plane, and multiple entity or virtual data load semantic shifting executors for the data plane. The purpose of it is to construct a physical or logical heterogeneous redundant processing unit in each plane. At the same time, the input agent is introduced to distribute information, and the mimic ruling is introduced to aid in output ruling. Mimic ruling result is then sent to the dynamic scheduling module through the feedback controller to manage, clean, or recover the executors.

12.1.3 DHR-Based Router Mimic Defense Model

Every software functional unit in the system can be described in the following model: there is a functionally equivalent heterogeneous executor pool $\overline{P} = (P_1, P_2, \ldots, P_n)$, and choose m executors $\overline{P}' = (P_1, P_2, \ldots, P_m)$ from the pool to work on. Input agent module distributes input information to m ($m \leq n$) executors, which compute the same input, get m results, conduct mimic ruling on the m results based on a certain strategy, and get the normalized output result that is exported through the output agent.

The DHR-based router architecture for defense model can effectively fend off known or unknown security threats in the mimic interface. Here are the three characteristics of this structure.

1. Heterogeneity

Heterogeneity means that two functionally equivalent executors are different in structural design; it describes the differentiation between two executors, which generally ensures that similar attacks on two executors will not cause completely consistent failure [6–8]. Attacks outside the mimic interface are not included here, for example, attacks based on functioning defects of the system. Generally speaking, vulnerabilities and backdoors of the router protocol executor are peculiar. For example, the SNMP software function vulnerability of Cisco 2600 series routers may be different from that of the 6500 series, and it is almost impossible for it to be the same as that of Juniper's router products. In this way, attacks based on a specific vulnerability could at most affect its directly related executors. DHR2 architecture is redundantly set with many heterogeneous functional executors, making it difficult for attackers' actions to enable the output vectors of feedback-control-based heterogeneous executors to result in a majority or completely consistent result under noncooperative conditions or to achieve mimic escape. In other words, any attack that cannot achieve mimic escape, if identified by the mimic ruling, will be cleaned offline by the negative feedback control mechanism through invocation of problematic executors, or its data will be recovered online through invocation of standby executors. However, the heterogeneity between given executors is usually unable to be measured or precisely controlled; the completely heterogeneous routing protocol executor really does not exist. A completely heterogeneous router protocol executor is nonsexist in reality. Therefore, in engineering, heterogeneity cannot be solely relied on to ensure system security.

2. Redundancy

Redundancy means that heterogeneous executor in current service set is a real or virtual object with diversified configuration. In the usual sense, heterogeneous redundancy could ensure relative correctness of the mimic ruling result. When faced with the same attack, the output response of the heterogeneous executor should not be the same. The addition of the number of heterogeneous executors in the current service set can nonlinearly increase the accuracy of mimic ruling's perception of threats and provide necessary conditions for the introduction of dynamicity. Of course, too large a redundancy would inevitably increase the cost of system implementation and deployment.

3. Dynamicity

Dynamicity refers to the changing of the heterogeneous redundant operating environment of the current service set based on the output state of the mimic ruler and the policy and schedule of the space-time dimension. It mainly reflects in the following aspects:

(1) Once the mimic ruler finds inconsistency in the multimode output vectors in the current service set, it will trigger the negative feedback controller. The controller will replace or migrate the executors in the service set progressively or iteratively according to the scheduling strategy or designate a certain executor to go offline for cleaning and restarting or reorganization and reconfiguration until the ruler returns to the stable and balanced status.

(2) Irregularly replacing or reconfiguring (reorganizing) the working executors in the service set through external control commands could help reduce the exposure time of specific components or structures per unit time, increase the unpredictability of the target object's defense behavior, and reduce risks of its vulnerabilities or backdoors to be taken advantage of. In particular, dynamicity could disrupt or clear attackers' early effort to launch latent or covert attacks, thus destroying attacks based on state transition mechanism.

(3) Dynamicity is also an expansion of diversity in the time dimension. Even if it does not perceive any threat, dynamicity can increase the uncertainty of attacks based on specific backdoor or scenario and lower the success rate of collaborative attacks against multiple targets under noncooperative conditions. If it detects a threat, it can replace, isolate, and clean the infected executor in time to block or disrupt the attacker's continuous control of the target system, trying to ensure the integrity of system functions and the privacy of sensitive data.

12.1.4 System Architecture Design

12.1.4.1 Overall Framework

The following issues need to be solved to build a router with an endogenous security mechanism based on the DHR2 model.

(1) Construction of heterogeneity of functional executors. Commercial companies normally will not invest human resources to develop two sets of heterogeneous software with the same function. Fortunately, as for routing protocols, there are already several mature open-source software packages, such as FRRouting [9], Quagga [10], and BIRD [11], which are designed by different teams using different languages and architectures and are absolutely impossible to have similar vulnerabilities [12, 13]. Besides, some mainstream router vendors have rolled out router emulators, for example, Cisco Simulator [14], Packet Tracer [15], Juniper Olive [16], and Huawei eNSP [17], and the open-source emulators such as GNS3 [18] and QEMU [19]. All these can be used to simulate a routing system with equivalent functions based on executable files without source codes, so they provide a wide material basis to build router heterogeneity. Furthermore, multiple versions of the same routing software are also heterogeneous to a certain degree, where new bugs will be brought in when fixing the old ones. The introduction of the mimic defense mechanism can enable simultaneous run of both old and new versions to overcome this problem. For example, 30% of the bugs in Quagga 0.99.9 are not found in Quagga 0.99.1 [12]. This method can speed up the maturity of the new version.

(2) Dynamic scheduling of functional executors. The dynamicity of the DHR2 model requires that the functional executors can be dynamically selected from the executors' pool and loaded, in which the abnormal executors can be cleaned offline and the visual structural representation of the working executors can be dynamically changed. The network function virtualization (NFV) technology can be used to instantiate the heterogeneous redundant executors and carry the functional executors on virtual hosts. In this way, you can conveniently and cost-effectively implement dynamic scheduling of executors and, at the same time, lower the cost of introducing heterogeneity and redundancy.

(3) Distribution agent on the message processing path and the insertion of the mimic ruling point. In the traditional router, the tight coupling of hardware and software, the customized internal communication interface, and the self-enclosed message processing make the implementation of mimic defense difficult and less economic. However, SDN technology [20] brings some good news to the DHR2 model. Its southbound APIs standardize the internal message interface between hardware and software in traditional routers and open them so that all the messages are carried through the standard OpenFlow interface [21], are dispatched and judged through the distribution agent, and mimic

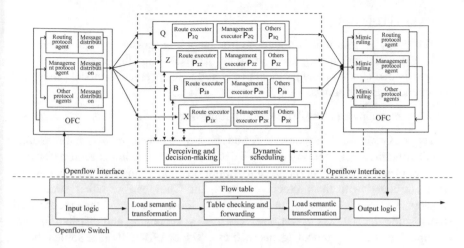

Fig. 12.3 Mimic router system architecture

ruling module inserted in the OpenFlow controller (OFC). At the same time, the mimic ruling can be performed through the routing table generated by the redundant heterogeneous routing executor to ensure the correctness of route computing result.

Based on the above considerations, the DHR2 router system architecture is designed as shown in Fig. 12.3. It includes two levels, a hardware level and a software level. The hardware level is the standard OpenFlow Switch (OFS), and the software level includes such units as OFC, proxy plug-in, mimic ruling, heterogeneous executor pool, dynamic scheduling, and the perceiving and decision-making unit, which are collectively called mimic plug-ins.

The OFC performs message distribution based on the protocol types of incoming messages, and multiple proxy plug-ins perform stateful or stateless processing based on respective protocol mechanisms and distribute messages to each executor. Executors process the messages through the protocol software and generate a flow table and corresponding output messages. Multiple output messages, after going through mimic ruling, will be sent to the path through the OFS. The routing table information generated by every executor will generate "trusted" entries through mimic ruling, which are then sent to the OFS. The mimic ruling result notifies the scheduling control component via the negative feedback mechanism to achieve the perception of running state of the executors, so as to perform executor scheduling according to the given scheduling strategies.

12.1.4.2   Function Unit Design

This section is a detailed introduction to the features and operating principles of important components.

1. Proxy plug-in unit

The proxy plug-in unit is a gateway for incoming messages (routing protocol packets and network management protocol packets). Thus the proxy plug-in unit can be divided into a routing protocol proxy and a management protocol proxy, and its functions are as follows.

(1) Dynamic copy and distribution of input messages

The proxy plug-in is responsible for duplicating the received external messages and distributing their copies to multiple heterogeneous function executors for processing. Copy and distribution are not simple operations; instead, they are somewhat correlated with the protocol of the agent. They can be either stateful or stateless. The stateful operation needs to modify the messages and then record their status, while the stateless operation refers to simple copying and distributing. For example, the routing protocol agents such as OSPF or ISIS will make some modifications of certain fields in the original messages and make them to adapt to the protocol state machine in each executor, while the management protocol agents such as SNMP or Telnet will only copy and distribute the load on the transport layer to the executors.

(2) Identification, filtering, and threat perception of unauthorized service messages

The messages processed by each proxy plug-in unit have different functional characteristics. If the proxy plug-in processes messages that are not in the scope of its own functions, it may bring certain security threats. Therefore, it is necessary to add on the proxy plug-in layer a certain security filtering list to scrutinize the incoming messages to block certain security threats out of the functional executors.

Meanwhile, the proxy plug-in, as the first pass of the system for incoming messages, can move forward the anti-attack function of the executor. Applying traditional security methods here cannot only block the complicated malicious traffic, thus reducing the threat to internal function executors, but also perceive threats through statistical analysis and intrusion detection and send out early warnings.

2. Mimic ruling unit

All outgoing messages need to go through the mimic ruling unit, and the unit takes the output vector of multiple heterogeneous executors as input. Through bit-level, load-level, behavior-level, or even content-level mimic ruling, the unit perceives abnormal internal function executors and triggers the corresponding post-processing programs through negative feedback. The perception and decision-making unit will specify according to security level its mimic ruling algorithm, for example, majority ruling, weighted ruling, or random ruling. Although mimic ruling can find out abnormal executors and screen them out, it cannot accurately tell what kind of abnormality happened to the executors. For example, a table entry in a router forwarding table can be changed if and only if the output results of most protocol executors are completely identical. Only one distinctive routing entry modification request issued by one executor will be blocked in the mimic ruling process. At the same time, the ruling unit extracts the multi-dimensional mimic ruling information and feeds it back to the perception and decision-making unit, providing the necessary information for the decision-making unit to evaluate the credibility of the executors.

The method of mimic ruling is related to the object it protects within the mimic interface. It is carried out on the control plane by comparing the routing tables generated by multiple executors. Due to the huge size of the routing table and the far lower frequency to modify the routing table than to read it, the value of the routing table can be gathered through hashing or checksum algorithm after the routing table is modified. It's much simpler to compute the ruling object of several bytes. For the management plane, the decision is made by comparing the differences in response messages generated by many executors for the same input stimulus. In semantically independent cases, external connection with multiple executors can be compared to make the ruling.

The basic assumption of the mimic ruling is that different executors do not have the same vulnerabilities, backdoors, or other hidden functions, that is, the same external trigger conditions do not cause a majority of heterogeneous executors to produce a completely consistent error output vector. However, there are always exceptions. For example, an attacker could trigger different abnormalities of multiple executors so that the multimode output vectors show the same error result at the same time. This is the consistent "super attack" to the dynamic redundant space with noncoordinating conditions to multiple targets, to which DHR could limit its chances of success but cannot solve it once and for all. In fact, for the attackers, they have to meet the attacking conditions: they have to get information of the vulnerability or backdoor of several internal executors, understand the working mechanism, and carry out a multi-targeted attack at the right time (especially across physical entity domains). It is quite hard to carry out the attack in the DHR architecture, because the dynamic operation of executors, mimic ruling of output results, and negative feedback modifying abnormalities will lead to ruling results that tend to converge uniformly, making it difficult to achieve the cross-domain (especially cross physical domain) coordinated attack. Therefore, the more redundant and heterogeneous the system is and the more diversified the ruling content is, the faster the feedback convergence is and the less chance collaborative attack has to succeed.

3. Dynamic scheduling unit

The main function of the dynamic scheduling unit is to manage the operation of executors in the heterogeneous executors' pool and its functional sub-pool. It schedules multiple heterogeneous function executors according to the scheduling policy specified by the decision-making unit to realize the dynamicity and diversity of executors. In this way, it will be more difficult for attackers to find or detect the target; interrupt the visibility or accessibility of unknown vulnerabilities, backdoors, virus Trojans, etc.; and destroy any attempt to attack multiple target executors in non-collaborative condition. The key part of a dynamic scheduling unit design is the execution strategy of executors. A random scheduling strategy based on the credibility of the executor is given below.

Each executor in the heterogeneous executors' pool has a credibility attribute value. The scheduling unit adopts a random scheduling method based on the credibility weight, which means that the higher the credibility of the executor, the higher its priority. Each executor is set with the same initial credibility value; however, the value will drastically decline if an abnormality is detected. Based on practical experience and engineering implementation constraint of seven executors, it is set that if four abnormalities in a row occur, the executor will be set unusable, while the longer the normal operation maintains, the more credible the executor will be. The growth rate of credibility is related to its current credit value and the interval of benchmark scheduling. The basic principle is to ensure that the credit value of the executor does not increase to the maximum in the next scheduling. Once an abnormality is identified, it's less possible for it to be used next time compared with other normal working executors. The credibility value is used as the weight in scheduling, and a weight-based random scheduling method is adopted. The larger the weight, the more probable for the executor to be used.

4. Heterogeneous executor pool

The heterogeneous executor pool is in store of heterogeneous executor units with equivalent functions. Its diversity increases the difficulty of attackers analyzing and exploiting vulnerabilities and backdoors, equipping the entire routing system with a powerful intrusion tolerance capacity. Functionally equivalent heterogeneous executors are divided into different isolated sub-nets. Heterogeneous executors on the same sub-net belong to different functional planes. The proxy plug-in ensures the consistency and integrity of executor data and state machine in different sub-nets.

In implementation, a heterogeneous executor pool is made up of multiple completely heterogeneous virtual executors constructed on three layers, namely, the underlying CPU type, operating system, and routing software, through virtualization technology based on open-source routing software. Open-source software choices include FRRouting, BIRD, and Quagga. The operating system can be constructed through Ubuntu, Debian, Centos, VyOS, etc. ×86, ×86–64, ARM32, and ARM64 can be simulated through virtualization methods so that the three layers (application, OS and CPU) are all made as heterogeneous as possible.

Although the heterogeneity of such executors is relatively coarse in granularity, it basically meets the requirements for vulnerability and backdoor defense due to large differences in the development team, operating system, and operating environment. It is a small probability event that common mode problems occur to the executors.

In the process of building such a pool, these heterogeneous executors can be packaged into a virtual machine, and a set of heterogeneous executors is formed. Commercial router vendors can consider purchasing the executor set to build heterogeneous redundant environment and their own mimic routers when it's not possible for them to develop their own executors.

5. Perception and decision-making unit

The perception and decision-making unit is mainly responsible for units including proxy plug-in, mimic ruling, heterogeneous executor pool, dynamic scheduling, etc. It defines the message distribution method of the proxy plug-in, the algorithm of the mimic ruling, and the strategic scheduling of the dynamic scheduling unit. It collects abnormalities and status information in the system operation and performs environment perception, based on which it also identifies abnormal executors by mimic ruling results and automatically modifies operating parameters through negative feedback to ensure an automatic change for active defense. At the same time, it carries out offline cleaning to executors with low credibility and restores to their initial state or triggers the traditional security mechanism for accurate inspection and cleanup.

Data cleaning and rewinding of abnormal executors are necessary tools to ensure a quick return to normal state. The cleaning process is divided into three steps:

(1) Clean the routing information generated by the associated executor. Firstly, make sure that the current routing table is generated by an abnormal executor, and then revoke the corresponding route forwarding table, computing again the new forwarding table and in the end complete the update.
(2) Revoke the routing information advertised by the associated executor. First, it is necessary to make sure what routing information the abnormal executor generates and sends to its neighbor and re-advertise the new routing information.
(3) Close the associated executor, restart booting and initialize memory, processes, and tasks, and reload the initial configuration file so that the entire system enters a completely new initial state. Status rewinding also makes the cleaned executor return from the initial state to a previous working state. Rewinding operation is done based on local buffered routing data pool, and different rewinding methods are available based on different routing protocols. For OSPF protocol, the external link state database (LSDB) information is synchronized to the local computer by means of state machine replication. For Border Gateway Protocol (BGP), information can be synchronized through the route refresh technology and smooth restart technology.

6. Load semantic transformation

The method to trigger backdoors with load characteristics is difficult to detect. Load semantic transformation unit uses its controller, message encapsulation, entry load semantic transformation, and exit load recovery to perform reversible transformation of the load data and to eliminate the instruction semantics hidden in the message load to open the backdoor to realize the protection of the network node with security defects. Besides, the load semantic transforming method can be dynamically selected according to management strategies, making attacks difficult to predict the transforming method and improving the security of the system.

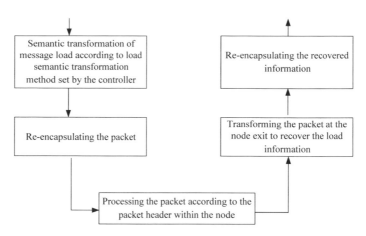

Fig. 12.4 Diagram of load semantic mimic concealing

Figure 12.4 is the diagram of the load semantic mimic concealing. The incoming message load into the node is semantically transformed according to the method set by the load semantic transformation controller. The methods include disturbing the code or other reversible transformation; re-encapsulating the transformed packet including the re-calculation of packet length, checksum, etc.; processing the packet according to the packet header within the node, specifically extracting the packet header from the table forwarding module and routing or forwarding the packet; inversely transforming the packet at the node exit to recover the load information; and re-encapsulating the recovered information. The transformation controller is in control of the transformation mechanism so that while mimetically concealing the data load semantics in communication, it can also dynamically change the transforming method. Therefore, chances of vulnerabilities, backdoors, or virus Trojans based on the load characteristics being triggered are reduced greatly.

12.1.5   Mimic Transformation of the Existing Network

The DHR-based router mimic defense architecture model is constructed into such an architecture through multiple virtualized heterogeneous executors. In fact, the mimic defense can also be realized through mimic transformation of the network based on diversity and heterogeneity of the routers. There are two alternatives to realize network mimic transformation.

The first is to change the self-constraint mechanism of the traditional router into mutual supervision mechanism through smooth upgrading of the router software. To be specific, it is to ensure the heterogeneity of neighbor routers when using routers for network deployment. Neighboring routers exchange information through new information channels. Each router completes its own and neighbor's route computing and sharing and chooses dynamically the computing result for its own and the neighboring router, which will go through the mimic ruling before generating the final forwarding table entries.

The second alternative is realized by deploying neighboring routers in a heterogeneous cluster in a traditional network by increasing corresponding routing supervision and mimic ruler. The method is to gather all the computing results generated by routers in the cluster and pass them on to a mimic ruler to generate a compliant computing result as the final forwarding table entry for distribution.

12.1.6 Feasibility and Security Analysis

Guided by the mimic defense mechanism, this section uses the software-defined network (SDN) and network function virtualization (NFV) technology to design a DHR implementation architecture of the router and gives out verification, testing, and assessment methods, and further discusses the mimic transformation methods of the Internet.

SDN and NFV increase the technical feasibility for the engineering and implementation of router mimic defense. First, mimic plug-ins are easy to be added. With the development of SDN and NFV, the dedicated devices gradually open up their internal structures and unify and standardize the module interfaces, making it more convenient to add mimic plug-ins. Second, the heterogeneity is easily satisfied. SDN decouples the dependency between hardware and software modules so that the modules can be provided by different manufacturers, enriching the species of executors, which is a good basis for heterogeneity. Third, the redundancy cost is controllable. NFV enables network functions to be realized virtually; thus the number of executors does not significantly affect the total cost, leaving room for increasing redundancy.

In terms of cost feasibility, the cost increase of the router mimic defense system mainly comes from the mimic plug-in, in which the input proxy module, mimic ruling module, dynamic scheduling module, and heterogeneous redundant executors are all implemented virtually and the load semantic transformation module is implemented through hardware logic. For low- and mid-range routers, the overall cost increase can be controlled within 10%, for high-end routers, not over 2%.

In terms of security performance, when equipped with the DHR mechanism, the router increases the difficulty for attackers to scan or detect vulnerabilities and makes internal vulnerabilities, backdoors, viruses, and Trojans harder to locate and even harder to be triggered. And even if they are triggered, it's less likely for them to achieve mimic escape. The DHR-based routers can tolerate the system continuing to provide credible service functions in the "virus-bearing" environment if there is no modification or it is unable to identify problems. This is decided by the DHR endogenous security mechanism and can greatly reduce the real-time demand for security maintenance and lower the cost of frequent upgrades. As the mimic plug-ins do not semantically explain and operate the passing information and are transparent to attackers, they are theoretically unreachable to attacks. Therefore, the introduction of mimic plug-ins won't bring extra risks to the system because of its own vulnerability.

Practice shows that the introduction of DHR mechanism in the router control plane is technically feasible and cost controllable. This neither affects the functionality or performance of the routers, nor significantly increases the router's capability to fight against known or unknown security threats.

12.2 Network Storage Verification System

12.2.1 Overall Plan

The network storage mimic verification system is a mimic transformation of the COTS file system. It is transformed by adopting mimic defense with a DHR-based storage architecture to process the hardware platform, local file system, and node operating system. It can increase the heterogeneity, diversity, dynamicity, and randomness of the system; solve the current problems facing the system, like similarity, unity, and staticity; and create dynamic, heterogeneous, and redundant storage environments, making it harder to take advantage of the system's unknown vulnerabilities, backdoors, viruses, Trojans, etc., to protect the system against unknown risks or unknown threats and to improve the storage security of the whole cluster.

Figure 12.5 is the overall scheme of the network storage mimic verification system.

The network storage mimic verification system is composed of a metadata server (MDS), an object-based storage device (OSD), an arbiter, a client, etc. The metadata server is used to manage metadata in the entire storage system, including the file control block inode and the directory tree, while the data server is for data segments. The client-generated files are stored in the metadata and data server, respectively, with the management info of files stored in the former and files which are strategically sharded or encrypted into segments stored in different data servers. The metadata server should consist of at least three functionally equivalent heterogeneous (or virtual) servers. The arbiter will forward the client request to at least three metadata servers, which work out results independently and return them to the arbiter. After performing mimic ruling, the arbiter selects a compliance result back to the client. It will trigger the background processing mechanism when abnormal metadata server or behaviors are detected.

Both the metadata and the data servers are independent computing systems, including hardware platforms, operating systems, local file systems, hypervisors, and virtualization software. By using multiple models of processors such as Intel, ARM, or Loongson and virtual monitors such as VMware, VirtualBox, Hyper-V, and KVM, the hardware platform could be transformed into multiple virtualized machines. Each of the virtual machines supports operating systems such as Ubuntu,

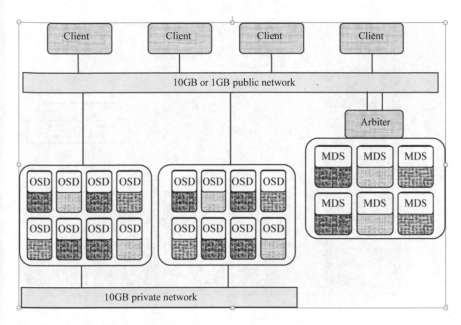

Fig. 12.5 Overall architecture of the verification system

Combinations of operating system etc.									
UFS	ZFS	Ext4	XFS	Btrfs	UFS	ZFS	Ext4	XFS	Btrfs
FreeBSD		Ubuntu, CentOS, Kylin, Oracle Linux			FreeBSD		Ubuntu, CentOS, Kylin, Oracle Linux		
VMware		VirtualBox		Hyper -V	VMware		KVM		
Windows					CentOS				

Fig. 12.6 Combinations of operating systems and local file systems

CentOS, Kylin, and FreeBSD, and file systems like Ext4, XFS, Btrfs, UFS, and ZFS. The multiple combination choices between the metadata server and data server help improve the heterogeneity of the system.

Combination choices of the operating system and local file system are shown in Fig. 12.6.

Figure 12.7 is the flow chart of the storage mimic verification system. Before operating an existing file (e.g., reading or writing), the client must first ask the metadata server to parse the path name in order to get the file's metadata (control information). When creating a file, the client also needs to ask the metadata server to create a file management structure (such as inode and insert it into the specified directory) and obtain its metadata. After that, according to the metadata and the location of data blocks needing to be read or written in the file, the client can calculate which data server the data blocks belong to and their locations in the server and then directly sends a request to the server to read and write the data blocks so as to reach the data blocks out of the system or write them into the system.

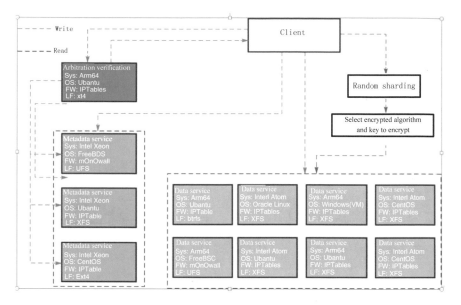

Fig. 12.7 Flow chart of the verification system

12.2.2   Arbiter

The arbiter is a kind of gateway located between the client and the metadata server. To the client, the arbiter is a metadata server that shields the real one in the system. The client sends requests to the arbiter and waits for the processing result. To the metadata server, the arbiter is the client whose requests need to be processed and results to be returned. The metadata servers behind the arbiter work independently. In fact, every metadata server thinks it is the only one in the system, unaware of the existence of others.

When there is a need to query the metadata, the client will pass the request to the arbiter, which will forward it to at least three heterogeneous metadata servers. The servers will generate a result, which will be passed to the client by the arbiter according to agreed policy (first-in, first-out/post correction, majority consistent, completely identical, etc.). If the policy of "first-in, first-out/post correction" is adopted, the first returned result won't go through arbitration and will be output directly to the client. After receiving results from all metadata services, the arbiter will perform a consistency check on them. The abnormal result will be returned for correction and sent back to the client, while post-processing procedure will start. If it is the result of other metadata servers that is abnormal, the information of the server node will be recorded in the warning log, and the status of the server will be degraded as standby and its service be migrated. A post-process program starts to perform reset and cleaning or to restructure the environment.

12.2.3   Metadata Server Cluster

The storage mimic verification system is equipped with multiple (both physical and virtual) metadata servers, called the metadata server cluster. Each of the servers in the cluster is different in its hardware platform, virtual machine monitor, operating system, and local file system to ensure that their security problems are inconsistent as much as possible. The servers can work in three states: operating, ready, and standby. Newly joined servers and those judged to be abnormal by the arbiter are in the standby state. All servers in the standby state will undergo a cleanup and post-processing procedure, after which they can enter the ready state.

The cleanup program will randomly select a software combination from the virtual machine monitors, operating systems, file systems, application software, etc. to reinstall the server.

The metadata server will be put automatically into the ready state for regular or routine cleanup after one cleanup cycle, no matter it is abnormal or not.

When the system first starts to operate, it will randomly and heterogeneously choose at least three metadata servers from those in the ready state and put them into the operating state. If one server is put into the standby state in the process, another server in ready mode will start up to ensure that there are at least three servers running at the same time.

12.2.4   Distributed Data Server

The network storage mimic principal verification system is configured with multiple data servers of different vendors to constitute a distributed heterogeneous data server cluster. The data servers in the cluster are of different structures, e.g., hardware platforms, operating systems, and local file systems, to ensure that the design flaws or security issues of the data servers are not completely consistent. The data servers are used to store data segments coming from the client. To improve the reliability of the storage system, the data server cluster will automatically generate a variable number of redundant segments and store them with or without encryption in different data servers. What any data server sees are data segments from different clients, of different size or even encrypted, so it is impossible to know the origin or panorama of the data.

The client and metadata servers in the system treat the data server cluster as a single object-based storage pool in which each object can store a data segment. An object takes the form of an independent local file in the data server. The object files are named in a unified manner, making the entire cluster a unified namespace. When it's needed to write data, the client or metadata server will conduct data slicing according to the metadata of the file it belongs to and use the Hash function SmartSection to calculate the storage locations of segments (target data server and directory) and names of the object files. Then the target data servers receive the segments, change them into local files, and save them in a designated directory with designated names. SmartSection, a pseudorandom data distribution function, is used to generate the storage server list: SmartSection Map. This approach has double advantages: first, it is fully distributed, and the client, storage server, or metadata server can easily and independently calculate the storage location of data segments; second, the map is rarely updated; this actually excludes any data exchange related to distribution. SmartSection solves the problem of data distribution and location at the same time.

The calculation of SmartSection is based on the OID of the data segment, that is, the inode of the data file + segment number of the data. In this way, the distribution of data segments does not rely on the application server on the front, nor does it rely on the metadata server. This approach increases the randomness and security of the storage system.

For reliability and security reasons, the verification system creates one or more copies of each data segment and automatically pastes them to different data servers. The copies are created by the system or data servers, with their names and locations being the same as those of the original. The only difference is the data server that stores them.

12.2.5 The Client

In the mimic verification system, the client is the user of the storage system; it saves files in the system and reads them when necessary.

Before saving the file data to the storage server, the client will randomly select a slicing method and segment size according to the file metadata provided by the metadata server, slice the file data, and store them in different data servers. The size of the segments is randomly selected, with a range of 1–4 MB. If the real length of the data is smaller than the segment, it will be added with random contents.

Before storing the file segments in the data server, the client will encrypt them so that all segments saved are cipher texts. The method of encryption is configurable, including:

(1) Uniform static encryption. All file segments saved in the storage server, regardless of their size and location in the storage directory, are subject to a uniform encryption method and key for original versions and subject to another uniform encryption method for their copies, as is shown in Fig. 12.8.
(2) Directory-based static encryption. It uses different encryption algorithms and keys on every level 1 directory or all levels of the directory. In this way, files would adopt different encryption methods if they are put under different levels in the directory, as is shown in Fig. 12.9.

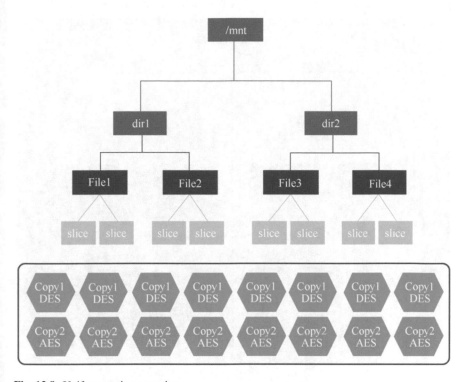

Fig. 12.8   Uniform static encryption

(3) Static encryption based on file size. Configure different encryption algorithms and keys according to the size of the file after being sliced.
(4) Dynamic encryption. After all the files saved in the system are sliced, the encryption algorithms and keys are dynamically selected for each segment.

The slicing and dynamic encryption process is shown in Fig. 12.10.

The client's data is saved in the data server after being sliced and encrypted. It's beyond the data server itself to know the meaning of each segment because the segments come from different clients and belong to different files and are encrypted.

12.2.6   *System Security Test and Result Analysis*

We conducted a security test on the mimic verification system and drew the following conclusions after analysis.

(1) The system has all the functions required for storage, including interfaces for the file, block, and object and also the virtualization-specific interface.

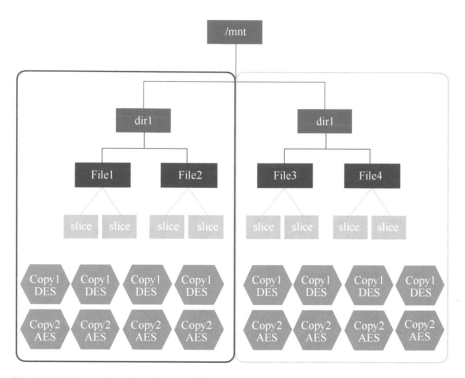

Fig. 12.9 Directory-based static encryption

(2) The system can monitor and compare all subsystems through the arbiter and is able to timely discover any abnormal access to the metadata and file data.

(3) The system is able to detect single executor attacks making use of the known or unknown vulnerabilities and preset backdoors. It can prevent theft of information, chaos, or paralyzation and replace automatically the abnormal executor under attack.

(4) The system can also detect multiple executor attacks making use of the known or unknown vulnerabilities, and preset backdoors. It can prevent theft of information, chaos, or paralyzation and implement preset protection method, which is to replace automatically and strategically the abnormal executor under attack according to the preset order.

(5) The system can discover attacks to individual or all executors from the data plane, the management plane, or the control plane and carry out preset protection method, which is the same as the method described in (4).

(6) The mimic defense mechanism is proven to be not significantly affecting the system functions or performance but improving greatly its ability to resist uncertain disturbances in a general sense.

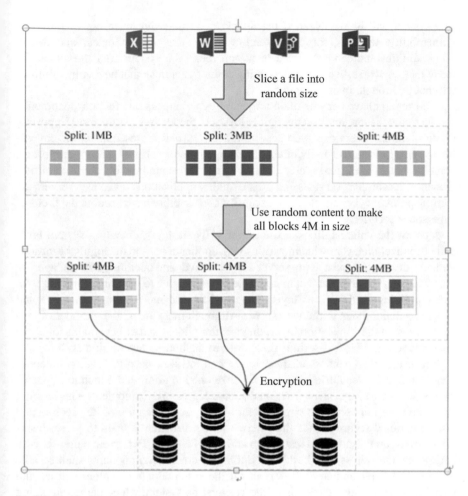

Fig. 12.10 The slicing and dynamic encryption process

12.3   Mimic-Structured Web Server Verification System

12.3.1   Threat Analysis

We know from Chap. 1 of the book that the inevitable existence of vulnerabilities and backdoors in both hardware and software is one of the fundamental reasons why the information system is under threat and that the core to improve security is the prevention and protection against all vulnerabilities (including backdoors and Trojans in a general sense). In the game of cyberspace attack and defense, the attacker always takes initiative, and the defender is always passively reacting.

The system will be paralyzed if the attacker can find and make use of only one vulnerability, whereas, for the defending party, they cannot detect and fix all vulnerabilities and can only trace back and mend the system after the attack. To make matters worse, even the patch fixing cannot guarantee that new vulnerabilities will not be brought in or generated.

Web servers have become main targets of cyber attacks due to their documents, services, and the tangible and intangible losses to the victims. As the most important mode of Internet services, web servers are the virtual representatives of governments, enterprises, and individuals. The government websites, corporate portals, personal homepages, blogs, etc. are all virtual representations of real-life entities. However, its complex services and mixed quality applications make web servers the starting point of cyber attacks, so its security and usability have become the focus of cyberspace security.

Most of the vulnerabilities at the server software layer, operating system layer and virtualization layer reside in various web applications, and the highly hazardous vulnerabilities often exist at the server software layer and operating system layer, so the two layers are often subject to new type of attacks due to their underpinning role. Vulnerabilities on those two layers cause wider and more serious damages and become a major threat to the web server. Although there are a large number of vulnerabilities on the application layer, they scatter widely and affect small scopes, so the damage caused is less serious than those to the bottom layer vulnerabilities.

Because of the openness of the Internet and the slow pace to develop independent technologies, countries that are less developed in information technology have no choice but to rely on international technologies and equipment. This current status quo has imposed a potential threat to the informatization of key sectors and also to national security since their network infrastructure is open to preset backdoors, transparent, subject to restriction and out of control. To prevent being set with backdoors, the web servers built with dubiously credible components shall be able to provide reliable and credible services if the system contains vulnerabilities and backdoors. This is also the basis of the verification system, which came into being as a result of the combination of the study of traditional web servers and their normal security threat with the basic idea of mimic defense.

12.3.2   Designing Idea

The idea of the mimic defense verification system in web servers is to construct a functionally equivalent, diversified, and dynamic dissimilar virtual web server pool according to the mimic defense principle. Technologies like redundant voting, dynamic executor scheduling, database instruction heterogenization, etc. are applied to break the attack chain and increase the difficulty to make use of system vulnerabilities and ensure the functionality and credibility of web services.

Figure 12.11 shows the logical structure of the verification system in terms of hardware, operating system, database, virtualization, virtual machine operating

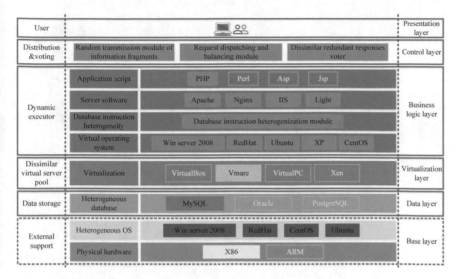

Fig. 12.11 The logical structure of the verification system

system, database instructions, server software, application scripts, and so on. Diversified designing is adopted on many layers of the software stack, providing a basis for the realization of mimic defense. It can be seen that diversity design is used at multiple levels of the software stack, and the diversity design provides the basis for the implementation of the mimic defense mechanism. The dynamicity, heterogeneity, and redundancy are manifested in:

(1) Heterogeneity. Configure different types of heterogeneous software and hardware on different layers. For example, Red Hat, Ubuntu, and others could be used on the virtual OS layer.
(2) Redundancy. Multiple software and hardware are deployed to simultaneously execute the request and vote on the result to realize the redundant operation.
(3) Dynamicity. According to the scheduling policy or feedback from threat perception, the heterogeneous executors which process a big part of services shall be replaced convergently in order to improve the uncertainty of the system running scenario.

12.3.3   System Architecture Design

The verification system of mimic construction in web server is designed according to the DHR model introduced in Chap. 7 (see Fig. 7.1). The system consists of modules including the request dispatching and balancing module, dynamically executing scheduler, dissimilar redundant response voter, Physical Web Server, dissimilar virtual web server pool, primary controller, and database instruction labelling module. The system architecture is as shown in Fig. 12.12.

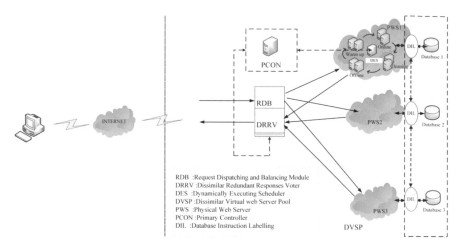

Fig. 12.12 Prototype architecture of mimic defense in web server

The request dispatching and balancing module (RDB) is based on the input and output proxy in Fig. 7.1. As the real entry of user requests, RDB dispatches requests, according to the resource strategy of maximum heterogeneity, to multiple separated but independently working web service executors (also objects for dynamic scheduling) in the dissimilar virtual web server pool. It is the prerequisite of a dynamic, diverse, and heterogeneous executor.

The dissimilar redundant response voter (DRRV), a crucial component of the mimic defense web server, is based on the multimode ruling in Fig. 7.1, which is the real exit of the server response. It could perform cross ruling of the responses of multiple heterogeneous executors to one request by the synchronous/asynchronous adaptive majority voting algorithm according to security level requirements and exclude inconsistent information to ensure the consistency of output results. The result will be sent to the negative feedback module of the primary controller.

The dynamically executing scheduler (DES) is a control unit for the state of executors in a dissimilar web service sub-pool. By switching the executor online/offline (level 2 dynamic scheduling), DES reduces the exposure time length of an executor and improves the security of the web service executor. By receiving the ruling result sent from the negative feedback control module of the primary controller, DES recognizes the ruling result and decides on the use of the event- or time-triggering strategy for the cleanup or rollback of the executors to ensure the integrity of the web service executor. By carrying out web server level-2 heterogeneity scheduling of the executors in the sub-pool, the DES enlarges the differences of executors participating in one and the same redundancy voting, and eventually the system is converged in a temporarily stable state.

The dissimilar virtual web server pool (DVSP) is based on multiple reconfigurable executors shown in Fig. 7.1. It is the real provider of web services and is composed of heterogeneous, diverse, and redundant web service executors. According to the heterogeneity maximization principle, all web service executors are aggregated into different physically separate and isolated sub-pools (also objects of heterogeneous combination). Multiple executors from different sub-pools non-collaboratively process the same web service request copied by the RDB module and return their responses to the DRRV. The heterogeneity and diversity of web service executors, and the independence and isolation between sub-pools, are the basis for the mimic-structured web servers to succeed.

The primary controller (PCON) is based on the negative feedback controller in Fig. 7.1. It monitors the running status of other functional modules to ensure that the system has sufficient resources. It also loosely couples each module unit in the system to prevent a single point of failure from causing functioning problems. The negative feedback control module is responsible for receiving the ruling result of the DRRV and transmitting the result to the DES.

The database instruction labelling (DIL) module consists of three sub-modules: SQL reserved word fingerprinting module, injection instruction filtering module, and database consistency offline voter. The SQL reserved word fingerprinting module processes the SQL reserved words from web applications by fingerprinting and characterizing the SQL instructions in web applications. The injection instruction filtering module filters the database read and write instructions according to the fingerprint and removes the illegal instructions the attackers injected. The database consistency offline voter votes to recover the inconsistent databases, that is, to restore and protect the failed databases to ensure their consistency.

12.3.4   Functional Unit Design

12.3.4.1   Request Dispatching and Balancing (RDB) Module

The main functions of this module are:

(1) Dynamic traffic replication: Multiple copies of a user request are made and sent to the back-end dissimilar virtual web server pool, which increases system uncertainty and difficulty for attackers to find vulnerabilities to ensure stable web service.
(2) Sub-net isolation: Request copies are forwarded to the heterogeneous resource pool in different sub-nets. The independently working virtual servers increase the robustness of the system by ensuring that one single point of failure won't affect the normal response of the system.

RDB isolates the back-end dissimilar virtual web server pools (DVSPs) through sub-net division, ensuring that the DVSPs are independent of each other and making it difficult for attackers to detect them. Acted as a reverse proxy, RDB receives user requests and makes three copies and forwards them to online web servers in different sub-nets. It improves system performance and adaptability by forwarding requests through multiple Network Interface Cards (NICs). The physical server which carries RDB supports at least four network adapters by VLAN or multiple

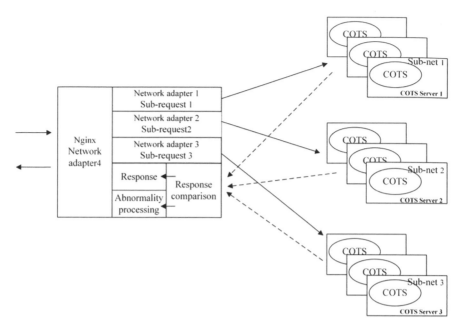

Fig. 12.13 Design model of RDB and DRRV

NICs technologies. Four independent adapters are used in this solution to realize the RDB, one for IP and port for external access, the others serving as reverse proxies to forward requests. In this way, the request traffic could go through different adapters to functionally equivalent heterogeneous web servers in different sub-nets, reducing traffic pressures to one single adapter. The difficulty in this design is the realization of dynamic replication and forwarding of requests and also the possible damage to system performance of the replication of one into multiple copies. RDB is the basis for implementing DRRV. Figure 12.13 shows the design model for RDB and DRRV.

RDB equips DRRV with a web server which is able to respond heterogeneously to one single request, ensuring the heterogeneity between servers processing the same request. It also adds more uncertainty to attackers in detecting real Banner information of the web servers.

As the most important parts of the web server verification system, RDB and DRRV serve as the key components to break the attacker's communication chain. RDB executes attacks on the back-end heterogeneous redundant web server executors and generates inconsistent results to most environment-dependent attacks. The DRRV discovers abnormality and interrupts attacks by voting on those results.

12.3.4.2 Dissimilar Redundant Response Voter

The function of the dissimilar redundant response voter (DRRV) is to vote on the multiple responses to the same request as per a specific algorithm and to return the result to the user. It helps generate an intrusion tolerant system to ensure the correctness of response data. It also enables a timely detection of the abnormal virtual server for the abnormal processing mechanism to step in.

The DRRV is a key technology in the mimic verification system of the web servers. It performs a mimic ruling to the heterogeneous web server responses by a given voting algorithm, such as a majority ruling, selects the right response, and sends it to the user; otherwise it encapsulates abnormalities ("warning" or "response") into special formats and sends then to the primary controller. Its main function is to prevent web pages from being tampered or key information from being disclosed.

12.3.4.3 Dynamically Executing Scheduler

The main function of the dynamically executing scheduler (DES) is to manage the operation of virtual servers in the dissimilar virtual web server pool, monitor the state of the virtual servers, schedule strategically, receive abnormalities forwarded by the primary controller, handle abnormal virtual servers timely, and interrupt the chain of attack based on vulnerabilities, backdoors, and Trojans. The DES starts and cleans the virtual servers in the dissimilar virtual web server pool according to the predetermined policy. It also performs the killing and cleaning task according to the abnormal information broadcast by the primary controller. Figure 12.14 is a sketch of the structure of DES, which consists of a monitoring system, an executor, and a distributor.

The key part of the DES is the virtual server scheduling strategy, a method of independent scheduling in the dissimilar virtual web server sub-pool, which is remotely assisted by the primary controller through the cross-platform message

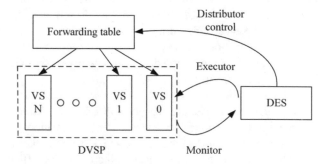

Fig. 12.14 Dynamically executing scheduler structure

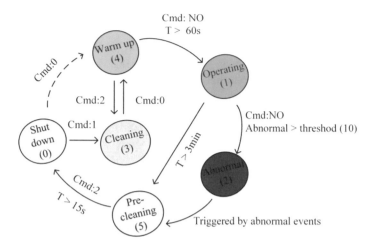

Fig. 12.15 Virtual server state switching scheme 1

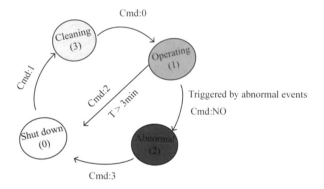

Fig. 12.16 Virtual server state switching scheme 2

transferring mechanism. This method can reduce the complexity of virtual server management and the independence of DES scheduling. Based on a visual programming script, the virtual server scheduling solution performs scheduling tasks, including start, stop, or snapshot recovery, of the virtual servers, makes a record of their running state, and stores them in the database. The two state switching modes of the virtual servers are shown in Figs. 12.15 and 12.16.

The main function of DES is to ensure the diversity and periodicity of the web servers and to carry out event-driven cleaning or rollback of the vulnerable servers. It shortens the time an attacker takes to detect a web server, increases the uncertainty of the detection result, disturbs the attacker's vision, and makes him unable to attack the target. More importantly, it significantly increases the difficulty of multi-target coordinated attack under noncooperative conditions.

12.3.4.4   Dissimilar Virtual Web Server Pool

The dissimilar virtual web server pool (DVSP) functions as a heterogeneous redundant virtual web service resource pool, which is characterized by heterogeneity, redundancy, diversity, and dynamicity. DVSP guarantees the diversity of output results of vulnerability scanning tools, increases the difficulty for attackers to analyze vulnerabilities or utilize backdoors, and enables the system to attain the ability of intrusive tolerance and self-recovery. Different dissimilar virtual web server subpools are divided in different sub-nets, and they cannot directly communicate with each other. There is a database proxy in each sub-net, where the database is shared. The databases in different sub-nets are synchronized in real time through the DRRV. The isolation and real-time synchronization of databases ensure the consistency and security of their data. Failure of any single database will not affect the operation of the overall system.

Heterogeneous web servers are the core of the mimic-structured web server system. Although they provide the same services, their heterogeneous applications, operating systems, and hardware reduce the occurrence of common mode failures to the heterogeneous platform servers. This is due to their effort to ensure the heterogeneity of vulnerabilities, backdoors, and Trojans in them. Therefore, it is not likely to trigger all vulnerabilities of the heterogeneous servers and generate an identical error. In the mimic-structured web server system, the scheduling selection algorithm based on a heterogeneous platform is very important. The criteria for selecting heterogeneous resources are as follows:

(1) Ensure that the system functionality set remain unchanged.
(2) Minimize the intersection of system vulnerabilities.

In a dissimilar virtual web server pool, there are multiple functionally equivalent heterogeneous web servers, with the online ones dynamically changing. The virtual server web service is scheduled according to a specific policy under the control of a dynamic web executor scheduler. Table 12.1 shows the diversified configuration of the software stack in a dissimilar virtual web server pool.

12.3.4.5   Primary Controller

The primary controller is mainly to monitor the running status of other functional modules, to ensure that there are sufficient resources in the system, and, at the same time, to loosely couple the modular units to system failure caused by any single point of failure. It also makes the system scalable and its management less complex. The primary controller collects the abnormality information detected by DRRV and DIL, processes it, and sends it to the DES; then the latter will carry out the virtual scheduling task in response to the abnormality information. In addition, the primary controller can monitor the operating status of the abovementioned components. The structure of the primary controller is shown in Fig. 12.17.

Table 12.1 Information list of a dissimilar virtual web server pool

No.	IP: PORT	Software stack	Snapshot name
1	192.168.10.211	Ubuntu12.04+Apache+PHP	SnapUbuntuApachePHP
2	192.168.10.212	Ubuntu 12.04+Nginx+PHP	SnapUbuntuNginxPHP
3	192.168.10.213	CentOS 6.6+Apache+PHP	SnapCentOSApachePHP
4	192.168.10.214	CentOS 6.6+Nginx+PHP	SnapCentOSNginxPHP
5	192.168.10.215	Windows XP+Apache+PHP	SnapWindowsXPApachePHP
6	192.168.10.216	Windows XP+Nginx+PHP	SnapWindowsXPNginxPHP
7	192.168.10.217	Windows 7+Apache+PHP	SnapWindows7ApachePHP
8	192.168.10.218	Windows 7+Nginx+PHP	SnapWindows7NginxPHP
9	192.168.10.64	Windows Server 2003+Lighttpd+PHP	SnapWinServer2003LighttpdPHP
10	192.168.10.65	Windows Server 2003+IIS+PHP	SnapWinServer2003IISPHP
11	192.168.10.66	Windows Server 2008+Apache+PHP	SnapWinServer2008ApachePHP
12	192.168.10.67	Windows Server 2008+IIS+PHP	SnapWinServer2008IISPHP
13	192.168.10.68	Windows Server 2008+Lighttpd+PHP	SnapWinServer2008LighttpdtpdPHP
14	192.168.10.69	Windows Server 2008+Nginx+PHP	SnapWinServer2008NginxPHP
15	192.168.10.81	Debian+Nginx+PHP	SnapDebianNginxPHP
16	192.168.10.82	Debian+Apache+PHP	SnapDebianApachePHP
17	192.168.10.83	Red Hat+Nginx+PHP	SnapRedHatNginxPHP
18	192.168.10.84	Red Hat+Apache+PHP	SnapRedHatApachePHP

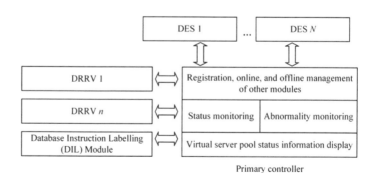

Fig. 12.17 Structure of dynamic web primary controller

12.3.4.6 Database Instruction Labelling (DIL) Module

The functions of the DIL module are:

(1) Characterize the SQL instructions of web applications.
(2) Filter the operation instructions (read or write) to the database according to the fingerprint, and delete illegal instructions injected by attackers.

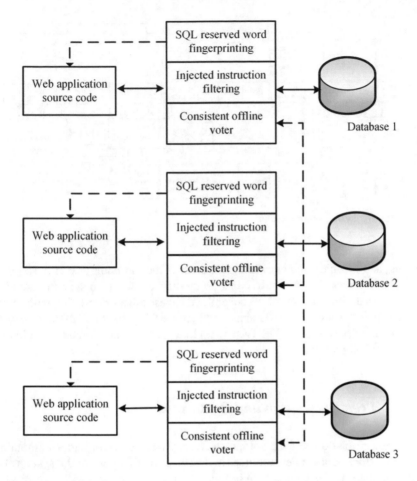

Fig. 12.18 DIL module functional structure

(3) Determine the abnormal database by voting and restore the faulty database.

The DIL module consists of three sub-modules: SQL reserved word fingerprinting module, injection instruction filtering module, and database consistency offline voter, as demonstrated in Fig. 12.18. The first sub-module processes the SQL reserved words in web applications by fingerprinting. The database redundancy voter functions to prevent malicious tampering with the database, ensuring the correctness of the database storage data, and the consistency of databases in the DVSP through synchronization.

The web applications in the same dissimilar virtual web server sub-pool are processed by the SQL reserved word fingerprinting module, so that the SQL reserved words are heterogeneous in the web applications. The injection instruction filtering module in the sub-pool filters and de-fingerprints the database access to the SQL reserved words. Differentiated SQL fingerprinting between sub-pools, SQL

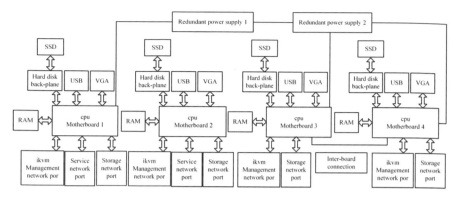

Fig. 12.19 Hardware platform design

filtering, and specific de-fingerprinting methods in the sub-pools ensure the security of database access. The database consistency offline voter is to synchronize databases across different sub-pools, recover the database table information with inconsistent voting, restore a faulty database, and ensure the correctness of data. In order to reduce false positives, the DRRV uses an interval-based abnormal threshold dual decision-making mechanism.

12.3.5 Prototype Design and Realization

The hardware platform of the mimic-structured web server provides computing, switching, storage, and other resources and basic device support for the system. The hardware platform design is shown in Fig. 12.19.

The case of the hardware is a standard 19-inch 2U rack-mount device, which accommodates multiple computing nodes, each equipped with an independent CPU, a memory, a hard disk, a network, and other components. The nodes are powered by one set of 1 + 1 redundant power supply with an overall power consumption of fewer than 1000 watts. It consists of three main types of hardware interfaces.

1. Service interface

The service interface provides services based on the HTTPS communication protocol and is an entry point for users to access the system. The motherboard No. 1 receives user requests through the service port, exchanges data with motherboard No. 2, and outputs web server response to users.

2. Management interface

The management interface distributes system resources and conducts system maintenance. The administrator carries out those tasks through the ikvm management interface on motherboards No. 1, 2, 3, and 4.

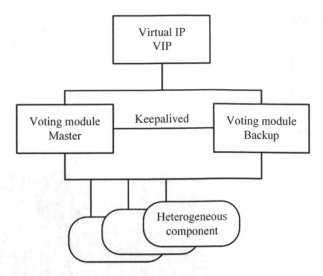

Fig. 12.20 High availability of the voting module

3. Storage interface

The storage interface is used mainly to store or back up the key data. The system stores key data through the storage NIC on the motherboards No.1, 2, 3, and 4. Motherboards No. 3 and 4 are double-click hot standby storage units.

The Keepalived tool is used in order to achieve high availability of the voting module and avoid a single point of failure, as shown in Fig. 12.20. When the primary voting module fails, the voting module on the other board takes over the virtual IP service.

GlusterFS is employed to achieve high availability of storage. It can make data consistent between two voting modules for the goal of data disaster tolerance and network redundancy. The high availability of storage is as shown in Fig. 12.21.

According to the above designing ideas and methods, two prototypes of the mimic-structured web server are implemented, as shown in Figs. 12.22 and 12.23.

12.3.6   Attack Difficulty Evaluation

As shown in Fig. 12.24b, the mimic-structured web server is equivalent to a DHR software protocol stack based on the I[P]O model. Figure 12.24a is a hypothetical attack chain, which is agreed to be reset once the attack is intercepted.

Figure 12.25 is the state transition diagram of the attack chain.

Figure 12.26 is a state transition diagram under successful interception.

Suppose P_i is the attack success rate of each stage under the non-mimic condition, and P_{A_i} is the rate of Executor A being successfully attacked in stage i. The voter compares the returned values of each executor. If the returned values are

Fig. 12.21 Diagram of high availability of storage

Fig. 12.22 Prototype 1

Fig. 12.23 Prototype 2

exactly the same, the value of V_i is "1"; otherwise it is "0." The success rate P_i' under the mimic condition is shown in Fig. 12.27.

Since attack difficulty varies at different stages, the possible results of attacks are as shown in Fig. 12.28, where the "escalation of privilege" of Step 4 is much more difficult than the "Trojan Reverse Connection" of Step 7; thus P_4 is far lower than P_7. Under the mimic defense conditions, the returned information of the three executors in Step 4 tend to be consistent, and V_4 is 1, while V_7 can only be "0" because of the heterogeneity of the system. Therefore, Step 7 is more vulnerable to attacks in the mimic defense conditions.

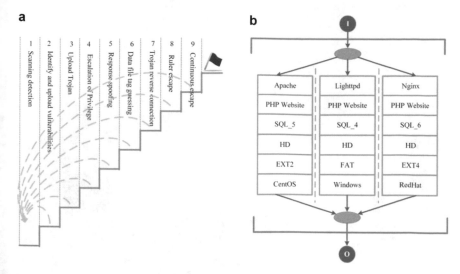

Fig. 12.24   DHR (**a**) attack chain and (**b**) software protocol stack

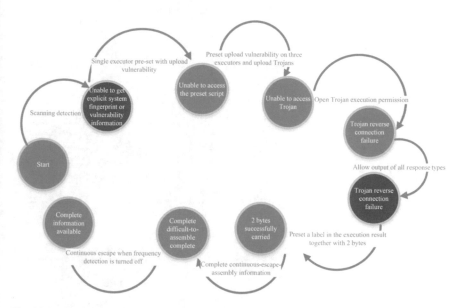

Fig. 12.25   State transition diagram of the attack chain

Next, suppose P'' is the success rate of the entire attack chain. As shown in Fig. 12.29, in the case of a black box, a complete attack may cause multiple alarms, each of which triggers the dynamic cleaning mechanism. Therefore, the attacker has to repeat the whole chain multiple times to complete the whole chain. The success rate of an attack is the product of all the success rates at each stage. Theoretically,

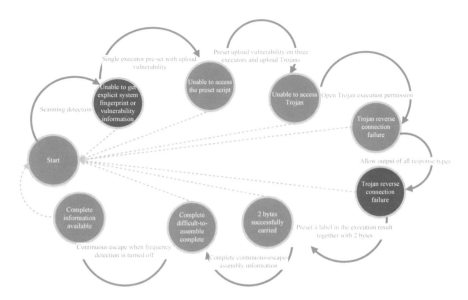

Fig. 12.26 State transition diagram under successful interception

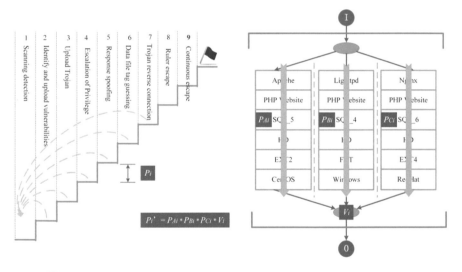

Fig. 12.27 Success rate of attacks under the mimic conditions

in the repeated attempts, if one step fails to produce a consistent output, the whole attack chain would fail.

It can be concluded that the more elaborately a structure is, i.e., the more dependent an attack is on the environment, the more likely it will be interrupted by the mimic defense. This is why the mimic defense can cope with unknown threats.

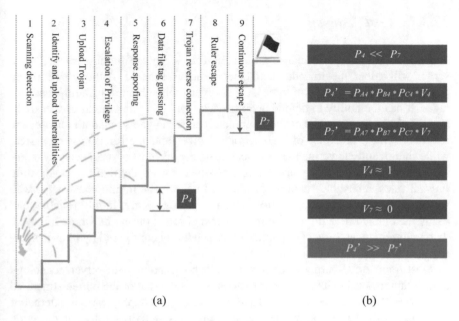

(a)   (b)

Fig. 12.28 Attack difficulty at different stages (**a**) Stage 4 and 7 in attack chain (**b**) Calculated attack difficulty at stage 4 and 7

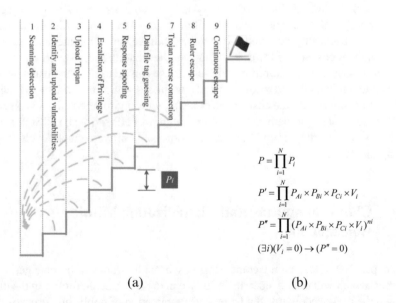

(a)   (b)

Fig. 12.29 Success rate of attacks (**a**) All stages in attack chain (**b**) Calculated success rate

12.3.7   Cost Analysis

When function and performance are guaranteed, a typical mimic-structured web server will adopt the tri-redundant heterogeneous scheme. Although its initial acquisition cost equals that of three COTS servers, this scheme has multiple advantages. It has the ability to defend against backdoors and vulnerabilities that traditional defense technologies cannot completely defend against, also it increases the reliability and availability of web service devices and the robustness of system while significantly lowering the technical difficulty and cost of daily security maintenance (including patching, upgrading antivirus software, configuring or adjusting firewall rules, analyzing running logs, etc.). In short, the mimic-constructed web service system is a holistic solution which is highly secure, highly reliable, and easily maintained. It is more effective and cost efficient during its whole life cycle than traditional web servers or methods independently configured by plug-in security instruments.

Most commercial-purpose web servers adopt hot standby multi-server technology to maintain its availability, the cost of which is close to that of the mimic-structured web server. To ensure security, you have to purchase and deploy various external or built-in security devices or systems, such as firewalls, antivirus, intrusion detection system, etc. All of the systems and devices added together will cost more than that of the mimic-structured web server but are far less effective in defending against unknown vulnerabilities and backdoors. Let's think about the follow-up maintenance; as the mimic defense technology is born with an ability to guard against unknown vulnerabilities and backdoors, it does not need to immediately respond to or mend the frequently occurred 0-day vulnerabilities, so its maintenance difficulty and manpower cost are far lower than those of the existing security solutions.

On the other hand, a simplified mimic defense can be used in the usual business environment for small- and medium-sized enterprises, which uses the bi-redundancy voting to further reduce the cost. In the cloud service environment, a single executor method (such as single-path redundant execution of PHP scripts or of SQL scripts) can be used to achieve mimic defense at some key software stack levels without raising the costs.

12.4   Cloud Computing and Virtualization Mimic Construction

In the past 10 years, cloud computing has brought tremendous changes to the Internet, which will only deepen its influence in the foreseeable future. In the digital age and the digital economy, the large-scale application of public and private clouds has made the scattered data storage and code computing more cost-effective than the hardware and software resource pooling application in the cloud computing scenario. "Users determine applications"; the cloud computing platform cannot only provide computing power including processor capacity, memory capacity, disk capacity, network port, and traffic but also provide customized or shared software service capabilities.

12.4.1 Basic Layers of Cloud Computing

The widely recognized cloud computing architecture is generally divided into three categories, namely, the Infrastructure as a Service (IaaS), Platform as a Service (PaaS), and Software as a Service (SaaS).

(1) Infrastructure as a Service (IaaS) mainly consists of computers, servers, communication devices, storage devices, etc. and provides users with IT infrastructural services, i.e., computing, storing, or network capabilities, thanks to the advancement of virtualization. Virtualization can be used to virtualize all kinds of computing devices into computing resources in the virtual resource pool, virtualize the storage devices into storage resources, and virtualize the network devices into network resources. When a user subscribes to these resources, the data center manager can directly package the subscribed portion and send it to the user.

(2) Platform as a Service (PaaS) in cloud computing can be seen as providing the same function the operating system/development tools play in the traditional computer architecture of "hardware + operating system/development tools + application software." PaaS is positioned to provide users with a complete set of support platforms for developing, running, and operating applications via the Internet. Just as in the PC software development model, programmers use development tools to develop and deploy applications on a computer with a Windows or Linux operating system. The best-known products of this kind are Microsoft's Windows Azure and Google's GAE.

(3) Software as a Service (SaaS) is a software application model that provides software services over the Internet. Users do not need to invest a lot of money in hardware, software, nor development teams. Instead, they only need to pay a certain rental fee to enjoy the corresponding services through the Internet, and the system maintenance and upgrading are at the cost of the operator.

12.4.2 Cloud Computing Architecture Layers

1. Display layer

   In general, the main function of the display layer is to present the needed content and service experience in a user-friendly manner. The following middleware technologies will be used to provide services:

HTML: Standard web page technology, mainly HTML4. However, HTML5 can better promote the development of web pages in many aspects, such as video and local storage.

JavaScript: A dynamic language for web pages. JavaScript-based AJAX can create more interactive dynamic pages.

CSS: Mainly used to control the appearance of web pages and to elegantly separate the content from its form.

Flash: the most commonly adopted RIA (Rich Internet Applications) technology. It can provide richer web-based applications than HTML and other technologies can currently provide.

Silverlight: very programmer-friendly for it allows the use of C# [5] in programming.

2. Middle layer

Supported by the lower-layer infrastructure resources, the middle layer provides multiple services, for example, caching and REST services. Those services can be used to support the display layer or directly used by the user. The following are the five main technologies:

REST: it could easily provide the caller with some of the services supported by the middleware layer.

Multi-tenancy: allowing a single application to serve multiple customers with good isolation and security. This technology could effectively reduce the acquisition and maintenance cost of the application.

Parallel processing: in order to process massive amounts of data, it is necessary to use a huge CPU cluster for large-scale parallel processing, for example, Google's MapReduce.

Application server: optimizes cloud computing to a certain degree based on the original application server. For example, the Jetty application server for Google App Engine.

Distributed cache: this technology could not only effectively reduce the pressure on the background server but also quicken the corresponding feedback speed. The most famous example is Memcached.

3. Infrastructure layer

The role of this layer is to prepare the required computing and storage resources for the middleware layer or the user. The four main technologies used in this layer are as follows:

Virtualization: it can be used to generate multiple virtual machines on a single physical server, with sufficient isolation between each of them. The purpose is not only to reduce the cost of server acquisition but also to reduce the operation and maintenance cost of the server as much as possible. The mature X86 virtualization technology includes VMware's ESX and open-source Xen.

Distributed storage: a complete set of distributed storage systems is needed to carry the large amount of data and also to ensure their manageability.

Relational database: it is usually expanded on the basis of the original relational database or optimized in management for wider application in the cloud environment.

NoSQL: it is needed to support the applications which some relational database cannot sustain in the mass data application scenario.

4. Management layer

The six main management functions are:

Account Management: it makes it easier for users to log in or for administrators to manage the account under prescribed security preconditions.

Service-Level Agreement (SLA) Monitoring: it's used to necessary to monitor the performance of virtual services and applications on different layers to meet the given SLA.

Billing Management: it offers an accurate bill to charge users based on resource statistics.

Security Management: comprehensive protection of IT resources such as data, applications, and accounts from criminals and malicious programs.

Load Balancing: it deals with unexpected situations by distributing traffic to multiple instances of an application or service.

Operation and Maintenance Management: it is mainly to make the operation and maintenance operations as professional and automated as possible to reduce the cost of the cloud computing center.

12.4.3 Virtualized DHR Construction

As the rapid-developing cloud service platform brings many benefits such as agile service, efficient resource utilization, and user customization, the security and credibility issues have become the focus. As the software and hardware are developed by different vendors using different software or hardware devices, components, and middleware or even IP cores, all the globalized design, production, tool and service chains, APP platforms migrating to the cloud and the ubiquitous development of open-source models provide high value targets and hard-won opportunity for deliberate acts exploiting the vulnerability and uncertain credibility of information systems. In particular, with the acceleration of the cloud-network-terminal process in different domains, at different levels and in different applications, the corresponding vulnerabilities, backdoors, and Trojans are also accelerating toward the cloud. The traditional border protection in the business environment and the authentication mode is no longer fit for the cloud computing architecture. The pooling of software and hardware resources and the resulted virtual server escape attack, making the cloud platform's service credibility and information security issues quite prominent.

However, from the perspective of mimic defense, the hardware and software resources at all levels in the cloud usually come from diversified markets and thus are naturally heterogeneous; the cloud resource allocation mechanism undoubtedly belongs to the category of redundancy deployment and usually takes the form of large-scale physical server clusters; the dynamicity and randomness are naturally introduced in the allocation or release mechanism of pooled resources and virtual hosts; the virtual host migration process naturally realizes functionally equivalent environmental changes. These are all the basic elements needed in mimic defense. Choose the appropriate "pass or fortress" to comprehensively utilize these intrinsic features and introduce the virtualized DHR structure in the pool resource allocation, process task management, virtual container migration, security management, SLA performance guarantee, file storage, etc. For example, using the intra-cloud transfer and migration to realize the switching of virtual service environments, or using the user environment assignment mechanism to implement the plural virtual host allocation and replacement across physical entities, or using the virtual input/output proxy and the arbiter to construct the virtualized mimic bracket. If the mimic defense can be deployed in a layered, coordinated, and virtualized manner in the cloud platform, and the traditional security technologies can be integrated into the virtual host environment, it's hopeful to enable the cloud services to break out of the security dilemma. There are three possible modes to adopt the mimic cloud. The first is a fully transparent mode, that is, the mimic structure is unperceivable to users. This mode requires the mimic cloud to be able to automatically create a virtual mimic environment. The second is the nontransparent mode, where users apply for virtual mimic components, such as heterogeneous executors, input/output proxy or an arbiter, and can provide diversified software versions for virtual hosts, as well as functional software for mimic brackets. This mode places the highest demand for the skills of the user. The third mode is the SaaS, where users directly apply to the mimic cloud for commercialized or customized software services, without worrying about the implementation of mimic technology. The user does not have to worry about the realization of mimic technology. This mode is suitable for the mature software service and is of medium difficulty for the professional development team to implement. The author believes that the mimic SaaS is likely to become the standard method for highly reliable, highly available, and highly credible cloud services.

12.5 Application Consideration for Software Design

The dynamic heterogeneous redundancy architecture can also be applied to high-reliability and high-security software designing. Although physical isolation of the processing space cannot be guaranteed, the following measures can still enhance the software's anti-attack capability and reduce reliability or security issues software designing flaws might bring.

12.5.1   Effect of Randomly Invoking Mobile Attack Surface

The dynamic invocation of those functionally equivalent heterogeneous redundancy modules through random parameters in the system (such as the current number of processes, processor occupancy, remaining storage space, etc.) could increase the uncertainty of software operation, thereby increasing the difficulty to take advantage of the design defects of these modules. For example, it is not certain for the software configured with multiple printer functions to adopt which kind of function. Even if those functions may have unknown vulnerabilities or are implanted with backdoors in their designing, the reachability of the attack packages will become uncertain because the random calling has turned the static attack surface into a moving one. Of course, the moving attack surface effect is achieved only when the software designer does not intentionally insert backdoors or malicious codes.

12.5.2   Guard Against Hidden Security Threats from Third Parties

The current integration of third-party software modules or commissioned third-party development has become a common practice in the industry. It is a serious challenge to ensure cyber service security even when the credibility of commissioned development components cannot be ensured. For example, Baidu received a report on February 28, 2017, that malicious codes are implanted when users downloaded software from two of Baidu-owned websites: http://www.skycn.net/ and http://soft.hao123.com/. Baidu then immediately investigated the case and found out the problem. Baidu made an announcement 4 days later on "Investigation into the Hidden Malicious Code found in Baidu's Websites," which says that the hao123 software downloader on the two websites was developed by a third-party team who turned out to have implanted a risky driver to hijack user traffic and gain more shares of income from Baidu Alliance. If the company employs the mimic structure, it could avoid the risk structurally by using multiple functionally equivalent third-party hardware and software components with dubious credibility. It may cost more in the first phase, but its advantages are quite obvious in the long run.

12.5.3   Typical Mimic Defense Effects

In theory, an invocation request can activate multiple heterogeneous redundancy modules simultaneously and, through a multimode ruling mechanism, select completely identical or majority identical results as the output of invocation. In this way, the design flaws or externally injected malicious codes cannot work, so they are suitable to be used in applications requiring high security and high availability. However, the logic "simultaneous activation" has to turn to physical "separate activation" at extra performance overhead. Fortunately, in the "era of low-cost multicore or many-core processor," this problem can be markedly weakened by scheduling multiple heterogeneous redundancy modules through multi-core or many-core parallel computing.

As to the problem of the unavoidable design defects or trapdoors in complicated software architecture when there is no deliberate insertion of backdoors, the mimic-structured application can be used to improve this problem from the structure level. It is a major advancement in security technology and has general significance for basic software, core software, tool software, and important applications and management software.

12.6   Commonality Induction of System-Level Applications

From the above applications, it is easy to see the following commonalities in the system-level application of mimic defense:

(1) Comply with the I[P]O model.
(2) Have one-to-many input splitters (can be assigned statically or dynamically).
(3) There are conditions for diversified heterogeneous redundant configuration of hardware and software components or virtual components.
(4) Have a multimode decision output mechanism capable of aggregating multiple outputs with independent processing space.
(5) All adopt the multimode ruling-based strategy scheduling and the negative feedback mechanism of multi-dimensional dynamic reconstruction.
(6) All are to guard against security threats brought about by unknown vulnerabilities, trapdoors, or backdoors from within the system.

References

1. Zeyu, S., Lili, J.: China's industrial software: a reflection on the "prism gate". China Inf. E Manuf. **11**, 24–33 (2013)
2. CERT. A briefing on the existence of preset backdoors on multiple router devices. http://www.cert.org.cn/publish/main/9/2014/20140429121938383684464/20140429121938383684464_html (12 Dec 2016)
3. Rekhter, Y., Li, T.: A Border Gateway Protocol 4 (BGP-4). RFC 4271, Internet Engineering Task Force, 2006, http://www.ietf.org/rfc/rfc4271.txt
4. Moy J. OSPF Version 2. RFC 2328, Internet Engineering Task Force, 1998, http://www.ietf.org/rfc/rfc2328.txt
5. Malkin G. RIP Version 2. RFC 2453, Internet Engineering Task Force, 1998, http://www.ietf.org/rfc/rfc2453.txt

6. Chun, B.G., Maniatis, P., Shenker, S.: Diverse replication for single-machine byzantine-fault tolerance. Usenix Technical Conference, Boston, pp. 287–292 (2008)
7. Junqueira, F., Bhgwan, R., Hevia, A., et al.: Surviving Internet catastrophes. Usenix Technical Conference, Anaheim, pp. 45–60 (2005)
8. Zhang, Y., Dao, S., Vin, H., et al.: Heterogeneous networking: a new survivability paradigm. Workshop on New Security Paradigms, pp. 33–39 (2001)
9. Xorp, Inc. http://xorp.org (22 Mar 2016)
10. Quagga. Quagga software routing suite. http://www.quagga.net (22 Mar 2016)
11. BIRD. The BIRD internet routing daemon. http://bird.network.cz (22 Mar 2016)
12. Keller, E., Yu, M., Caesar, M., et al.: Virtually eliminating router bugs. ACM Conference on Emerging Networking Experiments and Technology, Rome, pp. 13–24 (2009)
13. Knight, J., Leveson, N.: A reply to the criticisms of the Knight & Leveson experiment. ACM SIGSOFT Softw. Eng. Notes. **15**(1), 24–35 (1990)
14. Cisco 7200 Simulator. Software to run Cisco IOS images on desktop PCs. http://www.ipflow.utc.fr/index.php/Cisco_7200_Simulator (22 Mar 2016)
15. Packet Tracer. http://www.cisco.com/Web/learning/netacad/course_catalog/PacketTracer.html (22 Mar 2016)
16. Olive. Software to run Juniper OS images on desktop PCs. http://juniper.cluepon.net/index.php/Olive (22 Mar 2016)
17. Huawei. eNSP:Enterprise Network Simulation Platform. http://support.huawei.com/enterprise/toolNewInfoAction!toToolDetail?contentId=TL1000000015&productLineId=7919710 (22 Mar 2016)
18. GNS3. GNS3 Technologies Inc. https://www.gns3.com (22 Mar 2016)
19. QEMU. http://wiki.qemu.org (22 Mar 2016)
20. Casado, M., Freedman, M.J., Pettit, J., et al.: Ethane: taking control of the enterprise. ACM SIGCOMM Comput. Commun. Rev. **37**(4), 1–12 (2007)
21. Goransson, P., Black, C.: The OpenFlow specification. Software Defined Networks, pp. 81–118 (2014)

Chapter 13
Testing and Evaluation of the Mimic Defense Principle Verification System

In March 2014, China National Digital Switching System Engineering and Technological Research Center (NDSC), together with No.32 Research Institute of China Electronic Technology Group Corporation (CECT), Zhejiang University (ZJU), Fudan University, Shanghai Jiao Tong University (SJTU), Shenzhen ZTE Corporation, Wuhan FiberHome Telecommunication Technologies Co. Ltd., Chengdu Maipu Communication Technology Co., Ltd., etc., undertook the research and development of the "cyberspace mimic defense principle verification system," an assignment issued by Shanghai Municipal Science and Technology Commission. The assignment aims to verify the effectiveness and applicability of the mimic defense principle in the field of information and communication networks. For this purpose, it was broken down into two subjects of study: one is the mimic defense principle verification in the router environment (or mimic-structured router) for specific purposes; the other is the mimic defense principle verification in the web server environment (or mimic-structured web server) for general purposes. In November 2015, the R&D teams completed their verification systems, respectively, and conducted self-tests. In December 2015, Shanghai Municipal Science and Technology Commission submitted an application to the Ministry of Science and Technology of the People's Republic of China for testing of the systems. In January 2016, the Ministry of Science and Technology officially approved the application, and then Shanghai Municipal Science and Technology Commission started the joint test and evaluation of the mimic defense principle verification systems. It is hereby declared that the data and information of this chapter mainly come from the records and documents generated during the test and evaluation processes.

J. Wu, *Cyberspace Mimic Defense*, Wireless Networks,
https://doi.org/10.1007/978-3-030-29844-9_13

13.1 Mimic Defense Principle Verification in the Router Environment

13.1.1 Design of Test Methods for Mimic-Structured Routers

Traditional router testing includes at least function and performance testing, as well as the testing of stability, reliability, conformance, and interoperability. Among them, the function testing is mainly used to evaluate the router's interface function, communication protocol function, packet forwarding function, routing information maintenance function, management control function, and security function. The performance testing mainly examines the router's throughput, time delay, packet loss rate, back-to-back frame count, system recovery time, and other indicators. Traditional router testing methods can evaluate routers against the functionality, performance, and reliability indicators, but don't work well in evaluating the capability of routers in resisting network attacks, and are also unable to verify the implementation of the mimic defense mechanism in routers.

To this end, the test team has designed a test method for the mimic defense mechanism and another test method for the defense effect. These two methods comprehensively utilize the open internal module interface, result comparison analysis, etc. to evaluate the defense capability of the tested object from the perspectives of implementation process and effect. According to this test idea, the test content can be divided into the following three parts [1]:

(1) Basic router performance test. It tests the mimic-constructed routers and determines whether the mimic-structured routers will influence the basic ability of computation and forwarding performance when added with relevant units and modules for implementing the mimic defense mechanism.
(2) Mimic defense mechanism. It tests the relevant modules that realize the mimic defense mechanism and checks whether they can operate normally according to the design requirements and can effectively realize the mimic defense mechanism.
(3) Defense effect test. It tests the system to see if the system can ① change the display property of inherent vulnerabilities or backdoors; ② disturb the lockability of vulnerabilities or backdoors and the reachability of attack links; and ③ dramatically increase the difficulty of utilizing its apparent vulnerabilities or backdoors on condition that the inherent vulnerabilities or backdoors still remain there.

The overall architecture of testing is shown in Fig. 13.1.

The test methods can be divided into two types: conformity test and open fitting test [2]. The former applies to the tested objects, which have standards or design specifications to follow, to check if they conform to the relevant standards, theoretical framework, or design requirements [3]. The latter is used to evaluate the effect of the mimic mechanism in cutting off the steps of an attack chain, for

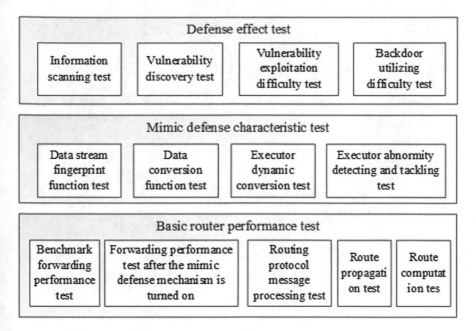

Fig. 13.1 Test architecture design

in which the next step will not start unless the preceding one succeeds. The open fitting test is done by making public the system implementation structure, setting the internal observation points, implanting backdoors, shutting down the mimic mechanism, etc.

The tests are based on the theory of mimic defense, with reference to the following standards:

(1) GB/T18018-2007 *Information Security Technology—Technical requirements for router security*
(2) YD/T1156-2001 *Test Specification for High-End Router*
(3) GB/T 20984-2007 *Information Security Technology—Risk assessment specification for information security*
(4) GB/T 14394-2008 *Computer software reliability and maintainability management*

The test environment settings are shown in Fig. 13.2. One port of the system under test is connected to the Internet to receive attacks from external networks. The other three ports are connected to the virtual network built with the Spirent TestCenter. The video server, mail server, terminal, and other service nodes, as well as virtual nodes used to carry out attacks, are virtualized in the virtual network. All the relevant tests of the mimic-structured router verification system are completed and analyzed in the topological environment. The test instruments include Spirent TestCenter, Wireshark [4], Nmap [5], Nessus [6], Metasploit [7], etc.

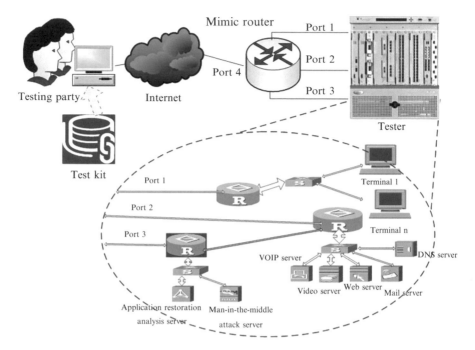

Fig. 13.2 Test topology environments

13.1.2 Basic Router Function and Performance Test

The purpose of the test in this section is to verify whether the functional conformity and operational consistency of the system are guaranteed after the mimic defense mechanism is introduced in the router, which is a prerequisite for security test. The basic performance test mainly includes routing protocol function test and forwarding performance test. Spirent TestCenter is used to complete the function and performance tests of the protocol.

13.1.2.1 Routing Protocol Functional Test

The test is to verify whether the routing protocol function implemented by the system conforms to the standard protocol specification. The object of the test is set as OSPF (open shortest path first) protocol [8]. The test method is to send specific data packets to the tested system through the Spirent TestCenter and carry out compliance testing in accordance with the OSPF protocol standard. The test results are shown in Table 13.1.

We can see from the test results that the OSPF protocol implemented by the mimic-structured router conforms to the RFC standard and that the introduction of the mimic defense mechanism does not affect the router's original routing protocol function.

Table 13.1 Function test results of the OSPF protocol

Test No.	Test item	Test results
1	OSPF header validity check	Passed
2	HELLO header validity check	Passed
3	HELLO messages sent via various ports	Passed
4	Receive HELLO message	Passed
5	Determination of master/slave relationships during database synchronization	Passed
6	Send and receive the database description message	Passed
7	Receive invalid database description messages	Passed
8	Synchronize various data bases of link-state advertisement	Passed
9	Send and receive link-state request messages, link-state update messages, and link-state confirmation messages	Passed
10	Link-state update message validity check	Passed
11	Route propagation	Passed
12	Shortest path first algorithm	Passed

13.1.2.2   Forwarding Performance Comparison Test

The test is to compare the changes of the system's forwarding performance before and after the introduction of the mimic defense mechanism so as to compare the influence of the mimic defense mechanism on the router's basic forwarding performance. The test was done in the following steps: Step 1, turn off the mimic defense function of the system (in this case, the system is equivalent to a traditional router), test the forwarding performance of the system, and take the result as a benchmark for comparison; Step 2, separately test the forwarding performance of the system under the condition of just turning on data transformation or only turning on dynamic scheduling function of the system; Step 3, test the forwarding performance of the system when turning on both data transformation and dynamic scheduling functions (i.e., all mimic functions of the system are turned on) simultaneously. Connect the system under test through the GE interface of the tester, and test the forwarding performance indicators of the system at the rate of 1000 MB/s. When both packet loss rate and delay jitter are all zero, the throughput rate and delay of the system are selected as the evaluation indexes of the forwarding performance. The frame sizes tested at each step include 64, 128, 256, 512, 1024, 1280, and 1518 bytes. The test results are shown in Table 13.2.

The test results show that when the mimic-structured router was partially or completely turned on the mimic defense function, its forwarding performance (throughput rate and delay) did not apparently decrease compared with the benchmark forwarding performance. Therefore, it can be concluded that the introduction of mimic defense mechanism has no obvious influence on the basic forwarding performance of the router.

Table 13.2 Comparison of test results of forwarding performance

Test condition	Average throughput rate (%)	Average delay/ms
Benchmark (turn off the mimic function)	85.7	0.35
Only data transformation is enabled	83.7	0.35
Only dynamic scheduling is enabled	84.2	0.38
Both data transformation and dynamic scheduling are enabled	85.4	0.39

In short, after the introduction of the mimic defense mechanism, the mimic-structured router can still ensure its basic functions, and its performance indicators do not show obvious decline.

13.1.3   Test of the Mimic Defense Mechanism and Result Analysis

In order to achieve the goal of mimic defense, the mimic router adds the data transformation mechanism, data stream fingerprint generation mechanism, dynamic executor scheduling mechanism, multimode ruling mechanism, and negative feedback control mechanism to its architecture. Before evaluating the defense effect of the mimic defense mechanism, it is necessary to first test whether the mimic-structured router has realized the alleged mimic defense mechanism and ensure that the mechanism can fulfill their expected functions according to the design requirements.

At the data level, regarding the two tests of data transformation mechanism and data stream fingerprint generation mechanism, it is necessary to make corresponding test data packets and set up multiple traffic monitoring points in and out of the system and judge whether the corresponding functions have been realized by comparing the data at different monitoring points. The functions of data transformation mechanism and data stream fingerprint generation mechanism are mainly realized in the input and output proxy module. The setting of data monitoring points is shown in Fig. 13.3.

13.1.3.1   Data Transformation Function Test

The test is to verify whether the system can perform data transformation of the input data message and perform inverse transformation before the data message leaves the system, thus to prevent utilizing the data stream on the date plane to trigger backdoors/vulnerabilities, which will lead to malfunction or failure of the router. Test procedure: send multiple data packets on a certain line to the system under test. Some of the packets shave specific load, but no packets are transmitted on other lines. Then record the passing packets at monitoring points on the data-sending line. Meanwhile, detect the passing packets at monitoring points on other lines. Once

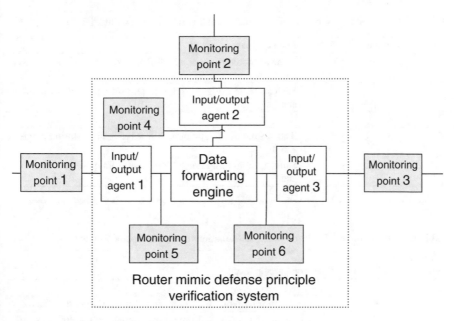

Fig. 13.3 Setting scheme of the data monitoring point

Table 13.3 Function test results of data transformation (monitoring point 1 input)

Monitoring point no.	Traffic	Load comparison result
1	Y	Same as monitoring point 3, different from monitoring points 5 and 6
2	N	–
3	Y	Same as monitoring point 1, different from monitoring points 5 and 6
4	N	–
5	Y	Same as monitoring point 6, different from monitoring points 1 and 3
6	Y	Same as monitoring point 5, different from monitoring points 1 and 3

packets are found passing, then compare these data packets with the intercepted packets at other monitoring points to judge if the date transformation function works well. Repeat the procedure three times, during each procedure the packets are sent through a different route. The test results are shown in Tables 13.3, 13.4, and 13.5.

The test results show that, when data packets entered the system under test, its load was transcoded by the input/output agent, and the data were converted reversely before being sent out of the tested system, so that the data were consistent with the input data load. The data were successfully transformed in the system under test.

Table 13.4 Functional test results of data transformation (monitoring point 2 input)

Monitoring point no.	Traffic	Load comparison result
1	Y	Same as monitoring point 2, different from monitoring points 4 and 5
2	Y	Same as monitoring point 1, different from monitoring points 4 and 5
3	N	–
4	Y	Same as monitoring point 5, different from monitoring points 1 and 2
5	Y	Same as monitoring point 4, different from monitoring points 1 and 2
6	N	–

Table 13.5 Function test results of data transformation (monitoring point 3 input)

Monitoring point no.	Traffic	Load comparison result
1	N	–
2	Y	Same as monitoring point 3, different from monitoring points 4 and 6
3	Y	Same as monitoring point 2, different from monitoring points 4 and 6
4	Y	Same as monitoring point 6, different from monitoring point 2 and 3
5	N	–
6	Y	Same as monitoring point 4, different from monitoring points 2 and 3

13.1.3.2   Data Stream Fingerprint Function Test

The test is to verify whether the system can normally add data stream fingerprint to the incoming data packets, which will then be deprived of the data stream fingerprint before leaving the mimic-structured router to prevent attackers from using the data plane function components to tamper with or transfer the packets. Test procedure: send multiple data packets to the tested system on a certain line, but no packets are sent on other lines, and then record all the data packets flowing through the monitoring points on the data-sending line. At the same time, check at other monitoring points the passing packets to see if they have been inserted with character string fingerprints, thereby to judge whether the data stream fingerprint function work normally. Repeat the procedure three times; during each procedure the test data are input on a different line. The test results are shown in Tables 13.6, 13.7, and 13.8.

The test results show that the input/output agents added fingerprint to each incoming data packet according to the algorithm and deleted the fingerprint from the data packets before they left the system. The data stream fingerprint function was normally implemented.

Table 13.6 Test results of data stream fingerprint function (monitoring point 1 input)

Monitoring point no.	Traffic	Character string fingerprint
1	Y	N
2	N	–
3	Y	N
4	N	–
5	Y	Y
6	Y	Y

Table 13.7 Test results of data stream fingerprint function (monitoring point 2 input)

Monitoring point no.	Traffic	Character string fingerprint
1	Y	N
2	Y	N
3	N	–
4	Y	Y
5	Y	Y
6	N	–

Table 13.8 Test results of data stream fingerprint function (monitoring point 3 input)

Monitoring point no.	Traffic	Character string fingerprint
1	N	–
2	Y	N
3	Y	N
4	Y	Y
5	N	–
6	Y	Y

13.1.3.3   Protocol Executor Random Display Test

The test is to verify whether the system under test can normally complete the selection and dynamic switch of the executors so as to carry out external display at random while completing normal routing computation and other functions. Test procedure: record every 10 min the number of the executor displaying itself after the switch function of the executor is started, which is the number of the worker executor. At the same time, inject the routing table entry into the system under test, input test data stream through the tester continuously into the tested system, and monitor whether the data flow is interrupted at the exit to observe if the updated data packets of the routing protocol are sent normally. The default switching time of the worker executor is a random value between 5 and 15 min. To facilitate the test, the switch time is set as 10 min, a fixed value; the routing protocol is OSPF, which is injected into four routes. The seven executors are numbered from A to G; the simultaneous online inspector executors are set to five. The five online executors will be

Table 13.9 Observation results of protocol executors displayed

Observation time/min	Worker executor no.	Inspector executor no.
5	E	A, B, C, D, E
15	C	B, C, D, E, F
25	F	C, D, E, F, G
35	A	A, D, E, F, G
45	B	A, B, E, F, G
55	A	A, B, C, F, G
65	G	A, B, C, D, G
75	D	A, B, C, D, E

selected from the seven executors in turn according to the number order. The observation shall start from the time when the system initialization is complete.

When the test started, the external display of the executors at each observation time point is shown in Table 13.9, where you can see that the worker executors displaying externally in each cycle of the system are different from those in the previous observation cycle and are scheduled in a random sequence. When routing is injected into the executors, the LSA notices sent by the OSPF protocol are also captured at each monitoring point. After injecting the test data stream, the traffic can be observed at the exit port without any interruption.

The test results indicate that the mimic-structured router can realize the dynamic random scheduling of the online executors, and the data forwarding and routing protocol operation are not affected by the scheduling operation.

13.1.3.4   Protocol Executor Routing Abnormity Monitoring and Handling Test

The purpose of this test is to verify whether the system can activate the feedback control function and switch the abnormal executors through the multimode ruling mechanism according to the set rules. Test procedure: start the switch function of the executor, simulate the neighboring router through the tester to inject the routing table entries into the tested system, and at the same time, input test data streams continuously to the system through the tester. Next, simulate the malicious invasion of the worker executor, modify the route by writing a static spurious route directly to the executor online rather than through the central controller, and observe the scheduling of the executor and the forwarding of the test flow. Repeat the procedure three times at different time points. The observation shall start from the time when the system initialization is complete.

It can be seen from the test results in Table 13.10 that after each injection of a false route into the online executor, the multimode ruling point of the system immediately detected the abnormity of the routing table and switched the worker executor. The routing table of the newly displayed executor did not contain the injected false routing table items. When executor B that had been switched offline was scheduled to the display state again, its routing table did not have the injected

Table 13.10 Dynamic conversion test results of the protocol executor

Time/min	Operation	Observation results
3	Observe the display of the executor	Executor B is *worker*
4	Inject a false route	The false route is injected into the routing table of executor B
5	Observe the display of the executor	Executor D is *worker*, and the routing table does not contain the false route
7	Inject a false route	The false route is injected into the routing table of executor D
9	Observe the display of the executor	Executor C is *worker*, and the routing table does not contain the false route
14	Observe the display of the executor	Executor C is *worker*, and the routing table does not contain the false route
15	Inject a false route	The false route is injected into the routing table of executor C
17	Observe the display of the executor	Executor F is worker, and the routing table does not contain the false route

spurious route since the tested system had cleaned its data. No flow interruption or flow interference occurred during the test.

The test results show that the system under test can timely detect the abnormal table entry from the routing table and schedule and switch the executor. After switching, the abnormal routing table entry will be cleared.

13.1.3.5   Endogenous Flow Interception Test

The test is to verify whether the multimode ruling point of the system can block the abnormal flow sent out positively by the abnormal executor. Test procedure: suppose the worker executor has been maliciously controlled by an attacker. The testing personnel sends commands to the worker through its NMS (network management system), so that the worker can actively send out a TFTP (trivial file transfer protocol) service request, simulating the process in which the router initiates a connection to the attacker and send data to it after vulnerability or backdoor is activated. Then set up monitoring points on both sides of the multimode ruling point to check if the request is intercepted. For comparison, in the second test, send the TFTP external connection command (with the same destination IP address) to the worker and three inspector executors at the same time, and observe at the monitoring points if the TFTP request is sent out. The monitoring point deployment scheme is shown in Fig. 13.4.

The test results are as follows: in the first test, the data message of TFTP request was detected at monitoring point 1, but it was not detected at monitoring point 2; in the second test, the data message of TFTP request was detected at both monitoring points 1 and 2.

It can be seen from the test results that, when only one executor makes outbound connection, the multimode ruling point treats it as an abnormity and intercepts the

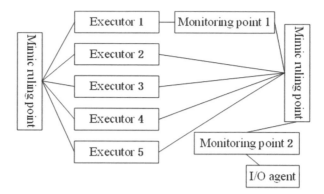

Fig. 13.4 Monitoring point deployment scheme for endogenous flow interception test

connection request packet. When most of the inspector executors send connection requests externally at the same time, the multimode ruling point will consider this as a normal behavior and release the worker's request.

The test results show that the multimode ruling mechanism of the system under test can effectively prevent individual executors from sending abnormal messages after their vulnerabilities or backdoors have been triggered.

13.1.4 Defense Effect Test and Result Analysis

Attacks targeting routers are mainly divided into two types [9]: one is to attack the defects of the routing protocol's own mechanism, making the routing protocol unable to converge or generate correct route output. The target of such attacks is the protocol mechanism, which is not within the defense scope of the system under test. Therefore, such attack scenarios are not tested. The other is to attack the vulnerabilities introduced during the router implementation or the backdoors implanted purposefully. Such attacks will gain control of targets, steal information, and disrupt or paralyze the target system. They shall be mainly guarded against in the tested system, and the defense effect test mainly focuses on such attack scenarios.

The vulnerabilities or backdoors introduced in the process of implementation can be divided into two types depending on their locations in the system under test. One is located in each executor, and the other is in a mimic plug-in (a functional part introduced to realize the mimic mechanism, including input/output agent, multimode ruling, dynamic scheduling, and other modules). For mimic plug-ins, the dynamic scheduling module is connected to the executor in one direction, and the attack message cannot reach it. The input/output agent takes responsibility for message matching and forwarding, and the multimode ruling handles the bitstream; neither of them processes data content or semantics. Therefore, the attacks cannot be launched against mimic plug-ins. In addition, as the function of mimic plug-in is single and simple, and the quantity of function codes is much less than the code

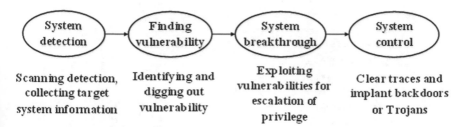

Fig. 13.5 Attack chain exploiting vulnerability of the router

quantity of the executor, the code auditing method can be used for security assessment and testing. Hence, this section focuses on the defense effect testing of vulnerabilities or backdoors in the executor.

13.1.4.1   Attack Models and Testing Scenarios

Considering the defense target of the system under test, the typical attack chain of an attacker using vulnerability to attack the router is shown in Fig. 13.5. The attacker first identifies the information, such as device model, operating system version, surrounding network topology, open port number, and enabled network service, of the target router by means of scanning detection. The attacker identifies the vulnerabilities on the target router based on the scanned and detected information, by which he can get escalation of privilege; after that, they can modify the router's ACL (access control list) rules to open the door for malicious traffic to enter the internal network; or he can modify the routing table and deploy hidden channels to perform sniffing and man-in-the-middle attacks on all the data passing through the router. Furthermore, they can set backdoors in the router system for long time and hidden control.

By analyzing the above attack chain, we can see that the success of the previous aspect of the attack chain is the premise of the subsequent aspects. In order to cover all the aspects of the attack chain and evaluate the effect of mimic mechanism on severing all aspects of the attack chain, this section uses the open coordinated test method to test and analyze the defense effect of a mimic router, that is, to test each aspect of the attack chain item by item on condition that the attacker has completed the preceding step of the aspect in the attack chain.

13.1.4.2   System Information Scanning Test

System information scanning is the first step of network attack. It detects the open port, the enabled services, and OS version of the target system to determine the methods and tools to be used in subsequent attacks. The test is to verify the system's ability to defend against the first step of the attack chain. Test procedure: use the Nmap tool to scan the system ten times, starting at a randomly selected time.

```
PROTOCOL STATE           SERVICE
1        open            icmp
2        open|filtered   igmp
4        open|filtered   ipv4
6        open            tcp
17       open            udp
41       open|filtered   ipv6
46       open|filtered   rsvp
47       open|filtered   gre
89       open|filtered   ospfigp
103      open|filtered   pim
112      open|filtered   vrrp


PORT      STATE          SERVICE VERSION
23/tcp    open           telnet
161/udp   open           snmp    SNMPv1 server; nil SNMPv3 server (public)
| snmp-sysdescr: ZXR10 ROS Version V4.6.03b GER Software,
| Version V2.6.03b41 Copyright (c) 2000-2005 by ZTE Corporation Compiled Jul
520/udp   open|filtered  route
1701/udp  open|filtered  L2TP
Network Distance: 1 hop
Service Info: Host: ZXR10
```

Fig. 13.6 System information scanning results

The scan from the beginning to the return of the scan results is seen as a scan process. In the scanning process, the system executors are switched through random scheduling.

Since the results of the ten scans are basically similar, we select the results of one of the scans for analysis. The scanning results are shown in Fig. 13.6. The scan lasted for 904 s. From the open port number and the service on the port, we knew that it was a routing device, and its model was ZXR10. By reading the schedule log of the target system, we learned that in the scanning process, there were three executors going online in turn, which were ZTE executor, Cisco executor, and Quagga executor. Therefore, the final test result got through Nmap scan is actually not the result of scanning of the same target, and the test conclusion based on such results is actually inaccurate. The results of each of the ten scans are inconsistent; in other words, the scan test results are unpredictable.

The working mechanism of the mimic-structured router enables it to run multiple executors at the same time and display externally the random scheduling and the information displayed by the system hops. It is difficult to accurately identify the relevant information of the target system only depending on one or two scans. To obtain accurate target information, the attacker will have to increase the number and frequency of scans. This will greatly increase the cost of the attack. Furthermore, if the scan frequency is too high, the attack is easy to be detected by security detection equipment or through log analysis. Therefore, in

Table 13.11 Vulnerability scanning results

Vulnerability level	Vulnerability quantity	Vulnerability details
Critical	0	–
High	1	SNMP Agent Default Community
Medium	0	–
Low	1	Unencrypted Telnet Server

the scanning detection stage, mimic defense mechanism can effectively confuse the attacker, which obviously obstructs the attacker from obtaining accurate information about the target system, increases the time spent in scanning detection, and significantly raises the difficulty for attackers to obtain information of the target system.

13.1.4.3 Mimic Interface Vulnerability Detection Test

In the previous test, the Nmap scanning detection did not confirm the specific information of the target system, but it knew which ports were open and which service protocols were running. The purpose of this test is, on the basis of the above scanned system information, to scan vulnerabilities in these open services and to test and verify the display property of the inherent vulnerabilities or backdoors in the mimic interface. The test was done by scanning several services of the target system ten times for vulnerabilities with the Nessus tool.

Only one of the test results displays a high-risk vulnerable point in the target system, as shown in Table 13.11. The high-risk vulnerable point is the SNMP community string [public]. This is because there is an executor in the mimic interface, where the community string is set to public. Due to the dynamic nature of the system, other executors do not use this easy-to-guess community string. As a result, only one of the ten scans discovered the vulnerability in the target system.

Vulnerability scanning is the second step of the attackers. The results of this step will determine the subsequent attack methods, tools, and the degree of difficulty. If the scan detects vulnerabilities in the target system, the attack will probably be successful. Through the random scheduling mechanism, the system displays multiple executors in the form of hops, making it hard for the vulnerability scanning to get the accurate vulnerability information of the changed target system within a short time. Therefore, the mimic defense mechanism can change the display features of vulnerabilities or weak points of the system if they haven't been eliminated yet.

13.1.4.4 Test of Difficulty in Vulnerability Exploitation Within Mimic Interface

The test is to evaluate the difficulty in verifying the utilization of the apparent vulnerabilities within the mimic interface. By skipping the preceding part of the attack chain, we directly assume that the attacker has been informed that there is an exploitable vulnerability in the mimic interface. On this basis, we use the vulnerability to launch attacks to verify the defense effect of the system. We selected the SNMP vulnerability of Cisco IOS for testing. The vulnerability exists in the Cisco executor of the system, and when exploited, it will enable TFTP, through which users can download the configuration file of the target system. The test is performed by first suspending the dynamic scheduling and multimode ruling mechanism within the mimic interface, that is, shut down the mimic defense mechanism of the system, and set the Cisco executor as worker. In this case, we think that the system is equivalent to a Cisco router and verify the exploitability of the vulnerability. And then, restore the dynamic scheduling and multimode ruling mechanism in the mimic interface, that is, start the mimic defense mechanism to verify the exploitability of the vulnerability again.

The test results are as follows: after turning off the mimic mechanism and using the Metasploit tool to penetrate into the system, we got results as shown in Fig. 13.7, which shows that the vulnerability was exploited successfully to obtain the target system's enable password as well as the corresponding system configuration file. We launched the above attacks every 180 s in 60 min, and all the results we got are the same.

When turning on the mimic mechanism and repeating the above process, we found the vulnerability could not be exploited successfully. This is because the

```
msf auxiliary(cisco_config_tftp) > exploit
[*] Starting TFTP server...
[*] Scanning for vulnerable targets...
[*] Trying to acquire configuration from 1.1.1.66...
[*] Scanned 1 of 1 hosts (100% complete)
[*] Providing some time for transfers to complete...
[*] Incoming file from 1.1.1.66 - 1.1.1.66.txt 860 bytes
[*] Saved configuration file to /home/1.1.1.66.txt
[+] 1.1.1.66:161 Unencrypted Enable Password: !qaz@wsx
[*] Collecting :!qaz@wsx
[+] 1.1.1.66:161 Username 'nmgdcy' with Password: n@Qos$mpls
[*] Collecting nmgdcy:n@Qos$mpls
[+] 1.1.1.66:161 SNMP Community (RO): nmgdcy@demaxiya
[*] Collecting :nmgdcy@demaxiya
[+] 1.1.1.66:161 SNMP Community (RW): public
[*] Collecting :public
[*] Shutting down the TFTP service...
[*] Auxiliary module execution completed
```

Fig. 13.7 Vulnerability exploitation results when the mimic mechanism is turned off

mimic-structured router adopts the multimode ruling mechanism, and the TFTP data stream triggered by the attacker through the vulnerability is an individual behavior (only one executor responded to the same TFTP input, but other executors did not). Therefore, it could not pass the ruling point, so the exploitation of the vulnerability failed.

To trigger and exploit vulnerabilities is the core step of attacks. Anyway, whether the vulnerabilities can be successfully triggered is the ultimate indicator to evaluate the security of a system. Therefore, the mimic defense mechanism can greatly increase the difficulty of exploitability of apparent vulnerabilities of the system if they haven't been eliminated.

13.1.4.5   Test of Difficulty in Utilizing Backdoors in the Mimic Interface

The test is to evaluate the difficulty of using backdoors in the mimic interface to test the defense ability of mimic defense mechanism against backdoor attack. Backdoors can be utilized to paralyze a given information system, i.e., to disrupt its availability, information acquisition, confidentiality, and integrity. Therefore, test cases are set around these three aspects, and before the test, implant a backdoor into the system executor. The test method is similar to the difficulty test of system vulnerability utilization. First, shut down the mimic defense mechanism of the system, and set the executor with backdoors as worker to verify the exploitability of the backdoor. Then enable the system mimic defense, and verify the exploitability of the backdoor again.

1. Disrupting backdoors

In order to verify the defense effect of the disrupting backdoor, we set a backdoor on the executor, which will be triggered if the received neighbor field of the OSPF HELLO message contains 0×5a5a5a02, and the executor will automatically add a default route to A.B.C.D and make other routes invalid. In this case, A.B.C.D is specified by the "designated router" field in the HELLO message.

When the mimic mechanism was closed, the data sent by the tester to router port 1 was correctly forwarded out through port 2. The testing party sent the backdoor trigger message through port 4 to trigger the backdoor. The data originally sent through port 2 was modified and sent out through port 3. It can be seen through SNMP that a default route has been added to the executor, and the next hop of the router is 192.168.3.100, as shown in Fig. 13.8. This indicates that the backdoor has been successfully triggered and exploited. Next, open the mimic mechanism, and repeat the above backdoor triggering operation. Through SNMP, we can still see that the backdoor executor generated default route, showing the backdoor of the executor had been triggered, but the data flow didn't change. This suggests that the backdoor in the mimic interface can be triggered but cannot been successfully exploited.

Analysis of the test results shows that the tampered backdoor of the routing table cannot be successfully utilized when the mimic defense mechanism is enabled. This

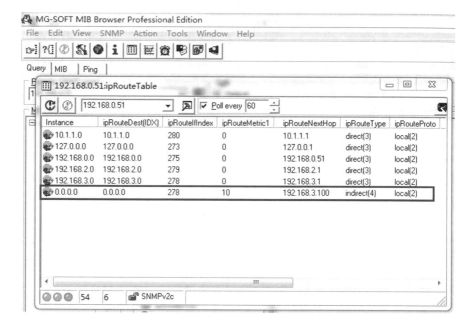

Fig. 13.8 Utilization results of routing table tampering backdoor when the mimic mechanism is turned off

is determined by the multimode ruling features of the mimic defense mechanism. The routing table modified by the attacker through the backdoor only exists on a single executor, which is inconsistent with the routing table of other online executors. For this reason, attacks cannot pass the ruling point, and the routing table tampering fails.

2. Theft backdoor

In order to verify the defense effect of the backdoor against theft of information, we set a backdoor on the executor, which will automatically send the internal routing table information to the specified IP address every 30 s as UDP messages. The IP address is specified by the SNMP trigger message.

When closing the mimic mechanism, the testing party sent the backdoor trigger message through port 4, and then the tester received the UDP message with the original destination port number 8993 at port3, as shown in Fig. 13.9. We found that the information carried by the payload was consistent with the routing table of the executor, indicating that the backdoor was successfully triggered and exploited. Then, turned on the mimic mechanism and repeated the above backdoor trigger operation, the tester did not receive the UDP message with original destination port number 8993 at port 3. This indicates that the mimic mechanism prevents the backdoor from being used successfully.

It can be seen from the test results that, when the mimic defense mechanism is enabled, the theft backdoor cannot be used successfully, just like the routing table

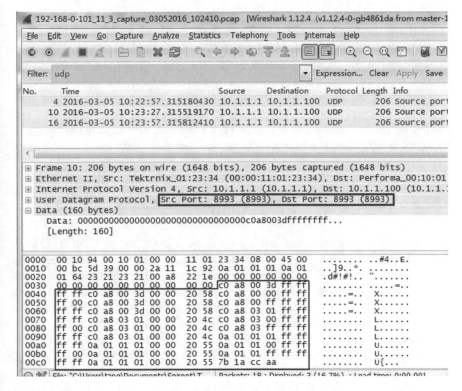

Fig. 13.9 Theft backdoor exploitation results when the mimic mechanism is turned off

tampering backdoor, due to the multimode ruling feature. Sending UDP messages by the executor with the backdoor is also an individual behavior, which will be found at the ruling point and cannot be sent out.

3. Paralyzing backdoor

To verify the defense effect against the paralyzing backdoor, we implanted the backdoor artificially in the data forwarding plane, which, after receiving the UDP message with the source and destination port number 66535 and the load of full F, will cause the system to paralyze and the forwarding function to fail.

When the mimic mechanism was turned off, the background traffic data sent by the tester to router interface 1 was correctly forwarded through port 2. After that, the backdoor trigger message was superimposed on port 1, and it was found that port 2 of the tester did not receive any message. When examining the forward unit through the serial port, we found that it did not respond to any commands, indicating that the backdoor was successfully triggered and exploited. Then we started the mimic mechanism and repeated the above backdoor trigger operation, and the tester was able to receive background traffic and backdoor trigger messages at port 2 all the time. This indicates that the mimic mechanism prevents the backdoor from being utilized.

It can be seen from the test results that after the mimic defense mechanism is enabled, the paralyzing backdoor cannot be successfully utilized. This is the result of the load semantic transformation function of the forwarding plane, which dynamically changes the data load entering the forwarding plane to disable the backdoor that relies on special bitstreams to trigger its hardware logic.

Through the analysis of the test results of the above three test cases, we can draw the following conclusion: by introducing dynamic heterogeneous redundant and multimode ruling mechanism, mimic defense can greatly increase the difficulty of backdoor attacks without eliminating them in the mimic interface.

13.1.5   Test Summary of Mimic-Structured Router

Routers are the most important basic core devices in the network space. Due to their closure, they often carry a lot of unknown vulnerabilities and backdoors and are vulnerable to attacks. The DHR architecture used to design the mimic-structured routers is a new way to enhance router security. The verification test results of the mimic-structured router principle show that the system can achieve the core function of mimic defense mechanism without affecting its basic functionality and performance and that it can effectively block the attacks based on unknown vulnerabilities, greatly increasing the backdoor attack difficulty and enhancing its own safety protection ability. All these have proved the effectiveness of the mimic defense mechanism. The test methods and means of the mimic-structured router principle verification system can be referred to in the formulation of test specifications or standards of mimic defense mechanism.

13.2   Mimic Defense Principle Verification in the Web Server Environment

13.2.1   Design of Test Methods for Mimic-Structured Web Servers

The purpose of the mimic principle verification test in the web server environment is to evaluate the effectiveness of mimic defense theory and the feasibility of its engineering implementation. So the focus of this test and assessment is not only on judging whether there are vulnerabilities, backdoors, or Trojans in the target object but also on checking whether the attacks based on these vulnerabilities, backdoors, and Trojans can succeed and whether the successful attacks can be used sustainably. In other words, whether the mimic defense can achieve the goal that "vulnerability may not be exploited, or if exploited, it may not be effective."

13.2.1.1 Test Process Design

The test team designed a test method for mimic-structured web server principle verification system. The method is carried out, respectively, in the basic function and compatibility test, mimic defense mechanism test, defense effect test, and web server performance test [10]. It includes the following four parts [11]:

1. Web server basic function test and compatibility test

After the mimic defense mechanism is introduced in the web server, the test examines whether the system's function compliance and page compatibility are guaranteed, whether they conform to the HTTP protocol, and whether they affect the normal use of users.

2. Mimic defense mechanism test

The test examines whether the relevant modules that realize the mimic defense mechanism can operate normally according to the design requirements and can effectively realize the mimic defense mechanism.

3. Defense effect test

The test examines whether the mimic-structured web server principle verification system still can provide secure and reliable services when vulnerabilities and backdoors exist and can fundamentally change the traditional network security defense methods, block the attack chain that employs the vulnerabilities or backdoors, and achieve the effective defense against unknown vulnerabilities and backdoors.

The mimic defense principle web server is a new type of web security defense means. To verify its effectiveness and guarantee the integrity and objectivity of the test results, the test shall follow the rules below:

(1) Developers of the system under test shall not conduct incremental development.
(2) The system under test does not adopt the existing security protection means, such as firewalls and intrusion detection system.
(3) Deploy or inject known vulnerabilities and backdoors to simulate the unknown ones without using the existing security protection means, because we are unable to construct a scenario to test the attacks based on unknown vulnerabilities and backdoors.

The test method completely simulates the vulnerability discovery technology and attack technology accessible to hackers and is an in-depth test of the security defense ability of the target web service system, with the tested equipment exposed in a more real-life environment. The controllable methods and means are used to detect target servers, web applications, and the vulnerabilities in network configuration, including black box testing, injection testing, and gray box testing based on compromise rules. Eventually, the testing personnel will simulate the hacker attack to verify the defense effect of the mimic web server principle verification system.

The test is done according to the theory of mimic defense, with reference to the following standards:

(1) GB/18336-2008 *Information Technology—Security Techniques—Evaluation criteria for IT security*
(2) GB/T 30279-2013 *Information Security Technology—Vulnerability Classification Guide*
(3) GB/T 20984-2007 *Information Security Technology—Risk assessment specification for information security*
(4) GB/T 14394-2008 *Computer software reliability and maintainability management*

4. Web server performance test

The test verifies whether the mimic defense mechanism affects the performance of the system, whether a performance bottleneck is introduced in the engineering implementation on condition that the effectiveness of security defense is ensured.

13.2.1.2   Test Environment Setting

To avoid errors in the test results caused by differences in the web server hardware, the uniform web server hardware configuration is adopted in the test, as shown in Table 13.12.

13.2.2 Basic Functional Test and Compatibility Test for Web Servers

The test in this section aims to verify whether the functional conformity and page compatibility of the system are guaranteed after the introduction of mimic defense mechanism in the web server, whether they conform to the HTTP protocol function,

Table 13.12 Web server hardware configuration table

Type	Configuration information
CPU model	Intel Xeon(R) CPU E5-2620 v3 @ 2.40 GHz
System board	GIGABYTE MD30-RS0
Graphics card	ASPEED, 16 MB
Disc drive	LSI MR9260-8i SCSI Disk Device
RAM	LSI MegaRAID SAS 9260-8i (32 GB)
Network card 1	Broadcom BCM57810 NetXtreme II 10 Gige
Network card 2	Intel(R) I210 Gigabit Network Connection

Table 13.13 Test content of HTTP communication protocol

No.	Name	Test items	Test results
1	GET request test	Use the HTTP protocol test tool to send a GET request to the web server and see if the server will return a response message (header and body)	Can receive the GET request and return the correct response information
2	POST request test	Similar to Sample 1, send POST request	Can receive the POST request and return the correct response information
3	PUT request test	Similar to Sample 1, send PUT request	Can receive the PUT request and return the correct response information
4	DELETE request test	Similar to Sample 1, send DELETE request	Can receive the DELETE request and return the correct response information
5	HEAD request test	Similar to Sample 1, send HEAD request	Can receive the HEAD request and return the correct response information
6	OPTION request test	Similar to Sample 1, send OPTIONS request	Can receive the OPTIONS request and return the correct response information
7	TRACE request test	Similar to Sample 1, send TRACE request	Can receive the TRACE request and return the correct response information
8	Persistent connection test	Send a request to the web page through the tester and observe the TCP connection	Support HTTP1.1 persistent connection
9	Non-persistent connection test	Send a request to the web page through the tester and observe the TCP connection	Support HTTP1.1 non-persistent connection

and whether they affect the normal use of users. This is the fundamental guarantee for web servers to provide services and is the basic premise of all system tests.

13.2.2.1 HTTP Protocol Function Test

HTTP protocol is the basic protocol that web servers follow [12], and its latest version is version 1.1. It defines a few different ways for the client to interact with the server, among which there are four basic ways of communication, namely, GET, POST, PUT, and DELETE [13]. In this test, in order to ensure the integrity of the test, we conducted a comprehensive HTTP 1.1 communication protocol test for the system in accordance with the HTTP/1.1 protocol [14], and the test items are shown in Table 13.13.

All the tests were successful. For the specific test flow and results, please refer to 4.2 HTTP1.1 Protocol Conformance Test, 4.6.6 Persistent Connection Test, and 4.6.7 Non-Persistent Connection Test in the *Joint Test Report on the Mimic defense Principle Verification System of Web Servers*.

The test results show that the web server's mimic defense principle verification system fully supports the HTTP protocol function, and the introduced mimic defense mechanism will not affect the normal protocol function of the web server.

Table 13.14 Compatibility test content and results

No.	Test items	Test environment	Test results
1	Page display integrity test	Operating system + browser: Linux+Chrome, Win8+Chrome, Win7+Chrome, Win10+Chrome, Linux+Firefox, Win8+Firefox, Win7+Firefox, Win XP+Firefox, Win7+IE, Win8+IE, Win XP+IE	The system's page display integrity is consistent with that of a normal web server
2	Web page function integrity test	Operating system + browser: Linux+Chrome, Win8+Chrome, Win7+Chrome, Win10+Chrome, Linux+Firefox, Win8+Firefox, Win7+Firefox, Win XP+Firefox, Win7+IE, Win8+IE, Win XP+IE	The function integrity of the system page is consistent with that of a normal web server
3	JS script compatibility test	Linux+Chrome, Win8+Chrome, Win7+Chrome, Linux+Firefox, Win8+Firefox, Win7+Firefox, Win XP+Firefox, Win7+IE, Win8+IE, Win XP+IE	The compatibility of the system's JS script is consistent with the common web server performance
4	Typeset layout test	Operating system + browser: Linux+Chrome, Win8+Chrome, Win7+Chrome, Win XP+Chrome, Linux+Firefox, Win8+Firefox, Win7+Firefox, Win XP+Firefox, Win7+IE, Win8+IE, Win XP+IE	The typeset layout of the system's web pages is consistent with that of ordinary web servers
5	Web page layouts at different resolutions	Operating system + browser + resolution: Linux+Chrome+w-800 h-600, Linux+Chrome+w-1280 h-720, Linux+Chrome+w-1855 h-1056, Linux+Firefox+ w-800 h-600, Linux+Firefox+w-1280 h-720, Linux+ Firefox+w-1855 h-1056, Win7+Firefox+w-800 h-600, Win7+Firefox+w-1280 h-720, Win7+Firefox+w-1920 h-1080, Win7+Firefox+w-1928 h-1044, Win7+IE+ w-800 h-600, Win7+IE+w-1280 h-720, Win7+IE+ w-1920 h-1080, Win7+IE+w-1928 h-1044	Under different resolutions, the web page layout of the system is consistent with that of ordinary web server

13.2.2.2   Page Compatibility Comparison Test

User experience is a key consideration for web servers. Due to the expansion of mainstream browsers (including IE, Firefox, Chrome, etc.) and the diversification of display devices, the web server system needs to be compatible with a variety of user environments. In this section, a test case is designed to compare the web page provided by the web server's mimic defense principle verification system and that of an ordinary server, and judge whether there are differences in page display integrity, page function, JS script compatibility, page typeset layout, and page layout under different resolutions. The test items are shown in Table 13.14.

All the tests were successful. For the specific test process and results, please refer to 4.6 Compatibility Conformance Test of *Joint Test Report on the Mimic Defense Principle Verification System of Web Servers*.

Table 13.15 Function test items of the mimic defense principle

No.	Test objects	Test methods	Test results
1	RDB module	Capture the front-end exit data, and observe their destination address and content	The packets output to each virtual server pool were the same
2	RRV	Change the web pages of one of the three online virtual server pools, and observe the changes	The website normally provided services, and the server control side returned an error message
3	DES	It is prescribed that the virtual server pool can have 5 DESs, all of which stay online. Change the web pages of one of them; observe the changes	Based on the abnormal event driving, the changed server was cleaned and restored
4	PCON	Switch between different scheduling strategies to see if the system can operate normally	The system ran normally
5	DIL module	Whether DIL has the instruction fingerprint labelling capability, the offline instruction judgment capability, and the capability to maintain database consistency	All the capabilities passed the test
6	Test of emergency response capability of PCON	A power outage occurs to a virtual server pool	The system ran normally

The test results show that the web server mimic defense principle verification system will not affect web page compatibility and user access after the introduction of the mimic defense mechanism.

13.2.3   Mimic Defense Mechanism Test and Result Analysis

The mimic defense principle verification system of web servers is composed of a request dispatching and balancing (RDB) module, a dissimilar redundant response voter (DRRV), a dynamically executing scheduler (DES), a primary controller (PCON), and a database instruction labelling (DIL) module. Before evaluating the defense effect of the system, we will first test whether the modules of the system have realized the designed mimic defense mechanism and make sure whether the mechanism can perform its functions according to the design requirements. The test items are shown in Table 13.15.

All the tests were passed. For the specific test process and results, please refer to 4.1 functional test of *Joint Test Report on the Mimic Defense Principle Verification System of Web Servers*.

The test results show that all the functions of each module in the mimic defense principle verification system of web servers conform to the mimic defense principles.

13.2.4   Defense Effect Test and Result Analysis

The defense effect test flow of the web server mimic defense principle verification system completely simulates the hacker attack flow. From the perspective of hackers, the testing personnel at this stage carried out the scanning detection test, operating system security test, as well as data security test, antivirus Trojan test, and web application attack test [15–17]. After that, they compared the test results of the mimic-structured web server principle verification system with those of the common servers and analyzed the mimic defense mechanism that played defensive function in the test, determined the state of system vulnerabilities or backdoors (online/offline) when the mimic defense mechanism worked, and evaluated the defense effect of the principle verification system of mimic-structured web servers.

All the tests were passed. For the specific test process and results, please refer to 4.3 Security Test of the *Joint Test Report on the Mimic Defense Principle Verification System of Web Servers* and the *Report on Internet Penetration Test of the Mimic Defense Principle Verification System of Web Servers*.

13.2.4.1   Scanning Detection Test

Attackers can obtain key information of the target system through scanning detection before making attack plans. If the web server system can resist their scanning detection, it will be much stronger to fend off attacks and disrupt their attack plans. The scanning detection test is divided into fingerprint scanning test and vulnerability scanning test, as shown in Table 13.16.

The test results show that the mimic defense principle verification system of web servers uses the heterogeneous, redundant, and dynamic characteristics to disturb the judgment of the attackers and plays a hidden role, thus forcing the attackers to face mounting obstacles to launching attacks.

13.2.4.2   Operating System Security Test

As the software stack at the lower layer, the operating system (OS) supports the stable operation and normal use of the upper layer software stacks. Any malicious attack on the OS of the web servers will bring about serious security threats. The OS security test can be divided into OS escalation of privilege test, OS control test, OS directory leak test, and OS paralysis test according to the common attack types. The test items are shown in Table 13.17.

The test results show that the web server's mimic defense principle verification system adopted a heterogeneous operating system; in the dissimilar virtual web server pool, there were almost no consistent operating system vulnerabilities between functionally equivalent heterogeneous web service executors; the DRRV technology was used to timely detect abnormalities, and cut off the output of the

Table 13.16 Test content of scanning detection

Test item	Test category	Defense mechanism				Vulnerability or backdoor state	Test results
		Heterogeneous	Redundant	Dynamic	Cleaning		
Scanning detection test	Fingerprint scanning test	√	√	√		Online	The test was passed. The heterogeneously connected executors resulted in inconsistent fingerprint information, and the output result couldn't go through the voting process
	Vulnerability scanning test	√	√	√		Online	The test was passed. The heterogeneously connected executors resulted in inconsistent vulnerability information, and the output result couldn't go through the voting process

Table 13.17 Contents of operating system security test

| Test item | Test category | Defense mechanism | | | | Vulnerability or backdoor state | Test results |
		Heterogeneous	Redundant	Dynamic	Cleaning		
OS security testing	OS escalation of privilege test	√	√		√	Offline	The test was passed. The executors' operating systems were inconsistent; not all executors had privilege vulnerability. Moreover, the methods of escalation of privilege were not consistent; the attack couldn't be successfully executed in all executors
	OS control test	√	√		√	Offline	The test was passed. The executors' operating systems were inconsistent; the system control program depends on the system environment. Moreover, the control methods were inconsistent; the attack couldn't be successfully executed in all executors
	Operating system file directory leak test	√		√		Offline	The test passed. The random data generated in the file directory led to inconsistent information of the file directory; the output result couldn't be voted on
	OS paralysis test	√	√		√	Offline	The test was passed. The executors' operating systems were inconsistent, not all executors had the paralysis vulnerability, and the paralyzing methods were inconsistent; the attack couldn't be successfully executed in all executors

target under attack; the cleaning mechanism was used to timely dispatch the executor to clear and isolate the attack target, thus blocking the persistent attack attempt. Hence, the mimic defense principle verification system of web servers has been proved to have the ability to resist the attack on the operating system layer.

13.2.4.3 Data Security Test

Data are an important target of the attacker. How to protect data security is also an important security issue faced by web server systems. The data security test can be divided into transmission node sniffing detection test, transmission node paralyzing test, website directory extraction test, SQL instruction fingerprint damage test [18], and voter logic escape test according to the ways of acquiring or destroying data. The test items are shown in Table 13.18.

The test results show that the web server mimic defense principle verification system adopts the randomization technology to construct the mimic defense characteristics at the data level, including dynamicity of data transmission, heterogeneity of data operation instruction, and heterogeneity of data storage, and employs the redundant voting technology to timely detect abnormalities and cut off the output of the target. In terms of the voter logic, the dynamic switch of the executors makes it impossible for the attacker to launch reliable and persistent cooperative escape attacks. Therefore, the mimic defense principle verification system of web servers has been proved to have the ability to resist the attacks on the data layer.

13.2.4.4 Anti-Trojan Test

Viruses or Trojans are powerful weapons of attackers; they are very harmful and evolve a wide range of variants. If the web server system can resist the attacks of the Trojans, it may resist a majority of attacks. The antivirus Trojan test is divided into Trojan connection test, Trojan execution test, test of resource depletion caused by malicious pop-up windows, server information leakage virus test, system paralysis test, and website content tampering test according to the types of attack of the Trojan horse. The test items are shown in Table 13.19.

The test results show that the mimic defense principle verification system of web servers adopts different software and hardware structures between heterogeneous web servers with equivalent functions in the dissimilar virtual web server pool (DVSP). And since the execution of Trojans depends heavily on the system environment, the same virus Trojan can hardly be triggered in the heterogeneous execution environment. The dissimilar redundant response voter (DRRV) was used to timely detect anomalies and cut off the output of the target. At the same time, the cleaning mechanism was used to dispatch the executors in time to clear and isolate the attacked target, thus blocking the continuous attempt of the attack. Therefore, the mimic defense principle verification system of web servers is capable of defending against Trojan attacks.

Table 13.18 Data security test content

| Test item | Test category | Defense mechanism | | | | Vulnerability or backdoor state | Test results |
		Heterogeneous	Redundant	Dynamic	Clean		
Data security test	Transport node sniffing test		√	√		Online	The test was passed. The data in the information fragment randomization transmission module was transmitted randomly through the node. A single node sniffer couldn't get all the information
	Transmission node paralyzing test		√	√		Online	The test was passed. The data in the information fragment randomization transmission module was transmitted randomly through the node. A single node paralysis did not affect information transmission
	Web directory extraction test	√		√		Online	The test was passed. The random data generated in website directory led to inconsistent website directory information. The output result couldn't go through the voting process
	SQL instruction fingerprint damage test	√		√		Online	The test was passed. The SQL instructions were dynamically and randomly switched, and the damaged executor was cleaned and restored
	Voter logic escape test		√	√		Online	The test was passed. The executors were dynamically switched, and could not launch reliable and persistent cooperative escape attacks.

Table 13.19 Anti-Trojan test content

Test item	Test category	Defense mechanism				Vulnerability or backdoor state	Test results
		Heterogeneous	Redundant	Dynamic	Cleaning		
Anti-Trojan test	Trojan connection test	√	√		√	Offline	The test was passed. The executors' operating systems were inconsistent, so the connection information couldn't go through the voting process
	Trojan execution test	√	√		√	Offline	The test was passed. The executors' operating systems were inconsistent, so the execution result couldn't go through the voting process
	Test of resource depletion caused by malicious pop-up windows	√	√		√	Offline	The test was passed. The executors' operating systems were inconsistent; the virus couldn't be executed successfully in all the executors
	Server information leakage virus test	√	√		√	Offline	The test was passed; the service information of the executors was inconsistent; the output result couldn't go through the voting process
	System paralyzing virus test	√	√		√	Offline	The test was passed. The executors' operating systems were inconsistent, and the virus could not be executed successfully in all the executors
	Website content tampering virus test	√	√		√	Offline	The test was passed; the executors' operating systems were inconsistent. The virus could not be executed successfully in all the executors

13.2.4.5   Web Application Attack Test

The web application—the most important content of a web server system—is directly exposed to users and becomes the most direct target of attacks. In addition, the web application more or less has security vulnerabilities due to the carelessness of developers. The web server system that can resist the attack on the web application will constitute the first security barrier. The web application attack test is divided into directory configuration vulnerability test, SQL injection test, parsing vulnerability test, file inclusion test, and DoS vulnerability test by the common attack types. The test items are shown in Table 13.20.

The test results show that the mimic defense principle verification system of web servers adopts different software and hardware structures between heterogeneous web servers with equivalent functions in the DVSP. In addition, since the running of web applications also relies heavily on the system environment, the heterogeneous environment at the bottom can block the vulnerability exploitation attack of the upper application. The DRRV was used to timely detect abnormities and cut off the output of the attacked target. At the same time, the cleaning mechanism was used to dispatch the executors in time to clear and isolate the attacked target, thus blocking the continuous attack attempt. Therefore, the mimic defense principle verification system of web servers is capable of defending against Trojan attacks.

13.2.5   Web Server Performance Test

The test is to verify the effectiveness of the mimic defense mechanism installed in the web server and then examine whether the system performance is greatly affected by the mechanism. This is an important indicator in measuring the practical application value of the mimic defense principle verification system of web servers. Spirent Avalanche 3100B tester was used.

There are four key indexes used to evaluate the performance of the web server: max concurrent TCP connection capacity (MCTCC), throughput, average response time (APT), and transactions per second (TPS) [19].

1. Max concurrent TCP connection capacity (MCTCC)

As indicated in IETF 2647, MCTCC is the maximum number of TCP connections that can be established concurrently between hosts through the gateway or between hosts and gateways [20].

2. Throughput

Throughput mainly reflects the data packet forwarding ability of the network equipment, which is usually represented as the data forwarding ability of network equipment without packet loss. Its unit is kbps [21].

Table 13.20 Web application attack test content

Testing item	Test category	Defense mechanism					Vulnerability or backdoor state	Test results
		Heterogeneous	Redundant	Dynamic	Cleaning			
Web application attack test	Directory configuration vulnerability test	√	√		√		Offline	The test was passed. The server software of the executors was inconsistent across the executors. Not all executors had directory configuration vulnerabilities, so the attack could not be successfully executed in all the executors
	SQL injection test	√	√	√	√		Online	The test was passed, the SQL fingerprints of the executors was inconsistent, and the attack could not be successfully executed in all the executors
	Parsing vulnerability test	√	√		√		Offline	The test was passed. The server software of the executors were inconsistent across the executors, and not all the executors had parsing vulnerabilities, so the attack could not be successfully executed in all the executors
	Uploading vulnerability test	√	√		√		Online	The test was passed. There was inconsistency at levels of the executor's operating system and server software. So it was impossible to launch attacks through the uploaded script
	File inclusion test	√	√		√		Online	The test was passed. There was inconsistency at levels of executor's operating system and server software. It was impossible to launch attack through the script with execution
	DoS vulnerability test	√	√		√		Offline	The test was passed. The server software between executors was inconsistent. Not all executors had DoS vulnerability, and the attack could not be successfully executed in all executors

Table 13.21 Benchmark web server performance test content

No.	Test item	Server configuration	Test results
1	Single web server access performance test	Virtual machine software: VMware CPU: dual cores and dual threads RAM: 2 GB Page: 1 KB static page (the 32 KB static web pages are for throughput test)	TPS: 5114 MCTCC: 186652 Throughput (kbps): 472144 RTT (ms): 0.58
2	Single web server access performance test	Virtual machine software: VMware CPU: Dual cores and dual threads RAM: 2 GB Page: 1 KB dynamic page (the 32 KB dynamic web pages are for throughput test)	TPS: 4362 MCTCC: 179230 Throughput (kbps): 465716 RTT (ms): 4.766
3	Single web virtual server + mount database access performance test	Virtual machine software: VMware Database software: MySQL CPU: dual cores and dual threads RAM: 2 GB Page: 1 KB dynamic page (the 32 KB dynamic web pages are for throughput test)	TPS: 708 MCTCC: 29984 Throughput (kbps): 135665 RTT (ms): 58.814

3. Average response time (ART)

ART is the time required by the server to respond to the client's request. This index is the average duration for the server to complete a client request after running steadily.

4. Transactions per second (TPS)

A transaction is a process in which a client sends a request to the server, and the server responds to it. This index represents the number of transactions processed by the server per second and determines the server's transactional load capacity.

For the specific test process and results, please refer to 4.5 Performance Test of the *Joint Test Report on the Mimic Defense Principle Verification System of Web Servers*.

13.2.5.1 Benchmark Web Server Performance Testing

The mimic defense principle verification system of web servers is built by the basic web servers, in which the RDB, DRRV, and other modules are developed on the basis of the reverse proxy server and its middleware, so it is necessary to design test cases as benchmark references for system performance comparison and module performance comparison. The test items are shown in Table 13.21.

Table 13.22 DIL module performance test content and results

No.	Test item	Server configuration	Test results
1	Single web virtual server + database agent + mount database access performance test	Virtual machine software: VMware Database software: MySQL Database agent: Amoeba CPU: Dual cores and dual threads RAM: 2 GB Pages: 1 KB dynamic pages (the 32 KB dynamic pages are for throughput test)	TPS: 612 MCTCC: 26375 Throughput (kbps): 110012 RTT (ms): 55.67
2	Single web virtual server + database agent + mount database + DIL module access performance test	Virtual machine software: VMware Database software: MySQL Database agent: Amoeba CPU: dual cores and dual threads RAM: 2 GB Pages: 1 KB dynamic pages (the 32 KB dynamic pages are for throughput test)	TPS: 480 MCTCC: 24002 Throughput (kbps): 109343 RTT (ms): 61.592

13.2.5.2 DIL Module Performance Test

The database and database agent are deployed in the mimic defense principle verification system of the web servers, and the DIL module of the system is developed on the basis of the database agent. Therefore, it is necessary to design test cases to test the effect of the database, database agent, and DIL module on the overall performance of the system. The test items are shown in Table 13.22.

The test results show that the application of DIL module changed the number of transactions per second apparently, down by 21.56%, indicating that the heterogeneous processing and execution of SQL instructions have caused a small portion of performance loss and that the DIL module still needs to be optimized in the design and implementation.

13.2.5.3 System Overall Performance Test

The overall performance of the mimic principle system of web virtual servers was tested on RDB+DRRV+DIL+ 3 web virtual servers as well as on a mount database and a database agent. The test items are shown in Table 13.23.

By comparing the test results, we found that the number of transactions per second and the response time of the mimic defense principle verification system of web servers changed remarkably, down by 20.26% and up by 96.23%, respectively. However, the DIL module performance test results show that the decrease of transactions per second is mainly caused by the application of DIL module; the extension of response time is mainly due to the application of RDB module and DRRV module. When compared to the unprotected benchmark virtual web server, the mimic defense principle verification system of the web server suffers a certain performance loss, but the millisecond response time does not affect the user experience, and it is within the acceptable range.

Table 13.23 System overall performance test content and results

No.	Test item	Server configuration	Test results
1	RDB module + DRRV module + three web virtual servers + database agent + mount database + DIL module access performance test	Virtual machine software: VMware Database software: MySQL Database agent: Amoeba CPU: dual cores and dual threads RAM: 2 GB Pages: 1 KB dynamic pages (the 32 KB dynamic pages are for throughput testing)	TPS: 488 MCTCC: 23139 Throughput (kbps): 109458 RTT (ms): 109.246

13.2.6   Summary of the Web Principle Verification System Test

All the test results show that the mimic defense principle verification system of web servers cannot only satisfy the standards for common web server functionality and performance but also resist the known risks and unknown threats based on vulnerabilities and backdoors in the DHR architecture, change the display form of vulnerabilities or backdoors, and block the response chain of most web server vulnerabilities or backdoor attacks. Its superposition and iteration effect can increase the attack difficulty nonlinearly [22, 23]. In addition, the dynamic and redundant defense makes it almost impossible to achieve reliable and continuous cooperative escape attacks in the DHR architecture. In a word, the system can achieve the mimic defense goal that "vulnerability may not be exploited, or if exploited, it may not be effective."

13.3   Test Conclusions and Prospects

The ministry of science and technology of China entrusted Shanghai Municipal Commission of Science and Technology to organize a joint test, verification, and evaluation of the mimic defense principle verification system. The joint test team consisted of more than ten organizations, such as the Institute of Information Engineering of the Chinese Academy of Sciences, the National Research Center for Information Technology Security, the China Academy of Information and Communications Technology (CAICT), No. 61 Research Institute of the Equipment Development Department of People's Republic of China Central Military Commission, Shanghai Jiao Tong University (SJTU), Zhejiang University (ZJU), Beijing Qihu Technology Co., Ltd., Qiming Xingchen Information Security Technology Co., Ltd., and Antiy Science & Technology Co., Ltd. The team formulated a detailed test and verification plan according to relevant national technical standards and specifications and the principle of mimic defense system. Under the plan, the team carried out the "black box" joint test and "injection" penetration test on the "mimic defense principle verification system of web servers ('mimic-structured web server')" and the "mimic defense principle verification system of routers ('mimic-structured router')," respectively. The test lasted for 6 months, during which 21 academicians and more than 110 peer experts participated in the evaluation work at different stages. The test evaluation expert group concludes that the testing was organized orderly and conducted rigorously under a reasonable plan in a number of methods and that the test results were recorded in detail, and the outcome of test is true, which fully proves that the mimic defense theory is effective, and its engineering implementation is feasible.

The test evaluation conclusions are as follows:

(1) The "mimic-structured web server and mimic-structured router" are successful applications of the theory and method of mimic defense. Equipped with the endogenous security mechanism based on heterogeneous redundancy and multi-dimensional dynamic reconfiguration, they both can provide standard-compliant functionality and performance of web servers and routers. Besides, they can independently and effectively deal with and resist the known risks and uncertain threats arising out of the vulnerabilities, backdoors, viruses, and Trojans in the targets, and their superposition or iteration effect can nonlinearly increase the difficulty of attacking on the targets. The test results completely conform to the expectation of the mimic defense theory.

(2) Mimic defense is a creatively active defense theory and method in cyberspace. It can systematically enforce general and apparent prevention and resistance against various security threats coming from unknown vulnerabilities, backdoors, viruses, and Trojans in the mimic interface without the support of traditional security methods and facilities. The "black box, gray box, and injection" tests plus the comprehensive verification analysis show that the existing scanning detection, vulnerability exploitation, backdoor setting or virus injection, Trojan implantation, as well as Advanced Persistent Threat (APT) do not work as expectedly in the mimic defense system, or if they occasionally succeed, their effect is not reliable. This theory and method are likely to reverse the current strategic decline of "attack and defense asymmetry" in cyberspace and subvert the defense concept of "the era of attacks from software and hardware code vulnerabilities."

(3) The mimic defense mechanism allows the parts, modules, or subsystems and other hardware and software components in the mimic interface to carry unknown vulnerabilities and backdoors and tolerates the existence of unknown viruses and Trojans in the operating environment within the interface. The "injection" penetration test results show that, under the action of heterogeneous redundancy and multi-dimensional dynamic reconfiguration mechanism, the attacks based on the vulnerabilities and backdoors of the software and hardware components in the mimic interface have difficulty realizing precise, reliable, and persistent cooperative escapes in the mimic ruling aspect, resulting in lower real-time requirements for system security protection and reduced frequency and cost of version upgrading in the full life cycle. In the global ecology, when equipped with the mimic defense technology, the "virus-bearing" commercial or open-source software and hardware will become manageable and controllable

information systems. Moreover, mimic defense can effectively curb the advantage of the "seller's market," which arises from its "backdoor engineering and hidden vulnerabilities," so it is a revolutionary approach to "change the rules of game in cyberspace."

(4) The test results also show that mimic defense is not just a special technology for security protection but a universal robust control architecture for information systems, providing a trinity functionality of "security protection, reliability assurance, and service provision." Mimic defense cannot only provide the desired service functions for highly reliable and highly credible application scenarios but also naturally accept existing or new technologies in the information field and integrate existing or future security technologies in the cyberspace. The technology is a new way to overcome the bottleneck of system architecture in response to the national "autonomy and control" strategy and is realistically indispensable to and strategically influencing the general objectives of cyber security and informatization in the global environment, which are like "two wings of a bird and two wheels of an engine."

(5) The theoretical expectation and test verification show that mimic defense also has security gains over non-mimic security issues, such as original design flaws with network protocols, service function implementation algorithms, statistical reuse mechanisms, or attacks by using social engineering methods and means, but its effect is uncertain.

Through the joint test and evaluation, we have verified the effectiveness of mimic defense theory and provided a test specification reference for the mimic defense network equipment and information systems and pushed forward the research and application of the mimic defense theory. Interested readers can pay attention to the reprint of this book, in which Chap. 10 has an added Sect. 10.4.6. The author believes this book will help product developers and manufacturers better understand the technical standards of mimic defense and enable independent quality inspection organizations to effectively examine the authenticity of mimic products to maintain the market order and protect the rights and interests of end users.

Although the principle of mimic defense can be applied universally, it shall be tailored for different sectors and requires continued technological and methodological innovations in defense and testing.

A newly innovated concept needs to be tested in the market before attaining market success. Mimic defense is no exception.

Finally, what should be specially pointed out is that the principles and technologies of mimic defense originating from reliability theory and methodology can be quantifiably tested and evaluated through reliability test and verification for they are calibratable, designable, measurable, and verifiable. In other words, just as stability triangle in Euclidean geometry can be proved by the fact that the sum of the three interior angles in a triangle is equal to 180°, the mimic defense system certainly can be measured by the "white-box" test since its functionality and performance depend on the dynamic heterogeneous redundancy effect. Compared with most other security defense technologies, the mimic defense technology has reached quite a new level.

References

1. Hailong, M., Yiming, J., Bing, B., et al.: Test and analysis of router mimic defense capability. J. Cyber Secur. **2**(1), 29–42 (2017)
2. Ning, L., Zhanhuai, L.: Research and practice of software testing strategy based on black box test. Appl. Res. Comput. **26**(3), 33–37 (2009)
3. Zhiyi, Z., Zhenyu, C., Baowen, X.: Research progress of test case evolution. J. Softw. **24**(4), 663–674 (2013)
4. Wireshark. https://www.wireshark.org (15 Oct 2016)
5. Lyon, G.: Nmap security scanner. http://namp.org (15 Oct 2016)
6. Nessus. Tenable Network Security. http://www.tenable.com/products/nessus-vulnerability-scanner (15 Oct 2016)
7. Metasploit. https://www.metasploit.com (15 Oct 2016)
8. Moy, J.: OSPF Version 2. STD 54, RFC 2328. https://doi.org/10.17487/RFC2328; http://www.rfc-editor.org/info/rfc2328 (15 Oct 2016)
9. Andrew, A.V., Konstantin, V.G., Janis, N.V.: Hacking Cisco Networks Exposed. Translated by Xu Hongfei, Sun Xuetao, Deng Qihao. Tsinghua University Press, Beijing (2008)
10. Weitao, L.: Testing and Analysis of Web Application System. Beijing JiaoTong University, Beijing (2011)
11. Zheng, Z., Bolin, M., Jiangxing, W.: Test and analysis of web server principle verification system based on mimic construction. J. Cyber Secur. **2**(1), 13–28 (2017)
12. Berners-Lee, T., Masinter, L., Mccahill, M.: Uniform Resource Locators (URL). RFC Editor (1994) RFC1738: http://www.rfcreader.com/#rfc1738
13. Wu, L.: Research on Transformation of WSP Protocol and HTTP Protocol Based on Message Type. National University of Defense Technology, Changsha (2004)
14. Fielding, R.: RFC 2616: hypertext transfer protocol-HTTP/1.1. http://www.w3.org/Protocols (15 Oct 2016)
15. Aroms, E.: NIST Special Publication 800–115. Technical guide to information security testing and assessment. CreateSpace (2012)
16. Okhravi, H., Hobson, T., Bigelow, D., et al.: Finding focus in the blur of moving-target techniques. IEEE Secur. Priv. **12**(2), 16–26 (2014)
17. Ron, D., Shamir, A.: Quantitative analysis of the full bitcoin transaction graph. In: Financial Cryptography and Data Security, pp. 6–24. Springer, Berlin (2013)
18. Zhuo, Z.: Research on SQL Injection Attack Technology and Preventive Measures. Shanghai Jiao Tong University, Shanghai (2007)
19. Avritzer, A., Weyuker, E.J.: The role of modeling in the performance testing of E-commerce applications. IEEE Trans. Softw. Eng. **30**(12), 1072–1083 (2005)
20. Newman, D.: Benchmarking Terminology for Firewall Performance. RFC Editor (1999) RFC2647: http://www.rfcreader.com/#rfc2647
21. Hongbing, S., Mo, C., Yibing, C., et al.: Research on IPv4/IPv6 translation gateway performance test. Comput. Eng. **32**(24), 93–95 (2006)
22. Qing, T., Zheng, Z., Weihua, Z., et al.: Design and implementation of mimic defense web server. J. Softw. **28**(4), 883–897 (2017)
23. Zheng, Z., Bolin, M., Jiangxing, W.: Testing and analysis of the mimic defense principle verification system of web serves. J. Cyber Secur. **2**(1), 13–28 (2017)

Chapter 14
Application Demonstration and Current Network Testing of Mimic Defense

14.1 Overview

In order to promote the development of the mimic defense technology, the Cybersecurity Bureau of the MIIT (Ministry of Industry and Information Technology) issued a *Notice on Launching a Pilot Program on Mimic Defense Technology* (hereinafter referred to as the *Notice*) on October 13, 2017. Under the *Notice*, the Henan Communications Administration (HCA) was required to (1) organize and guide China Unicom Henan Branch (hereinafter referred to as CUHB) and Giant Technology Co., Ltd. (hereinafter referred to as Gianet) to deploy mimic-structured domain name servers, mimic-structured web servers, and mimic-structured routers in their current networks; (2) assist CUHB and Gianet in their cooperation with the cyber mimic defense (CMD) research team to study and develop a plan for the pilot work; and (3) prepare a summary report of the pilot program based on the evaluation of the effectiveness of CMD in enhancing the security of existing networks and defending against unknown cyber attacks. From January 2018 onwards, the abovementioned devices have been put into application successively across some SOEs, such as the ICBC and the SGCC, for demonstration purposes, and selected as the serial target facilities in the First "Qiangwang" International Elite Challenge on Cyber Mimic Defense held in Nanjing, China.

In order to implement the requirements of the *Notice* and successfully complete the pilot program of mimic network device application, the research team invited three professional domestic security testing teams to conduct an expert test on a variety of mimic structure network products. Among them, the China Academy of Information and Communications Technology (hereinafter referred to as the CAICT) founded in 1957 is a scientific research institute directly under the MIIT. Since its inception, CAICT has been providing strong support for the industry's major strategies, plans, policies, standards, testing, and certification, acting as a key facilitator in the leapfrog development of the communications industry and the innovation of the IT sector in China. The team members engaged in this test are

© Springer Nature Switzerland AG 2020
J. Wu, *Cyberspace Mimic Defense*, Wireless Networks,
https://doi.org/10.1007/978-3-030-29844-9_14

veterans in the Ministry of Industry and Information Technology of the People's Republic of China (MIIT) inspections and evaluation of security of all kinds of communications network units and are leading members in drafting and compiling of the serial state standards for cyber security defense of the communications industry, as well as in formulating the MIIT security defense evaluation forms over the years, and can accurately and authoritatively interpret the assessment requirements and standards and the evaluation forms.

Nanjing Cyber Peace Technology Co., Ltd. (hereinafter referred to as NJCP) was established in October 2013. It set up Beijing Cyber Peace Network Security Technology Co., Ltd., a wholly owned subsidiary, in 2015, and received tens of millions of yuan in its Pre-A round financing in 2016. The company's core team members came from Tsinghua University (TSU) and top 500 world-class IT players, with 85% of its employees holding a bachelor degree or above and 30% a master's degree or above. NJCP has rich experience in cyber security Attack/Defense combat. Its founding team set up Blue-Lotus, the most prestigious cyber security Attack/Defense combat team in China, which entered into the finals of DEFCON CTF (Network Security World Cup) for five consecutive years. In 2016, Blue-Lotus became the second in the world and second in Asia in DEFCON CTF. The members engaged in the test are NJCP's Attack/Defense Lab team members who are experienced in penetration testing technology. They got remote guidance from the leading players of the XCTF-TSU joint lab team and Blue-Lotus team during implementation of the program.

As a leading professional information security provider in China, Beijing Topsec Network Security Technology Co., Ltd. (hereinafter referred to as Topsec) has been committed to providing specialized information security products and services for corporate users. As early as 2004, the company established the "Topsec Security Operation and Maintenance Center" to provide security O&M outsourcing services for corporate users, being the first commercialized security O&M service organization in China. Topsec established two security O&M centers, respectively, in 2007 and 2009 in conjunction with China Unicom Beijing and China Telecom Group (CTG), making full use of their superior resources to provide secure O&M services to corporate customers. Currently, Topsec has provided secure O&M services for more than 5800 companies. In 2012, the "Topsec Cloud Security Center" was set up to offer convenient and efficient cloud security services for corporate users based on a cloud security service platform.

14.2 Application Demonstration of the Mimic-Structured Router

According to the unified deployment of HCA, mimic-structured routers (referred to as mimic routers) were officially put online at Gianet on April 8, 2018 for demonstration application. The mimic routers were deployed incrementally at Gianet to extend the routing domain and share the load of the existing service support network of Gianet to carry the highly credible, reliable, and secure service delivery.

14.2.1   Status Quo of the Pilot Network

14.2.1.1   Threat Analysis

The position of a router in the network and its routing forwarding function make the router an excellent entry point for attackers to act. Once controlled by an attacker, it will cause incalculable hazards to cyber security. Anyone who controls the router can easily get access to user privacy data, monitor user surfing behaviors, obtain user account and password, tamper with key user data, push and spread false information, disrupt network data flow, paralyze network information interactions, even directly launch network attacks, etc.

The internal network of Gianet's service platform is built upon conventional routers, where all service data are routed and forwarded on the data paths defined by the network nodes. The security of network devices will determine the correctness of the service data paths. Gianet has introduced security measures such as IDS, firewalls, and WAF to improve the security of the cloud service host but is in short of security defense means to protect the routing switching devices, although the firewalls can filter out to a certain extent malicious access to and attacks on these devices. However, in dealing with unknown threats, especially attacks exploiting 0day vulnerabilities or backdoors, the security defense ability is by no means guaranteed. Once a vulnerability or backdoor of a type of routing switching device is successfully exploited, it will cause fatal impacts on user-entrusted data, business and services, etc.

Therefore, blocking or dissolving the attack chains based on the inherent security flaws of the routing switching nodes has become one of the challenges that Gianet must solve to provide trusted services. The mimic router pilot program will be mainly focused on the mimic defense of the routing switching nodes in Gianet's underlying network.

14.2.1.2   Application Scenario

As one of the domestic providers of IDC and cloud services, Gianet offers professional one-stop products and services to small and medium-sized enterprises (SMEs). A computer room at Gianet is a high-standard multi-line BGP computer room. To provide highly efficient and reliable cloud service capabilities, the BGP line is interconnected with mainstream network operators to avoid the latency caused by identification programs of different lines in a single-line computer room and secure high-speed user access from different lines. The computer room is connected directly to the Zhengzhou backbone network through an export bandwidth of

200G and then accesses the multi-line backbone networks such as China Telecom, China Unicom, China Mobile, and CERNET. Gianet's AS number is 37943, with a stock of 260,000 IP addresses, including the segments of 116.255.0.0/16, 122.114.0.0/16, 203.171.0.0/16, and so on. To ensure the stability and high speed of its network, Gianet adopts the dual-point backup mode in all its devices. Its internal network consists of three layers, namely, the core layer, the convergence layer, and the access layer:

(1) The core layer is virtualized as a backbone router via the stacking technology, and BGP sessions are established with multiple operators through the BGP for efficient service provision. It enhances the reliability of uplink bandwidth and connection for every operator through multi-link convergence.
(2) The convergence layer is interconnected with the core layer egress router through OSPF. It is used to aggregate the cloud hosts and IDC addresses and traffic, advertise routes to the core router, and introduce traffic to the subordinate hosts and servers from the core layer.
(3) The access layer applies VXLAN technology to connect cloud hosts and servers.

The mimic router put online this time is essentially a routing/switching platform with the ES functionality, supporting mainstream routing protocols and various network services, and defending against known or unknown vulnerabilities through the uncertainty effect generated by its own mimic structure. Given Gianet's architecture and deployment features, if the mimic router is deployed at the exit of the entire network, it will be difficult to complete the deployment in a short time since it requires the cooperation of multiple external network operators. Due to the tight schedule, we plan to deploy the mimic routers at the convergence layer of Gianet's intranet, extend the routing domain over its existing service support network and carry its service data in a load-sharing mode. In this way, we can deploy the mimic routers incrementally.

14.2.1.3   Product Plan

A router is a highly complex information system. A commercial router requires routing protocols, configuration management protocols, and dozens of value-added services of various types. At present, there are hardly small enterprises among the providers of commercial routers, reflecting to some extent the complexity of the router technology.

The R&D of mimic router products actually revolves around two key issues. The first is how to build a commercial router that meets the industry's network access requirements and supports a large number of different types of routing protocols, management protocols, and value-added services. The second is how to introduce the mimic structure into the router design so as to obtain GRC capabilities, including the mimic interface (MI) setting of the special closed system, the plural heterogeneity design of an executor, and the like.

Comprehensive analysis suggests that the SDN/NFV-based technology is a good choice. The philosophy of SDN architecture that seperates control from forwarding is employed to strip off the data plane, and the NFV technology is used to support various business and service demands of the commercial routers, so that generalized routing, management, and service functions can be provided by way of cheap and cheerful COTS virtual routing function software. The COTS switch boards are also used to support the high-speed data forwarding requirements. The product architecture is shown in Fig. 14.1. In fact, the security threats to a router mainly come from the routing control plane. Therefore, the R&D of mimic router products follows the overall technical mindset of incremental evolution. Above all, at the routing control plane, the mainstream routing is mimicked through protocols backed by the existing research foundation. Other protocols as well as services are supported by mature virtualized commercial routers, and then gradually added to the control and management plane. The data forwarding plane relies on the COTS switch boards to support the service and statistics functions as well as the carrier-grade hardware reliability.

The overall system plan is shown in Fig. 14.2, where the portion circled in the red box indicates a typical DHR architecture. The routing protocol adopts a mimic structure to ensure correct routing calculation. The service traffic taking a small proportion is undertaken through the virtualized routing software stacking load, while the forwarding traffic taking a big proportion is subject to line-speed processing through hardware table look-up and forwarding (TLU&F). The isolation of mimic and non-mimic processing scenarios ensures the security of configuration management.

In respect of the heterogeneous router executors, the source code cross-compilation heterogeneous redundancy (HR) is applied, as shown in Fig. 14.3. With cross-compilation, the source code is compiled into routing software running on different architectures, and the heterogeneity of the instruction set is employed to

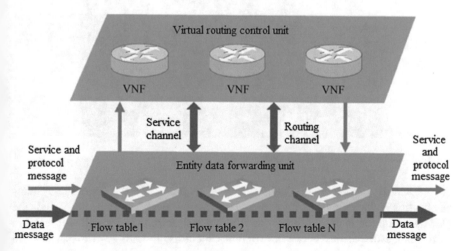

Fig. 14.1 Product architecture of mimic router products

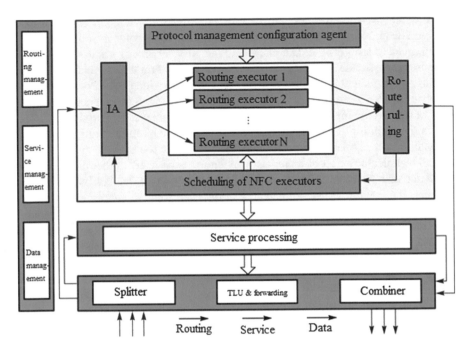

Fig. 14.2 Overall plan for the mimic router

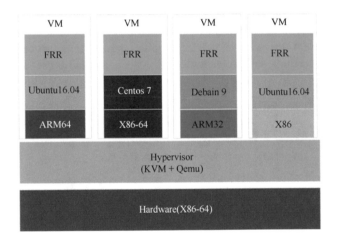

Fig. 14.3 Source code cross-compilation HR plan

defend against the backdoors and vulnerabilities present in the operating systems and routing software of the virtual machine, where the ARM architecture needs the *QEMU* simulator as the translator for different instruction sets, which will result in performance disparities between different executors, though such disparities are acceptable for the mimic routers regarding their routing calculations.

Fig. 14.4 Appearance of
MR1810E-N4

To ensure that each executor can start a session with the neighboring router and perform routing calculation simultaneously, it is necessary to select an executor to establish a real connection with the neighboring router for route advertisement, as well as to establish a false connection with the rest executors through the agent. In the ruling mechanism, the isogenous HR mechanism ensures that the output of the routing table can be normalized. When all the executors initiate a route-update or when the timer times out, the weighted ruling mechanism will select the compliant routing table entries and issue them. When the ruling mechanism finds any inconsistent entry output in the executors, it will clear and correct the erroneous routing information sent out by the executor if the executor establishing the real connection has an erroneous output.

At the negative feedback scheduling (NFS) level, in the initial state, the system selects five executors from the head of the executor queue and runs them online to participate in the ruling. Once the mimic ruling (MR) detects an executor that responds abnormally, it will put the executor offline, clean it, reload it online, and resynchronize it through the routing cache mechanism. In addition, due to the random operation command, the system will implement forcible switchover within the mimic interface to ensure the uncertainty of the CMD environment.

The R&D of the MR1810E-N4 mimic router is based on the COTS mega-components. The overall unit specification is 4U 19″ rack-mounted, with a redundant power backup and carrier-class reliability. The product appearance is shown in Fig. 14.4.

14.2.1.4 Application Deployment

As a router at the convergence layer, the mimic router is connected upstream with the core router of Gianet to exchange routing information through the OSPF protocol and is connected downstream to the cloud service platform to migrate at least one service of the network segment C out from the cloud environment for mimic router testing. The deployment diagram is shown in Fig. 14.5.

In the deployment of a mimic router, only minor adjustments to the network configuration are required at the pilot site to achieve the incremental deployment,

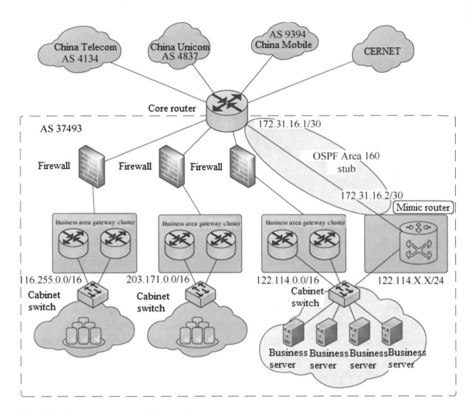

Fig. 14.5 Access scheme of the mimic router

while the cost of the current network transformation remains low. At the upstream of the mimic router, the interoperability with its routing protocols can be realized by modifying the configuration of one router; at the downstream of the mimic router, the existing server hosting service can be migrated to the mimic router by modifying the gateway of the service host, so that the existing business will be carried by the mimic router. Therefore, in the deployment process, two network engineers can complete the packaged work in 3 days, including the device racking, networking, power-on, configuration, protocol interworking of the mimic router, and the smooth migration of the business.

In summary, the mimic router introduces a DHR architecture to generate Endogenous Security (ES) performance, while ensuring the overall functionality and performance of the system. The external interfaces, interaction protocols, and interoperation interfaces of the overall unit are consistent with those of the conventional router, and so is its current network deployment approach: It can be completed by simply modifying the configuration of the neighbor routers and host instead of the transformation at the network physical level and the software code level. Therefore, a mimic router can be deployed at almost zero cost in any applicable scenarios of a conventional router, without any special or adaptive transformation of the application scenarios.

14.2.1.5   Cost Analysis

The MR1810E-N4 mimic router is designed according to the network access indicators of high-end routers, it defends the core site against key security threats to the routers, that is, the CMD mechanism protects the router's configuration information from being maliciously modified while the routing table from being deliberately falsified, thus ensuring the correctness of the router's routing function. Other security threats are beyond the CMD scope.

According to the product plan displayed in Fig. 14.2, the data TLU&F function is performed by a COTS switch board, the business service is borne by the COTS virtual routing software, and multiple HR routing software carried by virtual machines are introduced only at the routing control level. Then, mimic components are introduced to construct a mimic router. A cost analysis is made as follows from two aspects: engineering and O&M.

1. Engineering cost analysis

In respect of engineering cost, the additional cost upon the introduction of the mimic mechanism consists of two parts: the R&D cost of the HR routing software and mimic components and the cost of the COTS system carrying this software.

(1) Manpower input costs

The R&D of the heterogeneous routing software is based on open source routing and virtualized OS software, which requires one engineer, with an R&D cycle of 1 month.

The software R&D of mimic components, including input agent distribution, output mimic ruling and dynamic scheduling, etc., requires four engineers, with an R&D cycle of 3 months.

The COTS board, COTS software, heterogeneous executors (HEs), and mimic components are integrated together, coupled with the development of software such as business adaptation, service migration, configuration management, log maintenance, and system O&M interface, which requires four engineers, with an R&D cycle of 3 months.

Therefore, the manpower investment in the entire R&D process is 25 person months.

(2) Increased cost of hardware

The increased hardware cost is relevant to the way the mimic components and the heterogeneous routing virtual machine software are deployed.

Resource sharing mode: The above software is deployed on the COTS computing board running the business module (CPU, 2×Intel Xeon E5-2680v3; RAM, 32GB; Disk, 1 TB). This deployment allows the heterogeneous executors to share the same resources with the business module, while the mimic components share the same resources with the master control module. There are two flaws in this deployment approach: First, the heterogeneous executors compete with the business module for resources. When the traffic processing load is heavy, it will hinder the route computing speed of the heterogeneous executors. Second, the reliability of all the heterogeneous executors depends on the relevant performance of the current COTS computing board, reducing the reliability and maintainability of the entire system. Although this deployment method does not increase the hardware cost, it will increase the overhead of the COTS computing board running the business module (with the CPU up by 5.6%*4, memory up by 5.4%*4, and hard disk up by 3%*4), as well as the processing overhead of the COTS computing board running the master control module (CPU up by 6.3%, memory up by 10%, and hard disk up by 1%). In the strict sense, the power loss is also increased.

Resource monopoly mode: The above software is deployed on an independent COTS computing board. This deployment allows the heterogeneous executors to be distributed across different computing boards, bypassing resource competition, and ensuring robust gains of the router mimic structure. However, the overall unit cost increases with the need to add a COTS computing board. Since the computational complexity of the heterogeneous executors and mimic components stems mainly from route computing and distribution scheduling ruling, the CPU occupation and memory usage are much lower than that of the service-bearing COTS computing board. Therefore, the performance indicators of the COTS computing board carrying the heterogeneous executors and mimic components are far outraced by the service-bearing computing board. In this case, the cost of the new COTS computing board (CPU, Intel Xeon E5-2620v3; RAM, 32GB; Disk, 1 TB) is less than 10% of the overall unit.

2. O&M increment upgrading

The O&M model of the mimic routers will change after the introduction of the mimic mechanism.

Log information processing: As the CMD mechanism can timely detect, identify, and block through mimic ruling the security threats introduced by known/unknown vulnerabilities, which are not available for a conventional router, it is necessary to develop certain interfaces through which you can incorporate such threat information into the existing risk management platform and train the O&M personnel so that they are able to identify and analyze the mimic threat information.

Patching or system upgrading: The O&M personnel of conventional routers must closely track the latest vulnerability patches and versions released by the vendor. They need to have professional knowledge to patch or upgrade the system in time. Failure to get the timely announcement from the vendor will put the routers in danger and expose the network to serious risks. Lack of professional competence may lead to a downtime or outage due to the incompetent patching or upgrading operation. What's worse, a malicious attacker may issue false information to the O&M personnel and make them download, update, and install the backdoor-implanted patch/upgrade package of the system, while a conventional router has no means to identify the existence of such preset backdoors, thus leaving the routers in a completely uncontrollable state. In contrast, a mimic router with endogenous security features can detect and prevent vulnerability/backdoor-based attacks without relying on the threat signatures, avoid the threats arising from failures to timely patch the system or upgrade its version, and save the effort of ceaseless patching and blocking vulnerabilities, so it greatly reduces the O&M workload resulting from system vulnerability patch upgrading, breakdown maintenance, and security policy maintenance. Also, since the endogenous security mechanism (ESM) does not rely on the operation skills of professionals, it helps lower the professional and quality requirements for the O&M personnel and drastically reduces the O&M costs.

Cost analysis in the three aspects (device cost increment, upgrading O&M amount, and safety defense increment) shows that the mimic router is an integrated product with lower costs but higher security gains.

14.2.1.6   Application Outcome

Since its entry into practical application on April 2, 2018, the mimic router has been running reliably and stably. The overall traffic is maintained at around 60 Mbps, with the average daily forwarding data of about 3.7 Tb, providing ongoing data routing and forwarding services for part of Gianet's cloud services.

14.2.2   Current Network Testing

14.2.2.1   Testing Purpose

To verify the effectiveness of CMD technology in improving router security defense capabilities, and the impact of the application deployment of the mimic router on the existing network services, as well as to evaluate the impact of the mimic router on the functionality and performance of the existing network services, we sequentially conducted the function testing, performance testing, and compatibility testing. To evaluate the security defense capabilities of the mimic router, we carried out the comparative testing, injected testing, and online testing.

14.2.2.2   Testing Plan

The testing of the mimic router at Gianet is divided into two phases: production function testing and professional safety testing. The former includes function testing, performance testing, and compatibility testing, while the latter includes security comparative testing, security injected testing, and analysis and evaluation.

Production function testing mainly tests whether the mimic router can perform configuration management and maintenance normally, dock with the protocols of other routers on the network, advertise correct/learned routing information, and forward Gianet's service data via normal routing, etc. This process is mainly used to test whether a mimic router has the same functionality and performance as a conventional router does. Comparison and injection are adopted in the professional security testing to test the security defense capabilities of the mimic router to see if the router can effectively address vulnerability/backdoor-based attacks in the mimic interface and the related security issues.

14.2.2.3 Testing and Evaluation Items

According to the items of the testing plan, coupled with the testing principles and benchmark function testing methods given in Chap. 10, we refined the items of testing and evaluation of the MR1810E-N4 mimic router, as shown in Table 14.1.

Function testing as well as compliance testing has been done on the mimic-structured router as described in Chap. 10. The testing results show that the introduction of a mimic mechanism has not changed the functional features of the router. This meets the testing specifications such as the mainstream standard routing protocols and configuration management protocols. A mimic router has the same functions as a conventional router does.

Comparative testing has been done on the mimic-structured router as described in Chap. 10 to compare the forwarding performance of the routers by putting on/off the mimic mechanism. The testing results show that the introduction of a mimic mechanism does not affect the forwarding performance of the router, while the throughput and latency of the mimic router meet the industry standards.

Security comparison testing has been done on the mimic-structured router as described in Chap. 10. The testing results show that, through dynamic scheduling and mimic disguise, the mimic router effectively reduces the accuracy of target information acquisition via scanning detection, thus changing the display properties of the system vulnerabilities.

Security injection testing has been done on the mimic-structured router as described in Chap. 10. The testing results show that a mimic router has the ability to resist unknown threats in the mimic interface and defend against unknown attacks based on backdoors/vulnerabilities.

14.2.2.4 Current Network Testing

On April 2, 2018, the mimic-structured router was officially launched online after passing the testing of consistency and interoperability with Gianet's deployment devices, carrying part of Gianet's cloud services. Gianet did not deploy any additional security measures for the mimic router, which opened a normal administrator account with a weak password to the Internet and received real attacks from the Internet through an open interface.

In its operation, the mimic router monitored a large number of targeted attacks, including yet not limited to scanning detection, password blasting, privilege escalation, routing tampering, and malicious resources loading. As of June 25, 2018, the mimic router had detected and blocked a total of 108,167 attacks, covering all the

Table 14.1 Testing and evaluation items

Evaluation target	Testing phase	Main item	Evaluation item
Compliance testing of the mimic router	Function testing	(1) Test whether the mimic router's network access function, the interface signal, and the indicator light are normal (2) Test whether the mimic router can be remotely configured, managed, and maintained (3) Test whether the mimic router can forward data via static routing	Evaluate whether the mimic router can work normally
	Performance testing	1) Test the performance of the mimic router	Evaluate whether the mimic router can forward data according to a given routing switching capability, and whether the throughput and latency meet the industry standards
	Compatibility testing	(1) Test whether the mimic router can establish a corresponding routing protocol session with Gianet's conventional router(s) (2) Test whether the mimic router can discover the local routing and correctly advertise it to a neighbor router (3) Test whether the mimic router can correctly learn the routing from a neighbor router (4) Test whether the mimic router can forward data according to the normal routing	Evaluate whether a mimic router can communicate with conventional routers and forward data according to dynamic protocol specifications
Security defense capability testing of the mimic router	Security comparative testing	(1) Test the security gain of the mimic router based on the existing defense means (2) Test the security defense capabilities of the mimic router without applying existing defense means such as the IDS, firewalls, and the systems of black/white list filtering and illegal information checking	Compare the defense capabilities of existing defense means and of mimic defense techniques through the security threats spotted by the mimic router
	Security injection testing	(1) Inject a backdoor to the router (2) Test the ability of the router to resist an attack exploiting the injected backdoor (3) Inject a vulnerability into the mimic router (4) Test the ability of the mimic router to resist attacks exploiting the injected vulnerability	Evaluate the defense capabilities of the mimic router in resisting unknown threats by testing the effect of its defense against injected backdoors and vulnerabilities without patching

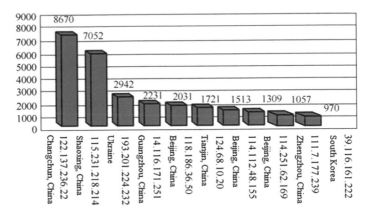

Fig. 14.6 Top 10 attackers (by threat number) and their locations

abovementioned categories. These threats came from 3750 IPs in different jurisdictions, and the IP addresses of the top 10 attackers (by threat number) and their locations are as shown in Fig. 14.6. All attack threats were spotted, recorded, and blocked by the mimic router.

14.2.2.5   Testing and Evaluation

Up to now, the mimic router has been detecting and blocking the ongoing threats from the Internet while ensuring normal data forwarding, safeguarding the reliable, stable, and secure operation of the devices. The testing and analysis indicate that the mimic router has high cost performance ratio in terms of credibility, availability, and reliability with a relatively small investment.

14.3   Mimic-Structured Web Server

14.3.1   Application Demonstration

14.3.1.1   Application of the Mimic-Structured Web Server (MSWS) in a Financial Enterprise

1. Threat analysis

A certain financial enterprise boasts strong comprehensive strength, stable operation and management, a sound service network, and stable outbound services. In addition, the enterprise has strong scientific and technological strength and well-established security measures. However, in the current cyberspace environment, various network devices, operating systems, and applications have been confirmed to have massive vulnerabilities and backdoors that can be exploited by attackers. The resulting uncontrollable security issues are likely to occur within the enterprise. The company's e-commerce website has the following security threats:

(1) Risks for the website to be tampered with.
(2) The website's mono-defense method cannot cope with increasingly sophisticated means of attack.

2. Application scenario.

The application target is the static resources of the financial enterprise's e-commerce website. The goal is to build a tamper-resistant mimic structure Web system to prevent static Web resources from being maliciously falsified, ensure the stable operation of the website, and promptly push the warning information to the company's early warning linkage platform.

3. Product plan

Modular access: It is not feasible for the mimicry of complex business logic to be compatible with all the original functions and methods in a short period of time. Therefore, for large-scale enterprise-class applications to serve such a financial player, it is necessary to divide the modules by functions to suit its in-house business logic and incorporate the mimic defense structure approach by way of the defense at the gateway and the core site. In the enterprise-wide application, we consider separating the dynamic data stream from the static data stream to first mimicry deploy the static resources module and then perform fine-grained business logic partitioning on the dynamic resources, thereby realizing the modular access of mimic defense.

Stable service: The network applications after the mimic defense transformation should also have the disaster tolerance capabilities of the enterprise-class applications. Therefore, the accessed mimic defense modules should also be subject to redundant backup. In application, this enterprise adopts a deployment method of dual-node file-level hot standby for the redundant backup of the mimic voter.

Compatibility with current security measures: Large-scale enterprises will inevitably add corresponding security protection measures to their service applications in the long run. Owing to its endogenous nonlinear effect, the mimic defense framework has unique advantages in compatibility with conventional security measures. Any enhanced security means increase the dissimilarity between executors in functional equivalence, enabling the mimic interface to get exponential defense gains. In addition, the marriage between the warning information generated by the mimic defense voter and the early warning linkage platform also enhances the overall security protection capability of the enterprise. Figure 14.7 illustrates the device appearance of the application deployment.

4. Application deployment

To introduce mimic defense into a financial enterprise, we have to solve the following problems one by one: construction of heterogeneous executors, access to the

Fig. 14.7 Mimic-
structured Web server

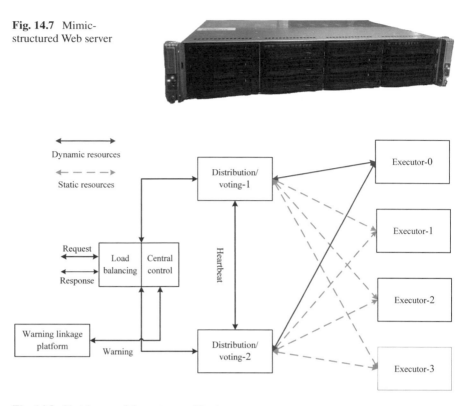

Fig. 14.8 Sketch map of the system architecture

mimic defense voter, compatibility with existing defense measures, and guarantee of the service stability. This application is mainly focused on protecting the static resources of the website to ensure the system of high availability and high security. To meet the security and compatibility requirements, the architecture of the MSWS is designed, as shown in Fig. 14.8. The website's dynamic and static resources are separated, with the latter subject to mimic transformation. Static resources are deployed on Executors 1, 2, and 3, while dynamic resources on Executor 0. The distribution and voting module distributes the requests for static and dynamic resources to the Executors 1, 2, and 3 as well as Executor 0, and then votes on the static resources. In order to ensure high availability and high reliability of the service, the distribution and voting module adopts the dual-node hot standby. If the master distribution and voting nodes turn abnormal and fail to serve normally, the backup distribution and voting nodes will immediately go online to provide services. To integrate the existing defense means of the financial enterprise, we transmit the threat log generated by the mimic defense voter to the enterprise's warning platform via its internal network to realize the warning linkage mechanism.

This application mainly aims to protect the static resources of the website to ensure normal access to its pages. Upon the mimic deployment, the website gets the following characteristics:

(1) Easy access: It can easily access the application environment.
(2) High security: The heterogeneous executors and redundant execution and ruling mechanisms improve system security.
(3) High availability and high reliability: The double-layer defense design of the heterogeneous executor's multi-machine hot standby and security devices' dual-machine hot standby ensures the stable operation of the system.
(4) Timely warning: Warning messages can be pushed to the financial enterprise's linkage warning platform.
(5) Easy management: It provides users with a convenient and efficient interface for visualized management.

5. Cost analysis

In view of the high reliability and availability requirements of the financial enterprise, this application deployment adds a dual-node distribution and voting server, where the two nodes act as the redundant hot standby; the dynamic and static resources are separated via virtualization technology, where no hardware device has been added. As a result, this application only involves one dual-node distribution and voting server, and the cost increase is not very high compared with the cost of the original website.

6. Application effect

After being applied to the network environment of the financial enterprise, the MSWS has been running stably and can detect faults and attacks in time and ensure the security and reliability of the static resources management on its e-commerce website.

14.3.1.2 Application of the MSWS on a Government Website

1. Threat analysis

The website has the following security threats:

(1) The site takes CentOS as the operating system and Tomcat as the Web service. Both types of software expose a lot of vulnerabilities.
(2) Although the website adopts security measures such as IDS, it fails to effectively address 0day vulnerabilities or backdoor attacks.

2. Application scenario

The majority of resources on the official website are static resources generated by the back-office server, and the primary station is separated from the back office. Besides, a backup server is set up to improve the reliability of the official website.

Fig. 14.9 The network
framework diagram of
target object

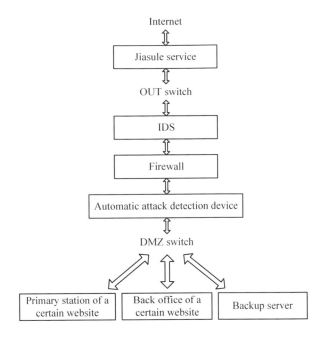

The primary Web server that provides static web pages is the access target of external users. With reference to the mimic defense deployment principle, the primary Web server should be considered as a defense target. To ensure the security of the official website, the accelerator *Jiasule (Joy of acceleration)* has been introduced to provide CDN services, which not only accelerates the content transmission but also avoids redundant traffic from entering the network service environment. After passing the *Jiasule* service area, the traffic first goes through the IDS intrusion detection system that blocks part of the intrusion behaviors, and then enters the DMZ area in accordance with the firewall rules. Before entering the primary station, the traffic must be subject to automatic detection by the automatic attack detection device. The network framework consisting of existing security devices and the Web server is shown in Fig. 14.9.

The official website had a daily access of about 350,000 times and a traffic of 5–10 GB. Under the original security device deployment, the number of daily attacks detected by *Jiasule* reached 10,000 times, mainly including Webshell, file injection, file inclusion, malicious scanning, code execution, malicious collection, SQL injection, etc. The deployed IDS intrusion detection system detected 82,136 malicious attacks in half a year. The specific attack types are shown in Table 14.2. According to the analysis of the testing results, Trojans, backdoors, SQL injection attacks, command injection, and attacks exploiting software vulnerabilities remain the main forms of attack. It can be seen that the official website has been constantly attacked by attackers, while the test results show no interception result of a 0day vulnerability attack.

Table 14.2 Testing results of the IDS intrusion detection system

Type of attack	Number of attacks
HTTP_Trojan backdoor_Webshell_PHP_eval_base64_decode Trojan	31,769
HTTP_SQL injection attack	13,859
HTTP_Linux command injection attack	10,094
HTTP_IIS parsing vulnerability	4448
HTTP_Acunetix_WVS_vulnerability scan	4185
HTTP_GNU_Bash remote arbitrary code execution	3356
HTTP_ Trojan backdoor_Webshell_china_chopper_20160620_asp control command	1732
HTTP_ directory traversal [../]	1700
HTTP_ Trojan backdoor_Webshell_PHP_fatalshell Trojan upload	1619
HTTP_XSS script injection	1566

3. Product plan

Only slight changes can turn the website into highly reliable and highly trusted mimic defense site under the existing conditions. Highly reliable: Use the existing backup server as the redundant executor of the MSWS; Highly trusted: Integrate existing security devices into the mimic architecture to significantly improve the target system's capabilities in sensing and defending against unknown attacks.

Design of the heterogeneous executors: In reference to the software and hardware configuration of the official network system, to maximize the heterogeneity between the executors, the software service layer and the executor operating system layer are isomerized, respectively. In order to eliminate the compatibility gap between different levels of software, a relatively stable combination of heterogeneous executors is determined through the testing of various combinations between different levels.

Design of synchronous data transmission: This function involves whether the file systems of the transmission systems are mutually supportive or compatible, and whether a transmitted file is accepted by the receiving system. At the same time, the stability of data synchronization needs to be considered.

Load balancing access: In order to reduce the service stability of the official website imposed by the demonstration application system, we distribute the traffic to the MSWS and the original official website server by 1:9 in load balancing.

The MSWS is based on the underlying platform of the Web application server, with the CMD technology as the core. The device appearance is shown in Fig. 14.10.

4. Application deployment

The official website has been subject to mimic transformation on the basis of its status quo. The deployment framework of the MSWS at the official website is shown in Fig. 14.11. During the operation, the original requested traffic is directly forwarded to the primary station of the official website by the DMZ switch, where

Fig. 14.10 Mimic-
structured Web server

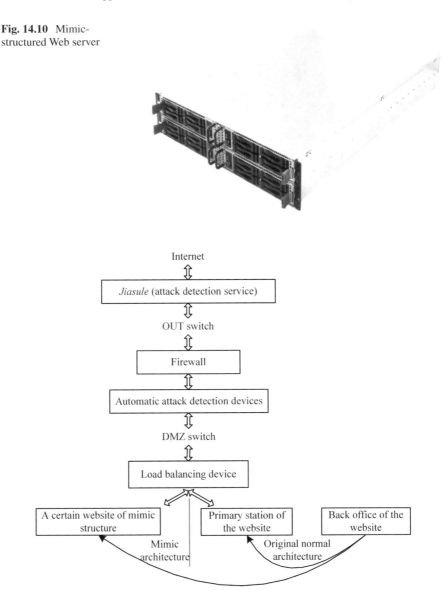

Fig. 14.11 Deployment framework of the MSWS (CMDA Web server)

the load balancing devices are responsible for equalizing the requested traffic respectively to the original normal-structure official website and the mimic-structure official website according to the preset ratio. Among them, the load balancing devices used are all domestic brands; the MSWS is responsible for carrying the upgraded mimic application of the official website, providing Web services and security defense capabilities.

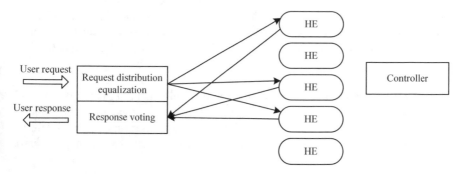

Fig. 14.12 Internal architecture of the MSWS (CMDA Web server)

In order to ensure that the newly introduced MSWS and other attached devices in the pilot program will not affect the normal service of the official website, the redundant features of the MSWS and the dual-machine hot standby configuration at the device level has realized a disaster tolerance to a degree equivalent to that of the original system at the Web application and device level, minimizing service losses in the event of cyber security/device failures at the official website.

The MSWS constructs the HR Web service executors (HRWSEs) based on the "request-response" service model. The internal architecture of the MSWS is shown in Fig. 14.12. When a request arrives, it is distributed to N HEs; the voting module receives and compares the similarities and differences between the N responses, generates compliant response outputs, and identifies abnormal executors based on the voting information. The HRWSEs not only implement the mimic defense function but also guarantee the high availability of Web services through their redundant executor configuration model.

The traffic is distributed to the MSWS and the primary station of the original official website through the load balancing device, which will automatically direct all request traffic to the primary station when the MSWS fails.

5. Cost analysis

A certain website requires a controllable cost of CMD deployment and O&M to provide considerable security gains for the central official website. Therefore, controlling the deployment and O&M cost and improving security become two key adjustment goals in the official website's mimic upgrading and adjustment process.

(1) Hardware purchase

Before the adjustment, the official website used three Sugon servers (Quad-Core AMD Opteron(tm) Processor 2378 2*4, 16G, 2 TB + 254 GB) to carry, operate, and maintain its Web services, but this type of server, purchased more than a dozen years ago and was already out of production, was not suitable to be the target of hardware cost comparison. Therefore, we chose Sugon D620 I620-G20 (Intel Xeon E5-2620 v4, 16GB*2, 300GB*3), the purchase cost of which was about 20,000 yuan per unit. The cost of purchasing an MSWS and deploy it in the official website was 70,560 yuan, a 17.6% increase in the hardware purchase cost compared to an MSWS with three Sugon I620-G20 servers (Xeon E5-2620 v4, 16GB*2, 300GB*3).

(2) Deployment cost

Some tasks should be done before the online deployment of the MSWS, including website mimic adjustment, website smooth migration, functional conformance testing, etc. For this purpose, we have invested 2 person months to complete the related work, a total investment of 2 person months*10,000 yuan per person month = 20,000 yuan.

(3) Operation and Maintenance (O&M)

The O&M management cost the MSWS requires in the pilot program is reflected in two aspects: One is the cost of environmental resources, mainly including electricity charges, rack space charges, etc. Through parameter comparison and calculation, the basic O&M cost of an MSWS is lower than that of three Sugon servers; the other is the cost of human resources or O&M manpower cost. Since a mimic device has ES features, it can avoid the threats arising from a failure of timely patching or upgrading of the system version and save the work of ceaselessly patching to block vulnerabilities, thus reducing the O&M manpower cost.

(4) Security defense

At present, a certain website has deployed firewall devices (32,000 yuan), cloud CDN service (200,000 yuan/year), and IDS devices (374,800 yuan) in its original environment, while the Sugon servers hosting the website are deployed in the DMZ area of the firewall. If the MSWS is successfully connected to the current network environment and has access to the database, it needs to be connected into the DMZ area of the firewall, just like the Sugon servers carrying the website. With its ESM, the MSWS itself has a good security defense function. Therefore, the security defense investment in the MSWS deployment is 0 yuan.

6. Application effect

The MSWS officially began to provide external services in April 2018 after passing the functional conformance and security tests and has been operating in a stable and normal state so far. According to the relevant data, the original defense measures of the website are relatively secure, where security incidents have not occurred, and the MSWS has not only achieved the same security defense effect as that of its original protective measures, but even found and intercepted more abnormal behaviors. Besides, throughout its life cycle, the MSWS is significantly more cost-effective compared to the existing security defense devices, being a CMD product solution featuring high security, high availability, low cost, and low false positives.

14.3.1.3 Application of the Mimic-Structured Web Virtual Host (MSWVH) in Gianet Fast Cloud (GFC)

1. Threat analysis

The Web-based GFC service has the following security threats:

(1) The majority of the websites in the Fast Cloud system are redeveloped based on open source versions, where the source code security audit is insufficient. These websites have a large number of loopholes and backdoors easy to be detected and exploited, thus vulnerable to attacks.
(2) The mono-defense WAF method deployed in the Fast Cloud system cannot address the increasingly dazzling attack means.
(3) The application layer has an extremely large attack surface, which is mainly threatened by PHP vulnerabilities and SQL injections.
(4) The websites in the Fast Cloud system are exposed to webpage tampering for driving traffic to hack sites, which in worse cases results in data loss and malicious content alteration, harming the interests and public image of customers.
(5) There is a lack of means to effectively deal with unknown threats such as 0day vulnerabilities.

2. Application scenarios

With full-featured business processing capabilities, GFC provides professional Web cloud service support for SMEs. The GFC system is a cloud service cluster built with the underlying physical devices, where virtual machines are created, and the VXLAN technology is applied to ensure that the networks of virtual machines are isolated from each other to prevent against cross-network attacks such as ARP. Then multiple virtual sites are created in each virtual machine, and each virtual site can provide Web access service for one of the users' websites and create independent system users as well as system permissions for secure isolation.

The virtual sites running Web services of the website are the targets for external users to access, so the virtual sites in the cloud environment constitute the attack surface to be protected. In order to ensure the security of the Fast Cloud service host, Gianet has introduced security defense measures such as IDS, firewalls, and WAF. The user access traffic first passes through the IDS intrusion detection system that blocks part of the intrusions, and then enters the cloud environment in accordance with the firewall rules. Before entering the virtual site, the traffic must be subject to WAF detection. WAF defense includes the systems of black/white list filtering and illegal information checking, as well as third-party *anquanbao* website vulnerability defense functions. The network framework composed of the existing security defense measures and the Web cloud service host (WCSH) is shown in Fig. 14.13.

It can be seen that the security defense measures currently used at Gianet's WCSH act to spot and block intrusions through conventional methods such as rule matching, black/white list matching, prior condition matching, etc. However, since security threats are multi-dimensional, although WAF based on packet filtering

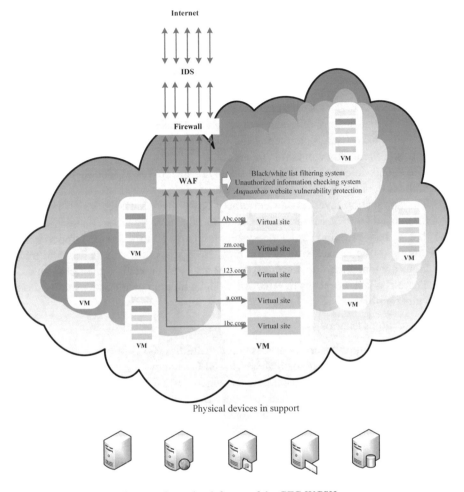

Fig. 14.13 Schematic diagram of security defenses of the GFC-WCSH

technology elevates the intrusion threshold for attackers via a large number of rules and thus blocks most types of attacks, those initiated through internal security vulnerabilities in Web services or possibly pre-implanted backdoors usually rely on normal services as their attack links and are not bound by the WAF rules.

In summary, the GFC-WCSH calls for effective defense technologies to enable itself to defend against unknown threats.

3. Product plan

A key concern in application demonstration is how to guarantee that the migration of 20,000 websites to the MSWVH will not affect the normal service of these user websites. As a solution, the double-layer disaster tolerance is achieved at both the system level and the web application level based on the endogenous redundancy of the CMD technology, ensuring that the double-layer disaster tolerance mechanism will be started immediately to ceaselessly provide normal services to users even if the system fails due to unknown factors in the migration process.

In view of the stability of the MSWVH in operation, the relevant contingency measures have been designed during the deployment process. To prevent the occurrence of force majeure or factors that cannot be checked out and solved in a short period of time, a disaster tolerance server is used to set up the dual-machine hot standby service for the MSWVH, which can migrate the business to the disaster tolerance server in real time to safeguard highly available Web services.

The MSWVH takes the physics devices of GFC as the underlying platform, with CMD technology as its core. The device appearance upon the MSWVH deployment is shown in Fig. 14.14.

4. Application deployment

To improve the security defense capabilities of GFCWCS to deal with unknown threats, especially 0day vulnerabilities and backdoor attacks, we deploy the MSWS and build the MSWVH on the GFCWCS platform. Based on the results of GFC's current major security threat analysis, the deployment of the current network follows the plan of "redundancy voting of static resources and heterogeneous monomers of dynamic content." As a platform to provide the public with Internet business functions, the MSWVH is required not only to provide business functions persistently and stably but also to prevent user information from leakage. In view of GFC's realities of providing WCS for SMEs via cloud service hosts, the corresponding MSWVH deployment plan is given as shown in Fig. 14.15. A large number of Web service attacks are launched toward upper-layer applications. In order to achieve a more remarkable defense effect at a lower cost, the MSWVH will bring the existing Web applications of the GFC service platform to mimic processing so as to build a lightweight MSWVH.

Fig. 14.14 The MSWVH

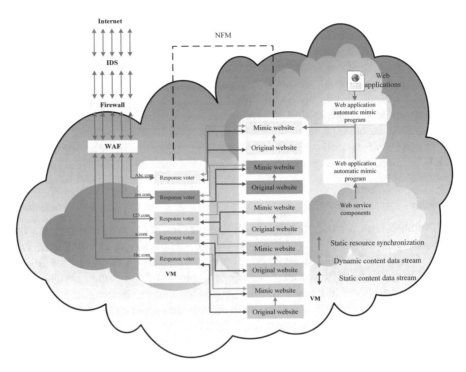

Fig. 14.15 Deployment plan for the MSWVH

The input agent (IA) and output agent (OA) have been added to the entrances and exits of existing web services. For a static resource request, the IA replicates the request and distributes it to the redundant static resource components. The static resource voter collects and compares the similarities and differences of the two responses, and obtains the unique correct response output against the voting algorithm. If the responses are inconsistent, the static file restoration method is applied to eliminate the threat; for a dynamic content request, the IA forwards the request to the mimic operation environment and outputs the correct execution results. If an execution abnormality is detected, it indicates a dynamic attack, and the real-time heterogeneous mechanism for dynamic codes will immediately eliminate the threat. The working principle is shown in Fig. 14.16.

The MSWVH will give full play to its features of flexible deployment and fast access, enabling convenient transplantation and expansion. While providing equivalent Web service functionality and performance, the lightweight application deployment is not costly, though still significantly improving the security defense capability as well as the high availability of Web services.

5. Cost analysis

As a cloud service provider, GFC seeks to provide substantial security gains for SME websites with minimal CMD deployment costs. Therefore, reducing

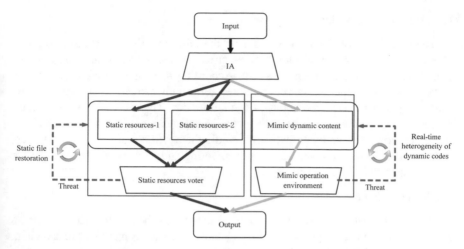

Fig. 14.16 Diagram of the working principle of the MSWVH

deployment costs and improving security are two equally important goals in the mimic upgrading transformation of GFC's virtual Web hosts.

(1) Hardware purchase

Hardware devices are service bearers. The newly purchased hardware devices for mimic upgrading are the most intuitive and direct deployment cost. Before the adjustment, GFC used two servers (Intel Xeon E5-2620 v4 *2, 256GB, 5 TB, about 55,000 yuan per unit) to carry Web services of 20,000 websites. After the voting software module is installed, the CPU payload increases no more than 10%. In view of the current CPU usage, there is no need to upgrade the CPU. After redundant storage of static resources, the disk occupation growth has not exceeded 20% (the original website template storage occupation of GFC's virtual Web host was 145.4 MB, which turned to 165.7 MB upon the mimic adjustment, an increase of 13.96%). During the pilot process, the MSWVH added a solid-state drive of 1 TB (3000 yuan per unit) to each of the original two servers at a storage increment of 20%, the solid-state drives cost 6000 yuan, and the average hardware upgrading cost for each website was only 0.3 yuan, an increase of 5.45%. Even if two servers (Intel Xeon E5-2620 v4 * 2, 256GB, 6 TB, about 58,000 yuan per unit) were purchased for the deployment of 20,000 MSWVHs, the required hardware acquisition cost will only be 116,000 yuan, or an average hardware cost of 5.8 yuan for each website.

(2) Deployment cost

Manpower cost is also necessary in the MSWVH deployment. GFC's technology team and CMD team invested a total of 10 person months *6000 yuan/person month = 60,000 yuan, which was averaged to 20,000 websites. Namely, the initial deployment cost for each website was only 3 yuan. As the MSWVH applications increases, the average initial deployment cost per site will be significantly reduced.

(3) Operation and Maintenance (O&M)

The O&M cost of the MSWVH during the pilot process consists of two parts: One is the basic O&M cost, including electricity charges, rack space charges, air-conditioning, refrigeration charges, etc. According to Gianet's measurement, the basic O&M cost of the two servers remained almost unchanged before and after the mimic adjustment, averaged about 340 yuan per day. The other is the O&M manpower cost. Once the 20,000 websites got adjusted and upgraded, the MSWVH can detect abnormalities and automatically handle them without artificial interference, reducing the workload of security troubleshooting and thus saving the demand for additional security O&M personnel.

(4) Security defense

No new security defense devices and systems have been purchased in GFC's mimic upgrading and adjustment this time, and the new investment herein is 0 yuan.

In summary, MSWVH is a low-cost cloud service product from the initial acquisition and deployment to the security management and O&M processes.

6. Application effect

On April 2, 2018, the MSWVH was officially launched after passing the test of functional conformance with the existing products. Up to now, over 20,000 websites have been deployed in the MSWVH format, all running normally and steadily. Although firewalls can defend against web threats of the majority types, some attacks have still managed to break through the plug-in firewall defenses and pose a threat to user websites, while the MSWVH that applies the CMDA can intercept and record these attacks with a minimal false negative rate and false positive rate. The MSWVH's plan designed for defense at the core site not only effectively solves the threat of malicious tampering of a user website but also detects attack behaviors unknown upon the existing defense means via voter abnormalities, and can classify the attack types simply according to the abnormal information. In future applications, the MSWVH will collaborate with on-site snapshots, file tracking, data recovery, and other forensics techniques to accurately locate unknown attacks based on automatic warning and handling of attack behaviors.

14.3.2   Current Network Testing

14.3.2.1   Testing of the MSWS

1. Testing purpose

In order to evaluate the defense effect and the operation status of the MSWS running on the official website in the pilot application, upon the completion of the functional conformance as well as security testing, we deployed the MSWS to the current network environment of the website and conducted online testing so as to provide data support for the application effect of the pilot work.

2. Testing plan

During the current network testing, the testing data are collected from the MSWS, the load balancer, and the official website platform, respectively. The hardware parameters and software configurations of the MSWS are respectively shown in Tables 14.3 and 14.4. The A10 Networks® Thunder® ADC series are selected as the load balancing devices, and their product line of high-performance and next-generation application delivery controllers will deliver better-established availability, acceleration, and security for the applications. The hardware parameters and software configurations of the devices are shown in Tables 14.5 and 14.6, respectively. The access traffic is distributed through the load balancer to the MSWS and the center's original platform server by 1:9. The current network testing scenario is shown in Fig. 14.17.

3. Testing and evaluation items

(1) Functional conformance testing

Once deployed at a certain government website, the MSWS was subject to a detailed and rigorous function testing and evaluation according to the comparative testing method stated in Chap. 10. The testing resorted to a combination of manual

Table 14.3 Hardware parameters of the MSWS

CPU	Intel Xeon E5-2620 v4 [2.10GHz,20MB tri-level cache]
RAM	32G DDR3
NIC	GbE Intel I350
Storage	1T 2.5 inch SATA
Power supply	675 W Hot-plugging power supply

Table 14.4 Software configurations of the MSWS

Physical host operating system	Windows Server 2012 R2, Centos 7		
VM software	VMWare 10.1		
Executor serial number	DUT2-1	DUT2-2	DUT2-3
VM operating system	Windows Server 2012 R2	SUSE11	Solaris10
Web server	Jetty	Resin	Tomcat
Web application	JAVA		
Mimic structure system control software	RDB and DDRV operating software		

Table 14.5 Hardware parameters of the load balancing A10 Networks Thunder ADC

Brand model number	Thunder840 ADC
CPU	Intel Communication Processor
RAM	8G ECC RAM
Network interface	1GE electrical port 5; 1/10GE fiber optical port (SPF+) 2
Storage	SSD
Power consumption	57 W/75 W

Table 14.6 Performance parameters of the load balancing A10 Networks Thunder ADC

(L4/L7)Application throughput (L4/L7)	5 Gbps/5 Gbps	SSL packet throughput	1 Gbps
4-layer CPS	200 K	SSL CPS	RSA(1K):2K;RSA(2K):500
4-layer HTTP RPS	1 million	DDoS prevention SYN/second	1.7 million
4-layer concurrent session	16 million	Application delivery partition (ADP)	32
7-layer CPS	50 K		

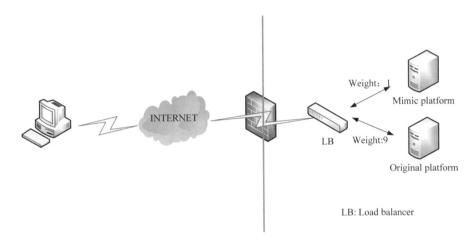

Fig. 14.17 Schematic diagram of the current network testing scenario

sampling access and automated web crawler scanning. The two access methods complement each other and can fully verify whether the mimic-structured website functionally conforms to the original website. The testing items are shown in Table 14.7.

In the manual sampling access testing, 24 test pages were extracted from 14 different functional modules of the website, where all the test pages succeeded in the output, with consistent test results compared to the original website. The automated web crawler scanning testing was performed on the mimic-structured website and the original website, respectively, where the total amount of links was 2437, including 2431 live links and 6 dead links, again indicating consistent results. Namely, the official website of a mimic structure functionally conformed to the original website.

(2) Security testing

Once deployed at a certain government website, the MSWS was subject to a detailed and rigorous security testing and evaluation according to the method stated in Chap. 10. There were five tested items: scanning detection, Trojan implantation, Webshell connection, webpage tampering, and system failure. In order to evaluate the effectiveness of CMD technology in preventing against known and unknown cyber attacks, we carried out a security test on a mimic structured website based on its pilot application. The corresponding items are shown in Table 14.8.

Table 14.7 Items of the functional conformance testing and evaluation

Evaluation target	Testing phase	Main item	Evaluation item
Business functionality of a mimic-structured website	Functional conformance testing	Test whether all the functional modules of the mimic-structured website can provide normal Web services, and whether they conforms to the original website in terms of business functionality	Evaluate whether the website of a mimic structure functionally conforms to the original website
		Test the total amount of live and dead links of the mimic-structured website and the original official website, and compare the amounts and the links	

Table 14.8 Items of the security testing and evaluation

Evaluation target	Testing phase	Main item	Evaluation item
Security defense capability testing of a mimic-structured website	Security penetration testing	(1) Scan a mimic-structured website (2) Scan a website under a single executor in a mimic structure	Compare the scanning results and evaluate the defense effect of the mimic-structured website
	Security injection testing	(1) Inject Trojans, Webshells, and malicious webpages into a website under a single executor in a mimic structure (2) Inject Trojans, Webshells, and malicious webpages into all the websites under all the executors in a mimic structure (3) Turn off the impact of a single executor or a Web server in the executor to test the impact of a system failure	Evaluate the security defense capabilities of the mimic-structured website by testing the effect of preventing against injected Trojans, backdoors (vulnerabilities) and malicious tampering without patching

The results of security penetration testing on a mimic-structured website show that the website has effectively resisted penetration scanning attacks through the voting mechanism, while the results of security injection testing on that website indicate that the mimic upgrading and adjustment has enabled the website to effectively defend against attacks such as Trojans, backdoors, loopholes, and malicious tampering, as well as to effectively tolerate system failures.

(3) Current network testing

The current network testing has been done on the basis of the functional conformance testing and security testing to test whether the MSWS is able to detect cyber attacks and system abnormalities in time. The testing data source is as follows:

(a) The mimic ruling log generated by the distribution ruling unit (DRU).
(b) The load balancer has the function of capturing network traffic in real time, and the related traffic data is stored as a traffic log.
(c) The O&M data of the website platform includes access logs and attack filtering logs.

Since the launch of the MSWS, the website has been working stably. Within 20 days from April 2 onwards, the total access to the MSWS was 632,892 times, an average of 3,164,4.4 times per day, with 12,736 abnormal visits spotted and recorded, which averaged at 636.8 visits per day or 26.53 visits per hour. Among them, the time periods recording the maximum and minimum abnormal access were 11:00 to 12:00 AM and 2:00 to 3:00 PM, respectively, which averaged at 58.15 and 20.9 visits per day accordingly. The number of threats recorded accounted for 2.01% of the total access. The data trend graphs for access traffic, daily abnormal access and hourly abnormal access are shown in Figs. 14.18, 14.19, and 14.20, respectively.

According to the analysis of access traffic and abnormal traffic, the MSWS is able to record its own overall running state and the abnormal access traffic passing through itself while carrying 10% of the data traffic of the website, and the running state remains stable and normal.

Based on the data records of CMD threat perception logs, we have analyzed sensitive information in abnormal access traffic, such as the access categories and sources. The abnormal access traffic recorded by the MSWS can be divided into six categories:

(1) Xss cross-site attack attempts, including abnormal access requests attempting to exploit functions in JavaScript, such as "publish/main/9/javascript:history. back()," etc.
(2) Webshell connection attempts, including abnormal access requests attempting at php, asp, jsp, and other types, such as "/plus/mytag_js.php," "/index.asp," etc.
(3) Xml utilization attempts, including abnormal access requests attempting to exploit external xml files, such as "rss/bulletin_2_0.xml," etc.
(4) Catalog browsing attempts, including abnormal access requests attempting at catalog blasting, such as "/yishi/," "/admin/," etc.
(5) Variable utilization attempts, including abnormal access requests attempting to operate on variables in the source code, such as "public/column/4664041?type =4&catId=4694378&action=list," etc.
(6) Other abnormal access, including abnormal access requests for scanning potential backup files in the website, such as "/robot.txt," "/www_cert_org_cn.rar," etc.

Among the 6 types of abnormal access, we have recorded 29 xss cross-site attack attempts, 237 Webshell connection attempts, 749 xml utilization attempts, 2839 catalog browsing attempts, 237 variable utilization attempts, and 6798 times of

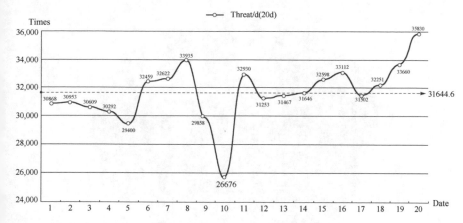

Fig. 14.18 Access traffic log

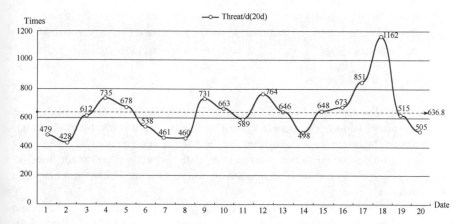

Fig. 14.19 Daily abnormal access traffic

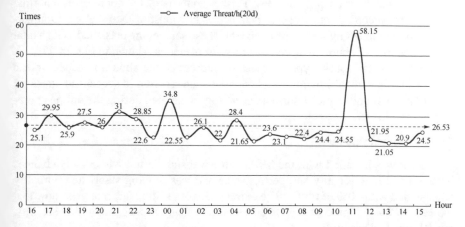

Fig. 14.20 Hourly abnormal access traffic

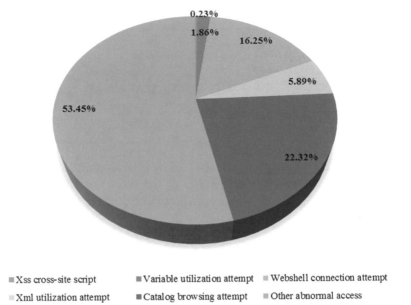

Fig. 14.21   Threat classification

other abnormal access. The MSWS threat classification is shown in Fig. 14.21. The first five types of abnormal access are remarkably out-proportioned by the last one, which has not been accurately analyzed yet. Further survey and analysis are necessary backed by on-site snapshot, file tracking, data restoration, and other forensic techniques, and this task will be applied in future product R&D, practice, and deployment.

After analyzing the sensitive entries in the six types of abnormal access traffic, we found a total of 7114 threat sensitive words: /robots.txt, 1,304 times; /rss/bulletin_2_0.xml, 672 times; /plus/mytag_js.php, 308 times; /index.php, 304 times; /index.aspx, 318 times; and /index.asp, 294 times. Further analysis of the sources of abnormal access in the recorded sensitive words indicates that most of them belong to fingerprint information left by web crawler access. These sources are classified, with a total of 43 categories of abnormal access sources discovered, as shown in Fig. 14.22.

Analysis of the access types and access sources of the abnormal access traffic recorded leads to the following conclusion: no data record of a successful attack on the MSWS is found in the abnormal access logs generated during the online operation of the MSWS.

4. Testing and evaluation

After it was installed with the MSWS, the website did not witness any changes in its service functionality and performance and was running stably and normally when added with 10% of traffic. The current network testing and analysis show that the MSWS has a high cost-effectiveness ratio in many aspects, including security, availability, and credibility.

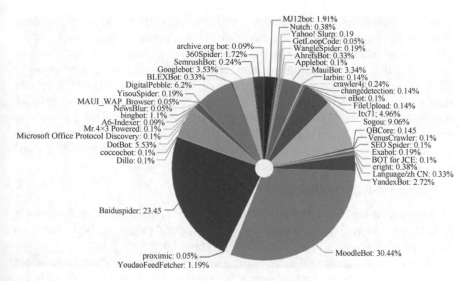

Fig. 14.22 Abnormal access sources

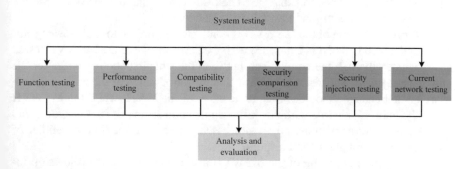

Fig. 14.23 Testing roadmap

14.3.2.2   Testing of the MSWVH

1. Testing purpose

Function testing, performance testing, and compatibility testing have been sequentially carried out to verify the effectiveness of CMD technology in GFC's deployment and its impact on the normal service functionality of the website, with comparative testing, injection testing, and current network testing carried out to evaluate the defense capabilities of the MSWVH.

2. Testing plan

On the basis of the guaranteed ongoing and stable operation of cloud services, the testing has been carried out in phases as shown in Fig. 14.23, sequentially function testing, performance testing, compatibility testing, security comparison testing, security injection testing, and analysis and evaluation. Function testing, performance testing, and compatibility testing mainly test whether the functionality of the MSWVH can guarantee the normal provision of website services; security comparison testing and security injection testing mainly test the defense capabilities of the MSWVH via comparison and injection methods; the current network testing mainly analyzes the operation status of the MSWVH to test and evaluate its current network security; while the analysis and evaluation phase is focused on analyzing the testing results accumulated in the previous phases, quantifying the pilot deployment situation and evaluating the engineering feasibility of CMD technology in the cloud environment. This process is also essential for the pilot work.

3. Testing and evaluation items

To assess the cost and potential defense benefits of CMD technology deployed in a cloud environment, we have identified the testing and evaluation items based on the target of each testing phase, as shown in Table 14.9.

The functions before and after the mimic Web deployment have been tested through comparative testing methods. The HTTP protocols and the mimic metafunctions of the MSWVH are tested and evaluated according to the relevant principles and methods stated in Chap. 13. The results show that the MSWVH has the same functions as the original Fast Cloud WVH does.

Performance testing of the MSWVH is conducted according to the relevant principles and methods stated in Chap. 13, with the performance of the MSWVH tested and evaluated through the comparison between the situations before and after the mimic deployment. The test results show that the introduction of the mimic mechanism has not changed the performance features of the WVH, while the MSWVH meets the indicators in line with the online application standards for Gianet's cloud virtual host in terms of transactions processed per second, maximum number of concurrencies, throughput, and response time.

The compatibility testing of the MSWVH also complies with the relevant principles and methods stated in Chap. 13, with the testing and evaluation done to judge whether there are disparities in terms of page display integrity, page layout, JS compatibility and webpage layout at different resolutions. The results show that the introduction of the CMD mechanism will affect neither webpage compatibility nor user access experience.

In line with the relevant principles and methods stated in Chap. 13, the security comparison testing of the MSWVH is conducted to test and evaluate the defense effectiveness of CMD technology over security technology applied to common cloud service systems. The results show that, with the voting mechanism and script isomerization technology, the MSWVH effectively resists attacks such as malicious tampering, Web application vulnerabilities, PHP Trojan injections, SQL injections, etc., improving the security defense capabilities of the cloud virtual host.

Injection testing of the MSWVH complies with the relevant principles and methods stated in Chap. 13, with the testing and evaluation done to find out that the MSWVH is able to withstand unknown threats as well as to defend against attacks based on vulnerabilities/backdoors.

Table 14.9 Testing and evaluation items

Evaluation target	Testing phase	Main item	Evaluation item
Compliance testing of the MSWVH	Function testing	1) Test whether the MSWVH can provide Web services normally 2) Test whether the MSWVH conforms to the HTTP 1.1 protocol consistency	Evaluate whether the MSWVH can provide Web services normally
	Performance testing	1) Test the number of transactions processed by the MSWVH per second 2) Test the maximum number of concurrencies of the MSWVH 3) Test the throughput of the MSWVH 4) Test the response time of the MSWVH	Evaluate whether the MSWVH can meet the indicators in line with the online application standards for Gianet's cloud virtual host in terms of transactions processed per second, maximum number of concurrencies, throughput, and response time
	Compatibility testing	1) Test the page display integrity of the MSWVH 2) Test the webpage functional integrity of the MSWVH 3) Test the web layout ability of the MSWVH 4) Test the capability of the MSWVH in switch between persistent connection and non-persistent connection	Evaluate whether the MSWVH can ensure normal Web service, webpage compatibility requirements
Security defense capability testing of the MSWVH	Security comparative testing	1) Test the security gain of the MSWVH based on the existing defense 2) Test the security defense capabilities of the MSWVH without applying existing defense means such as the IDS, firewalls, and the systems of black/white list filtering and illegal information checking	Compare the defense capabilities of existing defense means and of mimic defense techniques through the security threats spotted by the MSWVH
	Security injected testing	1) Inject a backdoor to the MSWVH 2) Test the ability of the MSWVH to resist an attack exploiting the injected backdoor 3) Inject a vulnerability into the MSWVH 4) Test the ability of the MSWVH to resist an attack exploiting the injected vulnerability.	Evaluate the defense capabilities of the MSWVH in resisting unknown threats by testing the effect of its defense against injected backdoors and vulnerabilities without patching
	Current network testing	Test the security of the MSWVH in the current network environment	Evaluate the security of the MSWVH by analyzing its operation status in the current network environment

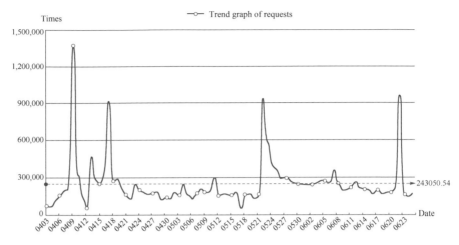

Fig. 14.24 Access traffic trend

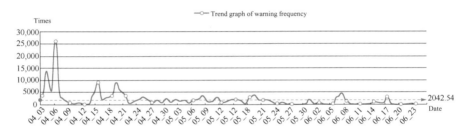

Fig. 14.25 Threat perception trend

4. Current network testing

From April 3 to June 25, the MSWVH received a total of 20,416,254 visits to 300 websites, with an average daily access of approximately 243,050 times. The access traffic trend is shown in Fig. 14.24, proving that the MSWVH can provide ongoing and stable external services.

For effective evaluation, 300 websites were randomly selected for operational monitoring analysis. During the operation, the system defended against 171,673 attacks in all, with an average of 2042 times per day. The number of attack threats that are automatically warned and disposed of accounted for 0.9% of the total traffic. The threat perception trend is shown in Fig. 14.25.

The analysis of access logs and threat perception logs mainly reveals the following four types of attack: scanning detection, SQL injection, overflow attack, and php_shell attack, as shown in Fig. 14.26, respectively, accounting for about 50%, 4%, 2%, and 35% of the total threats resisted by the MSWVH according to statistics. In addition, 9% of the total attack threats resisted by the MSWVH are unknown

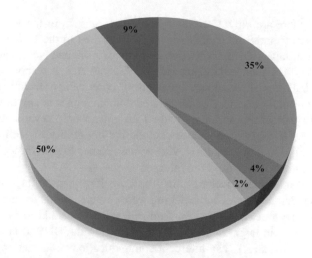

■ PHP_shell attack   ■ SQL injection attack   ■ Overflow attack   ■ Scanning detection   ■ Others

Fig. 14.26   Threat classification

attacks. This part of attack behaviors cannot be analyzed and classified by simple features, which should be further analyzed with forensic techniques including network data forensics and system data forensics.

5. Testing and evaluation

Up to now, the number of websites deploying the MSWVH has exceeded 20,000, and all the websites are running normally and steadily. According to testing and analysis, the MSWVH shows a high cost-effectiveness ratio in terms of security, availability and reliability upon a relatively small input.

14.4   Mimic-Structured Domain Name Server (MSDN Server)

14.4.1   Application Demonstration

14.4.1.1   Threat Analysis

The domain name system (DNS) is the core infrastructure of the Internet, responsible for the translation of domain names to IP addresses. Due to historical reasons, the domain name service system has been suffering from an inappropriate management order, unbalanced regional development, incomplete rules and regulations, and increasingly prominent security problems, which seriously affects the national network sovereignty maintenance and social development stability.

Currently, there are only 13 domain name root servers in the world, with the master root server and 9 secondary root servers located in the United States, 3 in Europe and 1 in Japan. The Internet Corporation for Assigned Names and Numbers (ICANN), an NPO under the US Department of Commerce, is responsible for managing the global Internet domain name system. Theoretically or technically, ICANN can close the resolution service of the related domain names at any time, making a certain country, organization, or individual instantaneously "evaporate" from the Internet space. Of course, we cannot exclude the possibility of taking the innate advantage of hijacking or tampering with domain names to gain or peek into sensitive data and information across the Internet.

In recent years, cyber attacks against the DNS have caused huge losses. First, impersonating domain names and phishing websites have become crucial means for criminals to carry out economic fraud and illegal propaganda to infringe on the privacy of citizens in large numbers. Next, exploiting the DNS to launch man-in-the-middle (MITM) attacks can intercept communications information of any government websites, emails, and business socials, or tamper with data and software in sensitive industries such as control scheduling, financial transactions, and logistics transportation, leaving tremendous hidden hazards to cyber security.

The domain name address resolution system itself has inevitable software/hardware security vulnerabilities and cannot block the backdoor/trapdoor/front door problems caused by malicious code implantation conducts. Besides, human science and technology cannot completely screen the unknown vulnerabilities/backdoors theoretically and engineeringly. Based on the vulnerability of the domain name protocols and the related services or defense systems, an attacker can exploit the communication protocol and the unknown software/hardware backdoors to hijack domain names by technical means such as tampering with domain name cache data or protocol packets for the impersonation of any website, including government websites, to implement malicious attacks, including but not limited to false information disclosure, unperceivable Trojan implantation, and confidential data theft.

The mimic domain name system (MDNS) faces three major threats, namely, impersonation (such as cache poisoning and domain name hijacking due to unknown vulnerabilities and backdoors), obliteration (such as .cn domain name records deleted from the root domain name server), and occoecation (e.g., the root domain name server fails to resolve domain name resolution requests from China). The root cause of the above threats lies in the backdoors of the system's own software/hardware as well as the vulnerabilities of the domain name resolution communication mechanism exploited by attackers, resulting in tampering of the mapping between the domain name and IP. To this end, the mimic interface of the mimic domain name server (MDN server) is set in the domain name resolution response packets that can affect the mapping between the domain name and IP, so that the server can rule the packets returned by each executor to detect and defend the mapping tampering attacks caused by the system's inherent software/hardware backdoors and the resolution communication mechanism's vulnerabilities.

In summary, the Internet domain name service system remains the "pain point" concerning the autonomy, security, and credibility of China's cyber infrastructure and is the Sword of Damocles hanging over China's state security, so it needs an immediate solution. However, cyber revolution has never been done at one kick. It is a matter of long-term deliberation and development, but time waits for no man, we must right now come up with a feasible and innovative solution.

14.4.1.2   Application Scenario

Established in October 2008, CUHB is the branch of China Unicom in Henan Province. It is responsible for its local construction and operation and is one of the major telecom service providers in Henan Province. As for CUHB's domain name service node in Luoyang, the original configurations included a total of 7 recursive domain name servers and 1 authoritative domain name server, they constitute the local cache recursive domain name service node in the metropolitan area network (MAN) of Luoyang. With the Anycast deployment, the seven servers in the node release the province-wide unified service IP addresses 202.102.224.68 and 202.102.227.68 in support of load balancing. When any one or more servers fail, the services can be automatically taken over by other servers within the node. If all the devices or switches in the node fail, the router will automatically switch to the Zhengzhou node to provide services. Currently, the server at the Luoyang node is designed with a peak load capacity of 30,000 QPS for a single server and 210,000 QPS for the entire node. The performance of the recursive domain name servers at the same node can no longer meet the design requirements for triple redundancy, and a system multiplication is required based on the MDN server.

The mimic authoritative domain name (MADN) server is planned to be deployed in Gianet's computer room in Zhengzhou. The MADN server itself is a DNS system with the function of authoritative domain name resolution, which can be deployed incrementally without changing the existing network topology, business services, and management methods. For the sake of stability, in the early stage the MADN server is responsible for new domain name resolution requests, and gradually takes over the resolution requests of the existing DNS.

14.4.1.3   Product Plan

As the underlying device providing Internet access to the public, the MDNS not only needs to resolve the authorized domain names in the local network but is responsible for providing users with recursive resolution capabilities of non-local authorized domain names. An MDNS consists of MADN servers and mimic recursive domain name (MRDN) servers and applies to various domain name resolution scenarios across the Internet, telecommunication networks, industrial control networks, and the IoT.

The overall architecture of the MDNS is shown in Fig. 14.27.

The network element function is described as follows:

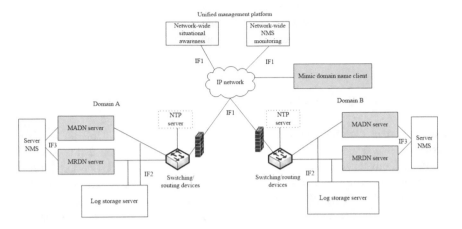

Fig. 14.27 Overall architecture of the MDNS

(1) Unified management platform: Unified monitoring and analysis of network-wide domain name services, including:

Network-wide NMS: Unified monitoring of DNS operation indicators across the entire network.
Network-wide situational awareness system: Data mining for DNS parsing logs, providing network-wide user behavior analysis and security situational awareness capabilities.

(2) Domain name service node: A domain name service node includes the following devices:

The MADN server: A server based on the mimic structure, with authorization for the applied domain, responsible for preserving the record information on the original domain name resources of the domains that have been authorized upon application.
The MRDN server: A server based on the mimic structure, responsible for accepting the request sent by clients (parsers), obtaining the query result required by the user by issuing a query request to the authoritative domain name servers at all levels, and finally returning the result to the client parser.
Parsed log storage server: Independent from the MADN/MRDN server, used to store DNS parsed logs.
Network management & monitoring system: Implementing network monitoring of DNS operation indicators.

The MDNS includes MADN servers, MRDN servers, and mimic domain name clients (domain name parsers). In order to defend against the attacks based on the system's own software/hardware vulnerabilities, the MDN server applies DHR design to the operating system layer and the domain name resolution application software layer. In order to defend against the communication mechanism vulnerabilities, the MDN server applies DHR design to the dimensions of time, geography, frequency, and physics. Based on the mimic structure, the MDN server introduces DHR executors to ensure that the executors can hardly generate common mode failures. The policy distribution and scheduling mechanism is introduced in the heterogeneous executor management process to enhance the uncertainty of the apparent structural characterization of target objects under the functionally equivalent conditions and increase nonlinearly the difficulty for attackers to detect or predict the defense behaviors; the DHR's convergent multi-dimensional dynamic reconfiguration mechanism, which is reconfigurable, reorganizable, reconstructable, redefinable, and virtualized, makes the defense scenario more targeted at the attacker behavioral outcomes, subverting inheritable and reproducible attack experience and making it impossible for an attack to produce a plannable and predictable task effect; the DHR closed-loop feedback control mechanism can provide non-specific surface defense functions not relying on threat feature acquisition, implement point defense based on specific awareness, and effectively block the coordinated escape in the space-time dimension through "trial and error" attacks on the heterogeneous executors of the target object.

The structure of the MDN server is shown in Fig. 14.28. In multiple processes such as network architecture, system logic, domain name resolution software, and security and maintenance management, it adopts distributed deployment, high robust control, executor heterogeneity, defense against domain name specific/non-specific attacks and DDOS attacks, security monitoring protection, security self-recovery, and other technologies to safeguard the secure and reliable operation of

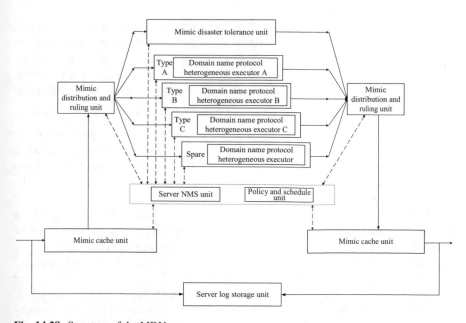

Fig. 14.28 Structure of the MDN server

General-class MDN server   DC-class MDN server   Carrier–class MDN server

Fig. 14.29 Appearance of the MSDN server MD-2100X-GS series

the DNS of the telecom operator. The MDN server uses an optimized high-performance commercial server as the underlying platform, which, mainly based on software, realizes the mimic structure and is able to provide highly reliable, available, and carrier-class domain name services.

The R&D of the MSDN server MD-2100X-GS is based on commercial standard servers with carrier-class reliability. The appearance of the series is shown in Fig. 14.29.

14.4.1.4 Application Deployment

The MDN server itself is a complete DNS with cache resolution and recursive functions. It can be inserted into the existing DNS based on the Anycast mechanism through incremental deployment without changing the current network topology, business services, and management methods. For the sake of prudency, the initial resolution requests may be offloaded, that is, the number of requests for offloading is $\frac{1}{N+1}$ (N is the number of DNS cache servers in the current network). When the system is abnormal, the health check function of the MDN server can be automatically used to exit the DNS resolution service without affecting the current network service. Later, it will gradually take over all the resolution requests of the existing DNS. In addition, all servers in the current DNS can be reused to maximize the returns on the existing investments.

The MDN server is deployed in an incremental manner without changing the existing domain name protocols and address resolution facilities. Since the current network topology, business services and management methods remain unchanged, the current network just needs a little transformation. To be specific, only about 2 person weeks of work for each set of MDN servers.

The MDNS can defend against attacks based on software/hardware vulnerabilities of the DNS. The incremental deployment of MDNSs can realize security defense, situational awareness, and emergency response of top-level MADN servers at the state level (.cn), as well as security defense and emergency recovery of the MRDN/MADN servers of carriers, service providers, and enterprises and institutions, without changing the existing domain name protocols and address resolution facilities.

14.4.1.5 Cost Analysis

The existing proactive and passive defense theories and methods are based on precise threat perception and follow the defense pattern of "threat perception, cognitive decision-making, and problem removal." They are almost defenseless against "known unknown" security risks or "unknown unknown" security threats, except for encrypted authentication measures when conditions permit. Although businesses are investing more in cyber security to resist attacks, conventional cyber security defense means have failed in ensuring a high level of security despite heavy spending. For example, JPMorgan Chase & Co. has a cyber security team of over 1000 staff members plus a budget of $250 million, it still got hacked in 2014. The CMD technology based on the GRC structure generates an ES effect without relying on "attached or external" defense measures, which can effectively suppress the security threats caused by known or unknown vulnerabilities, backdoors, and Trojans in the mimic interface. Thus, the function of unknown threat defense that cannot be realized through conventional defense methods can be achieved by the mimic defense technology at a reasonable cost, so mimic defense will become the main trend of cyber security defense.

CUHN and Gianet demand the maximum security gain for the DNS at minimal CMD deployment costs. Therefore, reducing deployment costs and improving security are two equally important goals in the pilot of the MDN server. The pilot deployment cost is hereinafter analyzed and evaluated from three aspects: hardware purchase, security defense, and deployment and O&M.

1. Hardware purchase

The newly purchased hardware devices for mimic upgrading constitute the most intuitive and direct deployment cost. Compared to conventional plans for domain name server deployment, the number of COTS servers required for a single set of carrier-class MDN servers has increased to 5 from 2 (35,000 yuan per unit), which is equivalent to an increase of about 6% in the total cost. Prior to the transformation, Gianet used DNSPOD for authoritative domain name resolution. The main hardware overhead introduced by the DC-class MDN server to provide domain name resolution services for 20,000 websites is to add 3 new COTS servers (20,000 yuan per unit) to each set. The hardware purchase cost is about 60,000 yuan, an increase of less than 16%. In addition, an MDN server is ten times better than a conventional one in performance, and the CMD team is researching and developing a cloud-based mimic domain name resolution service for a large number of users, which will greatly reduce the cost amortization among users.

2. Security defense

No new security defense devices and systems have been purchased in GFC's mimic upgrading and adjustment this time, and the new investment herein is 0 yuan. The operational data shows that the existing defense means can only defend against attacks with known features, while the mimic mechanism can defend against various known or unknown attacks based on the vulnerabilities in the mimic interface.

3. Deployment and O&M

The maintenance and management cost of the MDN server during the pilot process consists of three parts. The first is the manpower cost for the initial deployment, where CHHN and Gianet each invested 1 person week, while the CMD team invested 2 person weeks, costing a total of 6000 yuan. The second is the basic cost of O&M, including electricity charges, rack space charges, and air-conditioning and refrigeration costs, etc. After calculation, CHHN and Gianet found a slight increase in electricity and rack space charges after the transformation. The third is the O&M manpower cost. After the MDN servers were installed, the O&M remained unchanged, and the added MDN servers did not entail patching, upgrading, troubleshooting, and security policy maintenance, so the manpower cost was greatly lowered.

Analysis of costs in hardware purchase, security defense, and deployment O&M indicates that the MDN server is a domain name server product with a lower input cost and a higher security gain, whether it is in initial acquisition and deployment or in day-to-day management and O&M.

14.4.1.6 Application Effect

The MSDN server provides highly credible, available, and reliable mimic domain name resolution services at a relatively low cost (up by about 6–16% for mimic adjustment). The MSDN server adopts a standard protocol interface and follows the existing management and maintenance mechanisms, with a small workload for current network transformation. Since its launch at tightly defended CUHN, the server has recorded an average daily resolution amount of 2.41 billion times, and the system has been in secure and reliable operation. At Gianet, the server's average daily resolution record is 300 million times. In the absence of any additional security facilities, 320,000 attacks have been detected and intercepted cumulatively, covering scanning detection and domain name configuration tampering, and the system is running safely and reliably. The pilot operation of the two sites (CUHN and Gianet) proves that the MSDN server can effectively cope with and defend against known risks and unknown threats based on the target object's vulnerabilities and virus Trojans, etc., and has a universal and outstanding preventive effect on all kinds of known and unknown security threats in the mimic interface.

14.4.2 Testing and Evaluation

To verify the effectiveness of CMD technology in the current network deployment, check the impact of its introduction on the normal functionality of domain name resolution service, and evaluate the performance effect of the MSDN server in the existing network, we sequentially conducted function testing, performance testing, and compatibility testing. To evaluate the security defense capabilities of the MDN server, we carried out comparison testing, injection testing, and current network testing.

We have identified the testing and evaluation items based on the target of each testing phase, as shown in Table 14.10.

The MSDN server security testing was conducted in accordance with the methods in Chap. 10 to assess its defense capabilities of resisting unknown threats. The function, performance, and compatibility of the MSDN server were tested and evaluated according to state as well as industry standards. The testing results show that the MSDN server fully meets the relevant standards and requirements.

Following is a description of mainly the testing of the MSDN server in current networks:

14.4.2.1 CUHN

The MDN server was officially put into operation at CUHN on January 23, 2018, and the system has been running in a stable and reliable manner from then on, with a total resolution amount of 204.8 billion times, an average of 2.41 billion times per day or 28,584 times per second, and a maximum of 39,438 times and a minimum of 17,605 times per second. The total amount of successful resolutions is 183.6 billion times, which averaged 2.16 billion times per day or 25,183 times per second, with a maximum of 35,253 times and a minimum of 15,406 times per second. The total amount of recursive resolutions is 24.6 billion times, which averaged 290 million times per day or 3456 times per second, with a maximum of 6896 times and a minimum of 1803 times per second. The system features a successful resolution rate of 99.96%.

The access traffic trend of the MDN server is shown in Figs. 14.30 and 14.31.

Since it was put online at the Luoyang node of CUHN, the MDN server has identified a large number of poisoning attacks, DDOS attacks, amplification attacks and other attacks, and has detected and blocked 837 security threats that conventional security measures failed to find out. These security threats originated from 114 IPs. CUHN deployed strict security defense facilities in front of the domain name server in accordance with the state-level cyber security requirements, such facilities effectively intercepted the attacks, like scanning detection, though, they failed to detect the attacks intercepted by the MDNS. According to operational data analysis, the IPs and locations of the top 10 attackers are shown in Fig. 14.32, where all attack threats have been spotted and blocked by the MDN server.

Table 14.10 Testing and evaluation items

Evaluation target	Testing phase	Main item	Evaluation item
Evaluate the domain name resolution capabilities of the MSDN server	Function testing	(1) Test whether the service functions are consistent before and after the mimic deployment (2) Test the mimic meta-functions of the MSDN server, such as request distribution equalization, response voting, etc. (3) Test the DNS protocol functionality of the MSDN server	Evaluate whether the MSDN server can ensure normal functional requirements for domain name resolution and mimic defense
	Performance testing	Test the performance of the MSDN server	Evaluate whether the MSDN server can ensure normal DNS service performance requirements
	Compatibility testing	(1) Protocol compatibility (2) Client compatibility (3) Interconnection with other domain name servers	Evaluate whether the MSDN server can ensure normal domain name resolution compatibility requirements
Evaluation of the security defense capabilities of the MSDN server	Security comparative testing	1) Test the security gain of the MSDN server based on the existing defense 2) Test the security defense capabilities of the MSDN server without applying existing defense means such as the IDS, firewalls, and the systems of black/white list filtering and illegal information checking	Compare the defense capabilities of existing defense means and of mimic defense techniques through the security threats spotted by the MSDN server
	Security injected testing	1) Inject a backdoor into the MSDN server to test the ability of the MSDN server to resist attacks exploiting the injected backdoor 2) Inject vulnerability into the MSDN server to test the ability of the MSDN server to resist attacks exploiting the injected vulnerability	Evaluate the defense capabilities of the MSDN server in resisting unknown threats by testing the effect of its defense against injected backdoors and vulnerabilities without patching
	Current network testing	Test the security of the MSDN server in the current network environment	Evaluate the security of the MSDN server by analyzing its operation status in the current network environment

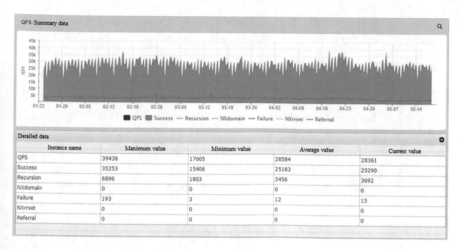

Instance name	Maximum value	Minimum value	Average value	Current value
QPS	39438	17605	28584	28361
Success	35253	15406	25183	25290
Recursion	6896	1803	3456	3692
NXdomain	0	0	0	0
Failure	193	3	12	15
NXrrset	0	0	0	0
Referral	0	0	0	0

Fig. 14.30 High-speed cache unit traffic flows of the MDN server

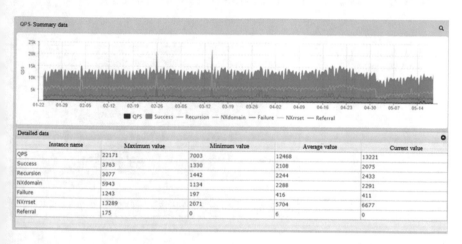

Instance name	Maximum value	Minimum value	Average value	Current value
QPS	22171	7003	12468	13221
Success	3763	1330	2108	2075
Recursion	3077	1442	2244	2433
NXdomain	5943	1134	2288	2291
Failure	1243	197	416	411
NXrrset	13289	2071	5704	6677
Referral	175	0	6	0

Fig. 14.31 Executor unit traffic flows of the MDN server

In order to ensure the secure and reliable operation of the DNS, CUHN restricted the geographical service partition of the domain name servers. The domain name server at the Luoyang node only serves the cities within the province, such as Luoyang, Puyang, and Sanmenxia. As a result, the source IP addresses of DDOS attacks are mainly located in these cities. The poisoning attackers may come from any corner of the Internet, while the statistical results tell that their IP addresses are mainly based in places such as Guangzhou, Hangzhou, Beijing, and Taipei.

Analysis of the attacks since the launch identifies a total of four types of attacks, namely, poisoning attacks, DDOS attacks, amplification attacks and other unidentified attacks, where the first two types account for the majority share. The numbers of classified attacks are shown in Fig. 14.33.

TOP10 IP

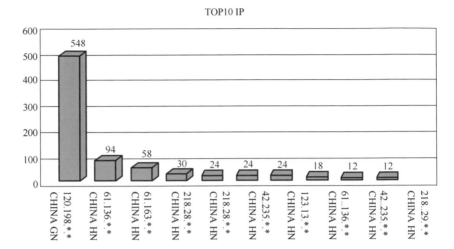

Fig. 14.32 IPs and locations of the top 10 attackers by the number of threats

Fig. 14.33 Statistics of classified attacks

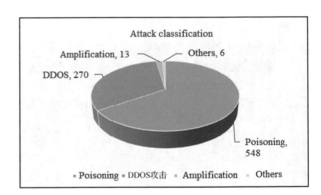

14.4.2.2   Gianet

The MDN server was applied online at Gianet on April 12, 2018, and has been acting as a stable and reliable system so far. The system's total amount of resolutions is 11.2 billion times, which averaged at 300 million times per day, with a successful resolution rate of 98.4%. In Gianet's pilot program, the relevant domain names of GFC were migrated gradually. Large-flow requests were used to verify the stability of the system in the early testing stage, they contributed to the large traffic in the first stage, and the test traffic was removed in the later stage.

Now the MADN server provides domain name resolution services for about 20,000 virtual websites of Gianet. As shown in Fig. 14.34, the current request rate of the MADN server is 3581 times per second.

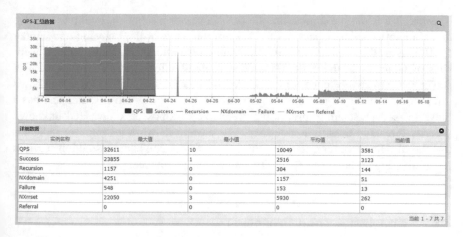

Fig. 14.34 Executor unit traffic flows of the MDN server

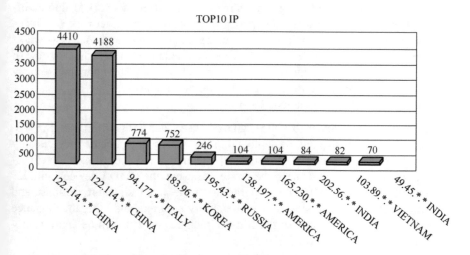

Fig. 14.35 IP addresses of the top 10 attackers (by threat number) and their locations

During its operation, the MDN server detected a large number of attacks (mainly scanning detection attacks), which totaled about 320,000 times, an average of 8687 times per day. To illustrate, we look at the data recorded on May 24, 2018: these threat attacks originated from 889 IPs scattered in multiple countries. The IP addresses of the top 10 attackers (by threat number) and their locations are shown in Fig. 14.35. All attack threats were spotted, recorded, and blocked by the MDN server.

Since it was put into operation, the MSDN server has been providing recursive domain name resolution services for 1.5 million users of the Luoyang node under CUHN, as well as authoritative domain name resolution services for 20,000 websites of Gianet, and the system has been running safely and stably. The MSDN server can effectively cope with and defend against known risks and unknown threats based on the target object's vulnerabilities and virus Trojans, etc., and has a universal and outstanding preventive effect on all kinds of known and unknown security threats in the mimic interface. The current network testing also indicates that the MSDN server can provide "three-in-one" domain name resolution services integrating high reliability, credibility, and cost performance.

14.5 Conclusions and Prospects

The research team invited three professional domestic security testing teams to conduct the current network test on a variety of mimic structure devices, which lasted a fortnight, with a dozen of testing experts involved in current network testing in different phases. The test evaluation expert group for the current network concludes that the testing was well organized under a comprehensive plan and done in a diverse number of methods, that the test results were recorded in detail, and the outcome of test is true, and that the mimic defense theory is effective in strengthening security of the current network and preventing unknown cyber attacks.

The testing analysis is concluded as follows:

(1) In terms of deployment cost, the CMDA devices such as the MS router, MSWS, and MSDN server can all be deployed incrementally, the price of which is only 2–20% higher over that of conventional system devices, indicating a lower system deployment cost. During the entire life cycle, the CMDA devices with ES features need neither frequent upgrading of the attack signature database and software/hardware versions nor vulnerability scanning and other maintenance and management efforts, so the life cycle aggregate cost will be significantly lower than that of conventional devices.

(2) In terms of defense effectiveness, mimic structure devices can effectively cope with and defend against known risks and unknown threats based on the target object's vulnerabilities and virus Trojans, etc., and has a universal and outstanding preventive effect on all kinds of known and unknown security threats in the mimic interface, being a three-in-one information system architecture that integrates "security defense, reliability guarantee, and service provision." In particular, such devices have the ability to perceive, learn about, and defend against unknown threats, thus providing unknown attack analysis with accurate data evidence that the conventional defense devices fail to present.

(3) In terms of the current network transformation, the mimic structure devices do not need much alteration since they can be connected with the existing hardware/software devices through standard protocol interfaces and run under the existing operation, administration, and maintenance system.

This test has verified the feasibility and effectiveness of the mimic defense products applied in the current network. The test records can serve as references for setting CMDA device test specifications in the existing networks and for setting the mimic defense standards. The pilot program has actually promoted the development and application of the mimic defense products.

Compared with the pure environment and controlled threats in laboratory testing, the real network environment is much more complicated and changeable. The current network testing not only provides a test environment to seek methods and key technologies for further optimization and refinement of the CMDA system but also points out the prospects for the subsequent development. It should be noted that CMD is the first defense technology that provides the ability to perceive, learn about, and defend against unknown threats in cyberspace and provides accurate and reproducible evidence for the technical analysis of unknown threats. CMD will possibly subvert the defense concept of "the era of attacks from software/hardware code vulnerabilities," and will have indispensable practical implications for and broad strategic influence on the pursuit of our general objective of cyber security and informatization in the global environment, which are like "two wings of a bird or two wheels of an engine."

Printed in the United States
By Bookmasters